OCEAN

OCEAN

THE WORLD'S LAST WILDERNESS REVEALED

LONDON, NEW YORK, MELBOURNE,
MUNICH AND DELHI

SENIOR EDITORS Peter Frances, Angeles Gavira Guerrero

SENIOR ART EDITOR Ina Stradins

PROJECT EDITOR Rob Houston

PROJECT ART EDITORS Peter Laws, Kenny Grant, Maxine Lea, Mark Lloyd

EDITORS Rebecca Warren, Miezan van Zyl, Ruth O'Rourke, Amber Tokeley

DESIGNERS Francis Wong, Matt Schofield, Steve Knowlden

CARTOGRAPHERS Roger Bullen, Paul Eames, David Roberts. Iowerth Watkins

INDEXERS Sue Butterworth, John Dear

DTP DESIGNERS Julian Dams, Laragh Kedwell

PROOF-READERS Polly Boyd, Ben Hoare

JACKET DESIGNERS Lee Ellwood, Sharon Spencer

SCHERMULY DESIGN COMPANY

SENIOR EDITOR Cathy Meeus

DESIGNERS Dave Ball, Lee Riches, Steve Woosnam-Savage

ART EDITOR Hugh Schermuly

CARTOGRAPHER Sally Geeve

EDITORS Gill Pitts, Paul Docherty

DESIGN ASSISTANT Tom Callingham

PICTURE RESEARCHER Louise Thomas

ILLUSTRATORS Mick Posen (the Art Agency), John Woodcock, John Plumer, Barry Croucher (the Art Agency), Planetary Visions

PRODUCTION CONTROLLER Joanna Bull

MANAGING EDITORS Sarah Larter, Liz Wheeler

MANAGING ART EDITOR Philip Ormerod

PUBLISHING DIRECTOR Jonathan Metcalf

ART DIRECTOR Bryn Walls

CONSULTANT John Sparks

First published in Great Britain in 2006
This edition published in 2008
by Dorling Kindersley Limited
80 Strand, London WC2R 0RL

A Penguin Company

A CIP catalogue record for this book is available from the British Library

ISBN 978 1 4053 3092 3

Colour reproduction by Colourscan, Singapore
Printed and bound in China by Leo Paper Products

See our complete catalogue at
www.dk.com

CONTENTS

OCEAN LIFE

ATLAS OF THE
OCEANS

ABOUT THIS BOOK

THIS BOOK IS DIVIDED INTO four chapters. An overview of the physical and chemical features of the oceans is given in the INTRODUCTION; OCEAN ENVIRONMENTS looks at the main zones of the oceans, while OCEAN LIFE examines the life-forms that inhabit them. The ATLAS OF THE OCEANS contains detailed maps of the oceans. Most chapters are divided into smaller sections.

INTRODUCTION

This opening chapter is divided into four sections. In OCEAN WATER, the properties of water itself are examined. OCEAN GEOLOGY covers the materials of the ocean floor and the way that it changes over time. CIRCULATION AND CLIMATE is about the interaction between oceans and the atmosphere and the large-scale movement of water, while TIDES AND WAVES looks at movements and disturbances of water on a smaller scale.

◀ OCEAN WATER
This section covers the properties of the water molecule, the chemistry of seawater, and the way that attributes such as temperature, pressure, and light transmission change with depth in ocean water.

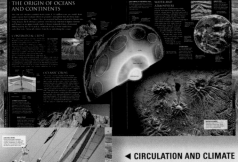

OCEAN GEOLOGY ▶
As well as describing the composition of the ocean floor, this section looks at the processes that shape it, tracing the origin of the oceans and their changing size and shape over geological time.

◀ CIRCULATION AND CLIMATE
This section describes the large-scale circulation of the oceans, both at depth and at the surface. It also looks at ocean climates and the many ways in which the oceans and the atmosphere influence one another.

TIDES AND WAVES ▶
The regular movements of the tides are described here, as well as the way that disturbances spread out across the surface in the form of waves.

OCEAN ENVIRONMENTS

This chapter looks at specific parts of the oceans. It is divided into sections on different zones, starting with COASTS AND THE SEASHORE and then moving to progressively deeper waters, first with SHALLOW SEAS and then THE OPEN OCEAN AND OCEAN FLOOR. A final section, POLAR OCEANS, looks at the frozen waters around the North and South poles. In each section, explanatory pages describe typical features and formative processes, while the succeeding pages contain profiles of actual features. The profiles are arranged by geographical location, starting with the Arctic Ocean and followed, in order, by the Atlantic, Pacific, Indian, and Southern oceans.

▲ EXPLANATORY PAGES
These pages describe general types of environments. The example above is taken from the shallow seas section.

OCEAN LIFE

This chapter contains two sections. The INTRODUCTION TO OCEAN LIFE covers the ecology and history of marine life and the way that marine organisms are classified. It is followed by a larger section, KINGDOMS OF OCEAN LIFE. This is divided into domains or kingdoms and, in the case of the plant and animal kingdoms, further divided into smaller groups. In each case, a general overview of the organisms that make up the group is followed by profiles of a selection of individual species. The section begins with the smallest forms of life, the bacteria and archaea, and ends with the animal kingdom.

DOMAIN	Eucarya
KINGDOM	Animalia
PHYLUM	Mollusca
CLASSES	8
SPECIES	50,000

colour-coded panel shows position of group being described (indicated with white outline) in the classification hierarchy

GROUP INTRODUCTION ▶
Pages such as the ones shown here describe groups of organisms in general. All introductions contain an account of the defining physical characteristics, usually followed by further information on behaviour, habitats, and classification.

ATLAS OF THE OCEANS

The final chapter of the book is an atlas of the world's oceans. It includes maps of the five major oceans. The pages that immediately follow each whole-ocean map contain more detailed maps of selected regions of that ocean. All the maps have been produced using data collected from a combination of satellite- and ship-borne instruments. They are labelled to show the names of the seas, undersea features (such as ridges, trenches, and seamounts), and prominent coastal features. They also show ocean depths and the boundaries between tectonic plates. At the end of the book there is a separate index of features identified on the maps.

▼ REGIONAL MAP
As well as maps, these pages also include profiles of individual seas or undersea features. The example shown here is from the section on the Pacific Ocean.

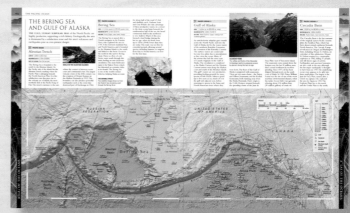

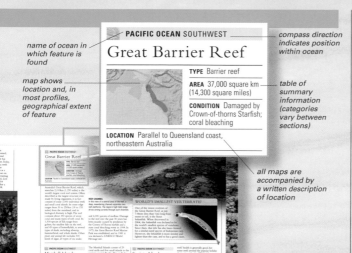

name of ocean in which feature is found

PACIFIC OCEAN SOUTHWEST

Great Barrier Reef

compass direction indicates position within ocean

map shows location and, in most profiles, geographical extent of feature

TYPE	Barrier reef
AREA	37,000 square km (14,300 square miles)
CONDITION	Damaged by Crown-of-thorns Starfish; coral bleaching
LOCATION	Parallel to Queensland coast, northeastern Australia

table of summary information (categories vary between sections)

all maps are accompanied by a written description of location

◄ FEATURE PROFILES
In most cases, explanatory pages are followed by profiles of actual features. For example, the profiles shown here describe coral reefs from around the world. Most profiles are illustrated with colour photographs.

◄ SPECIES PROFILES
All species profiles contain a text description and, in most cases, a colour photograph and distribution map.

CLASS GASTROPODA

Flamingo Tongue
Cyphoma gibbosum

name of group to which species belongs

common name of species is followed by scientific name

LENGTH	3–4cm (1–1½ in)
HABITAT	Coral reefs at about 15m (50ft)
DISTRIBUTION	Western Atlantic, from North Carolina to Brazil; Gulf of Mexico, Caribbean Sea

table of summary information (varies between categories)

all distribution maps are accompanied by a written summary of the range of the species

shaded area of map shows known natural range of species

CONSULTANT

John Sparks is a curator in the Department of Ichthyology at the American Museum of Natural History and an adjunct professor at Columbia University in New York City.

CONTRIBUTORS

Richard Beatty Glossary

Kim Bryan Introduction to Ocean Life, Bacteria and Archaea, Protists, Fungi, Molluscs, Arthropods, Red Crab Migration

David Burnie Animal Life, Reptiles, Birds, Mammals

Robert Dinwiddie Ocean Water, Circulation and Climate, Tides and Waves, Coasts and the Seashore, Shallow Seas, Polar Oceans, Ocean Yacht Racing, Shutting Down the Atlantic Conveyor, Hurricane Katrina, Global Warming and Sea-level Rise, Coastal Defences, The *Titanic* Disaster

Frances Dipper Introduction to Ocean Life, Sponges, Cnidarians, Segmented Worms, Flatworms, Ribbon Worms, Bryozoans, Echinoderms, Small Bottom-living Phyla, Planktonic Phyla, Tunicates and Lancelets, Jawless Fish, Cartilaginous Fish, Bony Fish

Philip Eales Ocean Geology, Atlas of the Oceans, Oceanography from Space, The Indian Ocean Tsunami, Ice-shelf Break-up

Monty Halls Diving Tourism

Sue Scott Shallow Seas, Red and Brown Seaweeds, Plant Life, Green Seaweeds, Green Algae, Mosses, Flowering Plants, Fishing

Mike Scott The Open Ocean and Ocean Floor, Exploration with Submersibles, Cold Water Reefs, Biodiversity Hot Spots, Whale Migration, Wind Farming in the Baltic

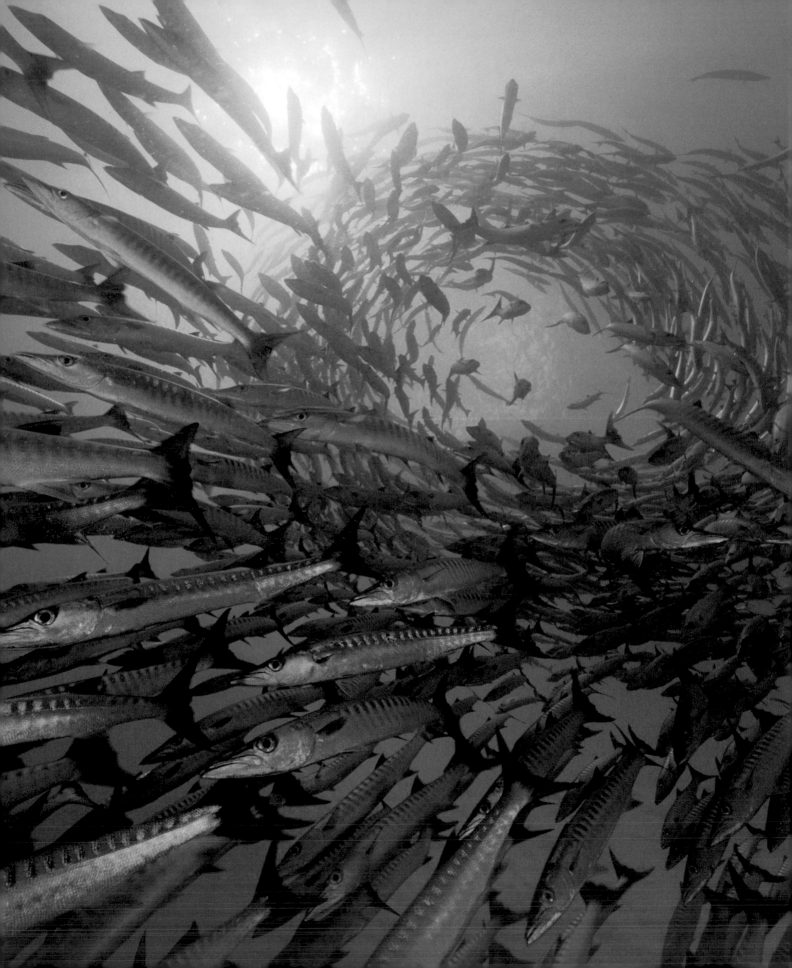

FOREWORD

We should call our planet Ocean. A small orb floating in the endless darkness of space, it is a beacon of life in the otherwise forbidding cold of the endless Universe. Against all odds, it is also the Petri dish from which all life known to us springs.

Without water, our planet would be just one of billions of lifeless rocks floating endlessly in the vastness of the inky black void. Even statisticians revel in the improbability that it exists at all, with such a rich abundance of life, much less that we as a species survive on its surface. Yet, despite the maze of improbability, we have somehow found our way to where we are today. Humans were enchanted by the sea even before the the Greek poet Homer wrote his epic tale of ocean adventure, the *Odyssey*. It is this fascination that has driven us to delve into this foreign realm in search of answers, but the sea has always been reluctant to give up its secrets easily. Even with the monumental achievements of past explorers, scientists, and oceanographers, we have barely ventured through its surface.

It is estimated that over 90 per cent of the world's biodiversity resides in its oceans. From the heartbeat-like pulsing of the jellyfish to the life-and-death battle between an octopus and a mantis shrimp, discoveries await us at every turn. And for every mystery solved, a dozen more present themselves. These are certainly exciting times as we dive into the planet's final frontier. Aided by new technology we can now explore beyond the two per cent or so of the oceans that previous generations observed. But even with the advent of modern technology it will take several more generations to achieve a knowledge base similar to the one we have about the land.

No matter how remote we feel we are from the oceans, every act each one of us takes in our everyday lives affects our planet's water cycle and in return affects us. All the water that falls on land, from the highest peaks to the flattest plains, ends up draining into the oceans. And although this has happened for countless millions of years, the growing ecological footprint of our species in the last century has affected the cycle in profound ways. From fertilizer overuse in landlocked areas, which creates life-choking algal blooms thousand of kilometers away, to everyday plastic items washing up in even the most remote areas of the globe, our actions affect the health of this, our sole life-support system.

This statement is not here to make us feel that we are doomed by our actions, but rather to illustrate that through improved knowledge of the ocean system and its inhabitants we can become impassioned to work towards curing our planet's faltering health. By taking simple steps, such as paying a little more attention to our daily routines, each one of us can have a significant positive impact on the future of our planet and on the world our children will inherit. In short, it would be much healthier for us to learn to dance nature's waltz than to try to change the music.

FABIEN COUSTEAU

MOVING EN MASSE
The fast, co-ordinated movement of a shoal of fish is one of the most spectacular sights in the oceans. These Blackfin Barracuda have formed a spiralling shoal in water around the Solomon Islands. Such shoals are often found in the same place several months, or even years, apart.

ICE AND SEA
An albatross is seen here against the backdrop of a huge iceberg near South Georgia in the south Atlantic. All icebergs in southern-hemisphere waters were once part of the vast Antarctic Ice Sheet, having entered the sea after breaking away from one of the ice shelves that extend from land out into the Southern Ocean.

OCEAN PARTNERSHIP
Relationships between animals in the oceans are often intricate. This Anemone Shrimp is using its long limbs and antennae to crawl over an anemone, removing waste and pieces of food. The shrimp is the chief beneficiary of this relationship, as the anemone not only provides food but also protects it from predators.

NORTHERN EXPOSURE

The world's shorelines can be inhospitable places to live. Common Guillemots nest in colonies, favouring rocky cliffs. These birds are clinging to a rock off the Scottish coast, while being battered by a fierce gale. During the storm, many of the birds were swept off the rock and some of their eggs were washed into the sea.

PROTECTION OF THE YOUNG

The Packhorse Lobster inhabits the continental shelves off Australia and New Zealand. The female shown here is carrying eggs under her abdomen. Up to two million eggs at a time can be stored in this way. Despite producing eggs in such prodigious numbers, these lobsters are threatened by overfishing and catches are now restricted.

CORAL REFUGE
The Spiny-headed Blenny is a common but inconspicuous
inhabitant of reefs in the Caribbean Sea. Just 2cm (¾in) long, it
is often found in holes that were originally made by worms or
molluscs but have since been abandoned. This one is surrounded
by star coral polyps and, like the corals, is feeding on plankton.

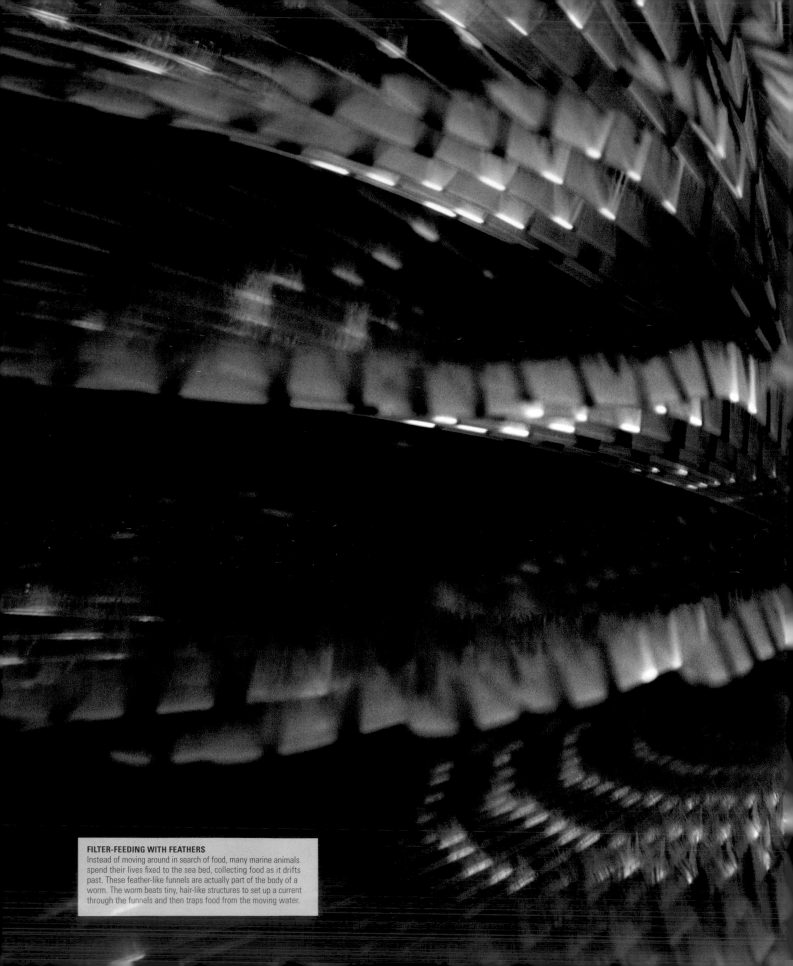

FILTER-FEEDING WITH FEATHERS
Instead of moving around in search of food, many marine animals spend their lives fixed to the sea bed, collecting food as it drifts past. These feather-like funnels are actually part of the body of a worm. The worm beats tiny, hair-like structures to set up a current through the funnels and then traps food from the moving water.

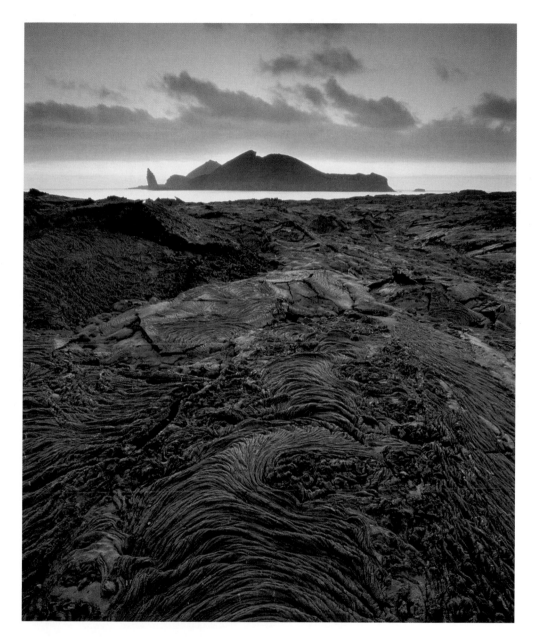

NEW COASTLINE
The shape of a coastline is determined by a balance of forces. The coastlines of the Galapagos Islands in the eastern Pacific are relatively new, having formed when the islands were created by volcanic eruptions. The lava seen here solidified about 100 years ago, but more recent eruptions have occurred on some of the group's younger islands.

THE FALL OF THE APOSTLES
On this part of the southern Australian coast, marine erosion is the dominant force. A line of limestone cliffs is slowly being worn back by the sea, leaving behind isolated stacks of rock. The stacks are collectively known as the Twelve Apostles, although when they were named there were only nine of them and there are now just eight.

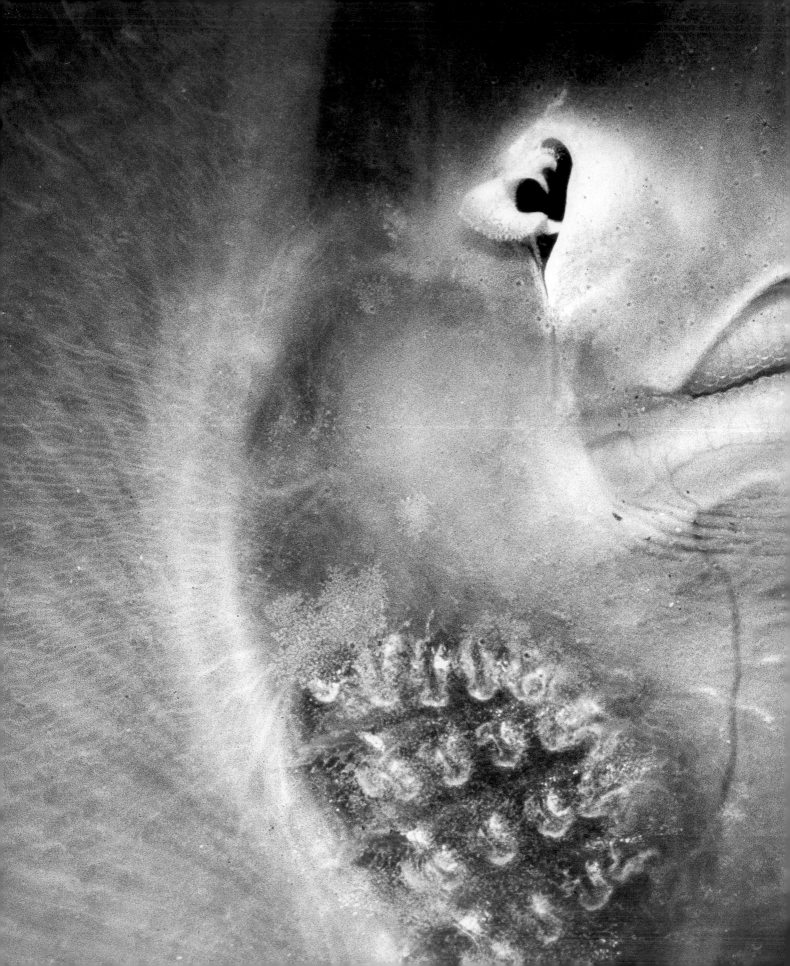

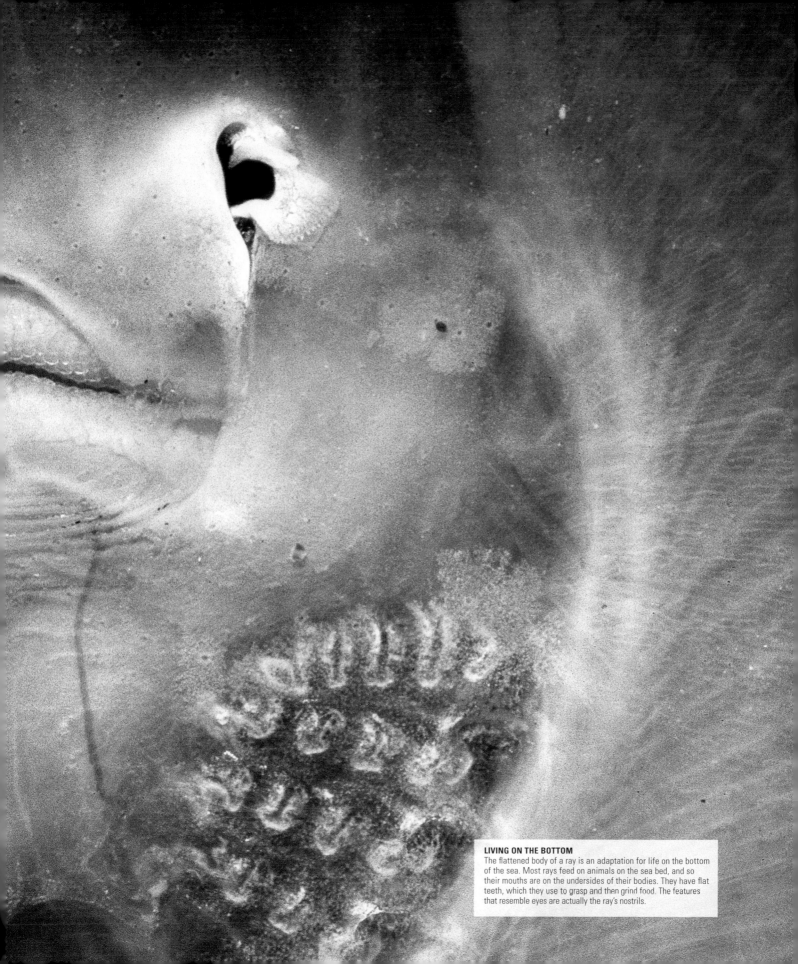

LIVING ON THE BOTTOM
The flattened body of a ray is an adaptation for life on the bottom of the sea. Most rays feed on animals on the sea bed, and so their mouths are on the undersides of their bodies. They have flat teeth, which they use to grasp and then grind food. The features that resemble eyes are actually the ray's nostrils.

AVOIDING A STING
Clownfish have a remarkable relationship with some anemones, feeding and sleeping among them. The anemones' stinging tentacles repel all other fish, but the clownfish avoid triggering the firing of the anemones' stinging cells with an undulating swimming action and by secreting chemicals that suppress the firing process.

LEARNING TO SWIM
Being able to swim is a useful skill for Polar Bears, which for part of the year track their prey across shifting sea ice in the Arctic Ocean and are sometimes seen in the water many miles from land. While underwater, they keep their eyes open but close their nostrils. They can remain submerged for up to two minutes.

ACROBATS OF THE OCEANS
The Common Dolphin is one of the oceans' most abundant cetaceans, a group of mammals that as well as dolphins includes whales and porpoises. It is a sociable animal, sometimes gathering in noisy groups of thousands of individuals. The fast-moving shoal seen here is feeding on sardines in False Bay off South Africa.

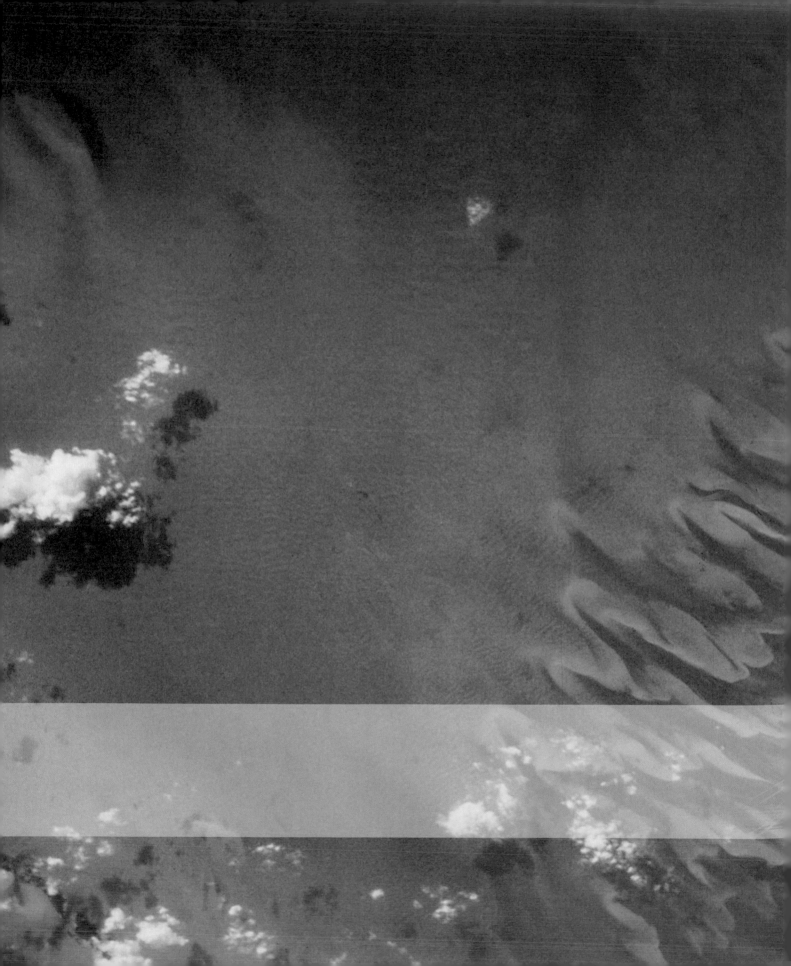

INTRODUCTION

THE EARTH'S OCEANS CONTAIN about 1.36 billion cubic kilometres (324 million cubic miles) of seawater. Dissolved in this are some 48 million billion tonnes (47 million billion tons) of salts, gases, and other substances. The base substance, water itself, has many unusual properties, such as its high surface tension and heat capacity, which are of tremendous significance to everything from the oceans' ability to support life to their stabilizing effect on the world's climate, and their ability to transmit waves. Also of significance is the variability of ocean water – the sea is not uniform but varies spatially and sometimes seasonally in attributes such as its temperature, pressure, dissolved-oxygen content, and level and quality of light illumination. These attributes are important in numerous key respects.

OCEAN WATER

CRASHING WAVE
This "barrel" wave is crashing onto the north shore of the island of Oahu, Hawaii. Inspiring sights such as this are only possible because of some of the unusual properties of water.

THE PROPERTIES OF WATER

THE MAIN CONSTITUENT OF THE OCEANS IS, of course, water. The presence of large amounts of liquid water on the Earth's surface over much of its history has resulted from a fortunate combination of factors. Among them are water's unusually high freezing and boiling points for a molecule of its size, and its relative chemical stability. Water also has other remarkable properties that contribute to the characteristics of oceans – from their ability to support life to effects on climate. Underlying these properties is water's molecular structure.

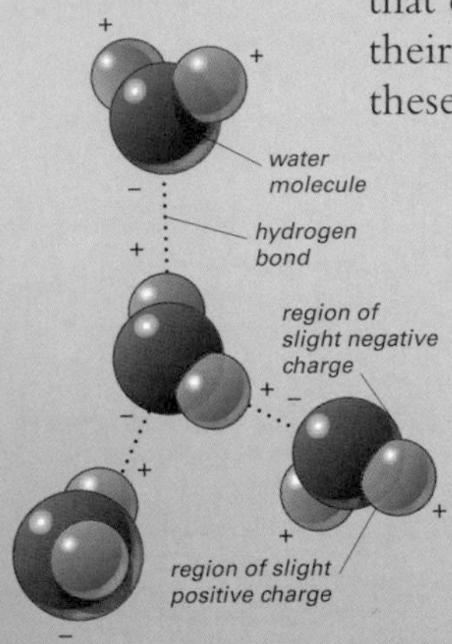

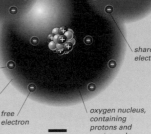

hydrogen atom consists of one proton and one electron

hydrogen nucleus, consisting of single proton, contains positive charge

shared electron

one of eight electrons in oxygen atom

oxygen atom

free electron

oxygen nucleus, containing protons and neutrons, has positive charge

water molecule

hydrogen bond

region of slight negative charge

region of slight positive charge

THE WATER MOLECULE

A molecule of water (H_2O) consists of two hydrogen (H) atoms bound to one atom of oxygen (O). Crucial to formation of the bonds between the oxygen and hydrogen atoms are four tiny negatively charged particles called electrons, which are shared between the atoms. In addition, six other electrons move around within different regions of the oxygen atom. This electron arrangement makes the H_2O molecule chemically stable but gives it an unusual shape. It also produces a small imbalance in the distribution of electrical charge within the molecule. An important result of this is that neighbouring water molecules are drawn to each other by forces called hydrogen bonds.

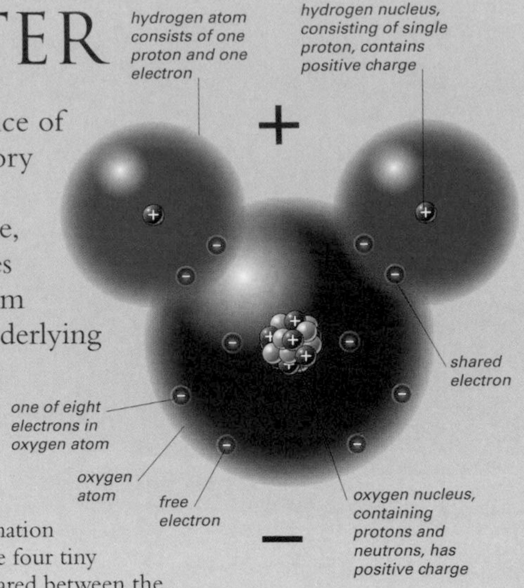

HYDROGEN BONDS
A hydrogen bond is an attractive electrostatic force between regions of slight positive and negative charge on neighbouring water molecules. Several bonds are visible here.

CHARGE IMBALANCE
The distribution of negative charges (electrons) and regions of positive charge in an H_2O molecule causes one side to carry a slight positive charge and the other side a slight negative charge.

SURFACE TENSION

One special property of liquid water that can be directly attributed to the attractive forces between its molecules is its high surface tension. In any aggregation of water molecules, the surface molecules tend to be drawn together and inwards towards the centre of the aggregation, forming a surface "skin" that is resistant to disruption. Surface tension can be thought of as the force that has to be exerted or countered to break through this skin. Water's high surface tension has various important effects. Perhaps the most crucial is that it is vital to certain processes within living organisms, for example water transport in plants and blood transport in animals. Surface tension also allows small insects such as sea skaters to walk and feed on the ocean surface, and it even plays a part in the formation of ocean waves (see p.76).

WALKING ON WATER
Certain insects, such as sea skaters and water striders (pictured below), exploit surface tension to walk, feed, and mate on the surface of the sea, lakes, or ponds.

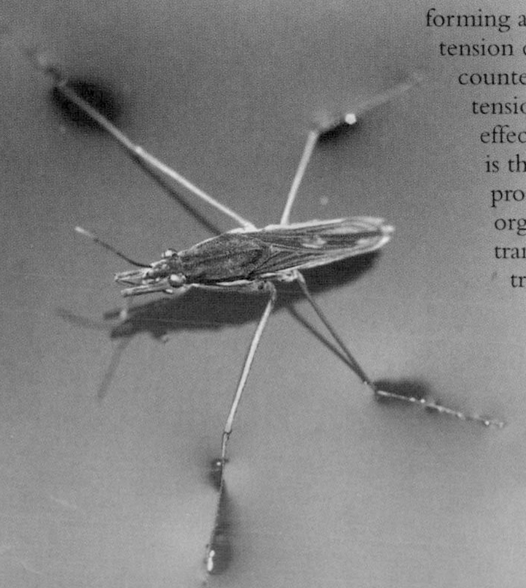

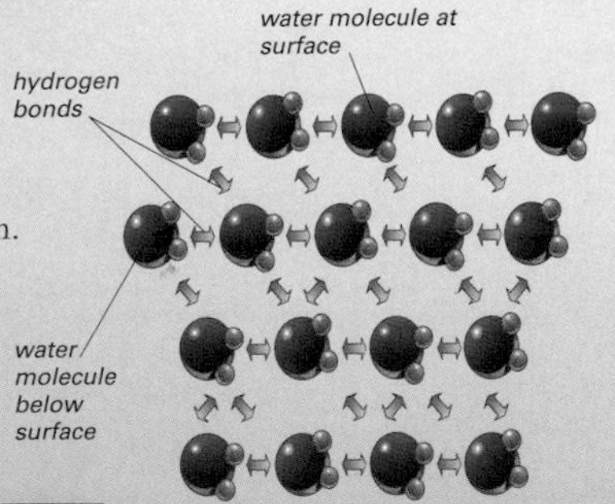

water molecule at surface

hydrogen bonds

water molecule below surface

CAUSE OF SURFACE TENSION
In a drop of water, molecules are pulled in all directions by hydrogen bonding with their neighbours. But at the surface, the only forces act inwards, or sideways, towards other surface molecules.

WATER DROPLETS
The shape of these droplets results from surface tension. The forces pulling their surface molecules together are stronger than the gravitational forces flattening them.

LAND AND SEA
Water's high heat capacity means that the Sun warms the sea more slowly than land. In this satellite-generated temperature map of south California during a heat wave, much of the land (red) is 50ºC (122ºF) or more, but the sea (left) is cool, at 10ºC (50ºF).

SPECIFIC HEAT CAPACITY

Specific heat capacity (SHC) is the energy (in joules) needed to raise the temperature of 1 gram of a substance by 1ºC. Listed below are the SHCs of 13 liquids, measured at room temperature unless otherwise stated.

SUBSTANCE	JOULES/GRAM ºC
Liquid ammonia at -40ºC (-40ºF)	4.7
Fresh water	**4.19**
Seawater at 2ºC (35ºF)	**3.93**
Glycerin	2.43
Ethanol (ethyl alcohol)	2.4
Acetone	2.13
Kerosene	2.01
Olive oil	1.97
Benzene	1.8
Turpentine	1.72
Freon 12 refrigerant at -40ºC (-40ºF)	0.88
Bromine	0.23
Mercury	0.14

HEAT CAPACITY

A second property of liquid water that can be attributed to hydrogen bonding is its unusually high heat capacity, which exceeds that of nearly all other known liquids (see table, left). When heat is added to water, most of the heat is used to break hydrogen bonds linking the molecules. Only a fraction of the energy increases the vibrations of the water molecules, which are detected as a rise in temperature. This means that areas of ocean can absorb and release huge amounts of heat energy with little change in temperature. It also means that movements of water – ocean currents – transfer enormous amounts of heat energy around the planet. This role of ocean currents is vital to Earth's climate (see p.66).

HEAT ON THE MOVE
In this temperature map of part of the northwest Atlantic, the water surface ranges from about 5ºC (41ºF) (blue) to 25ºC (77ºF) (red). A warm current, the Gulf Stream is visible in red.

WATER TWISTER
The effects of surface tension can cause moving sheets, jets, and streams of water to assume or hold together in some surprising forms, as in this slightly spiral-shaped water jet.

THREE STATES OF WATER

The temperatures at which water changes between its three states – melting point (ice to liquid water) and boiling point (liquid water to water vapour) – are both high compared with substances having similarly sized molecules. For ice to melt and water to vaporize, high levels of energy are needed to break all the hydrogen bonds. Water is also unusual in that its solid form is slightly less dense than its liquid form, so ice floats in liquid water. The reason for this is that the molecules in ice are loosely packed, whereas those in liquid water move around in snugly packed groups. The fact that ice floats on liquid water is important because it allows the existence of large areas of polar sea-ice (see p.66). These affect heat flow between ocean and atmosphere and help stabilize ocean temperatures and the Earth's climate.

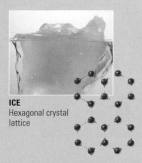

ICE
Hexagonal crystal lattice

LIQUID WATER
Small clumps of bonded molecules

WATER VAPOUR
Widely spaced unbonded molecules

SOLID, LIQUID, AND GAS
In ice, hydrogen bonds hold the water molecules together in a rigid structure. In liquid water, the bonds hold the molecules in small, moving clumps. In water vapour, there are no hydrogen bonds.

SIDE BY SIDE
Water is the only natural substance found in all three states at the Earth's surface. Sometimes, ice, liquid water, and condensing water vapour can be seen side by side, as here at the fiord in Spitsbergen.

THE CHEMISTRY OF SEAWATER

THE OCEANS CONTAIN MILLIONS OF DISSOLVED chemical substances. Most of these are present in exceedingly small concentrations. Those present in significant concentrations include sea salt, which is not a single substance but a mixture of charged particles called ions. Other constituents include gases such as oxygen and carbon dioxide. One reason the oceans contain so many dissolved substances is that water is an excellent solvent.

THE SALTY SEA

The salt in the oceans exists in the form of charged particles, called ions, some positively charged and some negatively charged. The most common of these are sodium and chloride ions, the components of ordinary table salt (sodium chloride). Together they make up about 85 per cent by mass of all the salt in the sea. Nearly all the rest is made up of the next four most common ions, which are sulphate, magnesium, calcium, and potassium. All these ions, together with several others present in smaller quantities, exist throughout the oceans in fixed proportions. Each is distributed extremely uniformly – this is in contrast to some other dissolved substances in seawater, which are unevenly distributed.

volcanic ash drifts down to sea

salts are leached from rocks into rivers and streams and flow to ocean

salt spray onto land

nutrients from soil wash into rivers and streams, and flow to ocean

BREAKDOWN OF SALT
If 10 litres (17½ pints) of seawater are evaporated, about 353g (12¾oz) of salts are obtained, of the types shown below.

10 litres (17½ pints) of seawater

other salts 8.9g (⅓oz)

gypsum 15.4g (½oz)

magnesium salts 56g (2oz)

sodium chloride (halite) 273g (10oz)

sodium ion (positive charge)

slow uplift of sedimentary rocks at continental margins, exposing salts, minerals, and ions at surface

uptake of nutrients by phytoplankton

nutrient upwelling

exchange of gases between phytoplankton and seawater

sinking and decomposition of dead organisms

WATER AS A SOLVENT
The charge imbalance on its molecules makes water a good solvent. When dissolving and holding sodium chloride in solution, the positive ends of the molecules face the chloride ions and the negative ends face the sodium ions.

sodium chloride crystal

water molecule

chloride ion (negative charge)

SOURCES AND SINKS

The ions that make up the salt in the oceans have arrived there through various processes. Some were dissolved out of rocks on land by the action of rainwater and carried to the sea in rivers. Others entered the sea in the emanations of hydrothermal vents (see p.188), in dust blown off the land, or came from volcanic ash. There are also "sinks" for every type of ion – processes that remove them from seawater. These range from salt spray onto land to the precipitation of various ions onto the seafloor as mineral deposits. Each type of ion has a characteristic residence time. This is the time that an ion remains in seawater before it is removed. The common ions in seawater have long residence times, ranging from a few hundred years to hundreds of millions of years.

RIVER DISCHARGE
River discharge is a mechanism by which ions of sea salt and nutrients enter the oceans. Here, the Noosa River empties into the sea on the coast of Queensland, Australia.

SOURCES, SINKS, AND EXCHANGES
Shown here are various sources, sinks, and exchange processes for the ions, salts, and minerals (yellow arrows), gases (pink arrows), and plant nutrients (turquoise arrows) in seawater.

KEY

gases

ions, salts, and minerals

plant nutrients

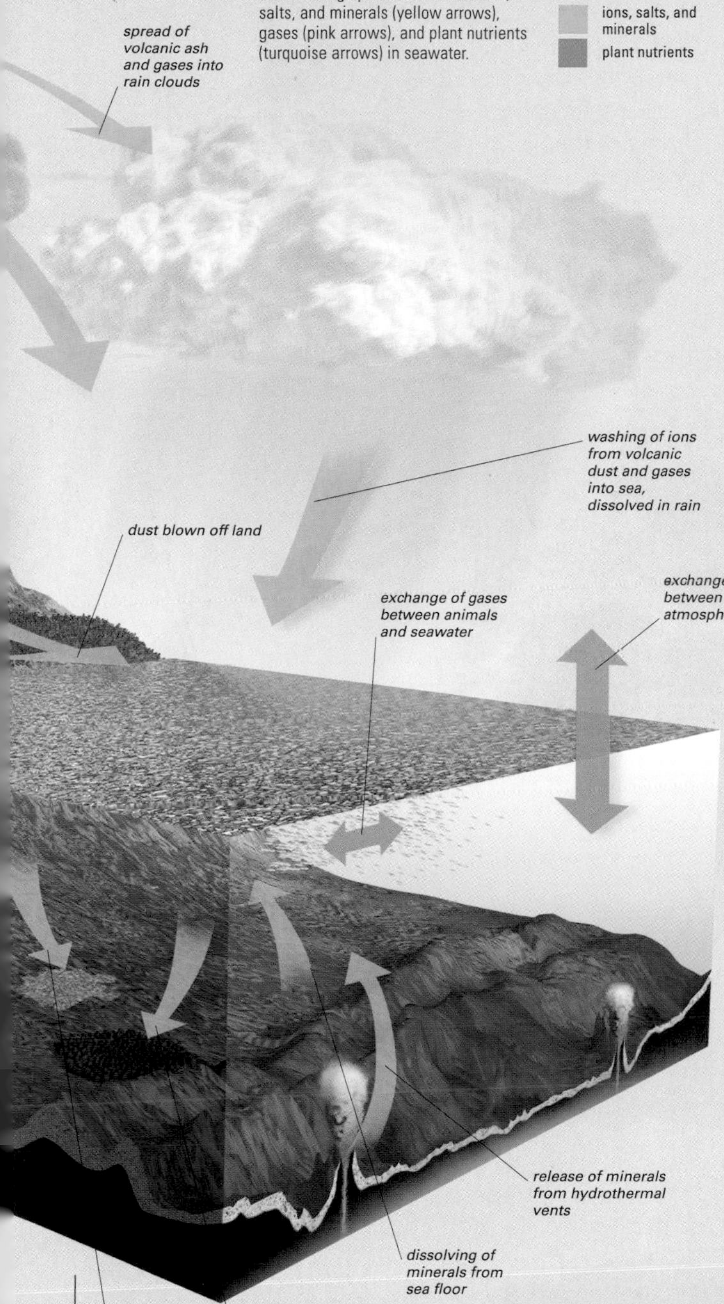

spread of volcanic ash and gases into rain clouds

washing of ions from volcanic dust and gases into sea, dissolved in rain

dust blown off land

exchange of gases between animals and seawater

exchange of gases between ocean and atmosphere

release of minerals from hydrothermal vents

dissolving of minerals from sea floor

precipitation of minerals onto sea floor

carbonates incorporated into sea-floor sediments from animal shells

SILICEOUS DIATOMS
These tiny forms of planktonic organisms have cell walls made of silicate. They can only grow if there are sufficient amounts of silica present in the water.

GASES IN SEAWATER

The main gases dissolved in seawater are nitrogen (N), oxygen (O_2), and carbon dioxide (CO_2). The levels of O_2 and CO_2 vary in response to the activities of photosynthesizing organisms (phytoplankton) and animals. The level of O_2 is generally highest near the surface, where the gas is absorbed from the air and also produced by photosynthesizers. Its concentration drops to a minimum at about 1,000m (3,300ft), where oxygen is consumed by bacterial oxidation of dead organic matter and by animals feeding on this matter. Deeper down, the O_2 level increases again. CO_2 levels are highest at depth and lowest at the surface, where the gas is taken up by photosynthesizers faster than it is produced by respiration.

CARBON SINK
Many marine animals, such as nautiluses (below), use carbonate (a compound of carbon and oxygen) in seawater to make their shells. After they die, the shells may form sediments and eventually rocks.

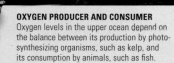

OXYGEN PRODUCER AND CONSUMER
Oxygen levels in the upper ocean depend on the balance between its production by photo-synthesizing organisms, such as kelp, and its consumption by animals, such as fish.

NUTRIENTS

Numerous substances present in small amounts in seawater are essential for marine organisms to grow. At the base of the oceanic food chain are phytoplankton – microscopic floating life-forms that obtain energy by photosynthesis. Phytoplankton need substances such as nitrates, iron, and phosphates in order to grow and multiply. If the supply of these nutrients dries up, their growth stops; conversely, blooms (rapid growth phases) occur if it increases. Although the sea receives some input of nutrients from sources such as rivers, the main supply comes from a continuous cycle within the ocean. As organisms die, they sink to the ocean floor, where their tissues decompose and release nutrients. Upwelling of seawater from the ocean floor (see p.60) recharges the surface waters with vital substances, where they are taken up by the phytoplankton, refuelling the chain.

PLANKTON BLOOM
This satellite image of the Skagerrak (a strait linking the North and Baltic seas) shows a bloom of phytoplankton, visible as a turquoise discoloration in the water.

TEMPERATURE AND SALINITY

OCEAN WATER IS NOT UNIFORM BUT VARIES in several physical attributes, including temperature, salinity, pressure, and density. These vary vertically (dividing the oceans into layers), horizontally (between tropical and temperate regions, for example), and seasonally. The basic variables, temperature and salinity, in turn produce variations in density that help drive deep-water ocean circulation.

MARCH 2003

TEMPERATURE

Temperature varies considerably over the upper areas of the oceans. In the tropics and subtropics, solar heating keeps the ocean surface warm throughout the year. Below the surface, the temperature declines steeply to about 8–10°C (46–50°F) at a depth of 1,000m (3,300ft). This region of steep decline is called a thermocline. Deeper still, temperature decreases more gradually to a uniform, near-freezing value of about 2°C (36°F) on the sea floor – this temperature subsists throughout the deep oceans. In mid-latitudes there is a much more marked seasonal variation in surface temperature. In high latitudes and polar oceans, the water is constantly cold, sometimes below 0°C (32°F).

MARCH 2001

PACIFIC VARIATION
The Pacific experiences long-term fluctuations (called El Niño Southern Oscillation) in the temperature patterns in its surface waters. These images contrast the patterns in March in two different years – red and white indicate warmer water, green and blue cooler water.

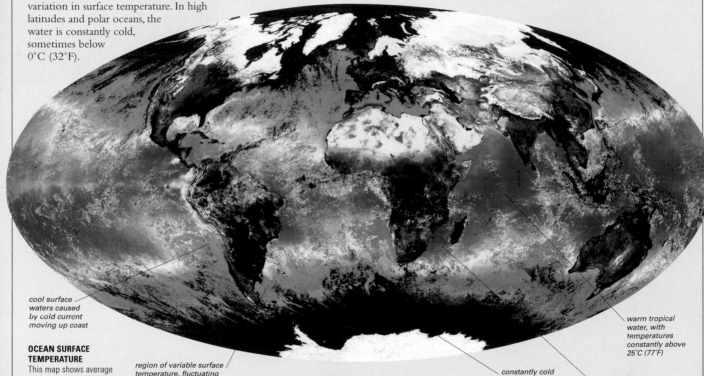

cool surface waters caused by cold current moving up coast

warm tropical water, with temperatures constantly above 25°C (77°F)

OCEAN SURFACE TEMPERATURE
This map shows average surface temperatures in March. Proximity to the equator is the main factor determining surface temperature, but ocean currents also play a role.

region of variable surface temperature, fluctuating seasonally from 7 to 20°C (45–68°F)

North America

constantly cold water with temperatures in the range 0–3°C (32–37°F)

warm surface waters caused by warm current moving down southeast coast

constantly cold water off Greenland

constantly warm pool of water in Caribbean Sea

South America

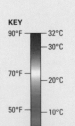

thermocline, where temperature declines rapidly with depth

KEY

90°F	32°C
	30°C
70°F	20°C
50°F	10°C
30°F	0°C

TEMPERATURE AND DEPTH
Shown in early summer, the vast bulk of ocean water in this part of the north Atlantic is uniformly cold (below 5°C/41°F). Only a thin surface layer from the tropics into mid-latitudes is warmed above this base level.

cold bottom water at a uniform temperature of 2°C (36°F)

SALINITY

Salinity is an expression of the amount of salt in a fixed mass of seawater. It is determined by measuring a seawater sample's electrical conductivity and averages about 35 grams of salt per kilogram of seawater. Salinity varies considerably over the surface of oceans – its value at any particular spot depends on what processes or factors are operating at that location that either add or remove water. Factors that add water, causing low salinity, include high rainfall, river input, or melting of sea-ice. Processes that remove water, causing high salinity, include high evaporative losses and sea-ice formation. At depth, salinity is near constant throughout the oceans. Between the surface and deep water is a region called a halocline, where salinity gradually increases or decreases with depth. Salinity affects the freezing point of seawater – the higher the salinity, the lower the freezing point.

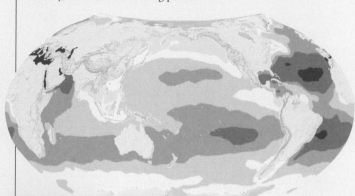

EASY FLOATING
In some enclosed seas where evaporative losses are high and there is little rainfall or river inflow, the sea-water can become so saline and dense that floating becomes easy. This is the case here in the Dead Sea.

GLOBAL SALINITY
Surface salinity is highest in the subtropics, where evaporative losses of water are high, or in enclosed or semi-enclosed basins (such as the Mediterranean). It is lowest in colder regions or where there are large inflows of river water.

KEY
37
36
35
34
33
32
31
30
29
under 29
parts per thousand (‰)

DENSITY

The density of any small portion of seawater depends primarily on its temperature and salinity. Any decrease in temperature or increase in salinity makes seawater denser – an exception being a temperature drop below 4°C (39°F), which actually makes it a little less dense. In any part of the ocean, the density of the water increases with depth, because dense water always sinks if there is less dense water below it.

Processes that change the density of seawater cause it to either rise or sink, and drive large-scale circulation in the oceans between the surface and deep water (see p.60). Most important is water carried towards Antarctica and the Arctic Ocean fringes. This becomes denser as it cools and through an increase in its salinity as a result of sea-ice formation. In these regions large quantities of cold, dense, salty water continually form and sink towards the ocean floor.

DENSITY LAYERS IN ATLANTIC
The oceans each contain distinct, named water masses that increase in density from the surface downwards. The denser, cooler masses sink and move slowly towards the Equator. The cold, high-density deep and bottom waters comprise 80 per cent of the total volume of the ocean.

PRESSURE

Scientists measure pressure in units called bars. At sea level, the weight of the atmosphere exerts a pressure of about one bar. Underwater, pressure increases at the rate of one bar for every 10m (33ft) increase in depth, due to the weight of the overlying water. This means that at 70m (230ft), for example, the total pressure is eight bars or eight times the surface pressure. This pressure increase poses a challenge to human exploration of the oceans. To inflate their lungs underwater, divers have to breathe pressurized air or other gas mixtures, but doing so can cause additional problems (arising from the dissolution of excess gas in body tissues). These problems limit the depths attainable.

DECOMPRESSION STOP
To avoid a condition called "bends" that can arise from decompressing too quickly, on their way to the surface scuba divers sometimes have to make timed stops to release excess gas.

NATURAL ADAPTATION
Elephant seals can dive to depths of up to 600m (2,000ft). They have evolved various adaptations for coping with the high pressure, including collapsible ribcages.

DISCOVERY

DECOMPRESSION

After working underwater for hours at a time, professional divers routinely undergo controlled decompression in a purpose-built pressure chamber. These facilities are also used to treat pressure-related diving illnesses and for research into diving physiology.

PRESSURE CHAMBER
The person being decompressed may have to breathe a special gas mixture while the ambient pressure is slowly reduced.

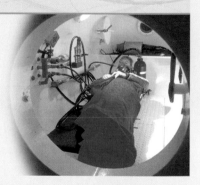

warm surface flow

Atlantic Central Water: warm, low-density surface waters in the tropics and subtropics

Atlantic Intermediate Water: cool layer of intermediate density, forms and sinks in north Atlantic, then moves south

Antarctic Intermediate Water: cool layer of intermediate density, sinks and moves north

Antarctic Bottom Water: coldest and densest layer, forms close to Antarctica, sinks then moves north

mid-ocean ridge

North Atlantic Deep Water: cold, dense water, forms and sinks in north Atlantic, then moves south

LIGHT AND SOUND

LIGHT AND SOUND BEHAVE VERY DIFFERENTLY in water than in air. Most light wavelengths are quickly absorbed by water, a fact that both explains why the sea is blue and why ocean life is concentrated near its surface – almost the entire marine food chain relies on light energy driving plant growth. Sound, in contrast, travels better in water, a fact exploited by animals such as dolphins.

LIGHT IN THE OCEAN

White light, such as sunlight, contains a mixture of light wavelengths, ranging from long (red) to short (violet). Ocean water strongly absorbs red, orange, and yellow light, so only some blue and a little green and violet light reach beyond a depth of about 40m (130ft). At 90m (300ft), most of even the blue light (the most penetrating) has been absorbed, while below 200m (650ft), the only light comes from bioluminescent organisms, which produce their own light (see p.224). Because they rely on light to photosynthesize, phytoplankton are restricted to the upper layers of the ocean, and this in turn affects the distribution of other marine organisms. Intriguingly, many bright red animals live at depths that are devoid of red light: their colour provides effective camouflage, since they appear black.

LIGHT PENETRATION
The red and orange components of sunlight are absorbed in the top 15m (50ft) of the ocean. Most other colours are absorbed in the next 40m (130ft). Wavelength is measured in nanometres (nm).

COLOUR RESTORATION
At a depth of 20m (65ft), most animals and plants look blue-green under ambient light conditions (top). Lighting up the scene with a photographic flash or torch reveals the true colours of the marine life (bottom).

FIREFLY SQUID
This squid produces a pattern of glowing spots (photophores). When viewed by a predator swimming below, the spots help camouflage its outline against the moonlit waters above.

FISH VISION

Fish have excellent vision, which helps them find food and avoid predators. Many can see in colour. The lens of a fish's eye is almost spherical and made of a material with a high refractive index. It can be moved backwards and forwards to focus light on the retina.

FISH EYE
The lens of a fish's eye bulges through the iris (the dark central part) almost touching the cornea (outer part). This helps to gather the maximum amount of light and gives a wide field of view.

INTRODUCTION

LOOKING UP
Seen from underwater, only a part of the surface of the sea appears lit up, while the rest looks dark. This is an effect of the way light waves are bent (refracted) when they enter the sea from the air.

SEA COLOURS

Seawater has no intrinsic colour – a glass of seawater is transparent. But on a clear, sunny day, the sea usually looks blue or turquoise. In part, this is due to the sea surface reflecting the sky, but the main reason is that most of the light coming off the surface has already penetrated it and been reflected back by particles in the water or by the sea bed. During its journey through the water, most of the light is absorbed, except for some blue and green light, which are the colours seen. Other factors can modify the sea's colour. In windy weather, the surface becomes flecked with white, caused by trapped bubbles of air, which reflect most of the light that hits them. Rain interferes with seawater's light-transmitting properties, so rainy, overcast weather generally produces dark, grey-green seas. Occasionally, living organisms, such as "blooms" of plankton can turn patches of the sea vivid colours.

VIVID GREEN FROM ALGAL BLOOM

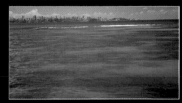
TROPICAL TURQUOISE

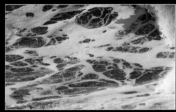
GREY FOAMY TEMPERATE SEA

OCEAN SHADES
A green sea (top) is sometimes caused by the presence of algae. Turquoise is the usual shade in clear tropical waters, while grey water flecked with white foam is typical of windy, overcast days.

PEOPLE

WALTER MUNK

The Austrian-American scientist Walter Munk (b.1917) pioneered the use of sound waves in oceanography. A professor at the Scripps Institute of Oceanography in San Diego, California, Munk demonstrated that by studying the patterns and speed of sound propagation underwater, information can be obtained about the large-scale structure of ocean basins.

UNDERWATER SOUNDS

The oceans are noisier than might be imagined. Sources of sound include ships, submarines, earthquakes, underwater landslides, and the sounds of icebergs breaking off glaciers and ice shelves. In addition, by transmitting sound waves or bouncing them off underwater objects (echolocation) whales and dolphins use sound for navigation, hunting, and communication. Sound waves travel faster and further underwater than they do in air. Their speed underwater is about 1,500m (5,000ft) per second and is increased by a rise in the pressure (depth) of the water and decreased by a drop in temperature. Combining these two effects, in most ocean regions, there is a layer of minimum sound velocity at a depth of about 1,000m (3,300ft). This layer is called the SOFAR (Sound Fixing and Ranging) channel. The properties of the SOFAR channel are exploited by people using underwater listening devices and, it has been theorized, by animals such as whales and dolphins.

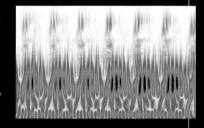

HUMPBACK WHALE SONG
The peaks and troughs in this spectrogram show the changes in frequency of a few seconds of repeated sound made by a Humpback Whale.

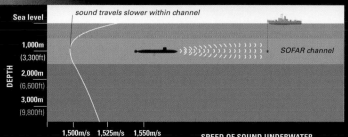

THE SOFAR CHANNEL
Low-frequency sounds generated in the SOFAR channel are "trapped" in it by inward refraction from the edges of the channel. As a result, sounds can travel very long distances in this ocean layer.

Sea level

sound travels slower within channel

1,000m (3,300ft)

SOFAR channel

DEPTH

2,000m (6,600ft)

3,000m (9,800ft)

1,500m/s (4,900ft/s)   1,525m/s (5,000ft/s)   1,550m/s (5,085ft/s)

SPEED OF SOUND UNDERWATER

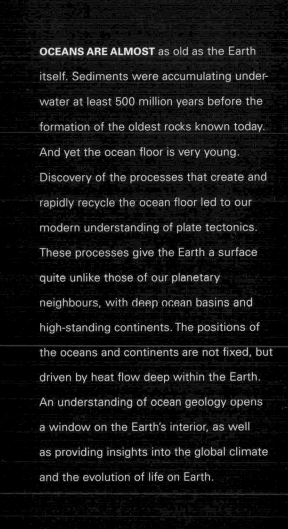

OCEANS ARE ALMOST as old as the Earth itself. Sediments were accumulating underwater at least 500 million years before the formation of the oldest rocks known today. And yet the ocean floor is very young. Discovery of the processes that create and rapidly recycle the ocean floor led to our modern understanding of plate tectonics. These processes give the Earth a surface quite unlike those of our planetary neighbours, with deep ocean basins and high-standing continents. The positions of the oceans and continents are not fixed, but driven by heat flow deep within the Earth. An understanding of ocean geology opens a window on the Earth's interior, as well as providing insights into the global climate and the evolution of life on Earth.

OCEAN GEOLOGY

OCEANIC LAVA
Steam mixes with surf as lava from Kilauea Crater reaches the Pacific Ocean on the south shore of Hawaii. Basaltic lavas such as this make up the oceanic part of the Earth's crust.

THE FORMATION OF THE EARTH

THE EARTH STARTED TO FORM MORE THAN 4,500 million years ago in a disc of gas, dust, and ice around the early Sun. This protoplanetary disc, as it is known, was held in orbit by the gravitational field of the young star. Gravitational attraction between dust particles in the disc produced small rocks, and collisions concentrated the rocks into several rings. The most densely populated rings went on to form the planets of the Solar System.

EARLY SOLAR SYSTEM
The early Solar System contained a disc of dust, ice, and gas, from which the rocky inner planets and gaseous outer planets formed.

small pieces of rock and ice pulled together by gravitational attraction

planetesimals start to form in protoplanetary disc around Sun

1 COLD ACCRETION
Under gravity, pieces of rock and ice coalesced. Material sharing the same orbit around the Sun clumped together.

2 PLANETESIMAL
Large concentrations of rock and ice formed planetesimals. Their gravity fields became stronger as their mass increased, smoothing out their irregular surfaces.

rocks accelerate towards primordial Earth

BIRTH OF THE EARTH

Initially, the rocks within each ring drifted together, due to their mutual gravitational attraction, in a process known as cold accretion. The largest bodies in each ring attracted the most material and grew to form small planetary bodies, or planetesimals, one of which was the early Earth. Planetesimals are loose collections of rock and ice, with a uniform structure. As the mass of a planetesimal grows larger, it exerts a stronger gravitational pull, becoming more tightly held together and attracting nearby rocks with greater force. Collisions between planetesimals broke them apart or grouped them together. In the inner Solar System, the planetesimals in each orbiting ring came together to form the rocky planets. The Earth was born in this way about 4,560 million years ago.

3 HEAVY BOMBARDMENT
Smaller rocks were pulled towards growing planetesimals with increasing force, producing more energetic high-speed impacts. This bombardment continued until most of the material in the vicinity of Earth's orbit was consumed.

impact of Mars-sized body leads to total melting

impacts generate surface heat and local melting

INTERNAL HEAT

The early Earth was mostly cold, mostly solid, and had a fairly uniform interior composition. Today it has a hot interior, part of it liquid, with well-defined zones of different composition. This transition may have its roots in several different energy sources. Localized surface melting would have occurred when the kinetic energy of incoming rocks was converted to heat during impacts. More significant heat sources would have been the decay of radioactive elements in the interior rocks and the heat released by the Earth's contraction under the force of its own gravity – a process that led to an event called the iron catastrophe (see below). Impact with a sufficiently large body might have released enough heat to melt the Earth's interior, and this may have happened more than once.

MOON FORMATION
Early in its history, the Earth was struck by a large planetesimal, creating the Moon, tilting the Earth's axis of rotation, and leaving it with a slightly eccentric orbit.

material ejected during collision later cooled and coalesced to form the Moon

---

THE IRON CATASTROPHE

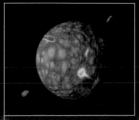

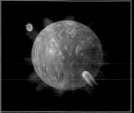

① As the Earth grew larger, the strength of its gravitational field increased, which in turn attracted more material.

② Eventually, the gravitational field was strong enough to cause the Earth to contract, converting gravitational potential energy into heat.

③ Enough heat was released to melt the iron contained in the Earth's rocks, allowing it to flow down to the centre of the Earth.

④ The sinking of large amounts of iron released further heat, enough to melt the entire interior of the planet in the event called the iron catastrophe.

CONVECTION AND DIFFERENTIATION

After the interior of the Earth melted, its heaviest constituents were able to sink to the centre and the lighter ones to rise towards the surface. One-third of the planet's mass pooled at the centre and formed a dense core consisting mainly of iron, the heaviest of the common elements making up the Earth. The core became the hottest part of the planet, up to 4,700°C (8,500°F), and a source of heat for the molten rocks above. Most materials expand as they are heated, becoming less dense and more buoyant. This is the basis of convection, which provided a mechanism for carrying heat and material from the interior of the Earth towards the surface. Vigorous convection cells carried hot, buoyant material upwards, where it lost heat by conduction near the surface, before sinking again. Lighter materials such as aluminium were left behind at the surface, forming a thin crust. In this way, the Earth became differentiated into layers of different chemical composition: a metallic core, a rocky mantle, and a buoyant crust. This occurred as early as 4,500 million years ago.

A LAYERED EARTH
The early Earth had a uniform composition but melting allowed chemical "zoning" to develop.

convection carries internal heat to surface

lighter materials rise up through semi-fluid mantle

heavy materials sink to form dense core

carbon dioxide

water vapour

nitrogen

ATMOSPHERE AND OCEAN
The lightest materials of all, gases and water, were expelled from the interior to form the outer atmospheric and ocean layers at an early stage in the Earth's history.

MANTLE SAMPLE

Our knowledge of the Earth's interior is based on indirect methods, such as studying seismic signals or gravity variations. While mantle rocks can be found on the surface, they have been altered by heating and other processes, so scientists are now trying to directly measure conditions in the mantle as part of the Integrated Ocean Drilling Program.

THE DEEPEST HOLE
The Japanese research ship *Chikyu* ("Earth") is designed to drill through 7,000m (4.35 miles) of ocean crust, in seas up to 2,500m (8,200ft) deep, to retrieve rock cores from the mantle.

THE EARTH TODAY

The Earth's interior is now split into three chemically distinct layers, which can be further split by changes in their physical properties due to temperature and pressure variations with depth. The core consists of an iron-nickel alloy, with some impurities, at a temperature of 3,500–4,700°C (6,300–8,500°F). Iron in the inner part of the core has solidified under the immense pressure, but the outer part is still a free-flowing liquid. The mantle of silicate rock surrounding the core has also solidified, but a form of convection called "solid-state creep" still takes place, with material in the lower mantle moving a few centimetres per year. The upper mantle, within about 350km (215 miles) of the surface, is a more easily deformed "plastic" region. Above it floats a thin crust enriched in lighter elements, with average thickness ranging from 8km (5 miles) beneath the oceans to 45km (28 miles) beneath the continents.

THE EARTH'S INTERNAL STRUCTURE
Although hot enough to melt surface rocks, the Earth's mantle is kept solid by the high pressure of overlying rock. Melting only occurs in discrete locations where extremely hot parts of the mantle are decompressed.

crust of silicate minerals, rich in aluminium

upper mantle

lower mantle

inner core of solid iron-nickel

outer core of liquid iron-nickel

mantle of silicate minerals, rich in magnesium

thin, dense oceanic crust

thick, light continental crust

THE ORIGIN OF OCEANS AND CONTINENTS

THE EARTH'S OCEANS FORMED MORE THAN 4,000 million years ago, mainly from water vapour that condensed from its primitive atmosphere but also from water brought from space by comets. Once its materials had been softened by density into vertical layers, the Earth had a uniform crust that was enriched in lighter elements and floated on an upper mantle made of denser materials. Later, a horizontal differentiation also developed as continents began to form, made from rocks that were chemically distinct from those underlying the oceans.

CONTINENTAL CRUST

The continents include a wide range of rock types, including granitic igneous rocks, sedimentary rocks, and the metamorphic rocks formed by the alteration of both. They contain a lot of quartz, a mineral absent in oceanic crust. The first continental rocks were the result of repeated melting, cooling, and remixing of oceanic crust, driven by volcanic activity above mantle convection cells, which were much more numerous and vigorous than today's. Each cycle left more of the heavier components in the upper mantle and concentrated more of the lighter components in the crust. The first micro-continents grew as lighter fragments of crust collided and fused. Thickening of the crust led to melting at its base and underplating with granitic igneous rocks. Weathering accelerated the process of continental rock formation, retaining the most resistant components, such as quartz, while washing solubles into the ocean.

THE OLDEST ROCKS
These sedimentary rocks on Baffin Island lie on the Canadian Shield. The stable continental shields contain the world's oldest rocks, granites dating back to 3,800 million years.

zircon crystals, among the earliest continental crust materials

ZIRCON

primitive continental crust thickens above sinking mantle flow, without mantle interference

sedimentary rocks

primitive oceanic crust

volcanic activity adds igneous rocks to surface above rising flows

OCEANIC CRUST

The oceanic crust has a higher density than the continental crust, making it less buoyant. Both types of crust can be thought of as floating on the "plastic" upper mantle, and the oceanic crust lies lower due to its lower buoyancy. It is relatively thin, with a depth no more than 11km (7 miles), compared with 25–70km (15–43 miles) beneath the continents. It consists mainly of basalt, an igneous rock that is low in silica compared with continental rocks, and richer in calcium than the mantle. Basalt lava is created when hot material in the upper mantle is decompressed, allowing it to melt and form liquid magma. The decompression occurs beneath rifts in the crust, such as those found at the mid-ocean ridges, and it is through these rifts that lava is extruded onto the surface to create new ocean crust.

basaltic lava — rift — basalt sheets (dykes) — sediment — ocean surface — ocean crust — lithosphere

gabbro — peridotite

Moho — magma rises to surface — upper mantle

asthenosphere

OCEAN-FLOOR STRUCTURE
Three layers of basalt in the crust (basaltic lava, dykes, and gabbro) are separated from the mantle by the Mohorovicic discontinuity (the Moho). The upper mantle is fused to the base of the crust to form the rigid lithosphere. The asthenosphere is the soft zone over which the plates of the lithosphere glide.

MANTLE ROCKS
Peridotite is the dominant rock type found in the mantle, consisting of silicates of magnesium, iron, and other metals. Sometimes it is brought to the surface when parts of the ocean floor are uplifted, as here in Newfoundland, Canada, or as fragments from volcanic activity.

DEVELOPMENT OF CONTINENTAL CRUST
Modification of the crust above rising mantle flows was delayed by the continuous intrusion of mantle basalt, resulting in the greenstone belts found today at the heart of each continental shield.

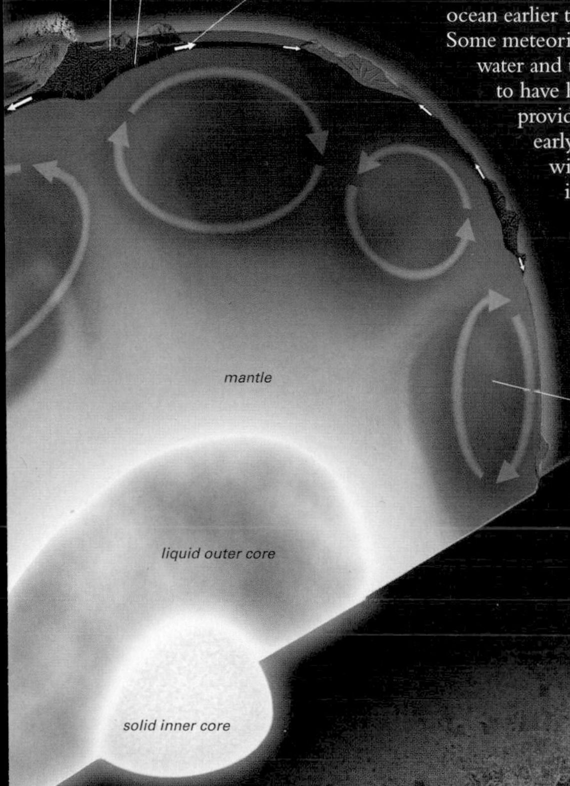

greenstone belts above rising mantle flow

basalt continuously intrudes from mantle

crust pulled apart by convective motion in mantle

mantle

liquid outer core

solid inner core

vigorous convection cells in upper mantle

WATER AND ATMOSPHERE

During the process of differentiation, volatile materials were expelled from the interior of the Earth by volcanic activity. The lightest gases, such as hydrogen and helium, would quickly have been lost to space, leaving a stable atmosphere of nitrogen, carbon dioxide, and water vapour. Some of the water vapour would have condensed to form liquid water, and it seems there was a significant ocean earlier than 4,000 million years ago. Some meteorites contain 15–20 per cent water and the early Earth is thought to have had the same composition, providing an ample source for the early ocean. More water arrived with impacting comets. It was in the ocean that free oxygen first appeared, with the arrival of photo-synthesizing life around 3,500 million years ago.

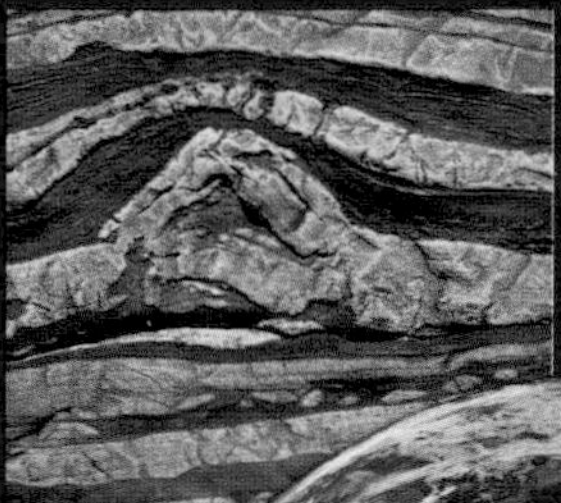

THE EARLY EARTH
The Earth had deep oceans from an early stage, with volcanoes and an increasing area of continental crust standing above the surface. The ocean became salty as weathering of surface rocks added minerals to the water.

ocean water from volcanic eruptions and comet impacts

traces of early meteorite and comet bombardment gradually erased

rifts occur when fragments of crust move apart

volcanic eruptions add gases and water vapour to atmosphere

BANDED IRON
Known as a banded-iron formation, this layered rock contains iron oxides that formed as the oxygen content of early oceans increased.

rivers erode and transport sediment

ANDEAN VOLCANOES
This radar image shows volcanoes formed from andesite lava, whose composition is intermediate between oceanic and continental rocks.

THE EVOLUTION OF THE OCEANS

spreading ridge

continent carried on plates

subduction zone

convection cell drives plate motion

EVER SINCE THE ATLANTIC COASTS OF SOUTH AMERICA and Africa were accurately charted, it has been apparent that they match like the pieces of a jigsaw puzzle. We now know that the continents move, that they were once joined together, and that today's oceans arose when the landmasses split apart. The evolving oceans have modified the global climate, and sea level has fluctuated in response to climate change and geological factors.

PLATE TECTONICS

The numerous convection cells (see p.43) that gave rise to the first fragments of continental crust gradually gave way to fewer, larger-scale convection cells as the mantle cooled. The continental fragments became consolidated into larger areas, and rifts formed at the thinnest parts of the ocean crust, splitting it into large plates. When the density of the oceanic and continental plates became sufficiently different, the oceanic crust sank where it met the more buoyant continental crust, creating subduction zones. Since then, the evolution of the oceans and continents has been dominated by plate tectonics (see pp.50–51). As the plates move, they carry the continents with them, with oceans opening and closing in between.

PLATE MOVEMENT
Crustal plates move around under the influence of convection cells, which reach deep down to the boundary between the outer core and the mantle.

1. CAMBRIAN (500 MYA)
The remains of the first supercontinent, Rodinia, were scattered, with the largest piece, Gondwana, lying in the south. The Iapetus Ocean separated Laurentia (North America) from Baltica (northern Europe). The Panthalassic Ocean occupied most of the northern hemisphere.

PANTHALASSIC OCEAN

LAURENTIA

SIBERIA

IAPETUS OCEAN

BALTICA

GONDWANA

"ancestral" North Atlantic lies between North America and Europe

scattered remnants of Rodinia

2. DEVONIAN (400 MYA)
The Rheic Ocean opened when a string of islands, which were to become western and southern Europe, broke away from Gondwana and moved towards Laurentia and Baltica, closing the Iapetus Ocean in the process.

SIBERIA

PANTHALASSIC OCEAN

EURAMERICA

AUSTRALIA

RHEIC OCEAN

GONDWANA

southern Europe joins Euramerica (Laurentia and Baltica) as Iapetus Ocean closes

first plants on land form vegetated areas

shallow continental -shelf seas

Ural Mountains

THROUGH THE AGES

As the Earth's plates have moved around, largely driven by the spreading ridges and subduction zones of the rapidly recycling oceanic crust (see p.50), continents have come together and moved apart – periodically grouping together to form "supercontinents". The German scientist Alfred Wegener proposed that 250 million years ago (mya) there was a supercontinent called Pangea, centred on the Equator and surrounded by one great ocean. It seems there was another grouping about 1,000 million years ago called Rodinia, and perhaps an earlier grouping before that. Each time that the continental landmasses have come together they have eventually been broken apart as deep rifts have opened up in their interiors, as is happening today in the Red Sea and Great Rift Valley of East Africa. Computer models of the crustal fragments and the locations of spreading and subduction have enabled fairly reliable reconstructions of the geography of earlier times back to 500 million years ago.

PANTHALASSIC OCEAN

PALEO-TETHYS SEA

PANGEA

SOUTH AMERICA

AFRICA

AUSTRALIA

GONDWANA

extensive deserts

southern ice cap covers most of South America, Africa, and Australia

3. CARBONIFEROUS (300 MYA)
As the supercontinent Pangea came together, continental masses stretched from pole to pole, almost encircling the Paleo–Tethys Sea to the east. Today's coal seams were laid down in swampy forests along the shores of equatorial shelf seas. An extensive ice cap built up as Gondwana moved over the South Pole.

KEY

 subduction zone

spreading ridge

outline of modern landmass

INTRODUCTION

PEOPLE

ALFRED WEGENER

Alfred Wegener (1880-1930) was a German scientist with interests in astronomy, meteorology, and geology. In 1915 he presented the theory of continental drift to explain the presence of identical rocks on opposite sides of the Atlantic Ocean and tropical plant fossils in the Arctic Circle. His ideas were not accepted until sea-floor spreading was discovered, providing a mechanism to explain his theory.

EPICONTINENTAL SEAS

At most times in the past, sea levels have been higher than they are today. This has given rise to shallow, tideless bodies of water called epicontinental seas covering extensive parts of the continental interiors. These were quite unlike the deep ocean basins and continental-shelf seas familiar to us today. The area of dry land was sometimes reduced to half its current extent by these seas, which were often very salty, low in oxygen, and devoid of life. They could isolate parts of continents, causing populations of living things to evolve separately. Epicontinental seas also affected the climate: their high salinity produced downwelling (see p.60) of dense water into adjacent equatorial oceans, in contrast to the polar downwelling that dominates the deep-ocean circulation today.

SHALLOW WATER
Conditions on the shore of North America's Western Interior Seaway 100 million years ago may have been similar to the shallow lagoons of the Bahama Islands today (right).

4. JURASSIC (150 MYA)
The Paleo–Tethys Sea closed as future parts of central Asia broke away from Gondwana and moved north, with the Tethys Ocean opening up behind them. The central Atlantic was opening, splitting Pangea into northern and southern components.

central Atlantic starts to open

rifting signals creation of floor of modern Pacific Ocean

polar ice cap lost

5. CRETACEOUS (100 MYA)
The break-up of Gondwana started with India, Africa, and Antarctica rifting apart. This also started the closure of the Tethys Ocean. The opening of the south Atlantic soon followed, Europe separated from North America, and the Arctic Ocean opened over the North Pole.

opening of north Atlantic splits apart Europe and North America

high sea levels

Western Interior Seaway

Gondwana breaks up

Turgai Seaway

Isthmus of Panama yet to close

remnants of Tethys Ocean

6. EOCENE (50 MYA)
India continued its rapid movement north, which would end with the uplift of the Himalayas when it hit Asia. Africa's convergence with Europe closed the western Tethys Ocean. Australia and South America both separated from Antarctica, allowing the establishment of the Circumpolar Current that isolated Antarctica from equatorial heat flow.

Antarctic ice cap begins to form

Australia moves north

INTRODUCTION

CURRENTS, CONTINENTS, AND CLIMATE

Along with the atmosphere, the oceans are the means by which heat is redistributed around the Earth. Most energy arriving from the Sun is absorbed as heat near the Equator. It is then redistributed to colder regions. About 40 per cent of the heat reaching the poles from the Equator comes via ocean currents. The pattern of circulation in the oceans therefore has a large influence on the Earth's climate (see pp.66–67). As continents, oceans, and currents have shifted through geological time, major climate changes have occurred. Conversely, warmer and colder periods affect sea level and the extent of seas. There is even evidence that the ocean froze to a depth of 2,000m (6,500ft) in places during a series of "snowball" events 750–580 million years ago, and possibly earlier, each event lasting up to 10 million years.

SNOWBALL EARTH
During snowball events, global glaciation would have left only the peaks of the highest mountains free of ice, as is the case today in Antarctica.

MESOZOIC CURRENTS
100 million years ago, ocean currents flowed through a continuous seaway from the Tethys Ocean in the east, through what is now the Mediterranean, the Central Atlantic between North and South America, and into the Pacific in the west.

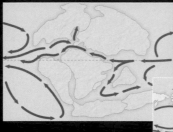

TODAY'S CIRCULATION
Today, equatorial ocean currents are blocked by landmasses, and the Circumpolar Current is the strongest current, blocking heat flow to the South Pole. The polar regions are colder.

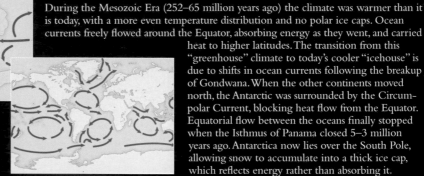

GREENHOUSE TO ICEHOUSE

During the Mesozoic Era (252–65 million years ago) the climate was warmer than it is today, with a more even temperature distribution and no polar ice caps. Ocean currents freely flowed around the Equator, absorbing energy as they went, and carried heat to higher latitudes. The transition from this "greenhouse" climate to today's cooler "icehouse" is due to shifts in ocean currents following the breakup of Gondwana. When the other continents moved north, the Antarctic was surrounded by the Circumpolar Current, blocking heat flow from the Equator. Equatorial flow between the oceans finally stopped when the Isthmus of Panama closed 5–3 million years ago. Antarctica now lies over the South Pole, allowing snow to accumulate into a thick ice cap, which reflects energy rather than absorbing it.

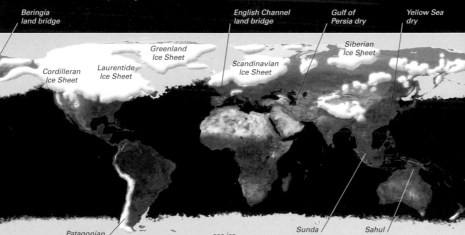

Beringia land bridge
English Channel land bridge
Gulf of Persia dry
Yellow Sea dry
Greenland Ice Sheet
Siberian Ice Sheet
Cordilleran Ice Sheet
Laurentide Ice Sheet
Scandinavian Ice Sheet
Patagonian Ice Sheet
sea ice
Sunda land bridge
Sahul land bridge
Antarctic Ice Sheet

LAST ICE AGE (21,000 YEARS AGO)
Today's climate swings between cold ice ages and warm interglacials on a 100,000-year cycle. During ice ages, northern-hemisphere ice caps expand, lowering global sea levels and revealing land bridges.

SEA-LEVEL CHANGE

Sea level has constantly changed through the Earth's history, being up to 400m (1,300ft) higher in the past. One of the factors controlling sea level is the global climate. Thermal expansion of ocean water increases global sea level by about 60cm (24in) for every 1°C (1.8°F) increase in temperature. More significant is the transfer of water between ice caps and the oceans during glacial cycles, which can account for a global change of 100–200m (330–655ft) over a few thousand years. The rate of sea-floor spreading can also affect global sea levels and has outweighed climatic factors at some times in the past. Faster-spreading ridges reduce the volume of the ocean basins as the younger, hotter crust rises higher, causing sea levels to rise (see p.88).

TEMPERATURE AND SEA LEVEL
Over the last 100 million years, climate has controlled sea levels, and these (in blue on graph) have dropped as the climate has cooled (temperature in yellow). At other times, low sea levels were due to reduced rates of sea-floor spreading.

MEDITERRANEAN BASIN HISTORY

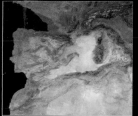
1 The Mediterranean was isolated from the Atlantic by the closure of the Strait of Gibraltar five million years ago, and evaporated to a salty desert.

2 21,000 years ago, sea levels were 120m (390ft) lower than they are today due to water being locked up in ice caps at the height of the last ice age.

3 10,000 years ago, ice-age meltwater started to flood the continental shelves exposed during the ice age, leaving today's familiar shoreline.

SEDIMENTARY BASINS

Most of the world's sedimentary rocks were laid down in water over continental shelves or in inland seas. The movements of the continents and changes in sea level have determined where this deposition occurred at particular times, and many former marine sedimentary basins are now far inland.

Oil and gas deposits are found in marine sedimentary rocks, the result of animal and plant remains decomposing and then being buried and compressed. About 20 per cent of the world's oil and gas production comes from offshore fields, but many offshore basins remain to be explored.

BASINS AND OILFIELDS
Sedimentary basins are found on the continental shelves and adjacent ocean floor, but also well inland where areas were once covered with water.

KEY
 onshore sedimentary deposits

 offshore sedimentary deposits

 Oil and gas deposits

ICE-AGE COAST
During ice ages, sea ice forms at lower altitudes than it does today. This scene may have been typical of the shores of western Europe 21,000 years ago.

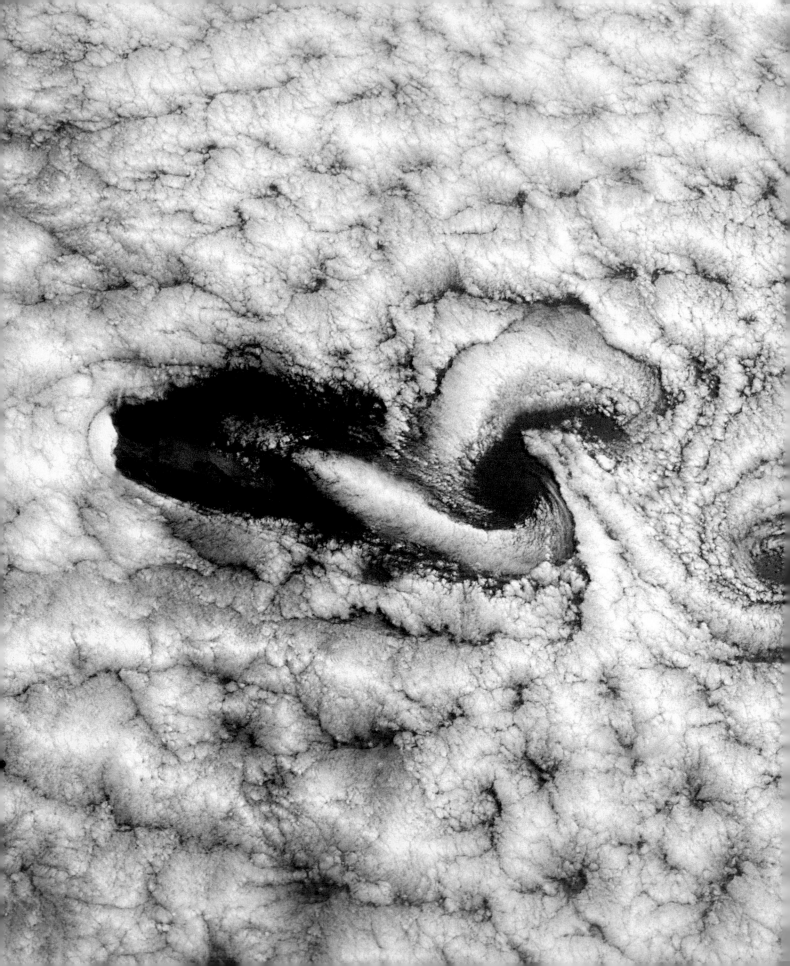

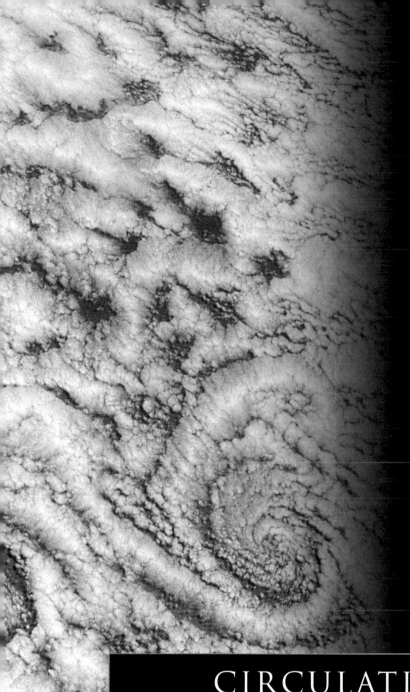

OCEAN WATER IS constantly in motion, and not simply in the form of waves. Throughout the oceans, there is a continuous circulation of seawater, both across the surface and more slowly at depth. Several related processes play a part in causing and maintaining these ocean currents. They include solar heating of the atmosphere, prevailing winds, the effect of the Earth's rotation, and processes that affect the temperature and salinity of surface waters. The various surface currents that are generated, some warm, some cold, have profound effects on climate in many parts of the world. Oceanic processes also play a part in the periodic climatic disturbances called El Niño and La Niña, and they help generate the extreme weather phenomena known as hurricanes and typhoons.

CIRCULATION AND CLIMATE

WHIRLING CLOUDS
These spiralling cloud patterns, caused when prevailing ocean winds encounter an island, are known as von Karman vortices. They were photographed from space over Guadalupe Island, near the Pacific coast of Mexico.

OCEAN WINDS

THE PATTERN OF AIR MOVEMENT over the oceans
results from solar heating of the atmosphere and the
Earth's rotation. This pattern of winds is modified by
linked areas of low and high pressure (cyclones and
anticyclones), which continually move over the oceans'
surface. Near coasts, additional onshore and offshore
breezes are common. These are caused by differences
in the capacity of sea and land to absorb heat.

ATMOSPHERIC CELLS

Solar heating causes the air in the Earth's atmosphere
to cycle around the globe in three sets of giant loops,
called atmospheric cells. Hadley cells are produced by
warm air rising near the Equator, cooling in the upper
atmosphere, and descending to the surface around
subtropical latitudes (30°N and S). Then the air moves
back towards the Equator. Ferrel cells are produced by
air rising around subpolar latitudes (60°N and S), cooling and falling in
the subtropics, and then moving towards the poles. Polar cells are caused
by air descending at the poles and moving towards the Equator.

CIRCULATION CELLS
*The atmospheric cells
produce north–south
airflows. These are
modified by Earth's
spin, producing winds
that blow diagonally.*

polar easterly
polar cell
air rises in
subpolar
latitudes
polar-front jet stream – narrow
ribbon of strong wind at high
altitude at top of front
Ferrel cell
air descends
in subtropical
latitudes
southwesterly
wind
Hadley
cell
air rises
at equator
direction of
Earth's spin
polar-front jet stream
northeasterly
trade
wind
subtropical
jet stream
southeasterly
trade wind
air descends
at pole
trade winds meet at Intertropical
Convergence Zone

THE CORIOLIS EFFECT

initial direction
of air
movemont
air
deflected
to right

The atmospheric cells cause air to move
in a north–south direction. This is
altered by the Coriolis effect, which is a
consequence of the Earth's spin. Because
the Earth turns continuously underneath the
airflow as it travels, the air appears to be
deflected from its straight north–south course.
It veers to the east when moving away from
the Equator, and to the
west when moving
towards it. The Coriolis
effect is an apparent,
not a true, force. No
actual force is exerted
on the wind.

Earth's
rotation
air
deflected
to left
initial direction
of air movement

AIR DEFLECTIONS
In the northern hemisphere,
the Coriolis effect causes
all air movements to be
deflected to the right of
their initial direction. In
the southern hemisphere,
they veer to the left.

SATELLITE IMAGING

The global pattern of ocean winds
is monitored by an instrument
called a scatterometer onboard the
NASA satellite QuikScat (right).
A scatterometer is a
radar device that can
measure both wind
speed and direction.

BREWING STORM
In this QuikScat image of an
Atlantic storm, orange areas
show the strongest winds.

PREVAILING WINDS

The winds produced by pressure differences and modified by the
Coriolis effect are called the prevailing winds. In the tropics and
subtropics, the air movements towards the Equator in Hadley cells
are deflected to the west. These are known as trade winds.
They comprise the northeastly trades in the northern
hemisphere, and southeasterly trades in the south.
At higher latitudes, the surface winds in Ferrel
cells deflect to the east, producing the westerlies.
In the southern hemisphere, these winds blow
from west to east without meeting land.
Those around latitudes of 40°S are known as
the Roaring Forties. In polar regions, winds
deflect to the west as they move away from
the poles. These are known as polar
northeasterlies and southeasterlies.

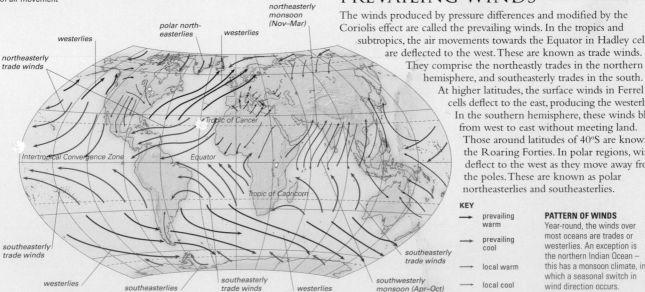

northeasterly
monsoon
(Nov–Mar)
polar north-
easterlies
westerlies
westerlies
northeasterly
trade winds
Tropic of Cancer
Intertropical Convergence Zone
Equator
Tropic of Capricorn
southeasterly
trade winds
southeasterly
trade winds
southwesterly
monsoon (Apr–Oct)
westerlies
southeasterlies
southeasterly
trade winds
westerlies

KEY

→	prevailing warm
→	prevailing cool
→	local warm
→	local cool

PATTERN OF WINDS
Year-round, the winds over
most oceans are trades or
westerlies. An exception is
the northern Indian Ocean –
this has a monsoon climate, in
which a seasonal switch in
wind direction occurs.

PRESSURE-SYSTEM WINDS

In any area of ocean where air sinks – often at subtropical latitudes – a zone of high atmospheric pressure, or anticyclone, develops. Where warm air rises, areas of low pressure, called cyclones or depressions, occur. These often develop near the Equator and subpolar latitudes. Cyclones and anticyclones create linked, circulating wind patterns, which continually move and change. In the northern hemisphere, there is a clockwise movement of air around an anticyclone, and an anticlockwise motion around a cyclone. This pattern is reversed in the southern hemisphere. Local pressure systems can affect the general pattern of prevailing winds. In particular, cyclones move swiftly over the ocean and can produce rapid changes in wind strength and direction.

CYCLONES AND ANTICYCLONES
Air moves from an area of high pressure towards one of low pressure, but the Coriolis effect modifies this, producing circular winds.

air ascends from cyclone

warm air rising

air descends into anticyclone

low pressure at centre

central area of high pressure

cold air sinks

air spirals around central area of low pressure

cold air flows towards area of low pressure

air moving from high to low pressure deflected by Coriolis efffect to form spiral

COASTAL BREEZES

Local winds, called onshore and offshore breezes, are generated near coasts, especially in sunny climes. Onshore breezes – sometimes called sea breezes – develop during the day. These are caused by the land heating up more quickly than the sea, as both absorb solar radiation. This occurs because the sea absorbs large quantities of heat energy with only a small rise in temperature, whereas the same amount of heat energy is likely to cause the land temperature to rise sharply (see p.33).

As the land warms up, it heats the air above it, causing the air to rise. Cooler air then blows in from the sea to take its place. In the evening, and at night, the opposite effect occurs. At nightfall, the land quickly cools down, but the sea remains warm and continues to heat the air above it. As this warm air rises, it sucks the cooler air off the land, and so generates an offshore breeze. This is sometimes called a "land breeze."

BREEZY COAST
On warm coasts, there is often a noticeable drop in temperature from midday as a cool sea breeze blows in off the water. The breeze typically reverses in the evening and at night.

warm air cools at high altitude

cold air sinks

air heats up and rises over land

cool air drawn in

ONSHORE BREEZE

DAY AND NIGHT
Land heats up faster than water during the day. Warm air rises over the land and draws in cold air from the sea. At night, the land cools more quickly, reversing the airflow.

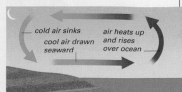

cold air sinks

cool air drawn seaward

air heats up and rises over ocean

OFFSHORE BREEZE

TRIMMING THE SAILS
A crew set their sails as they set off on the Sydney-to-Hobart yacht race. The course crosses the often stormy Bass Strait between Australia and Tasmania.

OCEAN YACHT RACING

Ocean yacht racing is the sport of competitive sailing, held over long distances and in open water. These races range from short but robust challenges lasting a few days, such as the famous Fastnet Race off southwest England and the annual Sydney-to-Hobart race, to long, multi-stage, round-the-world races, which can involve up to 6 months at sea. Some races are multi-handers, with crews of up to 18 per yacht, others are single-handers. The participants are typically highly experienced sailors. In multi-handed races, there will be a skipper/tactician, a navigator, and general crew whose responsibilities include sail changing and trimming. Solo racers have to do everything themselves. One race, the Global Challenge, is unusual in that its crew consists of ordinary people, some with little previous sailing experience, who have paid to take part under the leadership of a professional skipper.

To make races as equable as possible, usually competing boats are identical or a handicapping system is used to adjust the times of different classes of boat. The use of computer technology is paramount in modern racing. Navigation is electronically assisted, and computers are employed to monitor and help optimize boat performance. Vast amounts of weather data are downloaded via the internet during a race. An important skill is to be able to interpret this data, so as to know, for example, where the most wind is likely to be in the area ahead. Otherwise, doing well in a race is mainly down to tactics and seamanship — for example, knowing how to get the best out of a boat in both strong and light winds, or judging when best to tack (change course when sailing upwind).

RACING AROUND THE GLOBE

Three of the most famous ocean yacht races are global circumnavigations. Each is held every four years. The Volvo Ocean Race goes "round the right way" in the Southern Ocean (west to east, the same direction as the prevailing winds and currents), and includes stops. The Global Challenge is similar but goes "round the wrong way". The Vendée Globe Challenge is a non-stop race for single-handers.

→ Vendée Globe
→ Global Challenge
→ Volvo 2005–06

Rotterdam
Göteborg
Portsmouth
Les Sables-d'Olonne
New York Boston
Baltimore/Annapolis Vigo La Rochelle
ATLANTIC OCEAN
PACIFIC OCEAN
PACIFIC OCEAN
INDIAN OCEAN
Rio de Janeiro
Buenos Aries
Cape Town
Melbourne
Wellington
SOUTHERN OCEAN

EQUIPPED TO WIN

MULTIHULL RACING

TRIMARAN *Most ocean yacht races are for monohull yachts, but a few are open to multihulls (catamarans and trimarans), while some are for multihulls only. Multihulls are faster than monohulls, and although easier to capsize, they stay afloat even when severely damaged. Here, the trimaran Foncia is sailing on just one of its three hulls during the 2005 Grand Prix de Fécamp, off Normandy, France.*

YACHT CREWS

ALL HANDS ON DECK *The crews for some races are large – 18 people on each yacht for the Global Challenge, in which Barclays Adventurer competed in 2004–05. For this race, the crew, who are recently trained amateurs, participate in all duties on the yacht, including trimming sails, navigating, helming, cooking, and so on.*

HIGH-TECH AIDS

CONTROL CENTRE
Technology is all-important in modern racing, including the use of electronic charts and global positioning systems. Here, Frenchman Marc Thiercelin prepares for the Vendée Globe 2005–06 in the control centre on his yacht Pro-Form.

TURNED TURTLE *In 1997, the lone British yachtsman Tony Bullimore capsized in the Southern Ocean and was trapped for five days in his upturned yacht before rescuers arrived.*

HANGING ON *The crew of the yacht Astra (shown here) risk being swept overboard, the greatest disaster that can befall a sailor. To assist recovery in this eventuality, crew members usually wear radio beacons. It is also usual to be attached to the boat via a safety harness in anything other than calm daylight conditions.*

RISKY BUSINESS

INTRODUCTION

SURFACE CURRENTS

FLOWING FOR ENORMOUS DISTANCES within the upper regions of the oceans are various wind-driven currents. Many join up to produce large circular fluxes of water, called gyres, around the surfaces of the main ocean basins. Surface currents affect only about 10 per cent of ocean water, but they are important to the world's climate (see p.66), because their overall effect is to transfer huge amounts of heat energy from the tropics to cooler parts of the globe. They also impact on shipping and the world's fishing industries.

WIND ON WATER

When wind blows over the sea, it causes the upper ocean to move, creating a current. However, the water does not move in the same direction as the wind. Instead, it moves off at an angle – to the right in the northern hemisphere and to the left in the southern hemisphere. This phenomenon was first explained in 1902 by a Swedish scientist, Walfrid Ekman, using a model of the effect of wind on water now called the Ekman spiral. The model assumes that the movement of water in each layer of the upper ocean is produced by a combination of frictional drag from the layer above (or, in the top layer, from wind drag) and the Coriolis effect (see p.54). The model predicts that, overall, a mass of water will be pushed at right angles to the wind direction, an effect known as Ekman transport.

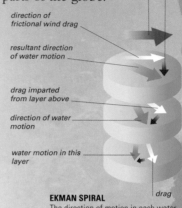

Coriolis deflection

wind

direction of frictional wind drag

resultant direction of water motion

drag imparted from layer above

direction of water motion

water motion in this layer

drag

EKMAN SPIRAL
The direction of motion in each water layer results from a combination of the drag from the layer above and a deflection caused by the Coriolis effect. This diagram shows the Ekman spiral in the northern hemisphere. In the southern hemisphere, deflection is to the left of wind direction.

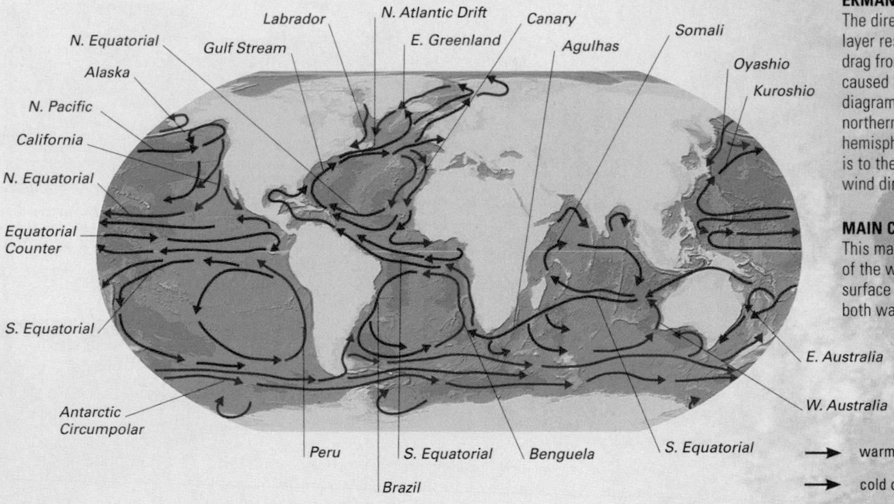

MAIN CURRENTS
This map shows all of the world's main surface currents, both warm and cold.

N. Equatorial
Alaska
N. Pacific
California
N. Equatorial
Equatorial Counter
S. Equatorial
Antarctic Circumpolar

Labrador
Gulf Stream
N. Atlantic Drift
E. Greenland
Canary
Agulhas
Somali
Oyashio
Kuroshio
E. Australia
W. Australia
Peru
S. Equatorial
Brazil
Benguela
S. Equatorial

→ warm current

→ cold current

OCEAN GYRES

The combination of prevailing winds (see p.54) and Ekman transport produces large-scale, circular systems of currents known as gyres. Altogether there are five ocean gyres – two in each of the Atlantic and Pacific oceans and one in the Indian Ocean. Each gyre consists of several named currents. Thus, the gyre in the north Pacific is made up of the Kuroshio current in the west, the California current in the east, and two other linked currents. Water tends to accumulate at the centre of these gyres – producing shallow "mounds" in the ocean.

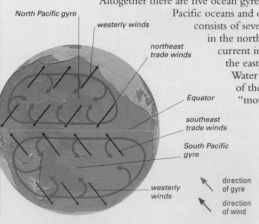

North Pacific gyre
westerly winds
northeast trade winds
Equator
southeast trade winds
South Pacific gyre
westerly winds

↗ direction of gyre

↗ direction of wind

GYRE CREATION
In the north Pacific, the combination of westerly and trade winds, always pushing water to the right (by Ekman transport) produces a clockwise gyre. In the south Pacific, where winds push water to the left, an anticlockwise gyre is created.

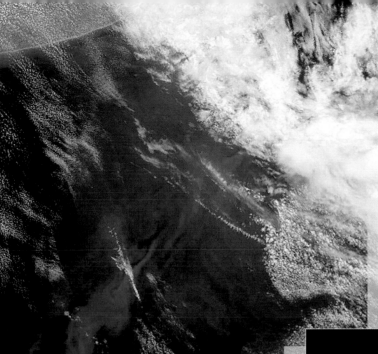

PEOPLE

BENJAMIN FRANKLIN

The American statesman and inventor Benjamin Franklin (1706–90) made one of the earliest studies of an ocean current, publishing a map of the Gulf Stream's course. He became interested in it after the British postal authorities asked him why American postal ships crossed the Atlantic faster than English ships. The answer was that American ships were utilizing an eastward extension of the Gulf Stream.

BOUNDARY CURRENTS

The currents at the edges of gyres are called boundary currents. Those on the western side of gyres are strong, narrow, and warm – they move heat energy away from the Equator. Examples of these currents are the Gulf Stream and the Brazil Current in the southwestern Atlantic. Eastern boundary currents are weaker, broader cold currents that move water back towards the tropics. Examples are the Benguela Current off southwest Africa and the California Current. At the gyre boundaries close to the Equator are warm, west-flowing equatorial currents. Other currents feed into or out of the main gyres. These include, for example, the warm North Atlantic Drift, an offshoot of the Gulf Stream, and cold currents that bring water down from the Arctic, such as the Oyashio and East Greenland currents.

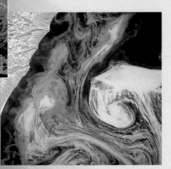

WARM CURRENT
Satellite devices can detect phytoplankton levels in the water, which can be related to temperature. Here, yellow and red indicate high levels of plankton and the warm Brazil Current.

COLD CURRENT
In this satellite view, sea-ice is visible flowing past the Kamchatka Peninsula in the cold Oyashio Current. Eddies within the current have produced spiral patterns in the sea-ice.

MEETING OF CURRENTS

In a few areas, warm and cold currents meet and interact. Examples include the meeting of the warm Gulf Stream with the cold Labrador Current off the eastern seaboard of the USA and Canada, and the meeting of the cold Oyashio Current with the warm Kuroshio Current to the north of Japan. At these confluences, the denser water in the cold current dives beneath the water in the warm current, usually producing some turbulence. This can trigger an upward flow of nutrient-rich waters from the sea floor, encouraging the growth of plankton, and producing good feeding grounds for fish, sea birds, and mammals.

OPPOSING CURRENTS
The warm Brazil Current on the left, and the colder Falklands Current on the right each carry differently coloured populations of plankton.

SEA SMOKE
Dolphins cavort amid steep waves. The "sea smoke" is created when water vapour is added to cold air drifting across the boundary between cold and warm currents.

INTRODUCTION

UNDERWATER CIRCULATION

THE WATERS THAT MAKE UP THE EARTH'S OCEANS circulate deep
below the surface. Subsurface currents are complex. Some are vertical,
moving water upwards and downwards to and from the surface,
processes called upwelling and downwelling. Surface and subsurface
currents are all linked in a global pattern of deep–water circulation.

DOWNWELLING

The most important causes of downwelling are thermohaline
processes ("thermo" means heat, and "haline" means salt), which
alter either the temperature or salinity of seawater. For example,
where warm, salty water is carried by a surface current into the
Arctic Ocean, it rapidly cools when it meets colder, less salty, polar
water. As it cools, its density increases, and it sinks down.
Downwelling also occurs on some coasts. For example, winds
blowing towards the Equator on the
western side of oceans push seawater towards land by
Ekman transport (see p.58). As it reaches the coast,
it is forced down. Finally, downwelling also
occurs beneath the mounds of water
that accumulate in the middle of
anticyclones (see p.55) and ocean
gyres (see p.58).

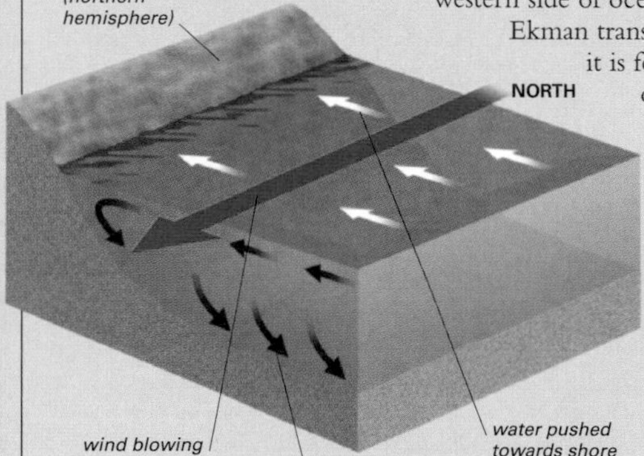

**NORTH ATLANTIC
DOWNWELLING ZONES**
At the important downwelling
sites shown here, warm
surface water meets colder
Arctic water, loses heat,
become denser, and sinks.

KEY

- - - → downwelling
──→ warm surface current
──→ loss of heat energy
──→ cold surface current
▬ downwelling zone

east-facing coast
(northern
hemisphere)

NORTH

wind blowing
towards the
Equator

water sinks near
coast

water pushed
towards shore
due to Ekman
transport

COASTAL DOWNWELLING
A wind blowing towards the
Equator on the western side
of an ocean, as here (left),
pushes seawater towards
the shore, where it sinks.

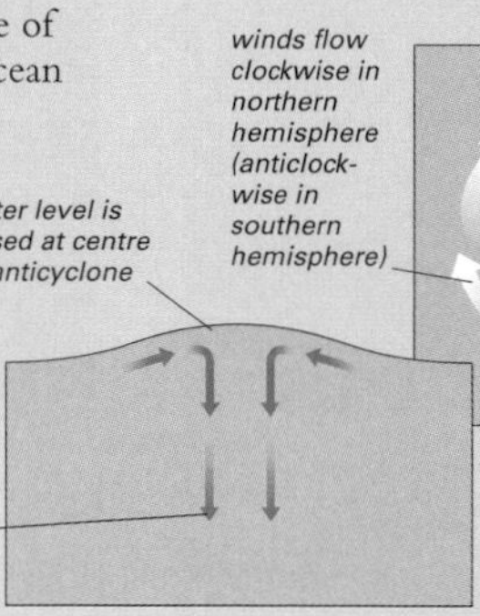

water level is
raised at centre
of anticyclone

water sinks
through effects
of gravity

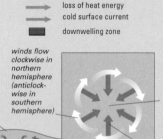

winds flow
clockwise in
northern
hemisphere
(anticlock-
wise in
southern
hemisphere)

accumulation
of water at
centre

Ekman
transport
pushes water
towards centre
of anticyclone

DOWNWELLING IN AN ANTICYCLONE
In an anticyclone, the circular
system of winds can push water in
to a central mound, where it sinks.

UPWELLING

Upwelling can occur in various situations, some of which are simply
the reverse of the conditions that cause downwelling. For instance,
winds blowing towards the Equator on the eastern sides of oceans
push seawater away from land by Ekman transport, so deeper water
must upwell near the coast to replace it. Water rises towards the surface
in the centre of cyclones (the opposite of anticyclones, see p.55), and
will also rise where surface waters tend to be pushed apart at boundaries
between ocean gyres, for example in some equatorial parts of the Pacific
and Atlantic. Some seawater upwells to replace sinking, denser water.
An example occurs around Antarctica, where upwelling replaces super-
dense, cold, salty water forming and sinking under developing sea-ice.

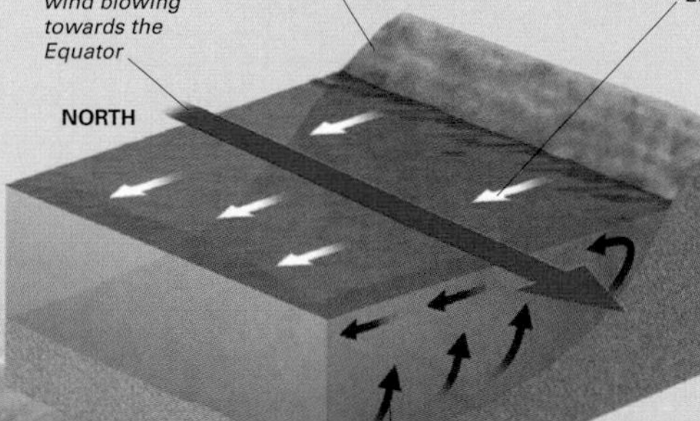

west-facing coast
(northern hemisphere)

wind blowing
towards the
Equator

NORTH

water moves away from
shore as a result of
Ekman transport

water moves upwards to
replace the water moving
offshore at the surface

**COASTAL
UPWELLING**
A wind blowing
towards the Equator
on the eastern side
of an ocean, as here,
pushes seawater
away from the shore,
causing upwelling
near the coast.

PLANKTON-HARVESTER
Where upwelling occurs, it brings
large amounts of nutrients up from
the sea floor. These encourage the
growth of plankton, attracting plankton-
grazers such as this Manta Ray as well as
smaller fish, whales, and other marine life.

DEEP-WATER CIRCULATION

Seawater circulates slowly through the deeper parts of the oceans, driven by water sinking in major downwelling zones, such as in the north Atlantic. Any specific mass of deep water has, at some time, sunk in one of these zones. Once it sinks, its properties, such as its salinity, remain stable for long periods – thus, every mass of deep water contains a "memory" of where it originally sank. By analyzing seawater samples from various parts of the deep oceans, it is possible to piece together the general pattern of deep-water flow. The indications are that there is a large-scale circulation involving all the oceans, called the global conveyor. A specific mass of seawater takes about 1,000 years to complete a lap of this circuit.

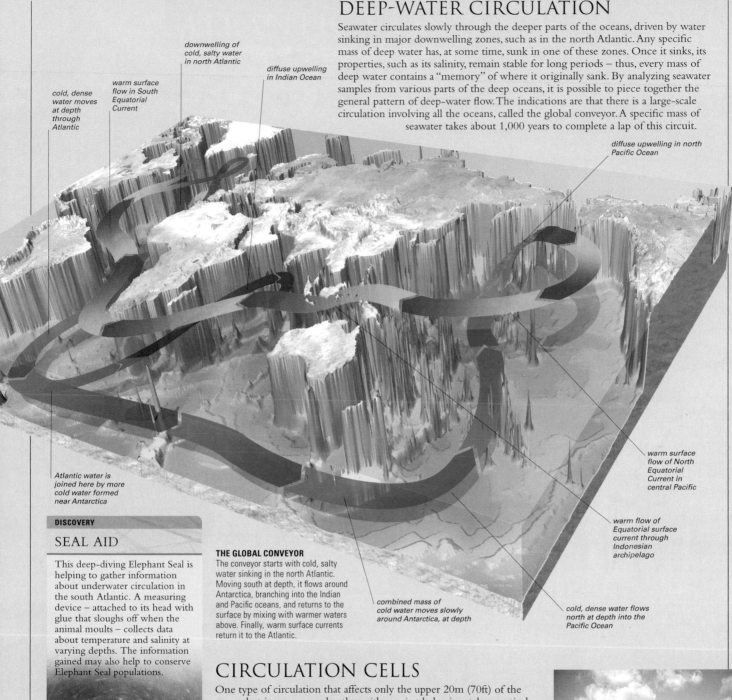

cold, dense water moves at depth through Atlantic

warm surface flow in South Equatorial Current

downwelling of cold, salty water in north Atlantic

diffuse upwelling in Indian Ocean

diffuse upwelling in north Pacific Ocean

warm surface flow of North Equatorial Current in central Pacific

warm flow of Equatorial surface current through Indonesian archipelago

cold, dense water flows north at depth into the Pacific Ocean

combined mass of cold water moves slowly around Antarctica, at depth

Atlantic water is joined here by more cold water formed near Antarctica

THE GLOBAL CONVEYOR
The conveyor starts with cold, salty water sinking in the north Atlantic. Moving south at depth, it flows around Antarctica, branching into the Indian and Pacific oceans, and returns to the surface by mixing with warmer waters above. Finally, warm surface currents return it to the Atlantic.

SEAL AID

This deep-diving Elephant Seal is helping to gather information about underwater circulation in the south Atlantic. A measuring device – attached to its head with glue that sloughs off when the animal moults – collects data about temperature and salinity at varying depths. The information gained may also help to conserve Elephant Seal populations.

CIRCULATION CELLS

One type of circulation that affects only the upper 20m (70ft) of the ocean, but is more complex than either a simple horizontal or vertical flow of water, is known as Langmuir circulation. This is wind-driven and consists of rows of long, cylinder-shaped cells of water, aligned in the direction in which the wind is blowing and each rotating in the opposite direction to its neighbour – alternate cells rotate clockwise and anticlockwise. Each cell is about 10-50m (30-165ft) wide and can be hundreds of metres long. On the sea surface, the areas between adjacent cells where seawater converges are visible as long white streaks of foam, or congregations of seaweed, called windrows. The whole pattern of circulation was first explained in 1938 by an American chemist called Irving Langmuir, after he had crossed the Atlantic in an ocean liner. It was subsequently named in his honour.

LANGMUIR WINDROWS
These long streaks of foam on the sea surface are the windrows of Langmuir circulation cells. The distance between windrow lines increases with the wind speed.

THE GLOBAL WATER CYCLE

THE WORLD'S OCEANS DO NOT FORM a self-contained system but continually exchange water with the atmosphere and landmasses through evaporation, cloud formation, precipitation, wind transport, and river flow. This complex of interconnected processes, which is ultimately driven by heat from the Sun, is called the global water cycle or hydrologic cycle. The cycle is made up of many smaller cycles, such as the formation and melting of sea-ice.

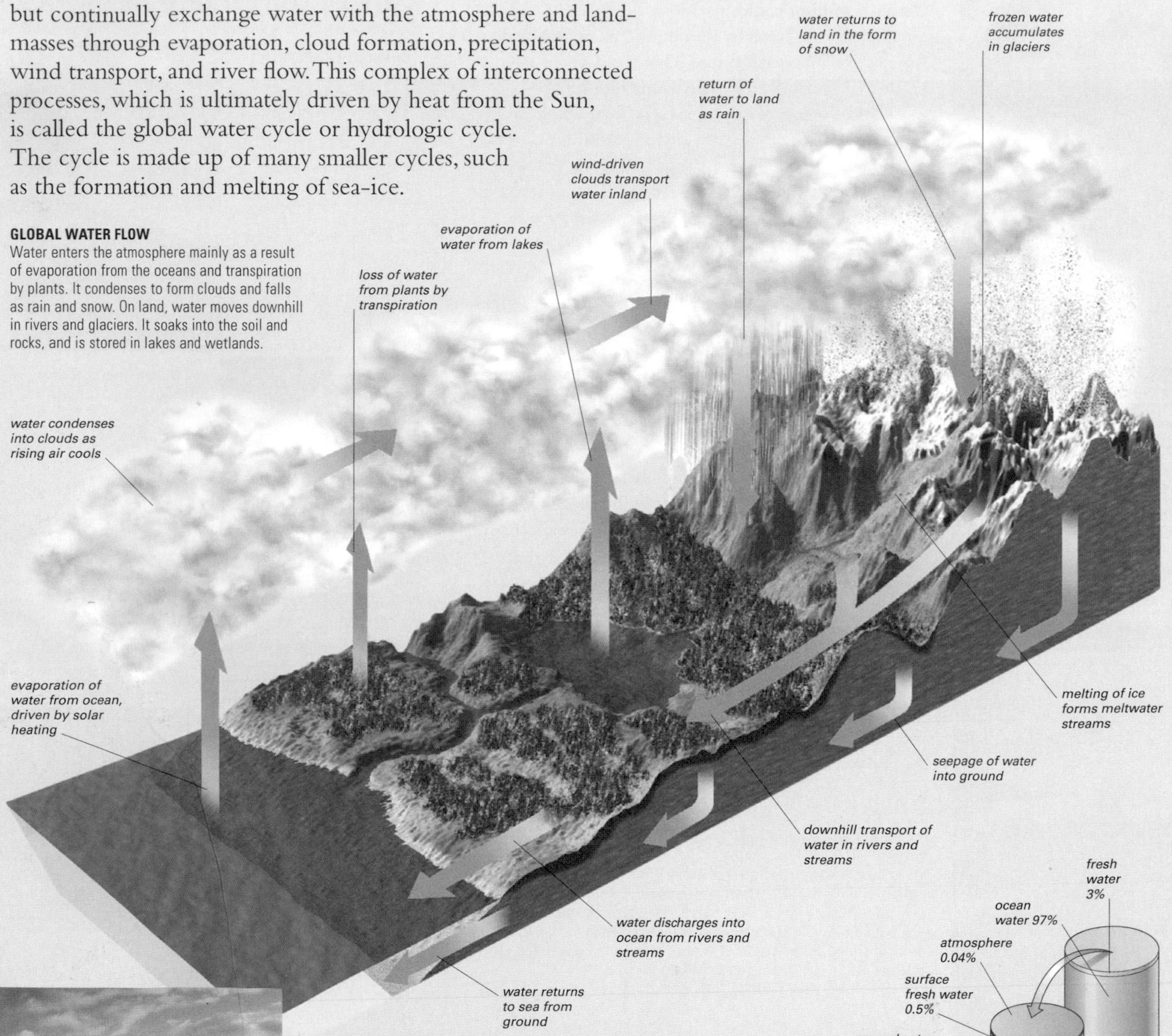

GLOBAL WATER FLOW

Water enters the atmosphere mainly as a result of evaporation from the oceans and transpiration by plants. It condenses to form clouds and falls as rain and snow. On land, water moves downhill in rivers and glaciers. It soaks into the soil and rocks, and is stored in lakes and wetlands.

water returns to land in the form of snow

frozen water accumulates in glaciers

return of water to land as rain

wind-driven clouds transport water inland

evaporation of water from lakes

loss of water from plants by transpiration

water condenses into clouds as rising air cools

evaporation of water from ocean, driven by solar heating

melting of ice forms meltwater streams

seepage of water into ground

downhill transport of water in rivers and streams

water discharges into ocean from rivers and streams

water returns to sea from ground

PLAYERS IN THE CYCLE

The sea, ice, mountains, and clouds all play a part in the global water cycle. This coastal scene is near Port Lockeroy in Antarctica.

EARTH'S WATER RESERVOIRS

Just over 1.4 billion cubic kilometres (or one-third of a billion cubic miles) of water exists on Earth. More than 97 per cent of this water is stored in the oceans as a component of salt water. The rest is fresh water. Of this, more than two-thirds is in the form of ice, locked up in the vast ice-sheets that cover Antarctica and most of Greenland, and in icebergs and sea-ice. Much of the rest is groundwater – contained in underground rocks – while a tiny amount (less than 1 part in 2,000) is water vapour in the atmosphere. Fresh liquid water on the Earth's land surface, in lakes, wetlands, and rivers, makes up just 0.5 per cent of all the world's fresh water, or 0.014 per cent of the total water. The Earth's different water reservoirs have not always had the same relative sizes that they have today. For instance, during the ice ages, a higher proportion was locked up in ice, with less in the oceans.

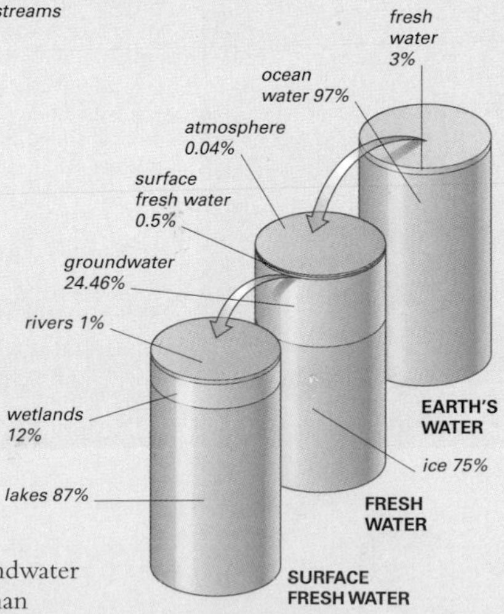

fresh water 3%

ocean water 97%

atmosphere 0.04%

surface fresh water 0.5%

groundwater 24.46%

rivers 1%

wetlands 12%

lakes 87%

EARTH'S WATER

ice 75%

FRESH WATER

SURFACE FRESH WATER

RELATIVE SIZES

Earth's ocean water (the bulk of the rear cylinder, above) hugely exceeds its reservoirs of fresh water, and the relative proportion of fresh water found on the land surface is tiny.

OCEAN EVAPORATION AND PRECIPITATION

A total of 425,000 cubic kilometres (102,000 cubic miles) of water evaporates from the oceans per year. Of this, 385,000 cubic kilometres (93,000 cubic miles) falls back into the sea as precipitation (rain, snow, sleet, and hail). The remainder is carried onto land as clouds and moisture. Evaporation and precipitation are not evenly spread over the surface of the oceans. Evaporation rates are greatest in the tropics and lowest near the poles. High rates of precipitation occur near the Equator and in bands between the latitudes of 45° and 70° in both hemispheres. Drier regions are found on the eastern sides of the oceans between the latitudes of approximately 15° and 40°.

PEOPLE

SENECA THE YOUNGER

In his book *Natural Questions*, the Roman statesman, dramatist, and philosopher Seneca the Younger (4 BC–AD65) pondered why ocean levels remain stable despite the continuous input of water from rivers and rain. He argued there must be mechanisms by which water is returned from the sea to the air and land and proposed an early version of the hydrologic cycle to explain this.

FRESHWATER INFLOW

The 40,000 cubic kilometres (9,000 cubic miles) of water lost from the oceans each year by evaporation and transport onto land is balanced by an equal amount returned from land in runoff. Just 20 rivers, including the Amazon and some large Siberian rivers, account for over 40 per cent of all input into the oceans. Inflows from the different river systems change over time as they are affected by human activity and climate change. For instance, global warming appears to have increased the flow from Siberian rivers into the Arctic Ocean, as water frozen in the tundra melts. These inflows lower the salinity of Arctic waters and may influence global patterns of ocean circulation (see p.63).

SIBERIAN RIVER LENA FLOODING
Climate change is thought to have contributed to severe flooding of the River Lena in recent years (below). A false-colour satellite image (left) shows the engorged river in black; red areas are ice.

EQUATORIAL RAINSTORM
In some areas near the Equator, as here in the tropical Pacific, annual rainfall is over 300cm (120in), compared to under 10cm (4in) in the driest ocean areas.

THE SEA-ICE CYCLE

In addition to the overall global water cycle, there is a local seasonal cycle in the amount of water locked up as sea-ice. In the polar oceans, the extent of sea-ice increases in winter and decreases in summer. This has important climatic consequences, because sea-ice formation releases, and its melting absorbs, latent heat to and from the atmosphere; and because the presence or absence of sea-ice modifies heat exchange between the oceans and atmosphere. In winter, sea-ice insulates the relatively warm polar oceans from the much colder air above, thus reducing heat loss. However, especially when covered with snow, sea-ice also has a high reflectivity (albedo) and reduces the absorption of solar radiation at the surface. Overall, the sea-ice cycle is thought to help stabilize air and sea temperatures in polar oceans. Also, because it affects surface salinity, sea-ice formation helps drive large-scale circulation of water through the world's oceans (see p.61).

SEA-ICE FORMING
As sea-ice forms, it releases heat to the atmosphere and increases the saltiness of the surrounding water (by rejecting salt). These processes affect climate and the circulation of seawater.

OCEANS AND CLIMATE

THE OCEANS HAVE A PROFOUND INFLUENCE on the world's climate, most strikingly in the way they absorb solar energy and redistribute it around the world in warm surface currents. Cold currents also produce local climatic effects, while alterations in currents are associated with climatic fluctuations such as El Niño (see p.68). The future behaviour of the oceans is crucial to the future course of climate change, as they are an important store for carbon dioxide, the principal greenhouse gas.

SOLAR HEATING
The surface layers of the oceans absorb over half the solar energy that reaches the Earth. Currents move this from the Equator towards the poles at a rate of about 1 billion megawatts.

WARM CURRENTS

Five or six major surface currents (see p.58) carry heat away from the tropics and subtropics towards the poles, giving some temperate regions a warmer climate than they would otherwise enjoy. A prime example is the effect of the warm Gulf Stream and its extension, the North Atlantic Drift, on Europe. The North Atlantic Drift carries heat originally absorbed in the Caribbean Sea and Gulf of Mexico right across the Atlantic, where it is released into the atmosphere close to the shores of France, the British Isles, Norway, Iceland, and other parts of northwestern Europe. As the prevailing westerly winds blow this warmed air over land, these countries benefit from a milder climate than equivalent regions at similar, or even lower latitudes, on the western side of the Atlantic. For example, winter temperatures are typically higher in Reykjavik, the capital of Iceland, than in New York. Similarly, in the northwest Pacific, the Kuroshio Current warms the southern part of Japan, while in the extreme southwest Pacific, the East Australian Current gives Tasmania a relatively mild climate.

BALMY BEACHFRONT
Penzance, in southwest England, has a mild climate that supports subtropical vegetation – the effect of the North Atlantic Drift is to raise temperatures here by about 5°C (9°F).

COLD CURRENTS

In some instances, the climatic effect of cold currents is simply to produce a cooler climate than would otherwise be the case. For instance, the west coast of the USA is cooled in summer by the cold California Current. Cold currents also affect patterns of rainfall and fog formation. In general, the various cold currents flowing towards the Equator on the eastern sides of oceans –combined with upwellings of cold water from depth in these regions – cool the air, reduce evaporative losses of water from the ocean, and cause downdraughts of drier air from higher in the atmosphere. Although cloud and fog often develop over the ocean in these areas (as what little moisture there is condenses over the cold water), these quickly disperse once the air moves over land. Thus, cold currents contribute to the development of deserts on land bordering the eastern sides of oceans, such as the Namib desert in southwestern Africa.

COAST OF NORTHERN CHILE
The cold Peru Current flows along the coast of northern Chile. It encourages the development of cloud and fog over the sea (visible above left in the satellite image) but also contributes to the extreme aridity of the coastal strip (left).

STRAYING NORTH

Most penguins live in Antarctica but, somewhat surprisingly, the world's most northerly-living penguins inhabit the Galapagos Islands, on the Equator. The islands have a cool climate – sea-surface temperatures in most years average 5°C (9°F) less than typical temperatures in the tropics, due to the cold Peru Current that flows up the west coast of South America.

CARBON IN THE OCEANS

FORAMINIFERAN SHELL

The oceans contain the Earth's largest store of carbon dioxide (CO_2) – the main greenhouse gas implicated in global warming. Huge amounts of carbon are held in the oceans, some in the form of CO_2 and related substances that readily convert to CO_2, and some in living organisms. The oceanic CO_2 is in balance with the atmospheric content of the same gas. For many years, the oceans have been alkaline, and acted as an important store for the excess CO_2 released by human activity. Biological and chemical processes turn some of this CO_2 into the calcium carbonate shells and skeletons of organisms, other organic matter, and carbonate sediments. However, the increasing CO_2 concentration is beginning to acidify the oceans, threatening shell and skeleton formation in marine organisms, as acid tends to dissolve carbonates. Further, some scientists fear that the rate at which the oceans can continue to absorb CO_2 will soon slow down, further aggravating global warming.

CARBON CONVERSION
CO_2 released from burning fossil fuels (right), after absorption into the oceans, can eventually end up in the shells of marine organisms in the form of carbonate.

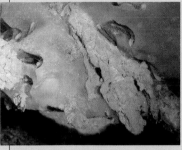

METHANE HYDRATE DEPOSIT
This substance is found as a solid on some areas of sea floor. There are concerns that ocean warming could release this into the atmosphere as methane gas, which traps more heat than CO_2.

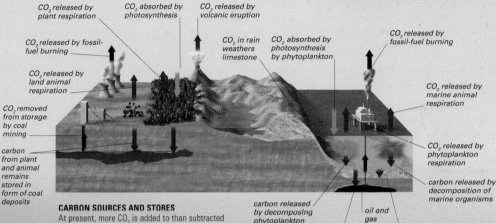

CO_2 released by plant respiration

CO_2 absorbed by photosynthesis

CO_2 released by volcanic eruption

CO_2 released by fossil-fuel burning

CO_2 released by land animal respiration

CO_2 in rain weathers limestone

CO_2 absorbed by photosynthesis by phytoplankton

CO_2 released by fossil-fuel burning

CO_2 removed from storage by coal mining

CO_2 released by marine animal respiration

carbon from plant and animal remains stored in form of coal deposits

CO_2 released by phytoplankton respiration

carbon released by decomposition of marine organisms

carbon released by decomposing phytoplankton

oil and gas

carbon in sediment turns into oil and gas

carbonate in sediment turns into limestone

CARBON SOURCES AND STORES
At present, more CO_2 is added to than subtracted from the atmosphere. Some of the excess is absorbed by the oceans, where some is held in solution and some incorporated into living organisms and sediments.

GOLDEN GATE FOG
The climate of San Francisco is influenced by exceptionally cold water, produced by upwelling, off the California coast. Fog is produced as westerly winds blow moist air over this cold water.

EL NIÑO AND LA NIÑA

EL NIÑO AND LA NIÑA ARE LARGE climatic disturbances caused by abnormalities in the pattern of sea surface temperature, ocean currents, and pressure systems. They are in the tropical Pacific Ocean. These disturbances have important repercussions for weather throughout the Pacific and beyond. Most scientists regard El Niño and La Niña as extreme phases of a complex global weather phenomenon called the El Niño–Southern Oscillation (ENSO).

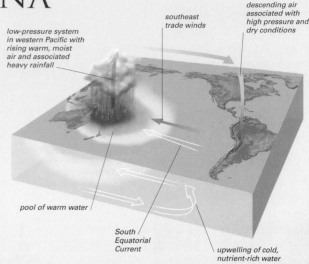

low-pressure system in western Pacific with rising warm, moist air and associated heavy rainfall

southeast trade winds

descending air associated with high pressure and dry conditions

pool of warm water

South Equatorial Current

upwelling of cold, nutrient-rich water

EL NIÑO EVENTS

The Spanish term El Niño means "little boy" or "Christ child". It originally denoted a warm current that was occasionally noticed around Christmas off Peru. Later it was restricted to unusually strong rises in temperature in the waters of the eastern Pacific, with a reduction in the upwelling of nutrient-rich waters that normally occur there. It is now used to mean a much wider shift in ocean and atmospheric conditions that affects the whole globe. El Niño events typically last from 12 to 18 months and occur cyclically, although somewhat unpredictably. On average, they occur about 30 times per century, with intervals that are sometimes as short as two years and sometimes as long as 10 years. Their underlying cause is not understood.

NORMAL PATTERN
A low-pressure system in the western Pacific draws southeast trade winds across from a high-pressure system over South America. These winds drive the South Equatorial Current, which maintains a pool of warm surface water in the western Pacific.

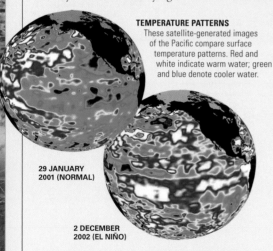

TEMPERATURE PATTERNS
These satellite-generated images of the Pacific compare surface temperature patterns. Red and white indicate warm water; green and blue denote cooler water.

29 JANUARY 2001 (NORMAL)

2 DECEMBER 2002 (EL NIÑO)

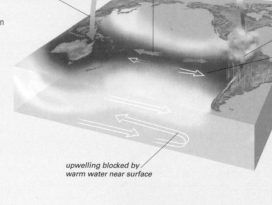

southeast trade winds reverse or weaken

descending air and high pressure brings warm, dry weather

low pressure and rising warm, moist air associated with heavy rainfall

warm water flows eastward, accumulating off South America

EL NIÑO PATTERN
During an El Niño event, the pressure systems that normally develop in the Pacific, and the southeast trade winds, weaken or reverse. The pool of warm surface water extends from the western Pacific into the central and eastern Pacific.

upwelling blocked by warm water near surface

EFFECTS OF EL NIÑO

An El Niño event causes wetter than normal conditions, and floods, in countries on the western side of South America, particularly Ecuador, Peru, and Bolivia. These conditions may also extend to the southeastern United States. In other parts of the world, it causes drier conditions. Drought and forest fires become more common in the western Pacific, particularly in Indonesia and parts of Australia, but also in East Africa and northern Brazil. The warmer waters in the eastern Pacific cause a reduction in the Peru Current and reduced upwelling near the coast of South America. This reduces the level of nutrients in the seawater, which has a negative impact on fish stocks. Other effects include a quieter Atlantic hurricane season and an increase in the extent of sea-ice around Antarctica. Japan, western Canada, and the western USA typically experience more storms and warmer weather than normal.

GIANT WAVES
During an El Niño event, storms become more frequent and violent in the central Pacific. These storms can produce gigantic waves, up to 10m (33ft) high, in Hawaii, as here on the island of Oahu.

width of rings directly related to amount of growth

1746–47 El Niño ring

EVIDENCE OF AN HISTORICAL EL NIÑO
Increased tree growth can be linked to high rainfall that occurred during historic El Niño events. One of the rings in this sample has been linked to an El Niño in 1746–47.

CORAL BLEACHING
This small circular coral reef has suffered severe bleaching (whitening). El Niño events are often associated with bleaching caused by unusually high sea surface temperatures.

MONITORING

The tropical Pacific is regularly monitored for temperature changes. The main monitoring methods are the use of satellites, which measure sea temperatures indirectly from slight variations in the shape of the ocean surface, and an array of instrumented weather buoys.

INSTRUMENTED BUOY

Buoys such as this one, strung in an array across the equatorial Pacific, are used to make regular measurements of water temperature at varying depths.

LA NIÑA EVENTS

La Niña is Spanish for the "little girl". A La Niña event is the reverse of an El Niño event. It is characterized by unusually cold ocean temperatures in the eastern and central equatorial Pacific, and by stronger winds and warmer seas to the north of Australia. La Niña conditions frequently, but not always, follow closely on an El Niño. Like El Niño, La Niña causes increased rainfall in some world regions and drought in others. India, Southeast Asia, and eastern Australia are lashed by rains, but southwestern USA generally experiences higher temperatures and low rainfall. Meanwhile, northwestern states of the USA experience colder, snowier winters. La Niña is also associated with an increase in Atlantic hurricane activity. Overall, the effects of a La Niña event often tend to be strongest during northern-hemisphere winters.

FOREST FIRES

In September 1988, during a La Niña event, forest fires raged in the western USA. This fire in Yellowstone National Park produced winds of 90kph (55mph).

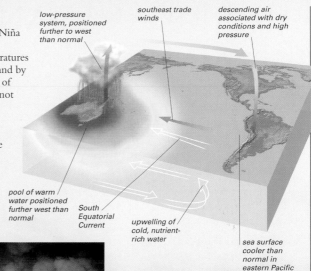

low-pressure system, positioned further to west than normal

southeast trade winds

descending air associated with dry conditions and high pressure

pool of warm water positioned further west than normal

South Equatorial Current

upwelling of cold, nutrient-rich water

sea surface cooler than normal in eastern Pacific

LA NIÑA PATTERN

During a La Niña event, the area of low pressure in the western Pacific is further west than normal, and the pool of warm surface water is also pushed west. Unusually cold surface temperatures develop in the eastern Pacific as the cold Peru Current strengthens off South America.

PERUVIAN FLOOD

Three men help a woman trying to cross the overflowing river Nepeña, near Chimbote, Peru, during the 1997–98 El Niño event. This El Niño ravaged Peru, causing $3.5 billion of damage.

HURRICANES AND TYPHOONS

HURRICANES AND TYPHOONS ARE NAMES USED IN different parts of the world for very similar weather phenomena. They are characterized by violent winds moving in a circular pattern over the ocean, dense bands of clouds, and rainfall. In the Atlantic they are known as hurricanes; those in the west Pacific are called typhoons. Similar phenomena elsewhere are called severe storms or cyclones. They start as a low-pressure system (depression) over warm oceans in the tropics, between latitudes 5° and 20°, and occur mainly in late summer.

KEY
→ hurricanes
→ severe cyclones
→ typhoons

DEVELOPMENT

All tropical cyclones develop from the effects of the Sun warming the surface of a broad area of ocean, and the air above it. This heating causes masses of warm, moist air to rise, creating a region of low pressure at the surface, and dense clouds above it. The low pressure sucks in more air, which spirals to the centre, creating a circular wind system. As it grows stronger, becoming a tropical storm, it is pushed westward by the prevailing trade winds. In the Atlantic, a storm attains hurricane status once its winds exceed 119kph (74mph). Eventually, most of these violent storms move away from the Equator – that is, to the north in the northern hemisphere. When one reaches land, it begins to lose energy, as it is no longer fed by heat from the ocean.

DISTRIBUTION
Severe tropical cyclones start as depressions over warm oceans in the tropics. They move across the ocean surface for several days, causing huge damage on reaching land. Their paths are shown in the map above.

THREE DEVELOPMENT STAGES

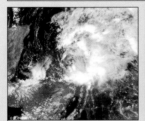 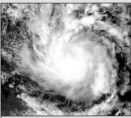 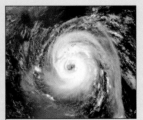

1 This swirling mass of warm, moist air rises over a tropical area of ocean and condenses to form clouds.

2 On reaching the tropical storm stage, a cyclone has a spiral form, with a dense central nucleus of clouds.

3 When at full typhoon or hurricane status, a cyclone has compacted and developed a clear central "eye".

STRUCTURE

A fully developed typhoon or hurricane is usually 300–600km (185–370 miles) in diameter and 10–15km (6–9 miles) high. At its centre is a calm region of low atmospheric pressure, called the eye. Within the rest of the cyclone, winds spiral in an anticlockwise direction in the northern hemisphere and clockwise in the southern hemisphere (the difference is due to the Coriolis effect, see p. 54). Within an area surrounding the eye, called the eyewall, the air spins upwards, forming dense clouds. The eye stays calm because the winds that spiral in towards it never reach the centre. Radiating out from the eye and eyewall are well-defined bands of clouds, called rainbands.

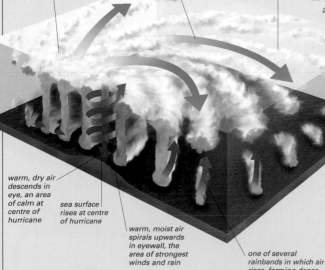

cap of cirrus clouds over cumulonimbus that forms bulk of clouds

high-level winds spiral outwards

spiral rainbands can extend for hundreds of kilometres from hurricane centre

warm, dry air descends in eye, an area of calm at centre of hurricane

sea surface rises at centre of hurricane

warm, moist air spirals upwards in eyewall, the area of strongest winds and rain

one of several rainbands in which air rises, forming dense cloud from which heavy rain pours

HURRICANE STRUCTURE
A hurricane consists of the central eye, which can be 8–200km (5–120 miles) across, the eyewall (a column of thick clouds, rain, and upward-spiralling winds), and the rainbands.

STORM CHASERS

The United States' National Oceanic and Atmospheric Administration (NOAA) monitors Atlantic hurricanes using specially equipped aircraft. They fly into hurricanes to drop instrument packages, which radio back data.

LOCKHEED WP-3D ORION
This turboprop aircraft, equipped with a sophisticated array of instruments, is one of those used in hurricane study.

HURRICANE CATEGORIES

A classification system called the Saffir–Simpson Scale divides hurricanes into five categories. It is used to estimate the damage and flooding to be expected along a coast impacted by the hurricane. Wind speed is the determining factor in the scale.

CATEGORY	WIND SPEED	HEIGHT OF SURGE
Tropical Storm	63–118kph (39–73mph)	1–1.5m (3–5ft)
Category 1 hurricane	119–153kph (74–95mph)	1.5–2m (5–6ft)
Category 2 hurricane	154–177kph (96–110mph)	2–3.5m (6–12ft)
Category 3 hurricane	178–209kph (111–130mph)	3.5–4.5m (12–15ft)
Category 4 hurricane	210–249kph (131–155mph)	4.5–6m (15–20ft)
Category 5 hurricane	over 249kph (155mph)	6–7.5m (20–25ft)

COASTAL EFFECTS

As it moves across the ocean, the low-pressure eye of a tropical cyclone sucks seawater up into a mound, which can be up to 3.5m (12ft) above sea level for a Category 2 hurricane or 7.5m (25ft) for a Category 5. When the cyclone hits land, the water in this mound surges over the coast in what is known as a storm surge. The surge may flood homes, wash boats inland, destroy roads and bridges, and seriously erode a section of coastline up to 150km (95 miles) wide. These effects compound the devastation caused by high winds, which can topple unstable buildings, uproot trees, damage coastal mangroves, and bring down power lines. Human deaths are not uncommon, so coastal areas threatened by a severe cyclone are normally evacuated in advance. Offshore, the water movements associated with a storm surge can devastate coral reefs. In the Caribbean, branching corals that live near the surface, such as elkhorn corals, are particularly vulnerable. Healthy reefs can recover from such damage, although it can take 10–50 years depending on the extent of injury.

CORAL DAMAGE
This colony of elkhorn coral was smashed by Hurricane Gilbert on Mexico's Caribbean coast in 1988.

DENSE CLOUDS
A swimmer watches heavy clouds associated with Hurricane Frances pass Cuba on their way towards Florida, 140km (85 miles) to the north, in 2004.

WATERSPOUT
Waterspouts are tornados (narrow, whirling masses of air) over the sea. They are quite commonly spawned around the edges of tropical cyclones.

INTRODUCTION

WATER WORLD
Two days after the passage of Hurricane Katrina, about 80 per cent of New Orleans was underwater, the result of breaches in the historic city's floodwater defences.

HURRICANE KATRINA

Hurricane Katrina, which hit the US Gulf Coast at the end of August 2005, was the deadliest hurricane to have affected the USA since 1928 and was by far the costliest hurricane in history. When it occurred, Katrina was also the fourth-strongest Atlantic hurricane ever recorded. By April 2006, the official death toll from Katrina had reached 1,605 with many hundreds of people still listed as missing. Estimates of the amount of property damage had reached $75 billion.

In addition to the deaths it caused, the hurricane brought extensive flooding to New Orleans. The city's vulnerability to flooding is well established. Much of the city lies at a lower level than the surface of Lake Pontchartrain, to its north, and these parts of the city are protected from the lake and from the Mississippi River, which runs through its centre, by reinforced embankments called levees. It had long been suspected that the levees would not stand up to the type of storm surge caused by a powerful hurricane such as Katrina, and this proved to be the case. When the storm struck, several of the levees failed. As the floodwaters receded, a massive clean-up was required.

Some climate scientists believe that global warming will increase the frequency of high-intensity tropical cyclones like Katrina. If this is the case, governments in affected countries will need to make provision for dealing with similar-scale catastrophes at regular intervals in future.

TRACK OF THE HURRICANE

The map below shows Katrina's path between 23 August 2005, when the storm first developed as a depression over the Bahamas, to 30 August, when it was downgraded to a tropical storm as it moved northward through the USA. Katrina crossed Florida on 25 August and hit the Gulf Coast on 29 August.

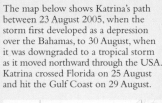

These satellite images show Katrina's appearance as it began to pass over Florida (top) and as it approached the US Gulf Coast and New Orleans (bottom).

USA

Tropical storm

Category 1

Category 3

Category 4

New Orleans

Gulf of Mexico ⊙ Tampa

Category 5

Category 4

Category 3

Category 1

Category 2

Tropical storm

Tropical depression

CUBA

PATH OF DESTRUCTION

LANDFALL IN FLORIDA

FORT LAUDERDALE, FLORIDA *At 6:30pm on 25 August 2005, Katrina made landfall on Florida's east coast, just north of Miami. With winds of 128kph (80mph), the coast was lashed by heavy rain and sandstorms. The hurricane caused two deaths, from falling trees, in the coastal city of Fort Lauderdale. Over the next 12 hours, the Florida death toll rose to 14.*

FLOODING OF NEW ORLEANS

BREACHED LEVEE *Most of the flooding in New Orleans stemmed from breaches in the levees that protect the city from nearby Lake Pontchartrain.*

RESCUING RESIDENTS *Three days after the hurricane struck, local volunteer teams were rescuing scores of stranded New Orleans residents by boat.*

COASTAL DEVASTATION

SMASHED HOMES *Some of the worst of the hurricane damage occurred in the city of Biloxi, on the coast of Mississippi. The high winds and a storm surge about 10m (33ft) high – the highest ever along this coastline – devastated homes, restaurants, beach-front casinos, and shrimp-fishing businesses. A total of 238 people died in Mississippi as a result of Hurricane Katrina.*

REPAIRING THE DAMAGE

STOPGAP REPAIR *Twelve days after Hurricane Katrina struck, helicopters were still patching up New Orleans' breached levees by dropping sandbags into them. Some of these levees were breached again, within a fortnight, when an even more powerful hurricane, Rita, hit the Gulf Coast of America.*

WAVES AND TIDES are two important physical phenomena that affect every area of the oceans but tend to be most noticeable, and have their main effects, on or near coasts. Ocean waves are mostly wind-generated and vary from tiny coastal ripples, to the regular, rolling swell of the open ocean, to monster breakers on world-famous surfing beaches. All waves transmit energy – when the waves reach land, this energy may be dissipated destructively, eroding coastlines, or constructively, building up features such as beaches. Tides are caused mainly by interactions between the Moon and Earth. As well as regular rises and falls in sea level, they can cause strong currents around coasts and, in some places, even more dramatic phenomena such as whirlpools and eddies.

WAVES AND TIDES

LAPPING WAVES
Waves lapping on the seashore meet a rocky stream at low tide near Kipahulu on the southeast coast of the Hawaiian island of Maui, in the Pacific.

OCEAN WAVES

WAVES ARE DISTURBANCES in the ocean that transmit energy from one place to another. The most familiar types of waves – the ones that cause boats to bob up and down on the open sea and dissipate as breakers on beaches – are generated by wind on the ocean surface. Other wave types include tsunamis, which are often caused by underwater earthquakes (see p.51), and internal waves, which travel underwater between water masses. Tides (see p.78) are also a type of wave.

WAVE PROPERTIES

A group of waves consists of several crests separated by troughs. The height of the waves is called the amplitude, the distance between successive wave crests is known as the wavelength, and the time between successive wave crests is the period. Waves are classified into types based on their periods. They range from ripples, which have periods of less than 0.5 seconds, up to tsunamis and tides, whose periods are measured in minutes and hours (their wavelengths range from hundreds to thousands of kilometres).

In between these extremes are chop and swell – the most familiar types of surface wave. Ocean waves behave like light rays: they are reflected or refracted by obstacles they encounter, such as islands. When different wave groups meet, they interfere – adding to, or cancelling, each other.

PARTICLE MOVEMENT
As waves pass over the surface, the particles of water do not move forward with the waves. Instead, they gyrate in little circles or loops. Underwater, the particles move in ever-smaller loops. At a depth below about half the distance between crests, they are quite still.

ROGUE WAVES
Interference between two or more large waves occasionally causes a giant or "rogue" wave. This one, recorded in the Atlantic Ocean in 1986, had an estimated height of 17m (56ft). It broke over the ship pictured, bending its foremast back by 20°.

WAVE GENERATION

Wind energy is imparted to the sea surface through friction and pressure, causing waves. As the wind gains strength, the surface develops gradually from flat and smooth through growing levels of roughness. First, ripples form, then larger waves, called chop. The waves continue to build, their maximum size depending on three factors: wind speed, wind duration, and the area over which the wind is blowing, called the fetch. When waves are as large as they can get under the current conditions of wind speed and size of fetch, the sea surface is said to be "fully developed". The overall state of a sea surface can be summarized by the significant wave height – defined as the average height of the highest one-third of the waves. For example, in a fully developed sea produced by winds of about 40kph (25mph), the significant wave height is typically about 2.5m (8ft).

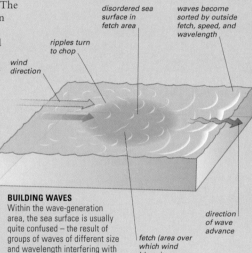

BUILDING WAVES
Within the wave-generation area, the sea surface is usually quite confused – the result of groups of waves of different size and wavelength interfering with each other. Outside this area, the waves become sorted by speed to produce a more regular pattern, called a swell.

CAPILLARY WAVES (RIPPLES)
These tiny waves are just a few millimetres high and have a wavelength of under 4cm (1½in).

CHOPPY SEA
In a choppy sea, the waves are 10–50cm (4–20in) high and have a wavelength of 3–12m (10–40ft).

FULLY DEVELOPED ROUGH SEA
Wind speeds over 60kph (40mph) can generate very rough seas with waves several metres high.

WAVE PROPAGATION

In the fetch, many different groups of waves of varying wavelength are generated and interfere. As they disperse away from the fetch, the waves become more regularly sized and spaced. This is because the speed of a wave in open water is closely related to its wavelength. The different groups of waves move at different speeds and so are naturally sorted by wavelength: the largest, fastest-moving waves at the fore, the smaller, slower-moving ones behind. This produces a regular wave pattern, or swell. Occasionally, groups of waves from separate storms interfere to produce unusually large "rogue" waves. As they propagate across the open ocean, wind-generated waves maintain a constant speed, which is unaffected by depth until they reach shallow water. Only with waves of extremely long wavelength – tsunamis – is the speed of propagation affected by water depth.

SWELL
A swell is a series of large, evenly spaced waves, often observed hundreds of kilometres away from the storm that spawned them. Wavelengths range from tens to hundreds of metres.

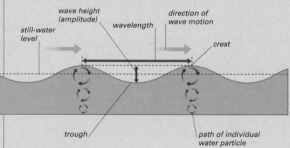

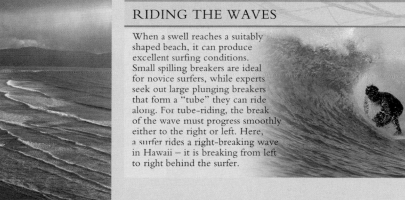

PLUNGING BREAKER
"Barrel" or "tube-forming" breakers like this occur when the waves reaching shore have large amounts of energy. The sea bed must be firm and quite steep.

ARRIVAL ON SHORE

As waves approach a shore, the motion they generate at depth begins to interact with the sea floor. This slows the waves down and causes the crests in a series of waves to bunch up – an effect called shoaling. The period of the waves does not change, but they gain height as the energy each contains is compressed into a shorter horizontal distance, and eventually break.

There are two main types of breaker. Spilling breakers occur on flatter shores: their crests break and cascade down the front as they draw near the shore, dissipating energy gradually. In a plunging breaker, which occurs on steeper shores, the crest curls and falls over the front of the advancing wave, and the whole wave then collapses at once. Waves can also refract as they reach a coastline. This concentrates wave energy onto headlands (see p.93) and shapes some types of beach (see p.106).

SHOALING AND BREAKING
Shoaling occurs as waves enter shallow water. The wave-length and speed both decrease, but the wave gains height. When the crest gets too steep, it curls and breaks.

water motion occurs offshore to depth of half the wavelength

wave shortens in length and decreases in speed, but increases in height

wave finally breaks

water motion caused by the wave begins to interact with the sea bed, and slow down

wave reaches critical ratio of wave height to wavelength and break

water carried up shore in swash zone

WAVE REFRACTION
When waves enter a bay enclosed by headlands, they are refracted (bent) as different parts of the wave-front encounter shallow water and slow down.

HUMAN IMPACT

RIDING THE WAVES

When a swell reaches a suitably shaped beach, it can produce excellent surfing conditions. Small spilling breakers are ideal for novice surfers, while experts seek out large plunging breakers that form a "tube" they can ride along. For tube-riding, the break of the wave must progress smoothly either to the right or left. Here, a surfer rides a right-breaking wave in Hawaii – it is breaking from left to right behind the surfer.

INTRODUCTION

TIDES

TIDES ARE REGULAR RISES AND FALLS in sea level, accompanied by horizontal flows of water, that are caused by gravitational interactions between the Moon, Sun, and Earth. They occur all over the world's oceans but are most noticeable near coasts. The basic daily pattern of high and low tides is caused by the Moon's influence on the Earth. Variations in the range between high and low tides over a monthly cycle are caused by the combined influence of the Sun and Moon.

HIGH AND LOW TIDES

Although the Moon is usually thought of as orbiting the Earth, in fact both bodies orbit around a common centre of mass – a point located inside the Earth. As the Earth and Moon move around this point, two forces are created at the Earth's surface: a gravitational pull towards the Moon, and an inertial or centrifugal force directed away from the Moon.

Earth

gravitational pull of Moon creates tidal bulge

Moon

inertial force creates second tidal bulge

centre of mass of Earth–Moon system

Earth's spin causes bulges to sweep over surface

bulges due to combined forces

These forces combine to produce two bulges in the Earth's oceans: one towards the Moon, and the other away from it. As the Earth spins on its axis, these bulges sweep over the planet's surface, producing high and low tides. The cycle repeats every 24 hours 50 minutes (one lunar day) rather than every 24 hours (one solar day), because during each cycle, the Moon moves round a little in its orbit.

DAILY TIDES
The two ocean bulges caused by the gravitational interaction between the Earth and Moon are shown (much exaggerated) here.

TIDAL PATTERNS

If no continents existed and the Moon orbited in the Earth's equatorial plane, the sweeping of the tidal bulges over the oceans would produce two equal daily rises and falls in sea level (a semidiurnal tide) everywhere on Earth. In practice, landmasses interfere with the movement of the tidal bulges, and the Moon's orbit tilts to the equatorial plane.

Consequently, many parts of the world experience tides that differ from the semi-diurnal pattern. A few have just one high and one low tide a day (called diurnal tides), and many experience high and low tides of unequal size (known as mixed semidiurnal tides). In addition, the tidal range, or difference in sea level between high and low water, varies considerably across the globe.

SEMIDIURNAL TIDE

DIURNAL TIDE

MIXED SEMIDIURNAL TIDE

KEY

diurnal

mixed

semidiurnal

------- small tidal range

········· medium tidal range

——— large tidal range

GLOBAL PATTERNS
This map shows the general pattern of tides (diurnal, semidiurnal, or mixed) and size of tidal range (average difference between high and low water) around the world.

ATLANTIC OCEAN

PACIFIC OCEAN

INDIAN OCEAN

SOUTHERN OCEAN

INTERTIDAL LIFE

Compared to permanently submerged plants and animals, organisms living in the intertidal zone have to cope with many extra stresses. They need to adapt, for instance, to the problem of becoming dried out (desiccated) when the tide is out. They may also have to endure extreme cold on frosty winter nights and even predation by land animals. Mussels, for example, often have to wait for hours between high tides to feed. At low tide, their shells close tightly to prevent desiccation and to protect against predators.

MONTHLY CYCLE

In addition to the daily cycle of high and low tides, there is a second, monthly, cycle. In this case, the Sun and Moon combine to drive the cycle. As with the Moon, the interaction between the Earth and Sun causes bulges in the Earth's oceans, though these are smaller than those caused by the Moon. Twice a month, at the times of new and full Moon, the Sun, Moon, and Earth are aligned, and the two sets of tidal bulges reinforce each other. The result is spring tides – high tides that are exceptionally high, and low tides that are exceptionally low. By contrast, at the times of first and last quarter Moon, the effects of the Sun and Moon partly cancel out, bringing tides with a smaller range, called neaps.

ALTERNATING SPRINGS AND NEAPS
Twice a month (top), the alignment of the Sun, Moon, and Earth creates spring tides. At other times (left), when the Sun and Moon lie at right angles, it creates neap tides. The alternation between springs and neaps can be seen in the 28-day tidal graph shown below.

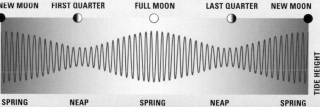

TIDAL CURRENTS

The vertical variation in sea level that occurs locally with tides can happen only through horizontal flows of water, called tidal currents. Over each daily tidal cycle, the currents generally run fastest about half-way between high and low tide at that location – at intermediate times they slow ("slack water") and then reverse direction.

The shape of a coast can have a crucial influence on current strength. Bottlenecks to water flow, such as narrow channels and promontories, are often associated with very powerful currents, called tidal races, that develop twice or four times a day. Where the flowing water meets underwater obstructions, phenomena such as whirlpools or vortices (spiralling, funnel-shaped disturbances), eddies (larger, flatter, circular currents), and standing waves may develop.

Other tide-related phenomena include tide rips – turbulence caused by converging currents – and overfalls, defined as a tidal current flowing opposite to the wind direction.

COOK STRAIT CURRENTS
These maps show the pattern of strong tidal currents in the Cook Strait, between the North and South islands of New Zealand, which occur twice a day, just over six hours apart. Water must funnel through a narrow channel in the Strait.

LOW TIDE AT BAMBURGH BEACH
Due to tides, large swathes of coast around the world are alternately covered and uncovered by the sea. This intertidal sandflat in Northumberland, England, has a tidal range averaging about 4m (13ft).

SURVIVING THE OLD SOW

The Old Sow has caused about a dozen fatalities from drowning over the past 200 years. Most of these involved mariners who strayed too close to the whirlpool in small rowing or sailing boats. In recent times, a few people in powered boats have had anxious experiences when their engines have stalled. Experienced mariners advise that if caught in a whirlpool, the priority is to keep the boat on an even keel and avoid getting swamped. Most objects floating in a stable position will eventually spin clear.

MINI-VORTEX
This mini-whirlpool, about 6m (20ft) wide and 50cm (16in) deep, would be called a "piglet" by experienced Old Sow watchers. Sometimes, several of these small vortices occur rather than a single large whirlpool.

The Old Sow Whirlpool

FEATURES
Tidal race, small whirlpools, occasional large whirlpool

TIMING
Four times daily

LOCATION Between Deer and Indian Islands, New Brunswick, Canada, and Moose Island, Maine, USA

Situated in Passamaquoddy Bay on the US–Canadian border, the Old Sow is one of the largest whirlpools in the world, and by far the largest in the Americas. Passamaquoddy Bay is at the lower end of the Bay of Fundy, which is famous for its strong tides. The Old Sow, when it appears, is located at a spot in the bay where various tidal streams flowing through the channels between

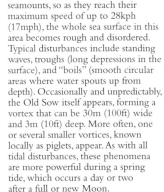

COASTAL SETTING
The Old Sow develops between Deer Island (top), Indian Island (right), and Moose Island (foreground).

different islands converge during the ebb tide or diverge during the flood tide. As they flow, these currents encounter numerous underwater obstructions, such as ledges and small seamounts, so as they reach their maximum speed of up to 28kph (17mph), the whole sea surface in this area becomes rough and disordered. Typical disturbances include standing waves, troughs (long depressions in the surface), and "boils" (smooth circular areas where water spouts up from depth). Occasionally and unpredictably, the Old Sow itself appears, forming a vortex that can be 30m (100ft) wide and 3m (10ft) deep. More often, one or several smaller vortices, known locally as piglets, appear. As with all tidal disturbances, these phenomena are more powerful during a spring tide, which occurs a day or two after a full or new Moon.

Lofoten Maelstrom

FEATURES
Tidal race and large, weak eddy

TIMING
Four times daily

LOCATION Between Lofoten Point and Mosken in the Lofoten Islands, off northwest Norway

Also known as the Moskenstraumen, the Lofoten Maelstrom is a complex pattern of sea-surface disturbances caused by tidal flows of water over a broad, submerged ledge of rock between two of the Lofoten Islands. These flows result from large sea-level differences that develop four times a day between the Norwegian Sea and the Vestfjord on the eastern side of the Lofoten Islands. The word maelstrom originates with the tidal phenomena in this area, and is derived from the Nordic word *male*, meaning "to grind". In Norse mythology, the Maelstrom was the result of a large salt-grinding millstone on the floor of the Norwegian Sea, which sucked water into its central hole as it turned. First described by the Greek explorer Pytheas in the 3rd century BC, the Lofoten Maelstrom is marked on many historical charts as an enormous and fearsome whirlpool. In 1997, a detailed study of tidal currents in the vicinity of the island of Mosken found that the reality is somewhat different. Although some strong tidal currents were measured, no obvious large whirlpool, with a vortex, was detected. Instead, the researchers found a weak eddy, about 6km (4 miles) in diameter, to the north of Mosken. This eddy develops twice a day during the flood tide, when it moves in a clockwise direction, and twice on the ebb tide, when it moves slightly further north and has an anticlockwise movement.

JULES VERNE

The French novelist Jules Verne (1828–1905) made reference to the Lofoten Maelstrom in his tale of undersea exploration, *Twenty Thousand Leagues Under the Sea*. At the end of the novel, Captain Nemo and his submarine, *Nautilus*, are sucked down into the whirlpool, "whose power of attraction extended to a distance of twelve miles", suffering an unknown fate.

TIDAL DISTURBANCE
For centuries, the Lofoten Maelstrom had a reputation as one of the world's most powerful tidal phenomena.

ATLANTIC OCEAN NORTHEAST

Saltstraumen

FEATURES
Tidal race and small whirlpools

TIMING
Four times daily

LOCATION Between Saltenfiord and Skjerstadfiord, northwest coast of Norway

The Saltstraumen tidal race occurs on the northwest coast of Norway and is generally acknowledged to be the strongest and most extreme tidal current in the world. It forms at a bottleneck between the Saltenfiord, an inlet from the Norwegian Sea, and the neighbouring Skjerstadfiord: its driving force is a difference in sea level of up to 3m (10ft) that develops four times a day between the two bodies of water. The channel at the centre of the bottleneck – Saltstraumen itself – is a 3km- (2-mile-) long strait between two headlands, with a width of just 150m (500ft) and a depth that varies from 20 to 100m (65 to 330ft). Twice a day, some 400 billion litres (88 billion gallons) of water roar through this strait on the flood tide, reaching maximum speeds of up to 40kph (25mph), as tidal forces act to fill the 50km- (30-mile-) long Skjerstadfiord. Twice a day, the waters flow out again through the same channel. The flows of water, and associated whirlpools, are equally strong during the ebb as the flood tide. Despite Saltstraumen's ferocity, the channel is regularly used by shipping. For short periods every day, the tidal flows slow almost to a halt, allowing large vessels to pass safely into and out of Skjerstadfiord. Smaller vessels do remain at risk from residual underwater currents during these periods of "slack water", but many experienced pilots still venture out. Saltstraumen offers both interesting opportunities for divers and excellent angling (see panel, below). Incoming tides carry large amounts of plankton through the channel, and fish of various sizes follow.

DANGEROUS WATERS
When the tidal race flows, the spin-off vortices, which can be 10m (33ft) across, are capable of pulling objects down to the rocky bottom of the channel.

DISCOVERY

LIFE BENEATH THE WHIRLPOOLS

It is possible to dive into and explore the Saltstraumen, although this can safely be attempted only when the tidal streams are at a minimum. Divers have discovered a rich and colourful marine life at the bottom of the channel, dominated by long strands of kelp and a variety of invertebrates, as well as fish such as lumpsuckers, coley, and Wolf-fish.

TEEMING WITH LIFE
Invertebrate life at the bottom of the channel includes colourful sponges and anemones.

ATLANTIC OCEAN NORTHEAST

Corryvreckan Whirlpool

FEATURES
Tidal race, standing waves, and whirlpools

TIMING
Twice daily

LOCATION Between the islands of Jura and Scarba, west coast of Scotland, UK

The most famous tidal phenomenon in the British Isles can be found in the Gulf of Corryvreckan. Twice a day on the flood tide, strong Atlantic currents and unusual underwater topography conspire to produce an intense tidal race. As the tide enters the narrow bottleneck at Corryvreckan, currents of up to 22kph (14mph) develop. Underwater, these currents encounter a variety of irregular features on the sea bed, including a conical obstruction known as the Pinnacle, which rises to within 30m (95ft) of the surface. The steep east face of this obstruction forces a plume of water to the surface, producing whirlpools and standing waves up to 4m (13ft) high, and the roar of the rushing water can be heard up to 5km (3 miles) away. Classified as non-navigable by the Royal Navy, the Corryvreckan has caused numerous emergencies and sinkings.

SPIN-OFF VORTEX
In the whirlpool area, massive upthrusts of water occur in pulses, producing vortices that spin away with the tidal flow.

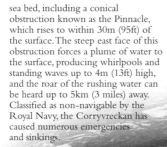

DISTURBED SEA
An area of disturbance begins to develop in the channel north of Jura, seen here with the island of Scarba lying behind it.

ATLANTIC OCEAN NORTHEAST

Slough-na-more Tidal Race

FEATURES
Tidal race with eddies and standing waves

TIMING
Four times daily

LOCATION Between Rathlin Island and Ballycastle Bay, County Antrim, Northern Ireland, UK

The Slough-na-more Tidal Race results from strong tidal flows of billions of litres of seawater between the Atlantic Ocean and the Irish Sea, via a narrow channel. During spring tides, the tidal stream can attain a speed of 13kph (8mph). Where it passes Rathlin Island, a complex of fast-moving currents, eddies, and standing waves is created. In contrast, the same sea area is usually calm during other phases of the tidal cycle. In 1915, the strength of the Slough-na-more Tidal Race forced the Irish steam coaster SS *Glentow* aground on the Irish coast, and the ship later broke up.

ATLANTIC OCEAN NORTHEAST

Needles Overfalls

FEATURES
Tidal race and overfalls

TIMING
Four times daily

LOCATION Needles Channel, northwestern coast of the Isle of Wight, England, UK

The Needles Channel is a 7km- (5-mile-) long stretch of water between a line of chalk sea stacks on one side (the Needles) and an underwater reef on the other. This stretch of water is affected by short, breaking waves (overfalls) at the time of the maximum ebb or flood tide. If the wind is blowing in the opposite direction to the tidal stream, these overfalls are greatly exacerbated, producing an extremely rough sea.

ATLANTIC OCEAN EAST

Garofalo Whirlpool

FEATURES
Tidal race, small whirlpools, and overfalls

TIMING
Four times daily

LOCATION Strait of Messina, between the northeast coast of Sicily and the south coast of Italy

The Strait of Messina separates the "toe" of Italy from the Mediterranean island of Sicily. It varies in width from 3 to 16km (2 to 10 miles) and is the site of numerous complex currents and small whirlpools that vary over the tidal cycle and hamper navigation through the Strait. In Italy, the small whirlpools that form are called *garofali*, but in the English-speaking world, the whole system of tidal disturbances is known as the Garofalo Whirlpool.

PACIFIC OCEAN NORTHEAST

Yellow Bluff Tide Rip

FEATURES
Tide rip, standing waves, and eddies

TIMING
Twice daily

LOCATION San Francisco Bay, California, USA

A tide rip is a stretch of rough, turbulent water caused by a tidal current converging with, or flowing across, another current. Thus it differs from a tidal race, which occurs where a tidal stream of water accelerates through a narrow opening in a coast. An example of a tide rip occurs at a place called Yellow Bluff in San Francisco Bay, not far from the bay's entrance, the famous Golden Gate.

Four times a day, strong movements of water occur through the Golden Gate – twice flowing into the bay on the flood tide and twice flowing out on the ebb tide. These currents can reach a speed of up to 8kph (5mph) during spring tides. Inside the bay, the pattern of currents becomes more complex, as they either split (during the flood tide) or converge (during the ebb tide) from different parts of the bay. The currents are also modified by the varying depth of the water around the shoreline, by the shoreline's shape, and by subsurface obstructions. At Yellow Bluff, disturbances to the sea surface are most noticeable during the ebb tide, when the tidal streams are converging, and are characterized by such phenomena as extremely rough, fast-moving water, standing waves, and eddies. The spot is popular with extreme kayakers, who challenge themselves against the strong currents and surf on the standing waves.

PACIFIC OCEAN NORTHEAST

Skookumchuck Narrows Tidal Race

FEATURES Tidal race, small whirlpools, and standing wave on flood tide

TIMING Four times daily; flood tide twice daily

LOCATION Skookumchuck Narrows, British Columbia, Canada

One of the world's most famous tidal races occurs at the Skookumchuck Narrows on British Columbia's Sunshine Coast, not far from Vancouver (*Skookum* is a Native American word for "strong" and *chuck* means "water"). Four times a day, there is a strong tidal rush of water through this 300m- (1,000ft-) wide channel, which connects two inlets into the coast – the Sechelt and Jervis inlets. A 3m (10ft) difference in sea level between low and high tide causes some 350 billion litres (77 billion gallons) of seawater to rush through the gap, creating turbulence and some small whirlpools. On the flood tide, when water is flowing into the Sechelt Inlet (but not the ebb tide, when it flows out), the tidal stream across an outcrop of bedrock in the channel creates a large standing wave – a mound of breaking water that remains stationary at a particular spot on the surface. At its peak, the flow rate is about 18 million litres (4 million gallons) per second, and current velocities can reach 28kph (19mph).

POWERFUL RAPIDS
Here, water is flowing right to left, from the Sechelt Inlet into Jervis Inlet. Six hours later, it flows back in the opposite direction.

HUMAN IMPACT

SURF-KAYAKING

Skookumchuck Narrows is a popular destination for enthusiasts of extreme surf-kayaking. The standing wave that arises there is up to 2.5m (8ft) high and 7m (23ft) wide and is regarded as one of the world's great white-water kayaking locations. When surf-kayaking, the object is to stay in the wave as long as possible, which requires strength and skill.

TROUBLED WATERS
A walkway hanging beneath the Oonaruto Bridge, which spans the Naruto Straits, provides an excellent view of the whirlpools below.

Naruto Whirlpool

FEATURES
Tidal race and whirlpools

TIMING
Four times daily

LOCATION Naruto Strait, between the islands of Shikoku and Awaji, Japan

The Naruto is a spectacular system of whirlpools that develops four times a day in a narrow channel separating the island of Shikoku (one of Japan's main islands) from Awaji Island, a much smaller island lying off Shikoku's northeastern coast. The channel, called the Naruto Strait, is one of several that join the Pacific Ocean to the Inland Sea, which is a large body of water lying between Shikoku and Japan's largest island, Honshu. Four times a day, billions of litres of water move into and out of the Inland Sea through this channel, generated by tidal variations in sea level between the Inland Sea and the Pacific Ocean of up to 1.5m (5ft).

The tidal flows can reach speeds of up to 15kph (9mph) during spring tides (that is, twice a month, around the time of a full or new Moon). They create vortices up to 20m (65ft) in diameter where they encounter a submarine ridge. These vortices are not stationary but tend to move with the current, persisting for 30 seconds or more before disappearing. The whirlpools can be viewed from Awaji Island, from sightseeing boats that regularly negotiate the rapids, or from a 1.3km- (⁴⁄₅-mile-) long bridge that spans the Naruto Strait.

HUMAN IMPACT

ARTISTIC INSPIRATION

The Naruto Whirlpool has existed since ancient times. It is mentioned many times in Japanese poetry and is possibly the only tidal phenomenon to feature in a well-known piece of art, namely *Whirlpool and Waves at Naruto, Awa Province*, by the 19th-century Japanese artist Utagawa Hiroshige (a fragment is shown below).

OCEAN ENVIRONMENTS

COASTS INCLUDE SOME of the most
beautiful, but also some of the most rapidly
changing, places on Earth. There are a great
many forms that they can take – from cliffs
composed of anything from limestone to
lava, to beaches, spits, and barrier islands,
river deltas, estuaries, and tidal flats.
Each coast has its own unique history of
formation, brought about by processes
such as land rise and fall, sea-level change,
glacial and volcanic action, and marine
erosion and deposition. On and around
these coasts, a variety of habitat types –
ranging from sandy coastal dunes to salt
marshes, coastal lagoons, and, in the tropics,
mangrove swamps – are shaped by the
interaction of tidal flows, breaking waves,
discharge of river sediment, and a variety of
biological and human-induced processes.

COASTS AND THE SEASHORE

SANDSTONE COAST
This dramatic Australian coastline consists of
eroded sandstone strata beautifully shaped
and sculpted by an azure sea.

COASTS AND SEA-LEVEL CHANGE

A COAST IS A ZONE WHERE THE LAND MEETS THE SEA — it extends from the shoreline inland to the first significant terrain change. Coastlines constantly alter in response to sea-level change, land-based processes, wave action, and tides. There are several different types of coasts including drowned and emergent coasts. The processes that shape each are to some extent influenced by sea-level change, the causes of which may be global (as in oceanic melting of the world's ice sheets) or local (involving local upward or downward movement of land causing local sea-level rise).

UPLIFTED TERRACE
This coastal region of New Zealand has experienced a localized sea-level fall in the recent geological past, as the land was significantly raised by an earthquake. What was beach is now flat cliff-top.

GLOBAL SEA-LEVEL CHANGE

The most important cause of a global change in sea level is an increase or decrease in the extent of the world's ice sheets and glaciers. This is related to Earth's climate. If it cools, more water becomes locked up as ice, so there is less in the oceans. If it heats up (global warming), the ice melts and increases the volume of ocean water. Another cause of global sea-level change, which is also affected by climate, is a rise or fall in ocean temperature. Warming lowers the density of water, so if the upper layers of the oceans heat up, they expand and increase the total volume of the oceans. Any changes in the size of the ocean basins, the ocean's containers, also impact globally on sea levels. For example, a change in activity at mid-ocean ridges can have such an effect and may be important in driving long-term sea-level change.

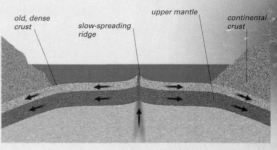

old, dense crust / slow-spreading ridge / upper mantle / continental crust

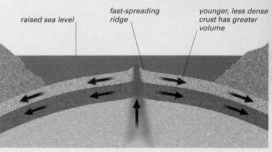

raised sea level / fast-spreading ridge / younger, less dense crust has greater volume

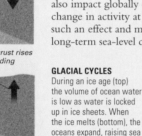

reduced ocean water / ice sheet / oceanic crust rises / continental crust depressed by ice

increased ocean water / continental crust rises due to unloading of ice

oceanic crust depressed

GLACIAL CYCLES
During an ice age (top) the volume of ocean water is low as water is locked up in ice sheets. When the ice melts (bottom), the oceans expand, raising sea levels globally.

OCEAN-BASIN CHANGE
A slow, global rise in sea level can occur when new crust is produced at a fast-spreading mid-ocean ridge. The relatively hot, buoyant new crust swells, pushing the ocean water upwards.

LOCAL SEA-LEVEL CHANGE

Local sea-level change occurs when a particular area of land rises or falls relative to the general sea level. One of the main causes is tectonic uplifting of land, which occurs in regions where oceanic crust is being forced beneath continental crust (a process often associated with earthquakes). Another cause is glacial rebound, which is a gradual rise of a specific area of land after an ice sheet that once weighed it down has melted.

During the last ice age, heavy ice sheets covered much of North America and Scandinavia. Since the ice melted, these regions have risen, and they continue to do so today at rates of up to a few centimetres a year. In contrast, other coastal areas are slowly sinking. Often, this occurs where a heavy load of coastal sediments is pushing the underlying bedrock down. A slow subsidence is occurring, for example, on the eastern coast of the USA. Many volcanic islands also start to subside soon after they form. This is due to the fact that the material from which they are created cools, compacts, and then contracts, while the sea floor under them warps downwards.

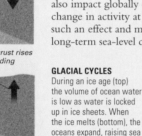

SINKING ISLANDS
These two volcanic Pacific islands, Rai'atea (top) and Bora-Bora, are subsiding. Locally, the current global rise in sea level is therefore slightly exacerbated.

DROWNED COASTS

A drowned (or submergent) coast is the result of a global or regional sea-level rise. There are two distinctive types — rias and fiords. In a ria coast, the sea-level rise has drowned a region of coastal river valleys, forming a series of wide estuaries, often separated by long peninsulas. In a fiord coast, the sea-level rise has drowned one or more deep, glacier-carved valleys. Both types are characteristically irregular and indented. Due to a significant global rise in sea level over the past 18,000 years, drowned coasts are common worldwide. Ria coasts are particularly prevalent in northwestern Europe, the eastern USA, and in Australasia. Large numbers of fiords are present in coastal Norway, Chile, Canada, and New Zealand.

RIA COAST
The coastline around Hobart, in Tasmania, Australia, was formed by a rise in sea level flooding a series of river valleys. Here, the Hobart Bridge spans one such drowned valley.

EMERGENT COASTS

Emergent coasts occur where land has uplifted faster than the sea has risen since the last ice age. The causes are either activity at the edge of a tectonic plate or glacial rebound. On emergent coasts, areas that were formerly sea floor may become exposed above the shoreline, while former beaches often end up well behind the shoreline, or even on cliff-tops. Sometimes, staircase-like structures called marine terraces are created by a combination of uplift and waves gradually cutting flat platforms at the bases of cliffs (wave-cut platforms). Emergent coasts are typically rocky, but sometimes they have a smooth shoreline. Examples of these coasts occur on the USA's Pacific Coast, in Scotland, Scandinavia, New Zealand, and Papua New Guinea.

PAST CHANGE

Scientists study past sea-level changes by examining rocks and fossils near shorelines. They also analyze ocean sediments to calculate past ocean temperatures and climatic properties. Over the past 500 million years, global sea levels have fluctuated by more than 300m (1,000ft). About 120,000 years ago, sea level was about 6–10m (20–35ft) higher than it is today, but some 18,000 years ago, it was about 120m (400ft) below today's level. Most of the rise since then occurred prior to 6,000 years ago. From some 3,000 years ago to the late 19th century, sea level rose at about 0.1–0.2mm (1/254–1/127in) per year. Since 1900, this has increased to 1–3mm (1/25–3/25in) per year.

FOSSIL MAMMOTH TOOTH

RAISED BEACH
In this bay in the Hebridean Islands, Scotland (left), the green areas behind the beach are former beaches that have been raised by glacial rebound since the end of the last ice age.

UPLIFTED CLIFFS
These marine cliffs in Crete, Greece (right), have been uplifted by tectonic activity, eroded, and finally tipped from the horizontal, also by tectonic activity.

THEN AND NOW
The red dotted line on this map shows where the east coast of North America was 15,000 years ago. At that time, mammoths roamed on what is now continental shelf – it is not uncommon for a mammoth tooth (above) to turn up in fishing trawls from these areas.

New York
Washington D.C.
NORTH AMERICA
ATLANTIC OCEAN
Miami

OCEAN ENVIRONMENTS

SURROUNDED BY WATER
The Italian city of Venice currently floods up to 200 times a year and is severely threatened by future sea-level rise, although a project to build a tidal barrier was launched in 2003.

GLOBAL WARMING AND SEA-LEVEL RISE

The measurement of global sea-level change is complex and, until satellite-based techniques were introduced in 1992, was somewhat imprecise. During the 1980s, a consensus emerged that sea level had been rising at 1–3mm (¹/₃₂–¹/₈ in) per year since 1900, whereas the new satellite techniques indicate a current average rise of 2.9mm (¹/₈ in) per year. Since 1900, there has also been a rise in the temperature of Earth's atmosphere and oceans (global warming) of 0.4–0.8°C (0.7–1.4°F). There are two plausible mechanisms by which the temperature rise might be linked to the sea-level rise. First, through melting of glaciers and ice sheets, which increases the amount of water in the oceans; and second, through the expansion of seawater as it warms. As there are no other convincing explanations of what might be causing the sea-level rise, the view of most scientists is that global warming is the cause.

Based on different models of the future course of global warming (which most scientists now believe is linked to human activity), it is possible to make various predictions of how sea level will change in the future. For example, the Intergovernmental Panel on Climate Change predicts that, by the end of the 21st century, there will be a further sea-level rise of 110–880mm (4½–35in). This rise will displace tens of millions of people living in low-lying coastal areas and have a devastating effect on some small island nations. Continued global warming will eventually melt the Greenland Ice Sheet, raising sea levels by a little over 6m (20ft), flooding most of the world's coastal cities.

SEA-LEVEL RISE IN SOUTHEASTERN USA

The maps below indicate the areas of the southeastern USA that would be threatened by sea-level rises of 1m (3ft), 2m (6½ft), and 6m (20ft). A 1m (3ft) rise is a little above the upper end of estimates for what can be expected this century. With this rise, parts of Florida and southern Louisiana would be inundated up to 30km (18 miles) from the present coastline. If the Greenland Ice Sheet were to melt, which would appear to be unlikely in this century but could happen within a few hundred years if global warming continues, a rise of a little over 6m (20ft) would submerge a large part of Florida, while Louisiana would be flooded as much as 80km (50 miles) in from the present coastline.

■ flooded area

1m (3ft) rise

2m (6½ ft) rise

6m (20ft) rise

EFFECTS OF GLOBAL WARMING

GLACIER RETREAT

PERUVIAN ANDES *Global warming is having a marked impact on the world's glaciers. The majority have shrunk since 1975, as their ice has melted faster than new ice has formed. These photographs from the same viewpoint show the extent of a glacier in the Cordillera Blanca, Peru, in 1980 (left) and 2002 (right).*

SUBMERGING ISLANDS

FUNAFUTI ATOLL *This atoll is part of Tuvalu, a group of small, low-lying Pacific islands whose future existence is threatened by sea-level rise.*

HIGH TIDE *Homes on Funafuti Atoll are already flooded by lagoon waters from time to time during exceptionally high tides.*

POPULATIONS AT RISK

CITY UNDER WATER *Dhaka, the capital of Bangladesh, together with about three-quarters of the country's land area, is less than 8m (27ft) above sea level. Much of the country would be flooded by melting of the Greenland Ice Sheet. A rise of 1.5m (5ft) would inundate one-sixth of Bangladesh and displace about 17 million people. The country is already severely affected by river floods.*

ANIMALS IN DANGER

STARVED TO DEATH *Polar Bears are one of the animal species most severely threatened by global warming. The bears use Arctic sea-ice as their summer hunting ground, and as the extent of sea-ice diminishes, so do their opportunities for hunting and feeding.*

OCEAN ENVIRONMENTS

COASTAL LANDSCAPES

A GREAT VARIETY OF LANDSCAPES ARE FOUND along the coastlines of the world's oceans. Coasts are shaped by processes such as sea-level change and wave erosion, as well as by land-based processes such as weathering, erosion and deposition by rivers, glacier advance and retreat, the flow of lava from volcanoes, and tectonic faulting. Some coastal features are made by living organisms, including the reefs built by corals and the harbours, coastal defences, and artificial islands built by humans.

CLASSIFICATION OF COASTS

Coasts can be classified as either primary or secondary. Primary coasts have formed as a result of land-based processes, such as the deposition of sediment from rivers (forming deltas), land erosion, volcanic action, or rifting and faulting in the Earth's crust. Coasts formed as a result of recent sea-level change, which include drowned coasts and emergent coasts (see pp.88–89), are also usually considered primary, as are coastlines consisting mainly of wind-deposited sand, glacial till, or the seaward ends of glaciers. Coasts are considered secondary if they have been heavily shaped by marine erosional or depositional processes, or by the activities of organisms, such as corals, mangroves, or, indeed, people. A few coasts – for example, emergent coasts that have undergone significant marine erosion – display both primary and secondary features and so fit into an intermediate category.

FRINGING REEF
This reef-fringed coast, around the south Pacific island of Bora Bora, is a secondary coast, as it has been modified by the activities of living organisms, notably corals.

ARTIFICIAL COAST
Singapore Harbour, in Southeast Asia, is an example of a coast that has been heavily shaped by human activity. Before human intervention, it was a mangrove-lined estuary.

VOLCANIC COAST
This land-eroded volcanic cone is in the Galapagos Islands. The entire coastline around these islands was formed by volcanic activity and so is a primary coast.

SEA ARCH
This spectacular arch in southern England is known as Durdle Door. A remnant of a once much larger headland, it is a classic feature of a marine-eroded coast.

WAVE-EROSION COASTS

Of all the different types of coastal landscape, perhaps the most familiar are wave-eroded cliffed coasts, a type of secondary coast. Wave erosion on these coasts occurs through two main mechanisms. First, waves hurl beach material against the cliffs, which abrades the rock. Second, each wave compresses air within cracks in the rocks, and on re-expansion the air shatters the rock. Where waves encounter headlands, refraction (bending) of the wave fronts tends to focus their erosive energy onto the headlands. At these headlands, distinctive features tend to develop in a classic sequence. First, deep notches and then sea caves form at the bases of cliffs on each side of the headland. Wave action gradually deepens and widens these caves until they cut through the headland to form an arch. Next, the roof of the arch collapses to leave an isolated rock pillar called a stack, and finally the stack is eroded down to a stump.

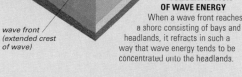

part of wave front opposite beach continues forward

beach

energy concentrated on headland as wave front refracts

erosion eventually divides headland into stacks

part of wave front opposite headland slows as it encounters shallower water

lobe of sediment

wave front (extended crest of wave)

CONCENTRATION OF WAVE ENERGY
When a wave front reaches a shore consisting of bays and headlands, it refracts in such a way that wave energy tends to be concentrated onto the headlands.

UNDERCUT CLIFF
Wave action has eroded a notch, and an adjoining platform, at the base of this cliff in the Caribbean.

SEA CAVE
This deep indentation and sea cave have been eroded into cliffs in the Algarve, Portugal.

SEA STACKS
The Old Harry Rocks are chalk sea stacks at a headland near Swanage in southern England.

MARINE-DEPOSITION COASTS

Marine depositional coasts are formed from sediment brought to a coast by rivers, eroded from headlands, or moved from offshore by waves. An important mechanism in their formation is longshore drift. When waves strike a shore obliquely, the movement of surf (swash) propels water and sediment up the shore at an angle, but backwash drags them back down at a right angle to the shore. Over time, water and sediment are moved along the shore. Where the water arrives at a lower-energy environment, the sediment settles and builds up to form various depositional features, including spits, baymouth bars, and barrier islands (long, thin islands parallel to the coast).

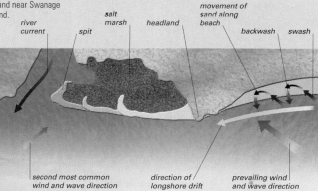

river current

spit

salt marsh

headland

movement of sand along beach

backwash

swash

second most common wind and wave direction

direction of longshore drift

prevailing wind and wave direction

SPIT FORMATION
On this coastline, sand and water is carried past the headland by longshore drift, but the sand settles at the mouth of an estuary where the waves are opposed by the sluggish outflow from a river. There it forms a slowly growing spit – a sandy peninsula with one end attached to the land.

BAYMOUTH BAR
Where a spit extends most or all of the way across the mouth of a bay or estuary, the result is called a baymouth bar. Here, a bar across the mouth of an estuary in Scotland, and an older spit, have created a sheltered coastal area of sandflats and salt marshes.

CLATSOP SPIT
This aerial view shows the impressive Clatsop Spit, at the mouth of the Columbia River, Oregon, USA. The spit extends for 4km (2½ miles) across the river mouth and is still growing.

Greenland Ice Coast

TYPE	Primary coast
FORMATION	Extension of ice-sheet to sea level in outlet glaciers
EXTENT	About 1,000km (620 miles)

LOCATION Parts of western and eastern coasts of Greenland

An ice coast forms where a glacier extends to the sea, so that a wall of ice is in direct contact with the water. This is a common feature around the highly indented margins of Greenland, mainly at the landward end of long fiords. Together, these ice walls form an interrupted ice coast, and they are the source of enormous numbers of icebergs, many of which escape the fiords and eventually reach the Atlantic. The ice coast extends along only a fraction of the total Greenland coastline, which is an astonishing 44,000km (27,500 miles) long.

ICE COAST NEAR CAPE YORK

Acadia Coastline

TYPE	Primary coast
FORMATION	Glaciation, then drowning by sea-level rise
EXTENT	66km (41 miles)

LOCATION Southeast of Bangor, Maine, northeastern USA

The coastline of Acadia in Maine is one of the most spectacular in the northeastern USA. It now forms the Acadia National Park, most of which is found within a single large island, Mount Desert Island, and some smaller associated islands. Sea-level rise since the last ice age has separated these islands from each other and from the mainland. The mountains that make up the basis of this coastline began to form 500 million years ago from seafloor sediments. Magma (molten rock) rising up from Earth's interior intruded into and consumed these sedimentary rocks, producing a mass of granite that was gradually eroded to form a ridge. About 2–3 million years ago, a huge ice sheet started to blanket the area, depressing the land and sculpting out a series of mountains

MOUNT DESERT ISLAND
The south-facing coast of Mount Desert Island consists of a series of fractured granitic steps that were produced by the action of glaciers some 100,000 years ago.

separated by U-shaped valleys. Since the ice sheet receded, the land has gradually rebounded upwards, but global sea-level rise has caused the Atlantic to overtake the rebound at a rate of 5cm (2in) per century. Today, waves and tidal currents are major agents of change at Acadia, gradually eroding the cliffs and depositing rock particles mixed with shell fragments at coves around the coastline.

STAIRWAY TO THE SEA
The tops of the columns form stepping stones that first lead up from the foot of the cliff to a mound and then progress downwards until they dip below the sea.

ATLANTIC OCEAN NORTHWEST

Hatteras Island

TYPE	Secondary coast
FORMATION	Deposition of sediment by waves and currents
EXTENT	112km (70 miles)

LOCATION Off the coast of North Carolina, northeastern USA

Hatteras Island is a classic barrier island of sandy composition. It runs parallel to the mainland and is long and narrow, with an average width of 450m (1,500ft), and it has been shaped by complex processes of deposition effected by ocean currents and waves. It is part of a series of barrier islands called the Outer Banks and has two distinct sections, which join at a promontory called Cape Hatteras. The dangerously turbulent waters in this area have resulted in hundreds of shipwrecks over the centuries.

HATTERAS SHORELINE
Hatteras Island is a typical barrier island, being low-lying with wave-straightened shorelines.

CAPE HATTERAS LIGHTHOUSE

Erosion and deposition often cause shorelines to migrate. In 1999, the Cape Hatteras Lighthouse was moved because the sea had begun to lap at its base, threatening its destruction.

NEW POSITION
The lighthouse is now located about 450m (1,500ft) back from the shoreline.

ATLANTIC OCEAN WEST

Les Pitons

TYPE	Primary coast
FORMATION	Volcanic lava-dome formation followed by volcano collapse and erosion
EXTENT	7km (4½ miles)

LOCATION Southwestern coast of St. Lucia, Lesser Antilles, eastern Caribbean

The southwestern coastline on the Caribbean island of St. Lucia is rocky, highly indented, and steeply shelving. A landmark here is Les Pitons ("The Peaks"), two steep-sided mountain spires, each more than 740m (2,430ft)

high. These are the eroded remnants of two lava domes (large masses of lava) that formed some 250,000 years ago on the flank of a huge volcano. The volcano later collapsed, leaving behind the peaks and other volcanic features in the area. The volcanic rocks on this coast are densely vegetated, except on the very steepest parts of Les Pitons themselves. Beneath the sea are some scattered coral reefs within a series of protected marine reserves. This region was declared a World Heritage Site in 2004.

TWIN PEAKS
In this view, Petit Piton is the nearer peak, while Gros Piton, which is slightly higher and much broader, is visible in the background.

ATLANTIC OCEAN NORTHEAST

Giant's Causeway

TYPE	Primary coast
FORMATION	Cooling of basaltic lava flow from an ancient volcanic eruption
EXTENT	1km (³⁄₅ mile)

LOCATION Northernmost point of County Antrim, Northern Ireland, UK

The Giant's Causeway is a tightly packed cluster of some 40,000 columns of basalt (a black volcanic rock). It is located at the foot of a sea cliff that rises 90m (300ft) on the northern coast of Northern Ireland. Although legend says the formation was created by a giant called Finn McCool, it in fact resulted from a volcanic eruption some 60 million years ago, one of a series that brought about the opening up of the North Atlantic. The eruption spewed up vast amounts of liquid basalt lava, which cooled to form the columns. They are up to 13m (42ft) tall and are mainly hexagonal, although some have four, five, seven, or eight sides.

ATLANTIC OCEAN NORTHEAST

Gruinard Bay

TYPE	Primary coast
FORMATION	Ice-sheet retreat and post-glacial rebound
EXTENT	13km (8 miles)

LOCATION West of Ullapool, northwestern Scotland, UK

Around Gruinard Bay in Scotland there is evidence of a phenomenon known as post-glacial rebound, in which a landmass, once pushed down by the huge weight of ice sheets during the last ice age, rises again. In some areas, such as Scotland and Scandinavia, this upward rebound has outstripped the sea-level rise caused by the ice sheets melting. At Gruinard Bay, rebound is indicated by its raised beaches – flat, grassy areas behind the present-day beaches. Over the last 11,000 years, this part of Scotland has been moving upwards relative to sea level at about 5cm (2in) per century.

RAISED BEACH
The green area beyond the present-day beach, well above the line of high tide, is the remnant of an ancient beach.

ATLANTIC OCEAN NORTHEAST

Devon Ria Coast

TYPE	Primary coast
FORMATION	Former river valleys drowned by sea-level rise
EXTENT	About 100km (60 miles)

LOCATION Between Plymouth and Torbay, southwestern coast of England, UK

Much of the south coast of the English county of Devon consists of the drowned valleys of the rivers Dart, Avon, Yealm, and Erme, and the Salcombe–Kingsbridge Estuary. The inlets, also known as rias, are separated by rugged cliffs and headlands. This beautiful coastal area was formed by the partial flooding of valleys, through which small rivers once flowed, as a result of global sea-level rise since the last ice age. The rise in sea level has been accentuated by the fact that the southern and eastern parts of Britain have been tipping downwards since the last ice age at a rate of up to 16cm (7in) per century, a trend that continues today.

SALCOMBE–KINGSBRIDGE ESTUARY
The highly scenic Salcombe Estuary is the largest of the five rias on the south Devon coast. Its protected waters provide ideal conditions for sailing.

ATLANTIC OCEAN NORTHEAST

White Cliffs of Dover

TYPE	Secondary coast
FORMATION	Marine erosion of a large mass of ancient chalk
EXTENT	17km (11 miles)

LOCATION Southeastern coast of England, to east and west of Dover, UK

One of England's most famous natural landmarks, the White Cliffs of Dover run along the northwestern side of the Strait of Dover, the narrowest part of the English Channel. They are complemented on the French side of the Strait by similar cliffs at Cap Blanc Nez. The chalk from which the cliffs are composed was formed between 100 million and 70 million years ago, when a large part of what is now northwestern Europe was underwater. The shells of tiny planktonic organisms that inhabited those seas gradually accumulated on the sea floor and became compressed into a layer of chalk that was several hundred metres thick. Subsequently, as the sea level fell during successive ice ages, this mass of chalk lay above the sea, and it later formed a land bridge between present-day England and France. However, about 8,500 years ago, the build-up of a large lake in an area now occupied by the southern North Sea caused a breach in the land-bridge. It eroded rapidly, causing flooding of the area that now forms the English Channel.

Today, the cliffs at Dover continue to be eroded at an average rate of a few centimetres per year. Occasionally a large chunk detaches from the cliff edge and falls to the ground. Many marine fossils have been discovered in the cliffs, ranging from sharks' teeth to sponges and corals.

HIGH CHALK CLIFFS
Up to 100m (330ft) high, these cliffs owe their remarkable appearance to the almost pure chalk of which they are composed.

ATLANTIC OCEAN NORTHEAST

Cape Creus

TYPE	Primary coast
FORMATION	Land-eroded rocky coastline of schists and other metamorphic and igneous rocks
EXTENT	10km (6 miles)

LOCATION Northeast of Girona, northeastern Catalonia, Spain

AN EASTERLY POINT OF THE CAPE

Cape Creus marks the point where the mountains of the Pyrenees meet the Mediterranean Sea. It has one of the most rugged coastlines in the entire Mediterranean region, with cliffs made of extremely rough-textured rocks, interspersed by small coves. Designated as a natural park in 1998, Cape Creus also boasts a varied underwater marine life, and is rich in invertebrate animals such as sponges, anemones, fan worms, and red corals. As such it is a popular diving location. The landscape is said to have inspired the Spanish surrealist artist Salvador Dali (1904–1989), and it features in many of his paintings, including *The Persistence of Memory*.

Western Algarve

TYPE Secondary coast	
FORMATION Erosional action of waves on ancient rock strata	
EXTENT 135km (84 miles)	

LOCATION Southern and southwestern coast of Portugal

The western Algarve coast extends from the city of Faro in southern Portugal to Cape St. Vincent, at the southwestern tip of the Iberian Peninsula, and then for a further 50km (30 miles) to the north. This coastline, which is bathed by the warm Gulf Stream, is notable for its picturesque, honey-coloured limestone cliffs, small bays and coves, sheltered beaches of fine sand, and emerald-green water. Many stretches of this coast show typical features of marine erosion at work, including caves at the feet of cliffs, grottoes, blowholes, arches forming through headlands, and sea stacks (isolated pillars of rock set off from headlands). Although limestone is a primary component of the landscape, other rocks, including sandstones and shales, form parts of the cliffs along scattered stretches of the coast. The strikingly beautiful scenery has made this coast a popular holiday destination.

BALANCED STACK
At Marinha Beach near Carvoeiro, wind and waves have produced distinctive rock formations, such as this eroded sea stack balanced on the shoreline.

 ATLANTIC OCEAN EAST

Amalfi Coast

TYPE Secondary coast	
FORMATION Marine erosion of folded and inclined limestone rock strata	
EXTENT 69km (43 miles)	

LOCATION Southern side of the Sorrento Peninsula, south of Naples, southern Italy

Stretching along the southern edge of the Sorrento Peninsula, south of Naples, the Amalfi Coast is famous for its steep cliffs punctuated by caves and grottoes, and for its picturesque coastal towns, some of which are built into the cliffs. The inclined layers of limestone rock that form the cliffs lie at the foot of the Lattari Mountains and were formed between 100 and 70 million years ago.

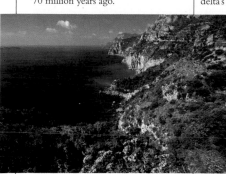

CLIFFS AT SANT ELIA POSITANO

 ATLANTIC OCEAN EAST

Nile Delta

TYPE Primary coast	
FORMATION Deposition of sediment at mouth of River Nile	
EXTENT 240km (150 miles)	

LOCATION North of Cairo, northern Egypt

The Nile Delta is one of the world's largest river deltas. As with all deltas, its shoreline is classed as a primary coast because it formed as a result of sediment deposition from a river, a land-based process. The flow of water that formed it and nourishes it has been reduced significantly by the Aswan Dam in Upper Egypt and by local water usage. The sand belt at the delta's seaward side, which prevents flooding, is currently eroding, and anticipated future rises in sea level pose a threat to its agriculture, freshwater lagoons, wildlife, and reserves of fresh water.

PROTECTIVE BELT
In this satellite view, the sand belt at the front of the Nile Delta is clearly visible. The protrusions through the sand belt mark the mouths of two Nile tributaries.

OCEAN ENVIRONMENTS

ATLANTIC OCEAN SOUTHEAST

Skeleton Coast

TYPE	Secondary coast
FORMATION	Wind-formed desert dunes
EXTENT	500km (310 miles)

LOCATION Northwest of Windhoek, on the northwestern coast of Namibia

The Skeleton Coast is an arid coastal wilderness in southwestern Africa, where the Namib Desert meets the South Atlantic. Its northern part is dominated by sand dunes that extend to the sea. Further south, these are replaced by low gravel plains. An important influence on this largely straight coastline is the Benguela Current, a surface current that flows in a northerly direction offshore, bringing cool waters from the direction of Antarctica. Prevailing southwesterly winds blow onto the coast from the Atlantic, but as they cross the cold offshore water, any moisture in the air condenses. This leads to an almost permanent fog bank and allows strange desert plants such as *Welwitschia mirabilis*, a species that survives for hundreds of years, to thrive. The coast is home to a large seal colony at Cape Fria in the north and includes many salt pans.

HUMAN IMPACT
SHIPWRECKS

The Skeleton Coast is aptly named. Its frequent fogs, onshore winds, and pounding surf have made it a graveyard for both ships and sailors. Behind the coast is a steep mountain escarpment, so before the days of rescue parties, the escape route for shipwrecked mariners was a long march along the coast through an arid desert.

WOODEN SKELETON
This wreck of a wooden vessel is one of many ships that have foundered on this treacherous coast.

HIGH DUNES AND POUNDING SURF
The coast's high dunes present ever-changing contours as they are blown by strong southwesterly winds. Below the dunes, waves pound the beaches.

INDIAN OCEAN NORTHWEST

Red Sea Coast

TYPE	Primary coast
FORMATION	Faulting and sinking of land
EXTENT	1,900km (1,200 miles)

LOCATION Coasts of Egypt, Sudan, Eritrea, and Saudi Arabia, from gulfs of Suez and Aqaba to Djibouti

The Red Sea was created as a result of a rifting process that has been gradually separating Africa from the Arabian Peninsula for the past 25 million years. Rifting is the splitting of a region of Earth's crust into two parts, which then move apart, creating a new tectonic plate boundary. This process begins when an upward flow of heat from the Earth's interior stretches the continental crust, causing it to thin, and eventually it may fracture, or fault. Sections of crust may sink, and if either end of the rift connects to the sea, flooding will occur, creating new coasts. On both sides of the Red Sea, there is evidence of the downward movement of blocks of crust, in the form of steep escarpments (lines of mountains). The Red Sea shoreline itself shelves steeply in many parts. On the land side, the coast is sparsely vegetated because of the region's hot, dry climate, but underwater there are many rich and spectacular coral reefs.

SEA MEETS DESERT
The steep Sarawat mountain escarpment that runs the length of the coast can be seen in the distance in this view of the Red Sea coast of the Sinai Peninsula.

Tigris-Euphrates Delta

INDIAN OCEAN NORTHWEST

TYPE	Primary coast
FORMATION	Sediment deposition from Tigris, Euphrates, and Karun
EXTENT	150 km (95 miles)

LOCATION Parts of southeastern Iraq, northeastern Kuwait, and southwestern Iran

The Tigris-Euphrates delta is a broad area of marshes and alluvial plain at the northern head of the Arabian Gulf, formed from sediment deposited by three major rivers. An important haven for wildlife, the delta has undergone substantial change over the past 30 years because of various damming and drainage schemes. These changes threaten coastal fisheries and endanger several species of waterfowl and mammals. Some restoration work on the delta began in 2003.

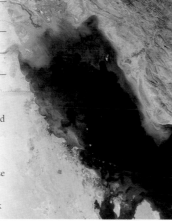

SATELLITE VIEW
The delta's seaward edge has advanced by about 250km (150 miles) in the past 3,000 years.

The Twelve Apostles

INDIAN OCEAN SOUTHEAST

TYPE	Secondary coast
FORMATION	Wave erosion of cliffs producing large sea stacks
EXTENT	3km (2 miles)

LOCATION Near Port Campbell, southwest of Melbourne, Victoria, southeastern Australia

ONGOING EROSION
The effects of wave erosion can clearly be seen at the bases of the remaining Apostles.

One of Australia's best-known geological landmarks is a group of large sea stacks formed through the erosion of 20-million-year-old limestone cliffs. Known as the Twelve Apostles, even though there were originally only nine of them, the stacks are up to 70m (230ft) tall. In 2005, one of the stacks collapsed, leaving just eight. Collapses such as this are quite common and are an integral part of the erosion process.

Krabi Coast

INDIAN OCEAN NORTHEAST

TYPE	Primary coast
FORMATION	Chemical erosion of limestone followed by drowning
EXTENT	160km (100 miles)

LOCATION Andaman Sea coast of southwestern Thailand

The area around Krabi on the western coast of southern Thailand is notable for its fantastic-looking formations of partially dissolved limestone, known as karst. This limestone was originally formed about 260 million years ago. At that time, a shallow sea covered what is now south Asia and slowly built up deposits of shells and coral that sediments washed in from the land subsequently buried. These formed layers of limestone, which were later thrust upwards and tipped over at an angle when India began to collide with mainland Asia some 50 million years ago. Around Krabi and Phang Nga Bay to its north, chemical erosion of these limestone strata by rainwater, followed by sea-level rise, has created thousands of craggy karst hills and islands. These include a number of isolated cone- and cylinder-shaped karst towers that rise out of the sea to heights of up to 210m (700ft) and groups of towers that sit on broad masses of limestone. Many of these karst formations are elongated in a northeast-southwest direction, reflecting the axis (or strike line) around which the original layers of limestone were tipped.

KOH TAPU ISLAND
Some of the karst formations along this coast have been weathered into unusual shapes, as in these examples at Koh Tapu Island in Phang Nga Bay to the north of Krabi.

Hong Kong Harbour

PACIFIC OCEAN WEST

TYPE	Secondary coast
FORMATION	Artificial coast built around various natural harbours and nearby islands
EXTENT	40km (25 miles)

LOCATION Southeast of Guangzhou, on the South China Sea coast of southeastern China

VICTORIA HARBOUR
This view shows Victoria Harbour with Hong Kong Island on the left and Kowloon on the right. Visited by more than 200,000 ships per year, the harbour is one of the world's busiest.

A number of natural harbours surround Hong Kong Island, which is the best-known part of the Hong Kong region of China. The largest, naturally deepest, and most sheltered of these harbours is Victoria Harbour, which has an area of over 42 square km (16 square miles) and is situated between Hong Kong Island and Kowloon Peninsula. Other smaller harbours include Aberdeen Harbour, which separates Hong Kong Island from a satellite island, Ap Lei Chau. The margins of all these harbour areas have been artificially modified by the construction of concrete piers, seawalls, jetties, and other structures. This coastline can be classified as a secondary coast because it has been modified by living organisms, in this case, humans. In the whole of the Hong Kong region, more than 100km (60 miles) of coastline have been artificially constructed or modified.

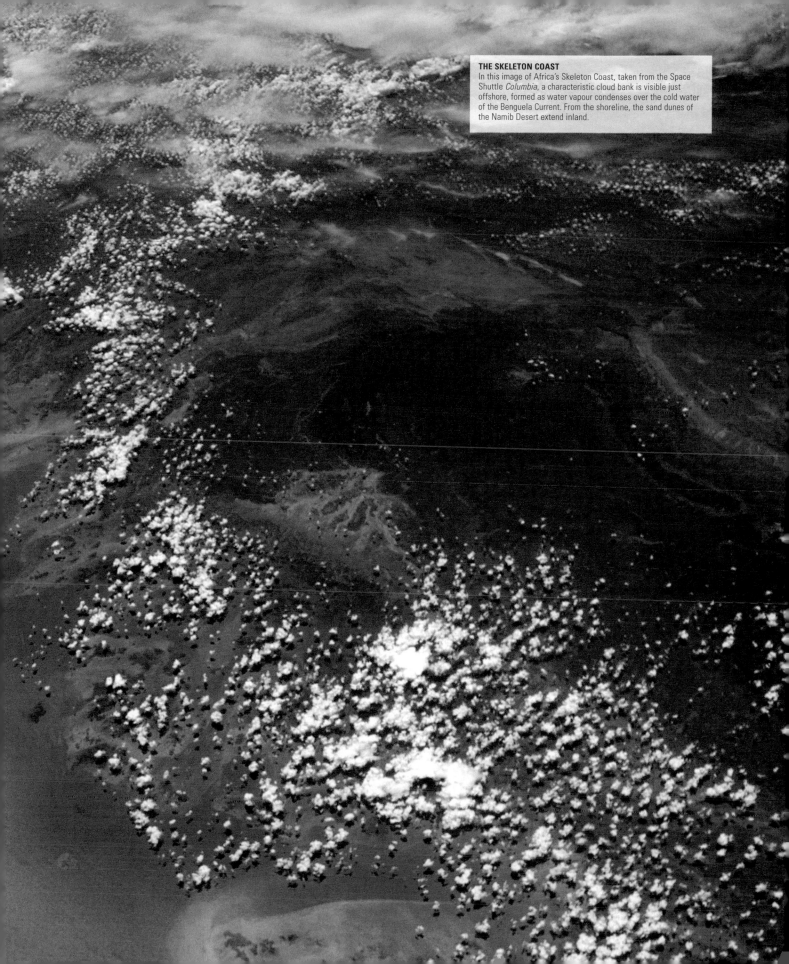

THE SKELETON COAST
In this image of Africa's Skeleton Coast, taken from the Space Shuttle *Columbia*, a characteristic cloud bank is visible just offshore, formed as water vapour condenses over the cold water of the Benguela Current. From the shoreline, the sand dunes of the Namib Desert extend inland.

PACIFIC OCEAN WEST

Ha Long Bay

TYPE	Primary coast
FORMATION	Chemical dissolution and drowning of limestone formations
EXTENT	120km (75 miles)

LOCATION On the Gulf of Tonkin, east of Hanoi, Northeastern Vietnam

Ha Long Bay is a distinctive region on the coast of Vietnam, within the Gulf of Tonkin. It consists of a body of water filled with nearly 2,000 islands composed of karst (limestone partially dissolved by rainwater). This landscape, which covers an area of just over 1,500 square km (585 square miles), was created by sea-level rise and flooding of a region with a high concentration of karst towers. Several of the islands are hollow and contain huge caves, and a few have been given distinctive names, such as Ga Choi ("Fighting Cocks") Island, Man's Head Island, and the Incense Burner, as a result of their unusual shapes. Most are uninhabited. The Bay's shallow waters are biologically highly productive and sustain hundreds of species of fish, molluscs, crustaceans, and other invertebrates, including corals. Designated a World Heritage Site in 1998, Ha Long Bay is currently under threat from water pollution as a result of mining activities, environmental degradation from urban development nearby, destruction of mangroves, and the removal of corals from reefs for sale to tourists.

TOWERING LIMESTONE
Several large karst islands, each topped with thick tropical vegetation, tower over a central area of Ha Long Bay. These islands rise up to 200m (660ft) above sea level.

PACIFIC OCEAN WEST

Huon Peninsula

TYPE	Primary coast
FORMATION	Uplift of fossil coral reefs as a result of tectonic plate movement
EXTENT	80km (50 miles)

LOCATION Eastern Papua New Guinea, north of Port Moresby

For hundreds of thousands of years, the Huon Peninsula has been forced upwards at a rate of about 25cm (10in) per century by movements of the Earth's crust at a tectonic plate boundary. This activity has pushed coastal coral reefs above the shoreline to form a series of terraced reefs on land. The oldest of these are hundreds of metres back from the coast. By studying them, scientists have learned much about changes in sea level and climate over the past 250,000 years.

AERIAL VIEW OF THE PENINSULA

PACIFIC OCEAN NORTHEAST

Puget Sound

TYPE	Primary coast
FORMATION	Glacier-carved coastal channels and bays
EXTENT	150km (90 miles)

LOCATION North and south of Seattle, Washington State, northwestern USA

Puget Sound, with its numerous channels and branches, was created primarily by glaciers. About 20,000 years ago, a glacier from present-day Canada advanced over the area, covering it in thick ice. Over the next 7,000 years, glaciers advanced and retreated several times. When they finally withdrew, they left behind many deeply gouged channels and thick layers of mud, sand, and gravel deposited by meltwater. Waves and weather have since reworked the deposits, moulding landforms and shoreline, and forming beaches, bluffs, spits, and other sedimentary features.

SOUND SETTLEMENTS
Much of the shoreline around Puget Sound has now been settled. The town of Tacoma is seen here, with Mount Rainier in the distance.

PEOPLE

GEORGE VANCOUVER

In 1792, the British sea captain George Vancouver (1757-98) became the first European to explore the area we now know as Puget Sound, as commander of the ship *Discovery*. He gave names to some 75 islands, mountains, and waterways in the area, and the city of Vancouver, Washington, was subsequently named after him. Vancouver named Puget Sound after a Lieutenant Puget, who took the first party ashore to explore its southern end.

PACIFIC OCEAN NORTHEAST

Big Sur

TYPE	Intermediate coast
FORMATION	Tectonic uplift combined with rapid wave erosion
EXTENT	145km (90 miles)

LOCATION Southeast of San Francisco, coast of California, USA

The Big Sur coastline of central California, where the rugged Santa Lucia Mountains descend steeply into the Pacific Ocean, is one of the most spectacular in the USA. Like much of the west coast of North America, Big Sur is an emergent shoreline, in that the coast has risen up faster than sea level since the end of the last ice age. This uplift has resulted from interactions at the nearby boundary between the Pacific and North American tectonic plates – this region is criss-crossed by a complex system of faults in the Earth's crust and is subjected to frequent earthquakes. At Big Sur a combination of tectonic uplift and relentless wave erosion has produced steep cliffs and partially formed marine terraces (platforms cut at the base of cliffs by waves and then lifted up). The coast is susceptible to landslides as a result of wave action, the weakening of the cliffs by faulting and fracturing, the destruction of vegetation by summer fires, and heavy winter rainfall.

RAISED PLATFORM
In this view of part of Big Sur, a grassed-over marine terrace (the green area) is visible above the present-day cliff, with a raised ancient cliff behind it.

PACIFIC OCEAN CENTRAL

Hawaiian Lava Coast

TYPE	Primary coast
FORMATION	Lava flow into the sea from an active volcano
EXTENT	20km (14 miles)

LOCATION Southeastern coast of the Big Island of Hawaii, USA

One of the fastest ways for a coast to change shape is as a result of lava flow to the sea. In southeastern Big Island, new coast has been added intermittently since 1969 as a result of lava flows from satellite craters of the active volcano Kilauea. Lava from the Pu'u O'o crater flows some 15km (9 miles) to the sea, where it cools and hardens to form land. This coastal landscape is a primitive scene of black beaches and dark cliffs made of rough, fractured lava. Plants begin to colonize newly formed areas of the coast within months of their formation.

STEAM PLUMES
As red-hot lava enters the sea, it solidifies amid huge plumes of steam. Newly forming shoreline sometimes collapses to reveal ripped-open lava tubes.

PACIFIC OCEAN SOUTHEAST

Chilean Fiordlands

TYPE	Primary coast
FORMATION	Deep glacier-carved valleys flooded by sea-level rise
EXTENT	1,500km (950 miles)

LOCATION Pacific coast of southern Chile from Puerto Montt to Punta Arenas

ICE-CHOKED FIORD
The calving ends of outlet glaciers, which choke the waters with icebergs, are found at the landward end of some fiords.

The Chilean fiordlands are a labyrinth of fiords, islands, inlets, straits, and twisting peninsulas, lying to the west of the snow-capped peaks of the southern Andes. The fiordlands extend for most of the length of southern Chile, as far south as Tierra del Fuego, and their total area is some 55,000 square km (21,500 square miles).

Some 10,000 years ago, this region was covered in glaciers, but these have largely retreated into large ice-filled areas within the mountains on the Chile–Argentina border called the Northern and Southern Patagonian Ice-fields. The glaciers left behind a network of long, deeply gouged valleys, which were filled by glacier meltwater and then flooded by the sea to form today's fiords. Rainfall here is heavy, and clear skies are rare because the moisture-laden Pacific air cools and forms clouds as it rises to cross the Andes. On the edges of the fiords, waterfalls cascade down steep granite walls, while hundreds of species of birds nest and feed around the often mist-shrouded coast and islands. Mammals that live along this coast include sea lions, elephant seals, and marine otters.

COASTAL DEFENCES

Coastal defence refers to various types of engineering techniques aimed at protecting coasts from the sea. The threats posed by the sea fall into two main categories. First is the danger of flooding of low-lying coastal areas during severe storms. Second is the continuous gradual erosion of some coasts. There are a number of different approaches to coastal defence. To prevent flooding of low-lying regions, one solution is to build a large-scale system of dams and tidal barriers. Another is to encourage the development of natural barriers, such as salt marshes, around coasts, and to conserve existing areas of this type. A third possibility is managed retreat. Instead of trying to hold back the sea, some areas of coast are allowed to flood. The idea is that, in time, the flooded land will turn into a marsh, providing natural protection.

To slow coastal erosion, various "hard" engineering techniques are commonly employed, such as the building of sea walls, breakwaters, or groynes. These methods can be effective for a while (they usually have to be rebuilt after a few decades), but are expensive and can increase erosion on neighbouring areas of coast by interfering with longshore movement of sediment. "Soft" engineering techniques are more environmentally friendly. They include the temporary solution of beach nourishment (see panel, right), which has to be repeated every few years, and encouraging the development of coastal dunes.

CRUMBLING COASTS

In 2000, the US Federal Emergency Management Agency estimated that as many as 87,000 houses in the USA are in danger of falling into the sea by the year 2060. Among them are the condemned houses, pictured below, on eroding cliffs at Governors Run in Chesapeake Bay, Maryland. In California, about 86 per cent of the coast is actively eroding. Similarly, stretches of the eastern coast of England are eroding at a rate of up to 1.8m (6ft) a year – the highest rate in Europe. Coastal defences can slow coastal erosion temporarily, but in the long run maintenance will become prohibitively expensive. In the end, the sea will triumph.

TYPES OF DEFENCES

HARD ENGINEERING

SEA WALL *A sea wall is designed to reflect wave energy. Modern walls have a curved top that prevents water spraying over the wall in storms. A wall protects the land behind it for some years but usually increases erosion of the beach in front of it.*

ROCK GROYNE *This consists of a pile of large rocks built out from the shore. The aim is to slow erosion by causing a local build-up of sand, but it can aggravate erosion nearby.*

SOFT ENGINEERING

DUNE STABILIZATION *Coastal dunes provide valuable protection against erosion if they can be stabilized and prevented from shifting. This is usually achieved by planting with grasses.*

BEACH NOURISHMENT *This involves adding large amounts of sand to a beach. Waves and tides spread the material along the coast, temporarily building up its natural defences.*

MODERN SOLUTION

GEOTUBE *A geotube is a long, cylindrical container, over 2.3m (8ft) in diameter, made of a durable textile or plastic and filled with a slurry of sand and water. Different types can be laid along the top of a beach, or inside a dune, or just offshore, where they reduce coastal erosion and protect beachfronts. This tube is part of the Barren Island Tidal Wetland project in Maryland, USA.*

DAMS AND STORM-SURGE BARRIERS *The Netherlands has invested in an extensive series of engineering works to protect a large region of the country from future marine flooding. Known as the Deltaworks, it includes many dams and moveable storm-surge barriers. The works were initiated in 1953 after a serious storm and floods killed a total of 1,835 people.*

LARGE-SCALE PROTECTION

OCEAN ENVIRONMENTS

BEACHES AND DUNES

BEACHES ARE DEPOSITS OF SEDIMENTARY MATERIAL, ranging in size from fine sand to rocks, that commonly occur on coasts above the low-tide line. Sources of beach material include sediment brought to a coast by rivers, or eroded from cliffs or the sea floor, or biological material such as shells. This material is continually moved on and off shore and around coasts, by waves and tides. Wind can also influence beach development and is instrumental in forming coastal dunes.

DISSIPATIVE BEACH AND DUNE
Dissipative beaches are usually made up of fine sand, and they slope at an angle of less than 5°.

BEACH ANATOMY

A typical beach has several zones. The foreshore is the area between the average high- and low-tide lines. On the seaward side of the foreshore is the nearshore, while behind it is the backshore; the latter is submerged only during the very highest tides and usually includes a flat-topped accumulation of beach material called a berm. The sloping area seaward of the berm, making up most of the foreshore, is the beach face. At the top end of the beach face there are sometimes a series of crescent-shaped troughs, called beach cusps. The swash zone is the part of the beach face that is alternately covered and uncovered with water as each wave arrives. Seaward of the swash zone, extending out to where the waves break, is the surf zone. The shape of a beach often alters as wave energy changes over the year.

SWASH
The surge of water and sediment up a beach when a wave arrives is called swash. If waves reach a beach at an angle, the combined effect of swash and backwash moves material along the beach.

TYPES OF BEACH

The level of wave energy, the direction the waves arrive from, and the geological make-up of a coast all affect the type of beach that will form. Dissipative beaches are gently sloping and absorb wave energy over a broad area, while reflective beaches are steeper and shorter, and consist of coarser sediment. If a cliffed coast contains a mixture of both easily eroded and erosion-resistant rock, headlands tend to form, with crescent-shaped beaches within the bays (embayed beaches) or smaller "pocket" beaches. Both of these tend to be "swash-aligned" – the waves arrive parallel to shore and do not transport sediment along the beach. Many long, straight beaches are "drift-aligned" – the waves arrive at an angle and sediment is moved along the beach by longshore drift.

RANGE OF BEACHES
This imaginary coast (right) shows several beach types, ranging from a tombolo (a sand deposit between the mainland and an island) to a drift-aligned beach.

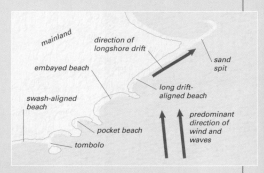

mainland
direction of longshore drift
embayed beach
sand spit
swash-aligned beach
long drift-aligned beach
pocket beach
tombolo
predominant direction of wind and waves

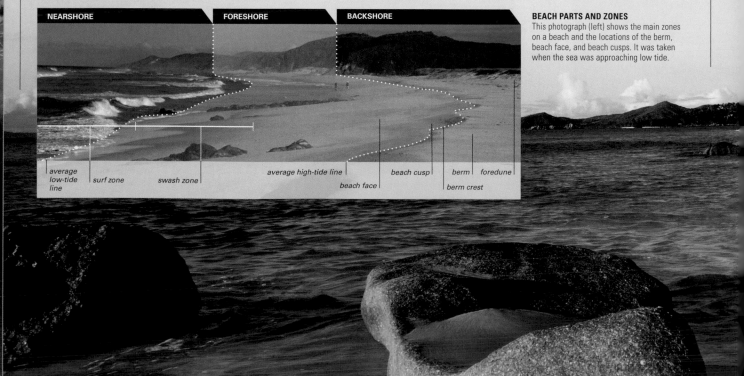

NEARSHORE   FORESHORE   BACKSHORE

average low-tide line | surf zone | swash zone | average high-tide line | beach face | beach cusp | berm crest | berm | foredune

BEACH PARTS AND ZONES
This photograph (left) shows the main zones on a beach and the locations of the berm, beach face, and beach cusps. It was taken when the sea was approaching low tide.

BEACH COMPOSITION

The composition of a beach at any particular location depends on the material available and on the energy of the arriving waves. Most beaches are composed of sand, gravel, or pebbles produced from rock erosion. Sand consists of grains of quartz and other minerals, such as feldspar and olivine, typically derived from igneous rocks such as granite and basalt. Other common beach-forming materials, seen particularly in the tropics, include the fragmented shells and skeletons of marine organisms. In general, higher wave energies are associated with coarser beach material, such as gravel or pebbles, rather than fine sand. Occasionally, large boulders are found on beaches – usually they have rolled down to the shore from local cliffs, but some boulders have ended up on beaches as a result of glacial transport or even backwash from tsunamis.

pebbles or medium gravel 8mm–1.5cm (³⁄₈–½ in) in diameter

very fine gravel 2–4mm (¹⁄₁₆–⅛ in) in diameter

very coarse sand 1–2mm (¹⁄₃₂–¹⁄₁₆ in) in diameter

medium sand 0.25–0.5mm (¹⁄₁₀₀–¹⁄₅₀ in) in diameter

fine sand 0.125–0.25mm (¹⁄₂₀₀–¹⁄₁₀₀ in) in diameter

coarse silt 0.03–0.06mm (¹⁄₈₅₀–¹⁄₄₀₀ in) in diameter

GRAIN SIZES
The silts, sands, and gravels that make up most beaches tend to become sorted by the action of waves, with material of different sizes deposited on different parts of the beach.

MARRAM GRASS
This grass is a common colonizer of embryo dunes. It develops deep roots that allow it to tap into deep groundwater stores. The roots bind the sand together, while the grass traps more blown sand, assisting in foredune development.

COASTAL DUNES

Coastal dunes are formed by wind blowing sand off the dry parts of a beach. Dunes develop in the area behind the backshore, which together with the upper beach face supplies the sand. For dunes to develop, this sand has to be continually replaced on the beach by wave action. The actual movement of sand to form dunes occurs through a jumping and bouncing motion along the ground called saltation. Some coastal areas have more than one set of vegetated dunes that run parallel to the shoreline. The dunes closest to shore are called foredunes, behind them is a primary dune ridge, secondary dune ridge, and so on. These anchored, vegetated dunes are important for the protection they provide against coastal erosion. On some coasts, non-vegetated, mobile dunes occur, which move in response to the prevailing winds. These can often be anchored by planting with grasses.

REFLECTIVE BEACH
This beach in the Seychelles is an example of a reflective beach because of its quite steeply shelving face. It has a distinct berm and berm crest.

Pink Sands Beach

TYPE Dissipative beach, protected by reefs	
COMPOSITION Sand mixed with broken shells and skeletons	
LENGTH 5km (3 miles)	

LOCATION Harbour Island, off Eleuthera, northeast of Nassau, Northern Bahamas

Pink Sands in the Bahamas is a gently sloping beach that faces eastwards onto the Atlantic Ocean. It is protected from ocean currents by an outlying reef. The pale pink colour of the sand comes from small, single-celled organisms called foraminiferans, in particular, the species *Homotrema rubrum*, also known as the Sea Strawberry. The shells of these organisms are bright red or pink due to the presence of an iron salt. In parts of the Bahamas they are abundant, living on the underside of reefs. When they die, they fall to the sea floor where they are broken by wave action and mixed with other debris, such as the white shells of snails and sea urchins, as well as mineral grains. This mixture is then finely pulverized and washed up on the shore as pink-coloured sand by wave action.

GENTLE SLOPE
Pink Sands is an example of a dissipative beach, on which waves break some distance from the shore, then slowly roll in, dissipating their energy across a broad surf zone.

Copacabana Beach

TYPE Embayed, dissipative beach	
COMPOSITION White sand	
LENGTH 4km (2½ miles)	

LOCATION Rio de Janeiro, southeastern Brazil

One of the most famous beaches in the world, Copacabana Beach is a wide, gently curving stretch of sand between two headlands. Behind the beach lies the city of of Rio de Janeiro, with green, luxuriant hills in the hinterland. The beach is crowded much of the year and is known for its beach sports and New Year's Eve firework displays. The sea area off the beach is not always recommended for swimming, due to strong currents.

COPACABANA LOOKING NORTH

St. Ninian's Tombolo

TYPE Tombolo	
COMPOSITION Yellow and white sand	
LENGTH 1km (⅗ mile)	

LOCATION West coast of southern Mainland, the main island of the Shetland Isles, off Scotland, UK

St. Ninian's Isle in the Shetland Isles provides a classic example of a tombolo, or ayre – a short spit of sedimentary material that connects an island to a nearby land mass or mainland. A tombolo is formed by waves curving around the back of an island so that they deposit sediment on a neighbouring land mass, at the point directly opposite the island. Over time, these sediments gradually build up into a tombolo, which typically projects at right angles to the coast and has a beach on each side. St. Ninian's Tombolo has been in existence for at least 1,000 years, and its permanence may be due to a cobble base underlying the sand. This tombolo tends to become lower and narrower during storms as a result of destructive wave action, while during calmer weather the waves build it up again with sand carried from offshore or the nearshore. The sediment that forms a tombolo may come from the mainland, the island, the sea floor, or a combination. Scientists have deduced that a tombolo will usually form when the ratio between an island's distance from shore and its length parallel to the shore is less than 1:5 (for St. Ninian's Isle this ratio is less than 1). If the ratio is greater than 1:5, a feature called a salient may form – a sand spit that reaches towards the island but does not quite reach it.

NARROW CONNECTION
A slim, sandy tombolo extends from the Shetland island of Mainland in the foreground, to St. Ninian's Isle.

North Jutland Dunes

TYPE	Coastal dunes
COMPOSITION	Yellow sand, Marram Grass
LENGTH	250km (155 miles)

LOCATION North and northwest coast of Jutland, Denmark

Much of the northern coastline of Denmark's Jutland Peninsula consists of sand dunes, which cover several thousand square kilometres of coast. These dunes are "active" in that they have a natural tendency to migrate

along the coast, carried by wind (sand drift) and wave erosion. In some areas, attempts have been made to restrict this dune drift, to prevent the sand inundating summer houses. Some early attempts were fruitless. For example, sand fences were built into the dunes during the Second World War, but the dunes have since moved behind them, leaving the fences on the beach. More recently, many dune areas have been stabilized more successfully by planting with grasses and conifer trees.

SHIFTING SANDS
Many sand dunes on the peninsula have Marram Grass growing in them, which helps constrain their movement.

Porthcurno Beach

TYPE	Pocket beach
COMPOSITION	Yellow-white sand, composed mainly of shell fragments
LENGTH	150m (500ft)

LOCATION Southwest of Penzance, Cornwall, southwestern England, UK

Porthcurno is a typical pocket beach located near Land's End at England's southwesternmost tip. Like all pocket beaches, it nestles between two headlands that protect the sandy cove from erosion by winter storms and strong currents. Pocket beaches are

common where cliffs made of different types of rock are subject to strong wave action. Rock that is especially hard and resistant to erosion forms headlands, while intervening areas of softer rock are worn down to form pocket beaches. Unlike other beaches, pocket beaches exchange little or no sand or other sediment with the adjacent shoreline, because the headlands prevent longshore drift. The sea at Porthcurno is a distinctive turquoise, possibly due to the reflective qualities of the sand, which is made mainly of shell fragments.

GRANITE HEADLANDS
The headlands on either side of the beach are formed from 300-million-year-old granite.

Chesil Beach

TYPE	Storm beach on tombolo
COMPOSITION	Shingle of flint and chert
LENGTH	29km (18 miles)

LOCATION West of Weymouth, Dorset, southern England

Chesil Beach forms the seaward side of the Chesil Bank, a remarkably long, narrow bank of sedimentary material that connects the coast of Dorset in southern England to the Isle of Portland. Behind the bank is a tidal lagoon called the Fleet. Running parallel to the coast, Chesil Bank looks like a barrier island. However, because it connects the mainland to an island, it is classified as a tombolo. How Chesil Bank and its beach originally formed is debated – the most widely accepted theory is that it originally formed offshore and was then gradually moved to its current location by waves and tides. The beach is classified as a storm beach, as it is affected by strong waves because it faces southwest towards the Atlantic and the prevailing winds. Like most storm beaches, it is steep, with a gradient of up to 45 degrees, and is made of shingle.

CHESIL BANK
The bank is about 200m (650ft) wide and 15m (50ft) high along its entire length. The beach (left) is on its seaward side.

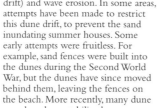

DISCOVERY

GRADED PEBBLES

Chesil Beach's pebbles change in size progressively from potato-sized at one end to pea-sized at the other. This reveals the differences in wave energy along its length – at one end, strong waves wash smaller pebbles offshore; at the other, weaker waves wash them onshore.

OCEAN ENVIRONMENTS

Cap Ferret

TYPE	Coastal dunes on a spit
COMPOSITION	Sand, grasses, forest
LENGTH	12km (7½ miles)

LOCATION Coast of Aquitaine, southwest of Bordeaux, southwestern France

Cap Ferret lies at the southern end of a long sand spit in western France. It separates the Arcachon Lagoon from the Atlantic Ocean and forms part of the spectacular Aquitaine coast, which at 230km (143 miles) is the longest sandy coast in Europe. This region is characterized by a series of straight, sandy beaches backed by longitudinal sand dunes, which are the highest dune formations in Europe. They include the highest individual European sand dune, the Dune du Pilat, which rises to about 115m (380ft) above sea level.

Behind the main dune area is a forest, originally planted in the 18th century to try to prevent the dunes shifting. Unfortunately, this coast is undergoing serious erosion, of more than 10m (33ft) a year in some places, mainly because excessive urban development has degraded the vegetation cover.

SAND MOUNTAINS
Along the coast to the north and south of Cap Ferret, mini-mountains of pale, rippling sand are backed by an extensive vegetation cover.

Banc d'Arguin

TYPE	Coastal dunes and tidal flats
COMPOSITION	Yellow sand
LENGTH	160km (100 miles)

LOCATION Between Nouakchott and Nouadhibou on the northwest coast of Mauritania, West Africa

The Banc d'Arguin National Park is a vast region of dunes, islands, and shallow tidal flats covering more than 12,000 square km (4,600 square miles) of the Mauritanian coast. The dunes, which consist mainly of windblown sand from the Sahara, are concentrated in the southern region of the Park. Banc d'Arguin contains a variety of plant life and is a major breeding or wintering site for many migratory birds, including flamingos, pelicans, and terns. It was declared a World Heritage Site in 1987.

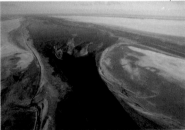

SAND BANKS AT BANC D'ARGUIN

Jeffreys Bay

TYPE	Series of gently sloping, dissipative beaches
COMPOSITION	Sand
LENGTH	24km (15 miles)

LOCATION Southwest of Port Elizabeth, eastern Cape Province, South Africa

Jeffreys Bay is famous both as a highly popular surfing spot and for the large numbers of beautiful seashells that wash up on its shores. It consists of a series of wide beaches strung out along a southeast-facing stretch of the South African coastline.

As a surfing destination, Jeffreys Bay is regularly ranked among the top five beaches in the world by those seeking the "perfect wave". The most acclaimed surfing spot or wave "break"

is known as Supertubes. Here, the combination of shoreline shape, bottom topography, and direction of wave propagation regularly generates waves that form huge, glassy-looking hollow tubes as they break. Other nearby wave breaks in Jeffreys Bay have been given such colourful names as Boneyards, Magna Tubes, and Kitchen Windows. Some of these waves can carry a skilled surfer several hundred metres along the beach on

a single ride. The same waves that attract surfers are also responsible for the vast numbers and wide variety of seashells that are washed up onto the beach with each tide.

Conchologists have identified the shells of over 400 species of marine animals, including various gastropods, chitons, and bivalves, making the bay the most biologically diverse natural coastline in South Africa. Dolphins, whales, and seals are also seen.

HEADING FOR SUPERTUBES
The waves at Supertubes may be 3m (10ft) high and invariably break right-to-left when viewed from the shore.

HUMAN IMPACT

HIDDEN DANGERS

Every surfing spot, including Jeffreys Bay, has dangers that would-be surfers should know about. The most important are rip currents. The enormous volume of seawater washed up on shore by the waves tends to pool at specific points on the beach and is then funnelled back out to sea in swift currents. These move rapidly away from the beach, straight out through the surf zone, and can sweep unsuspecting swimmers out to sea. They can be escaped by swimming parallel to the shore. At Jeffreys Bay, there have also been rare reports of surfers being bitten by sharks, most often by the Sandtiger or Ragged-tooth Shark.

Anjuna Beach

TYPE	Series of embayed and pocket beaches
COMPOSITION	Yellow sand
LENGTH	2.5km (1½ miles)

LOCATION On the Arabian Sea coast, northwest of Panaji, southwestern India

Anjuna Beach is one of the most scenic and popular of the renowned string of beaches that lie on the coast of the Indian State of Goa. The beach has an undulating shape and is broken up into several sections by rocky outcrops that jut into the sea. By reducing rip currents and cross-currents, these outcrops help to make Anjuna one of the safest bathing beaches on the Goa coast. During the monsoon season, from June to September, much of the beach sand is stripped away and carried offshore by heavy wave action, but after the monsoons, calmer seas restore the sand deposits.

PICTURESQUE SETTING
With its calm seas and sand crescents backed by swaying palms and low, rocky hills, Anjuna has been a favoured holiday destination since the 1960s.

Cox's Bazar

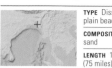

TYPE	Dissipative coastal plain beach
COMPOSITION	Yellow sand
LENGTH	120km (75 miles)

LOCATION South of Chittagong, southeastern Bangladesh

Cox's Bazar lies on a northeastern stretch of the Bay of Bengal and is the second-longest unbroken natural beach in the world – Ninety Mile Beach (see p.112) in Australia is the longest. It fronts a range of dunes and, at its southern end, a spit of land. The dunes, spit, and beach have been built up over hundreds of years through a combination of wave action and deposition of sediment from the Bay of Bengal. This is a gently sloping beach that offers safe swimming and surfing and is also popular among collectors of conch shells.

SATELLITE VIEW OF COX'S BAZAR

Shell Beach

TYPE	Embayed beach
COMPOSITION	Shells of a species of cockle
LENGTH	110 km (70 miles)

LOCATION Northwest of Perth, Western Australia

Shell Beach, in Western Australia's Shark Bay, has a unique composition, consisting almost entirely of the white shells of *Fragum erugatum*, a species of cockle (a bivalve). The beach lies in a partially enclosed area of Shark Bay known as L'Haridon Bight. This cockle thrives here because its predators cannot cope with the high salinity of the seawater. On the foreshore of Shell Beach, the layer of shells reaches a depth of 8–9m (26–30ft) and also forms the sea floor, stretching for hundreds of metres from the shoreline. On the upper parts of the beach, away from the water line, many of the shells have become cemented together, in some areas leading to the formation of large, solid conglomerations. These are mined to make decorative wall blocks.

SHELL BANK
Individual shells in the beach are about 1cm (½in) wide. Accumulations of these shells over about 4,000 years has led to the formation of a long bank along the seashore.

OCEAN ENVIRONMENTS

PACIFIC OCEAN SOUTHWEST
Ninety Mile Beach

TYPE	Dissipative coastal plain beach
COMPOSITION	Yellow sand
LENGTH	151km (94 miles)

LOCATION Southeast of Melbourne, Victoria, southeastern Australia,

Australia's Ninety Mile Beach, on the coast of Victoria, has a solid claim to be the world's longest uninterrupted natural beach. The beach runs in a southwest to northeasterly direction and fronts a series of dunes. Waves generally break too close to the beach for good surfing, and strong rip currents make the conditions hazardous for swimmers. In its northeastern part, several large lakes and shallow lagoons, known as the Gippsland Lakes, lie behind the dunes. Beneath the sea, vast plains of sand stretch in every direction and are home to a large variety of small invertebrate life, including crustaceans, worms, and burrowing molluscs.

AERIAL VIEW
Facing out onto the Bass Strait, Ninety Mile Beach is subject to strong waves during the winter months.

PACIFIC OCEAN SOUTHWEST
Moeraki Beach

TYPE	Embayed beach
COMPOSITION	Dark sand and large boulders
LENGTH	8km (5 miles)

LOCATION Northeast of Dunedin, southeastern New Zealand

The beach north of Moeraki on New Zealand's South Island is strewn with large, near-spherical boulders. Their origin is unclear, but the most widely accepted scientific view is that they are mineral concretions that formed 60 million years ago in mudstones – layers of softer sedimentary rock on the sea floor. These mudstones were later uplifted and now form a cliff at the back of the beach. There, gradual erosion exposes and releases the boulders, which eventually roll down onto the beach.

SUPERSIZED BOULDERS
The boulders are up to 3m (10ft) in diameter and some weigh several tons. Some are half-buried in the sand.

PACIFIC OCEAN CENTRAL
Punalu'u Beach

TYPE	Pocket beach
COMPOSITION	Black sand
LENGTH	1km (³/₅ miles)

LOCATION Northeast of Naalehu, Big Island, southeastern Hawaii

Punalu'u Beach on Hawaii's Big Island is a steeply shelving pocket beach. It is best known for its dramatic-looking black sand, which is composed of grains of the volcanic rock basalt. The sand has been produced by wave action on local cliffs of black basaltic lava. Punalu'u, in common with about half the land area of Hawaii, lies on the flank of Mauna Loa, the world's most massive volcano. Lava produced by the volcano dominates the local landscape – although no lava has reached Punalu'u from Mauna Loa or the nearby active volcano Kilauea for several hundred years.

The beach is a popular location for swimming and snorkelling, but underwater springs that eject cold water into the sea close to the beach can be discomforting. Punalu'u Beach is also visited by Green Turtles, which come to eat seaweed off rocks at the edge of the beach and bask on the warm, heat-absorbing sand.

PACIFIC OCEAN NORTHEAST
Columbia Bay

TYPE	Series of embayed beaches
COMPOSITION	Gravel and rocks
LENGTH	50km (31 miles)

LOCATION Southwest of Valdez, southern Alaska, USA

Many beaches in southern Alaska, and other beaches at high latitudes in the northern hemisphere, consist of gravel, small rocks, and boulders. These materials come from coarse glacial till – mixtures of clay, silt, sand, gravel, and rocks that were carried to a location by ancient glaciers, remaining there when the glaciers melted. The till has usually been reworked by wave action, with the lighter material (clay, silt, and sand) washed away and the heavier gravel and rocks sorted by size and deposited in different areas along the shoreline. Such is the case in Columbia Bay, a region within Alaska's Prince William Sound. Many of the beaches in this area have old tidal lines visible above the present ones, the result of a huge earthquake in 1964 that raised the land by 2.4m (8ft).

BEACH AND BAY
The backshore area visible here, which has been colonized by plants, was foreshore prior to the 1964 earthquake.

BLACK AND BLUE
The sand on Punalu'u is almost perfectly black, contrasting with the deep blue Pacific waters. Removal of the sand is prohibited.

PACIFIC OCEAN NORTHEAST

Oregon National Dunes

TYPE Coastal dunes

COMPOSITION Yellow sand, grasses, conifers

LENGTH 64km (40 miles)

LOCATION Southwest of Portland, Oregon, northwestern USA

Oregon National Dunes is the largest area of coastal sand dunes in North America, extending along the coast of Oregon between the Sislaw and Coos rivers. These dunes have been created through the combined effects of coastal erosion and wind transport of sand over millions of years and extend up to 4km (2½ miles) inland, rising to 150m (500ft) above sea level. A continuum of dry and wet conditions extends through the dune area. Close to the beach are low foredunes of sand and driftwood

stabilized by Marram Grass. Behind these are hummocks where sand collects around vegetation. Water accumulates around the hummocks seasonally, giving them the appearance of floating islands. Behind the hummocks are further distinct regions, ranging from densely vegetated areas that become marsh-like in winter to completely barren, wind-sculpted high dunes. The dunes are a popular location for various recreational activities, including riding all-terrain vehicles (ATVs) and dune buggies.

DUNE DESTABILIZATION

The use of dune buggies and all-terrain vehicles (ATVs), especially when raced in large numbers, may destroy the grass on the dunes, making them susceptible to wind scour. This may in turn lead to self-propagating breaches in the dune ridges. To protect the dunes, ATV usage is restricted.

SEA OF DUNES
The wind moulds the sand of the dunes into wave shapes, with crests at right angles to the wind direction.

PACIFIC OCEAN NORTHEAST

Dungeness Spit

TYPE Sand spit

COMPOSITION Sand

LENGTH 9km (5½ miles)

LOCATION Northwest of Seattle, Washington State, northwestern USA

Dungeness Spit, thought to be the world's longest natural sand spit, juts out from the Olympic Peninsula in Washington State, USA. It is part of

the Dungeness National Wildlife Refuge and is as little as 15m (50ft) wide in places. In addition to its great length, the spit has a complex shape, the result of seasonal changes in wind and wave direction. During part of the year, these bring sandy sediments from the northwest, and at other times from the northeast. The resulting pattern of sedimentation has created a large sheltered coastal area, providing refuge for many shorebirds and waterfowl, which nest along the beach, and for Pacific Harbour Seals. The tidal flats nourish a variety of shellfish, and the inner bay is an important nursery habitat for several salmon species.

GROWING SPIT
The spit grows at about 4.5m (15ft) a year. It provides shelter for a large inner bay and an area of tidal flats.

PACIFIC OCEAN EAST

Tamarindo Beach

TYPE Embayed beach

COMPOSITION Yellow sand

LENGTH 3km (2 miles)

LOCATION Northwest of San José, northwestern Costa Rica

Tamarindo Beach is a curved, gently shelving crescent of sand situated between two mangrove-lined estuaries and backed by dry tropical forest.

COASTAL SETTING
In this view, the main part of Tamarindo beach is in the background, with the entrance to an estuary that curves round behind the beach on the right.

It faces directly onto the Pacific, with its enormous fetch (wave-generation area), and so benefits from a strong year-round incoming swell, making the beach a popular surfing location. To the north and south of the main beach are two further beaches that together form the Las Baulas National Marine Park. These are important nesting sites for the Leatherback Turtle from October to March.

ESTUARIES AND LAGOONS

ESTUARIES AND COASTAL LAGOONS ARE BOTH semi-enclosed, coastal bodies of water. An estuary typically connects to the open sea, is quite narrow, and receives a significant input of fresh water from one or more rivers. This fresh water mixes with the salt water to a varying degree, depending on river input and tides. Many estuaries are simply the seaward, tidally affected ends of large rivers. Coastal lagoons are usually linked to the sea only by one or more narrow channels, through which water flows in and out; sometimes these channels open only at high tide.

ESTUARY FORMATION

Estuaries form in four main ways. For example, the sea level may rise and flood an existing river valley on a coastal plain, such as in Chesapeake Bay, in the USA. Alternatively, the sea level can rise to flood a glacier-carved valley, forming a fiord. Estuaries formed in this way are deeper than other types, but have shallow sills at their mouths that partially block inflowing seawater. Coastal wave action can also create an estuary, by building a sand spit or bar across the open end of a bay fed by a stream or river (see p.93). Other estuaries result from movement at tectonic faults (lines of weakness) in the Earth's crust, where downward slippage can result in a surface depression. This becomes an estuary if seawater later floods in.

FORMATION PROCESSES
An estuary can form when sea-level rise causes the seaward end of a river valley to flood (top) or inundates a glacier-carved valley to create a fiord (middle), or when a spit extends across a bay (bottom).

CONGO RIVER ESTUARY
Formed by flooding of a river valley, this estuary is the world's second largest (after the Amazon) in terms of discharge rate.

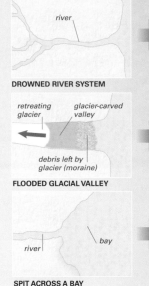

DROWNED RIVER SYSTEM
river
estuary
delta

FLOODED GLACIAL VALLEY
retreating glacier
glacier-carved valley
debris left by glacier (moraine)
sill
estuary (fiord)

SPIT ACROSS A BAY
river
bay
sand spit
bay
longshore current

TYPES OF ESTUARY

The way in which fresh and salt water mixes in an estuary determines its classification. A strong river inflow usually means minimal mixing – the less dense fresh water flows over the denser salt water, which forms a wedge-shaped intrusion into the bottom of the estuary. This is a salt-wedge (river-dominated) estuary. In partially mixed and fully mixed (tide-dominated) estuaries, there is considerable mixing, producing turbulence and increased salinity in the fresh water. In each case, this is balanced by a strong, tidally influenced influx of salt water from the sea: this influx brings sediments from offshore, which are deposited as mud in the estuary.

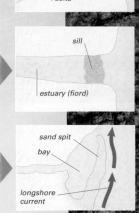

strong flow of fresh water
minimal mixing of salt and fresh water
slightly salty water flows out
fresh water
wedge of sea water
small tidal countercurrent

SALT-WEDGE ESTUARY
In a salt-wedge estuary (left), there is a strong flow of fresh river water over a wedge of salt water, with little mixing between the two layers.

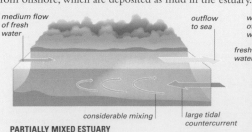

PARTIALLY MIXED ESTUARY
In this type of estuary, there is considerable mixing between fresh and salt water. The saltiness of the water increases with depth in all parts of the lower estuary.

medium flow of fresh water
outflow to sea
considerable mixing
large tidal countercurrent

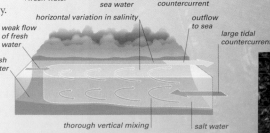

FULLY MIXED ESTUARY
In a fully mixed (or tide-dominated) estuary, the fresh and salt water are well-mixed vertically, but there is some horizontal variation in saltiness.

weak flow of fresh water
horizontal variation in salinity
outflow to sea
large tidal countercurrent
fresh water
thorough vertical mixing
salt water

CARVED BY GLACIERS
A fiord is an estuary formed by the sea flooding a deep valley originally carved out by a glacier. Norway's Geiranger Fiord is 20km (12 miles) long, and reaches a depth of 200m (660ft).

ESTUARINE ENVIRONMENTS

Estuaries are unique coastal environments. Being typically long and funnel-shaped, tides don't just rise here – they rush in, creating strong currents and, sometimes, wall-like waves called tidal bores. The high rate of sedimentation means that mud accumulates, so tidal mudflats and salt marshes (see pp.124–25) or in the tropics, mangrove swamps (see pp.130–31), develop. Despite the effects of tides and currents, the high turbidity that reduces plant photosynthesis, and fluctuations in salinity and temperature, most estuaries are biologically highly productive. This is partly owing to the high concentration of nutrients in river water, and because estuaries are well oxygenated. Although only a limited range of organisms, such as mussels, cope with living in estuaries, populations are often huge.

COMMON EUROPEAN OYSTER

RICH FOOD SOURCE
Estuaries attract waders and other shorebirds because of the high concentrations of small animals (such as worms and shrimp) that live in the mud deposits. These lapwings and an egret are congregating to feed in the Thames estuary, UK.

ESTUARY DWELLER
Various species of starfish tolerate the estuarine environment, where they feed on mussels, crustaceans, and worms. This Common Starfish is in an estuary in Brittany, France.

COASTAL LAGOONS

Coastal lagoons occur worldwide, and are different from the lagoons found at the centres of coral atolls (see p.152). Calmer and usually shallower than estuaries, most lagoons are connected to the sea by tidal channels. Although fresh water does not usually flow into coastal lagoons, some do receive a significant river inflow. So, as well as saltwater lagoons, there are also some partly, or predominantly, freshwater lagoons.
  In hot climates, some lagoons are hypersaline (saltier than ocean water), due to high evaporative losses. Although some coastal lagoons are severely polluted, the cleaner ones are often well stocked with fish, crustaceans, and other marine life, and frequently attract large numbers of shorebirds. Some provide feeding or breeding areas for sea turtles and whales.

LAGOON AND CHANNELS
Matagorda Bay is a lagoon on the coast of Texas, USA, separated from the Gulf of Mexico by a long, narrow peninsula. Two channels, located near the southwest corner of the lagoon, connect it to the gulf.

OCEAN ENVIRONMENTS

ATLANTIC OCEAN NORTHWEST
St. Lawrence Estuary

TYPE Salt-wedge (river-dominated) estuary

AREA Approximately 25,000 square km (10,000 square miles)

LOCATION Quebec, eastern Canada

The St. Lawrence Estuary is one of the world's largest estuaries. Some 800km (500 miles) long, it discharges about 12 million litres (2.6 million gallons) of water into the Gulf of St. Lawrence each second. The estuary is rich in marine life. In its wide middle and lower reaches, the icy Labrador Current flows 300m (1,000ft) below the surface in the opposite direction to the main estuarine flow. In one section, near the mouth of a fiord that branches off the estuary, the Current's nutrient-rich waters rise abruptly and mix with warmer waters above. This upwelling of nutrients encourages plankton growth, providing the base of a food chain that involves many species of fish and birds, and a small population of Beluga Whales.

WINTER SCENE
In winter, much of the estuary becomes iced over. A stretch of the estuary is seen here at low tide, shortly after sunrise.

ATLANTIC OCEAN NORTHWEST
Chesapeake Bay

TYPE Partially mixed estuary

AREA 8,200 square km (3,200 square miles)

LOCATION Surrounded by Maryland and parts of eastern Virginia, USA

Chesapeake Bay is the largest estuary in the USA. Its main course, fed by the Susquehanna River, is over 300km (185 miles) in length. It has numerous sub-estuaries, and more than 150 rivers and streams drain into it. This body of water was created by sea-level rise drowning the valley of the Susquehanna and its tributaries over the last 15,000 years. Once famous for its seafood, such as oysters, clams, and crabs, the bay is now far less productive, though it still yields more fish and shellfish than any other estuary in the USA. Industrial and farm waste running into the bay causes frequent algal blooms, which block sunlight from parts of its bed. The resulting loss of vegetation has lowered oxygen levels in some areas, severely affecting animal life. The depletion of oysters, which naturally filter water, has had a particularly harmful effect on the bay's water quality.

BAY BRIDGE
A major bridge in the upper bay connects Maryland's rural eastern shore to its urban western shore.

DISCOVERY
IMPACT CRATER

In the 1990s, drilling of the sea bed at Chesapeake Bay led to the discovery of a meteorite impact crater 85km (53 miles) wide under its southern region. The 35-million-year-old crater helped shape today's estuary.

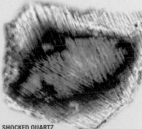

SHOCKED QUARTZ
Evidence for the crater included the discovery of grains of shocked quartz, which forms when intense pressure alters its crystalline structure.

MAIN CHANNEL FLOWING THROUGH DELTA

ATLANTIC OCEAN WEST
Mississippi Estuary

TYPE Salt-wedge (river-dominated) estuary

AREA 60 square km (25 square miles)

LOCATION Southeastern Mississippi Delta, southeastern Louisiana, USA

The Mississippi Estuary is about 50km (30 miles) long and lies at the seaward end of the Mississippi River, where the river flows through its own delta. The estuary consists of a main channel and several sub-channels. Together, these discharge an average of some 18 million litres (4 million gallons) of water per second into the Gulf of Mexico. The main channel is a classic example of a salt-wedge estuary – its surface waters contain little salt, but they flow over a wedge of salt water, which extends at depth for several kilometres up the estuary.

ATLANTIC OCEAN WEST

Laguna Madre

| **TYPE** |
| Hypersaline coastal lagoon |
| **AREA** |
| 3,660 square km (14,400 square miles) |

LOCATION Southern Texas, USA, and northeastern Mexico, along the coast of the Gulf of Mexico

The Laguna Madre is a shallow lagoon extending about 456km (285 miles) along the coast of the Gulf of Mexico. Its northern part, in Texas, is separated from the Gulf by a long, thin barrier island, Padre Island. The southern part, in Mexico, is similarly cut off by a string of barrier islands. The entire lagoon connects with the Gulf only via a few narrow channels, and it is less than 1m (3ft) deep in most parts. It is saltier than seawater because it receives no input of river water and lies in a hot, dry region, leading to high rates of evaporation. Seagrass meadows and several species of crustaceans and fish thrive in the lagoon, which also supports many wintering shorebirds and waterfowl. Threats to its health include coastal development, dredging, overfishing, agricultural pesticides, and algal blooms.

FLY-FISHING FOR REDFISH
The sale of licences for fly-fishing – for trout and redfish – in the Laguna provides funds for protecting its water quality and wildlife.

TWO LAGOONS
In this aerial view, Lagoa dos Patos is the pale central area. Below it, the darker Mirim Lagoon extends to the Brazil–Uruguay border.

ATLANTIC OCEAN SOUTHWEST

Lagoa dos Patos

| **TYPE** |
| Tidal coastal lagoon |
| **AREA** |
| 10,000 square km (3,900 square miles) |

LOCATION South of the city of Porto Alegre, southern Brazil

Lagoa dos Patos ("Lagoon of Ducks") is the world's largest coastal lagoon. Its name is said to have been given to it by Jesuit settlers in the 16th century, who bred waterfowl on its shores. It is a shallow, tidal body of water, 250km (155 miles) long and up to 56km (35 miles) wide. A sand bar separates it from the Atlantic, with which it connects at its southern end via a short, narrow channel that disgorges a large plume of sediment into the ocean. Marine animals use this channel to access the lagoon; sea turtles are found in the lagoon in spring and summer.

At its northern end, the lagoon receives an inflow of fresh water from the Guaíba Estuary, formed from the confluence of the Rio Jacui and three smaller rivers. Along its inner side are a number of distinctive wave-like "cusps" that have been caused by the accumulation and erosion of sediments driven by tidal action and winds. The salinity of the lagoon varies. It consists mainly of fresh water at times of high rainfall, but there is considerable saltwater intrusion at its southern end at times of drought. Lagoa dos Patos is one of Brazil's most vital fishing grounds. However, run-off from rice fields and pastureland, industrial effluents, and increasing population have led to concerns for the lagoon's ecosystem.

ATLANTIC OCEAN SOUTHWEST

Amazon Estuary

TYPE Salt-wedge (river-dominated) estuary

AREA Approximately 20,000 square km (7,800 square miles)

LOCATION Northern Brazil

The Amazon Estuary is a stretch of the River Amazon that extends more than 300km (190 miles) inland from the river's mouth to an area southwest of the city of Macapà. Varying in width from 25 to 300km (15 to 190 miles), the estuary is partly filled by numerous low-lying, forested islands.

The Amazon Estuary has by far the largest water output of any estuary in the world, discharging an average of 175 million litres (40 million gallons) per second into the Atlantic. The sheer magnitude of this discharge means that, almost uniquely among estuaries, there is very little saltwater intrusion into it. Instead, nearly all of the mixing between the river's discharge and seawater occurs outside the estuary, on an area of continental shelf. Despite the relative lack of seawater intrusion, the whole estuary is significantly affected by twice-daily tides, which cause inundation (by river water) of most of the islands in the estuary.

MARAJO ISLAND
The Amazon Estuary is so enormous that the biggest of the forested islands lying within it, Marajo Island, has its own river system.

HUMAN IMPACT

POROROCA SURF

Tidal bores, locally called *pororocas*, occur on large spring tides in several of northern Brazil's river estuaries. Some of these bores attain heights of 3m (10ft) and can be surfed for several kilometres. This sport is rather hazardous, however, because the waters through which the *pororocas* surge are home to dangerous snakes, fish, and crocodiles.

ATLANTIC OCEAN SOUTHWEST

River Plate

TYPE Salt-wedge (river-dominated) estuary

AREA 35,000 square km (13,500 square miles)

LOCATION On the Argentina–Uruguay border, east of Buenos Aires and southwest of Montevideo

The River Plate, or Rio de la Plata, is not a river but a large, funnel-shaped estuary formed by the confluence of the rivers Uruguay and Paraná. These rivers and their tributaries drain about one-fifth of the land area of South America. At 290km (180 miles) long and 220km (136 miles) wide at its mouth, the Plate discharges about 25 million litres (5.5 million gallons) of water per second into the Atlantic Ocean. As well as transporting this vast amount of water, the estuary receives about 57 million cubic metres (2 billion cubic feet) of silt each year from its input rivers. This mud accumulates in great shoals, so that the water depth in most of the estuary is less than 3m (10ft). Constant dredging is therefore needed to maintain deep-water channels to the ports of Buenos Aires, which lies near the head of the estuary, and Montevideo, which is close to its mouth. Surface salinity varies uniformly through the estuary, from close to zero in its upper parts to a value just below average ocean salinity near its mouth. At depth, a wedge of salt water penetrates deep into the estuary. Biologically, the River Plate is highly productive, yielding large annual masses of plankton, which support large numbers of fish and dense beds of clams. It is also a habitat for the La Plata Dolphin, a rare, long-beaked species of river dolphin.

SATELLITE VIEW
The main flow of river water over the sediments on the estuary bed is visible here, as well as the River Paraná at top left and River Uruguay at top centre.

Curonian Lagoon

TYPE
Freshwater coastal
lagoon

AREA
1,580 square km
(610 square miles)

LOCATION On the Baltic Sea coasts of Lithuania and
the Kaliningrad Oblast (part of Russia)

The Curonian Lagoon is a non-tidal
lagoon on the southeastern edge of
the Baltic Sea, with an average depth
of just 3.8m (12ft). The River Neman
flows into the lagoon's northern
(Lithuanian) section, which discharges
into the Baltic via a narrow channel,
the Klaipeda Strait. While most of the
lagoon consists of fresh water, seawater
sometimes enters its northern part via
the Klaipeda Strait following storms. In
the past, the lagoon has suffered heavy
pollution from sewage and industrial
effluents, but attempts are now being
made to address this problem.

The lagoon is separated from the
Baltic by the narrow, curved Curonian
Spit, which is 98km (60 miles) long.
The spit is notable for its mature
pinewoods and drifting barchans
(sand dunes), some reaching a height
of 60m (200ft), which extend for
31km (20 miles) along the spit. The
sandy beaches on the spit, together
with vistas over the lagoon, woods,
and drifting dunes, make it a tourist
attraction, and in 2000 the entire spit
was designated a UNESCO World
Heritage site.

DUNES AND LAGOON
This quiet corner of the northern part of the
lagoon is backed by the Curonian Spit's high
dunes. Migrating birds use the lagoon and
nearby Neman Delta for vital rest breaks.

READS ISLAND
This low-lying island,
in the upper part of the
estuary, is a breeding
ground for avocets and
other rare birds and is
managed as a nature
reserve. The view here
is looking downstream.

Humber Estuary

TYPE Fully mixed
(tide-dominated) estuary

AREA Approximately
200 square km
(80 square miles)

LOCATION West and southeast of Kingston-upon-
Hull, eastern England, UK

This large estuary on Great Britain's
eastern coastline is formed from the
confluence of the rivers Ouse and
Trent. It discharges about 250,000
litres (60,000 gallons)
of water per second
into the North Sea,
the largest input from
any British river into
this sea. After the end
of the last ice age,
when sea levels were
much lower, the
Humber was a river
that flowed up to 50km (30 miles)
past the present coastline before
reaching the sea.

About 100,000 cubic metres
(3.6 million cubic feet) of sediment
are deposited in the estuary every
year, mainly from offshore by tidal
action. Shifting shoals formed by this
sediment can obstruct shipping. The
estuary's intertidal areas are productive
ecosystems that support a wide range
of molluscs, worms, crustaceans, and
other invertebrates. These are vital
sources of food for birds, especially
waders. The estuary also supports
a colony of Grey Seals, and many
lampreys pass through it every year.

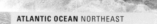

Hardanger Fiord

TYPE Highly stratified
estuary; fiord

AREA Approximately
1,500 square km
(600 square miles)

LOCATION Southeast of Bergen, southwestern
Norway

Like all fiords, the Hardanger Fiord in
Norway is much deeper than a typical
coastal-plain estuary, with a maximum
depth of some 800m (2,600ft). Near
its mouth is a sill just 150m (500ft)
deep. At 183km (114 miles) long, it is
the third longest fiord in the world.

UPPER FIORD
The fiord's narrow upper parts are fed by
several spectacular waterfalls, such as the
Vøringsfossen, which treefalls 182m (600ft).

Hardanger Fiord was formed about
10,000 years ago, when a large glacier
that had carved out and occupied
a deep U-shaped valley in the area
began to melt and retreat. As it did
so, seawater flooded into the valley
to create the fiord. Today, the fiord
continues to receive a large input
of fresh water from glacier melt.
Throughout much of its length, the
fiord is stratified into a lower layer of
salt water, which moves into the fiord
during flood tide, and an upper layer
of fresher water that flows outwards
to the sea on the ebb tide.

Eastern Scheldt
Estuary

TYPE
Former estuary, now
a sea-arm

AREA
365 square km
(140 square miles)

LOCATION Southwest of Rotterdam, southwestern
Netherlands

The Eastern Scheldt Estuary is a tidal
body of water 40km (25 miles) long,
with a salinity similar to that of
seawater. Since the late 1980s, it has
been cut off from its input of fresh
water from the River Scheldt by dams,
leading to its reclassification as a sea-
arm rather than an estuary.

It has also been defended against
seawater flooding by a storm-surge
barrier (see p.104). This was originally
to have been a fixed dam to prevent
any ingress of seawater at all, but there
were fears that, with a dam of this
type, the estuary would gradually lose
its salinity, producing an adverse effect
on its fauna and flora – in particular,
there were concerns that it would end
the large-scale mussel and oyster
farming in the area and degrade the
tidal flats and salt marshes that form
an important habitat for birds. The
government of the
Netherlands therefore
commissioned a
moveable barrier, the
construction of which
was completed in 1986.

STORM BARRIER GATES
The gates are usually
raised, allowing tidal water
in and out of the Eastern
Scheldt Estuary. They are
lowered about twice a year,
during stormy weather.

OCEAN ENVIRONMENTS

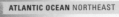

Gironde Estuary

TYPE Fully mixed (tide-dominated) estuary

AREA Approximately 500 square km (200 square miles)

LOCATION North of Bordeaux, western France

The Gironde Estuary, formed by the confluence of the rivers Garonne and Dordogne, is the largest estuary in Europe at almost 80km (50 miles) long and up to 11km (7 miles) wide. The estuary's average discharge rate into the Atlantic is 1 million litres (220,000 gallons) per second. It has a large tidal range, of up to 5m (16ft) during periods of spring tide, and the strong tidal currents in the estuary, as well as numerous sand banks, tend to hamper navigation. One of the Gironde's most impressive features is its tidal bore – a large, wall-like wave at the leading edge of the incoming tide – known locally as the Mascaret. Occurring with each flood tide at the time of spring tides (that is, twice daily for a few days every fortnight), the bore surges from the Gironde upstream into its narrower tributaries. On the Garonne, the Mascaret sometimes forms a barrelling wave, which can reach a height of 1.5m (5ft) and tends to break and reform.

The Gironde is an important artery of the Bordeaux wine region and a rich source of eels and a wide variety of shellfish, which feature on local restaurant menus. Wild sturgeon (the source of caviar) were once also plentiful in the estuary, and although their numbers have declined due to overfishing, they are still farmed in small numbers.

THE MASCARET
When it reaches the River Dordogne, the Mascaret, or Gironde tidal bore, turns into a series of waves, which may travel up to 30km (20 miles) upstream.

Venetian Lagoon

TYPE Saltwater coastal lagoon

AREA 550 square km (210 square miles)

LOCATION On the Adriatic coast of northeastern Italy

The Venetian Lagoon is a very shallow, crescent-shaped coastal lagoon off the northern part of the Adriatic Sea. It is the largest Italian wetland and a major Mediterranean coastal ecosystem.

In addition to Venice, which sits on a small island at the centre, the lagoon contains many other islands, most of which were marshy but have now been drained. Its average depth is just 70cm (28 inches), so most boats cross the lagoon only via dredged navigation channels, and four-fifths of its area consists of salt marshes and mudflats. It takes in both riverine fresh water and seawater, and its tides have a range of up to 1m (3ft). During periods of spring tide, Venice is regularly flooded (see p.90). While its inhabitants have coped with this problem for centuries, land subsidence and rising sea levels now pose a major threat to the city and its art treasures. Marine life in the lagoon includes many species of fish (from anchovies to eels, mullet, and sea bass) and invertebrates. Sea birds, waterfowl, and waders proliferate on the many uninhabited islands. Efforts are now being made to reduce industrial and agricultural pollution, including attempts to capture pollutants by means of shrubs planted along the edges of the lagoon.

WATERY GEM
In the centre of this photograph, taken from the International Space Station, is the fish-shaped main island of Venice. Below it is one of the lagoon's three protective barrier islands.

JAMES ISLAND

Gambia Estuary

TYPE Salt-wedge (river-dominated) estuary

AREA Approximately 1,000 square km (400 square miles)

LOCATION East of Banjul, Gambia, West Africa

The Gambia Estuary is the western half of the River Gambia, which runs 1,130km (700 miles) through West Africa. The estuary is tidal throughout and discharges about 2 million litres (450,000 gallons) per second into the Atlantic during the rainy season, but only 2,000 litres (450 gallons) in the dry season. It contains abundant stocks of fish and shellfish, including various species of barracuda, catfish, and shrimp. James Island, near the estuary's mouth, was a former slave-collection point and is now a UNESCO World Heritage Site.

Ebrié Lagoon

TYPE
Coastal lagoon of variable salinity

AREA
520 square km
(200 square miles)

LOCATION West of Abidjan, Ivory Coast, West Africa

The Ebrié Lagoon is one of three long, narrow lagoons that line the shores of the West African state of Ivory Coast. With a length of 120km (62 miles) and an average width of 4km (2½ miles), it is the largest lagoon in West Africa. Its average depth is 5m (16ft). At its eastern end, it

TIAGBA VILLAGE
In the village of Tiagba, on the outskirts of a small island in the Ebrié Lagoon, the buildings are raised up on wooden piles.

connects to the Atlantic via a narrow artificial channel, the Vridi Canal, which was opened in 1951. Abidjan, the largest city in Ivory Coast, stands on several converging peninsulas and islands in an eastern part of the lagoon; other communities situated on or in the lagoon include Jacqueville and the village of Tiagba (see below). The Komoé River provides the main input of fresh water. In winter the lagoon becomes salty, but it turns to fresh water during the summer rainy reason. The levels of pollution in the lagoon have been moderately high for some years due to discharge of untreated industrial effluents and sewage from the nearby urban areas.

Kerala Backwaters

TYPE Chain of coastal saltwater lagoons

AREA Approximately 1,000 square km (400 square miles)

LOCATION Southeast of Cochin, Kerala State, southwestern India

The backwaters of Kerala in southern India are a labyrinth of lagoons and small lakes, linked by 1,500km (900 miles) of canals. The lagoons are

VEMBANAD LAKE
Vembanad, the largest Kerala coastal lagoon, is listed as a Wetland of International Importance under the Ramsar Convention.

shielded from the sea by low barrier islands that formed across the mouths of the many rivers flowing down from the surrounding hills. During the summer monsoon rains, the lagoons overflow and discharge sediments into the sea, but towards the end of the rains, the seawater rushes in, altering salinity levels. The aquatic life in the backwaters, which includes crabs, frogs, otters, and turtles, is well adapted to this seasonal variation.

LAKES AND LAGOON
In this satellite view, the Coorong Lagoon is the narrow blue strip behind the yellow sand dunes. Above are the lakes Alexandrina (left) and Albert (right).

PELICANS IN DECLINE

The Coorong is home to a large breeding colony of Australian Pelicans, which inhabit a string of islands in the centre of the lagoon. Since the 1980s, however, their numbers have fallen significantly due to reduced flows of fresh water into the Coorong from the Murray River. The resultant higher salt levels in the lagoon have reduced the growth of an aquatic weed that is a major part of the food chain.

AUSTRALIAN PELICANS
This pelican, one of seven species worldwide, is widespread in Australia, where it lives on freshwater, brackish, and saltwater wetlands.

Coorong Lagoon

TYPE
Saltwater coastal lagoon

AREA
200 square km
(80 square miles)

LOCATION Southeast of Adelaide on the southeastern coast of South Australia

The Coorong Lagoon is a wetland that lies close to the coast of South Australia. It is famous as a haven for birds, ranging from swans and pelicans

to ducks, cranes, ibis, terns, geese, and waders such as sandpipers and stilts. The lagoon is separated from the Indian Ocean by the Younghusband Peninsula, a narrow spit of land covered by sand dunes and scrubby vegetation. The lagoon is about 150km (93 miles) long, with a width that varies from 5km (3 miles) to just 100m (330ft). At its northwestern end, the lagoon meets the outflow from Australia's largest river, the Murray, after the river has passed through Lake Alexandrina. In this region, called the Murray Mouth, both river and lagoon meet the sea, and the Coorong can receive both fresh and salty water.

The lagoon was once freely connected to the lake, from which it received a much larger supply of fresh water. In 1940, however, barrages were built between the lagoon and the lake to prevent seawater from reaching the lake and the lower reaches of the Murray River.

The salinity of the lagoon's waters increases naturally with distance from the sea due to evaporative losses. However, reduced water flows from the Murray, due to a combination of barrage construction and extraction of water for irrigation projects, has caused a gradual further increase in salinity throughout the lagoon. There

is ample evidence that this has adversely affected the lagoon's ecosystem. In particular, several species of plant have become less abundant or disappeared, many fish species have declined, and migratory bird numbers have fallen. Further, the reduced flow from the Murray may result in the eventual closure of the channel joining the lagoon to the ocean, which would prevent migration of fish and other animals between the two.

INDIAN OCEAN SOUTHEAST

Northern Spencer Gulf Estuary

TYPE
Inverse estuary

AREA
Approximately
5,000 square km
(2,000 square miles)

LOCATION Northwest of Adelaide, South Australia

The estuary in the north of Australia's Spencer Gulf is classified as an inverse estuary, owing to its unusual pattern of salt distribution and water circulation.

DEEP GULF
Spencer Gulf is the larger wedge-shaped coastal indent visible in this satellite image. The desert around its head helps produce the estuary's unusual circulation pattern.

In a reverse of the usual pattern, this estuary's waters become saltier towards it head, away from its mouth. This is because its head is surrounded by hot desert and loses more water to evaporation than enters it from rivers. The head's high salinity means that it draws in from the mouth ocean water of lower salinity than the water drawn in by a typical estuary. The estuary is surrounded by extensive tidal flats, seagrass banks, and mangroves.

PACIFIC OCEAN WEST

Pearl River Estuary

TYPE
Salt-wedge (river-dominated) estuary

AREA
1,200 square km
(450 square miles)

LOCATION Northwest of Hong Kong, Guangdong, southeastern China

The bell-shaped Pearl River Estuary receives and carries most of the outflow from the Pearl River, the common name for a complex system of rivers in the southern Chinese province of Guangdong. The estuary is nearly 60km (37 miles) long, and its width increases from 20km (12 miles) at its head to about 50km (30 miles) at its mouth. To the north and west of the estuary is a delta, formed from the confluence of the Xi Jiang and other rivers of the Pearl River system. Together, these rivers discharge an average of

GUANGZHOU
Formerly known as Canton, this large and busy port city lies on a northerly extension of the Pearl River Estuary.

10 million litres (2.2 million gallons) of water per second into the South China Sea. Mostly less than 9m (30ft) deep, but containing some deeper dredged channels, the Pearl River Estuary has a tidal range of 1–2m (3–6ft). It drains water from some of the most densely populated areas of China, and so is severely polluted with sewage and industrial waste. About 560 million tonnes (550 million tons) of domestic waste and 2 billion tonnes (1.95 billion tons) of industrial effluent enter the estuary each year. Over the past 20 years, this pollution has led to frequently occurring algal blooms that threaten local fishing and aquaculture. Pollution is also a threat to the 1,400 Chinese White Dolphins that live in the estuary.

PACIFIC OCEAN WEST

Yangtze Estuary

TYPE
Partially mixed estuary

AREA
2,500 square km
(1,000 square miles)

LOCATION Northwest of Shanghai, eastern China

The Yangtze Estuary is the lower, tide-affected part of the Yangtze (or Changjiang) – the longest river in Asia and the third longest in the world. The estuary occupies 700km (430 miles) of the river's 6,300-km (3,900-mile) length. Near its mouth, it splits into three smaller rivers and numerous streams that run through a delta. Here, silt deposition continually creates new land, which is used for agriculture.

The estuary carries an average of 30 million litres (6.6 million gallons) of water per second into the East China Sea; its average depth is 7m (23ft), and the average tidal range at its mouth is 2.7m (9ft). It supports large numbers of fish and birds, although fish stocks have declined over the past 20 years due to overfishing and pollution. The estuary's waters may be fresh, brackish, or salty, depending on the season.

In winter, salt water intrudes a significant distance upstream, making the water unfit for drinking and irrigation. Recently, this intrusion has occurred more frequently due to reduced river flow – a reduction that is likely to be exacerbated by the Three Gorges Dam project further upstream. Reduced flows may worsen the acute water shortage in the city of Shanghai on the estuary's southern shore, as well as affect the dispersion and dilution of pollutants around the estuary. Silt deposition in the delta is also likely to fall, reducing the rate of new land creation.

WUHAN BRIDGE
The double-decker Wuhan bridge, completed in 1957, was the first bridge built over the lower part of the Yangtze.

PACIFIC OCEAN SOUTHWEST

Doubtful Sound

TYPE	Highly stratified estuary; fiord
AREA	70 square km (30 square miles)

LOCATION West of Dunedin, southwestern South Island, New Zealand

Doubtful Sound is one of 14 major fiords that were formed 15,000 years ago in a scenic part of New Zealand's South Island. Some 40km (25 miles) long and opening onto the Tasman Sea, it is surrounded by steep hills from which hundreds of small waterfalls descend during the rainy season. Its name originated in 1770 during the first voyage to New Zealand by the English explorer Captain James Cook (1728–79). He called the fiord Doubtful Harbour because he was sceptical of being able to sail out again if he entered it. Doubtful Sound is the second-longest and the deepest of the New Zealand fiords, with a maximum depth of 421m (1,380ft). It receives fresh water from a hydroelectric power station at its head and from a huge 6,000m (236in) of rainfall annually. Like all fiords, it contains fresh water in its top few metres and a much denser, colder, saltier layer below. There is little mixing between the two. Doubtful Sound is home to Bottlenose Dolphins, New Zealand Fur Seals, and many species of fish, starfish, sponges, and sea anemones.

SOUND VIEW
This view of the head of Doubtful Sound, looking towards the open ocean, is from the hills of the south-central region of New Zealand's South Island.

PACIFIC OCEAN NORTHEAST

San Francisco Bay

TYPE	Partially mixed tectonic estuary
AREA	4,160 square km (1,600 square miles)

LOCATION Central California, western USA

San Francisco Bay, the largest estuary on North America's west coast, consists of four smaller, interconnected bays. One of these, Suisun Bay, receives fresh water drained from about 40 per cent of California's land area. This water flows into San Pablo Bay and then Central Bay, where it mixes with salt water that has entered at depth from the Pacific Ocean via the Golden Gate channel. From Central Bay, there is little flow of fresh water to the largest body of water, South San Francisco Bay, but there is

OAKLAND BAY BRIDGE
Thick fog surrounds the lower half of the San Francisco–Oakland Bay Bridge, one of five bridges that cross the bay.

some surface outflow of brackish water to the Pacific. San Francisco Bay is a tectonic estuary – one caused by movement at tectonic faults (lines of weakness) in the Earth's crust, of which there are several in the area, notably the San Andreas Fault.

During the past 150 years, human activity has resulted in the loss of 90 per cent of the bay's surrounding marshy wetland, a greatly reduced flow of fresh water (which has been diverted for agricultural purposes), and contamination by sewage and effluent. Nevertheless, the bay remains an important ecological habitat. Its waters are home to large numbers of economically valuable marine species, such as Dungeness Crab and Pacific Halibut, and millions of geese and ducks annually use the bay as a refuge.

PACIFIC OCEAN EAST

Laguna San Ignacio

TYPE	Hypersaline coastal lagoon
AREA	360 square km (140 square miles)

LOCATION On the Pacific coast of the Baja California Peninsula, Mexico, southeast of Mexicali

The Laguna San Ignacio is a coastal lagoon in northwestern Mexico best known as a sanctuary and breeding ground for Pacific Gray Whales. Latin America's largest wildlife sanctuary, it is also an important feeding habitat for four endangered species of sea turtle. The lagoon, which is 40km (25 miles) long and on average 9km (6 miles) wide, receives only occasional inflows of fresh water, and its evaporative losses are high. Its salinity is therefore significantly higher at its head than at its mouth, where it connects to the sea. Apart from whale watching, the main human activities in the area are small-scale fisheries and oyster cultivation. In 1993, the lagoon was designated a World Heritage Site.

LAGOON BEACH
Waves break on the shore at San Ignacio Lagoon, which is surrounded by a landscape of sparse desert scrub.

HUMAN IMPACT

WHALE WATCHING

The Laguna San Ignacio is a popular whale-watching site. Between January and March, large numbers of Gray Whales can be found there. The whales, which often approach boats, use the upper part of the lagoon for giving birth, while the lower lagoon is where males and females look for mates. Females swim with their calves in the middle part of the lagoon.

OCEAN ENVIRONMENTS

SALT MARSHES AND TIDAL FLATS

A SALT MARSH IS A VEGETATED AREA OF COAST that is partly flooded by the sea at high tide and completely flooded by the highest spring tides. Many areas of salt marsh are bordered by tidal flats. These are broad areas of mud or sand, mainly without vegetation, that are uncovered at low tide and covered as the tide rises. Salt marshes and tidal flats are depositories for large amounts of organic material, derived from decaying plants and animals. This provides the base for an extensive food chain.

FORMATION AND FEATURES

Tidal flats occur on low-energy sheltered coasts, such as estuaries and enclosed bays, where sediment held in the water settles out and builds up. The most extensive flats occur where there is a high tidal range. Tidal flats may consist either of sand (sandflats) or mud (mudflats), or a mixture of these. Mudflats contain a higher concentration of the decaying remains of dead organisms than sandflats and are also the first stage in the development of salt marshes. These develop on the landward side of mudflats. As various salt-tolerant plants grow, their roots trap sediment and stabilize the mud. As the vegetated flat builds up, different types of plants become established. The result is a salt marsh, consisting of blocks of flat, low-growing vegetated areas of mud, broken up by sinuous channels.

DISTRIBUTION
Salt marshes and tidal flats occur only north of the latitude of 32°N and south of 38°S. In latitudes nearer the Equator, they are replaced by mangrove swamps.

BAY OF FUNDY
In this small sub-estuary of Canada's Bay of Fundy, an area of salt marsh is visible in the background. In the foreground is a broad intertidal area of mud and gravel.

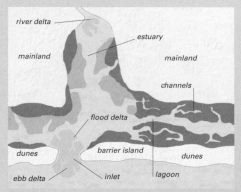

river delta
estuary
mainland
mainland
channels
flood delta
dunes
barrier island
dunes
ebb delta
inlet
lagoon

COASTAL SETTING
Salt marshes commonly develop in coastal lagoons or in estuarine areas that are sheltered from the sea by spits or barrier islands. The channels transport salt water, plankton, nutrients, sediment, and plant detritus into and out of the marsh.

KEY
salt marsh
tidal flats

ZONES AND EVOLUTION

Salt marshes have two main zones. The parts flooded by every high tide are called low marsh, while the areas that are only occasionally flooded are termed high marsh. Each zone is colonized by distinct species of salt-tolerant plants. Each species, of which there are many, has developed special mechanisms to deal with the high levels of salt they are exposed to: some possess salt-excreting glands, for instance, while others have storage systems for collecting the salt until they can dilute it with water. Salt marshes and adjoining mudflats usually evolve over time. As sediment builds up, the mud surface in the marsh, the adjoining flats, and the bay or estuary as a whole tends to rise. As it does so, areas of low marsh become high marsh and areas of mudflat are colonized by plants, turning into low marsh.

SEA LAVENDER
Sea lavenders are common high-marsh colonizers. They bloom in summer, producing purple or lavender flowers.

SALT-MARSH ZONES
The low marsh is the part flooded once or twice a day at high tide, while the high marsh is the area above the mean high-tide level – it is flooded only occasionally, by the highest spring tides. Each zone has distinctive vegetation.

SALT-MARSH CORDGRASS
Also called Smooth Cordgrass, this species is the dominant low-marsh plant throughout the Atlantic coast of North America. Stands of this grass grow to 2m (7ft) high.

pool
highest spring tide
upper high marsh
mean high tide
lower high marsh
mean sea level
upland
high marsh
low marsh
mudflat

ALGAE-COVERED MUDFLATS
Some mudflats, such as these in Alaska, become
heavily encrusted with green algae. The algae
is often itself colonized by large numbers of
tiny marine snails.

CONSERVING SALT MARSHES

Salt marshes are threatened
worldwide through being built
on, converted to farmland, or
even used as waste dumps. Over
half of the original salt marshes in
the USA, for example, have been
destroyed. This is regrettable as
salt marshes are valuable wildlife
habitats and centres of biodiversity.

MARSH HOUSING DEVELOPMENT
This coastal development in Myrtle Beach,
South Carolina, USA, has been built on top of
a drained salt marsh. However, the adjoining
area of marsh has been carefully preserved.

ANIMAL LIFE

Measured by the amount of organic matter (the base material for food
chains) that they produce, salt marshes are extremely productive habitats.
Most of this material comes from decaying plant material. When plants die,
they are partially decomposed by bacteria and fungi, and the resulting detritus
is consumed by animals such as worms, mussels, snails, crabs, shrimps, and
amphipods living in the marsh, and zooplankton living in the salt water. These
in turn provide food for larger animals. Salt marshes provide nursery areas for
many species of fish, and feeding and nesting sites for birds such as egrets, herons,
harriers, and terns. Tidal flats are
home to many types of crustaceans,
worms, and molluscs, which either
feed on the surface or burrow beneath
it. These in turn provide food for
enormous numbers of wading birds.

NATTERJACK TOAD
This toad, found in parts of
western and northern Europe,
inhabits upper salt marsh
habitats (just below the high
marsh), where it uses shallow
ponds to breed.

GREAT EGRET
A common inhabitant of salt marshes in the
USA and parts of east Asia, the Great Egret
feeds on small fish, invertebrates, and small
mice. It breeds in some marshes.

LUGWORM CASTS
Lugworms live in burrows some 20–40cm
(8–16in) deep in tidal flats. They feed by
taking in sand or mud, digesting any organic
matter, and excreting the rest as a cast.

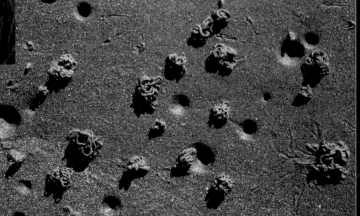

MARSH AT LOW TIDE
Patches of salt marsh surround the basin, together with tidal flats that can extend for up to 5km (3 miles) from the shore at low tide.

ATLANTIC OCEAN NORTHWEST

Minas Basin

TYPE Tidal sandflats and mudflats, and salt marshes

AREA 1,250 square km (490 square miles)

LOCATION Eastern part of Bay of Fundy, Nova Scotia, Canada

The Minas Basin is a semi-enclosed inlet of the Bay of Fundy. It consists of a triangular area of tidal mudflats and sandflats, surrounded by patches of salt marsh, most of which have been dyked and drained for agriculture. Twice a day, the sea fills and empties the basin, rising and falling by over 12m (40ft), which is the largest tidal range in the world. No other coastal marine area has such a large proportion of its floor exposed at low tide. Sediments in the basin, which are brought in and deposited by tides, range from coarse sand to fine silt and clay. The tidal flats formed by these sediments contain high densities of a marine amphipod, the Bay of Fundy Mud-shrimp, which provides food for huge numbers of migrating shorebirds, including sandpipers and plovers. The numbers peak from July to October, and for some species exceed 1 per cent of the world population.

SEMIPALMATED SANDPIPER
Half a million Semipalmated Sandpipers stop off in the Minas Basin each year on their way from North America's Arctic regions to South America.

RACE POINT
A typical area of salt marsh can be seen here behind dunes at Race Point, at the northern extreme of Cape Cod.

ATLANTIC OCEAN NORTHWEST

Cape Cod Salt Marshes

TYPE Salt marshes

AREA 80 square km (30 square miles)

LOCATION Cape Cod, eastern Massachusetts, USA

Salt marshes are the dominant type of coastal wetland around Cape Cod, although about a third of the region's marshes have been lost or severely degraded within the past 100 years. These salt marshes occur behind barrier beaches or spits and within estuarine systems, and have developed over the past 3,000 years in response to sea-level rise. They mainly consist of high marsh, where the dominant plant species is Saltmeadow Cordgrass, with some scattered areas of low intertidal marsh, dominated by Smooth Cordgrass. The low marsh areas are flooded twice daily and the high marsh twice a month, during the highest spring tides. The largest individual marsh is the Great Salt Marsh to the west of the town of Barnstable. With deep channels running through it, this is a popular area to explore by kayak.

The marshes around Cape Cod serve as a breeding and foraging habitat for a diversity of brackish and freshwater animals. Among these are two rare and protected bird species, the Northern Harrier and Least Tern, and two endangered reptiles, the Diamond-backed Terrapin and Eastern Box Turtle. Restoring degraded salt marshes on Cape Cod is regarded as a top priority for many regional and national conservation organizations. Restoration will allow these wetlands to regain their function as a barrier protecting the coastline from storm surges and as a natural sponge that filters pollutants and excess nutrients from the water run-off in the region.

ATLANTIC OCEAN NORTHWEST

South Carolina Low Country

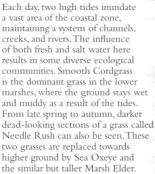

TYPE Salt marshes and tidal mudflats

AREA 1,600 square km (630 square miles)

LOCATION South Carolina coast, southwest and northeast of Charleston, USA

The Low Country contains one of the most extensive systems of salt marsh and tidal flats in the United States. Its size results from the broad, gently sloping, sandy coast of the US eastern seaboard, coupled with a moderately high tidal range of 1.5–2m (5–7ft).

Each day, two high tides inundate a vast area of the coastal zone, maintaining a system of channels, creeks, and rivers. The influence of both fresh and salt water here results in some diverse ecological communities. Smooth Cordgrass is the dominant grass in the lower marshes, where the ground stays wet and muddy as a result of the tides. From late spring to autumn, darker dead-looking sections of a grass called Needle Rush can also be seen. These two grasses are replaced towards higher ground by Sea Oxeye and the similar but taller Marsh Elder.

In the lower marshes and the bordering tidal flats, mud snails, crabs, shrimps, worms, and other tiny inhabitants burrow into the mud, while attached and clinging to the stalks of the grasses are Ribbed Mussels and Marsh Periwinkles. Among the fish living in the silty tidal wash are croaker, menhaden, and mullet. Birds living here include Marsh Wrens and Clapper Rails.

CORDGRASS MEADOWS
A tidal channel weaves its way through stands of Smooth Cordgrass, the dominant plant species in the lower marsh areas.

SALT-MARSH MOSAIC
The edges of the Wadden Sea are a mosaic of marsh patches broken up by shallow tidal channels.

ATLANTIC OCEAN NORTHEAST

Wadden Sea

TYPE Tidal mudflats and sandflats, salt marshes, and islands

AREA 10,000 square km (4,000 square miles)

LOCATION North Sea coast from Esbjerg, Denmark, along northern Germany, to Den Helder, Netherlands

The Wadden Sea is not a sea in the conventional sense but an extensive body of shallow water and associated tidal flats, salt marshes, and low lying islands in northwestern Europe.

Straddling the shores of Denmark, Germany, and the Netherlands, the Wadden Sea has been formed by storm surges and sea-level rise inundating an area of coast, combined with the deposition of fine silt by rivers. It is an important nursery for North Sea fish species such as plaice and Common Sole, and its extensive mudflats are home to enormous numbers of molluscs and worms. The salt marshes provide a habitat for more than 1,500 species of insects and are important feeding and breeding grounds for many species of birds. Unfortunately, these marshes are threatened by agriculture, dyke building for coastal protection, and tourism.

ATLANTIC OCEAN NORTHEAST

Morecambe Bay

TYPE Tidal mudflats and sandflats, and salt marshes

AREA 310 square km (120 square miles)

LOCATION Northwest England, UK

Formed from the confluence of five estuaries, those of the rivers Kent, Keer, Leven, Lune, and Wyre, Morecambe Bay is the largest continuous area of tidal flats in the UK. Broad, shallow, and funnel-shaped, the bay has a large tidal range, of up to 10.5m (35ft). During periods of spring tides, the sea can ebb as far as 12km (7 miles) back from the high-water mark. The flood tide comes up the bay faster than a person can run, and parts of the bay are also affected by quicksands, posing dangers for anyone who does not know the area well.

The bay's extensive mudflats support a rich and diverse range of invertebrate animals, including cockles and mussels, snails, shrimps, and lugworms, as well as one of the largest populations of shorebirds in the UK. The bay regularly hosts 170,000 wintering waders, with several species present in internationally significant numbers, including oystercatchers, curlews, Dunlins, and Knots. The tidal flats are surrounded by extensive salt marshes, which make up about 5 per cent of the total salt marsh in the UK

and support a number of rare plants. Much of this marsh area is grazed by sheep and cattle.

The bay is an important location for commercial fishing; the fish species most commonly caught include bass, cod, whitebait, and plaice. However, Morecambe Bay has not escaped the problems of pollution common to many coastal areas of northwestern Europe. Oil, chemicals, and plastic are among the more common pollutants of this ecosystem.

MORECAMBE MUDFLATS
The ebbing tide reveals half of the bay's total area as undulating expanses of mud and sand, meandering channels, and tidal pools.

HUMAN IMPACT

COCKLING

Morecambe Bay has many rich cockle beds. The cocklers use planks of wood called jumbos to soften the sand, which helps draw the cockles to the surface. Because of the fast-moving tides, cockling has to be carried out with an eye to safety. In February 2004, at least 21 Chinese migrant workers tragically drowned after being cut off by the tides.

The Wash

TYPE	Salt marshes, tidal sandflats, and mudflats
AREA	250 square km (100 square miles)

LOCATION Northeast of Peterborough, England, UK

The Wash is a large, square-mouthed, shallow estuary on the eastern coast of England, surrounded by extensive areas of tidal sandflats, some mudflats, and salt marshes. It is fed by four main rivers: the Great Ouse, Nene, Welland, and Witham. The sandflats of the Wash range from extensive fine sands to drying banks of coarse sand and are home to large communities of bivalve molluscs, crustaceans, and polychaete worms. The extensive salt marshes

TERRINGTON MARSHES
Located close to the mouth of the River Nene, these marshes form part of the Wash National Nature Reserve.

comprise the largest single area of this habitat in Britain and are growing in extent. The main plant species making up these salt marshes, which are traditionally used as grazing lands by farmers, are cordgrass and glasswort in roughly equal amounts.

The Wash is one of the most important sites in the UK for wild birds, its sheltered tidal flats providing a vast feeding ground for migrating birds, such as geese, duck, and waders.

These come to spend the winter in the Wash in huge numbers, with an average total of about 300,000 birds, from as far away as Greenland and Siberia. In addition, the Wash is an important breeding area for Common Terns and a feeding area for Marsh Harriers. It has been declared a Special Protection Area (SPA) under EU law.

In 2000, parts of the artificial coastal defences on the western side of the Wash were deliberately breached

to increase the area of salt marsh in the region. This has taken pressure off other nearby sea defences, because the newly establishing area of salt marsh soaks up wave energy, acting as a natural sea defence. This is a relatively novel approach to coastal management that employs "soft engineering" techniques to defend against the erosive power of the sea. It also has the added environmental advantage of providing additional habitat for wildlife.

EDGE OF THE MARSHES
The dominant plant species in the non-exploited areas of salt marsh, such as at La Turballe in the northern part of the marshes, include sea-blite, cordgrass, and glasswort.

Guérande Salt Marshes

TYPE	Salt marshes, artificial salt pans, and tidal mudflats
AREA	50 square km (20 square miles)

LOCATION Northwest of St. Nazaire, Atlantic coast of France

The region of salt marshes close to the medieval town of Guérande is most famous for its salt production but is also a noted ecological site, important for its role as a feeding and resting site

for large numbers of birds. The salt marshes came to exist in their present state through a combination of geology, climatic factors, and human intervention. Around the coast near Guérande, a system of spits and coastal dunes developed thousands of years ago, cutting off an area of shallow water, which was nevertheless subject to tides – seawater could flow in through two inlets in the dune belts. Over the centuries, marshes and tidal flats developed in this basin. During the past 1,000 years or so, these have been artificially converted into a mosaic of salt pans, separated by clay walls, although some areas remain unexploited. During the flood tide, seawater is allowed to flow through

channels into the pans, and during the warm summer months, when the rate of evaporation is high, sea salt is skimmed from the surface of the pans by an army of salt-farmers (*paludriers*).

The areas of marsh surrounding the salt pans are made up of various salt-tolerant plants. More than 70 different species of birds nest and breed in the area, and many species spend the winter here in large numbers. For many years, the salt-farmers and the French ornithological society, the LPO, have jointly organized exhibitions and guided tours in the Guérande Salt Marshes, which are themed on the economics of salt production, the ecology of the marshes, and their need for protection.

SALT HARVEST

The Guérande region has had salt pans for over 1,000 years. Today, about 200 salt-farmers work in the area, one of the few places in France where salt continues to be produced in a traditional manual way. The average annual harvest is about 10,000 tons of natural, mineral-rich sea salt, which is sold unrefined, with nothing added and nothing removed. The salt has a light grey colour because of its content of fine clay from the salt pans.

Saemangeum Wetlands

TYPE
Mudflats, sandflats, and
salt marshes

AREA
400 square km
(155 square miles)

LOCATION South of Seoul, on the west coast of South Korea

Situated at the confluence of the Mangyeung and Dongjin river estuaries, on South Korea's Yellow Sea coast, the Saemangeum Wetlands is a shorebird staging site of great importance. Its tidal flats and shallows support many bird species, some of which are considered to be globally threatened. In 2006, the status of this

SPOON-BILLED SANDPIPER
This extremely rare species is one of the shorebirds most threatened by the reclamation project.

wetland – as well as the thousands of migratory birds that depend on it as a key feeding area – came under threat due to the completion of a 33km- (22-mile-) long sea wall at the mouth of the two estuaries. The sea wall is part of a reclamation project

LOW TIDE AT DONGJIN ESTUARY
The area around the estuary consists of tidal flats and scattered salt marsh intersected by channels that fill at high tide.

that involves cultivating the tidal flats, and converting the remaining estuarine areas into freshwater reservoirs. The project is going ahead despite the fears of conservation groups that it will result in irreversible environmental damage.

Yatsu–Higata Tidal Flat

TYPE
Tidal mudflat

AREA
0.4 square km
(1/6 square mile)

LOCATION Narashino City, at the northern part of Tokyo Bay, Japan

Yatsu–Higata is a tiny rectangular mudflat at the northern end of Tokyo Bay, and is unusual because it is almost completely surrounded by a dense urban area. Once open shoreline, Yatsu–Higata now sits 1km (3/5 mile) inland. Twice daily, it experiences a tidal inflow and outflow of water from Tokyo Bay via two concrete channels. When the tide comes in, the mudflat fills with about 1m (3ft) of water. When it flows out, a variety of resident and migrant shorebirds congregate to feed on the lugworms, crabs, and other marine animals that live within the fine silt that remains. Yatsu–Higata is an important stopover point for migrating birds flying from Siberia to Australia and Southeast Asia.

Alaskan Mudflats

TYPE
Tidal mudflats

AREA
10,000 square km
(4,000 square miles)

LOCATION Various coastal inlets of southern and western Alaska, USA

Many areas on the coast of southern and western Alaska are fringed by mudflats that appear at low tide. They are formed of a finely ground silt that in some areas is several hundred metres deep. This silt has originated from the action of Alaska's numerous glaciers, which have been grinding away on the surrounding mountains for thousands of years. As these glaciers melt, the silt is carried to the coast in meltwater and deposited as

sediment when it reaches the sea. Because tidal ranges around Alaska are generally high, the total area of mudflats exposed at low tide is huge. These mudflats are an important stopover for migrating shorebirds. Various species of burrowing worm and bivalve mollusc are an important source of food for these waders and for the waterfowl that feed on the mudflats through the winter. Harbour Seals also use the mudflats as rest areas.

The mudflats are dangerous for human visitors, because in some areas they behave like quicksand. Even mud that at first seems firm enough to support a person may in reality be treacherous. A number of people have become stuck and some have even drowned.

DRYING MUDFLATS
These mudflats are at the edge of a large delta on the southwest coast of Alaska, formed by the Yukon and Kuskokwim rivers.

DIGGING FOR CLAMS

Brown bears are occasional visitors to some Alaskan mudflats, where they dig for Pacific Razor Clams buried in the mud. They probably find the clams by looking for the small holes they leave on the surface as they burrow down. Extracting them is tricky, as when disturbed, they burrow down further.

ALASKAN BROWN BEAR
This large adult bear is digging on the coast of Katmai National Park, at the eastern end of the Alaskan Peninsula.

OCEAN ENVIRONMENTS

MANGROVE SWAMPS

A MANGROVE SWAMP IS A COLLECTION of salt-tolerant evergreen trees, thriving in an intertidal environment in the tropics or subtropics. Mangrove swamps line about eight per cent of the world's coastlines, where they filter pollutants from river runoff and help prevent the silting up of adjacent marine habitats. They also protect coastlines against erosion and provide a home for fish, invertebrates, and many other animals.

AERIAL ROOTS
Many mangrove species have aerial roots. These prop the tree up and take in oxygen, which is usually not available in the mud that most mangroves grow in.

FORMATION

Mangrove swamps develop in coastal areas protected from direct wave action. These areas often fringe estuaries and coastal lagoons (see p. 114). Most mangroves develop in fine muds or sandy sediments that form in these environments. As the lower parts of the mangrove roots develop in the sediment, aerial roots form a tangled network above it. This traps silt and other material carried there by rivers and tides. Land is built up, and then colonized by other types of vegetation.

DISTRIBUTION
Mangrove swamps occur only between latitudes 32°N and 38°S. Salt marshes and tidal flats (see p.124) replace them elsewhere.

PLANTS

Some 54 species of trees and shrubs are classified as "true" mangroves, occurring only in mangrove habitats. Each has evolved special adaptations to the conditions they grow in, such as salty water. For example, some mangroves can excrete salt in their leaves. On most mangrove shorelines, there are two or three zones, each dominated by different mangrove species. In the Americas, just four main species are found. The area closest to the sea is dominated by red mangroves. Land-ward of this are black mangroves – the roots of this and some other species develop pencil-like breathing tubes, called pneumatophores. White and button mangroves grow further landward. The mangrove swamps on the coasts of the rest of the tropics contain greater species richness.

PNEUMATOPHORES
These vertical tubes grow up out of the sand or mud as extensions of horizontal roots. When exposed to the air, they take in oxygen.

RED MANGROVE
This mangrove species can grow in deep water by means of its numerous prop roots, which often have a reddish tint. It also has a particularly high salt tolerance.

ANIMAL LIFE

Mangrove swamps are rich centres of biodiversity. Mangrove trees produce enormous amounts of leaf litter, as well as twigs and bits of bark, which drop into the water. Some of this immediately becomes food for animals such as crabs, but most is broken down by bacteria and fungi, which turn it into food for fish and prawns. These in turn produce waste, which along with the even smaller mangrove litter, is consumed by molluscs, amphipods, marine worms, small crustaceans, and brittlestars. Some of these become food for larger fish, and the various fish species provide food for larger animals.

Across the world, mangrove swamps are home to an enormous number and diversity of birds and several endangered species of crocodiles. Other types of animals found in great numbers and diversity in mangrove swamps include frogs, snakes, insects, and mammals ranging from swamp rats to tigers.

BANDED ARCHERFISH
This little fish inhabits mangrove swamps in the Indian and Pacific oceans. It is known as an archer-fish because it feeds mainly on flying insects, which it knocks out of the air and into the water by spitting at them.

MANGROVE BRITTLESTAR
This scavenger is one of the few echinoderms found in mangrove swamps. It is highly mobile, using its long arms to pull itself along.

JABIRU STORK
This large stork inhabits mangrove swamps and other wetlands through-out the tropical Americas, feeding on a range of prey, including snakes.

SHELTER FROM PREDATORS
Cardinalfish, sheltering here in a mangrove swamp in Papua New Guinea, are one of the many types of small tropical fish that use mangrove roots for protection from predators.

HUMAN IMPACT

SHRIMP FARMING

Over half the world's mangrove swamps have been destroyed in recent decades and have been built on or turned into commercial enterprises such as shrimp farms (including the one shown here, in Vietnam). Unfortunately, intensive shrimp farming often has devastating environmental effects. Typically, the effluent from shrimp ponds pollutes nearby coastal waters, destroying more mangroves as well as coral reefs along the coastline.

MANGROVE-LINED CHANNEL
Here, parallel stands of Red Mangrove line a shallow offshoot channel of Florida Bay in the southern part of the Everglades National Park.

ATLANTIC OCEAN WEST

Everglades

PRINCIPAL SPECIES
Red, Black, and White Mangroves

AREA Mangroves only: 1,500 square km (600 square miles)

LOCATION Southwestern Florida, USA

Mangroves occupy a large, roughly triangular area at the southwestern tip of southern Florida, where a maze of islands along the coast is intersected by mangrove-lined channels. Here, where the salt water of the Gulf of Mexico and Florida Bay meets fresh water that has travelled from Lake Okeechobee in central Florida, is the largest area of mangrove swamps in North America.

ANHINGA
This diving bird hunts fish, frogs, and baby alligators in the Everglades mangroves.

The dominant species along the edges of the sea and the numerous channels is the Red Mangrove – water within the channels is normally stained brown from tannin contained in the leaves of this species. In addition to their role in stabilizing shorelines with

their large prop roots, Red Mangroves are crucial to the Everglades ecosystem, acting as a nursery for many species of fish, as well as shrimp, mussels, sponges, crabs, and other invertebrates. The other principal mangrove species in the Everglades are the Black Mangrove and White Mangrove. Both of these grow closer to the shore than Red Mangroves, so they are in contact with seawater only at high tide. The Everglades swamps provide a feeding

---

ATLANTIC OCEAN WEST

Alvarado Mangrove Coast

PRINCIPAL SPECIES
Red, White, and Black Mangroves

AREA 1,500 square km (600 square miles)

LOCATION Around Veracruz, southern Mexico, on Bay of Campeche, in southwestern Gulf of Mexico

The Alvarado Mangroves Ecoregion in southern Mexico is an extensive area of mangrove swamps mixed in with other habitats such as reed beds and palm forests. The mangroves grow on flat coastal land interspersed with brackish lagoons fed by several small rivers. The swamps are brimming with life, from rays gliding in the calm waters to snails climbing the mangrove roots, whose tangled network protects many fish and invertebrates from predators. Bird life in and around the swamps includes the Keel-billed Toucan, Reddish Egret, Wood Stork, and several species of heron and kingfisher, while the mammalian inhabitants include spider monkeys and West Indian Manatees. Some large areas of mangrove in the region have been destroyed, and those that remain are under pressure from logging, agricultural expansion, oil extraction, and frequent oil spills.

ATLANTIC OCEAN WEST

Sian Ka'an Biosphere Reserve

PRINCIPAL SPECIES
Red, Black, White, and Button Mangroves

AREA 1,000 square km (400 square miles)

LOCATION Eastern coast of Yucatan Peninsula, eastern Mexico, 150 km (90 miles) south of Cancun

Stretching for 120km (75 miles) along Mexico's Caribbean coast, the Sian Ka'an Biosphere Reserve contains a mixture of mangrove swamps, lagoons, and freshwater marshes; it was declared a World Heritage Site by UNESCO in 1987. The mangroves are protected from the energy of the Caribbean Sea by a barrier reef growing along the coast. However, the reserve's terrestrial part is between 20 and 75 per cent flooded, depending on season. Sian Ka'an's mangrove systems are some of the most biologically productive in the world and their health is critical for the survival of many species in the western Caribbean region. Hidden between the massive mangrove roots live oysters, sponges, sea squirts, sea anemones, hydroids, and crustaceans. Bird species found here include Roseate Spoonbills, pelicans, Greater

Flamingos, Jabiru Storks, and 15 species of heron. The swamps are also home to West Indian Manatees and two endangered crocodiles: the American Crocodile and Morelet's Crocodile. The explosion of tourism in the nearby resort of Cancun poses several threats to the area. Unregulated development has increased pollution and altered the distribution and use of water in Sian Ka'an, compromising the health of the mangroves.

BOAT TOUR
Because a large part of Sian Ka'an is flooded for much of the year, there are few roads into the area, so much of it can be reached and explored only by boat.

and nesting site for several mammals, including swamp rats, and numerous bird species, such as herons, egrets, gallinules, Anhingas, and Brown Pelicans. While much of the region has an abundant alligator population, the swamps are the sole remaining stronghold in the US for the rare and endangered American Crocodile. Also occasionally spotted in the channels between the mangroves are West Indian Manatees (sea cows).

The Florida mangroves are sometimes damaged by the hurricanes that hit the region, Hurricane Andrew in 1992 being an example. Hurricanes damage mangroves in two ways: strong winds may defoliate them, and storm surges harm them by depositing large quantities of silt on their roots. Fortunately, mangrove forests are resilient ecosystems, and they usually regenerate fully from hurricane damage within a few years.

CICHLID INVASION

Since 1983, the Mayan cichlid, an exotic fish species from Central America, has been spreading rapidly through the mangrove swamps and other wetland areas of the Everglades. No-one yet knows what effect it may have on the region's ecosystem. There are worries that it may displace native fish species; alternatively, it could be occupying a new "niche" that no other fish species has filled.

ATLANTIC OCEAN WEST
Zapata Swamp

PRINCIPAL SPECIES
Red, Black, White, and Button Mangroves

AREA
2,500 square km (1,000 square miles)

LOCATION Western Cuba, 160 km (100 miles) southeast of Havana

The Zapata Swamp is a mosaic of mangrove swamps and freshwater and saltwater marshes that form the largest and best-preserved wetland in the Caribbean. The swamp was designated a Biosphere Reserve in 1999 and forms a vital preserve for Cuban wildlife, a spawning area for commercially valuable fish, and a crucial wintering territory for millions of migratory birds from North America. More than

GREATER FLAMINGOS
Large numbers of these colourful birds live in the swamp, feeding off algae, shrimps, molluscs, and insect larvae that inhabit the mud at the bottom of the shallow waters.

900 plant species have been recognized in the swamp, and all but three of the 25 bird species endemic to Cuba breed there. Altogether, about 170 bird species have been identified in the swamp, including the Common Black-hawk, the Greater Flamingo, and the world's smallest bird, the Bee Hummingbird. It also contains the remaining few thousand Cuban Crocodiles. Mammalian residents include the Cuban Hutia, a gopher-like rodent, and the West Indian Manatee. The Manjuari, or Cuban Gar, is an unusual fish found only in the swamp. Adjacent to the swamp is the Bay of Pigs, where millions of land crabs breed each spring.

ATLANTIC OCEAN WEST
Belize Coast Mangroves

PRINCIPAL SPECIES
Red, Black, White, and Button Mangroves

AREA
2,800 square km (1,100 square miles)

LOCATION Eastern Belize, on the western margins of the Caribbean Sea

The mangrove swamps here are a nursery ground for many fish species associated with the huge Belize Barrier Reef. By filtering run-off from rivers and trapping sediment, the mangroves also protect the clarity of the coastal waters, helping the coral reef to survive. Numerous cays – small islands composed largely of coral or sand – along the coast are covered with mangroves and form a habitat for birds. In all, more than 250 bird species share the swamps with West Indian Manatees and a variety of reptiles, including boa constrictors, American Crocodiles, and iguanas.

MANGROVE ROOTS
A tangled maze of mangrove roots extends beneath the water's surface all along this coast, providing refuge for a variety of juvenile fish.

INDIAN OCEAN WEST

Madagascar Mangroves

PRINCIPAL SPECIES Grey, Yellow, Long-fruited Stilt, and Large-leafed Orange Mangroves

AREA 3,300 square km (1,300 square miles)

LOCATION Scattered areas around the coast of Madagascar, off the eastern coast of Africa

Mangroves occur in a wide range of environmental conditions on Madagascar, fostered by a high tidal range, extensive low-lying coastal areas, and a constant supply of fresh river water, which brings a high silt load. They occupy about 1,000km (600 miles) of the island's coastline and are often associated with coral reefs, which protect them from ocean swell. The mangroves, in turn, capture river sediment that otherwise would threaten both reefs and seagrass beds. Up to nine different mangrove species have been recorded in Madagascar, although only six are widespread. Several of Madagascar's endemic birds, including the Madagascar Heron, Madagascar Teal, and Madagascar Fish-eagle, use the mangroves and associated wetland habitats. Dugongs (relatives of manatees) glide through the waters, feeding on sea grasses, while huge quantities of invertebrates and fish swim freely among the finger-like roots of the mangroves. These provide an abundance of food for animals such as the Nile Crocodile, sharks, and aquatic and wading birds, such as herons, spoonbills, and egrets. Many of the fish and bird species here are found nowhere else in the world. Unfortunately, the mangroves are threatened by urban development, overfishing, and the development of land for rice and shrimp farming.

MANGROVE MAZE
This area of coastal mangroves, bisected by numerous channels, is located on the northeastern coast of Madagascar, at the mouth of the River Ambodibonara.

INDIAN OCEAN NORTH

Pichavaram Mangrove Wetland

PRINCIPAL SPECIES Grey, Milky, Stilted, Small-fruited Orange, and Yellow Mangroves

AREA 12 square km (5 square miles)

LOCATION South of Chennai (Madras), Tamil Nadu, southeastern India

The Pichavaram Mangrove Wetland lies on a delta between the Vellar and Coleroon estuaries in southeastern India. It consists of a number of small and large mangrove-covered islets intersected by numerous channels and creeks. Fishing villages, croplands, and aquaculture ponds surround the area. This small, carefully preserved wetland is thought to have saved many lives during the 2004 Indian Ocean tsunami. When the tsunami struck, six villages that were physically protected by the mangroves incurred no damage, while other, unprotected villages were totally devastated. The wetland may have reduced the tsunami's impact partly by slowing the onward rush of the sea through frictional effects and partly by absorbing water into its numerous canals and creeks.

INDIAN OCEAN NORTH

Sundarbans Mangrove Forest

PRINCIPAL SPECIES Sundri, Milky Mangrove, Yellow Mangrove, Indian Mangrove, Keora

AREA 8,000 square km (3,200 square miles)

LOCATION Southwestern Bangladesh and northeastern India, between Calcutta and Chittagong

This forest, a World Heritage Site since 1997, is the largest continuous mangrove ecosystem in the world. It is part of a huge delta formed by sediments from the rivers Ganges, Brahmaputra, and Meghna. The region contains thousands of mangrove-covered islands intersected by an intricate network of waterways. The Bengal Tiger swims here from island to island, hunting prey such as spotted deer and wild boar. Other inhabitants include Fishing Cats, Rhesus Macaque monkeys, Water Monitor lizards, hermit crabs, and various species of shark and dolphin. Habitat destruction threatens this region: more than half of the original mangroves have been cut down.

GHARIAL

One extremely endangered inhabitant of the wetlands and rivers of Bangladesh is the Gharial, a crocodilian. Once quite common in the Sundarbans, their numbers have dwindled due to accidental capture in fishing nets and other factors. Gharials are probably heading for regional extinction, although captive breeding programmes in India and Nepal aim to save the species.

SATELLITE VIEW
In this satellite view of part of the Ganges–Brahmaputra–Meghna delta, the Sundarbans Mangroves form the area that appears dark red. On the right is the Bay of Bengal.

PACIFIC OCEAN WEST

Kinabatangan Mangroves

PRINCIPAL SPECIES Stilt Mangrove, Long-fruited Stilt Mangrove, Grey Mangrove, Nipa Palm

AREA 1,000 square km (400 square miles)

LOCATION Southeast of Kota Kinabalu, eastern Sabah, Malaysia

Mangrove swamps occupy a coastal region of the Kinabatangan river delta, within eastern Sabah in the northern part of the island of Borneo. The mangrove swamps in this area form a complex mosaic with other types of lowland forest (including palm forest) and open reed marsh. They are home to dozens of species of saltwater fish, invertebrates such as shrimps and crabs, otters, and some 200 species of birds including various species of fish eagle, egret, kingfisher, and heron.

Irrawaddy Dolphins are also occasionally spotted in the region, while other spectacular inhabitants include Borneo's indigenous Proboscis Monkey and the Saltwater Crocodile (the world's largest crocodile species), which was almost hunted to extinction but whose numbers are now recovering. Over the past 30 years, there has been extensive clearance of mangroves in the Kinabatangan delta for purposes of timber and charcoal production. The mangroves have either been replaced by oil palms or the cleared land has been developed for shrimp farming. Inevitably, the wildlife has suffered, but the government of Sabah is now engaged in a large-scale mangrove replanting operation.

MANGROVE MONKEY
A female Proboscis Monkey, able both to swim and walk upright, is seen here with an infant, leaping across a waterway in the Kinabatangan mangroves. Her long tail helps to stabilize her movement through the air.

AERIAL ROOTS AT LOW TIDE
The mangroves are anchored in the soft mud by a dense network of roots that also provide a habitat for many animals.

PACIFIC OCEAN WEST

New Guinea Mangroves

PRINCIPAL SPECIES Grey, Long-fruited Stilt, Tall-stilted, and Cannonball Mangroves

AREA 10,000 square km (4,000 square miles)

LOCATION Scattered areas around the island of New Guinea in the western Pacific

Mangrove swamps occur in extensive stretches on New Guinea's coastline. The longest and deepest stretches are found on the south side of the island, around the mouths of large rivers such as the Digul, Fly, and Kikori rivers. Mangrove communities here are the most diverse in the world – more than 30 different species of mangroves have

been found in a single swamp – and they form a vital habitat for a variety of animals living on the water's edge. Underwater, over 200 different fish species, ranging from cardinal fish and Mangrove Jacks to seahorses and anchovies, have been recorded in either their adult or juvenile stages. Mudskippers (species of fish that can leave the water and climb trees), snails, and crabs climb the mangrove roots, while Saltwater Crocodiles patrol the channels between the mangrove stands. Although there are many species of fish and mangrove in these swamps, terrestrial animal diversity is

relatively low. Two endemic species of bat and a species of monitor lizard are found here.

Ten bird species are endemic, including the New Guinea Flightless Rail, two species of lory, the Papuan Swiftlet, Red-breasted Paradise-Kingfisher, and Red-billed Brush-turkey. Although largely intact, the mangrove regions in the western part of New Guinea have recently come under threat of pollution from the rapidly expanding oil and gas industries.

SEAHORSE
This small seahorse is adopting the yellow colour of fallen mangrove leaves.

PACIFIC OCEAN EAST

Darien Mangroves

PRINCIPAL SPECIES Red, Black, Button, White, Mora, and Tea Mangroves

AREA 900 square km (360 square miles)

LOCATION Southeast of Panama City on the Pacific coast of eastern Panama

The Darien mangrove swamps lie around estuaries in eastern Panama in the Darien National Park, adjacent to the Gulf of Panama. Here, the roots of mangroves create a haven for molluscs, crustaceans, and many fish species. Shrimp are particularly abundant – the larvae hatch offshore, migrate to the mangrove "nursery" for a few months, and then return to sea as adults. Some of the mangrove swamps in this region have been converted to shrimp ponds or farmland. Other threats include urbanization and pollution.

BLACK MANGROVE
These Black Mangroves are in the Punta Patiño Nature Reserve, a private reserve owned by a non-profit environmental group.

NEW GUINEA MANGROVES
This young Saltwater Crocodile is feeding among mangrove roots. Fully grown, this species is the largest of all crocodiles, growing up to 7m (23ft) long. Despite its name, it prefers fresh water, and adults compete fiercely for control of prime channels in swamps, often forcing juveniles into marginal rivers or out to the open sea.

THE SHALLOW SEAS that cover the continental shelves around the Earth's landmasses nurture an extraordinary diversity of life. Energy from the Sun and nutrients from the land and sea ensure good conditions for plant growth, on which all marine life depends. The Moon is also a key player. Its gravitational pull drives the tides, which uncover the seashores each day, creating tidal currents that distribute plant nutrients and bring food to waiting animals. Each area of sea bed provides a specific habitat for marine life that has adapted to the local conditions. The shallow seas comprise those parts of the oceans with which we are most familiar; yet, we are only just beginning to understand the complexity of life there and its importance to the overall health of the planet.

SHALLOW SEAS

CORAL REEFS
From polar seas to the tropics, the reflective surface of the sea hides a realm populated with unfamiliar life forms. Here in the tropics, animals look like plants, and plants hide inside the tissues of corals.

CONTINENTAL SHELVES

CONTINENTAL SHELVES ARE ESSENTIALLY the flooded edges of continents, inundated by sea-level rise after the last ice age. The shelf sea bed is now approximately 200m (600ft) below the surface, and its width varies, occasionally extending to hundreds of kilometres. The shelf sea bed and water quality are influenced by land processes. Rivers bring fresh water and nutrients, making shelf waters very productive ecologically, while river-borne material settles on the sea bed as sediment. The continental shelf has a huge diversity of marine life and habitats, but it is also the area of the sea that suffers most from pollutants.

FIORDS

Fiords are deep, sheltered sea inlets originally gouged out by glaciers and then flooded by the sea. They often extend many kilometres inland and are made up of deep basins, separated from the open sea by shallow sills.

This basin-and-sill structure has a huge influence on marine life. In this sheltered environment, still, dark salt water lies beneath peaty fresh water. This mimics the marine conditions off the continental shelf, and animals normally confined to much deeper water, such as cold-water corals, inhabit water shallow enough for divers to explore.

FERTILE FRINGES

The coastal fringes have the greatest diversity of life in the oceans. Light penetration is highly variable, from turbid basins to clear tropical waters. In many places, enough light reaches the shallow sea bed for good growth of photosynthetic organisms. Seaweeds, seagrasses, and phytoplankton thrive here, fed by solar energy, nutrients from land, and sediments stirred up by winds and currents. The coastal fringes are much more productive than the open oceans. Combined with diverse habitats, this results in complex marine communities, making rich feeding and nursery grounds for animals from deeper water. In higher latitudes, seasonal variations in the Sun's strength stimulate an annual cycle of plankton and seaweed growth. In the tropics, where seasons are less pronounced, seagrasses and seaweeds grow year-round.

SHALLOW SEAWEED
Seaweeds grow best on shallow, sunlit rocks, thrive in strong water movement, and provide food and shelter for many small animals.

PRODUCTIVE PLAINS

Much of the continental shelf is covered with deep sediments. Sand, gravel, and pebbles are deposited in shallow water, while fine mud is carried into deeper water offshore. An important part of shelf sediments is biogenic (made from the remains of living organisms). It consists of carbonates (chemical compounds containing carbon) derived from, for example, coral skeletons, and microscopic plankton.

At first sight, sediment plains appear barren. However, many different animals live hidden beneath the surface, either permanently or emerging from burrows and tubes to feed and reproduce. Shifting sand and gravel is a difficult place to live, but more stable sediments occur on deeper sea beds. Varying particle size makes it suitable for constructing burrows and tubes, and it can contain huge numbers of animals, providing a rich food source. These animal communities are all sustained by plankton falling from the continental-shelf surface waters, and by the products of decomposition of seagrasses and seaweeds.

SEDIMENT PREDATORS
Fish and starfish are top predators on sediments, eating the many different animals on the surface or buried beneath. Fish catch a wide range of creatures, while starfish capture slower-moving prey.

SHELF FISHERIES

The waters and sea bed of the continental shelf support most of the world's major fisheries. In coastal waters, there is planktonic food for larvae and cover for juveniles, and this is where 90 per cent of the world's total seawater catch reproduces. Demersal fish (living on or just above the sea bed) such as cod and haddock feed on sea-bed life. Pelagic (open water) shoaling fish such as sardines and herring feed on zooplankton, and are important food for larger fish such as mackerel and sharks, as well as for cetaceans and seabirds. Commercially important invertebrates such as prawns are caught in shelf waters. Worldwide, coastal communities are sustained by small-scale, inshore fisheries, which catch a wide range of marine life.

JUVENILE SHELTER
These baby cod are feeding in Horse Mussel beds, before moving offshore as adults.

QUEEN SCALLOP
Scallops feed by filtering seawater, and can be collected by diving, or farmed, with no damage to the marine environment.

SHELF DEPOSITS
The Mississippi River flows into the sea through a network of channels. As its silt-laden waters reach the sea, sediments fall out along the continental shelf.

GEOLOGY OF THE CONTINENTAL SHELF

Shelf deposits can be extremely thick. For example, those off eastern North America are up to 15km (9 miles) deep, and have been accumulating and compacting for millions of years. A cross-section here reveals ancient sediments other than those deposited by rivers and glaciers, including carbonates, evaporites, and volcanic materials. Carbonates are largely produced by marine life in shallow tropical seas. Evaporites are salts resulting from seawater evaporation in shallow basins or on arid coastlines. Evaporite deposits create domes in overlying sedimentary rocks, trapping oil and gas.

HUMAN IMPACT

COASTAL POLLUTION

For many years, coastal seas have been used as a convenient dump for human waste. Even the most remote seashores are now littered with plastic. More insidious is invisible pollution: nutrients and pathogens from sewage; heavy metals, organohalogens, and other toxins from industrial and agricultural effluents; radioactive waste from power stations; and hydrocarbons from effluents, oil spills, and other sources.

DREDGED TREASURE
Metals such as gold, tin, rare earth elements, and aggregates for the building industry are extracted by dredging continental shelves.

ROCKY SEA BEDS

FROM THE WARM TROPICS TO COLD POLAR SEAS, many distinctive communities of marine life develop on the rocky floors of shallow seas. Underwater rocks provide points of attachment for both seaweeds and marine animals and are often covered with life. Seaweeds thrive in the sunlit shallows and provide a sheltered environment for animal communities. Firmly attached animals extend arms and tentacles to catch planktonic food from water currents, or pump water through their bodies to filter out nutrients. Mobile animals graze seaweeds or prey on fixed animals or each other. The life on a rocky reef depends on many environmental factors.

ROCKY SLOPE
These underwater rocks in British Columbia, Canada, are covered with marine life. A sunstar and leather star search for prey among pink soft corals and sponges, while urchins graze below.

THE SEAWEED ZONE

Seaweeds rely on sunlight for growth, and thrive only on the shallowest rocks. The depth in which they can grow depends on water clarity, from a few metres in turbid seas, to more than 100m (330ft) in the clearest waters. In colder waters, huge forests of kelp and other large brown seaweeds dominate the shallows, with smaller seaweeds in deeper water. Large seaweeds are often scarce on rocks in the tropics – instead the Sun's energy is harnessed by tiny unicellular algae inside coral tissues. Seaweeds harbour a plethora of associated animals. Some live permanently in the seaweed zone, while others use it as a breeding ground or nursery before moving into deeper water.

FOOD SOURCE
Energy from sunlight captured by seaweeds is used by grazing animals. Here, green seaweeds cover rocks in Orkney, Scotland.

BALLAN WRASSE
In summer, adult ballan wrasses lay eggs in nests built of seaweed, secured in rock crevices. Young wrasses are often patterned, providing camouflage.

ROCK GRAZERS

Sea urchins are highly successful marine invertebrates, well defended by sharp spines. They graze the seabed, eating virtually everything except hard-shelled animals and coralline seaweed crusts. They have a profound effect on sea-bed communities. If urchins are abundant, they can seriously reduce the diversity of life on the sea bed, leaving urchin "barrens". Conversely, where urchins are sparse, they can increase diversity, by clearing spaces for new life to settle.

ANIMAL-DOMINATED DEEPS

In deeper water, light levels are too low for most seaweeds, although encrusting red seaweeds need little light and grow further down. Much of the plant-like growth in deeper water actually consists of fixed animals, which are most abundant in places with strong tidal currents. For mobile animals living here, the sea bed is a minefield of toxic substances, released by fixed animals to deter predators. Below 50m (160ft), water movement from waves is much less, and fragile animals such as sponges and sea fans can grow to a large size. Here, and in places more sheltered from water movement, a smothering layer of fine silt continually settles on the rock surfaces, restricting the animal life to forms that can hold themselves above the rock or can remove the silt. On the most heavily silted rocks, animals may grow only on vertical or overhanging surfaces.

ROCKY-BOTTOM PREDATOR
Stonefish have a textured skin and irregular shape, making them difficult to spot. A huge mouth engulfs prey while the dorsal spines contain venom that can be fatal.

protruding eye used when hiding in sediment

dorsal spines with poison glands

camouflage skin colour and texture

large mouth

tail fin

VERTICAL ROCK

Underwater cliffs are often more heavily colonized with invertebrates than gently sloping rocks. In shallow water exposed to strong waves, various mobile sea-bed animals, particularly grazing sea urchins and predatory starfish, find it harder to cling to vertical and overhanging surfaces, and are knocked off by waves in rough weather. Vertical walls receive less sunlight, and are harder places for seaweed spores to settle, so there is less competition from seaweeds here than on horizontal rock. At sheltered sites, upward-facing rock is often covered with silt, and has few animals, but vertical and overhanging rock, by contrast, is silt-free and may have abundant life. Ledges and crevices in underwater cliffs provide safe refuges for fish and crustaceans.

JEWEL ANEMONES
Multicoloured jewel anemones carpet vertical, wave-exposed rocks, with tentacles outstretched to catch food from the currents.

CREVICES AND CAVES

Irregularities in underwater rock features can provide additional habitats for marine life. Crevices and small caves provide shelter for nocturnal fish that hide during the day and are active at night. Elongated fish are well shaped to live in crevices, while fish that are active by day need holes to hide in at night and when predators approach. Deep, dead-ended caves contain a range of habitats, from sunlit, wave-exposed entrances to dark, still inner waters and sheltered sediments. Prawns and squat lobsters occupy cave ledges, while animals that actively pump water to feed live in the quiet water inside the cave and coat the walls. Flashlight fish hiding in caves during the day signal to each other with light produced by bacteria in organs beneath their eyes. Small crevices are important because they form a refuge for small animals from sea urchins.

TAKING REFUGE
The flattened body of this spiny squat lobster enables it to retreat far into narrow crevices if threatened, and the spines help to wedge it in small spaces.

STORMS AND SCOUR

Shallow rocky reefs take the full force of waves during storms, but rock-living animals and seaweeds on open, exposed coasts are firmly attached and are generally well-adapted to cope with pounding waves. Larger seaweeds and animals will be torn from shallower rocks, making space for new life to settle, while many seaweeds and colonial animals can regrow from holdfasts or basal parts. However, few animals or plants survive on rolling boulders or on bedrock scoured by nearby sand and pebbles. Where rock meets sand, there is often a band of bare, sandblasted rock. Just above, tough-shelled animals such as keelworms survive, together with patches of hard encrusting calcareous red seaweeds. Above this, fast-growing colonial animals such as sponges and barnacles can colonize in the intervals between storms.

KEELWORMS
Keelworms have a hard, calcareous shell that protects their bodies from sand scour.

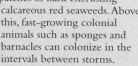

CORALLINE ALGAE
Like a coating of hard pink paint, encrusting coralline algae can withstand considerable scouring from nearby sand and pebbles.

SANDY SEA BEDS

MOST OF THE CONTINENTAL SHELF is covered with thick sediments, accumulated from millennia of land and coast erosion. The calcareous remains of marine life are continually added to the mix. Unlike deep-sea sediments (see pp.180–81), shelf sediments are stirred up by waves during storms, re-suspending nutrients and profoundly affecting marine life and productivity. Sediments are largely the domain of animals, as seagrasses and seaweeds grow only in limited, shallow areas. Buried beneath the surface of a sandy sea bed, there may be vast numbers of animals hiding from, or waiting for, prey.

GRAVEL AND SAND

The coarsest sediments from coastal and land erosion are usually deposited inshore by rivers and glaciers as they enter the sea. Frequently shifted by waves and tides, clean, coarse sand and gravel make a difficult habitat; typical inhabitants include tough-shelled molluscs, sea cucumbers, burrowing urchins, and crabs. A wider range of organisms live in the more stable sand and gravel, where purple-pink beds of maerl can be found. This unattached, calcareous seaweed (see p.245) is made up of coral-like nodules. The open structure of live maerl twiglets is ideal for sheltering tiny animals, newly settled from the plankton, while the dead maerl gravel underneath supports burrowing animals. Beds of seagrass and green seaweeds thrive in shallow sand, harbouring a wide range of life. Embedded shells and stones provide anchors for various seaweed species. Many fish have adapted to life on sandy sea beds,

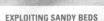
coarse bristles (chaetae) on sides

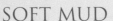
felt-like dorsal chaetae

SANDY HABITAT
A marine segmented worm, the Sea Mouse lives in muddy sand.

the most familiar being flatfish. Shallow-water anglerfish wave their fishing lures to tempt prey within striking distance of their huge mouths, while garden eels live permanently in sand burrows, partly emerging to eat plankton. Sand eels and cleaver wrasse dive into the sand to avoid predators.

GRAVEL DWELLER
This Flame Shell lives in a nest of gravel, pebbles, and shells. It pumps seawater through the nest, extracting food with its sticky, acidic tentacles.

EXPLOITING SANDY BEDS
Stingrays are among the many animals that hide in the sand of the sea bed; this Southern Stingray does so both to escape predators and to ambush prey.

MIXED SEDIMENTS

Most sediments on the continental shelf are a mix of coarse and fine materials. An important part of these are calcareous fragments, derived from hard-shelled animals. Mixed sediments offer a wider range of building materials for tubes and burrows than sand or mud and are easier to traverse, so a far greater variety of animals live here. Seaweeds and hydroids cover the bed, attached to shells and pebbles. Visible life includes tube worms, brittlestars, and burrowing anemones; most of these withdraw into the sediment if threatened. Below the surface, hidden

LIFE ON THE SEDIMENT
Its mouth fringed by tentacles, this half-buried sea cucumber (left) and a hermit crab inhabit these mixed sediments.

animals, including bivalves and crustaceans, provide a rich source of food for animals that can find and excavate it, such as starfish, crabs, and rays.

SOFT MUD

In sheltered waters in enclosed bays, estuaries, and fiords, and in the deeper parts of the continental shelf, the finest particles of sediment settle as soft mud. Easily stirred up, the fine particles smother newly settled larvae and clog gills. There is little oxygen just below the mud surface, so buried animals must find ways to obtain oxygen from seawater. Despite these challenges, mud can be very productive. Bacteria and diatoms are often abundant on the mud surface, providing food for hoovering animals such as echiuran worms. Stable burrows are more easily built in mud than in sand or gravel. Animals such as sea pens and burrowing anemones anchor themselves in the mud, raising sticky polyps and tentacles to catch the raining plankton or to ensnare a passing fish or crustacean.

ANCHORED IN MUD
This sea pen's branches are covered with small polyps that feed on the plankton.

OCEAN ENVIRONMENTS

SEA-BED STABILIZERS

This sea bed owes its luxuriant growth, including hydroids, soft corals, and brittlestars, to the many Flame Shells and Horse Mussels hidden under the surface. These molluscs bind the shifting sediments with strong threads, creating a stable, complex surface that many other animals can colonize. Flame Shell nests join together to form extensive reefs, with holes for water exchange, so many other organisms can live inside and beneath the nests.

SEDENTARY HABIT
This Norway Lobster lives in a U-shaped burrow with two exits and is mainly nocturnal.

SEA WRECKS

The complex shape and hard surfaces of shipwrecks such as the one shown here (the *Eagle,* off Florida) attract sedentary invertebrates and fish. A new wreck may take some time to become colonized, depending on the material from which it is made. Small hydroids, barnacles, and keelworms often settle first, paving the way for other animals and seaweeds to grow on their hard shells. Filter feeders thrive in enhanced currents on the super-structure, while the spaces inside offer hiding places for fish and octopuses.

BENEATH THE SURFACE

Wave-disturbed sand and gravel creates a mobile, well-oxygenated environment. Animals that live here, such as crustaceans, and echinoderms, move through the shifting sand without building permanent homes. Small animals that disturb sediments in this way, or by ingesting and defaecating it, are called bioturbators and are important recyclers of nutrients. Less-disturbed sediments are inhabited by sediment stabilizers. These sedentary animals, many living in permanent burrows or tubes, can cope with oxygen depletion and being covered over. Some strengthen their burrows by lining them with substances such as mucus and draw in seawater to supply food and oxygen. Others filter seawater or hoover sediment by extending their siphons to the surface. Microscopic creatures (the meiofauna) live in between the sand grains.

SEAGRASS BEDS AND KELP FORESTS

SEAGRASS BEDS AND KELP FORESTS are very different habitats, but both are highly productive and contribute significantly to the total primary production of inshore waters. Seagrasses are the only fully marine flowering plants. They thrive in shallow, sunlit water on sheltered, sandy sea beds primarily in warm water. Kelps are large brown seaweeds that grow as dense forests on rocks of the lower shore and subtidal zone, preferring cold water. Both of these ecosystems have a complex structure and provide shelter for a wide range of associated animals and seaweeds, some of them found nowhere else.

COLD-WATER KELP
Kelps are large brown seaweeds that live mainly in shallow subtidal zones.

ENDANGERED GRAZERS

Seagrasses are the primary food of Green Turtles, and the only food of manatees and Dugongs. Globally, these animals are now endangered or vulnerable, threatened by the destruction of their feeding grounds. The coastal areas in which seagrass beds are found are often vulnerable to pollution. Runoff of nutrients and sediments from land affect water clarity, and are probably the biggest threat worldwide.

SEAGRASS BEDS

Seagrasses are the only fully marine flowering plants (angiosperms), and grow best in shallow, sandy lagoons or enclosed bays, where the water clarity is good. They are also tolerant of variable salinity. Unlike seaweeds, seagrasses have roots, which they use to absorb nutrients from within the sediments, thus recycling nutrients that would otherwise be locked up below the surface. Their intertwined rhizomes and roots help to stabilize the sand, protecting against erosion and encouraging the build-up of sediments. The productivity and complex physical structure of seagrasses attracts a considerable diversity of associated species, some of which are only found in seagrass beds. A variety of seaweeds and sedentary animals, including species of hydroids, bryozoans, and ascidians, grow on the leaves. Seagrasses are also a critically important food for animals such as manatees, Dugongs, Green Turtles, and for many aquatic birds.

SEAGRASS MEADOWS
Seagrass meadows help to protect shallow sandy sea beds against erosion.

NATURAL CAMOUFLAGE
This Greater Pipefish's elongated shape and drab colour makes it hard to spot among seagrass leaves.

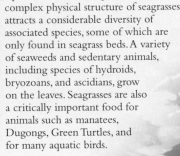

KELP FORESTS

The term "kelp" was originally used to refer to the residue resulting from burning brown seaweeds, which was used in soap-making. It is now used more generally to refer to the many kinds of large brown seaweeds of the order Laminariales. Kelp forests grow best in colder waters, on shallow rocks with good water movement. The top edge of some kelp beds is visible on the lowest tides. Kelps grow densely on rock slopes down to around 10–20m (30–70ft) deep, depending on water clarity. In deeper water, there is less light for photosynthesis and kelps grow more sparsely; in most coastal waters they cannot survive below 25m (80ft). In exceptionally clear water, kelps can grow at depths of 50m (160ft). Many kelp species have gas-filled floats, which hold the fronds up to the light and away from grazers. Within the kelp forest, waves are dampened and many organisms live in its shelter. Although kelp habitats support rich marine communities, only about 10 per cent of kelp is eaten directly by animals; the rest enters the food chain as detritus or dissolved organic matter.

DISTRIBUTION MAP
Seagrass beds flourish in the tropics, while kelp forests thrive in cold, nutrient-rich waters, extending into the polar regions.

■ kelp forests
■ seagrass beds

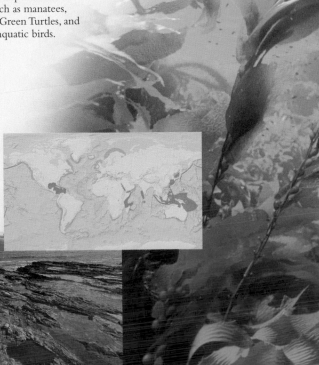

COASTAL DEFENCES
A band of Giant Kelp can help to protect coasts from severe storms by absorbing wave energy.

KELP COMMUNITIES

Many kelps are tree-like in shape, with a branched holdfast for attachment and a long stem (stipe), sometimes with floats, supporting a palm-like frond. This makes a kelp forest a multi-layered environment in which different organisms live at different levels. Small spaces in the holdfast can harbour hundreds of small animals from predators. Some kelps have rough stipes covered with red seaweeds, although sea urchins and limpets may graze these in calm weather and in deeper water. Actively growing kelp fronds exude slime, which deters most animals from settling, but as growth slows later in the season, the fronds may become covered with a few species, particularly bryozoans, hydroids, and tube worms. These animals reduce the light reaching the fronds, and some kelps shed their fronds to get rid of unwanted settlers before growing new ones. The sea floor beneath the kelps may be covered with marine growth, or relatively barren if heavily grazed by sea

KELP ANEMONE
This large anemone is unusually mobile, and crawls or drifts up onto seaweed fronds to catch floating prey.

BLUE-RAYED LIMPET
At the end of the growing season, these limpets move down into the holdfast to avoid being discarded with the old frond.

NURSERIES AND REFUGES

Seagrass beds and kelp forests are important refuges for young fish that need to hide from predators until they reach maturity. Many fish, such as the lumpsucker and Swell Shark, do not live among seagrasses or kelps as adults, but come into these habitats to spawn, giving their young a greater chance of survival. Small fish need small prey, and they find an abundance of food in the form of tiny worms, crustaceans, and molluscs among the seagrasses and in the sediment beneath, or in the undergrowth of kelp forests. These young fish are often unlike their parents, usually camouflaged in shades of green and brown to avoid detection. Some herbivorous fish from surrounding reefs come into seagrass beds only at night. Seagrass beds are important nurseries for some commercial invertebrates, including shrimp and cuttlefish.

LUMPSUCKER
This baby lumpsucker is very vulnerable. However, it is well camouflaged on kelp fronds, to which it attaches itself with a sucker.

DENSE KELP FOREST
Giant Kelp is the world's biggest seaweed. Its stipes can be more than 30m (100ft) long, and it can grow as fast as 50cm (20in) per day.

ATLANTIC OCEAN WEST

Laguna de Términos

COASTAL TYPE	Shallow lagoon
WATER TYPE	Tropical
PRIMARY VEGETATION	Seagrasses, seaweeds, and mangroves

LOCATION In the southwest of the Yucatán Peninsula, Campeche State, Mexico

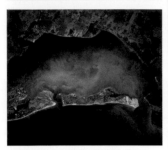

SATELLITE VIEW, WITH LAGOON AT TOP

Two channels connect this sediment-laden lagoon to the Gulf of Mexico, while three rivers feed in fresh water, producing a pronounced change in salinity. The seagrasses *Thalassia testudinum*, *Syringodium filiforme*, and *Halodule wrightii* cover 29 per cent of the lagoon. With 448 recorded animal species, Términos is the most species-rich of Mexico's four large lagoons.

ATLANTIC OCEAN NORTHEAST

Sound of Barra

COASTAL TYPE	Island chain with sounds
WATER TYPE	Cool
PRIMARY VEGETATION	Seagrasses, maerl, and kelp

LOCATION Between South Uist, Eriskay, and Barra, Outer Hebrides, Scotland, UK

Strong tidal currents flow through the Sound of Barra, and its clear, shallow waters and sandy sea floor provide an ideal habitat for the eelgrass *Zostera marina*. The eelgrass beds, together with beds of maerl (see p.245), are home to many species of small animals. Such rich, current-swept communities in this part of Scotland are threatened by the building of rock causeways across the sounds, which cut off the nutrient-bearing currents that are essential for healthy growth. Eelgrass also grows in nearby brackish lagoons, together with the tasselweed *Ruppia maritima*, which is regarded by some scientists as a type of seagrass. Forests of the kelp *Laminaria hyperborea* grow on rocks at the edges of the sound, and these are home to abundant sea squirts and sponges.

EELGRASS STANDS
Healthy stands of eelgrass now thrive in the current-swept sound. Almost 90 per cent of western Europe's eelgrass was lost to a wasting disease in the 1930s.

ATLANTIC OCEAN NORTHEAST

Falmouth Bay

COASTAL TYPE	Rocky with inlets
WATER TYPE	Cool
PRIMARY VEGETATION	*Laminaria hyperborea* kelp, eelgrass

LOCATION Southwest Cornwall, England, UK

The coastline of Falmouth Bay includes two drowned river valleys (rias), the Fal and Helford, which are now long, sheltered sea inlets. Because of their rich marine life, these inlets,

together with part of Falmouth Bay, have been designated as a European marine Special Area of Conservation. Beds of the eelgrass *Zostera marina* and maerl (see p.245) in the inlets are home to a wide variety of animals, including the rare Couch's Goby.

On the wave-exposed rocky coasts outside the inlets, the kelp *Laminaria hyperborea* grows in dense forests that support many associated seaweeds and animals. This kelp has a stiff stipe, which raises the frond off the sea bed and means that the forest has developed well vertically. On the rock beneath the kelp, there is competition for space among anemones, sponges, and smaller seaweeds, while other animals hide in kelp holdfasts. The kelp stipes have a rough surface and provide effective attachment points for red seaweeds, bryozoans, soft corals, and other types of encrusting animals. On the fronds, tiny Blue-rayed Limpets graze, and colourful sea slugs eat small hydroids and lacy bryozoans.

In deeper water, the kelp *Laminaria ochroleuca* grows, close to its northern limit in Europe. This kelp is similar to *Laminaria hyperborea*, but has a smooth stipe on which little can grow. Two other kelps are found in the area, the Sugar Kelp (*Laminaria saccharina*), which has a crinkled frond, and Furbelows (*Saccorhiza polyschides*), which has a large, hollow holdfast and grows up to 4m (13ft) long in just one season.

CORNISH KELP FOREST
This forest of *Laminaria hyperborea* kelp has many different plants and animals living on the rocks beneath it, and on the kelp itself.

ATLANTIC OCEAN SOUTHEAST

Saldanha Bay

COAST TYPE	Rocky and sandy bay with lagoon
WATER TYPE	Cool currents
PRIMARY VEGETATION	Kelp and eelgrass

LOCATION Western Cape, South Africa

The cold Benguela Current flowing northward along the west coast of South Africa brings nutrient-rich water that is ideal for kelp growth, and Sea Bamboo (*Ecklonia maxima*)

SOUTH ATLANTIC KELP FOREST
On the west coast of South Africa, Sea Bamboo is the largest of the local kelps. It can grow as tall as 15m (50ft).

is abundant in Saldanha Bay. The smaller Split-fan Kelp (*Laminaria pallida*) becomes dominant in deeper water. South Africa is famous for its diversity of limpets, and the kelp limpet *Cymbula compressa* is found only on Sea Bamboo. Its shell fits neatly around the stipe, where it grazes. The highly endangered limpet *Siphonaria compressa* occurs only in the bay's Langebaan Lagoon, grazing on the endemic eelgrass *Zostera capensis*.

INDIAN OCEAN WEST

Gazi Bay

COASTAL TYPE Shallow bay and fringing reef

WATER TYPE Tropical

PRIMARY VEGETATION Seagrasses and mangroves

LOCATION 50km (30 miles) south of Mombasa, Kenya

Gazi Bay's shallow, subtidal mud and sand flats are sheltered by fringing coral reefs. Twelve species of seagrass grow on the mudflats, and these seagrass beds cover about half of the bay's 15 square km (6 square miles). Mangrove-lined creeks flow into the bay, and this unusual proximity of mangrove, seagrass, and coral reef systems has led to scientific studies on how they interact. The seagrass beds proved to be important in trapping particles washed into the bay from the creeks. Most were trapped within 2km (1¼ miles) of the mangroves. The seagrass beds provide food directly for prawn larvae, zooplankton, shrimps, and oysters, and they are the main feeding grounds of all the fish in the bay, making them very important to the health of the local fisheries.

MANGROVES AND SEAGRASS
Unusually for mangrove-lined creeks, the water here is clear, and stands of seagrasses are able to flourish in the waterways leading into Gazi Bay.

INDIAN OCEAN EAST

Lombok

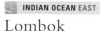

COASTAL TYPE Semi-sheltered bays on rocky coast

WATER TYPE Tropical

PRIMARY VEGETATION Seagrasses

LOCATION Lesser Sunda Islands, Indonesia

At least 30,000 square km (11,600 square miles) of sea bed around Indonesia is covered by seagrasses. In the warm, shallow lagoons and bays, 12 species of seagrass flourish. Gerupuk Bay in the south of the island of Lombok contains 11 of the 12 Indonesian seagrass species, with *Enhalus acoroides* and *Thalassodendron ciliatum* forming dense stands. Analyses of the gut contents of fish that live among seagrass in Lombok's waters revealed that crustaceans were the dominant food source. However, a species of *Tozeuma* shrimp found there avoids the attention of predators by having an elongated body coloured green with small white spots, a perfect camouflage against seagrass leaves. At low tide, local people use sharp iron stakes to dig for intertidal organisms, and this damages the seagrass leaves and roots, thereby threatening the survival of the beds.

HUMAN IMPACT

THREAT FROM TOURISM

The islands of Southeast Asia contain the greatest diversity of seagrasses in the world, but human activity threatens them in many places. Tourism is a means of bringing a much-needed boost to many local economies and this necessitates the building of hotels and other tourist facilities in previously unspoilt areas.

HOTEL DEVELOPMENT
Future tourist development in the region may threaten the Lombok seagrass beds as a result of pollution and loss of habitat through the building of beach facilities, such as marinas.

BAY OF PLENTY
The seagrass beds in Lombok's bays are a source of seaweeds, sea urchins, sea cucumbers, molluscs, octopus, and Milkfish for the area's inhabitants.

OCEAN ENVIRONMENTS

PACIFIC OCEAN WEST

Sea of Japan/East Sea

COASTAL TYPE Mainly rocky	
WATER TYPE Warm to cold	
PRIMARY VEGETATION Kelp and seagrasses	

LOCATION Off the west coast of the island of Hokkaido, northern Japan

The Sea of Japan/East Sea is influenced by the warm Tsushima Current from the south and the cold Liman Current from the north, so its marine flora is a rich mix of temperate and cold-water species because of the wide range of water temperatures in different parts of the coast. The mixing of these currents also provides plentiful nutrients for plant growth. Seagrass diversity is moderate, but eelgrasses are particularly well represented with seven species, several of them endemic to the area. Kelps are also diverse, with species of *Undaria*, *Laminaria*, and *Agarum* thriving in the colder waters in the north. Kelp is highly nutritious, and Hokkaido is the traditional centre of kelp harvesting.

INVASIVE KELP
Since 1981, Asian Kelp (*Undaria pinnatifida*) has spread from its indigenous sites in Japan, China, and Korea to four continents.

FEMALE RED PIGFISH IN KELP FOREST

PACIFIC OCEAN SOUTHWEST

Poor Knights Islands

COASTAL TYPE Offshore islands	
WATER TYPE Temperate	
PRIMARY VEGETATION Kelp and other brown seaweeds	

LOCATION Off the east coast of Northland, North Island, New Zealand

In 1981, a marine reserve was set up around the Poor Knights Islands, extending 800m (2,600ft) out from the shore. The area is popular with divers for its caves and kelp forests. In the most exposed places, the kelp *Lessonia variegata* is predominant, while at more sheltered sites *Ecklonia radiata* is more abundant, together with the large brown seaweed *Carpophyllum flexuosum*. Large numbers of sea urchins dominate in some places.

INDIAN OCEAN EAST

Shark Bay

COASTAL TYPE Shallow, semi-enclosed bay	
WATER TYPE Tropical; high salinity	
PRIMARY VEGETATION Seagrasses	

LOCATION Inlet of the Indian Ocean, north of Perth, Western Australia

Shark Bay is a UNESCO World Heritage Site, and it contains one of the largest, most diverse seagrass beds in the world. Its 12 species of seagrass, which include *Amphibolis antarctica* and *Posidonia australis*, dominate the subtidal zone to depths of about 12m (40ft). The vast seagrass beds provide

SEAGRASS BANKS
Shark Bay has one of the world's largest seagrass beds, covering about 4,000 square km (1,500 square miles).

food for one of the world's largest populations of Dugongs (see p.423), which are preyed on by sharks. The adjacent Hamelin Pool is too salty for seagrasses, but it is well known for the growth of stromatolites (see p.232).

POSIDONIA AUSTRALIS

PACIFIC OCEAN NORTHEAST

Izembek Lagoon

COASTAL TYPE Rocky coast and lagoon	
WATER TYPE Cold; low salinity	
PRIMARY VEGETATION Eelgrass and kelp	

LOCATION On the northern side of the Alaskan Peninsula, Alaska, USA

Izembek Lagoon covers 388 square km (150 square miles) of the Izembek State Game Refuge and is the site of one of the world's largest eelgrass beds. The eelgrass *Zostera marina* grows in dense beds here, both subtidally and on intertidal flats, where it is grazed by wading birds at low tide. Over half a million geese, ducks, and shorebirds stop over at the lagoon during migration to refuel on the eelgrass. On the rocky, open coasts outside the lagoon, kelp forests thrive in the cold water. The commonest forest-forming kelp here is Bull Kelp (*Nereocystis luetkeana*), which can grow to 40m (130ft) in length. Bull Kelp is an annual, which means that it reaches maturity within a single year. It grows quickly, at a rate of up to 13cm (5in) per day. The huge fronds, which have many long, strap-shaped blades, are supported by gas-filled bladders (pneumatocysts) that are up to 15cm (6in) in diameter.

FEEDING GROUNDS
After raising their young farther north, thousands of Brent Geese graze on eelgrass in Izembek Lagoon in the autumn before flying south to Baja California, Mexico.

Monterey Bay Kelp Forest

COASTAL TYPE
Rocky and sandy

WATER TYPE
Cool to warm

PRIMARY VEGETATION
Kelp

LOCATION South of San Francisco, California, USA

The California coast is famous for its beds of Giant Kelp (*Macrocystis pyrifera*), the largest seaweed on the planet (see p.240). It forms dense forests just offshore, and in Monterey Bay it outcompetes Bull Kelp for sunlight in many places, but the latter dominates in more exposed areas. Inshore of these giant species, other smaller kelps thrive. The kelp forests provide a unique habitat. Sea Otters (see p.406), which live among the kelp forests and eat sea urchins, are thought to be important in controlling the urchins, which graze on the kelp. Seagrasses of the genus *Phyllospadix* are also found in Monterey Bay. Unusually for seagrasses, they can attach to rock, and grow in the surf zone or in intertidal pools on rocky coasts. Each year over 140,000 tons of Giant Kelp are harvested in California for the extraction of alginates, which are used in the textile, food, and medical industries.

SUNLIT FOREST
In exceptional circumstances, Giant Kelp can be 80m (265ft) long. The forests are at their thickest in late summer, and decline during the dark winter months.

CORAL REEFS

CORAL REEFS ARE SOLID STRUCTURES built from the remains of small marine organisms, principally a group of colony-forming animals called stony (or hard) corals. Reefs cover about 300,000 square km (100,000 square miles) of the world's shallow marine areas, growing gradually as the organisms that form their living surfaces multiply, spread, and die, adding their limestone skeletons to the reef. Coral reefs are among the most complex and beautiful of Earth's ecosystems, and are home to a fantastic variety of animals and other organisms; but they are also among the most heavily utilized and economically valuable. Today, the world's reefs are under pressure from numerous threats to their health.

TYPES OF REEFS

Coral reefs fall into three main types: fringing reefs, barrier reefs, and atolls. The most common are fringing reefs. These occur adjacent to land, with little or no separation from the shore, and develop through upward growth of reef-forming corals on an area of continental shelf. Barrier reefs are broader and separated from land by a stretch of water, called a lagoon, which can be many kilometres wide and tens of metres deep.

Atolls are large, ring-shaped reefs, enclosing a central lagoon; most atolls are found well away from large landmasses, such as in the South Pacific. Parts of the reef structure in both atolls and barrier reefs often protrude above sea level as low-lying coral islands – these develop from wave action depositing coral fragments broken off from the reef itself.

Two other types of reef are patch reefs – small structures found within the lagoons of other reef types – and bank reefs, comprising various reef structures that have no obvious link to a coastline.

FRINGING REEF
A fringing reef directly borders the shore of an island or large landmass, with no deep lagoon.

BARRIER REEF
A barrier reef is separated from the coast by a lagoon. In this aerial view, the light blue area is the reef and the distant dark blue area is the lagoon.

ATOLL
An atoll is a ring of coral reefs or coral islands enclosing a central lagoon. It may be elliptical or irregular in shape.

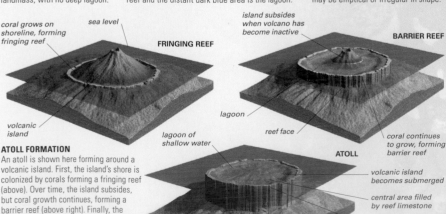

ATOLL FORMATION
An atoll is shown here forming around a volcanic island. First, the island's shore is colonized by corals forming a fringing reef (above). Over time, the island subsides, but coral growth continues, forming a barrier reef (above right). Finally, the island disappears, but the coral maintains growth, forming an atoll (right). Atolls can also form as a result of sea-level rise.

coral grows on shoreline, forming fringing reef

sea level

FRINGING REEF

volcanic island

island subsides when volcano has become inactive

BARRIER REEF

lagoon

reef face

coral continues to grow, forming barrier reef

ATOLL

lagoon of shallow water

volcanic island becomes submerged

central area filled by reef limestone

coral continues to grow where waves bring food

CORAL DIVERSITY
In this seascape off a Fijian island, groups of shoaling Sea Goldies hover over diverse species of coral, sponges, and other reef organisms.

REEF FORMATION

The individual animals that make up corals are called polyps. The polyps of the main group of reef-building corals, stony corals, secrete limestone, building on the substrate underneath. The polyps also form colonies that create community skeletons in a variety of shapes. An important contributor to the life of these corals is the presence within the polyps of tiny organisms called zooxanthellae, which provide much of the polyps' nutritional needs. Other organisms that add their skeletal remains to the reef include molluscs and echinoderms.

Grazing and boring organisms also contribute, by breaking coral skeletons into sand, which fill gaps in the developing reef. Algae and other encrusting organisms help bind the sand and coral fragments together. Most reefs do not grow continuously but experience spurts of growth interspersed with quieter periods, which are sometimes associated with recovery from storm damage.

STONY CORAL
This group of branching hard corals is growing in a depth of about 5m (16ft) off the coast of eastern Indonesia. Individual stony corals can grow up to several centimetres a year.

OPEN POLYPS
At the centre of each polyp is an opening, the mouth, which leads to an internal gut. The tissue around the gut secretes limestone, which builds the reef.

DISTRIBUTION OF REEFS

Stony corals can grow only in clear, sunlit, shallow water where the temperature is at least 18°C (64°F), and preferably 25-29°C (77-84°F). They grow best where the average salinity of the water is 36ppt (parts per thousand) and there is little wave action or sedimentation from river runoff. These conditions occur only in some tropical and subtropical areas. The highest concentration of coral reefs is found in the Indo-Pacific region, which stretches from the Red Sea to the central Pacific. A smaller concentration of reefs occurs around the Caribbean Sea. In addition to warm-water reefs, awareness is growing about other corals that do not depend on sunlight, and form deep, cold-water reefs – some of them outside the tropics (see p.178).

WARM-WATER REEF AREAS
The conditions needed for the growth of warm-water coral reefs are found mainly within tropical areas of the Indian, Pacific, and Atlantic oceans. The reefs are chiefly in the western parts of these oceans, where the waters are warmer than in the eastern areas.

COLD-WATER CORAL
This species, *Lophelia pertusa*, is one of a few of the reef-forming corals that grow in cold water, at depths up to 500m (1,650ft).

CORAL BLEACHING

Bleaching refers to colour loss in reef-building corals and occurs when the tiny organisms called zooxanthellae, which give corals their colours, are ejected from coral polyps or lose their pigment. In extreme cases, it can lead to the coral's death. Various stresses can cause bleaching, including pollution and ocean temperature rises. In recent decades, several mass bleaching events have been recorded, affecting corals over wide areas of ocean.

OCEAN ENVIRONMENTS

PARTS OF A REEF

Distinct zones exist on coral reefs, each with characteristic levels of light intensity, wave action, and other parameters. Each zone's characteristics determine the organisms that live there. The reef slope, or forereef, is the part that faces the sea. The upper parts of the reef slope are dominated by branching coral colonies and intermediate depths by massive forms. These are the areas of the reef with the greatest diversity of species.

At the top of the reef slope is the reef crest. This takes the brunt of the wave action and is subject to high light levels. Shoreward of the reef crest is the reef flat, a shallow, relatively flat expanse of limestone, sand, and coral fragments that may become exposed at high tide. The number of corals decreases towards the shore. Barrier reefs and atolls have a final zone, the lagoon area.

SPECIES DIVERSITY

In addition to reef-building corals, the warm, sunny waters of a reef are populated by a huge variety of other animals as well as seeweeds. The richest and healthiest reefs are home to thousands of species of fish and other marine vertebrates, such as turtles, while all the major groups of invertebrate animals are also represented.

These include sponges, worms, anemones, and non-reef-building corals (such as sea fans), crustaceans, molluscs (which include snails, clams, and octopuses), and echinoderms (sea urchins and relatives). Every nook and cranny of a reef is used by some animal as a hiding place and shelter. All the organisms in the reef are part of a complex web of relationships. Many organisms are also involved in mutualistic partnerships with other organisms, in which both species benefit.

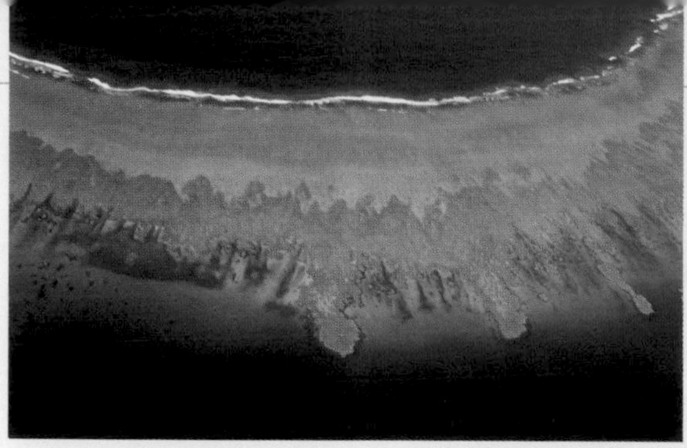

QUEEN ANGEL FISH
One of hundreds of fish species found on the Caribbean reefs, this juvenile angelfish feeds on small crustaceans and algae.

TUBE SPONGES
Different species of sponge are found in many parts of the reef, including caves and cavities, as well as on the open reef slope.

REEF CREST
In front of the reef crest (the uppermost, seaward part of a reef), spurs of coral sometimes grow out into the sea separated by grooves.

sea urchin

crinoid

elkhorn coral

staghorn coral

maze coral

tube sponge

sea fan

star coral

lettuce coral

REEF ZONES
The structure of a typical fringing reef, including forereef, reef crest, and reef flat, and some of the sea life that inhabits it, are shown here. The forereef has three zones, which are dominated by different coral forms: branching coral, massive coral, and platy coral. Individual corals are not shown to scale.

plate-like star coral

finger coral

sea whip

PLATY CORAL ZONE
Corals in this deep, dark part of the forereef expand horizontally to capture maximum sunlight, forming plate-like colonies.

beach

small brain coral

SEA URCHIN
Sea urchins graze on algae and are important in preventing algal overgrowth on coral reefs.

seagrass

golf ball coral

sea anemone

SAND AND ALGAL ZONE
This area is dominated by sand and seagrass, which may harbour small marine life.

REEF FLAT
The animals living here must be able to endure high temperatures and salinity.

REEF CREST
The corals inhabiting this zone are invariably robust, as they must withstand energetic wave action.

BRANCHING CORAL ZONE
This zone is just below the reef crest and is dominated by corals with branching forms, such as staghorn coral.

MASSIVE CORAL ZONE
This central part of the forereef is usually dominated by massive corals – that is, colonies with rounded shapes.

SUBMARINE STUDY
Here researchers record the frequency of algal species on a reef in the Hawaiian Islands, using a camera, a frame for delineating areas of reef, and underwater writing implements.

THE IMPORTANCE OF REEFS

Coral reefs are of inestimable value for many reasons. First, they provide a protective barrier around islands and coasts: without the reefs, these would erode away into the ocean. Second, reefs are highly productive, creating more living biomass than any other marine ecosystem and providing an important food source for many coastal peoples. Third, they support more species per square unit area than any other marine environment. In addition to known coral-reef species, scientists estimate that there may be several million undiscovered species of organisms living in and around coral reefs. This biodiversity may be vital in finding new medicines for the 21st century – many reef organisms contain biochemically potent substances that are being studied as possible cures for arthritis, cancer, and other diseases. Finally, because of their outstanding beauty, reefs contribute to local economies through tourism, particularly attracting snorkellers and scuba-diving enthusiasts (see p.474).

REEF FISHING
Small-scale fishing using hand nets, often transported to a suitable site by canoe, is common throughout the Indian and Pacific oceans, as shown here off Pantar Island in eastern Indonesia.

GOLDEN CRINOID
Crinoids, or feather stars, are related to starfish. They usually live in a hole or other shelter on the reef, extending their elegant arms to catch food.

crinoid arm

algae

CORAL POISONING

One of the most destructive fishing practices, liable to kill corals over wide areas of reef, involves the use of poison to help catch tropical fish for the aquarium trade. This is practised in parts of Southeast Asia such as the Philippines. The young boy photographed below, swimming at a depth of about 20m (70ft), carries a catch bag, net, and a squirt bottle containing a solution of sodium cyanide. The cyanide is used to immobilize selected reef fish, making them easier to catch, but kills all the living corals that it comes in contact with, taking a terrible toll on the health of the reef.

VULNERABLE REEFS

Many types of stress can damage reefs and are doing so today on a massive scale. Much of the harm is caused by human activity, including the effects of coastal pollution, uncontrolled development of coasts, and diving tourism. Other problems include the collection of corals and reef organisms for the aquarium and jewellery trades, uncontrolled mining of reefs for building materials, and destructive fishing practices. Natural disturbances include tropical storms and mass die-offs of animals that help to maintain reef health. Coral bleaching, linked to rises in sea temperatures (see p.153), is particularly worrying. Coral reefs can recover from periodic natural traumas but if they are subjected to multiple and sustained stresses, they perish. It has recently been estimated that two-thirds of the world's warm-water reefs are at risk of disappearing in the near future.

Bermuda Platform

TYPE	Atoll with fringing and patch reefs
AREA	370 square km (150 square miles)
CONDITION	Localized areas of damage

LOCATION Northwest Atlantic, extending west and north of the islands of Bermuda

The Bermuda Platform is the elliptical, flattened summit of a huge volcanic submarine mountain (seamount) in the northwest Atlantic. Its surface lies 14–18m (45–60ft) below sea level and is covered in a thick layer of limestone, formed over millions of years from the

BOILER REEFS
These small reefs, close to the surface, are called "boilers" after their frothy appearance when waves break on them.

remains of corals and other organisms growing on the platform. Along the platform's southern and eastern edges, limestone sand has gradually built up to form the Bermuda islands. Coral reefs are present around the other edges of the platform, forming an atoll, while patch reefs grow on its central surface. The diversity of reef flora and fauna here is less than that associated with the reefs in the Caribbean Sea to the south.

Nevertheless, 21 different species of stony coral, 17 species of soft (non-reef-building) coral, including many spectacular Purple Sea Fans, and about 120 different species of fish have been recorded here.

Lighthouse Reef

TYPE	Atoll with patch reefs
AREA	300 square km (120 square miles)
CONDITION	Generally healthy

LOCATION Western Caribbean, 80km (60 miles) east of central Belize

Lighthouse Reef is an atoll lying 55km (35 miles) east of the huge Belize barrier reef, off the coast of central Belize. It is roughly oval-shaped, about 38km (23 miles) long, and 8km (5 miles) wide on average.

Florida Reef Tract

TYPE	Barrier reef, patch reefs
AREA	1,000 square km (400 square miles)
CONDITION	Degraded; some recent recovery

LOCATION From offshore Miami Beach to south of Key West, Florida, USA

This system of coral reefs is 260km (160 miles) long and curves to the east and south of the Florida Keys. Some geologists classify it as a barrier reef, others as a barrier-like collection of bank reefs. It is the largest area of coral reefs in the USA and has a high biodiversity, being home to about 60 species of stony coral, 1,300 species of mollusc, and 500 species of fish. The reefs' health has declined over

the past 30 years, mainly due to human impact. Live coral cover has decreased, coral diseases have become extensive, inhabitants that were once common (such as the Queen Conch) have virtually disappeared, and the area of reef encroached on by mats

of algae has expanded. Causes of this degradation include overfishing, fertilizer run-off from south Florida, sedimentation onto the reefs as a result of dredging, and sewage pollution from boats. Other contributing factors include hurricane damage, declines in algae-grazing sea urchins, and direct damage from dive-boat anchors and ship groundings. Steps are being taken to reverse the decline, with some signs of success.

CARYSFORT REEF
Carysfort Reef, part of the Florida Reef Tract, lies close to Key Largo and is the site of many ancient shipwrecks.

Bahama Banks

TYPE	Fringing reefs, patch reefs, barrier reef
AREA	3,150 square km (1,200 square miles)
CONDITION	Generally healthy

LOCATION Bahamas, southeast of Florida, USA, and northeast of Cuba

The Bahamas is an archipelago of some 700 islands scattered over two limestone platforms, the Little Bahama and Great Bahama Banks, in the West Indies. The platforms have been accumulating for at least 70 million years – the Great Bahama Bank is over 4,500m (15,000ft) thick – yet their surfaces remain 10–25m (33–80ft) below sea level. Many of the islands have fringing coral reefs; there are also many patch reefs on the Banks and a

barrier reef near the island of Andros. The reefs are home to a range of corals and coral reef-dwelling animals that is typical for the western tropical Atlantic. Although local declines in coral cover and occasional outbreaks of coral disease have been recorded, the reefs are generally healthy. There has been concern about overgrowth of algae, but for now the algae are being kept in check by a thriving population of parrotfish, which graze the reefs.

HARD AND SOFT CORALS
This diverse group of corals, including a large Purple Sea Fan, was photographed off the island of New Providence.

Like all atolls, it is bounded by a ring-like outer structure of coral formations, many of which break the surface. These form a natural barrier against the sea and surround a lagoon, which sits on top of a mass of limestone. The lagoon is relatively deep but contains numerous patch reefs along with six small, sandy, low-lying islands, or cays (one containing a dive centre). At its centre is Lighthouse Reef's most remarkable feature – a large, almost circular sinkhole in the limestone, known as the Great Blue Hole. Approximately 145m (480ft) deep, this feature formed some 18,000 years ago during the last ice age, when much of Lighthouse Reef was above sea level. At that time, freshwater erosion

produced a complex of air-filled caves and tunnels in the limestone. At some point, the ceiling of one of the caves collapsed, producing what is now the entrance to the Blue Hole. Later, as sea level rose, the cave complex flooded, and it is now accessible only by adventurous scuba divers.

Apart from the Blue Hole, the atoll boasts large areas of healthy, abundant, mainly unexplored coral formations. As well as patch reefs within the atoll, around its margins are many spectacular coral-encrusted walls (drop-offs) that descend to depths of several hundred metres. Lighthouse Reef exhibits a biological diversity typical of the region; it is home to some 200 fish species and 60 species of stony coral.

HUMAN IMPACT

DIVING THE GREAT BLUE HOLE

The Great Blue Hole is one of the world's most exciting dive sites. It is not recommended for the faint-hearted (as sharks are commonly encountered) or for novice divers (because perfect buoyancy control is needed). At 38m (125ft) depth, an array of impressive ancient stalactites can be seen hanging from the slanting walls of the hole. The entrance to a system of caves and tunnels lies a few metres further down.

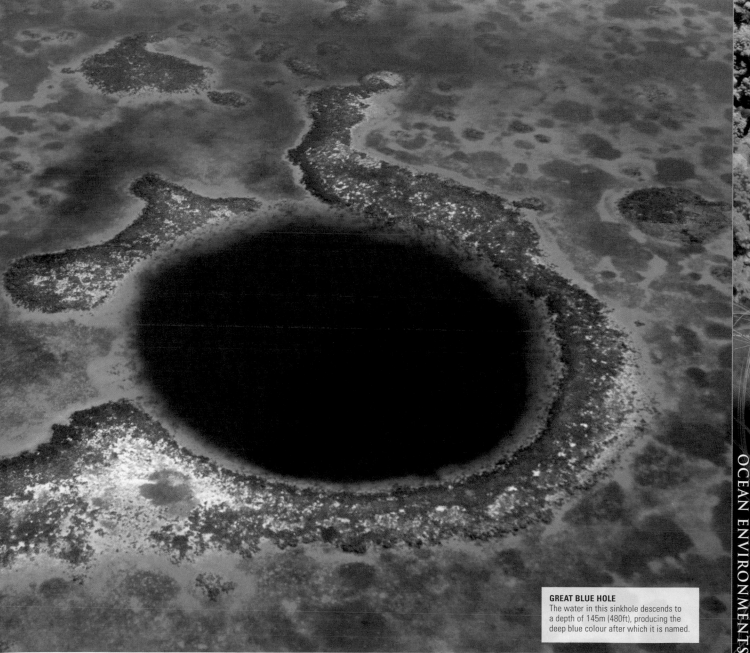

GREAT BLUE HOLE
The water in this sinkhole descends to a depth of 145m (480ft), producing the deep blue colour after which it is named.

INDIAN OCEAN NORTHWEST

Red Sea Reefs

TYPE	Fringing, patch, and barrier reefs; atolls
AREA	16,500 square km (6,300 square miles)
CONDITION	Generally good; localized damage

LOCATION Red Sea coasts of Egypt, Israel, Jordan, Saudi Arabia, Sudan, Eritrea, and Yemen

The Red Sea contains arguably the richest, most biologically diverse, and most spectacular coral reefs outside Southeast Asia. The coral reefs in the northern and southern areas of the sea differ considerably. In much of the northern section, the coasts shelve extremely steeply and there are few offshore islands. The coral reefs here are mainly narrow fringing reefs, with reef flats typically only a few metres wide, and slopes that plunge steeply towards the sea floor. In the south, off Eritrea and southwestern Saudi Arabia, is a much wider area of shallow continental shelf. Many of the reefs

in this area surround offshore islands, and there are fewer steep drop-offs. The southern Red Sea also receives a continuous inflow of water from the Gulf of Aden to its south that is high in nutrients and plankton, making the waters more turbid, or cloudy, which restricts reef development. Live coral cover throughout the Red Sea reefs is generally high, at about 60–70 per cent, as is the diversity of stony and soft corals, fish (including the famous Red Sea Lionfish), and other reef organisms. More than 260 different species of stony coral have been identified in the central Red Sea.

Although the Red Sea reefs are mainly healthy, intense diving tourism, especially in parts of Egypt, has caused severe local damage. Coral predation by the Crown-of-thorns Starfish has also been a problem, and there is a threat of oil spillages from tankers heading towards the Suez Canal.

GULF OF AQABA REEF
Groups of little red fish of the genus *Anthias* fluttering around hard coral heads, or colonies, are a familiar sight on Red Sea reefs.

INDIAN OCEAN NORTHWEST

Aldabra Atoll

TYPE	Atoll
AREA	155 square km (60 square miles)
CONDITION	Excellent, although it has suffered some coral bleaching

LOCATION Western extremity of the Republic of Seychelles archipelago, northwest of Madagascar

At 34km (20 miles) long and 14.5km (9 miles) wide, Aldabra is the largest raised coral atoll in the world. The term "raised" refers to the fact that the

limestone structures forming its rim, which originated from coral reefs, have grown into four islands that protrude as much as 8m (27ft) above sea level. Situated on top of an ancient volcanic pinnacle, the islands enclose a shallow lagoon, which partially empties and then fills again twice a day with the tides. Because of its remote location, and its status as a Special Nature Reserve and (since 1982) UNESCO World Heritage Site, Aldabra has escaped the worst of the stresses that human activities have placed on most of the world's coral reefs. Although, in common with many Indian Ocean locations, the atoll was affected by

a severe coral bleaching in 1997–98, its external reefs are in a near-pristine state. They are rich in marine life, featuring large schools of reef fish, Green and Hawksbill Turtles, forests of yellow, pink, and purple sea-fans, groupers, hammerhead sharks, and barracuda. The atoll's inner lagoon contains numerous healthy patch reefs, is fringed by mangrove swamps, and is inhabited by turtles, parrotfish, and Eagle Rays.

On land, Aldabra is famous for its giant tortoises, rare exotic birds such as the Flightless Rail, and giant Robber Crabs, which have claws big enough to crack open coconuts.

MUSHROOM ROCK
Strong tidal flows of ocean water into and out of Aldabra's lagoon have sculpted some raised clumps of old reef into mushroom-shaped islets known as champignons.

INDIAN OCEAN WEST

Bazaruto Archipelago

TYPE	Fringing reefs, patch reefs
AREA	150 square km (60 square miles)
CONDITION	Generally good; some damage

LOCATION Southeastern coast of Mozambique, northeast of Maputo

The Bazaruto Archipelago is a chain of sparsely populated islands on the coast of Mozambique, formed where sand was deposited over hundreds of thousands of years by the Limpopo River. A Marine National Park, established in 2001, covers most of the archipelago, protecting its impressive fringing reefs and kaleidoscopic range of marine life. More than 2,000 fish species, 100 species of stony corals, and 27 dazzling soft-coral species, including unusual "green tree" corals, are found on Bazaruto's reefs, as well as Eagle Rays, Manta Rays, and five species of turtle. The archipelago is also a refuge for one of the remaining populations of dugongs (see p.423) in the western Indian Ocean.

REEF SAFARI
A peaceful way of visiting the shallow, crystal-clear waters around the Bazaruto reefs is on a dhow, as part of a reef safari.

INDIAN OCEAN CENTRAL
Diego Garcia Atoll

TYPE	Atoll
AREA	44 square km (17 square miles)
CONDITION	Generally good; recovering from coral bleaching in 1998

LOCATION Chagos Archipelago, south of Sri Lanka, central Indian Ocean

This atoll, best known as a US military base, is also home to one of the world's largest populations of breeding sea birds. The atoll is unusual in that its coral limestone–based rim, raised above sea level, is a single island that runs almost completely round the central lagoon, which is 20–30m (65–100ft) deep. The reefs within the lagoon and around the atoll's edges are home to 220 species of stony coral.

WESTERN SIDE OF DIEGO GARCIA

INDIAN OCEAN CENTRAL
Maldives

TYPE	Atolls, fringing reefs
AREA	9,000 square km (3,500 square miles)
CONDITION	Recovering from coral bleaching

LOCATION Off southern India, southwest of Sri Lanka, in the Indian Ocean

The Maldives are a group of 26 atolls, many of them very large, in the Indian Ocean. The majority are composed of numerous separate reefs and coralline islets (some 1,200 in all), arranged in ring-like structures. Within most of the atoll lagoons, which are 18–55m (60–180ft) in depth, there are usually many patch reefs and numerous structures called faros, which are rare outside the Maldives. These look like mini-atolls and consist of roughly elliptical reefs with a central lagoon.

Most of the Maldivian atolls are themselves arranged in a large, elliptical ring, some 800km (500 miles) long and 100km (60 miles) wide. The reefs that fringe all the Maldivian atolls, islets, and faros contain more than 200 species of colourful stony coral, more than 1,000 different fish species, and are abundant in other marine life. Groupers, snappers, and sharks, for example, are frequently encountered.

In 1998, a severe coral bleaching event killed up to 90 per cent of the corals in some areas, and had a strong negative impact on diving and snorkelling tourism. By early 2006, however, some areas were effectively recolonizing with corals.

HUMAN IMPACT
ATOLL CITY

Male, the Maldives' capital city, covers the entire surface area of a coral island that forms part of an atoll rim. Its reef has been mined to provide building materials for artificially extending the island.

The partly dismantled reef leaves the island poorly protected from storms, so a sea wall has been built around much of its perimeter, preventing major damage during the 2004 Indian Ocean tsunami.

INDIAN OCEAN NORTHEAST
Andaman Sea Reefs

TYPE	Fringing reefs
AREA	5,000 square km (2,000 square miles)
CONDITION	Some areas poor due to coral bleaching, diver damage

LOCATION Andaman Sea coasts: Thailand, Myanmar, Andaman and Nicobar Islands, Malaysia, Sumatra

Most Andaman Sea reefs are fringing reefs around islands off the coasts of Thailand and Myanmar or, in the northwest, off the eastern coasts of the Andaman and Nicobar islands – the site of the largest continuous area of reefs in south Asia. About 200 coral species and more than 500 fish species have been recorded here. The reefs and islands are also important feeding and breeding grounds for endangered sea turtles. A coral bleaching event in 1998 badly damaged reefs around the Andaman and Nicobar islands, which until then were some of the most pristine anywhere, but hardly touched the Thailand reefs. The 2004 Indian Ocean tsunami caused relatively little damage. Other threats to these reefs include collection of marine life for aquariums, destructive fishing techniques, siltation caused by poorly managed deforestation on some of the islands, and anchor damage from dive boats.

SOFT CORAL COLONIES
These soft corals and glass fish, which are almost transparent, were photographed off southwest Thailand.

OCEAN ENVIRONMENTS

Shiraho Reef

TYPE	Fringing reef
AREA	10 square km (4 square miles)
CONDITION	Reasonable; recovering from severe coral bleaching in 1998

LOCATION Southeast coast of Ishigaki Island, at the southwestern extremity of Japanese archipelago

Shiraho Reef, off Ishigaki Island, part of the Japanese archipelago, came to notice in the 1980s as an outstanding example of biodiversity, with some 120 species of coral and 300 fish species concentrated in a few square kilometres. The reef also contains the world's largest colony of rare Blue Ridge Coral (*Heliopora coerulea*). For decades, environmentalists have battled to save the reef from the building of a new airport for Ishigaki. A proposal to construct the airport on top of the reef was dropped, but concern remains over plans to build it on land, as soil run-off and sedimentation in the reef area is likely to have an adverse effect.

BLUE RIDGE CORAL
Despite its name, the colour of this coral varies from violet through blue, turquoise, and green to yellow-brown. Its branching vertical plates can form massive colonies.

Tubbataha Reefs

TYPE	Atolls
AREA	330 square km (130 square miles)
CONDITION	Good; recovering from coral bleaching in 1998

LOCATION Central Sulu Sea, between the Philippines and northern Borneo

The Tubbataha Reefs lie around two atolls in the centre of the Sulu Sea and are famous for the many large pelagic (open ocean) marine animals attracted to them – such as sharks, Manta Rays, turtles, and barracuda. The steeply shelving reefs here are also rich in smaller life, including many species of crustaceans, colourful nudibranchs (sea slugs), and more than 350 species of stony and soft coral.

In the early 1990s, the Tubbataha Reefs were rated by scuba divers among the top ten dive sites in the world. However, during the 1980s they suffered considerable damage from destructive fishing practices and the establishment of a seaweed farm.

CORAL DROP-OFF
In this photograph of a steeply shelving reef slope, several species of soft coral are visible, together with a shoal of Longfin Bannerfish.

In 1988, the Philippines government intervened, declaring the area a National Marine Park, and since 1993 it has also been a UNESCO World Heritage Site. Today, the condition of the Tubbataha reefs has much improved, due to the enforcement of measures such as a prohibition on fishing and a ban on boats anchoring on the reefs (visiting craft must use mooring buoys). Live coral cover in most sites in 2004 showed significant increases after a coral bleaching episode in 1998.

Nusa Tenggara

TYPE	Fringing reefs, barrier reefs
AREA	5,000 square km (2,000 square miles)
CONDITION	Damaged by fishing practices

LOCATION Southern Indonesia, from Lombok in the west to Timor in the east

Nusa Tenggara is a chain of around 500 coral-fringed islands in southern Indonesia. The northern islands are volcanic in origin, while the southern islands consist mainly of uplifted coral limestone. Many of the reefs have been only rarely explored. However, what surveys have been carried out indicate an extremely high diversity of marine life in this region. For example, a single large reef can contain more than 1,200 species of fish (more than in all the seas in Europe combined), and 500 different species of reef-building coral. Common animals here include Eagle Rays, Manta Rays, Humphead Parrotfish, and various species of octopuses and nudibranchs (sea slugs). Major threats to the reefs in Nusa Tenggara include pollution from land-based sources, sediment pollution from logging, removal of fish from the reefs for the aquarium trade, and reef destruction by blast fishing.

REEF FLAT OFF PANTAR ISLAND
This shallow reef area, featuring numerous species of stony coral and a starfish, is in central Nusa Tenggara.

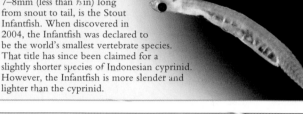

Great Barrier Reef

TYPE	Barrier reef
AREA	37,000 square km (14,300 square miles)
CONDITION	Damaged by Crown-of-thorns Starfish; coral bleaching

LOCATION Parallel to Queensland coast, northeastern Australia

Australia's Great Barrier Reef, which stretches 2,010km (1,250 miles), is the world's largest coral reef system. Often described as the largest structure ever made by living organisms, it in fact consists of some 3,000 individual reefs and small coral islands. Its outer edge ranges from 30 to 250km (18 to 155 miles) from the mainland, and its biological diversity is high. The reef contains about 350 species of stony coral and many of soft coral. Its 1,500 species of fish range from gobies, the smallest fish on the reef, and 45 species of butterflyfish, to several shark species, including silvertip, hammerhead, and whale sharks. The reef is also home to 500 species of algae, 20 species of sea

REEF CHANNEL
In this view of a central area of the reef, a deep, meandering channel separates two reef platforms. The region's high tidal range drives strong currents through such channels.

snake, and 4,000 species of mollusc. Damage to the reef over the past 30 years has resulted mainly from predation by the Crown-of-thorns Starfish and a mass coral bleaching event in 1998. In 1975, the Great Barrier Reef Marine Park was established, and in 1981 it was declared a UNESCO World Heritage site.

THE WORLD'S SMALLEST VERTEBRATE?

One of the tiniest residents of the Great Barrier Reef, at just 7–8mm (less than ⅓in) long from snout to tail, is the Stout Infantfish. When discovered in 2004, the Infantfish was declared to be the world's smallest vertebrate species. That title has since been claimed for a slightly shorter species of Indonesian cyprinid. However, the Infantfish is more slender and lighter than the cyprinid.

Marshall Islands

TYPE	Atolls
AREA	6,200 square km (2,400 square miles)
CONDITION	Generally good; some local degradation

LOCATION Micronesia, southwest of Hawaii, western Pacific

The Marshall Islands consist of 29 coral atolls and five small islands in the western Pacific. The atolls lie on top of ancient volcanic peaks that are thought to have erupted from the ocean floor 50–60 million years ago. They include Kwajalein, the largest atoll in the Pacific at 2,500 square km (1,000 square miles), and Bikini and Enewetak atolls, which were used by the USA for testing nuclear weapons between 1946 and 1962. Human pressures on these two remote, evacuated atolls have been minimal during the past 50 years, and marine life around them now thrives; for example, 250 species of coral and up to 1,000 species of fish have been recorded at Bikini.

MAJURO ATOLL
As with many Pacific atolls, the rim of Majuro Atoll consists partly of shallow submerged reef and partly of small, low-lying islands.

Society Islands

TYPE	Fringing reefs, barrier reefs, atolls
AREA	1,500 square km (600 square miles)
CONDITION	Good, but significant local damage

LOCATION French Polynesia, northeast of New Zealand, south-central Pacific

The Society Islands comprise a chain of volcanic and coral islands in the South Pacific, including islands with barrier reefs (such as Rai'atea), islands with both fringing and barrier reefs (such as Tahiti), and atolls or near-atolls (such as Maupihaa and Maupiti). The reefs' biological diversity is moderate compared with the reefs of Southeast Asia, although more than 160 coral species, 800 species of reef fish, 1,000 species of mollusc, and 30 species of echinoderm have been

recorded. The reefs' health is generally good, but some reefs around the busy holiday destination islands of Tahiti, Moorea, and Bora-Bora have been severely affected by construction, sewage, and sediment run-off.

MOOREA
A wide fringing reef almost completely surrounds the shoreline of mountainous Moorea, part of which is visible in this view.

Hawaiian Archipelago

TYPE	Fringing reefs, atolls, submerged reefs
AREA	1,180 square km (450 square miles)
CONDITION	Generally good; some degradation

LOCATION North-central Pacific

The Hawaiian Archipelago consists of the exposed peaks of a huge undersea mountain range. These mountains have formed over tens of millions of years as the Pacific Plate moves

northwest over a hotspot in the Earth's mantle. Coral reefs fringe some coastal areas of the younger, substantial islands at the southeastern end of the chain, such as Oahu and Molokai. To the northwest, located on the submerged summits of older, sunken islands, are several near-atolls (such as the French Frigate Shoals) and atolls (such as Midway Atoll). These reefs are highly isolated from all other coral reefs in the world, and although their overall biological diversity is relatively low, many new species have evolved on them. About a quarter of the animals and plants found in the Hawaiian Islands and a few nearby reefs are found nowhere else.

FRENCH FRIGATE SHOALS
Reef fish, including Longfin Bannerfish, Milletseed Butterflyfish, and Bluestripe Snappers, swim around a table coral.

THE GREAT BARRIER REEF
The warm, clear waters of the reef support an astonishing variety of life. Here, Fairy Basslets can be seen shoaling over vividly coloured soft corals. Bright coloration can serve several purposes for reef fish, including helping members of a species to recognize each other and acting as a warning to predators.

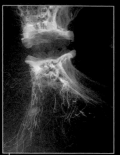

LION'S MANE JELLYFISH
This daunting giant of the plankton can grow up to 2m (6ft) across, with 60m (200ft) tentacles.

THE PELAGIC ZONE

THE PELAGIC ZONE IS THE WATER COLUMN ABOVE the continental shelf (although the term is also used to refer to the water column of the open ocean). It is a vast environment, and temperature and salinity variations within it result in distinct water masses. These are separated by "fronts" and characterized locally by different plankton. Coastal and shelf waters are more productive than the open ocean. When calm, the water stratifies, cutting off the surface plankton from essential nutrients in the layers below. Storms cause the layers to mix, stimulating phytoplankton blooms. High latitudes have seasonal plankton cycles; in warmer waters, seasonal upwelling of nutrient-rich deeper water triggers phytoplankton growth.

MICROSCOPIC PRODUCTIVITY

Much of the primary productivity in the world's oceans and seas occurs over the continental shelves. Tiny phytoplankton floating in the surface waters harness the Sun's energy through photosynthesis to produce living cells. Some of the tiniest algae (picoplankton) are thought to supply a considerable amount of primary production. As well as sunlight, nutrients and trace metals are needed for phytoplankton growth. These are often in short supply in the open ocean, but shelf waters benefit from a continual input from rivers, mixing by waves and, on some coasts, the upswelling of nutrient-rich water.

PRIMARY PRODUCTION
This satellite map shows variations in primary production, indicated by the concentration of the pigment chlorophyll a in the oceans and the amount of vegetation on land.

CHLOROPHYLL A CONCENTRATION
LOW   HIGH

VEGETATION INDEX
MIN   MAX

THE PLANKTON CYCLE

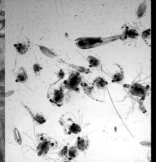

DRIFTING ZOOPLANKTON
Continental shelf zooplankton contains many larvae of sea bed animals which then drift away to new areas.

In temperate and polar seas, optimal phytoplankton growth occurs in both spring and summer. There are long daylight hours and maximum nutrient levels after winter storms have mixed the water column and resuspended dissolved nutrients from the sea bed. The well-known spring blooms can rapidly turn clear seawater into pea soup, or a variety of other colours depending on the organism. Typically, there is a succession of phytoplankton species with short blooms. Responding to abundant food and increasing temperatures, tiny zooplankton begin grazing the phytoplankton and reproducing. Bottom-living coastal animals release clouds of larvae to feed in the nutritious broth, before taking up life on the sea bed. Spawning fish also contribute a mass of eggs and larvae. Eventually, the phytoplankton is grazed down, nutrients are exhausted, and productivity drops off, in an annual cycle that will be renewed again next spring.

HUMAN IMPACT

RED TIDE

A rapid increase in a population of marine algae is called a bloom. This bloom on the Scottish coast, known as a red tide, was caused by the dinoflagellate *Noctiluca scintillans*. Sometimes blooms poison marine life. Often, the sheer numbers of organisms clog fish gills, suffocating them. Dense blooms occur naturally, but man-made pollution from nutrient runoff into the sea may also feed these blooms, making them more frequent and extensive.

RIDING THE CURRENTS

From tiny algae to giant jellyfish, the animals and plants of the plankton either float passively or swim weakly. This is mainly to keep them up in the sunlit surface waters, where most production occurs; these drifters must go wherever the currents take them. On most continental shelves, there is a residual drift in a particular direction, although wind-driven surface currents, where most of the plankton live, can move in any direction for short periods. Some animals go on long migrations to spawn, relying on residual currents to bring their larvae back to areas suitable for their growth into adults; for example, Conger Eel larvae take around two years to drift back from their spawning grounds far off the continental shelf. The larvae of the majoritiy of coastal animals, including those of barnacles, mussels, hydroids, and echinoderms, spend much shorter periods in the plankton – just long enough to disperse to new areas of coast. However, the plankton is a dangerous place, full of hungry mouths and tentacles, and though millions of eggs and larvea are released, the vast majority of planktonic feeders will die; only a lucky few find a suitable place to settle and grow.

COLONIZED ROPE
This rope was colonized over the course of a year by sea squirts, feather stars, fan worms, and anemones, their planktonic larvae having been transported by ocean currents.

MIGRATORY SHOALS
Pelagic fish such as these mackerel move around the ocean in response to temperature changes. They are among the pelagic zone's larger predators.

ACTIVE SWIMMERS

The animals of the plankton, especially small crustaceans such as copepods and krill, are eaten by fish, mainly small, shoaling species such as herrings, sandeels, sardines, and anchovies. Most of these fish live permanently in midwater, using the sea bed only to spawn or to avoid predators. They are strong swimmers (nekton), using speed to catch prey and evade predators. They can travel long distances against residual currents to feed and also to reach their spawning grounds. Small, shoaling fish are, in turn, food for larger predators, such as squid, tuna, cetaceans, and sharks. Whale Sharks, Basking Sharks, and Baleen Whales are among the largest of the marine animals, yet they feed feed directly on plankton, consuming vast quantities.

NEKTONIC INVERTEBRATE
Squid are the only invertebrates that swim strongly enough to be classed as nekton. They catch a variety of prey including fish and planktonic crustaceans.

PELAGIC FISHERIES

Continental shelf waters support massive quantities of pelagic fish, ultimately sustained by abundant plankton. The most important fisheries are for herring, sardines, anchovies, pilchards, mackerel, capelin and jackfish. Squid are also fished commercially. Many fish stocks are under severe pressure as boats and nets get bigger and the technology to pinpoint shoals becomes ever more sophisticated. Pelagic fish and squid are caught in drift nets which hang about 10m (32¾ft) down from the surface. In the north Pacific, some 170,000km (105,000 miles) of drift net is available to major fisheries; unfortunately these nets also trap cetaceans, turtles, and diving birds. Drifting longlines are used for tuna and swordfish; these also catch juvenile fish, sharks, turtles, and seabirds. Midwater trawls capture vast quantities of shoaling fish such as herring, mackerel, and sardines. Small-scale fisheries for a wide variety of other pelagic species are important in sustaining local coastal communities worldwide.

FOOD CHAIN THREAT
Sand eels are food for sea-birds (such as this Arctic Tern), seals, cetaceans, and larger fish. Despite their importance at the base of many food chains, vast quantities are taken by fisheries for feeding to livestock and farmed fish, and are burned as fuel oil.

THE STEEL-BLUE WAVES of the open ocean conceal an extraordinary landscape, where the continents plunge down to a vast, undulating, muddy plain. Here, the ocean water column supports layer upon layer of life, from the surface zone, powered by sunlight, to the crushing pressure of the darkest depths. In places, the abyssal plain is broken by underwater volcanoes or by mountain ranges high enough to rival the Himalayas. Springs of super-hot water emerge from these mountainsides, supporting living communities unlike any others on the planet. Elsewhere, the Earth's vast tectonic plates collide, ripping trenches in the ocean floor and stimulating powerful earthquakes. Yet fewer people have explored these mysterious depths than have flown in Space.

THE OPEN OCEAN
AND OCEAN FLOOR

BENEATH THE WAVES
In the deepest ocean, an underwater Mount Everest could be hidden beneath these waves – and still leave space for Mount Washington on top. As a result, we have better maps of the Moon than of the deep sea bed.

ZONES OF THE OPEN OCEAN

CONDITIONS IN THE OCEAN vary greatly with depth. Light and temperature changes occur quickly, while pressure increases incrementally. Although many of these changes are continuous, the ocean can be divided into a series of distinct depth zones, each of which produces very different conditions for living things.

THE SURFACE LAYER

The top metre of the ocean is the richest in nutrients. This upper layer is sometimes called the neuston, although this term is also used for the animals that live there, such as jellyfish. Amino acids, fatty acids, and proteins excreted by plants and animals float up into this surface layer, as do oils from the decomposing bodies of dead animals. These produce a rich supply of nutrients for phytoplankton.

The top metre of seawater is the interface where gas exchange takes place between the ocean and the atmosphere. This is vitally important to all life on Earth, as half of the oxygen animals need for survival comes from the ocean. Not surprisingly, phyto-plankton gather in this surface zone in daylight, as do the animals that feed on them. Sadly, this zone is also highly susceptible to chemical pollution and floating litter, which can be deadly for marine life.

NOCTURNAL AND DIURNAL DISTRIBUTION
Only a small proportion of marine life inhabits the deep zone; the majority live above 1,000m (3,300ft). The sunlit zone is dangerous for animals – many stay in the twilight zone by day and only go upwards at night. The sunlit zone is much emptier by day.

DAY
- 10% sunlit zone
- 75% twilight zone
- 15% deep zone

NIGHT
- 40% sunlit zone
- 50% twilight zone
- 10% deep zone

HUMAN IMPACT

FREE DIVING

When divers breathe compressed air underwater, excess nitrogen dissolves in their blood, and they risk the bends if they surface too fast. Free divers avoid this by holding their breath underwater. Pressure squeezes their lungs, but the surrounding blood vessels swell to protect them, and blood nitrogen levels stay safe. Trained free divers can hold their breath long enough to reach 200m (660ft), using aids to help them descend and ascend.

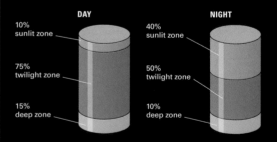

OCEAN ZONES

SUNLIT ZONE 0–200m
Seawater rapidly absorbs sunlight so only one per cent of light reaches 200m (660ft) below the surface. Phytoplankton use the light to photosynthesize, forming the base of food chains. This zone drives all ocean life.

TWILIGHT ZONE 200–1,000m
Too dark for photosynthesis, but with just enough light to hunt by, many animals move from this zone into the sunlit zone at night.

DARK ZONE 1,000–4,000m
Almost no light penetrates below 1,000m (3,300ft). From here to the greatest depths, it is dark, so no plants can grow, and virtually the only source of food is the "snow" of waste from above. Temperatures down here are a universally chilly 2–4°C (35–39°F), and the pressures so extreme that only highly adapted animals can survive.

The dark zone is defined as continuing down to the abyssal plain, below 4,000m (13,100ft). Technically, all the water below 1,000m (3,300ft) is a dark zone, where the only light comes from bioluminescent animals (see p.224). However, for convenience, the waters below the dark zone can be further subdivided.

ABYSSAL ZONE 4,000–6,000m
Beyond the continental slope, the sea bed flattens out. In many areas, it forms vast plains at depths below 4,000m (13,100ft). Some areas drop deeper to a sea floor that undulates down to depths of 6,000m (19,700ft). Around 30 per cent of the total sea-bed area lies between these depths. Animals living here move up and down through a narrow column above the sea bed, called the abyssal zone.

HADAL ZONE 6,000–11,000m
The sea floor plunges below 6,000m (19,700ft) in only a few deep ocean trenches. This hadal zone makes up less than 2 per cent of the total sea-floor area. Only two human beings have ever visited this zone (see p.183), and the pressures are so high that only a few unmanned submersibles are able to operate here.

We therefore know very little about what lives in these great depths, although a fish has been dredged from a depth of 8,370m (27,500ft) and shrimp-like amphipods have been photographed 10,911m (35,797ft) down. Some animals here may require the high pressure at these depths in order to survive.

SCALE

Empire State Building (443m/1,453ft)

Light Temperature Pressure People and Machin

- 200m (660ft) 20°C 1 atm sea level
  - 50m (164ft) deepest safe scuba div
  - 209m (685ft) deep freedive
  - 330m (1,082ft) scuba record by French diver
  - 600m (1,960ft) nuclear-powered submarine
- 1,000m (3,300ft) 5°C
- 2,000m (6,500ft) 2–4°C
- 3,000m (9,800ft)
- 4,000m (13,100ft) 2–4°C 100 atm
- 5,000m (16,400ft)
  - 4,500m (14,800ft) Alvin submersible
- 6,000m (19,700ft) 2–4°C 400 atm
- 7,000m (23,000ft)
  - 00m (21,000ft) inkai submersible
- 11,000m (36,100ft) 600 atm

deep trenches plunging to 11,000m (36,100ft) cover just 2 per cent of the sea bed.

0,911m (35,797ft) rieste submersible

Marine Life

- 201m (659ft) *Dolphin*
- 350m (1,148ft) *King Penguin*
- 680m (2,230ft) *Great White Shark*
- 1,000m (3,300ft) *Sperm Whale*
- 1,200m (3,937ft) *Leatherback Turtle*
- 1,580m (5,183ft) *Elephant Seal*

The column below shows the average depth (yellow band) and greatest depth (red band) of the oceans and some of the world's seas.

North Sea average depth **94m** (308ft)

Baltic Sea greatest depth **449m** (1,473ft)

North Sea greatest depth **700m** (2,296ft)

Arctic Ocean average depth **990m** (3,248ft)

Mediterranean Sea average depth **1,500m** (4,921ft)

Caribbean Sea average depth **1,512m** (4,960ft)

almost one third of the total sea-bed area is made up of abyssal plains at around 4,500m (14,800ft).

Atlantic Ocean average depth **3,330m** (10,925ft)

Indian Ocean average depth **3,890m** (12,762ft)

Pacific Ocean average depth **4,280m** (14,041ft)

Southern Ocean average depth **4,500m** (14,763ft)

Mediterranean Sea greatest depth **5,095m** (16,715ft) (Hellenic Trough)

Arctic Ocean greatest depth **5,601m** (18,377ft) (Molloy Deep)

Southern Ocean greatest depth **7,152m** (23,466ft)

Caribbean Sea greatest depth **7,685m** (25,213ft) (Caymen Trench)

Indian Ocean greatest depth **7,725m** (25,344ft) (Java Trench)

Atlantic Ocean greatest depth **8,962m** (29,404ft) (Puerto Rico Trench)

Beyond the abyssal plains, undulating, rocky sea bed stretches down to around 6,000m (19,700ft). Only the ocean trenches reach deeper.

Pacific Ocean greatest depth **10,920m** (35,829ft) (Marianna Trench)

ZONES OF LIFE
The different zones of life in the deep ocean are shown here, together with the depths reached by humans and a selection of marine animals. Most life is concentrated above 1,000m (3,300ft), where there is some light.

CRYSTAL WATERS
Crystal-clear tropical waters look idyllic, but the clarity indicates that there are few nutrients and therefore few phytoplankton in the water. As a result, feeding for animals is quite poor.

THE SUNLIT ZONE

The sunlit zone is the range in which there is enough sunlight for photosynthesis. The ocean absorbs different wavelengths of sunlight to differing extents (see p.38). Nearly all red light is absorbed within 10m (30ft), so red animals look black below this depth. Green light penetrates much deeper in clear water, down to around 100m (330ft), and blue light to twice that. Due to the presence of chlorophyll, phytoplankton preferentially absorb the red and blue portions of the light spectrum (for photosynthesis) and reflect green light. They can photosynthesize down to about 200m (660ft) in clear water.

In cloudy water, the sunlit zone is shallower, because light is absorbed more quickly. The accumulations of phyto- and zooplankton in fertile waters absorb sunlight, reducing the depth of the sunlight zone. Phytoplankton must stay in the sunlit layer during daylight to photosynthesize. Zooplankton follow them there to feed, along with animals that feed on zooplankton. This zone is dangerous because light makes animals conspicuous to their hunters.

FEEDING IN THE SUNLIT ZONE
The phytoplankton of the sunlit zone are the food of zooplankton. Larger animals, such as these shrimp, in turn feed on zooplankton. Phytoplankton are at the bottom of most ocean food chains.

LIVING IN THE SUNLIT ZONE

Phytoplankton must remain in the sunlit zone if they are to catch enough sunlight for photosynthesis. This zone is the warmest and richest in the nutrients needed for growth. It would be counterproductive to expend huge amounts of energy to stay in this zone, so phytoplankton have developed a wide range of mechanisms to help them hang there effortlessly. Buoyancy bubbles, droplets of oil, or stores of light fats keep some species afloat. Others are covered in spines, which increase their surface area and help buoy them up. Some phytoplankton form colonial chains, which produce more drag in water and slows the rate at which they sink. One group, called dinoflagellates, have thread-like flagellae that let them swim weakly. In this highly productive zone, phytoplankton produce half of the oxygen in the atmosphere. In temperate regions, phyto-plankton proliferate in summer, sometimes forming dense blooms.

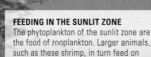

DIATOM
Diatoms are a very prolific type of phytoplankton. Some grow colonially, attached to rocks in chains or mats. Each year, six billion tonnes of phytoplankton grow in the oceans worldwide.

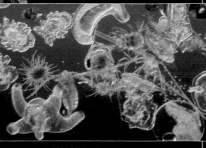

ZOOPLANKTON
This sample of zooplankton, collected in a net, includes an echinoderm (bottom left), a radiolarian, and a crab larva (centre), with a fish egg (bottom right).

PLANKTON AND NEKTON

In spring, as phytoplankton blooms begin to develop, zooplankton start multiplying. They follow the phytoplankton into the sunlit zone to feed. Most are herbivores that feed on phytoplankton; some are carnivores that hunt other zooplankton. Many are classed as meroplankton – the young of animals like crabs, lobsters, barnacles, and some fish – which have a planktonic larval stage and use the currents to spread. By taking advantage of the summer phytoplankton feast, they avoid competing for food with adults of their own kind. While plankton drift with the currents, many free-swimming animals (collectively called nekton) gather to feed on them: fish, squid, marine mammals, and turtles. These, in turn, are food for predatory fish and sea birds. Some larger animals, such as basking sharks, also feed on zooplankton and nekton.

SARGASSUMFISH
Here, two Sargassumfish are hiding in *Sargussum* seaweed, floating on the surface of the Sargasso Sea.

COPEPOD
Copepods are herbivores. They make up 70 per cent of the total zooplankton population, with thousands in a cubic metre.

THE TWILIGHT ZONE

In the twilight zone, there is just enough light for animals to see – and be seen. As a result, predators and prey are in constant battle. Many species are almost totally translucent, to avoid casting even a faint shadow. Others are reflective, to disguise themselves against the light from above, or have wafer-thin bodies that reduce their silhouette. To cope with dim light, many animals in this zone have large eyes.

The main source of food here is detritus. Many animals therefore migrate upwards into the sunlit zone, where food is plentiful, at night, returning to the twilight zone as the Sun rises. Millions of tonnes of animals, equivalent to around 30 per cent of the total marine biomass, make this daily trek – by far the largest migration of life on Earth. The length of the journey is a matter of scale. Small planktonic animals measuring less than 1mm ($^1/_{25}$in) in length may only migrate through 20m (70ft), but some larger shrimp travel 600m (2,000ft) each way, every day.

GIANT FILTER FEEDER
More than 11m (36ft) long, Basking Sharks like this one scoop up shoals of plankton, then filter them from the water with the white gill rakers inside their jaws.

THE DARK AND ABYSSAL ZONES

The waters below the twilight zone are all dark, cold, subject to high pressure, and impoverished in food. For animals adapted to these deep zones, pressure is not a problem: their liquid-filled bodies are almost incompressible, compared to the gas-filled bodies of surface-living birds and mammals, which are much more easily compressible and subject to the effects of pressure. Most fish use gases in their swim bladders to maintain buoyancy, and these are susceptible to pressure change. Many deep-water fish therefore have no functional swim bladders.

For most deep-water species, lack of food is the biggest problem: only about five per cent of the energy that plants produce at the surface filters down to these depths. Animals of the deep are typically slow-moving, slow-growing, and long-lived. They conserve energy by waiting for food to come to them. Many therefore have massive mouths and powerful teeth. Others use ploys to catch prey: anglerfish dangle lures, and some species even harbour luminescent bacteria or use chemical processes to make these lures and other structures glow.

SQUID OF THE OPEN OCEAN
Squid live in several zones of the ocean, from the sunlit zone, where this Big-fin Reef Squid is found, to the deep zone. Deep-sea squid are difficult to photograph and are often photographed only as dead specimens.

SPOOKFISH
The Brownsnout Spookfish is found at depths of up to 1,000m (3,300ft), living on the boundary of the dark zone. Its bones are so thin that it is almost transparent, and its large eyes look upwards to spot predators attacking from above. It feeds mainly on copepods, and gives birth to live young which float in the plankton.

mucus on body attract bacteria, which protect it from heat

red tentacles around head gather food and provide sensory information

HEAT-TOLERANT WORM
This polychaete worm (a type of segmented worm) was discovered by the *Alvin* submersible in 1979 – and named *Alvinella pompejana* in its honour. It is the most heat-tolerant animal on Earth, living near water emerging from hydrothermal vents at 300°C (570°F).

DISCOVERY

THE *CHALLENGER*

Many oceanographic discoveries were made by HMS *Challenger*, a converted British warship that made a 110,900km-(68,900-mile) voyage around the oceans from 1872 to 1876, collecting depth soundings as she went.

In March 1875, near Guam, she dropped a sounding line to a depth of 8,184m (26,850ft) and collected clay to prove this was the sea bed. By good luck, she was over the Mariana Trench, close to the deepest spot in the ocean, now appropriately called the Challenger Deep.

DEEP SQUEEZE
A polystyrene cup attached to the outside of a submersible resurfaces at a fraction of its original size, illustrating the effects of pressure in the deep ocean.

ALVIN SUBMERSIBLE
Alvin is designed to withstand the extreme pressure of the deep zone and has enabled scientists to make many important discoveries during over 4,000 dives.

THE HADAL ZONE

Few deep-water species have been observed in their natural environment of the hadal zone and even fewer photographed. Many species are known only from samples dredged up in nets, and most photographs are of dead specimens (including the Fangtooth on the right). Sometimes deep-sea animals can be studied in aquaria, but many species cannot survive the temperature and pressure changes when brought to the surface.

Although many of the animals here hunt each other, the food chain must begin with a supply of food from above. Whereas animals on the sea bed can patrol large areas to find food particles accumulated there over weeks and months, animals in midwater must grab food particles in the short time when they float downwards past them, which is much trickier. Only a small proportion of the detritus from above is harvested in midwater, so food is always scarce.

Scientists observing this zone often see the same species repeatedly. The environment of this zone is remarkably uniform worldwide and there are few physical or ecological barriers to block the movement of species. Many deep-water species therefore are widely distributed, and several are found in every ocean. As a result, species diversity is low: only around 1,000 of the known 29,000 fish species live at this depth.

FANGTOOTH FISH
The Fangtooth has been recorded at depths of 4,992m (16,380ft). Like many deep-water fish, it has a large head and massive teeth. Sensory organs along its body detect prey movement in the dark.

OCEAN EXPLORER
The *Ocean Explorer* is an ROV that can dive to 2,865m (9,400ft). Controlled by the mother ship through a cable, it can take photographs of the sea floor.

LOUIS S. ST-LAURENT

EXPLORATION WITH SUBMERSIBLES

By the mid-20th century, most of the Earth's land surface had been explored, but it wasn't until the 1960s, when a variety of submersibles were developed, that the amazing sights in the deep oceans began to be revealed. Submersibles are small underwater vehicles that can be manned or unmanned. They are designed to perform a variety of often highly specialized tasks.

The most famous manned submersible is *Alvin*, operated by Woods Hole Oceanographic Institute (see p.182–83). Built in 1962, *Alvin* made its first dive in 1964. By the end of 2005, it had completed 4,162 dives and spent the equivalent of 1,200 days underwater. A typical dive lasts about seven hours at an average depth of 2,000m (6,560ft). In 1966, it helped locate an H-bomb lost in the Mediterranean Sea off Spain. In 1977, the crew discovered the first hydrothermal vents. *Alvin* is still limited to a maximum depth of 4,500m (14,800ft). A new generation of submersibles is being launched (see panel, right), including a replacement for *Alvin*, scheduled for launch in 2009. Newer designs may no longer depend on ballast and buoyancy tanks to control their descent into the depths, but may instead use technologies have been developed for flight.

INTO THE DEEP

The only way for humans to explore deep water is in a pressurized diving bell or submersible, or to use an unmanned Remotely Operated Vehicle (ROV) to bring back photographs. *Kaiko*, a Japanese ROV, has reached depths of 11,000m (36,100ft) (see p.183). In 2006, trials of a Hybrid (H)ROV that is designed to operate on the very deepest parts of the sea floor will commence. It will be either free-swimming or tethered and will be able to drill cores, take sonar surveys, and record images.

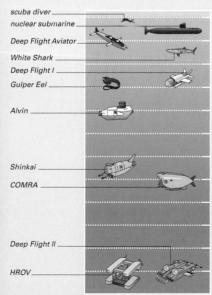

scuba diver	Sea Level
nuclear submarine	
Deep Flight Aviator	1,000m (3,300ft)
White Shark	
Deep Flight I	2,000m (6,500ft)
Gulper Eel	3,000m (9,800ft)
Alvin	4,000m (13,100ft)
	5,000m (16,400ft)
Shinkai	6,000m (19,700ft)
COMRA	7,000m (23,000ft)
	8,000m (26,300ft)
	9,000m (29,500ft)
Deep Flight II	10,000m (32,800ft)
HROV	11,000m (36,100ft)

TYPES OF SUBMERSIBLE

VIEWING MID-OCEAN

JOHNSON SEA-LINK *The acrylic sphere of the Johnson Sea-Link observation dome gives scientists a wide view of the animals that live in midwater. This highly manouevrable submersible has a depth limit of about 1,000m (3,300ft). It is 7m (23ft) long and can take a crew of four.*

EXPLORING THE SEA FLOOR

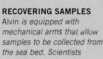

BLACK SMOKER *In April 1979, a team onboard Alvin were diving on the East Pacific Rise when they saw an astonishing sight: a tall spire of rock spewing out black fluid like billowing smoke. They had discovered the first black smoker.*

RECOVERING SAMPLES *Alvin is equipped with mechanical arms that allow samples to be collected from the sea bed. Scientists aboard the mother ship are ready to analyze material collected from the depths.*

DEEP DIVER

SHINKAI 6500 *Launched in 1989 by the Japan Marine Science and Technology Centre, Shinkai 6500 has dived to 6,527m (21,414ft), a record for an untethered manned submersible. In 2006 the China Ocean Minerals and R and D Association (COMRA) will test a manned craft that will reach a depth of 7,000m (23,000ft).*

FLYING UNDERWATER

DEEP FLIGHT I *The American engineer Graham Hawkes thinks future submersibles will "fly" underwater, like his Deep Flight Aviator and his latest prototype Deep Flight I.*

THE NEXT GENERATION *Hawkes believes that future submersibles, such as the proposed Deep Flight II, will use short, inverted wings to provide "negative lift" and pull the craft down quickly.*

OCEAN ENVIRONMENTS

SEAMOUNTS AND GUYOTS

SEAMOUNTS ARE TOTALLY submerged, undersea mountains that rise at least 1,000m (3,300ft) from the sea floor; smaller ones are called sea knolls. Guyots are seamounts that once rose above sea level – as a result, they have a flat top caused by erosion. Often isolated in deep ocean, seamounts and guyots provide a habitat for marine life adapted to shallower water. The obstruction of a seamount forces nutrient-rich, deep-sea currents to rise closer to the surface, forming eddies above the seamount. These trap nutrients and support plankton, which in turn attract shoals of fish.

PEOPLE

HENRY GUYOT

Arnold Henry Guyot (1807-1884) was the first professor of geology at Princeton University. He set up a system of weather observatories that led to the formation of the US Weather Bureau. Guyots were named in his honour by a later Princeton geology professor, Harry Hass. Hass discovered guyots using echo-sounding equipment during the Second World War.

GEOLOGICAL ORIGINS

Seamounts start as undersea volcanoes, where a rift in the sea bed allows volcanic eruptions. Many arise at rifts on the crest of mid-ocean ridges, formed by the movement of tectonic plates (see p. 185). Because these rifts are generally linear, seamounts tend to be elliptical or elongated in shape. They are made of volcanic basalt rock, but a thin layer of marine sediment accumulates over time. Seamounts often occur in chains or elongated groups, either because there are several weak spots along a rift, or because a series of seamounts originated sequentially at a single, stationary volcanic hotspot. Sometimes volcanic eruptions break above the ocean surface to form island chains, and these may continue out to sea as a line of guyots, or tablemounts. Newly formed volcanic rock is easily eroded so, over time, the above-water peak of the volcanic island is eroded down to a flat top. Then, as the ocean plates carry it away from the zone of volcanic activity, the flat-topped guyot sinks beneath the surface.

EVOLUTION OF A GUYOT

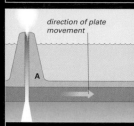

direction of plate movement

1 A guyot (A) begins life when a volcano erupts above a "hotspot", creating a small volcanic island.

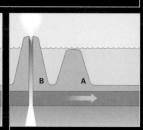

2 Over millennia, erosion reduces the island to a flat top at sea level, while it (A) moves away from the hotspot. A new island (B) forms

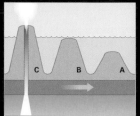

3 As the island moves further, it sinks and forms a guyot. New islands (B and C) erupt from the hotspot.

SEAMOUNT FORMATION
A seamount forms from an underwater volcanic eruption. Erosion here is slower than on land, so it remains conical.

WORLD DISTRIBUTION

There may be 100,000 seamounts and guyots in the oceans but few have been mapped or explored and the total number is unknown. Seamounts may occur either singly or in clusters or chains, reflecting zones of past volcanic activity. The Pacific, with its Ring of Fire, is the most volcanically active ocean, containing over 30,000 seamounts and guyots. Pacific chains typically form in a northwesterly direction, matching the direction of plate movement, with 10 to 100 seamounts in each chain, sometimes connected by an undersea ridge. In the Atlantic and Indian oceans, by contrast, seamounts mostly occur singly.

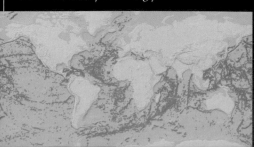

DISTRIBUTION MAP OF SEAMOUNTS
Some seamounts and guyots arise over volcanic hotspots, often in chains. Others form singly along mid-ocean ridges. Total numbers are unknown.

UPWELLINGS

The open ocean is mainly barren, because cold, nutrient-rich currents are confined to deep water, far beneath the reach of plankton. Seamounts – which stand up to 4,000m (13,000ft) proud of the sea bed – form a major obstruction to these currents, diverting them and pushing them upwards. This brings an upwelling of nutrients into the sunlit zone, and allows phytoplankton to flourish. As these nutrient-rich currents rush over the top of the seamount, they split in two and sweep around it. This makes the water above the seamount rotate, encircling a cylindrical column of still water that extends high above the height of the seamount. This "virtual" cylinder is called a Taylor Column. Above a seamount, it forms an area of back-eddies and still water in which nutrients accumulate and plankton get trapped. This creates a zone of incredible richness and productivity above the seamount – an "oasis" in the nutrient desert of the open ocean.

trapped nutrients and plankton

spiral flow

still water

upwelling

seamount

flow splits

deep-water current

WATER COLUMNS
The currents spiralling around and over a seamount create a column of still water above it. Plankton thrive on the nutrients trapped there.

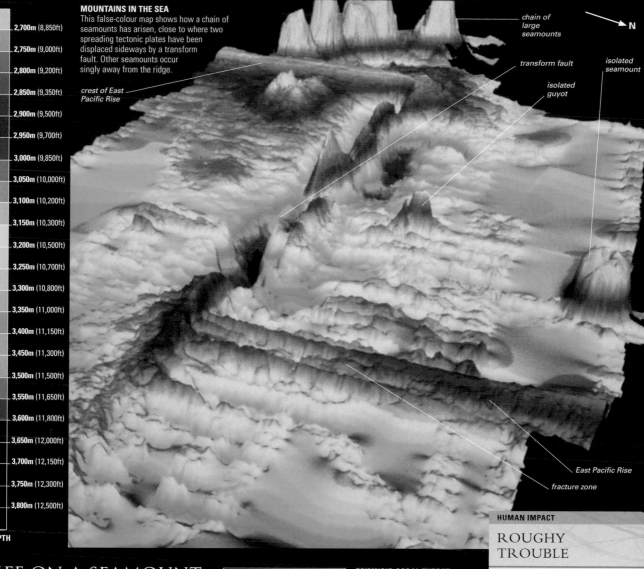

DEPTH

2,700m (8,850ft)	
2,750m (9,000ft)	
2,800m (9,200ft)	
2,850m (9,350ft)	
2,900m (9,500ft)	
2,950m (9,700ft)	
3,000m (9,850ft)	
3,050m (10,000ft)	
3,100m (10,200ft)	
3,150m (10,300ft)	
3,200m (10,500ft)	
3,250m (10,700ft)	
3,300m (10,800ft)	
3,350m (11,000ft)	
3,400m (11,150ft)	
3,450m (11,300ft)	
3,500m (11,500ft)	
3,550m (11,650ft)	
3,600m (11,800ft)	
3,650m (12,000ft)	
3,700m (12,150ft)	
3,750m (12,300ft)	
3,800m (12,500ft)	

MOUNTAINS IN THE SEA
This false-colour map shows how a chain of seamounts has arisen, close to where two spreading tectonic plates have been displaced sideways by a transform fault. Other seamounts occur singly away from the ridge.

crest of East Pacific Rise

chain of large seamounts

N

transform fault

isolated seamount

isolated guyot

East Pacific Rise

fracture zone

LIFE ON A SEAMOUNT

Some seamounts were first detected when fishermen discovered large shoals of fish in the area. The nutrient-rich waters trapped above seamounts support dense concentrations of phytoplankton as well as the zooplankton that feed on them. Free-swimming animals are attracted by this feast, including fish at densities found nowhere else in the open ocean. Predators such as sharks and seals also gather to feed. Seamount rock is colonized by suspension feeders – animals that catch plankton and detritus as it floats past. Only about one in a thousand seamounts has been explored underwater. However, in studies of 25 seamounts in the Tasman and Coral seas, 850 species (some previously thought extinct) were recorded. Seamounts are important biodiversity hotspots, with up to one-third of species found there restricted to a single seamount or group of seamounts.

PRIMNOID CORAL THREAT
Scientists fear some Primnoid coral species may be wiped out by bottom-trawl fishing before they have even been named.

SEAMOUNT FEEDER
Found in the tropics, this octocoral is a colony of soft corals. The feeding polyps, lined up along the branches, catch food from currents sweeping over the seamount.

FEEDING FROM THE CURRENTS
This squat lobster, or Pinch bug, is a scavenger living on rock faces. Currents welling over the Bowie Seamount in the northeast Pacific supply rich pickings.

ROUGHY TROUBLE

Fishermen thought they had found a bonanza in the 1980s when they discovered the huge shoals of fish that live over seamounts. For example, they could catch 100 tonnes of Orange Roughy (see below and p.354) in a single day. However, roughy, which can live for almost 150 years, is slow-growing, and does not produce eggs until it is 20–30 years old. Such heavy fishing cannot be sustained. World catches have declined hugely, and the roughy is now in danger.

OCEAN ENVIRONMENTS

THE CONTINENTAL SLOPE AND RISE

THE CONTINENTAL SLOPE AND RISE are areas of sloping sea floor that lead from the continental shelf to the abyssal plain. Beyond a point on the shelf called the shelf break, the sea bed begins to drop more steeply. This is the continental slope, which leads into the open ocean. It sweeps down to 3,000–4,500m (9,800–14,800ft), where the sea bed flattens out. In places, the slope is broken by submarine canyons. Sediments wash down these canyons, and accumulate at the base of the slope in a gentler gradient, forming the continental rise.

CONTINENTAL MARGIN
A typical continental margin is shown here, including the transition from a shoreline to the abyssal plain via the continental shelf, slope, and rise. The continental slope is about 140km (87 miles) wide, and the continental rise is about 100km (60 miles) wide. The vertical scale has been exaggerated: The continental slope actually has a gentle gradient, of about 1 in 50 (2 per cent); and the rise is even gentler, at about 1 in 100 (1 per cent).

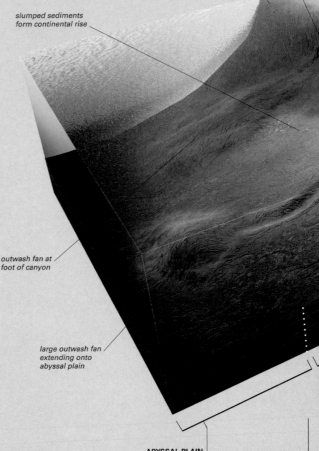

submarine canyon

shelf break – around 200m (660ft) below surface

slumped sediments form continental rise

outwash fan at foot of canyon

large outwash fan extending onto abyssal plain

CONTINENTAL SLOPE

The rock of the continental slope is blanketed by sediments washed from the land that have accumulated over millions of years. Crustaceans, echinoderms, and many other animals live in, or on, these sediments. The slope is dissected by deep canyons. These have been cut by an abrasive mix of sediment and water, called turbidity currents, which flow down the gorges at a rate of 80–100kph (50–60mph). Some submarine canyons are massive: the Grand Bahama Canyon in the Caribbean has cliffs rising 4,285m (14,060ft) from the canyon floor. Many canyons are seaward extensions of great rivers. At the canyon end, the sediment is deposited as a spreading outwash fan, extending far out onto the abyssal plain.

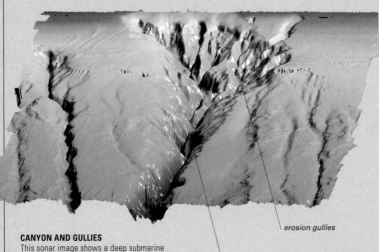

CANYON AND GULLIES
This sonar image shows a deep submarine canyon in the continental slope off Sodwana Bay, in KwaZulu Natal, South Africa.

erosion gullies

submarine canyon

ABYSSAL PLAIN
This flat plain is formed by a deep accumulation of sediments. It typically lies at a depth of 4,500m (15,000ft).

LIFE ON THE CONTINENTAL SLOPE

Like the shelf, the continental slope is enriched by nutrients washed off the land. This helps support both midwater (pelagic) and bottom-dwelling (demersal) fish. Fish stocks over most continental-shelf regions have declined dramatically in recent decades, as a result of over-exploitation and poor management, driving more fishermen to seek deeper-water species over the continental slope. Unfortunately for fisheries, although deep-water species are long-lived, they breed slowly, and stocks take a long time to recover. So, many fisheries are now in serious decline.

CATCHING SABLEFISH
Sablefish are caught with longlines, 1.2km (⅔ mile) long, which reach down towards the continental slope.

SABLEFISH
Sablefish breed slowly, and it takes 14 years to replace each fish caught. Fish farms (right) may be a better option.

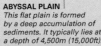

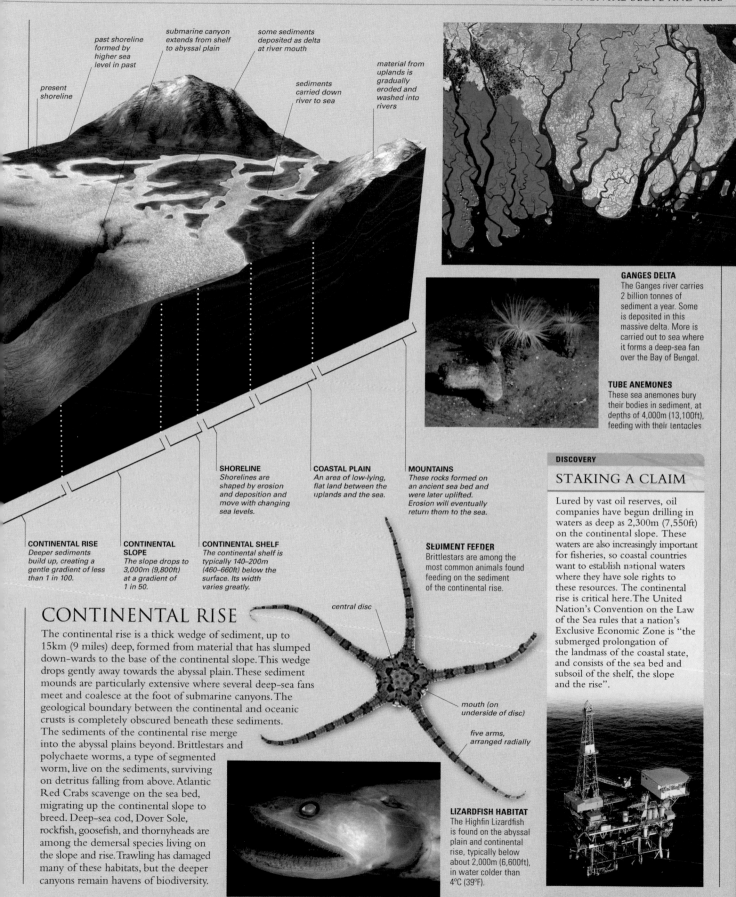

present
shoreline

past shoreline
formed by
higher sea
level in past

submarine canyon
extends from shelf
to abyssal plain

some sediments
deposited as delta
at river mouth

sediments
carried down
river to sea

material from
uplands is
gradually
eroded and
washed into
rivers

SHORELINE
*Shorelines are
shaped by erosion
and deposition and
move with changing
sea levels.*

COASTAL PLAIN
*An area of low-lying,
flat land between the
uplands and the sea.*

MOUNTAINS
*These rocks formed on
an ancient sea bed and
were later uplifted.
Erosion will eventually
return them to the sea.*

CONTINENTAL RISE
*Deeper sediments
build up, creating a
gentle gradient of less
than 1 in 100.*

**CONTINENTAL
SLOPE**
*The slope drops to
3,000m (9,800ft)
at a gradient of
1 in 50.*

CONTINENTAL SHELF
*The continental shelf is
typically 140–200m
(460–660ft) below the
surface. Its width
varies greatly.*

GANGES DELTA
The Ganges river carries
2 billion tonnes of
sediment a year. Some
is deposited in this
massive delta. More is
carried out to sea where
it forms a deep-sea fan
over the Bay of Bengal.

TUBE ANEMONES
These sea anemones bury
their bodies in sediment, at
depths of 4,000m (13,100ft),
feeding with their tentacles

STAKING A CLAIM

Lured by vast oil reserves, oil
companies have begun drilling in
waters as deep as 2,300m (7,550ft)
on the continental slope. These
waters are also increasingly important
for fisheries, so coastal countries
want to establish national waters
where they have sole rights to
these resources. The continental
rise is critical here. The United
Nation's Convention on the Law
of the Sea rules that a nation's
Exclusive Economic Zone is "the
submerged prolongation of
the landmass of the coastal state,
and consists of the sea bed and
subsoil of the shelf, the slope
and the rise".

SEDIMENT FEEDER
Brittlestars are among the
most common animals found
feeding on the sediment
of the continental rise.

CONTINENTAL RISE

The continental rise is a thick wedge of sediment, up to
15km (9 miles) deep, formed from material that has slumped
down-wards to the base of the continental slope. This wedge
drops gently away towards the abyssal plain. These sediment
mounds are particularly extensive where several deep-sea fans
meet and coalesce at the foot of submarine canyons. The
geological boundary between the continental and oceanic
crusts is completely obscured beneath these sediments.
The sediments of the continental rise merge
into the abyssal plains beyond. Brittlestars and
polychaete worms, a type of segmented
worm, live on the sediments, surviving
on detritus falling from above. Atlantic
Red Crabs scavenge on the sea bed,
migrating up the continental slope to
breed. Deep-sea cod, Dover Sole,
rockfish, goosefish, and thornyheads are
among the demersal species living on
the slope and rise. Trawling has damaged
many of these habitats, but the deeper
canyons remain havens of biodiversity.

central disc

mouth (on
underside of disc)

five arms,
arranged radially

LIZARDFISH HABITAT
The Highfin Lizardfish
is found on the abyssal
plain and continental
rise, typically below
about 2,000m (6,600ft),
in water colder than
4°C (39°F).

OCEAN ENVIRONMENTS

COLD-WATER COMMUNITY
A squat lobster shelters among the polyps of the cold-water stony coral *Lophelia pertusa*, or Tuft Coral, in a Norwegian fiord.

COLD-WATER REEFS

Deep-sea corals were first discovered in 1869, but it took the advent of sonar and deep-sea submersibles to reveal the size and abundance of the reefs that they build. Although less well studied than their tropical counterparts, these cold-water reefs are just as rich in life. The stony corals that form deep-water reefs flourish in water temperatures of 4–13°C (39–55°F). Unlike tropical corals, they can live in total darkness because they do not rely on zooxanthellae (p.153) living inside them to produce nourishment by photosynthesis in sunlight. Instead, they survive by filtering food from the water. Norwegian scientists have suggested recently that hydrocarbons (compounds of hydrogen and carbon, such as methane) seeping from the sea floor could provide an additional food source at some sites. Methane may provide energy for bacteria at the bottom of a food chain, which are then filtered from the water by the coral polyps.

One of the biggest reefs – covering 100 square km (38 square miles) – was discovered during an oil-related survey of the Atlantic Frontier, northwest of Scotland in 1998. *Lophelia pertusa* is the main reef-forming coral at these reefs, called the Darwin Mounds, which lie at a depth of 1,000m (3,300ft). Lophelia reefs occur at similar depths on many seamounts in the Atlantic, and also in shallow cold water such as in Norway's fiords. Several other coral species form cold-water reefs elsewhere in the world. For example, in the Pacific, the main reef species on seamounts and oceanic banks around Tasmania and New Zealand are *Goniocorella dumosa* and *Solenosmilia variabilis*. Over 1,300 species of animals have been recorded on deep-sea reefs, and they may be important nursery grounds for commercial fish species.

LOCATION OF DEEP-SEA REEFS

The map below shows the global distribution of cold-water reefs. Some of these reefs are small, while others cover up to 2,000 square km (770 square miles), although the map dots exaggerate their extent. The many reefs detected in the north Atlantic probably reflects the intensity of surveying there, particularly in the search for oil. More detailed surveys of other oceans are likely to reveal the existence of further deep-sea reefs.

PACIFIC OCEAN
ATLANTIC OCEAN
PACIFIC OCEAN
INDIAN OCEAN
SOUTHERN OCEAN
SOUTHERN OCEAN

LIFE IN COLD WATER

DEEP-SEA CORALS

LOPHELIA REEF *This Lophelia reef lies deep in the Atlantic off the west coast of Ireland, where it can be studied only by means of a submersible. Fortunately, Lophelia can also be viewed in waters as shallow as 39m (128ft) in some Norwegian fiords.*

GONIOCORELLA CORAL *This deep-sea coral thicket is made mostly of Goniocorella dumosa, a species that is restricted to the southern hemisphere. It forms reefs at depths to 1,500m (5,000ft).*

SQUAT LOBSTER *This tiny squat lobster is sitting on Madrepora oculata coral polyps, 390m (1,290ft) down in the Bay of Biscay, north of Spain.*

CHIROSTYLUS CRABS *Many animals live among the coral. These long-limbed crabs are crawling over a black coral in the northeast Atlantic.*

ASSOCIATED MARINE LIFE

THREATS FROM DEEP-SEA TRAWLING

DAMAGED REEF *Fishing gear has snagged on this reef west of Ireland, tearing off chunks of living reef that could be up to 8,500 years old. In 2005, the European Union banned fishing near the Darwin Mounds.*

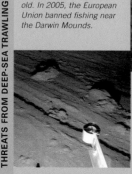

TRAWL MARK *Even before scientific surveys discovered them, many deep-water reefs had been severely damaged by trawls dragged across the sea bed to catch bottom-living fish. The scarred sea bed shown here is at a depth of 885m (2,900ft).*

OCEAN ENVIRONMENTS

OCEAN-FLOOR SEDIMENTS

OVER VAST AREAS OF THE SEA BED, THE UNDERLYING landforms are hidden beneath deep layers of sediments. Made up of silts, muds, or sands that have built up over 200 million years, they now form a blanket that is several kilometres thick in places. The sediments have various origins. One group, terrigenous sediments, come from land, mainly from fragments of eroded rock that are carried down rivers into the sea, then down the continental slope to form the continental rise and abyssal plain beyond. Other sediments are biogenic, formed from the hard remains of dead animals and plants. A few, called authigenic sediments, are made up of chemicals precipitated from seawater. There are even cosmogenic sediments, which come from outer space as particles in space dust and meteors. All accumulate to form extensive, flat plains. Various animals feed here and burrow into the sediments for shelter.

SEDIMENT THICKNESS

| 0 | 500m (1,650ft) | 1,000m (3,300ft) | 5km (3 miles) | 10km (6¼ miles) | 20km (12½ miles) |

MAPPING SEDIMENTS
Ocean sediment depths can be measured and mapped using echo-sounding. Some areas (white on the map above) are still unsurveyed. Sediment is thickest near land. Glaciers also carry many sediments into the oceans.

DEEP-SEA SEDIMENTS

The average thickness of sediments on the ocean floor is 450m (1,500ft), but in the Atlantic Ocean and around Antarctica, sediments can be up to 1,000m (3,300ft) deep. Closer to the continents – along the continental rise – sediments washed from the land accumulate more rapidly, and can be up to 15km (9 miles) deep. In the open ocean, further from the source of terrigenous sediments, the build-up rate is very slow: from a millimetre to a few centimetres in a thousand years. That is slower than the rate at which dust builds up on furniture in an average house. The accumulated sediments tell scientists a great deal about the last 200 million years of the Earth's history. Their form and arrangement provide a vivid snapshot of sea-floor spreading, the evolving varieties of ocean life, alterations in the Earth's magnetic field, and changes in ocean currents and climate.

WHITE CLIFFS OF DOVER
These chalk cliffs originated on the sea bed from a biogenic ooze, formed from algal scales (coccoliths) that built up to form layers hundreds of metres thick. They are now raised above sea level.

PTEROPOD OOZE

Pteropods are small winged snails that float in midwater. When they die, their internal shells of aragonite (calcium carbonate) sink to the sea bed, contributing to biogenic oozes. The presence of pteropod remains in samples collected from deep in the ooze reveal changes over millennia in water temperatures and sea levels.

SEDIMENTS DERIVED FROM THE LAND

Most terrigenous sediments come from the weathering of rock on land and are swept into the oceans, mainly by rivers but also by glaciers, ice sheets, and wind. Coastal erosion adds to these sediments. Often, they are washed down through submarine canyons to the deeper ocean. Sometimes, the route from land to sea is more indirect: volcanic eruptions eject material into the upper atmosphere before it falls as "rain" into the ocean.

In the deepest ocean floors, below about 4,000m (13,000ft), the main sediment is red clay, composed mostly of fine-grained silts that have washed off the continents and accumulated incredibly slowly – about 1mm or ¹⁄₃₂in per thousand years. These clays may include up to 30 per cent of fine, biogenic particles and have four main mineral components – chlorite, illite, kaolinite, and montmorillonite. Clay types depend on origin and climate. For example, chlorite dominates in polar regions, kaolinite in the tropics, and montmorillonite is produced by volcanic activity.

DUST STORM RESULTS IN SILT
Winds from arid regions, such as North Africa (shown in this satellite image) carry dust far out to sea, where it sinks to form silts.

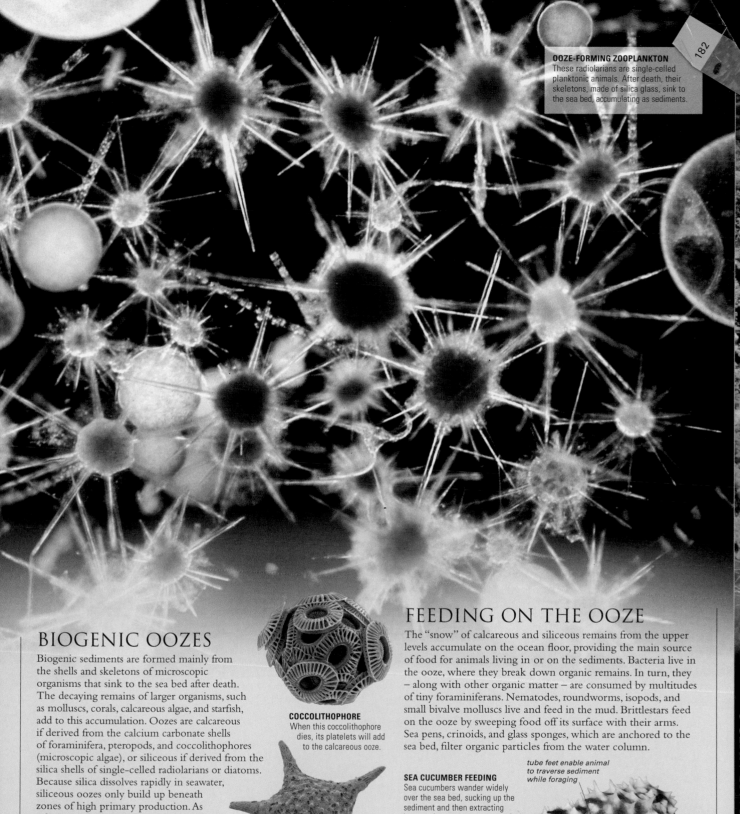

OOZE-FORMING ZOOPLANKTON
These radiolarians are single-celled planktonic animals. After death, their skeletons, made of silica glass, sink to the sea bed, accumulating as sediments.

BIOGENIC OOZES

Biogenic sediments are formed mainly from the shells and skeletons of microscopic organisms that sink to the sea bed after death. The decaying remains of larger organisms, such as molluscs, corals, calcareous algae, and starfish, add to this accumulation. Oozes are calcareous if derived from the calcium carbonate shells of foraminifera, pteropods, and coccolithophores (microscopic algae), or siliceous if derived from the silica shells of single-celled radiolarians or diatoms. Because silica dissolves rapidly in seawater, siliceous oozes only build up beneath zones of high primary production. As calcareous shells and skeletons sink, they reach a depth (around 4,500m/15,000ft) where the water becomes more acidic; this, combined with pressure, means calcareous remains are dissolved rapidly in seawater at depth. Calcareous oozes therefore occur only above this "calcium carbonate compensation depth", beneath which the sea bed consists mainly of terrigenous red clays.

COCCOLITHOPHORE
When this coccolithophore dies, its platelets will add to the calcareous ooze.

FORAMINIFERA
The tiny shells of dead foraminiferans add to the biogenic oozes.

FEEDING ON THE OOZE

The "snow" of calcareous and siliceous remains from the upper levels accumulate on the ocean floor, providing the main source of food for animals living in or on the sediments. Bacteria live in the ooze, where they break down organic remains. In turn, they – along with other organic matter – are consumed by multitudes of tiny foraminiferans. Nematodes, roundworms, isopods, and small bivalve molluscs live and feed in the mud. Brittlestars feed on the ooze by sweeping food off its surface with their arms. Sea pens, crinoids, and glass sponges, which are anchored to the sea bed, filter organic particles from the water column.

SEA CUCUMBER FEEDING
Sea cucumbers wander widely over the sea bed, sucking up the sediment and then extracting its organic content.

tube feet enable animal to traverse sediment while foraging

OCEAN ENVIRONMENTS

ABYSSAL PLAINS, TRENCHES, AND MID-OCEAN RIDGES

OVER VAST AREAS, THE SEA BED IS COVERED BY A flat expanse of accumulated sediments. Life here is sparse, relying entirely on food falling from above. In places, the abyssal plains are disrupted by more dramatic features, created by the movement of tectonic plates. Where plates diverge, magma wells up through the gap to create mid-ocean ridges. New sea bed is constantly being formed at these ridges. At the other extreme, where plates collide, one plate is dragged downwards, opening up a trench. These trenches are the deepest places on Earth.

ABYSSAL PLAINS

Over large areas of the ocean floor, sediments have built up a blanket several kilometres thick, obscuring the underlying topography. This produces vast flat or gently undulating abyssal plains at a typical depth of 4,500m (14,800ft). These are most common in the Atlantic, where the Sohm Plain alone covers 900,000 square km (350,000 square miles). Abyssal plains lie at different depths, with barriers between them, and this leads to submarine waterfalls, where water spills over the barrier and down into the plain below, at rates of up to 8kph (5mph). Occasional abyssal storms also occur, stimulated, in a way not yet fully understood, by instabilities at the ocean surface resulting from atmospheric conditions. Originally thought to be a world without seasons, recent studies have shown that life here responds to pulses of food from above, for instance when the summer bloom of plankton dies and sinks. Most animals in this zone are scavengers with a body temperature close to that of the surrounding water. They move and grow slowly, reproduce infrequently, and live longer than their relatives at the surface.

ABYSSAL FLOOR
A recent study off the east coast of North America revealed 798 species buried in a small sediment sample from the sea bed.

MANGANESE NODULE
In places, the abyssal plain is littered with potato-sized nodules of manganese, often contaminated with other valuable metals such as nickel, cooper, and cobalt.

fin near tail

mouth barbel

body

SEA-BED SCAVENGERS
Hagfish feed on animal corpses that fall to the abyssal plain. Blind and jawless, these primitive fish are attracted by smell. They bore into corpses, using their horny teeth, and secrete clouds of mucus to deter other scavengers.

THE DEATH OF A WHALE

Occasionally, a dead whale sinks to the abyssal plain and provides a feast. Scientists have counted 12,000 animals of 43 species feeding on the bones of a single whale. It may take them 11 years to strip the flesh from a Blue Whale. Later, bacteria invade and decompose the remaining bones. This process leaches out sulphides that sustain a complex community of crustaceans, clams, limpets,

OCEAN TRENCHES

Ocean trenches are created by a process called subduction. Where oceanic and continental tectonic plates collide, the denser but thinner oceanic plate is forced down beneath the thicker but less dense continental plate, and plunges to its destruction in the mantle deep below. Where two oceanic plates collide, the older plate is subducted beneath the younger. The buckling where the plates collide causes a deep depression at the point of impact – an ocean trench. These are the deepest places on the ocean floor.

Trenches are typically V-shaped, with steeper slopes on the continental side. The Pacific is the region of most active subduction, with 17 of the 20 major ocean trench systems. The Atlantic has two major trenches, the Puerto Rico and South Sandwich trenches, and the Java Trench is the only major trench in the Indian Ocean. The deepest trench on Earth is the Mariana Trench, located in the Pacific Ocean, near the Mariana Islands.

MARIANA TRENCH
The Mariana Trench is roughly 2,500m (1,600 miles) long and 70km (40 miles) wide. This section of the trench is just off the coast of Japan, which appears as a dark green mass here.

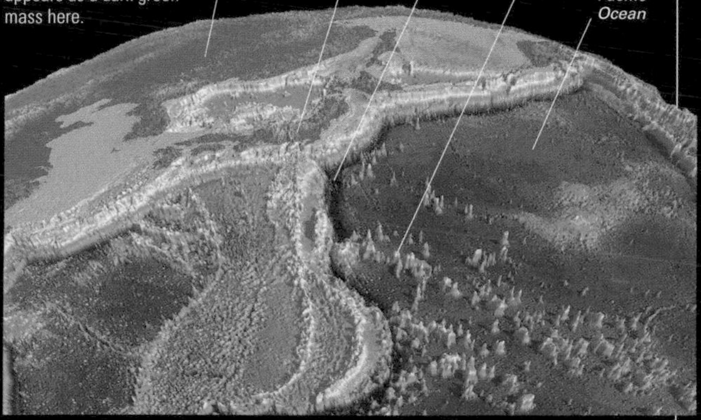

Japan
Mariana Trench
China
seamounts
Pacific Ocean

EXPLORING THE ABYSSAL PLAIN
Alvin has downward-facing portholes to study the abyssal plain. Its scientists deploy baited traps to attract deep-water life, such as Six-gilled Sharks.

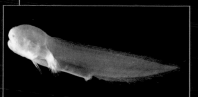

GELATINOUS BLINDFISH
A small number of these curious fish have been collected from the sea bed in the Atlantic, Pacific, and Indian oceans, at depths of at least 3,000m (10,000ft). Like many deep-water fish, they are almost transparent, with tiny eyes.

LIFE IN THE OCEAN TRENCHES

Animals have been found at great depths in the ocean trenches. The depth record belongs to a cuskeel, appropriately named *Abyssobrotula galatheae*, which was dredged from 8,400m (27,500ft) in the Puerto Rico Trench by the Danish research ship *Galathea*. This fish has an elongated body up to 20cm (8in) long, a swollen snout, and small eyes. It can cope with pressure more than 800 times that at sea level.

A segmented worm was found by *Galathea* in another deep trench, and sponges have been collected from 8,000m (26,000ft) in the Kuril–Kamchatka Trench. The unmanned Japanese submersible *Kaiko* photographed a shrimp-like amphipod called *Hirondella gigas* at 10,911m (35,797ft). In 1998, samples of this scavenging crustacean were collected in baited traps, and found to have a bacterium in their bodies that can survive only at above 525 times atmospheric pressure.

Kaiko also collected sediment samples from the Challenger Deep, which contained 432 different species of foraminiferans, and a range of bacteria that can grow only at pressures above 600 atmospheres.

DISCOVERY

THE *TRIESTE* EXPEDITION

In 1960, two oceanographers, Don Walsh and Jacques Picard, dived to 10,911m (35,797ft) in the Challenger Deep section of the Mariana Trench in the bathyscaphe *Trieste* – still the greatest depth reached by humans. It took five hours to descend to that depth, and after just 20 minutes hanging there, the crew began their return to the surface.

OCEAN ENVIRONMENTS

THE RING OF FIRE

All around the margins of the Pacific Ocean, tectonic plates are colliding. This produces a belt of intense volcanic and earthquake activity encircling the Pacific, known as the Ring of Fire. It extends for 30,000km (18,600 miles) in a series of arcs, from New Zealand, through Japan, and down the west coast of the Americas to Patagonia. About three-quarters of the Pacific lies over a single oceanic plate, the Pacific Plate, which is colliding around its edges with the Eurasian, North American, Indian, and Australian plates, as well as their associated minor plates. As the oceanic Pacific Plate plunges beneath the lighter continental plates, massive slabs of rock shatter explosively along faults, producing frequent earthquakes, while chains of volcanoes erupt along the lines of weakness.

Many of these volcanoes rise above sea level as islands, often arranged in characteristic arc-shaped chains that mirror the curved edge of the tectonic plate far below. Some of these islands are being eroded rapidly by the waves, and may one day become guyots (see p.174). Several deep ocean trenches are also aligned around this ring, where the sea bed is being dragged down by subduction, forcing deep gashes to open up in the sea floor. As they collide, the continental plates are compressed and thrust upwards, producing mountain chains such as the Andes.

SAFE HAVEN

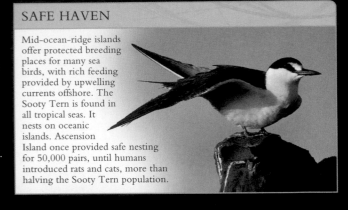

Mid-ocean-ridge islands offer protected breeding places for many sea birds, with rich feeding provided by upwelling currents offshore. The Sooty Tern is found in all tropical seas. It nests on oceanic islands. Ascension Island once provided safe nesting for 50,000 pairs, until humans introduced rats and cats, more than halving the Sooty Tern population.

THE RIDGE ON LAND
For most of its vast length, the Mid-Atlantic Ridge is hidden deep beneath the ocean. However, at Iceland, where both the Eurasian and North American plates are separating, it rises above the surface.

HOTSPOTS OF VOLCANIC ACTIVITY
Red on this map shows areas of volcanic activity around the Pacific Ocean, highlighting the Ring of Fire. These volcanoes form on continental plates as oceanic plates are thrust below.

MOUNT ST. HELENS
Mount St. Helens in Washington State, USA is part of the Ring of Fire. It erupted in May 1980, blowing the whole top off the volcano. Here, in 2004, a new lava cone has begun to grow, producing steam.

OCEAN ENVIRONMENTS

MID-OCEAN RIDGES

New sea bed is produced wherever tectonic plates diverge. As plates move apart, they create a rift. Magma wells up through this rift from deep in the Earth's mantle, forming volcanoes and creating an underwater mountain chain, called a mid-ocean ridge. The lava cools as it meets the seawater, and solidifies in vertical basalt dykes or fields of pillow lava (see p.44).

Mid-ocean ridges are assembly lines along which new ocean floor is being produced. The ridges and lava fields remain visible for some time before sediments accumulate over them. Sometimes the volcanoes extend above sea level, producing islands such as Iceland. Some mid-ocean ridges spread slowly, allowing deep rift valleys to form down their centres – others are much faster-spreading but lack rift valleys. Sometimes the ridges are disrupted sideways by transform faults.

As the new sea bed spreads outwards, tensions are created, making it crack. Water seeps into these cracks and re-emerges from hydrothermal vents (see p.188). The oceanic ridge system is the third largest feature on the Earth's surface, after the oceans and continents.

PILLOW LAVA
Under the high pressure of the deep ocean, lava oozes slowly from the mid-ocean crests. When it meets cold seawater, it cools rapidly to form globular masses, called pillow lavas due to their shape. About 3.5 square km (1.4 square miles) of new sea floor is formed each year along mid-ocean ridges.

ASCENSION ISLAND
Ascension Island arises where the Mid-Atlantic Ridge protrudes above sea level in the south Atlantic. It covers 90 square km (35 square miles) and ascends to 859m (2,817ft) on Green Mountain. Sooty Terns and sea turtles breed around its shores.

OCEAN WANDERERS
Macquarie Island, on the Macquarie Ridge, provides a nesting site for the Black-browed Albatross. Outside of the breeding season, it wanders the Southern Ocean.

RIDGES OF THE WORLD

The longest mid-ocean ridge occurs where the Eurasian and African plates are diverging from the North and South American plates. The Mid-Atlantic Ridge runs along this boundary for 16,000km (10,000 miles), from the Arctic Ocean to beyond the southern tip of Africa, equidistant between the continents on either side of the Atlantic and rising 2,000–4,000m (6,000–13,000ft) above the sea floor.

A chain of volcanoes runs down its length, most famously in Iceland, where an eruption in 1963 created a new volcanic island, Surtsey. Ascension Island and the Azores lie on the ridge, while St. Helena and Tristan da Cunha arise from isolated volcanoes, displaced from it. A valley, 25km (15 miles) wide, extends along the ridge crest. In the Pacific the main ridge system is the East Pacific Rise. This is the Earth's fastest spreading system separating at 13–16cm (5–6in) per year. A series of mid-ocean ridges encircle Antarctica, along the divergent boundaries between the Antarctic Plate and its neighbours, and the Carlsberg Ridge runs down the centre of the Indian Ocean.

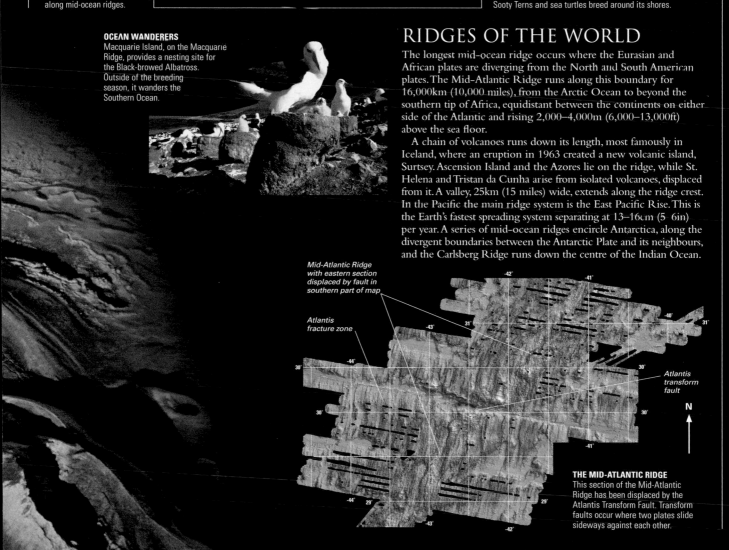

Mid-Atlantic Ridge with eastern section displaced by fault in southern part of map

Atlantis fracture zone

Atlantis transform fault

THE MID-ATLANTIC RIDGE
This section of the Mid-Atlantic Ridge has been displaced by the Atlantis Transform Fault. Transform faults occur where two plates slide sideways against each other.

OCEAN ENVIRONMENTS

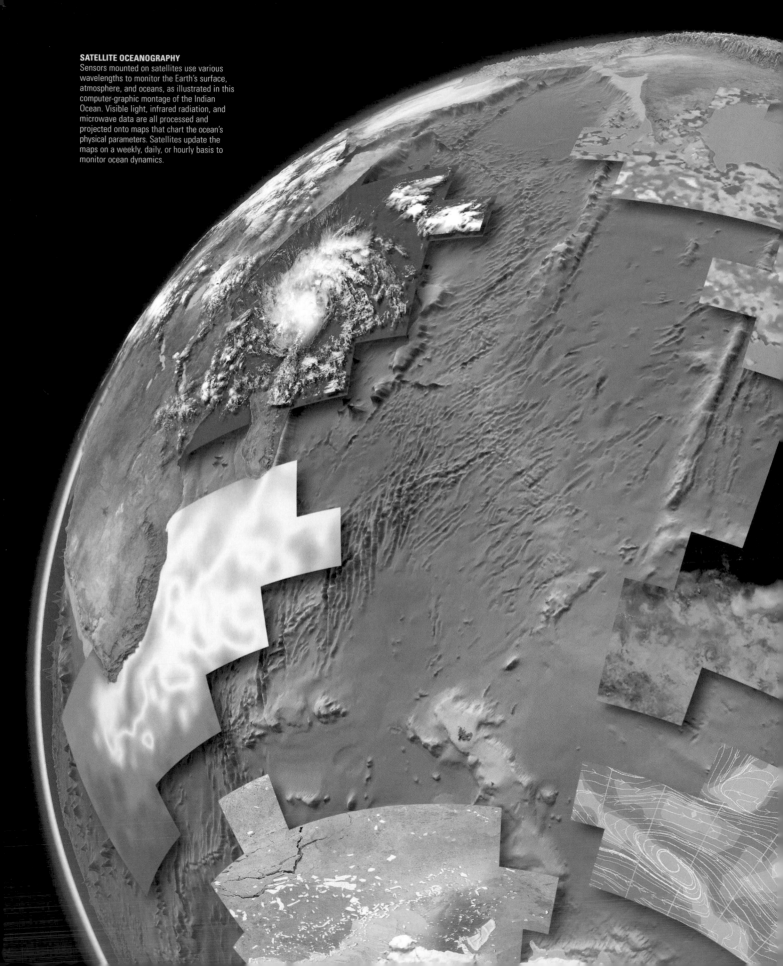

SATELLITE OCEANOGRAPHY
Sensors mounted on satellites use various wavelengths to monitor the Earth's surface, atmosphere, and oceans, as illustrated in this computer-graphic montage of the Indian Ocean. Visible light, infrared radiation, and microwave data are all processed and projected onto maps that chart the ocean's physical parameters. Satellites update the maps on a weekly, daily, or hourly basis to monitor ocean dynamics.

OCEANOGRAPHY FROM SPACE

The world's oceans are too vast to be adequately studied using ships alone. Even if all of the depth soundings that were taken during the 20th century were to be plotted, the resultant map would provide only sparse information on the sea floor and would even be blank in large areas. The advent of satellite remote sensing in the 1960s brought a revolution in oceanography. For the first time, it was possible to take a picture showing an entire ocean basin. Hurricane tracking and warning was one of the first benefits to accrue from early weather satellites. Eventually, a large range of sensors were developed to probe the physical attributes of the ocean surface and the atmosphere above.

Ocean colour, temperature, and surface roughness are among the parameters that can be systematically monitored in some detail. Satellite-derived information is a vital component of practical applications such as weather forecasting, commercial fishing, oil prospecting, and ship routing. In some cases, 30 years of continuous observations have been built up, helping scientists to track seasonal and long-term changes in the ocean environment and understand its effects on the global climate.

MEASURING OCEAN DEPTH FROM SPACE

Satellites cannot directly measure the depth of the sea floor, but it can be derived from the height of the sea surface. The sea is not flat. Water piles up above gravity anomalies caused by ocean-floor features such as seamounts, producing variations in the surface that are much larger than those produced by tides, winds, and currents. By comparing the height of the sea surface against a reference height, the depth of the sea floor can be estimated.

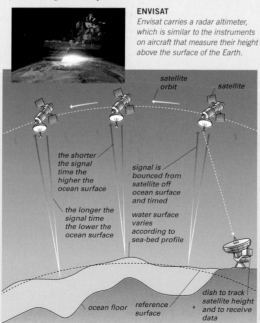

ENVISAT Envisat carries a radar altimeter, which is similar to the instruments on aircraft that measure their height above the surface of the Earth.

satellite orbit

satellite

the shorter the signal time the higher the ocean surface

signal is bounced from satellite off ocean surface and timed

the longer the signal time the lower the ocean surface

water surface varies according to sea-bed profile

dish to track satellite height and to receive data

ocean floor

reference surface

FEATURES STUDIED FROM SPACE

WEATHER

CLOUDS Cloud cover is detected using visible-light cameras, and cloud-top height data is derived from infrared radiometers on satellites such as Meteosat. These systems are used to track storms and forecast the weather.

RAINFALL The Tropical Rainfall Measuring Mission uses a microwave radiometer to see through clouds and detect the presence of liquid water in the atmosphere. Rainfall measures are used in computer models of the climate and ocean.

PLANT LIFE

CHLOROPHYLL Ocean colour cameras use wavelengths of visible light to measure the concentration of chlorophyll, which is present in phytoplankton. This information is used for water-quality assessment, finding fish, and in various aspects of marine biology.

WIND SPEED

MICROWAVE SCATTEROMETER Surface wind speed and direction are measured by satellites such as QuickScat, which bounces radio beams off the surface of the ocean. Wind-induced ocean waves modify the return signal, and the data can be used for meteorology and climate research.

ICE COVER

SYNTHETIC APERTURE RADAR Imaging radar systems, such as the one carried by Radarsat, penetrate clouds and can operate through the dark of the extended polar night to monitor ice shelves, sea-ice, and icebergs all year round.

TEMPERATURE

SURFACE TEMPERATURE Infrared radiometers can measure the temperature of the sea surface precisely. Shifts in ocean currents, cold-water upwelling, and ocean fronts can be monitored for ocean and climate research.

THERMAL GLIDER A new generation of instrument platforms is being developed to sample the vast subsurface volume of the world's oceans. Autonomous Underwater Vehicles, or "sea gliders", can undertake long cruises, surfacing every day to return their data via satellite communication links.

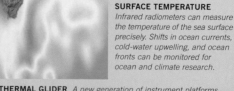

OCEAN ENVIRONMENTS

VENTS AND SEEPS

HYDROTHERMAL VENTS ARE SIMILAR to hot springs on land. Located near ocean ridges and rifts, at an average depth of 2,100m (7,000ft), they spew out mineral-rich, superheated seawater. Some have tall chimneys, formed from dissolved minerals that precipitate when the hot vent water meets cold, deep-ocean water. The mix of heat and chemicals supports animal communities around the vents – the first life known to exist entirely without the energy of sunlight. Elsewhere, slower, cooler emissions of chemicals called hydrocarbons occur from sites known as cold seeps.

DISCOVERING WHITE SMOKERS

The first hydrothermal vents that scientists observed from *Alvin*, a submersible, in 1977, were black smokers. Scientists then explored other sites near mid-ocean ridges and found more vent systems. Some looked different: their fluids were white, cooler, and emerged more slowly from shorter chimneys. These were called white smokers (see right).

DISTRIBUTION OF VENTS AND RIDGES
Since their discovery in 1977, hydrothermal vents have been found in the Pacific and Indian oceans, in the mid-Atlantic, and even in the Arctic, always near mid-ocean ridges and rifts.

HYDROTHERMAL VENTS

Hydrothermal vents always form close to mid-ocean ridges and rifts (see p.185), where new ocean crust is forming and spreading, and where magma from the Earth's mantle lies relatively close to the surface. Seawater seeps into rock cracks opened up by the spreading sea floor. It penetrates several kilometres into the newly formed crust, close to the hot magma below. This heats the water to 350–400°C (660–750°F). The high pressure at these depths stops it from boiling, and it becomes superheated, dissolving minerals from the rocks that it is passing through, including sulphur which forms hydrogen sulphide. The hot water rises back up through cracks and erupts out of the vents as a hot, shimmering haze, complete with its load of minerals.

BLACK AND WHITE SMOKERS

As superheated water erupts from a hydrothermal vent, it meets the colder water of the ocean depths. This causes hydrogen sulphide in the vent water to react with the metals dissolved in it, including iron, copper, and zinc, which then come out of solution in the form of sulphide particles. Sometimes these form pools on the sea bed. However, if the water is particularly hot, it spouts up a little before being chilled by the surrounding seawater, and the metal sulphides form a cloud of black, smoke-like particles. Some of these minerals form a crust around the "smoke" plumes, building up into chimneys that can reach tens of metres in height. Such vents are called black smokers. More recently, a different form of vent has been discovered. In these, the black sulphides come out of solution as solids well beneath the sea floor, but other minerals remain in the vent water. Silica and a white mineral called anhydrite form the "smoke" from these chimneys, which, because of their colour, are called white smokers.

THE FORMATION OF A SMOKER
Water, heated by magma deep beneath the sea bed, dissolves minerals from the rocks. When it erupts through vents, the water is chilled by the surrounding sea. This makes minerals precipitate as smoky clouds, which can be white or black; other minerals are deposited to form chimneys.

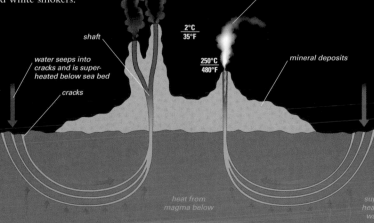

black cloud of metal sulphide particles

white cloud of silica and anhydrite particles

375°C
710°F

2°C
35°F

250°C
480°F

mineral deposits

shaft

water seeps into cracks and is super-heated below sea bed

cracks

heat from magma below

super-heated water

SMOKING CHIMNEYS
The minerals from black smokers, like this one, can increase the height of a chimney by an incredible 30cm (12in) a day. However, the chimneys are fragile, and they collapse when they get too high.

LIFE WITHOUT SUNLIGHT

The first biologists to explore hydrothermal vents were amazed at the life they saw. Masses of limpets, shrimps, sea anemones, and tube worms cluster close to the vents, beside unusually large clams and mussels. White crabs and a few fish, such as the Eelpout, scrabble among them. Not every vent system is the same: in the Atlantic, there are no tube worms, clams, or mussels, but lots of white shrimps. Some animals that live in darkness depend on sunlit waters for their food supply but vent animals are remarkable in that they do not need sunlight for energy. White mats of bacteria around vents are the key. They oxidize sulphides from the vent water to make energy, and are the vent animals' food source. Some animals have the bacteria living inside their bodies.

VENT FISH
This fish, called an Eelpout, feeds on mussels, shrimps, and crabs living around vents.

GHOSTLY CRAB
The Hydrothermal Vent Crab is one of many vent creatures. Each year, about 35 new species living around vents are being described by scientists.

DIFFERENT ANIMAL COMMUNITIES
Animal communities vary between vent systems. Vents on the Mid-Atlantic Ridge are inhabited by swarms of Rift Shrimps (shown here), feeding on sulphide-fixing bacteria, but there are no giant clams.

COLD SEEPS

The discovery of hydrothermal vents proved that not all deep-sea life depends on sunlight for energy. Soon, other sea-bed communities were found that could survive in the dark. In the Gulf of Mexico, diverse animal colonies live in shallow waters near where oil companies drill for petroleum. Here, seeps of methane and other hydrocarbons (compounds containing carbon and hydrogen) ooze up from rocks beneath the sea. Mats of bacteria feed on these cold seeps, providing energy for a food chain that includes soft corals, tube worms, crabs, and fish. Other animal communities in deep-sea trenches off the coasts of Japan and Oregon, USA, rely on methane, which is released by tectonic activity. Cold-seep communities may be more common than first thought at depths below 550m (1,800ft), although there is often no obvious seepage. Such communities may instead rely on chemical-rich sediments exposed by undersea landslides or currents.

OCEAN SMOKER
This black smoker, seen from *Alvin*, is similar to the one that scientists first observed in 1977, spewing out dark fluids from deep in the ocean crust.

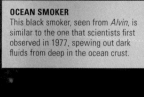

LIFE ON A SEEP
Mussels containing methane-fixing bacteria live alongside tube worms, soft corals, crabs, and an eelpout at this cold seep, 3,000m (9,800 feet) down on the sea bed near Florida.

WORM WITHOUT A MOUTH

The Vent Tube worm (below) can be 2m (6ft) long and as thick as a human arm. It has no apparent way of feeding. However, its body sac contains an organ called a trophosome, filled with grape-like clusters of bacteria. The worm's crimson plumes collect sulphides from vent water, and the bacteria use these to produce organic material, which the worm absorbs as food.

THE TWO POLAR OCEANS are the Arctic Ocean in the northern hemisphere and the Southern Ocean, which surrounds the continent of Antarctica, in the southern hemisphere. They differ from other oceans in several respects, not least in the sheer quantity of ice that floats on them. This includes sea-ice, which is frozen seawater, and icebergs and ice shelves, which are frozen fresh water. The polar oceans contain fewer temperature layers than other oceans, being uniformly cold, and they have different circulation patterns, which are partly wind-driven but also influenced by such factors as river inflow (in the Arctic Ocean) and sea-ice formation. The edges of the sea-ice are biologically productive zones, where plankton blooms occur in summer, attracting many fish, birds, and mammals.

POLAR OCEANS

PENGUINS UNDER THE ICE
These Emperor Penguins are swimming in a break between sea-ice off the coast of Antarctica. They can dive to 600m (2,000ft), staying down for up to 20 minutes.

ICE SHELVES

AN ICE SHELF IS A HUGE FLOATING ice platform, formed where a glacier, or group of glaciers, extends from a continental ice sheet over the sea. The landward side of an ice shelf is fixed to the shore, where there is a continuous inflow of ice from glaciers or ice streams that flow down from the ice sheet. At its front edge, there is usually an ice cliff, from which massive chunks of ice break off (calve) periodically, forming icebergs. Ice shelves are almost entirely an Antarctic phenomenon, with only a few small ones in the Arctic.

ANTARCTIC ICE SHELVES

Ice shelves surround about 44 per cent of the continent of Antarctica and cover an area of some 1.5 million square km (600,000 square miles). The largest is the Ross Ice Shelf, also called the Great Ice Barrier, discovered by Sir James Clark Ross (see panel, above). It is as large as mainland France, with an area of about 500,000 square km (190,000 square miles) and is fed by seven different ice streams. The second largest, the Ronne–Filchner Ice Shelf, covers about 430,000 square km (160,000 square miles). About 15 or so other ice shelves are dotted around the edge of the continent. Since 1995, a few of the smaller ice shelves around the Antarctic Peninsula, including parts of the Larsen Ice Shelf, have disintegrated, most probably as a result of ocean warming (see p.487).

ICE CLIFF
This massive ice cliff was photographed at the seaward edge of the Riiser-Larsen Ice Shelf. In front of it, Emperor Penguins queue to enter the water at Atka Bay, on the Weddell Sea.

Fimbul Ice Shelf
Lazerov Ice Shelf
Ekstrom Ice Shelf
Riiser-Larsen Ice Shelf
Weddell Sea
Larsen Ice Shelf
Brunt Ice Shelf
Amery Ice Shelf
Filchner Ice Shelf
Worche Ice Shelf
Ronne Ice Shelf
West Ice Shelf
George VI Sound
ANTARCTICA
Abbot Ice Shelf
Shackleton Ice Shelf
Ross Ice Shelf
Getz Ice Shelf
Ross Sea
Sulzberger Ice Shelf
Voyeykov Ice Shelf
Cook Ice Shelf

ICE-SHELF LOCATIONS
The two largest ice shelves – the Ross and Ronne–Filchner ice shelves – sit on either side of west Antarctica.

STRUCTURE AND BEHAVIOUR

Every ice shelf is anchored to the sea floor (ending at a point called the grounding line) and has a front part, that floats. The front part is usually 100–1,000m (330–3,300ft) thick, although only about one-ninth protrudes above water. The back of an ice shelf is fixed while the front part moves with the tides, creating stresses that can lead to the formation of cracks. Overall, there is a gradual movement of ice from the rear to the front of an ice shelf, from where large tabular icebergs occasionally calve. There is sometimes also a slow upward migration of ice, due to seawater freezing to the bottom of a shelf and the ice on the upper surface melting and evaporating in summer. Even deposits from the sea floor under an ice shelf are sometimes brought to the surface by this mechanism.

CALVING SHELF
The front part of an ice shelf will sometimes break up and the pieces drift off as tabular icebergs. Each piece visible here has a surface area of several square kilometres.

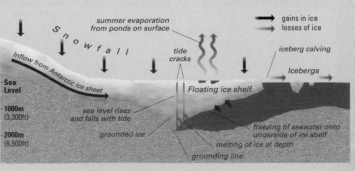

GAINS AND LOSSES
An ice shelf gains ice from glaciers flowing into its landward end, from new snowfall, and from seawater freezing to its undersurface. It loses ice by iceberg calving, by some summer melting of its upper surface and through evaporation, and by melting of part of on its undersurface.

summer evaporation from ponds on surface
gains in ice
losses of ice
s n o w f a l l
Inflow from Antarctic ice sheet
tide cracks
iceberg calving
Sea Level
Icebergs
-1000m (3,300ft)
sea level rises and falls with tide
Floating ice shelf
-2000m (6,500ft)
grounded ice
freezing of seawater onto underside of ice shelf
melting of ice at depth
grounding line

SURFACE AND INTERIOR

The upper surfaces of Antarctic ice shelves are inhospitable places. For most of the year, cold air streams called katabatic winds blow down from the Antarctic Ice Sheet and over the ice shelves. The surface of the ice is not flat, but is shaped by the winds into a series of ridges and troughs, called sastrugi. These are typically covered in a snow blanket. In some areas, the surface is littered with rocks from the input glacier or glaciers, or even with material that has been carried upwards from the sea floor by vertical movement. In summer, small ponds form on some ice shelves and provide a home for various types of microscopic organisms. Internally, an ice shelf usually contains some tide-induced cracks and crevasses.

CAVE INSIDE AN ICE SHELF
In summer, the internal cracks and crevasses in an ice shelf may enlarge to form caves as some of the ice melts.

BENEATH THE ICE SHELVES

Underneath the Antarctic ice shelves are extensive bodies of water that are some of the least explored regions on Earth. Seawater is thought to circulate constantly here, caused partly by new ice formation underneath and around the ice shelves. As new ice forms, it "rejects" salt, making the surrounding seawater denser. This causes the seawater to sink, and helps drive the circulation. Recent attempts have been made to explore these areas, using robotic submarines to take measurements. Little is known about the organisms that live here, although in 2005 a community of clams and bacterial mats was found on the sea floor under the Larsen B Ice Shelf after it broke up (see p.487).

LIFE UNDER THE ICE
Organisms such as starfish and worms live in shallow water around the edge of Antarctica, and possibly under the ice shelves.

annually reforming fast ice

new ice

marine ice is found beneath sea level

ice shelf

melting zone

ice platelets rise as density decreases

low-salinity water

high-salinity water

SEAWATER CIRCULATION
A continuous circulation of seawater is thought to occur under large ice shelves, driven by sea-ice formation on its undersurface and partial melting at depth.

grounding line

ice pump driven by salt rejection

ICEBERGS

ICEBERGS ARE HUGE, FLOATING chunks of ice that have broken off, or been calved, from the edges of large glaciers and ice shelves. These chunks range from car-sized objects to vast slabs of ice that are bigger than some countries. It is estimated that each year 40,000 to 50,000 substantial icebergs are calved from the glaciers of Greenland. A smaller number of gigantic icebergs break off the ice shelves around Antarctica. Surface currents carry icebergs away from their points of origin into the open ocean, where they drift and slowly melt. They can last for years and are a considerable danger to shipping.

ICEBERG PROPORTIONS
Because pure ice is 90 per cent as dense as seawater, an iceberg made entirely of ice will have only 10 per cent of its mass visible above water.

ICEBERG PROPERTIES

Icebergs consist principally of frozen fresh water, with no salt content. This is because they originate not from seawater but from glaciers or ice shelves (floating glaciers), and glaciers themselves come from compacted snow. Typically, an iceberg has a temperature of about -15 to -20°C (-4 to 5°F) at its core and 0°C (32°F) at its surface. In addition to ice, some icebergs contain rock debris. This is material that has fallen onto the parent glacier from surrounding mountains, or frozen to the glacier's edges, and eventually becomes incorporated into the ice. An iceberg's rock load affects its buoyancy. An iceberg with a high rock content may float up to 93 per cent submerged.

SIZES AND COLOURS

Icebergs include pieces of ice that are hundreds of square kilometres in area, down to ones the size of houses (bergy bits) or cars (growlers). Tabular icebergs may rise to a height of up to 60m (200ft) above the sea surface and extend underwater to a depth of up to 300m (1,000ft). Most icebergs appear white because of the light-reflecting properties of air bubbles trapped in the ice. Those made of dense, bubble-free ice absorb all but the shortest (blue) light wavelengths and so have a vivid blue tint. Occasionally, icebergs roll over and expose a previously submerged section to view, which appears aqua green because of algae growing in the ice.

RANGE OF SHAPES
Icebergs come in a range of shapes including tabular (flat-topped), domed, pinnacled or pyramidal, wedge-shaped, and various irregular shapes, as shown here.

TABULAR

PINNACLED

IRREGULAR

DOMED

NORTH ATLANTIC ICEBERGS

Most icebergs seen in the north Atlantic begin as snow falling on Greenland. This snow eventually becomes ice, which over thousands of years is transported from the Greenland ice sheet down to the sea as glaciers. Icebergs calved from the glaciers on the west coast of Greenland (and many from the east coast) move into Baffin Bay. The Labrador Current carries these icebergs southeast, past Newfoundland, into the north Atlantic. There, most of the icebergs rapidly melt, but a few reach as far south as 40°N – around the same latitude as New York and Lisbon.

ARCTIC OCEAN

Ellesmere Island

GREENLAND

Greenland Sea

Arctic Circle

ICELAND

Humboldt

Hayes

Baffin Bay

Jakobshavn

CANADA

NORTH ATLANTIC OCEAN

ORIGINS AND DISTRIBUTION
Most north Atlantic icebergs are calved by glaciers in west Greenland, such as the Jakobshavn and Hayes glaciers.

ICEBERG DETECTION

Because of their threat to shipping, north Atlantic icebergs are monitored by the US Coast Guard. Information on iceberg sightings, obtained by aircraft and ships, is fed into a computer along with ocean-current and wind data. The future movements of the icebergs are then predicted so that ships can be warned. The southernmost iceberg ever spotted in the Atlantic was only 250km (155 miles) from Bermuda at 32°N.

ICELANDIC ICEBERGS
These massive icebergs were calved from Breidamerkurjökull, a glacier in Iceland, and form a surreal tourist attraction, drifting in the glacial lagoon.

SOUTHERN OCEAN ICEBERGS

All Southern Ocean icebergs have broken off one of the ice shelves that surround Antarctica (see p.192). Most start off as extremely large, tabular icebergs – satellite monitoring of their drift tracks has provided useful information about Southern Ocean currents. After calving, these icebergs drift westward around Antarctica in a coastal current (the East Wind Drift). A few are carried in an eastward direction by the Antarctic Circumpolar Current. In extreme cases, they drift further, reaching as far north as 42°S in the Atlantic Ocean. The largest Southern Ocean iceberg ever recorded measured 290km (180 miles) long and 40km (25 miles) wide – about the size of Jamaica.

ATLANTIC OCEAN

AFRICA

SOUTH AMERICA

Antarctic Circle

INDIAN OCEAN

Minimum extent of sea-ice

ANTARCTICA

PACIFIC OCEAN

Maximum extent of sea-ice

Limit of icebergs

AUSTRALASIA

DISTRIBUTION
The approximate limit of iceberg drift from Antarctica is shown by the red dotted line. Most Southern Ocean icebergs remain close to the Antarctic Circle at 67°S.

ICE RAFTING

Icebergs that contain rock debris gradually release this material as they melt, and the debris sinks to the sea floor. Thus rock fragments can be transported from Greenland, for example, to the bottom of the north Atlantic. The process is called ice rafting. By examining sediment samples taken from the ocean floor, scientists can often identify rock fragments that have been transported in this way. Such studies can provide clues about past patterns of iceberg calving and iceberg distribution. For example, they have shown that there were short cold periods during the last ice age, called Heinrich Events, when vast armadas of icebergs were calved and crossed the Atlantic eastward from the coast of Labrador.

DIRTY ICEBERG
The fact that this iceberg contains considerable amounts of rock and dust is plain from its "dirty" appearance. This rock will end up on the sea floor as ice-rafted material.

OCEAN ENVIRONMENTS

WRECK OF THE TITANIC
The *Titanic*'s bow section, of which the upper deck and railings are seen here, is mostly intact, although deeply embedded in the sea floor.

THE TITANIC DISASTER

The sinking of the ocean liner *Titanic* on 15 April 1912 in the north Atlantic ranks as one of the worst peacetime maritime disasters in history. It is also arguably the most famous sinking of all time, partly because the ship had been considered unsinkable. A total of about 1,520 people died in the disaster, while just over 700 survived.

The exact sequence of blunders by which the *Titanic* came to collide with an iceberg has never been fully explained. It is known that during the 12 hours preceding the disaster, messages were sent from other ships that large icebergs lay in the *Titanic*'s path. However, these messages may not have reached the ship's bridge. When the collision occurred, the iceberg did not hit the *Titanic* full-on, but brushed the starboard side. However, this was enough to buckle the hull and dislodge rivets below the waterline, creating leaks into five of the ship's hull compartments. Although lifeboats were deployed, there were not enough to hold everyone. Furthermore, some were launched before they were full. As a result, about 1,500 people were still on the ship when it sank. Most are thought to have died of hypothermia in the ice-cold waters.

In 1985, the wreck of the *Titanic* was located by an American–French team, by means of an underwater vehicle with a video camera and lights attached. A notable discovery was that the ship had split in two before sinking – the bow and stern were found lying 600m (2,000ft) apart and facing in opposite directions.

FIRST AND LAST VOYAGE

The *Titanic* left England on 10 April 1912, bound for New York. After crossing the English Channel, the ship took on additional passengers in France, and also stopped in Ireland the next day, before continuing on its journey. Three days later, on 14 April, the ship's captain altered course slightly to the south, possibly in response to iceberg warnings received over the radio. However, at 11:40pm, lookouts spotted a large iceberg directly in front of the ship. Despite a frantic avoiding manoeuvre, the *Titanic* hit the iceberg and by 2:20am, the ship had sunk.

WAS THIS THE ICEBERG? *This photograph, taken six days later in the vicinity of the disaster, shows an iceberg that closely accorded with descriptions provided by survivors.*

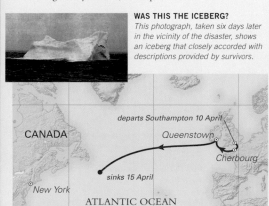

departs Southampton 10 April

CANADA

Queenstown

Cherbourg

sinks 15 April

New York

ATLANTIC OCEAN

THE HISTORY OF THE *TITANIC*

SETTING SAIL

LEAVING SOUTHAMPTON *At the time of her launch, the Titanic was the world's largest passenger liner and also the most opulent. When the ship left Southampton docks on 10 April 1912, she was carrying about 900 crew and 1,300 passengers, including some of the world's richest and most prominent people.*

MEDIA SENSATION *The sinking of the Titanic caused shock around the world. This Chicago newspaper dates from 16 April 1912.*

DISASTER

THE UNSINKABLE SINKS *In the early morning of 15 April, about 2.5 hours after colliding with an iceberg, the Titanic's stern rose out of the water as the ship sank.*

DISCOVERY OF THE WRECK

BOB BALLARD *Along with French scientist Jean-Louis Michel, American oceanographer Bob Ballard led the team that discovered the wreck of the Titanic on 1 September 1985, at a depth of 3,800m (12,500ft).*

LIFEBOAT WINDLASS *This piece of deck machinery was barely recognizable under a covering of rusticles (nodules containing a mixture of iron compounds and microbes that feed on wrought iron).*

ARTEFACTS

BANKNOTES *Banknotes in surprisingly good condition have been retrieved, including this $5 bill found in the purser's bag.*

CHINA DISHES *Rows of dishes were found lying on the sea floor. Such diverse items as books, watches, and wireless messages have also been retrieved, along with a bronze cherub and hundreds of other objects.*

OCEAN ENVIRONMENTS

SEA-ICE

SEA-ICE IS SEAWATER THAT HAS FROZEN at the ocean surface and floats on the liquid seawater underneath. It includes pack ice – ice that is not attached to the shoreline and drifts with wind and currents – and fast ice, which is frozen to a coast. Sea-ice formation and melting influences the large-scale circulation of water in the oceans. It has important stabilizing effects on the world's climate, since it helps control the movement of heat energy between the polar oceans and atmosphere. Sea-ice strongly reflects solar radiation, so in summer it reduces heating of the polar oceans. In winter, it acts as an insulator, reducing heat loss. Today, scientists are concerned about shrinking sea-ice in the Arctic because of its possible effects on climate and wildlife.

FORMATION

Seawater starts to freeze when it reaches a temperature of -1.8°C (28.8°F), slightly cooler than the freezing point of fresh water. Sea-ice formation starts with the appearance of tiny needle-like ice crystals (frazil ice) in the water. Salt in seawater cannot be incorporated into ice, and the crystals expel salt. The developing sea-ice gradually turns into a thick slush and then, under typical wave conditions, into a mosaic of ice platelets called pancake ice. Subsequently, it consolidates into a thick, solid sheet, through processes such as "rafting" (in which the ice fractures and one piece overrides another) and "ridging" (where lines of broken ice are forced up by pressure). Where ridging occurs, each ridge has a corresponding structure, a keel, that forms on the underside of the ice. Newly formed, compacted sheet ice is called first-year ice and may be up to 30cm (12in) thick. It continues to thicken through the winter. Any ice that remains through to the next winter is called multi-year ice.

TESTING THE ICE
Pancake ice, consisting of ice platelets, can be up to 10cm (4in) thick. Waves and wind have caused these platelets to collide, hence their curled-up edges.

HOW ICE FORMS
The stages of sea-ice formation vary according to whether the sea surface is calm or affected by waves. A typical sequence in an area of moderate wave action is shown below.

GREASE ICE
Fine ice spicules, called frazils, appear in the water. These coagulate into a viscous soup of ice crystals, called grease ice.

PANCAKE ICE
Wave action causes the grease ice to break into slushy balls of ice, called shuga. These clump into platter shapes called pancakes.

FIRST-YEAR ICE
The ice pancakes congeal, consolidate, and thicken through processes such as rafting and ridging to form a continuous sheet of ice.

MULTI-YEAR ICE
Further thickening, for a year or more, produces multi-year ice. This has a rough surface and may be several metres thick.

EXTENT AND THICKNESS

The extent of sea-ice in the polar oceans varies over an annual cycle. About 85 per cent of the winter ice that forms in the Southern Ocean melts in summer, and on average this ice only reaches a thickness of a metre or two. In the Arctic, much of the ice lasts for several seasons, and this multi-year ice attains a greater thickness – on average 2–3m (7–10ft). In winter, pack ice covers most of the Arctic Ocean. In summer, it usually halves in area. In recent years, the summer retreat has been more pronounced, raising fears that the Arctic ice coverage may disappear altogether over the next 30 to 40 years.

ARCTIC SEA-ICE COVERAGE
Coverage varies from a winter high of 15 million square km (6 million square miles) to a summer low of less than 7 million square km (3 million square miles).

☐ year-round ice

☐ winter sea-ice

Map labels: Arctic Circle; ARCTIC OCEAN

DISCOVERY

USS NAUTILUS

In 1958, a US submarine, the USS *Nautilus*, crossed the Arctic Ocean underneath its cover of sea-ice, passing the North Pole on 3 August. The crossing proved that there is no sizeable land mass in the middle of the Arctic Ocean. The submarine traversed the Arctic from the Beaufort Sea to the Greenland Sea in four days at a depth of about 150m (500ft).

GAPS IN THE ICE

Even in parts of the polar oceans that are more or less permanently ice-covered, gaps and breaks sometimes appear or persist in the ice. These openings vary greatly in size and extent and have different names. Fractures are extremely narrow ruptures that are usually not navigable by boats of any size. An ice lead is a long, straight, narrow passageway that opens up spontaneously in sea-ice, making it navigable by surface vessels and some marine mammals. Polynyas are persistent regions of open water, up to several hundred square kilometres in area and often roughly circular in shape. They sometimes develop where there is upwelling of warmer water in a localized area, or near coasts where the wind blows new sea-ice away from the shore as it forms.

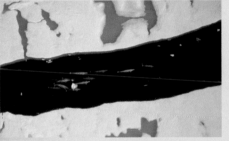

ICE LEAD
An ice lead forms when an area of sea-ice shears. Stresses from winds and water currents are thought to be the cause. Here a group of beluga whales swims along a lead.

LIFE AROUND THE ICE

Life thrives around sea-ice. One reason for this is that as ice forms, salt is expelled into the seawater, causing it to become denser and sink. This forces nutrient-laden water to the surface. In summer, the combination of nutrients and sunlight encourage the growth of phytoplankton, which provide a rich food source. These organisms form the base of a food chain for fish, mammals, and birds. In the Arctic, sea-ice provides a resting and birthing place for seals and walruses and a hunting and breeding ground for Polar Bears and Arctic Foxes. In the Antarctic, it supports seals and penguins. Breaks in the ice are vital to this wildlife. Seals, penguins, and whales rely on them for access to the air, while Polar Bears hunt near them. Decreases in Arctic sea-ice would drastically shrink some habitats, pushing them towards extinction.

ANTARCTIC KRILL
These crustaceans form an important part of the food chain in the Southern Ocean, where they congregate in dense masses.

HUMAN IMPACT

ICE BREAKERS

Ice breakers are ships designed for moving through ice-covered environments. An ice breaker has a reinforced hull and a bow shape that causes the ship to ride over sea-ice and crush it as it moves forward. The shape of the vessel clears the ice debris to the sides and underneath the hull, allowing steady forward progress. A modern ice breaker can advance through sea-ice up to 1.8m (6ft) thick.

WEDDELL SEAL
The Weddell Seal, found only in the Antarctic, is one of nine seal species that inhabit polar oceans. Weddell Seals never stray far from sea-ice.

POLAR OCEAN CIRCULATION

THE ARCTIC AND SOUTHERN OCEANS each have their own unique patterns of water flow, which link in with the rest of the global ocean circulation. These flows are driven partly by wind and partly by various factors that influence the temperature and salinity of the surface waters in these oceans – including seasonal variations in air temperature and sea-ice coverage, and large inflows of fresh water from rivers. Although driven by similar influences, the significantly different water-flow patterns of these two oceans are largely due to the fact that the Arctic Ocean is encircled by land, whereas the Southern Ocean surrounds a frozen continent.

CIRCULATION AND FEEDING
The Southern Ocean meets warmer water at the Antarctic Convergence, creating a biologically rich feeding area for whales, including these Humpbacks.

ARCTIC SURFACE CIRCULATION

The upper 50m (170ft) of the Arctic Ocean is affected by currents that keep it in constant motion. There are two main components to this circulation (see p.428–29). In a large area north of Alaska, there is a slow, circular motion of water called the Beaufort Gyre. This clockwise movement is wind-generated and completes one rotation every four years. The second component, the Transpolar Current, is driven by a vast quantity of water discharged into the Arctic Ocean from Siberian rivers.

MOUTH OF THE LENA RIVER
The Lena flows across Siberia and discharges 420 cubic km (100 cubic miles) of water into the Arctic Ocean every year.

ARCTIC DEEP-WATER CIRCULATION

In the deeper waters of the Arctic, there is a slow circulation of cold, dense water. This circulation is restricted by the structure of the Arctic Ocean, which consists of a central deep basin (the Arctic Basin), bisected by several underwater ridges and surrounded on most sides by shallow continental shelf. Only on the Atlantic side is there a connection between the deep waters of the Arctic and deep ocean waters to the south. On the opposite side, the connection with the Pacific is via the shallow and narrow Bering Strait. What little circulation of water occurs in the Arctic Basin mostly involves influxes of Atlantic water at various depths to the north of Russia, and outflows around Greenland.

ARCTIC BASIN CIRCULATION
Atlantic water enters to the north of Russia and Pacific water via the Bering Strait. As it cools, some of the Atlantic water becomes denser and dips far below the sea-ice cover, where it flows around slowly. The main outflow is to the east of Greenland.

→ atlantic water
→ pacific water

MISTY SEAS
The seas around the Antarctic Convergence are prone to mists. Here a cruise liner approaches a channel just south of the Convergence.

THE ANTARCTIC CONVERGENCE

The Antarctic Convergence is a region of the Southern Ocean encircling Antarctica, located roughly at latitude 55°S (but deviating from this in places), where cold, northward-flowing waters from Antarctica sink beneath the relatively warmer waters to the north.

At the Convergence, there is a sudden change in surface ocean temperature of 3-5°C (5-9°F) as well as alterations in the chemical composition of seawater. As a result, the Convergence forms a barrier to the movement of animal species, and the groups of marine animals found on either side of it are quite different. This is a turbulent area. The meeting of different water masses brings dissolved nutrients from the sea bed to the ocean surface. This acts as a fertilizer, encouraging the growth of plankton during the southern-hemisphere summer.

PHYTOPLANKTON
In summer, massive blooms of phytoplankton occur around the Antarctic Convergence, forming the base of a productive food chain.

ALBATROSSES
These Black-Browed and Grey-Headed albatrosses inhabit the biologically productive Southern Ocean.

SOUTHERN OCEAN CIRCULATION

In the Southern Ocean, surface waters move under the influence of two wind-driven currents. Off the coast of Antarctica, the Antarctic Coastal Current carries water from east to west around Antarctica. Several hundred kilometres north, the Antarctic Circumpolar Current (ACC) moves water in the opposite direction, from west to east, and pushes the Antarctic waters northwards. The ACC is a major ocean current that connects the Pacific, Atlantic, and Indian oceans and isolates Antarctica from the warmer ocean currents to the north. In the Southern Ocean, an important movement of water also occurs at depth. In an area near Antarctica, masses of dense, salty water forms as salt is rejected from seawater as it freezes. This cold water sinks and moves north into the southern Atlantic.

PEOPLE

FRIDJTOF NANSEN

The Norwegian explorer and scientist Fridjtof Nansen (1861-1930) is most famous for his Arctic voyage of 1893-1895 on a specially built wooden ship called the *Fram*. Nansen deliberately allowed the *Fram* to drift across the Arctic Ocean locked in ice, and in doing so proved the existence of the surface current now called the Transpolar Current. In 1895, setting off with one companion from the *Fram*, Nansen walked and skied to within 640km (400 miles) of the North Pole, closer than anyone else up to that time.

OCEAN LIFE

BY FAR THE LARGEST HABITAT on Earth, the oceans are more accurately seen as a great range of environments as disparate as mangrove swamps and deep-sea vents. Living organisms have found a place to take hold in every ocean environment, even the deepest trenches, more than 10km (6 miles) beneath the surface. Ocean life teems with greatest abundance and variety in the sunlit surface waters. Here, microscopic plants and plant-like organisms, the phytoplankton, fuel productive communities of organisms right up to top predators, such as Killer Whales. Life began in the oceans, and they were the site of many ground-breaking steps in evolution. Tracing the history of this evolution puts into context the astonishing variety of today's marine life.

INTRODUCTION TO OCEAN LIFE

KELP FOREST COMMUNITY
Ocean life develops into one of a number of characteristic communities, according to physical conditions. Here in cool, shallow water, a canopy of kelp towers above an undergrowth of encrusting red seaweed, while an eagle ray and smaller fish find shelter among the kelp fronds.

CLASSIFICATION

BY CLASSIFYING ORGANISMS AND FITTING them into a universally accepted framework, scientists have created a massive reference system that accommodates all forms of life. Over 2 million organisms have been described, but only about 16 per cent live in the oceans. The marine proportion is likely to increase, as many new species continue to be discovered annually, particularly in the deep ocean.

WHAT IS A SPECIES?

A species is the basic unit of classification. One commonly accepted definition of a species is a population of organisms that have so many features in common that they form a distinct group that interbreeds, producing fertile offspring in natural conditions. This definition cannot be applied to fossil species. There are scores of other species definitions, some of which incorporate both fossil and living species. In the end, a species can often be defined rather subjectively.

LINNAEAN HIERARCHY

Linnaeus used a hierarchy of ranked categories of increasing exclusiveness. Today's expanded system includes many ranks, from domain down to species. Below is an example of a series of ranked categories, illustrating those that classify the Common Dolphin.

DOMAIN Eucarya
Includes all Eukaryotes – organisms that have complex cells with distinct nuclei. Only bacteria and archaea fall outside this domain.

KINGDOM Animalia
Includes all animals – multicellular Eukaryotes that need to eat food for energy. All animals are mobile for at least part of their lives.

PHYLUM Chordata
Includes all chordates – animals possessing a notochord. In most cases, the notochord is replaced before birth by the backbone.

CLASS Mammalia
Includes all mammals – air-breathing chordates that feed their young on milk. The jaw is made up of a single bone.

ORDER Cetacea
Includes all cetaceans (whales and dolphins) – marine mammals that have a tail with boneless, horizontal flukes for propulsion.

FAMILY Delphinidae
Includes all dolphins (a subgroup of toothed cetaceans) with beaks and 50–100 vertebrae. The skull lacks a crest.

GENUS Delphinus
Includes a few colourful, oceanic dolphins with 40–50 teeth on each side of the jaw. These dolphins form large social groups.

SPECIES Delphinus delphis
Specifies a single type of dolphin with a V-shaped black cape under the dorsal fin and criss-cross hour-glass patterning on its sides.

PRINCIPLES OF CLASSIFICATION

Classification helps us make sense of the natural world by grouping organisms on the basis of features that they share. It gives scientists a clear and accurate understanding of the diversity of life, and because everyone uses the same system, the knowledge is accessible on a worldwide basis. The hierarchical system devised by the Swedish scientist Carolus Linnaeus (see panel, left) in the 18th century, still forms the basis of today's classification. Each species is identified with a unique two-part scientific name (made up of the genus and species name) then filed in a series of ever-larger groupings. However, as our knowledge increases, it is often necessary to revise the groups. Sometimes, this leads to subdivision of categories, for example phylum Arthropoda has been split into the subphyla Crustacea and Hexapoda. Frequently, a new species is identified when it is shown to be distinct from other populations.

THE EVIDENCE

In the past, scientists could identify and classify organisms only by studying anatomy, by looking at form, function, and embryological development (animals only), and by examining the fossil record. Recently, scientists have also been able to investigate organisms by looking at their proteins and their DNA. DNA is a complex molecule whose sequential structure is unique to each organism. The relatedness of organisms can be determined by comparing these DNA molecules for shared features. This molecular evidence has led to many revisions of classification.

DETAILED ANATOMY
By making a detailed anatomical examination of material in museum collections, scientists can distinguish between similar organisms and classify them according to shared characters.

CLADISTICS

By the 1950s, although most people used the same system of classification, the criteria they used for placing organisms in categories were often neither measurable nor repeatable. The idea emerged to analyse many characters using an automatic, computer-like process, not only to classify organisms, but also to trace their evolution. This process became known as cladistics, and it is a widely used technique today. A cladistic analysis examines a wide selection of characters shared by a study group of organisms. It finds the most likely pattern of evolutionary changes that link the organisms, involving the least number of steps (evolutionary branching points). It then arranges the organisms in a tree diagram (cladogram) that reflects their relationships. A cladogram is made up of nested groups called clades. A clade encompasses all the descendants of the group's common ancestor.

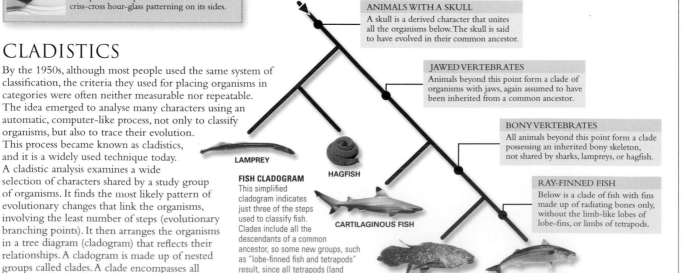

ANIMALS WITH A SKULL
A skull is a derived character that unites all the organisms below. The skull is said to have evolved in their common ancestor.

JAWED VERTEBRATES
Animals beyond this point form a clade of organisms with jaws, again assumed to have been inherited from a common ancestor.

BONY VERTEBRATES
All animals beyond this point form a clade possessing an inherited bony skeleton, not shared by sharks, lampreys, or hagfish.

RAY-FINNED FISH
Below is a clade of fish with fins made up of radiating bones only, without the limb-like lobes of lobe-fins, or limbs of tetrapods.

LAMPREY

HAGFISH

FISH CLADOGRAM
This simplified cladogram indicates just three of the steps used to classify fish. Clades include all the descendants of a common ancestor, so some new groups, such as "lobe-finned fish and tetrapods" result, since all tetrapods (land vertebrates) descend from lobe-fins.

CARTILAGINOUS FISH

LOBE-FINNED FISH AND TETRAPODS

RAY-FINNED FISH

MARINE LIFE

THE CLASSIFICATION FRAMEWORK USED in this book is shown on the following three pages. In this framework, all living things are divided into three domains. Within domains, only the marine groups are shown, although the numbers of classes and species cited include all organisms within the group whether they are marine or not. Some groupings, such as protists, are shown in dotted lines because although they are useful categories, they are not true, taxonomic groups. Others, such as small, bottom-living phyla and planktonic phyla, are ecological groupings and do not reflect taxonomy or evolutionary history.

BACTERIA

DOMAIN Bacteria   KINGDOMS 10   SPECIES many millions

ARCHAEA

DOMAIN Archaea   KINGDOMS 3   SPECIES many millions

EUKARYOTES

DOMAIN Eucarya   KINGDOMS At least 15   SPECIES 2 million

EUKARYOTES

THIS DOMAIN INCLUDES ALL ORGANISMS that have cells with a nucleus and other complex structures not seen in prokaryotes (bacteria, archaea). The eukaryotes comprise protists, seaweeds, plants, fungi, and animals.

Protists

KINGDOMS At least 10   SPECIES More than 100,000

UNTIL RECENTLY, THESE TINY, SINGLE-CELLED organisms formed a single kingdom, but apart from size, there is little to unite them. Their classification is in flux, but in this book, they are presented as a mixture of kingdoms and informal groups.

Diatoms
Informal group   CLASSES 2   SPECIES 10,000

Dinoflagellates
KINGDOM Dinoflagellata   CLASSES 4   SPECIES 4,000

Golden Algae
KINGDOM Chrysophyta   CLASSES 3   SPECIES 720

Radiolarians
KINGDOM Radiolaria   CLASSES 4   SPECIES 4,100

Foraminiferans
KINGDOM Foraminifera   CLASSES 2–3   SPECIES 10,000

Ciliates
KINGDOM Ciliata   CLASSES 3   SPECIES 10,000

Coccolithophorids
Informal group   CLASSES 1   SPECIES 200

+ SEVERAL MORE INFORMAL GROUPS AND KINGDOMS

Red seaweeds
KINGDOM Rhodophyta   CLASSES 1 or more   SPECIES 5,500

Brown seaweeds
KINGDOM Phaeophyta   CLASSES 1   SPECIES 2,000

Plants
KINGDOM Plantae   DIVISIONS 6   SPECIES 283,000

GREEN PLANTS HERE COMPRISE SIX DIVISIONS, with all vascular plants grouped in division Tracheophyta. The Hepatophyta (liverworts), Anthoceraphyta (hornworts), and Lycophyta (clubmosses) are not shown because they have no marine species.

GREEN SEAWEEDS AND ALGAE
DIVISION Chlorophyta   CLASSES At least 4   SPECIES 16,000

GREEN ALGAE
CLASS Prasinophyceae   ORDERS 1   SPECIES 200

GREEN SEAWEEDS
CLASSES Ulvophyceae, Cladophorophyceae, Bryopsidophyceae, Dasycladophyceae   SPECIES 1,200

+ SIX MORE CLASSES OF GREEN ALGAE

MOSSES
DIVISION Bryophyta   CLASSES 3   SPECIES 10,000

VASCULAR PLANTS
DIVISION Trachaeophyta   CLASSES 8   SPECIES 250,000

FLOWERING PLANTS
CLASS Angiospermae   ORDERS 30   SPECIES 235,000

+ SEVEN NON-MARINE CLASSES

+ THREE NON-MARINE DIVISIONS

Fungi
KINGDOM Fungi   PHYLA 4   SPECIES 600,000

Animals
KINGDOM Animalia   PHYLA About 30   SPECIES Over 1.5 million

THE FOLLOWING LIST OF ANIMAL PHYLA progresses from organisms with simple body plans and systems, such as sponges, to the most complex phylum, chordates, which contains humans. Each phylum represents a distinct body plan.

SPONGES
PHYLUM Porifera   CLASSES 3   SPECIES 15,000

CNIDARIANS
PHYLUM Cnidaria   CLASSES 4   SPECIES 9,000

CORALS AND ANEMONES
CLASS Anthozoa   ORDERS 10   SPECIES 6,000

JELLYFISH
CLASS Scyphozoa   ORDERS 4   SPECIES 200

BOX JELLYFISH
CLASS Cubozoa   ORDERS 1   SPECIES 16

HYDROIDS
CLASS Hydrozoa   ORDERS 7   SPECIES 2,700

PLANKTONIC PHYLA

THE FOLLOWING THREE PHYLA FLOAT with the ocean currents in the plankton and are grouped here on this basis. The Ctenophora and Chaetognatha contain so few species that they are known as minor phyla.

COMB JELLIES
PHYLUM Ctenophora   CLASSES 2   SPECIES 100

ARROW WORMS
PHYLUM Chaetognatha   CLASSES 2   SPECIES 70

ROTIFERANS
PHYLUM Rotifera   CLASSES 3   SPECIES 2,000

FLATWORMS
PHYLUM Platyhelminthes	CLASSES 4	SPECIES 20,000

RIBBON WORMS
PHYLUM Nemertea	CLASSES 2	SPECIES 900

SEGMENTED WORMS
PHYLUM Annelida	CLASSES 3	SPECIES 15,000

BOTTOM-LIVING PHYLA

MEMBERS OF THE FOLLOWING PHYLA all live in or on the ocean floor. The list is not comprehensive – the following phyla are among those not included: Entoprocta, Acanthocephala, Placozoa, and Loricifera.

GIANT TUBE WORMS
PHYLUM Vestimentifera	CLASSES 1	SPECIES 10

POGONOPHORAN WORMS
PHYLUM Pogonophora	CLASSES 1	SPECIES 110

SPOON WORMS
PHYLUM Echiura	CLASSES 2	SPECIES 140

LAMP SHELLS
PHYLUM Brachiopoda	CLASSES 2	SPECIES 300

HORSESHOE WORMS
PHYLUM Phoronida	CLASSES 1	SPECIES 12

PEANUT WORMS
PHYLUM Sipuncula	CLASSES 2	SPECIES 320

CYCLIOPHORANS
PHYLUM Cycliophora	CLASSES 1	SPECIES 1

GASTROTRICHS
PHYLUM Gastrotricha	CLASSES 1	SPECIES 430

ROUND WORMS
PHYLUM Nematoda	CLASSES 2	SPECIES 20,000

PRIAPULA WORMS
PHYLUM Priapulida	CLASSES 1	SPECIES 17

MUD DRAGONS
PHYLUM Kinorhyncha	CLASSES 2	SPECIES 150

WATER BEARS
PHYLUM Tardigrada	CLASSES 2	SPECIES 400

PTEROBRANCH WORMS
PHYLUM Pterobranchia	CLASSES 1	SPECIES 24

ACORN WORMS
PHYLUM Enteropneusta	CLASSES 1	SPECIES 70

MOLLUSCS
PHYLUM Mollusca	CLASSES 8	SPECIES 50,000

CAUDOFOVEATES
CLASS Caudofoveata	ORDERS 1	SPECIES 70

SOLENOGASTRES
CLASS Solenogaster	ORDERS 4	SPECIES 180

MONOPLACOPHORANS
CLASS Monoplacophora	ORDERS 1	SPECIES 8

TUSK SHELLS
CLASS Scaphopoda	ORDERS 2	SPECIES 350

BIVALVES
CLASS Bivalvia	ORDERS 13	SPECIES 14,000

GASTROPODS
CLASS Gastropoda	ORDERS 16	SPECIES 35,000

CEPHALOPODS
CLASS Cephalopoda	ORDERS 4	SPECIES 650

CHITONS
CLASS Polyplacophora	ORDERS 3	SPECIES 500

ARTHROPODS
PHYLUM Arthropoda	CLASSES 17	SPECIES 1.1 million

CRUSTACEANS
SUBPHYLUM Crustacea	CLASSES 6	SPECIES 50,000

WATER FLEAS AND RELATIVES
CLASS Branchiopoda	ORDERS 4	SPECIES 900

BARNACLES AND COPEPODS
CLASS Maxillopoda	ORDERS 23	SPECIES 13,000

MUSSEL SHRIMPS
CLASS Ostracoda	ORDERS 5	SPECIES 7,000

MALACOSTRACANS
CLASS Malacostraca	ORDERS 16	SPECIES 30,000

MANTIS SHRIMPS
ORDER Stomatopoda	FAMILIES 4	SPECIES 350

ISOPODS
ORDER Isopoda	FAMILIES 94	SPECIES 10,000

AMPHIPODS
ORDER Amphipoda	FAMILIES 119	SPECIES 10,000

KRILL
ORDER Euphausiacea	FAMILIES 2	SPECIES 90

LOBSTERS, CRABS, AND SHRIMPS
ORDER Decapoda	FAMILIES 105	SPECIES 8,500

+ 11 MORE MINOR ORDERS

SEA SPIDERS
SUBPHYLUM Pycnogonida	CLASSES 1	SPECIES 1,000

CHELICERATES
SUBPHYLUM Chelicerata	CLASSES 13	SPECIES 70,000

SPIDERS, SCORPIONS, TICKS AND MITES
CLASS Arachnida	ORDERS 12	SPECIES 70,000

HORSESHOE CRABS
CLASS Merostomata	ORDERS 1	SPECIES 4

INSECTS
SUBPHYLUM Insecta	CLASSES 6	SPECIES 950,000

+ 1 NON-MARINE SUBPHYLUM: MILLIPEDES AND CENTIPEDES (MYRIAPODA)

ECHINODERMS
PHYLUM Echinodermata	CLASSES 6	SPECIES 7,000

SEA LILIES AND FEATHER STARS
CLASS Crinoidea	ORDERS 5	SPECIES 630

STARFISH
CLASS Asteroidea	ORDERS 5	SPECIES 1,500

BRITTLESTARS
CLASS Ophiuroidea	ORDERS 3	SPECIES 2,000

SEA URCHINS
CLASS Echinoidea	ORDERS 15	SPECIES 940

SEA CUCUMBERS
KINGDOM Holothuroidea	ORDERS 6	SPECIES 1,150

SEA DAISIES
KINGDOM Concentricycloidea	ORDERS 1	SPECIES 2

CHORDATES

PHYLUM Chordata	SUBPHYLA 3	SPECIES 51,500

THE VERTEBRATES DOMINATE PHYLUM CHORDATA. The remaining two, much smaller, subphyla are united with vertebrates by the presence of the rod-like notochord, which becomes the backbone before birth in vertebrates.

TUNICATES (SEA SQUIRTS AND SALPS)

SUBPHYLUM Urochordata	CLASSES 4	SPECIES 2,000

LANCELETS

SUBPHYLUM Cephalochordata	CLASSES 1	SPECIES 25

VERTEBRATES

SUBPHYLUM Vertebrata	CLASSES 7	SPECIES Over 49,500

MANY AUTHORITIES EXCLUDE HAGFISH from the vertebrates due to their poorly developed vertebral column. They include hagfish within a larger group called the Craniata, meaning animals possessing a skull (cranium).

FISH

"FISH" IS AN INFORMAL TERM for five distinct groups of animals. The bony fish do not form a self-contained group, because all land vertebrates (tetrapods) descend from lobe-finned fish. Bony fish are a natural group only if tetrapods are placed within the lobe-fin group.

JAWLESS FISH

HAGFISH

CLASS Myxinoidea	ORDERS 1	SPECIES 50

LAMPREYS

CLASS Petromyzontida	ORDERS 1	SPECIES 38

CARTILAGINOUS FISH

CLASS Chondrichthyes	ORDERS 10	SPECIES 1,114

SHARKS AND RAYS

SUBCLASS Elasmobranchii	ORDERS 9	SPECIES 1,080

SHARKS

ORDERS 8	FAMILIES 29	SPECIES 480

SKATES AND RAYS

ORDERS 1	FAMILIES 12	SPECIES 600

CHIMAERAS

SUBCLASS Holocephali	SPECIES 1	SPECIES 34

BONY FISH

CLASS Osteichthyes	ORDERS 47	SPECIES 28,000

LOBE-FINNED FISH

SUBCLASS Sarcopterygii	ORDERS 3	SPECIES 8

RAY-FINNED FISH

SUBCLASS Actinopterygii	ORDERS 44	SPECIES 28,000

STURGEONS AND PADDLEFISHES		SALMONS	
ORDER Acipenseriformes	SPECIES 25	ORDER Salmoniformes	SPECIES 66
TARPONS AND TENPOUNDERS		LIGHTFISH AND DRAGONFISH	
ORDER Elopiformes	SPECIES 7	ORDER Stomiiformes	SPECIES 391
BONEFISH		GRINNERS	
ORDER Albuliformes	SPECIES 5	ORDER Aulopiformes	SPECIES 228
EELS		LANTERNFISH AND RELATIVES	
ORDER Anguilliformes	SPECIES 737	ORDER Myctophiformes	SPECIES 241
SWALLOWERS AND GULPERS		VELIFERS, TUBE-EYES, RIBBONFISH	
ORDER Saccopharyngiformes	SPECIES 28	ORDER Lampriformes	SPECIES 18
HERRINGS AND RELATIVES		COD FISH AND RELATIVES	
ORDER Clupeiformes	SPECIES 397	ORDER Gadiformes	SPECIES 475
MILKFISH		TOADFISH AND MIDSHIPMEN	
ORDER Gonorhynchiformes	SPECIES 27	ORDER Batrachoidiformes	SPECIES 69
CATFISH AND KNIFEFISH		CUSK EELS	
ORDER Siluriformes	SPECIES 2,867	ORDER Ophidiiformes	SPECIES 354
SMELTS AND RELATIVES		ANGLERFISH	
ORDER Argentiniformes	SPECIES 227	ORDER Lophiiformes	SPECIES 300

CLINGFISH		PIPEFISH AND SEAHORSES	
ORDER Gobiesociformes	SPECIES 120	ORDER Syngnathiformes	SPECIES 240
NEEDLEFISH		SCORPIONFISH AND FLATHEADS	
ORDER Beloniformes	SPECIES 186	ORDER Scorpaeniformes	SPECIES 1,326
SILVERSIDES		PERCH-LIKE FISH	
ORDER Atheriniformes	SPECIES 312	ORDER Perciformes	SPECIES 9,500
SQUIRRELFISH AND RELATIVES		FLATFISH	
ORDER Beryciformes	SPECIES 121	ORDER Pleuronectiformes	SPECIES 572
DORIES AND RELATIVES		PUFFERS AND FILEFISH	
ORDER Zeiformes	SPECIES 42	ORDER Tetraodontiformes	SPECIES 353
STICKLEBACKS AND SEAMOTHS			
ORDER Gasterosteiformes	SPECIES 16		

+ 17 MORE ORDERS

REPTILES

CLASS Reptilia	ORDERS 4	SPECIES 7,700

TURTLES

ORDER Chelonia	FAMILIES 12	SPECIES 300

SNAKES AND LIZARDS

ORDER Squamata	FAMILIES 44	SPECIES 7,400

CROCODILES

ORDER Crocodylia	FAMILIES 3	SPECIES 23

+ 1 NON-MARINE ORDER: THE TUATARAS (SPHENODONTIDA)

BIRDS

CLASS Aves	ORDERS 29	SPECIES 9,500

IN THIS CLASSIFICATION, the birds have been divided into 29 orders. Some scientists consider birds to be grouped within the reptiles.

WATERFOWL (DUCKS, GEESE, AND SWANS)

ORDER Anatidiformes	FAMILIES 2	SPECIES 149

PENGUINS

ORDER Sphenisciformes	FAMILIES 1	SPECIES 17

DIVERS

ORDER Gaviiformes	FAMILIES 1	SPECIES 5

ALBATROSSES AND PETRELS

ORDER Procellariiformes	FAMILIES 4	SPECIES 108

GREBES

ORDER Podicipediformes	FAMILIES 1	SPECIES 22

PELICANS AND RELATIVES

ORDER Pelicaniformes	FAMILIES 5	SPECIES 65

HERONS AND RELATIVES

ORDER Ciconiiformes	FAMILIES 7	SPECIES 119

BIRDS OF PREY

ORDER Falconiformes	FAMILIES 3	SPECIES 307

WADERS, GULLS, AND AUKS

ORDER Charadriiformes	FAMILIES 14	SPECIES 343

KINGFISHERS AND RELATIVES

ORDER Coraciiformes	FAMILIES 9	SPECIES 191

+ 19 NON-MARINE ORDERS

MAMMALS

CLASS Mammalia	ORDERS 26	SPECIES 5,000

THREE PARTLY OR WHOLLY MARINE mammal orders are listed here. The pinnipeds (seals, sea lions, and walruses), until recently classified as order Pinnipedia, do not form a natural group, and have been placed within order Carnivora (cats, dogs, bears, otters, and relatives). The 26 mammal orders includes new orders formerly classified as marsupials.

CARNIVORES

ORDER Carnivora	FAMILIES 9	SPECIES 249

WHALES AND DOLPHINS

ORDER Cetacea	FAMILIES 12	SPECIES 85

SEA COWS

ORDER Sirenia	FAMILIES 2	SPECIES 4

+ 23 MORE NON-MARINE ORDERS

OCEAN LIFE

RED SEA REEF
The Red Sea is one of the world's top 18 coral hot spots. Its colourful reefs are home to an abundance of marine life, including the venomous Red Lionfish.

BIODIVERSITY HOT SPOTS

Many people have heard of biodiversity hot spots, particularly in the context of documentaries about ocean life. These sites are very popular with filmmakers for the variety of life they exhibit. However, the term is a slight misnomer. Strictly speaking, such sites are "species diversity hot spots", places where the largest number of species are concentrated in a small area. Identifying such hot spots helps conservationists to decide where protected areas should be set up. However, places where species diversity is low, such as the ocean trenches, are also important because of the remarkable animals that live there.

The problem is that too little is known about the subject for scientists to be sure where the highest species diversity occurs in the ocean, beyond the shallow layer accessible to human divers. Assessments have been made of the richest coral reefs (see below), and of regions with the most seagrasses (see p.146). Another study has found hot spots for sea turtles, tuna, sharks, and other large predators, mainly near islands, shelf breaks, and seamounts. Recent research, including a 2006 study of Saba Bank in the Caribbean (see right) by the Conservation International Marine Rapid Assessment Programme, has added to the pool of knowledge. New Zealand and the United States are among the few countries that have tried to locate and assess species diversity hot spots in their territorial waters. However, far too little is known about the deep ocean for all marine hot spots to be identified.

CORAL REEF HOT SPOTS

Tropical coral reefs are popular with divers because they are colourful, shallow, and easy to reach. As a result, we know far more about life on coral reefs than many other ocean habitats. In 2002, a team of researchers led by Dr Callum Roberts from the University of York, UK, gathered data on 3,235 different species of reef fish, corals, snails, and lobsters. More than a quarter of the fish and snails and half of the lobsters were localized species found in just a few coral reefs.

The study pinpointed 18 coral reef hot spots (shown in red below). These sites cover 35 per cent of the world's total coral reef area but are home to more than 60 per cent of rare and localized reef species, so they are a high conservation priority. Richest of all are the reefs of southern Japan, followed by those in western Australia and the Gulf of Guinea, West Africa. The famous Great Barrier Reef, off eastern Australia, came fourth.

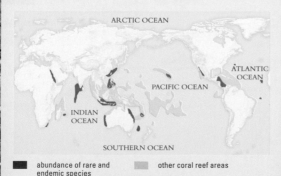

ARCTIC OCEAN
ATLANTIC OCEAN
PACIFIC OCEAN
INDIAN OCEAN
SOUTHERN OCEAN

■ abundance of rare and endemic species   ■ other coral reef areas

TYPES OF HOT SPOT

HIDDEN HOT SPOT

LOCH CARRON The northwest Highlands of Scotland may be scenic, but very little biodiversity is found in the harsh, rocky landscape that surrounds Loch Carron.

BENEATH THE SURFACE Underwater, however, Loch Carron is as full of life as any tropical coral reef. Animals include soft corals, Dahlia Anemones, and brittlestars.

CARIBBEAN TREASURE CHEST

SABA BANK During nine dives in 2006, a single diver found 50 species of sponge in Saba Bank off the Netherlands Antilles in the Caribbean.

NEW TO SCIENCE The Saba Bank study discovered a seven-spined goby living on the sea bed. It is a new species, and probably a new genus.

FISH HAVEN Soldierfish and snappers gather at a seamount in the Indian Ocean. The upwelling of nutrient-rich currents around seamounts makes them "oases" of the ocean.

SEAMOUNT COMMUNITIES

GUADALUPE SEAMOUNT Up to a third of species of seaweed, plants, and animals on isolated seamounts may be unique (or endemic) to that seamount, having evolved there over millions of years.

BURIED RICHES

SAMPLING SEDIMENTS One deep-sea sediment study found 798 species in 233 mud cores collected from a total area the size of an average room. Based on this, scientists estimate there could be up to 10 million species in the sea bed.

OCEAN LIFE

CYCLES OF LIFE AND ENERGY

ALL LIFE DEPENDS ON ORGANISMS that harness energy from either chemicals or the Sun to produce food. These organisms, whether phytoplankton, seaweeds, or bacteria, are called primary producers and form the first link of a food chain. This first link is just one point in a cycle that processes chemical energy and nutrients through the entire community of life in an ecosystem, into the physical environment, and back again.

ENERGY FLOW

As each organism in an ecosystem is eaten in turn by the next organism in the food chain, food energy flows from prey to consumer. The primary producers – the organisms such as diatoms and bacteria at the beginning of the food chain – are eaten by organisms called primary consumers, which are eaten by secondary consumers, and so on to top predators – animals not preyed upon by anything else. In land-based ecosystems, the total mass of organisms at each succeeding food-chain level decreases, leaving very few top predators. However, in marine ecosystems with phytoplankton as producers, the mass is greatest at the primary consumer level. This is possible because phytoplankton grow so rapidly that they provide great turnover despite having little mass.

FOOD-ENERGY PYRAMID
At each level of a food chain, energy is lost as heat, so less is available to the next consumer. The diminishing energy at each level can be represented by a pyramid (below) and accounts for the scarcity of top predators.

top predators
predators
consumers
primary producers

TOTAL ENERGY

BIOMASS PYRAMID
The biomass pyramid (below) for a system with plankton producers is partly inverted, because the producers have low total mass. Despite this, the rapid reproduction of the plankton keeps the food chain supplied.

top predators
predators
consumers
primary producers

TOTAL BIOMASS

RECYCLING

All living things need a supply of chemical nutrients, such as nitrates, phosphates, and silicates, to grow and reproduce. They are taken up by primary producers then passed along the food chain. Although some nutrients are available from seawater, most are derived ultimately from the sea floor. When an organism dies, any parts that are not eaten by other animals gradually sink to the sea floor, where they are broken down by bacteria and other decomposers. Faecal matter also ends up on the sea bed and is processed by detritus feeders or decomposers. Eventually, the nutrients are released into the environment in their mineral, non-living forms. They may then remain at depth, or they may be returned to surface waters by circulating water currents within an ocean basin (see upwelling, opposite).

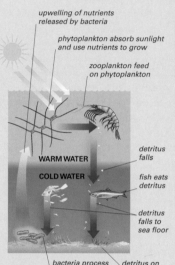

upwelling of nutrients released by bacteria

phytoplankton absorb sunlight and use nutrients to grow

zooplankton feed on phytoplankton

WARM WATER

COLD WATER

detritus falls

fish eats detritus

detritus falls to sea floor

bacteria process detritus

detritus on sea floor

NUTRIENT CYCLE
Small particles of organic matter, or detritus, are found in the water column. They may be eaten by scavengers or broken down still further by bacteria present in the water. However, many of them rain down on the ocean floor where they decompose, releasing nutrients. The nutrient cycle is completed by upwelling water currents that then carry the nutrients back to the surface where they can be utilized by the phytoplankton.

HUMAN IMPACT

OVER-HARVESTING

These fishermen are harvesting cod in the Pacific Ocean off Alaska. There has been a drastic decline in cod populations because the number being caught for human consumption far exceeds the reproductive rate of the fish. Despite the imposition of quotas by various governments in recent years, cod numbers have failed to recover.

COD FISHING IN ALASKA
Alaskan fishermen catch between 250,000 and 300,000 tons of cod annually, using various methods, including trawls and longlines.

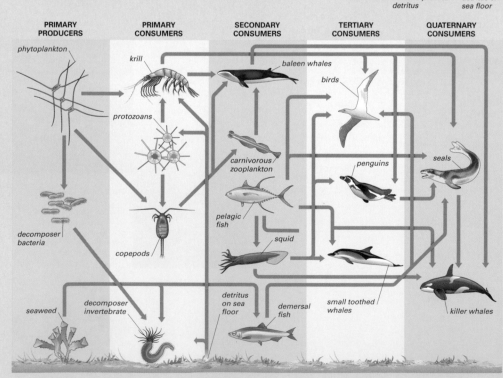

PRIMARY PRODUCERS	PRIMARY CONSUMERS	SECONDARY CONSUMERS	TERTIARY CONSUMERS	QUATERNARY CONSUMERS

phytoplankton
krill
baleen whales
birds
protozoans
carnivorous zooplankton
penguins
seals
pelagic fish
decomposer bacteria
copepods
squid
small toothed whales
killer whales
seaweed
decomposer invertebrate
detritus on sea floor
demersal fish

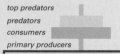

FOOD WEB
Many food chains have been combined to form this complex food web, extending from primary producers to quaternary consumers (top predators) for a Southern Ocean ecosystem. Each arrow shows the flow of food energy from prey to predator, grazer, or decomposer. It shows how organisms depend on one another for food. Some animals feed on organisms from several different levels of the food chain, adding to its complexity. Food webs are delicately balanced and easily upset by human interference.

PRODUCTIVITY

Throughout the world's oceans, the abundance of marine life varies dramatically. The ocean is more productive in some places and at some times than others. The amount of sunlight is a major influence on productivity and changes with latitude and time of year. The supply of nutrient-rich water from the sea floor and light for photosynthesis is affected by changing water movements and day length, affecting plankton levels. Temperature also affects productivity as it influences the rate of photosynthesis.

CLEAR, TROPICAL OCEAN
Tropical waters are not mixed seasonally, so few nutrients are returned to the surface, and little plankton growth is possible. Here, a solitary turtle cruises in crystal-clear surface waters near Hawaii.

RICH, MURKY TEMPERATE SEA
In coastal and temperate areas, water turbulence circulates nutrient-rich water that supports a variety of algae, such as this kelp forest in the Mediterranean.

UPWELLING

The open-ocean surface water can become impoverished, as nutrients are constantly absorbed by phytoplankton and fall with detritus to the sea floor. Nutrient-rich water can be restored to the surface on a large scale by vertical ocean currents in a process called upwelling (see p.60).

Near land, coastal upwelling is caused by surface currents, such as the Humboldt Current off South America (see p.58). In the equatorial waters of the Pacific and Atlantic, mid-ocean upwelling occurs when water masses are driven north and south by the trade winds, and cooler, nutrient-rich water rises to take their place. Polar upwelling can happen where winter storms cause intense water movement. When upwelling occurs and there is sufficient sunlight, phytoplankton multiply rapidly to support a vast number of organisms, creating the most productive ocean waters in the world.

NUTRIENT-RICH WATERS
Where there is upwelling, large numbers of small fish gather to feed on the plankton. They, in turn, attract larger predators like these Copper Sharks feeding on sardines off the coast of South Africa.

SWIMMING AND DRIFTING

MOST OF THE OCEAN'S LIVING SPACE IS NOT ON THE SEA BED but in the water column and out in the open ocean – areas known as the pelagic zone. Salt water provides support, as well as the nutrients that allow many plants and animals to live in the water column without ever going near the sea bed. Some animals live at the interface between ocean and air, or alternate between both environments, because it is more energy-efficient. The water surface, water column, and sea bed are all interconnected, and many animals move between these habitats.

PLANKTON

The sunlit, surface layers of the ocean are home to a huge array of tiny plants and animals (plankton) that drift with the water currents. Phytoplankton consist of plants or plant-like protists (see p.234) that can photosynthesize and make their own food. Along with fixed seaweeds, phytoplankton forms the basis of ocean food webs.

Zooplankton consists of animals, most of which are very small and feed on the phytoplankton. However, jellyfish can grow to a huge size. Many deep-sea forms have strange shapes and soft bodies that are very delicate. Some zooplankton, such as arrow worms, comb jellies, and copepods, live permanently in the plankton, hunting and grazing (holoplankton), while others are simply the larval and dispersal stages of animals, including crabs, worms, and cnidarians (meroplankton) that will spend part or all of their adult lives on the sea bed.

Many planktonic organisms have elegant spines, long legs, or feathery appendages that help them float. Tropical zooplankton generally have more of these than their temperate or polar equivalents because warm water tends to be less dense and viscous, and so provides less support.

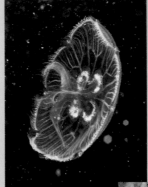

TEMPORARY PLANKTON
Most temporary zooplankton are the larvae of animals that, as adults, live on the sea bed. The Common Jellyfish, however, has a planktonic adult stage (shown above), and a fixed, asexual, juvenile stage (right).

PLANKTONIC LARVA
The eggs of the Common Shore Crab hatch into floating, spiny zoea larva.

NEKTON

Fish and most other free-living marine animals can all swim, even if only for short distances, over the sea bed. However, some animals spend their whole lives swimming in the open ocean and are collectively called nekton. This group includes many fish and all whales, dolphins, and other marine mammals, turtles, sea snakes, and cephalopods. There are also some representatives from other groups such as swimming crabs and shrimp. Most nektonic animals are streamlined, and there is a remarkable similarity in shape between some dolphins and open-ocean nektonic fish such as tuna.

TYPICAL NEKTON FEATURE
Most nektonic animals, including Dusky Dolphins, are vertebrates, (having either bones or cartilage).

THE OCEAN–AIR INTERFACE

Some animals live at the interface between air and water, either floating at the surface or alternating between the two environments. Oceanic birds such as albatrosses, petrels, gannets, and tropic birds spend their whole lives out at sea. They eat, sleep, preen, and even mate on the ocean surface. Large rafts of such sea birds are particularly vulnerable to oil spillages. Other diving seabirds, such as terns and puffins, alternate between hunting at sea and resting on land. Just as these birds plunge down into the water to catch fish, so some sharks lunge out of the water to catch birds and turtles. Flying fish leap into the air to escape their predators. Some planktonic animals live permanently at the water surface with part of their body projecting into the air. The By-the-Wind Sailor is a small, colonial cnidarian that is supported by a gas-filled float and transported by wind blowing against its vertical sail. Drifting with it on a raft of mucous bubbles is the Violet Seasnail, which also feeds on it. There are even surface-living insects, of the genus *Halobates*, that drift the oceans.

DRIFTING AT THE INTERFACE
The large gas-filled float of the Portuguese Man-of-War supports the whole colony at the water surface.

FLYING AND DIVING
The Brown Pelican is one of several species that dive or dip down from the air into the water to catch fish. It uses its capacious beak as a scoop.

FLOATING COMMUNITY
Ocean Sunfish often drift at the ocean's surface and will investigate any floating objects for potential food, such as jellyfish and planktonic crustaceans.

HUMAN IMPACT

LURING FISH

Fish-attracting devices (FADs) have increased catches in many areas by making fish stocks easier to exploit. However, they make no contribution to biological productivity as they simply gather fish together, and so may contribute to overexploitation. Artificial reefs also attract fish, but provide safe breeding sites, too.

FISH-ATTRACTING DEVICE
Even simple FADs, such as this floating buoy in Hawaii, will attract fish. Juvenile jacks and endemic Hawaiian damselfish can be seen sheltering under this one.

DRIFTING HOMES

Many pelagic fish species are attracted to floating objects that provide shelter from predators, currents, and even sunlight. Floating logs and seaweed also provide a meeting point. Fishermen have exploited this tendency by using Fish Attracting Devices (FADs, see panel, right) to concentrate fish in one area. These vary from simple rafts with hanging coconut palm leaves to complex technological devices.

Mini-ecosystems often develop on and around large drifting logs. Seaweeds and goose barnacles settle, providing shelter and food for crabs, worms, and fish. Shipworms bore into the wood, and their tunnels provide further refuge. Occasionally reptiles, insects, and plant seeds survive and drift on logs, and may eventually be washed ashore to colonize new places, including new volcanic islands.

SARGASSO HAVEN
Floating *Sargassum* seaweed provides a safe haven for the Sargassumfish. More than 50 animal species have been recorded in this habitat.

BOTTOM-LIVING

ANIMALS LIVING ON THE OCEAN FLOOR or within its sand and mud, either moving over it or firmly attached, are called benthic animals. On land, plants provide a structural habitat within which animals live. In the ocean, this is rarely the case, except in shallow, sunlit areas dominated by kelp, seaweeds, or seagrasses. Instead, wherever areas of hard sea bed provide a stable foundation, a growth of benthic animals develops, fixed to the sea bed and often resembling plants. A sea bed of shifting sediments is no place for fixed animals. Here, a community of burrowers develops instead.

FIXED ANIMALS

Many benthic animals such as sponges, sea squirts, corals, and hydroids spend their entire adult lives fixed to the sea bed, unable to move around. On land, animals must move about in search of food, whether they are grazers, predators, or scavengers. In the ocean, water currents carry an abundant supply of food in the form of plankton and floating dead organic matter. Fixed animals can take advantage of this by simply catching, trapping, or filtering their food directly from the water, without having to move from place to place. When it is time to reproduce, they simply shed eggs and sperm into the water, where the eggs are fertilized and grow into planktonic larvae. Sometimes, they retain their larvae or eggs, and release them only when the young are well developed. Water currents distribute the offspring to new areas, where they can settle and grow.

REEF-FORMING TUBE WORM
In some Scottish sea lochs, the chalky cases of tube worms form substantial reefs.

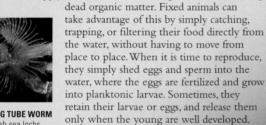

SEA BED IN THE SUN
Seaweeds anchor in the tidal zone of rocky shores and on rocky reefs, such as this one in the Canary Islands. On sunlit, temperate sea beds, it is seaweeds that provide the community structure.

BENEATH THE SEAWEED
Below the seaweed-dominated zone around northern European coasts, on sea beds too deep and dark for photosynthesis, Dead Man's Fingers, sponges, and tube worms typically grow attached to subtidal rocks.

MOBILE ANIMALS

Dense growths of seaweeds or fixed animals provide shelter and food for many mobile animals. Grazers, such as sea urchins, crawl through the undergrowth, eating both seaweeds and fixed animals. Meanwhile, crabs, lobsters, and starfish scramble and swim around, hunting and scavenging for food. Sea slugs are specialist predators, each species feeding on one, or a few, types of bryozoans, hydroids, or sponges. Sea slugs therefore live in close association with their prey and rarely stray far. Kelp holdfasts provide a safe haven for small, mobile animals such as worms.

FISH IN DISGUISE
Scorpionfish live on the sea bed among the seaweeds and fixed animals. Their intricate skin-flaps blend in with this habitat.

BURROWING AND BORING

Much of the sea floor is covered in soft sediments, such as sand and mud. Living on the surface of the sediment is both difficult and dangerous, and most animals burrow below or build tubes in which to live and hide. Bivalves and segmented worms cope especially well in this habitat, and many different species can be found in sediments all over the world. Safe under the sediment surface, a bivalve draws in oxygen-rich water and plankton through one of its two long siphons, expelling waste through the other. It never has to come out to feed or breathe. Piddocks and shipworms bore into rocks and wood, then use their siphons in a similar way. They are not completely safe here – predatory Moon Snails dig through the sediment and bore into bivalve shells, eating the contents. Ragworms are also active predators, hunting through the sediment for other worms and crustaceans. Some worms build flexible tubes from sand grains, their own secretions, or both. The tubes stick out of the sand, and they feed by extending feathery or sticky tentacles from the tube to catch plankton. If danger threatens, they can withdraw rapidly. A similar strategy is adopted by tube anemones and sea pens.

REPLACING SIPHONS
The siphon tops of buried bivalve molluscs are sometimes nipped off by flatfish but can regrow.

BORING INTO ROCK
The Boring Sponge uses chemicals to dissolve tunnels in calcareous shells and rocks, creating a living space for itself.

FIXED TO THE BOTTOM
Christmas Tree Worms live fixed to the bottom in hard tubes that they cement into coral reefs. They feed by filtering plankton from the water, using their beautiful double spiral of tentacles.

SYMBIOSIS

Bottom-living is a challenge for marine organisms. A safe crevice on a coral reef, for instance, is valuable, but fiercely fought over. The solution to finding a home is often to enter an intimate relationship with a different organism – a situation called symbiosis. When only one partner benefits, the relationship is called commensal, and often involves one animal providing a home for the other. Small Pea Crabs live inside mussels, gaining shelter and food, while the mussel merely tolerates their presence. Symbiosis in which both partners benefit is called mutualism. Many tropical gobies live in such relationships with blind or nearly-blind prawns. The prawn digs and maintains a sandy burrow that accommodates both, while its keen-eyed partner goby acts as a lookout. Some anemones adhere to the shells of hermit crabs, gaining from the crab's mobility and access to its food scraps. The crab is protected, in return, by the anemone's stinging tentacles. The third type of symbiosis is parasitism, in which one partner, the host, is harmed. The crustacean *Sacculina* spreads fungus-like strands through its host crab's body to extract nutrients, weakening or killing the crab.

MUTUAL RELATIONSHIP
The Banded Coral Shrimp earns its place in the moray eel's well-defended crevice by cleaning the teeth of its host.

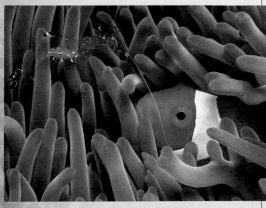

A HOME IN EXCHANGE FOR CLEANING
Large reef anemones often provide a haven for clownfish and tiny cleaner shrimps. The anemone benefits from the house-keeping activities of its guests.

OCEAN LIFE

ZONES OF OCEAN LIFE

NO PART OF THE OCEAN IS DEVOID of organisms, from polar seas to the tropics and from coasts and the seashore to the deepest depths. The sea bed and the water column above it both support a huge variety of life. However, marine organisms are distributed unevenly both horizontally and vertically. As on land, climate (mainly temperature) and food play a large part in determining distributions and biodiversity. In the harsh environment at the poles, there is less coastal life than in the warm tropics, but beneath the surface, Antarctic seas support rich marine communities. Although there is life at every depth, most creatures can only survive within particular depth zones at particular pressures, temperatures, and light regimes.

GEOGRAPHICAL ZONES

Seawater temperatures are much more stable than those on land because water loses and gains heat more slowly than does air. However, the distribution of marine coastal and continental shelf communities still follows a global pattern, with distinct polar, temperate, and tropical ecosystems. Coastal salt marsh in temperate parts is replaced in the tropics by mangroves. Kelp forests only grow in cool waters but extend into the tropics in places where cold water upwells from the deep, such as off the coast of Oman on the Arabian peninsula. Planktonic species and bottom-living species with planktonic larvae might be expected to occur anywhere that ocean currents take them. However, a boundary between water masses with different physical characteristics may present as effective a barrier in an ocean as mountains do on land. Below a certain depth, there are fewer such barriers, and conditions are stable and similar worldwide, so deep-sea animals often have very wide distributions.

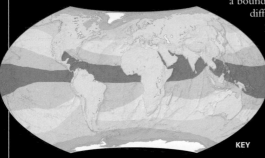

CLIMATIC ZONES
The shape and tilt of our planet results in differences in the amount of solar radiation reaching land and ocean at different latitudes. This produces large-scale climatic zones that ring the Earth.

KEY

■ equatorial	■ temperate
■ tropical	■ subpolar
■ subtropical	■ polar

ENDEMIC SPECIES

Some marine organisms, especially pelagic species, have a wide global distribution, since there are few barriers to their dispersal. Others live in restricted geographical ranges and are said to be endemic to a particular sea, island, or country. The most remote patches of habitat, such as small oceanic islands, tend to have the most endemic species. This is because animals in their dispersive stages, such as eggs and larvae, may survive only for short periods and so never reach distant shores. The Red Sea holds many endemic fish species. It is connected to the Indian Ocean only by a narrow channel and so is effectively isolated. Endemic fish are often those that cannot or do not swim far. Anemonefish for example, lay their eggs on rocks under their anemones and the young search for new anemones on the same reef. Flightless marine birds such as penguins are likewise restricted in their ability to colonize new areas.

GALAPAGOS PENGUIN
This penguin species lives only around the Galapagos islands. The cold, upwelling Cromwell Current keeps them cool in spite of the tropical climate. They are isolated by the surrounding warm waters, so cannot disperse beyond their home islands.

MALDIVES ANEMONEFISH
This endemic fish is not a strong swimmer. It does not have planktonic larvae and lives only in the Maldives and Sri Lanka in the Indian Ocean. Its host anemone has a wider distribution, because its larvae disperse on ocean currents.

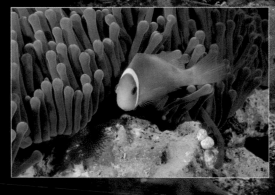

HUMAN IMPACT

SHIFTING ZONES

The northern and southern geographical limits of many shallow-water marine species are dictated by water temperature. Most species breed and disperse only within certain temperature limits. Climate change is slowly raising water temperatures and in the northern hemisphere, records have shown that some warm-water species are extending their ranges further north. Similarly, some cold-water species may be expected to retreat further north.

TROPICAL INVADER
Warm- water Triggerfish occur as far north as southern Britain but do not breed there. Continued ocean warming may change this.

DEPTH ZONES

As depth increases so does pressure, while light, temperature, and food supply decrease. These changes impose limits on the types of marine organisms that can survive and prosper at different depths. The areas on and over continental shelves around the world are rich in life as they are well-supplied with nutrients from river discharge and stirred-up sediments. Shoaling fish, such as herring, feed on plankton sustained by the nutrients. Most commercial fisheries are over continental shelves. Below the continental shelf, no phytoplankton or seaweeds grow. Pelagic animals either eat each other or make daily feeding migrations into the upper layers. Rocky areas support a diverse fauna including coldwater coral reefs, sponge reefs, and hydrothermal vent communities. Fine sediments cover the immense, flat abyssal plains at the foot of the continental slope. While microorganisms abound, large animals are relatively scarce.

seaweeds

0m

sponge

50m
(160ft)

starfish

100m
(330ft)

150m
(490ft)

200m
(660ft)

SEA BED IN THE SUNLIT ZONE

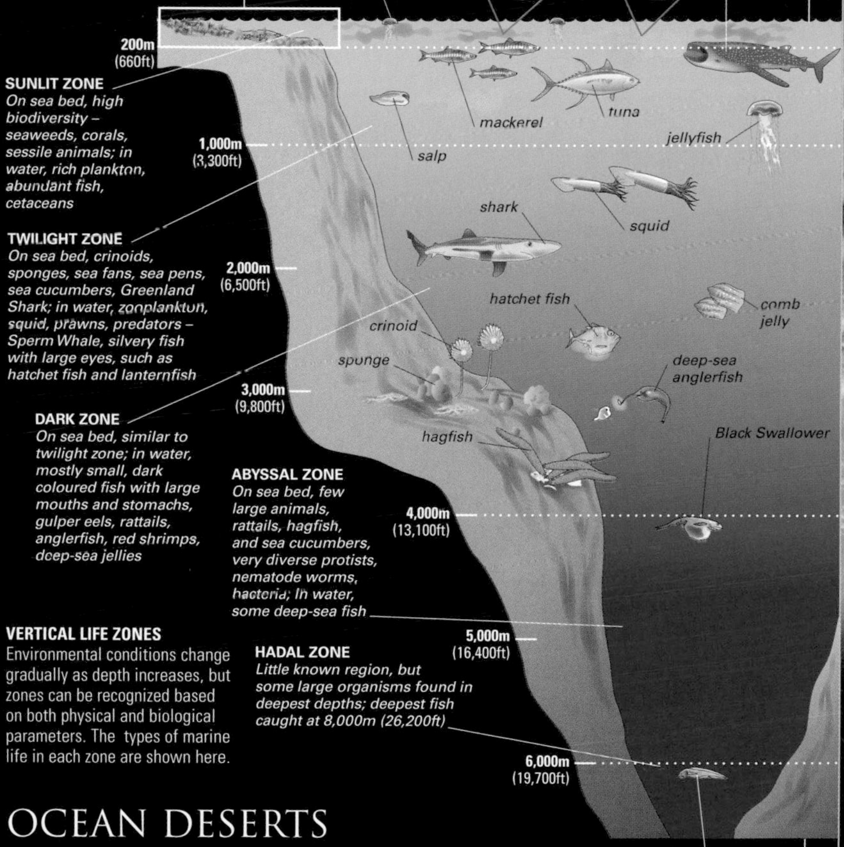

phytoplankton   zooplankton

Portuguese
Man-of-War

Whale
Shark

200m
(660ft)

SUNLIT ZONE
On sea bed, high
biodiversity –
seaweeds, corals,
sessile animals; in
water, rich plankton,
abundant fish,
cetaceans

1,000m
(3,300ft)

mackerel   tuna

jellyfish

salp

TWILIGHT ZONE
On sea bed, crinoids,
sponges, sea fans, sea pens,
sea cucumbers, Greenland
Shark; in water, zooplankton,
squid, prawns, predators –
Sperm Whale, silvery fish
with large eyes, such as
hatchet fish and lanternfish

2,000m
(6,500ft)

shark

squid

hatchet fish

comb
jelly

crinoid

sponge

deep-sea
anglerfish

3,000m
(9,800ft)

DARK ZONE
On sea bed, similar to
twilight zone; in water,
mostly small, dark
coloured fish with large
mouths and stomachs,
gulper eels, rattails,
anglerfish, red shrimps,
deep-sea jellies

hagfish

Black Swallower

ABYSSAL ZONE
On sea bed, few
large animals,
rattails, hagfish,
and sea cucumbers,
very diverse protists,
nematode worms,
bacteria; In water,
some deep-sea fish

4,000m
(13,100ft)

VERTICAL LIFE ZONES
Environmental conditions change
gradually as depth increases, but
zones can be recognized based
on both physical and biological
parameters. The types of marine
life in each zone are shown here.

5,000m
(16,400ft)

HADAL ZONE
Little known region, but
some large organisms found in
deepest depths; deepest fish
caught at 8,000m (26,200ft)

6,000m
(19,700ft)

OASIS BENEATH THE ICE
Seal breathing holes in sea ice create
oases on the sea bed beneath, where
benthic organisms enjoy the benefits
of a greater supply of light and nutrients.

cusk eel

OCEAN DESERTS

Some areas of ocean are similar to deserts on land and support few species. Clear, blue surface water over the deep oceans often supports only small amounts of plankton, because it is very poor in the nutrients and minerals needed by phytoplankton to grow. This is especially the case in areas where there are few storms to stir the water and bring nutrients up from deep water. The nutrient iron can be a limiting factor, and experiments in which areas were seeded with iron have shown greatly increased phytoplankton production. The ocean floor in abyssal depths can support only a few large animals and was once considered to be a virtual desert. However, recent work on deep-sea sediments has shown the opposite. If all the bacteria and tiny animals living between the sediment particles are counted, then this habitat is as diverse as a tropical rainforest.

BARREN POLAR SHORE
This polar shore in Greenland
supports little life due to the
grinding action of winter ice, though
below the reach of the ice, rich
communities may develop.

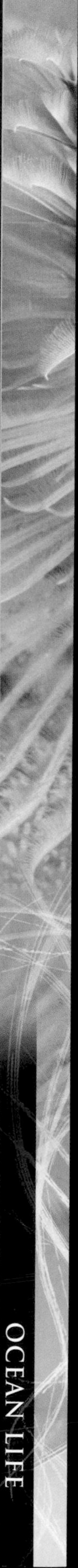

OCEAN LIFE

OCEAN MIGRATIONS

FEEDING AND BREEDING ARE THE MAIN REASONS that animals migrate. They move from one place to another, often at the same time of day or year, and usually follow the same, well-defined routes. Migratory species include many of the larger marine animals, such as whales and turtles, but smaller creatures, such as squid and plankton, also make spectacular journeys in order to survive and reproduce. Animal migration in the oceans is more complicated than on land because animals can move both horizontally and vertically through the water column.

TYPES OF MIGRATION

The driving force behind any animal migration is survival. Individuals must eat to live, and some will travel long distances to find food. Such journeys often coincide with peak production times of plankton and other food sources in particular places, such as sites of seasonal upwellings. A shorter, more regular feeding migration is made daily by plankton and active swimmers such as squid (see p.221).

A species' survival depends on reproductive success. Gathering together and breeding in a few places at the same time optimizes conditions for offspring survival. Breeding grounds where food is abundant and conditions favourable are used repeatedly, with individuals often returning to their birthplace to breed. Some shore animals also migrate up and down the beach, following the outgoing tide to feed and avoiding immersion by returning before the tide turns.

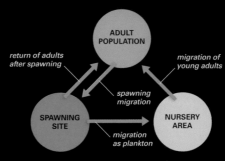

MARINE MIGRATORY CYCLE
Some marine organisms migrate to a specific spawning site to release their eggs. The eggs hatch into larvae that join the plankton and drift in the currents to another nursery area, where they feed and mature before joining the adult population.

DISCOVERY

TRACKING

Until recently, any journey made by a marine animal such as the Leatherback Turtle (shown below), was poorly understood because tracking devices used on land were inappropriate for use in water. This changed when satellite-tracking devices became available. Attaching one to a turtle does not impede or harm it in any way but it can still pose problems. Yet turtles are threatened in the wild, so knowing where a female goes after laying her eggs is vital to conservation work.

TRACKING TURTLES
A tracking device is being attached to this Leatherback Turtle on Juno Beach, Florida, USA. In case an opportunity does not arise to remove it manually, parts of the harness are designed to gradually disintegrate.

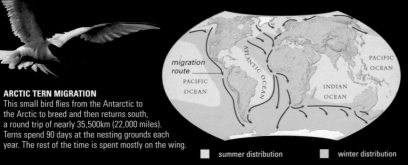

ARCTIC TERN MIGRATION
This small bird flies from the Antarctic to the Arctic to breed and then returns south, a round trip of nearly 35,500km (22,000 miles). Terns spend 90 days at the nesting grounds each year. The rest of the time is spent mostly on the wing.

■ summer distribution   ■ winter distribution

LOBSTER MIGRATION
Caribbean Spiny Lobsters migrate in single file across the sea floor in winter, seeking warmer water, and return to shallow water in summer.

MIGRATING BETWEEN SALT AND FRESH WATER

Although some marine species can cope with a great range of salinity and temperature, only a few move between fresh and salt water at particular stages of their lives. Some, such as salmon, start and finish their lives in fresh water and spend the rest of their time in the ocean. Such fish are described as anadromous. Eels, on the other hand, start and finish their lives in the sea, but spend 10 to 14 years in fresh water while maturing. Fish such as these are termed catadromous.

At maturity, both of these fish return to their birthplace to breed, after which they die. Changing from fresh to salt water or vice versa would be fatal to most fish but various physiological adaptations, including the way their kidneys function, allow both anadromous and catadromous fish to make the transition without experiencing any ill effect.

SALMON RETURNS TO FRESHWATER SPAWNING GROUNDS

1 At three to five years of age, a Coho Salmon is ready to return to the river where it was born to spawn. Some mature at only two years, returning to their home river as "jacks".

2 In migrating upstream to its spawning grounds, the salmon may swim 3,500km (2,175 miles) against the water current, negotiating several waterfalls and rapids.

3 As soon as the female has deposited her eggs in the gravel on the river bed, the male swims over them and releases his sperm, optimizing the chance of fertilization.

4 Newly hatched salmon live among the gravel until they absorb their yolk sacs and become fry. They then begin their journey downstream to the sea.

OCEAN LIFE

HORIZONTAL AND VERTICAL TRAVEL
Some animals migrate horizontally and
vertically. Here, Longfin Squid migrate
to spawn in May. They also move up and
down the water column each day to feed.

NAVIGATION

While satellite tracking provides
information on migration routes
and confirms that many
individuals travel the same path,
how animals navigate over vast
distances is still poorly understood.

Salmon return to their spawning
grounds by smelling the unique
chemical composition of the
water, but they first have to get
close enough to pick up this scent.
Some aquatic species may use
water currents to guide them,
while others are said to use the
Earth's magnetic field to navigate.

Animal migration is amazingly
accurate, both in terms of
direction and timing. Oceanic
birds, turtles, and mammals can
also navigate using the Sun, the
stars, and familiar landmarks.

BELUGA WHALES MIGRATING
Belugas, like these in Lancaster Sound,
Canada, live in the Arctic and subarctic, but
some migrate to warmer waters in summer.

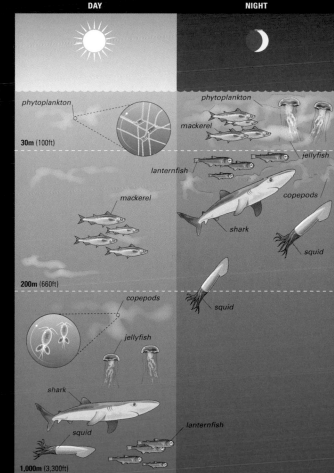

DAY — NIGHT

phytoplankton
30m (100ft)
phytoplankton
mackerel
jellyfish
lanternfish
mackerel
copepods
shark
squid
200m (660ft)
copepods
squid
jellyfish
shark
lanternfish
squid
1,000m (3,300ft)

VERTICAL MIGRATIONS

In temperate and tropical regions,
zooplankton migrates to the ocean
surface at night and then moves
down again during the hours of
daylight. In a single day, this vertical
movement may range from 400–
1,000m (1,310–3,300ft), depending
on the size and type of animal
involved. In polar regions, where
darkness lasts for several months,
zooplankton migrates up and
down on a seasonal basis, being at
the surface during summer and at
depth in winter.

It is thought that zooplankton
rises to feed on the phytoplankton
that lives in the surface waters, but
then retreats to depth for safety, or
possibly because it expends less
energy in cooler water. Maturing
planktonic larvae of animals such
as crabs will eventually migrate to
the sea floor and become benthic.

EARTH'S GREATEST MASS MIGRATION
During the day, when many animals remain
at depth out of sight of predators, the
phytoplankton utilizes the Sun's energy to
produce food. At night, the biomass of the
surface waters (the sunlit zone) increases by
as much as 30 per cent, as zooplankton come
up to feed on phytoplankton and are, in turn,
eaten by various fish and other animals. This
regular movement of animals up and down
the water column is the greatest mass
migration on Earth.

LIVING AT DEPTH

THE DEEP-SEA ENVIRONMENT APPEARS INHOSPITABLE – cold, dark, and with little food. However, it is remarkably stable: temperatures remain between 2 and 4°C (35 and 39°F) all year round, salinity is constant, and the perpetual darkness is overcome by novel communication methods (see pp.228–29). Although deep-sea pressures are immense, most marine animals are unaffected, since they have no air spaces, while animals living below about 1,500m (5,000ft), show subtle adaptations. Species diversity of large animals decreases with depth, but there is a huge diversity of small organisms living within deep-sea sediments.

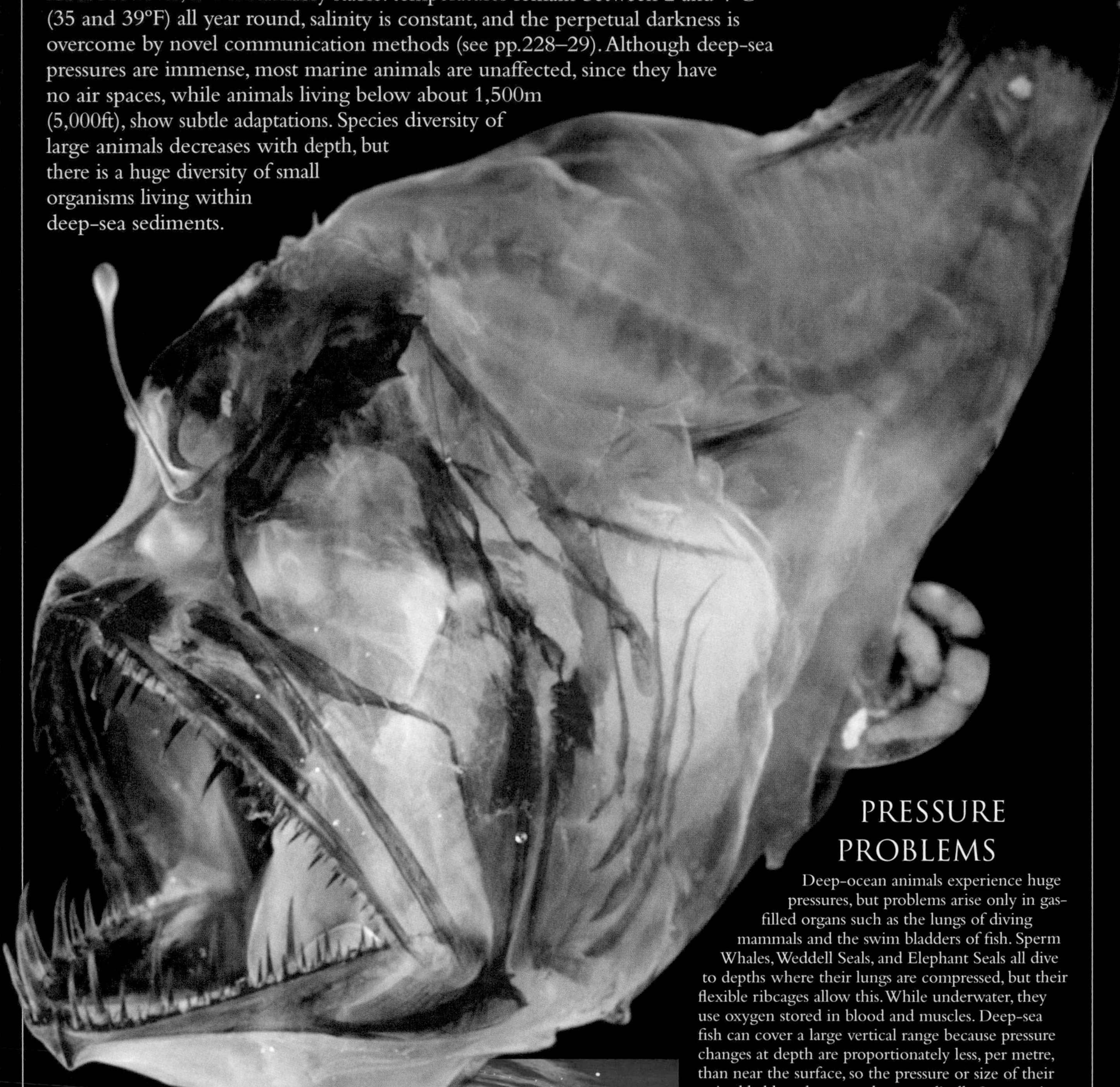

PRESSURE PROBLEMS

Deep-ocean animals experience huge pressures, but problems arise only in gas-filled organs such as the lungs of diving mammals and the swim bladders of fish. Sperm Whales, Weddell Seals, and Elephant Seals all dive to depths where their lungs are compressed, but their flexible ribcages allow this. While underwater, they use oxygen stored in blood and muscles. Deep-sea fish can cover a large vertical range because pressure changes at depth are proportionally less, per metre, than near the surface, so the pressure or size of their swim bladders does not change radically. In oceanic trenches, the pressure is so great that it affects the operation of biological molecules, such as proteins.

Pressure-loving bacteria in this habitat have specialized proteins – they cannot grow or reproduce when brought to the surface.

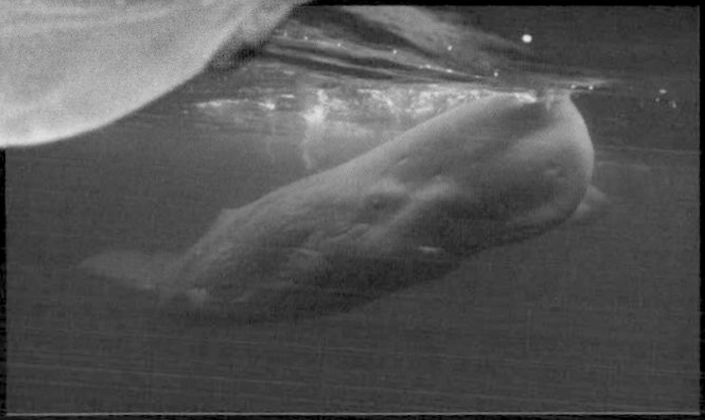

DEEP-SEA ADAPTATIONS
Anglerfish have a lightweight skeleton and muscles for neutral buoyancy. This specimen's muscles have been "cleared" to show the bone, which is stained red.

SPERM WHALE
Sperm Whales can dive to at least 1,000m (3,300ft), where the pressure is 100 times greater than at the surface.

FINDING FOOD

The major problem of living at depth is finding enough food. With the exception of communities based around hydrothermal vents and cold seeps (see pp.188–89), animals living in the deep ocean and on the deep-ocean floor are ultimately reliant on food production in the sunlit layer, thousands of metres above. In the depths, it is too dark for plant plankton to live and to provide food. Sometimes, large mammal or fish carcasses reach the sea bed, but most food arrives as tiny food fragments, slowly sinking from above. Much is eaten before it reaches the sea floor, but much is also added in the form of skins, shed from mid-water crustaceans and salps. Bacteria grow on such material, helping it to clump together and so fall more rapidly.

mouth surrounded by modified tube feet

tube foot, used to move across sea floor

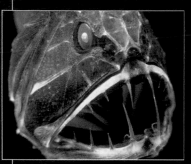

MIDWATER FEEDER
The Fangtooth lives at midwater depths of about 500–2,000m (1,600–6,500ft). Food is scarce, so its large mouth and sharp teeth help it to catch all available prey.

SEA-BED CONSUMER
Sea cucumbers vacuum up organic remains from the sea floor. At high latitudes, more food rains down in spring, following surface phytoplankton blooms; these rains may trigger sea cucumbers to reproduce.

OBSERVING DEEP-SEA LIFE

Before the advent of modern research submersibles, few biologists had the opportunity of seeing deep-sea animals alive and in the wild. Dredged and netted specimens are often damaged, and little can be learned from them about the animal's way of life. Modern submersibles have an excellent field of view, are equipped with sophisticated cameras and collecting equipment, and can operate to depths of 1,000m (3,300ft) or even 6,000m (20,000ft).

A WINDOW ON DEEP-SEA LIFE
Deep Rover is a two-person submersible capable of diving to 1,000m (3,300ft). Launched from a semi submersible platform, the occupants can see all the way round through the acrylic hull.

SCAVENGING GIANTS

Many deep-sea animals are smaller than their relatives in shallow water. This is an evolutionary response to the difficulties of finding food in the deep ocean. However, some scavengers survive by growing much larger than their shallow-water counterparts. For example, amphipod and isopod crustaceans, which measure only 1cm (½in) or so long, are common in shallow water, where they scavenge on rotting seaweed and other debris. Carrion in the deep sea is sparse, but it comes in big, tough lumps such as whale carcasses. Some deep-sea amphipods grow to a length of 10–15cm (4–6in), more than ten times larger than shallow-water species, and so are able to tackle such a bonanza. In the low temperature of the deep ocean, these animals move and grow slowly and reproduce infrequently, but live much longer than their shallow-water counterparts. Sea urchins, hydroids, seapens, and other animals also have giant deep-sea forms. Similar giants are found in cold Antarctic waters.

DEEP-SEA GIANT
The widespread deep-sea scavenger amphipod *Eurythenes* grows to over 8cm (3in).

STAYING ALOFT

Huge areas of the deep-sea floor are covered in soft sediments many metres thick, called oozes (see p.181). Sea-bed animals need ways of staying above these sediments so that they can feed and breathe effectively. Many sedentary filter-feeding animals, such as sea lilies, sea pens, and some sponges, have long stalks, enabling them to keep their feeding structures above the sediment. Some sea cucumbers have developed stilt-like tube feet that help them to walk over the sediment surface, instead of having to plough through it. Similarly, the Tripodfish props itself up on its fin tips. One species of sea cucumber, *Paelopatides grisea*, has an unusually flattened shape that allows it to lift itself off the sea bed with slow undulations of its body.

SEA LILIES ANCHORED IN THE OOZE
To catch food, sea lilies reach up into the current on stalks up to 60cm (2ft) high. The stalk extends deep into the sediment to provide an anchor.

BIOLUMINESCENCE

BIOLUMINESCENCE IS A COLD LIGHT produced by living organisms. On land, only a few nocturnal animals, such as fireflies, produce light but, in the ocean, thousands of species do so. Deep-water fish and squid use bioluminescence extensively but there are many other light producers, such as species of bacteria, dinoflagellates, sea pens, jellyfish, molluscs, crustaceans, and echinoderms. Evidence suggests that marine organisms use bioluminescence for defence (as camouflage or distraction), for finding and luring prey, and for recognizing and signalling to potential mates.

USING LIGHT TO COMMUNICATE
Many bioluminescent marine organisms use their light in communication. This bristlemouth fish can signal to its own kind with its specific photophore pattern.

LIGHT PRODUCTION

Bioluminescence is produced by a chemical reaction in special cells known as photocytes, usually contained within light organs called photophores. A light-producing compound called luciferin is oxidized with the help of an enzyme called luciferase, releasing energy in the form of a cold light. Most bioluminescent light is blue-green, but some animals can produce green, yellow, or more rarely, red light.

A range of light-producing structures is found in different animals. The hydroid *Obelia* has single photocytes scattered in its tissues, while certain fish and squid have complex photophores with lenses and filters. Some animals, including flashlight and eyelight fish, some anglerfish, ponyfish, and some squid, adopt a different strategy. They culture symbiotic, bioluminescent bacteria in special organs. The bacteria produce their light and are, in return, fed nutrients by their host and given a safe place in which to live.

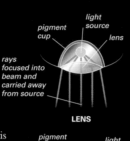

light source
pigment cup
lens
rays focused into beam and carried away from source

LENS

pigment cup
light source
light pipe

LIGHT PIPE

pigment cup
light source
deep-red pigment filter
filter allows only deep-red light to pass

COLOUR FILTER

PHOTOPHORE TYPES
Photophores often feature a pigment cup and a lens that directs the light into a parallel beam. With a light pipe, light can be channelled from the photophore, which might be buried in the animal's body. Colour filters in front of the light source fine-tune the colour of the emitted light.

HUNTING WITH A SPOTLIGHT
The Dragonfish produces a beam of red light, from a photophore beneath its eye, to spotlight its prey. Red light is invisible to most deep-sea animals.

LIGHT DISGUISE

Animals using bioluminescence to attract prey, or to signal to each other, risk alerting their own predators to their presence. However, lights can also be used for camouflage. Hatchetfish live at depths where some surface light is still dimly visible. To prevent their silhouette being seen from below, they manipulate the light they emit from photophores along their belly, to mimic the intensity and direction of the light coming from above. Bioluminescence is also used to confuse potential predators. Flashlight fish turn their cheek lights on and off. Some squid, shrimps, and worms eject luminous secretions or break off luminous body parts that act as decoys, while they escape.

MANIPULATING LIGHT
The silvery, vertical flanks of Hatchetfish reflect downwelling light, and their photophores shine downwards, camouflaging their silhouette from below.

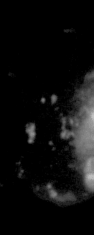

body covered with tiny, flashing photophores

organs producing downward-directed beams of light

LUMINOUS SMOKESCREEN
A Firefly Squid presents a predator with a myriad of confusing pin-prick lights emitted from its body. It can also secrete a cloud of luminous particles into the water to act as a smokescreen, allowing it to escape.

squid's ink is bioluminescent

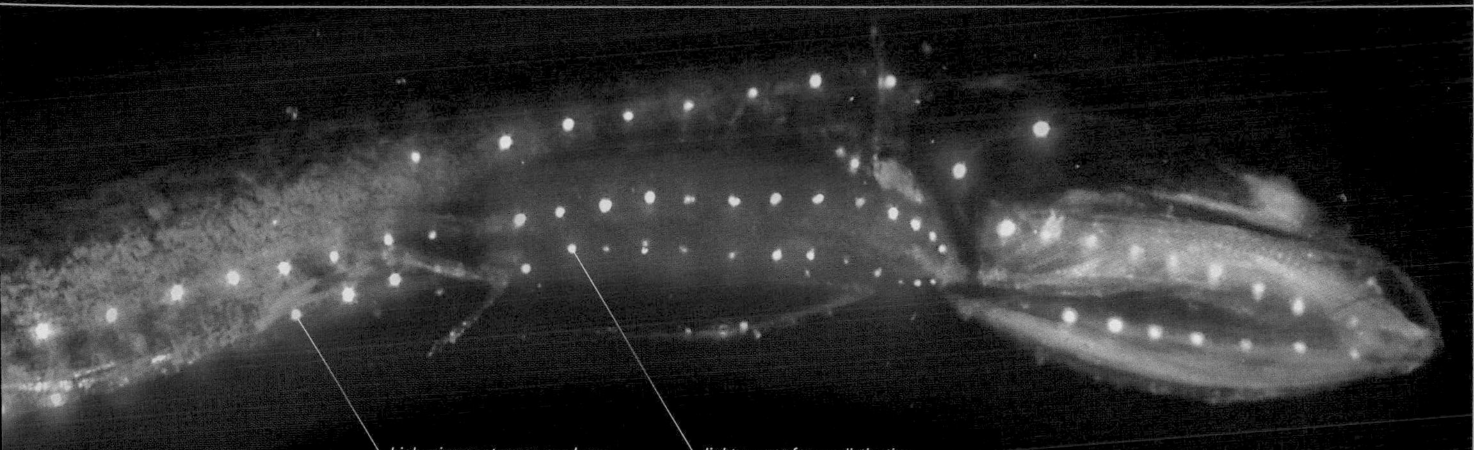

bioluminescent organ produces light and directs it downwards; the light merges with light downwelling from the sky and conceals the animal from predators below

light organs form a distinctive pattern recognized by other bristlemouths

PREDATORS

light-producing bacteria cause lure to glow

In the unlit regions of the deep ocean, many hunters try to attract prey, rather than go in search of it. After all, hunting by sight and chasing prey is difficult where the only available light is from bioluminescence. An obvious way of attracting prey is to use a luminous lure, and anglerfish are especially good at this. Anglerfish in the genus *Linophryne* have a head lure, like a fishing rod, lit by luminous bacteria, and a chin barbel with tiny photophores that produce their own light. Midwater fish often have thin skeletons and weak muscles to improve their buoyancy, so luring prey is an energy-efficient way for them to hunt. *Stauroteuthis syrtensis,* an unusual deep-sea octopus with glowing suckers, sets a deadly trap. Its eight tentacles are connected into a web, and its modified suckers, which have lost the ability to grasp, are bioluminescent. Although this species has never been seen hunting, its prey (which are primarily copepods) is probably lured towards the raised, light-emitting arms, and then enfolded and eaten.

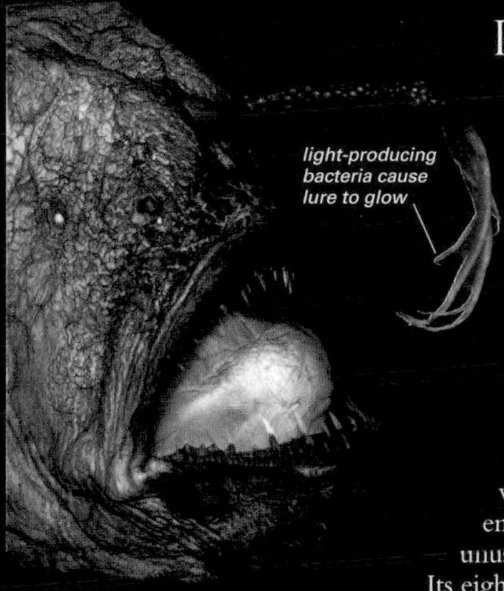

LUMINOUS LURE
Fish are attracted to the luminous lure of deep-sea anglerfish and are quickly snapped up. Most anglerfish are brown or black so that they do not light themselves up.

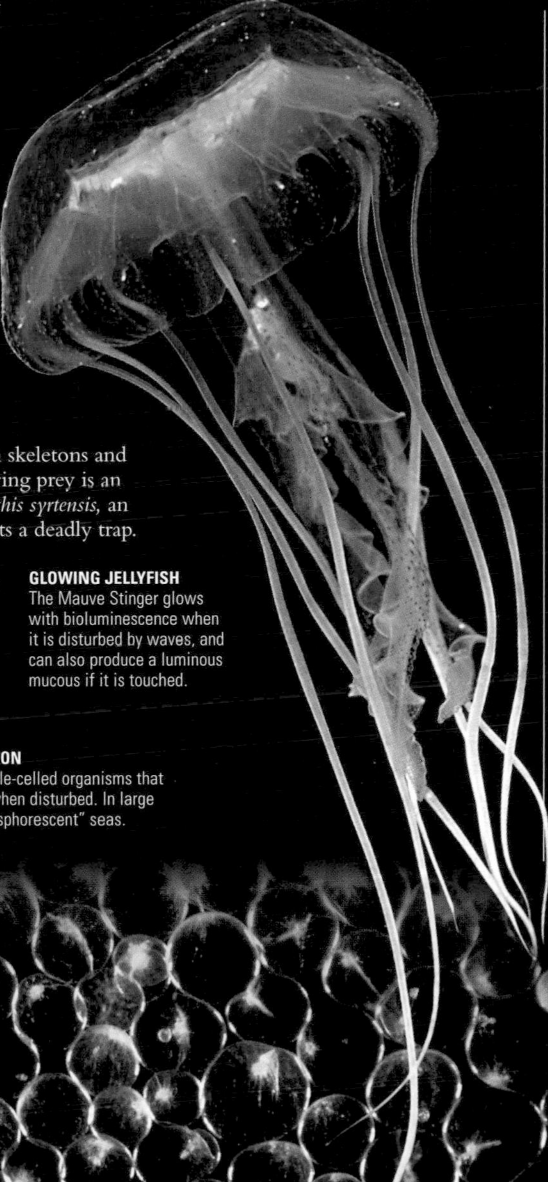

GLOWING JELLYFISH
The Mauve Stinger glows with bioluminescence when it is disturbed by waves, and can also produce a luminous mucous if it is touched.

PHOSPHORESCENCE

On a still, warm night, especially in the tropics, moving boats leave a glittering trail of light in their wake and divers can create swirling pinpricks of light by simply moving about. This phenomenon is caused by bioluminescent plankton, mostly dinoflagellates. Their light is often informally called phophorescence, because it is emitted when they are disturbed, but decays after a few seconds.

Biological phosphorescence is thought to be an anti-predation device. When dinoflagellates are attacked by planktonic copepods, they flash. This alerts nearby shrimps and fish to the copepods' presence, and the copepods themselves may then become prey. Some dinoflagellates, such as *Gonyaulax polyedra,* only produce light at night, so they do not waste energy on light production when it cannot be seen.

Deep-sea jellyfish may use a similar anti-predator strategy. The jellyfish light up only when disturbed by vibrations, which indicate an approaching predator. Often, a series of erratic flashes travels over the entire body surface. Such lights may serve to distract the predator.

BIOLUMINESCENT PLANKTON
Dinoflagellates are tiny, single-celled organisms that emit bright flashes of light when disturbed. In large numbers, they produce "phosphorescent" seas.

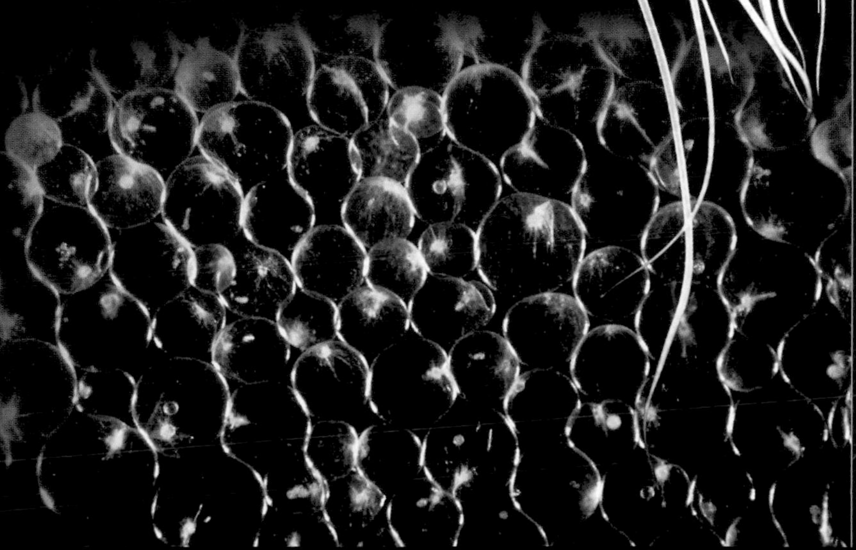

OCEAN LIFE

THE HISTORY OF OCEAN LIFE

LIFE HAS BEEN PRESENT IN THE OCEANS for over 3,500 million years. The great diversity of today's marine life represents only a minute proportion of all species that have ever lived. Evidence of early life is hard to find, but it is seen in a few ancient sedimentary rocks. The fossil record has many gaps, but it is the only record of what past life looked like. Fortunately, many marine organisms have shells, carapaces, or other hard body parts, such as bones and teeth. They are more likely to be preserved than entirely soft-bodied creatures, although in exceptional circumstances these have also been fossilized. Using fossils, and information from the sediments in which they are preserved, scientists can reconstruct the history of marine life.

PEOPLE

A.I. OPARIN

In 1924, Russian biochemist Aleksandr Oparin (1894–1980) theorized that life originated in the oceans. He suggested that simple substances in ancient seas harnessed sunlight to generate organic compounds found in cells. These compounds eventually evolved into a living cell.

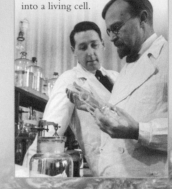

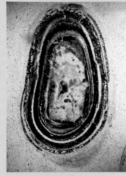

EARLY MICROFOSSILS
This micrograph of a section of chert (a form of silica) from the Gunflint Formation, Canada, includes 2,000 million-year-old microfossil remains. These microfossils contain the oldest and best-preserved fossil cells known.

3,800–2,200 MILLION YEARS AGO
THE ORIGIN OF LIFE

When the Earth formed, it was totally unsuitable for life. The atmosphere changed, however, and the oceans formed and cooled (see pp.44–45), so that by 3,800 million years ago, conditions allowed biochemical reactions to take place. It is thought that simple, water-soluble organic compounds called amino acids accumulated in the water, eventually forming chains and creating proteins. These combined with other organic compounds, including self-replicating DNA, to form the first living cells.

The Earth's atmosphere was further developed by mats of algae and cyanobacteria called stromatolites, whose fossil record stretches from over 3,500 million years ago to the present day. Stromatolites could perform photosynthesis, and their growth eventually flooded the atmosphere with oxygen. Cyanobacteria are single-celled organisms with DNA but no nucleus or complex cell organelles. It was not until 2,200 million years ago that cells with nuclei and complex organelles (eukaryote cells) appeared.

620–542 MYA PRECAMBRIAN LIFE

Ancient life, though soft-bodied, fossilizes under certain conditions, offering rare glimpses of early multicellular life. About 620 million years ago, a community of soft-bodied animals known as the Ediacaran fauna left their body impressions and trackways in a shallow sea bed. The sea bed now forms the sandstone of the Ediacaran Hills in Australia, where the fauna was discovered in the 1940s.

The ancient sea was inhabited by strange, multicellular animals. Some resembled worms and jellyfish, but others were thin, flat, and unfamiliar, making it difficult to know if they are related to existing animals or a separate, extinct, evolutionary line. These animals are the only link between the single-celled organisms that preceded them and the rapid diversification of life that followed. Ediacaran fauna are also found in Namibia, Sweden, Eastern Europe, Canada, and the UK.

EDIACARAN FOSSILS
These are typical examples of Ediacaran fossils preserved as impressions in rock. Mawsonite (left) is believed to be a complex animal burrow; Spriggina (below) may be an arthropod, or a new life-form.

550–530 MYA
CAMBRIAN EXPLOSION

Over 20 million years around the start of the Cambrian Period, many life-forms made a sudden appearance. Indeed, most of today's major animal groups (phyla) abruptly appear in the fossil record. The Cambrian Explosion of evolution may have been caused by the creation of new ecological niches as the coastline increased, due to the break-up of the Rodinia supercontinent. Further niches arose as a rise in sea level produced large expanses of warm, shallow water. The Cambrian seas were dominated by arthropods, chiefly trilobites, but there were also foraminiferans, sponges, corals, bivalves, and brachiopods. All readily fossilize, as they each have some sort of mineralized "skeleton".

FIRST REEFS
The Cambrian reefs were built by extinct sponges called archaeocyathids. They resembled tube sponges (above), having a similar shape and a calcareous skeleton.

ARTHROPOD TRAILBLAZERS
Trilobites evolved a multitude of different body forms and remained a ubiquitous arthropod group for the next 100 million years. They became extinct during the Permian Period.

BRACHIOPODS
Brachiopods may resemble bivalve molluscs, but they are unrelated life-forms and were among the first animals to appear in the Cambrian Period. Over 3,000 genera have been described. Only 300 species survive today.

LIVING MARINE STROMATOLITES
Built by Earth's oldest type of organism, stromatolites are now found in only a few places such as here, in the hyper-saline water of Hamelin Pool, Australia.

418–354 MYA THE AGE OF FISH

The earliest vertebrate fossils known are jawless fish that lived some 468 million years ago. Jawed fish appeared in the Silurian Period, following the development of massive coral and sponge reefs that provided them with a multitude of habitats in which to diversify. The now-extinct acanthodians, with their prominent spines on the leading edges of their fins, were among the earliest of these. Having hinged jaws allowed fish to feed more efficiently, and paired fins gave them the speed and manoeuvrability to hunt. The following Devonian Period (418–354 million years ago) saw an evolutionary radiation that could be called the "Age of Fish". Armoured fish called placoderms dominated Devonian seas, some reaching lengths of 6m (20ft). Ray-finned fish, sharks, and lobe-finned fish also appeared at this time and have survived to the present day, although marine lobe-finned fish are known only from the Coelacanth. Lobe-finned fish are important in the fossil record because one group gave rise to early tetrapods (limbed vertebrates).

EARLY JAWLESS FISH
Jawless fish first evolved in the ocean, later spreading into brackish and freshwater habitats. The bony head shield and dorsally situated eyes of this *Cephalaspis* suggest it is a bottom-dweller.

EVOLUTIONARY INSIGHT
This lobe-finned fish, *Tiktaalik roseae*, has gills and scales like a fish, but has tetrapod-like limbs and joints. This "missing link" helps to reveal how animals moved from the oceans onto land.

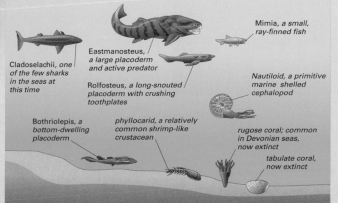

DEVONIAN COMMUNITY
The Devonian reef fauna (right) from Gogo, Australia, is typical of the time. It is dominated by a wide variety of armoured placoderms, but ray- and lobe-finned fish, and a shark, have also been found there.

Cladoselachii, one of the few sharks in the seas at this time

Eastmanosteus, a large placoderm and active predator

Mimia, a small, ray-finned fish

Rolfosteus, a long-snouted placoderm with crushing toothplates

Nautiloid, a primitive marine shelled cephalopod

Bothriolepis, a bottom-dwelling placoderm

phyllocarid, a relatively common shrimp-like crustacean

rugose coral; common in Devonian seas, now extinct

tabulate coral, now extinct

252–65 MYA GIANT MARINE REPTILES

During Triassic, Jurassic, and Cretaceous times, evolution of reptiles, similar to that of the dinosaurs on land, occurred in the oceans. Between 252 and 227 million years ago, three groups appeared – turtle-like placodonts, lizard-like nothosaurs, and dolphin-like ichthyosaurs. Of these, only ichthyosaurs survived until the Jurassic.

The Jurassic oceans teemed with life. Modern fish groups were well represented, as were ammonites, molluscs, squid, and modern corals. A variety of ichthyosaurs evolved, some giant forms reaching 9m (30ft) in length, but they soon died out and were replaced by modern sharks.

The gap left by the extinction of the placodonts and nothosaurs was filled by long-necked plesiosaurs. Those with a short body and tail and a small head lived in shallow water, while larger forms, called pliosaurs, probably lived in deep water. It is also likely that some of the flying reptiles, called pterosaurs, lived on coastal cliffs and survived by eating fish caught at the water's surface. During the Cretaceous Period, reptiles remained the largest marine carnivores, (plesiosaurs now co-existing with mosasaurs, distant relatives of monitor lizards,) but none survived the mass extinction that occurred 65 million years ago.

FOSSILIZED ICHTHYOSAUR
The dolphin-like features of this ichthyosaur are evident from its fossilized remains. The powerful tail was half-moon shaped, but here only the down-turned backbone is preserved.

dorsal vertebrae with attachment points for long ribs

rib

long neck comprising 30 vertebrae

PLESIOSAUR
Cryptoclidus eurymerus is a mid-Jurassic plesiosaur. It has a small head, long neck, and short tail, which is typical of shallow-water forms. Its sharp teeth indicate that it ate small fish or shrimp-like crustaceans.

pointed, interlocking teeth trap prey

plate-like, flattened collar bone lends support on land

large bones of pelvic girdle

paddle-like hind flipper

TIMELINE OF EARTH HISTORY

Million years ago		4,100 MYA	4,000 MYA	3,500 MYA	3,000 MYA	2,500 MYA	2,200 MYA
	CRYPTOZOIC AEON 4,500–542 MYA						

first organic molecules

first stromatolites

first microfossils

first eukaryotes and multicellular algae

50–14 MYA RETURN TO WATER

Following the mass extinction that saw the demise of marine reptiles, some mammals that had evolved on land began returning to the water. Around 50 million years ago, the oceans started to resemble modern oceans in terms of their geographical positions and fauna. The ancestors of whales, however, were unlike their modern counterparts. The earliest whale, *Pakicetus*, was probably a close relative of the hoofed mammals (ungulates), but it is known only from its skull. *Ambulocetus*, which means "walking whale", is another early form. It had few adaptations for living in water and probably still spent much time on land.

The productivity of the oceans increased, whales diversified, and other marine mammals appeared. Whales similar to today's toothed whales appeared first, and a few million years later, baleen whales evolved. By 24 million years ago, baleen whales had reached today's giant sizes, suggesting that plankton were present in vast numbers for them to feed on. Only 14 million years ago, pinnipeds and sirenians (Dugongs and manatees) evolved. It is thought that pinnipeds arose from a family of carnivores not unlike otters. Their present-day forms are seals, sea lions, and Walruses.

ANCIENT WHALE SKELETON
This skeleton has been exposed in a desert in Sacaco, Peru. Whales evolved over the last 50 million years, so this area must have been an ancient sea at some point in this period.

SKULL WITH A BLOWHOLE
The nostrils of *Prosqualodon davidi* are positioned on top of the head, forming a blowhole. This feature proves that this fossil skull is from a primitive whale.

TODAY: LIFE IN MODERN OCEANS

We know much more about life in today's oceans because, as well having entire organisms to study, we can also observe life cycles, locomotion, and behaviour. Each of the five oceans supports a wide variety of life. Some species are very specialized and are restricted to a small area, while others are migratory or generalists and have a wider distribution. Sometimes, closely related species live in the same habitat in different oceans, separated by land or other physical barriers (see right).

By studying living organisms and the characteristics of the water they live in, scientists can also better understand ancient ocean environments and organisms. The deep ocean is still poorly known, but it contains an ecosystem that could be crucial to our understanding of life – black smokers (see p.188). Isolated from sunlight and from the surrounding water by a steep thermal gradient, it is possible that this is the type of environment in which life first evolved 3,500 million years ago.

GREY REEF SHARK
Like its close relative the Caribbean Reef Shark (below), this shark lives in warm, shallow waters, near coral atolls and in adjacent lagoons. It is found in the Indian and Pacific oceans, but it is cut off from the Atlantic.

CARIBBEAN REEF SHARK
Like the Grey Reef Shark (above), this species lives in shallow water near coral reefs. Its range is isolated from the Indo-Pacific by the deep, cold ocean around South Africa, so it is restricted to warm parts of the Atlantic, from the Caribbean to Uruguay.

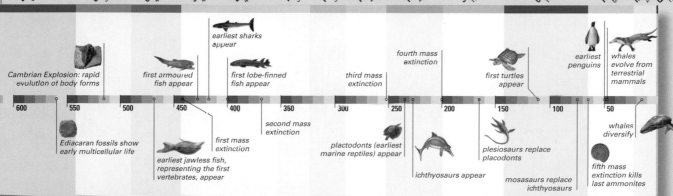

Timeline periods: EDIACARAN 635–542 MYA · CAMBRIAN 542–490 MYA · ORDOVICIAN 490–443 MYA · SILURIAN 443–418 MYA · DEVONIAN 418–354 MYA · MISSISSIPPIAN 354–323 MYA · PENNSYLVANIAN 323–290 MYA · PERMIAN 290–252 MYA · TRIASSIC 252–199.5 MYA · JURASSIC 199.5–142 MYA · CRETACEOUS 142–65 MYA · PALAEOGENE 65–23.3 MYA · NEOGENE 23.3–1.8 MYA · QUATERNARY 1.8–PRESENT

Cambrian Explosion: rapid evolution of body forms

earliest sharks appear

first armoured fish appear

first lobe-finned fish appear

fourth mass extinction

third mass extinction

first turtles appear

earliest penguins

whales evolve from terrestrial mammals

Ediacaran fossils show early multicellular life

first mass extinction

second mass extinction

plactodonts (earliest marine reptiles) appear

plesiosaurs replace placodonts

whales diversify

earliest jawless fish, representing the first vertebrates, appear

ichthyosaurs appear

mosasaurs replace ichthyosaurs

fifth mass extinction kills last ammonites

635 · 600 · 550 · 500 · 450 · 400 · 350 · 300 · 250 · 200 · 150 · 100 · 50

2,000 MYA · **1,500** MYA · **1,000** MYA · 700 MYA · 635 MYA · **PRESENT DAY**

PHANEROZOIC AEON 542 MYA–PRESENT

first fossil evidence of mineralized skeletons

beginning of the Ediacaran period, which soon features the first multicellular life

MASS EXTINCTIONS

The history of life is punctuated by five mass extinctions – catastrophic events in which many life forms died out. The first occurred 443 million years ago, when prominent marine invertebrates disappeared from the fossil record. About 368 million years ago, global cooling and an oxygen shortage in shallow seas caused about 21 per cent of marine families to disappear, including corals, brachiopods, bivalves, fish, and ancient sponges. At the end of the Permian Period, 252 million years ago, the cooling and shrinking of oceans killed over half of all marine life. Another mass-extinction event at the end of the Triassic Period, 199.5 million years ago, caused major losses of cephalopods, especially the ammonites.

The fifth extinction, 65 million years ago, caused the demise of the dinosaurs; in the oceans, it caused the giant marine reptiles to disappear. The next mass extinction is likely to be a result of human activity.

VOLCANIC ARMAGEDDON
Volcanic activity in the western Ghats of India is now thought to have been a factor in the most recent mass extinction. The eruptions would have caused destruction and climate change on a global scale.

AMMONITE FOSSIL
This ammonite species is one of the few to survive the late-Triassic mass extinction event.

LIFE ON EARTH was once thought to fall into
five great kingdoms – the animals, plants,
and fungi, and the two microscopic kingdoms,
the protists and the bacteria. Scientists are
now looking at life ever more closely, and
each discovery expands our perspective
on life's vast variety. Many experts now
consider that the familiar life-forms, plants
and animals, represent just two of 30 or
more kingdoms. The oceans are the
ancestral home of life and are still home to
all major groups of animals. Although plants
are far more diverse on land, their place is
taken in the oceans by a range of seaweeds
and microorganisms. The following section
showcases the entire range of ocean life.
It is organized into kingdoms and further
divided into the smaller units used by
scientists to order and understand nature.

KINGDOMS OF OCEAN LIFE

SUCCESS IN WATER
This Violet-Spotted Reef Lobster is a flamboyant
example of one of the marine success stories of
the animal kingdom – the varied and abundant
crustaceans. The crustaceans as a group
includes crabs, prawns, shrimps, and some
of the commonest members of the zooplankton.

BACTERIA AND ARCHAEA

DOMAINS	Bacteria
	Archaea
KINGDOMS	13
SPECIES	many millions

THE SMALLEST ORGANISMS ON EARTH are the bacteria and their relatives, the archaea. Bacteria occupy virtually all oceanic habitats, whereas most archaea are confined to extreme environments, such as deep-sea vents. Bacteria and archaea play vital roles in the recycling of matter. Many are decomposers of dead organisms on the ocean floor. Others are remarkable in being able to obtain their energy from minerals in the complete absence of light.

CARL WOESE

Born in New York in 1928, Carl Woese is the microbiologist who is responsible for the current division of living organisms into three domains, Archaea, Bacteria, and Eucarya, on the basis of his research into the RNA (a chemical related to DNA, called ribonucleic acid) found in ribosomes. Although Woese put forward his new classification in 1976, it was not until the 1980s that his hypothesis was accepted.

ANATOMY

Bacteria and archaea are single-celled organisms that are far smaller than any other, even protists. Most have a cell wall, which, in bacteria, is made from a substance called peptidoglycan. None has a nucleus or any of the other cell structures of more complex organisms (eukaryotes). Some bacteria and archaea can move by rotating threads called flagella; others have no means of propulsion.

Scientists separated the Archaea and Bacteria groups on the basis of chemical differences in their cell make-up. All living cells contain tiny granules (ribosomes), which help to make proteins, but those in archaea are differently shaped to those in bacteria. The oily substances that make up their cell membranes are also different. Additionally, archaea have special molecules associated with their DNA that protect them in the harsh environments in which they live.

Scientists now think that their chemical differences are sufficiently important to rank Archaea as a distinct evolutionary branch of life. Initially considered to be primitive, the archaea are now thought to be closer to the ancestors of eukaryotes than are the bacteria.

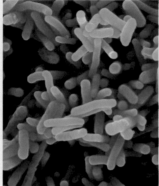

THRIVING IN THE RIGHT CONDITIONS
The bacterium *Nitrosomonas* forms colonies wherever there is enough ammonia and oxygen in the water.

HEAT-LOVING ARCHAEA
Most archaea can adapt to extreme conditions. This heat-loving example, GRI, was ejected from the sea floor in an undersea eruption.

HABITATS

Bacteria are found throughout the ocean environment, because nearly all habitats provide them with the materials necessary to obtain energy. Most bacteria obtain energy by breaking down organic matter. Much of this matter accumulates on the ocean floor and provides excellent conditions for the decomposer bacteria. However, bacteria are also found in smaller numbers in the water column, feeding on suspended matter.

A few kinds of bacteria, such as cyanobacteria, can photosynthesize and so live nearer to the surface, in brightly lit waters. Some form colonies and build huge structures, called stromatolites, near the shore.

Most archaea can live in extreme and harsh conditions, such as high temperatures, highly acidic water, high salinity, or low oxygen levels. For example, archaea live around deep-sea vents and obtain their energy from chemical reactions of methane and sulphide compounds ejected by the vents. Others survive in the very high concentrations of salt on some sea shores.

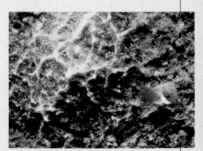

LIVING ON THE SEA BED
Bacterial mats form on the sea bed where oxygen supply is low. This mat of *Beggiatoa* sp. is at the mouth of the Mississippi, USA.

HYPER-SALINE CONDITIONS
The hyper-saline water of Hamelin Pool, west Australia, is ideal for stromatolites. The rocks are formed by the cyanobacteria cementing sediment particles together.

DOMAIN BACTERIA
Oscillatoria willei

SIZE Filament length 0.13mm
DISTRIBUTION Tropical waters

Once known as blue-green algae, cyanobacteria are bacteria that are able to use photosynthesis to make foods in a similar way to plants. *Oscillatoria willei* and other related cyanobacteria occur in rows of similarly sized cells that form filaments called trichomes. Many trichomes are enveloped in a firm casing, but in *Oscillatoria* the casing is thin or may be absent altogether, which allows the filaments to glide quickly forwards, backwards, or even to rotate. Some species of *Oscillatoria* can fix nitrogen but, unlike *Trichodesmium* (below), they may not have cells specialized for the purpose.

Fragments of filaments, called hormogonia, which consist of dozens of cells, sometimes break off and glide away to establish new colonies. These bacteria may cause skin irritations in humans who come into contact with them in tropical waters.

DOMAIN BACTERIA
Trichodesmium erythraeum

SIZE 1–10mm per colony
DISTRIBUTION Tropical and subtropical seas worldwide

Individual filamentous colonies of the cyanobacteria *Trichodesmium erythaeum* are just visible to the naked eye, and these bacteria have traditionally been known as sea sawdust by mariners. Under warm conditions, the bacteria is able to multiply extremely rapidly to create massive blooms that may have such an extent that they are visible from space.

This is a prolific nitrogen-fixing bacterium that harnesses about half of the nitrogen passing through

oceanic food chains. The bacteria form in long, multicelled filaments, in which some cells carry out nitrogen fixation, while others are specialized for photosynthesis. These tasks must be separated because the oxygen by-product that results from photosynthesis would interfere with the nitrogen-fixing process, so they cannot both occur in the same cell.

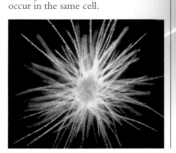

DOMAIN BACTERIA
Vibrio fischeri

SIZE 0.003mm cell length
DISTRIBUTION Worldwide

Many marine organisms, particularly those in the deep sea, make use of bioluminescence, the biochemical emission of light. Many of these creatures depend on bacteria, such as the rod-shaped *Vibrio fischeri,* to generate the light, and in these cases

the bacteria live within the body of their host in a mutually beneficial relationship. The bacteria produce light using a chemical reaction that takes place inside their cells. *Vibrio fischeri* also occurs as a free-living organism, moving through water by means of a flagellum and feeding on dead organic matter. The distinctive, comma-shaped cells seen in *Vibrio fischeri,* below, are characteristic of the genus. Other *Vibrio* species (which are not luminescent) are responsible for the potentially fatal disease cholera.

DOMAIN BACTERIA
Calothrix crustacea

SIZE Filament length 0.15mm
DISTRIBUTION Worldwide

Forming single filaments or small bundles, bacteria of the genus *Calothrix* are widespread in oceans everywhere. Unlike those of *Oscillatoria* and *Trichodesmium* (left), the filaments of *Calothrix crustacea* have a broad base and a pointed tip that ends in a transparent hair. The filament has a firm or jelly-like coating, which is often made up of concentric layers that may be colourless or yellow-

brown. Unusually, the filament grows in much the same way as a plant root, its growth being confined to a special region just behind the tip, called a meristem. Sometimes, the filament sheds the tapering tip above the growth region, enabling *Calothrix* to reproduce asexually by casting off fragments called hormogonia from the meristem. These fragments are able to form new filaments far away from the parent. These kinds of cyanobacteria often form slimy coatings on coastal rocks and seaweeds. At least one species of *Calothrix* is known to make up the photosynthetic part of some rocky shore lichens, such as *Lichina pygmaea* (p.257).

EYE LIGHTS

Eyelightfish (such as the one shown below) have light-emitting organs called photophores under each eye. The light is produced by colonies of *Vibrio fischeri* living in the photophores. The light organs display distinct patterns, and may be used as an aid to recognition and communication between fish of this species. The ability to emit light may also play a part in prey capture and the avoidance of predators.

DOMAIN ARCHAEA
Halobacterium salinarium

SIZE 0.001–0.006mm
DISTRIBUTION Dead Sea and other hypersaline areas of the world

Archaea that have adapted to live in waters with exceptionally high salt concentrations are called halophiles. One example of this type of organism is *Halobacterium salinarium*, which is rod-shaped, produces pink pigments called carotenoids, and forms extensive areas of pink scum on salt flats. The cell membranes of halophiles contain substances that make them more stable than other types of cell membrane, preventing them from falling apart in the high salt concentrations in which they live. Their cell walls are also modified, for the same reason. These bacteria obtain nourishment from organic matter in the water. In addition, their pigments absorb some light energy, which the bacteria then use for fuelling processes within the cells.

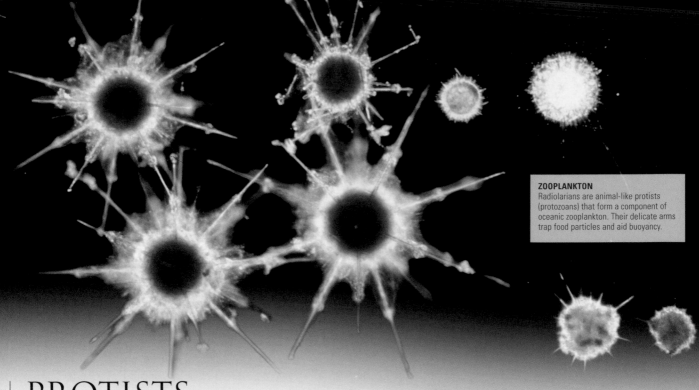

ZOOPLANKTON
Radiolarians are animal-like protists (protozoans) that form a component of oceanic zooplankton. Their delicate arms trap food particles and aid buoyancy.

PROTISTS

DOMAIN	Eucarya
KINGDOMS	At least 10
SPECIES	More than 100,000

PROTISTS ARE MICROSCOPIC ORGANISMS, many of which drift around with ocean currents and form a major part of the plankton. They are remarkably diverse. Many protists are single-celled and, in the past, they have been grouped together on this basis. Despite their extremely small size, they play a crucial role in the marine food chain, providing energy to all the other animals in the ocean.

ANATOMY

Protist body form is extremely varied. Some protists, such as diatoms and foraminiferans, have rigid outer skeletons and a fixed shape, while many unicellular protozoans, which are bounded only by a thin cell membrane, can continually change form. Radiolarians are spherical, and most have many long, delicate arms to aid buoyancy as they float passively. In contrast, flagellates have a whip-like thread on their body, which they can use to cause jerky movements. Diatoms and many dinoflagellates contain structures called chloroplasts, which turn them green. They produce food by photosynthesis and are collectively referred to as phytoplankton. Other protists, which feed on other organisms or detritus, are part of the zooplankton. Some protists are even luminescent.

DIFFERENT SHAPES
There are 10,000 species of diatoms. Each species has a differently shaped silica skeleton.

HABITATS

Protists live in every ocean in the world. Phytoplankton need lots of sunlight to photosynthesize, so they are found only in the surface layers where light penetrates. The zooplankton migrate up and down the water column, depending on the time of day, in a process called diel migration. They come to the surface at night to feed and then sink to deeper levels during the day (see p.221). Some protists live inside other life-forms and are essential to their wellbeing. Single-celled algae called zooxanthellae, for example, are found in both reef corals and some species of anemone; their relationship with their host is called mutualism because both organisms benefit. Protists also live on the sea bed and even in soft, deep-ocean sediment, and these are considered particularly vital to marine ecosystems.

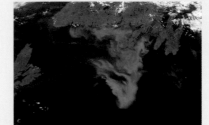

ALGAL BLOOM OFF NEWFOUNDLAND
When conditions are particularly favourable, phytoplankton multiply rapidly, causing an algal bloom, as shown here off the coast of Newfoundland, Canada.

SINGLE-CELLED COMPLEXITY
Foraminiferans have a rigid "shell" called a test. The test's shape varies from species to species, but it may have a complex spiral structure (left), reminiscent of a snail's shell. Foraminiferans' soft bodies protrude through their tests into long, branching extensions (pseudopodia), which they use to trap food.

PROTIST OR PROTOCTIST?

Until recently, scientists placed most single-celled eukaryotes, including unicellular algae, protozoans, and small, fungus-like organisms, in the taxonomic group Protista. Other scientists prefer to use the term Protoctista, which also includes the red and brown seaweeds. However, neither is a natural group, because many of the organisms they contain are not closely related.

GROUP DIATOMS

Ethmodiscus rex

DIAMETER 2–3mm (¹/₁₆–¹/₈in)

HABITAT Warm, nutrient-poor water

DISTRIBUTION Open ocean worldwide

Of the 10,000 species of diatom alive today, *Ethmodiscus rex* is the largest. It is a single cell with a rigid cell wall, called a test, which is impregnated with silica and covered in regular rows of pits. The test is made up of two disc-shaped halves, called valves, which fit tightly together. Because each

diatom has a unique test, *Ethmodiscus rex* can be easily identified in the fossil record. It is found in rocks that date from the Pliocene and the fossils can be up to 5 million years old. The cells need to remain near the water surface in order to utilize the Sun's energy for food, which they do by transforming the products of photosynthesis into oily substances that increase their buoyancy. *Ethmodiscus rex* can reproduce sexually but, if conditions are favourable, it multiplies rapidly, simply by dividing into two. Over a 10-day period, one individual that divides three times a day can have over 1.5 billion descendants.

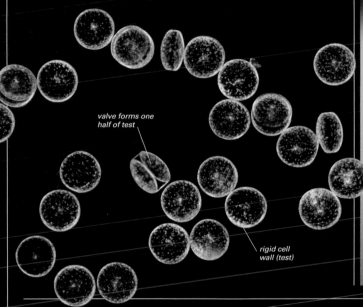

valve forms one half of test

rigid cell wall (test)

GROUP DIATOMS

Coscinodiscus granii

LENGTH Up to 0.1mm

HABITAT Surface waters

DISTRIBUTION Widespread in northern hemisphere

In bright light, the individual cells of *Coscinodiscus granii* appear golden-brown. This is because the numerous

chloroplasts inside the cell, which are visible through the transparent silica test, contain orange-brown pigments for photosynthesis, rather than the more familiar green chlorophyll. Like *Ethmodiscus rex* (left), *Coscinodiscus granii* is disc-shaped and radially symmetrical (that is, if it is cut in half, the two halves are mirror images of each other). During replication (asexual reproduction), the two halves (valves) of the test separate so that each daughter cell inherits one valve from its parent and creates the other valve itself. In contrast, a cell that is created as a result of sexual reproduction produces both halves of its test.

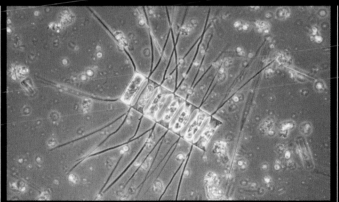

GROUP DIATOMS

Thalassiosira nordenskioeldii

LENGTH 0.01–0.05mm

HABITAT Cold water

DISTRIBUTION Northern hemisphere

This diatom is similar in structure to *Coscinodiscus granii* (above right), but differs in forming colonies in which rows of individual cells are joined together at regular intervals by threads of a substance called chitin. Like many diatoms that live in cold waters, the numbers of this species fluctuate greatly over the year. In spring, they multiply rapidly, in what is referred to as a bloom, to take advantage of the warmer, more favourable conditions. In autumn, their numbers fall as the water temperature decreases.

GROUP DIATOMS

Chaetoceros danicus

LENGTH 0.005–0.02mm

HABITAT Surface waters

DISTRIBUTION Worldwide

First described in 1844, *Chaetoceros* is one of the largest and most diverse genera of marine diatoms, containing well over 100 species. *Chaetoceros danicus* is a colonial form, and groups of seven cells are not uncommon (as shown here). It is easily recognized

because it has highly distinctive long, stiff hairs, called setae, which project perpendicularly from the margins of its test. and have prominent secondary spines along their length. Chloroplasts, which contain pigments used in photosynthesis, are numerous and found inside both the cell and the setae. The setae are easily broken and if large quantities lodge in the gills of a fish, they may kill it. The secondary spines anchor the setae to the sensitive gill tissue causing irritation, and the fish reacts by producing mucus. Eventually, it dies from suffocation.

KINGDOM DINOFLAGELLATA

Ceratium tripos

LENGTH 0.2–0.35mm

HABITAT Surface waters

DISTRIBUTION Worldwide

The unique three-pronged shape of the dinoflagellate *Ceratium tripos* makes it easy to identify among the phytoplankton, where it is one of the dominant organisms. Although this species is usually solitary, several individuals may be seen together, attached to each other by the single apical horn. This occurs when a cell divides and the daughter cells remain linked in short chains. *Ceratium tripos* is sometimes parasitized by other protists.

apical horn

lateral horns aid floatation

KINGDOM DINOFLAGELLATA

Protoperidinium depressum

LENGTH Up to 0.3mm

HABITAT Open water

DISTRIBUTION Worldwide

Like most dinoflagellates, *Protoperidinium depressum* is a single-celled organism with a complex armoured test. It is widest in the mid-region where there is a prominent groove, called the cingulum, which encircles the body and houses a whip-like flagellum. Movement of this and a second flagellum set at right angles to it causes the typical spiral motion of the cell. This ability to move allows *Protoperidinium depressum* to predate other small organisms. Also capable of bioluminescence, it has been mistaken for *Noctiluca scintillans* (see below).

KINGDOM DINOFLAGELLATA

Dinophysis acuta

LENGTH Up to 0.95mm (¹/₃₂ in)

HABITAT Open water

DISTRIBUTION Cold and temperate waters worldwide

This species is one of the largest belonging to the genus *Dinophysis*. It is plant-like in having numerous yellow-green chloroplasts, which it uses to manufacture food by photosynthesis, but it is also animal-like in being able to move using its two whip-like flagella for propulsion. The cells are oblong in shape, rounded posteriorly, and have a robust, laterally compressed test that is made up of a number of different plates. Like diatoms, most dinoflagellates have unique tests by which they can be identified even as fossils. All species of *Dinophysis* are toxic and have been responsible for a number of die-offs in shellfish when blooms occur. Under certain conditions, such as a long period of calm weather, they multiply very rapidly by simple division, sometimes becoming so numerous that the water changes colour. This phenomenon is referred to as a red tide. A huge amount of toxins and okadaic acid builds up in the water and passes up the food chain via shellfish in which the toxins accumulate. If the contaminated shellfish is then eaten by humans, it causes sickness and severe diarrhoea. Sexual reproduction in *Dinophysis acuta* is not well understood but it is thought to involve cells that in the past were described as a different species, *Dinophysis dens*.

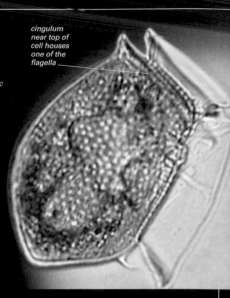

cingulum near top of cell houses one of the flagella

KINGDOM DINOFLAGELLATA

Noctiluca scintillans

DIAMETER Up to 2mm (¹/₁₆ in)

HABITAT Suface waters

DISTRIBUTION Worldwide

Also known as Sea Sparkle, *Noctiluca scintillans* is a large dinoflagellate that lives near the surface of the ocean where it feeds on other planktonic organisms. It has a flattened spherical body but no protective test. It is colourless, although the presence of photosynthetic organisms within the cell may give it a pink or greenish tinge. Usually only one of the two flagella is visible. The flagellum is not used in locomotion but, instead, sweeps food into the oral cavity and removes waste matter. To control its buoyancy, *Noctiluca scintillans* can adjust the concentration of its cell contents. This species, which is bioluminescent in some areas (*noctiluca* means "shining lantern"), may also form red tides (see opposite) and has been linked to fish and invertebrate deaths. Little is known of its complex life cycle. Reproduction can be either sexual or asexual by simple division.

BIOLUMINESCENCE

Floating just below the surface of the water at night, dinoflagellates, and in particular *Noctiluca scintillans*, are the commonest cause of bioluminescence in the open ocean. Millions of *Noctiluca scintillans* cells twinkle in the waves, hence the common name Sea Sparkle. The blue-green light is emitted from small organelles within the cells and is generated by a chemical reaction. Unlike many bioluminescent fish, it does not depend on light-emitting bacteria.

KINGDOM DINOFLAGELLATA

Gymnodinium pulchellum

DIAMETER 0.025mm

HABITAT Surface waters

DISTRIBUTION Temperate and tropical waters above continental shelves, and Mediterranean

Some red-tide organisms such as *Gymnodinium pulchellum* produce toxins that affect the nervous system and the clotting properties of the blood, causing high mortality among fish as well as invertebrates. The cause of red tides is not well understood but some scientists think they may be influenced by coastal pollution providing nutrients that might otherwise be in short supply and so normally limit the population size. Rapid reproduction by simple cell division results in huge numbers of *Gymnodinium pulchellum* being present in the water, turning it a characteristic brown-red colour, as shown here in the seas around Hong Kong. Unlike many other types of dinoflagellate, this species lacks a test and also produces food by photosynthesis.

KINGDOM CHRYSOPHYTA

Dictyocha fibula

LENGTH 0.045mm

HABITAT Surface waters

DISTRIBUTION Atlantic, Mediterranean, Baltic Sea, and eastern Pacific off coast of Chile

The golden-yellow pigments visible in this image of *Dictyocha fibula* indicate that it belongs to the large and complex group of golden algae known as Chrysophyta. The word Dictyocha means "net" and refers to the large windows in the silica test.

Only three species of *Dictyocha* are alive today. They are all that remains of a group of organisms that flourished more than 5 million years ago and whose fossils are abundant in some Miocene deposits.

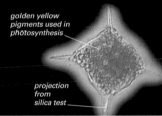

golden yellow pigments used in photosynthesis

projection from silica test

KINGDOM RADIOLARIA

Cladococcus viminalis

DIAMETER 0.08mm

HABITAT Surface waters

DISTRIBUTION Mediterranean

Radiolarians produce extremely complex silica tests of spines and pores that are laid down in a well-defined geometric pattern. The spines aid buoyancy and the pores provide outlets for cell material, called pseudopodia, which engulf any food that becomes trapped on the spines and carry it to the centre of the cell to be digested. *Cladococcus viminalis* is a polycystine radiolarian, which are the most commonly fossilized radiolarians and are frequently found in chalk and limestone rocks.

KINGDOM FORAMINIFERA

Hastigerina pelagica

LENGTH 6mm (¼in)

HABITAT Warm waters at depth of 200m (660ft)

DISTRIBUTION Subtropical and tropical waters of North Atlantic and western Indian Ocean

Foraminiferans are unicellular organisms that are found only in marine habitats. *Hastigerina pelagica* is one of the larger forms. It is often pinkish-red in colour and has a calcareous test with several globular-shaped chambers from which radiate calcite spines covered with cytoplasmic strands (pseudopodia) for collecting food. *Hastcrigina pelagica* is unique in that it surrounds its test with a gelatinous capsule of tiny frothy bubbles, which is thought to aid buoyancy. Dinoflagellates sometimes live on the surface of the capsule and up to 79 have been counted on a single individual, though 6–10 is more usual. The relationship between the two organisms is not clearly understood as *Hasterigina pelagica* is carnivorous, yet the dinoflagellates are unharmed.

globular-shaped chamber of calcareous test

calcite spines aid buoyancy

KINGDOM CILIATA

Strombidium sulcatum

DIAMETER 0.045mm

HABITAT Surface waters

DISTRIBUTION Atlantic, Pacific, and Indian oceans

Organisms such as *Strombidium sulcatum* are classified as ciliates because the cell membrane has many hair-like projections, called cilia, that are used in locomotion.

In *Strombidium sulcatum*, the cilia are restricted to a collar at one end of its spherical body. There is no test.

Among protists, ciliates have the most complex cells, with two nuclei. The macronucleus is needed for growth and reproduction, and the micronucleus is essential for sexual reproduction. *Strombidium sulcatum* reproduces asexually by splitting in two, but during sexual reproduction two cells join together for a considerable period and swap parts of their micronuclei. When the genetic composition of the joined micronuclei has become identical, the cells separate again. Further changes occur and both cells then split into two. This form of sexual reproduction is called conjugation.

GROUP COCCOLITHOPHORIDS

Emiliania huxleyi

DIAMETER 0.006mm

HABITAT Surface waters

DISTRIBUTION Atlantic, Pacific, and Indian oceans

Emiliana huxleyi has a golden-yellow, free-swimming form and a resting stage, which is spherical and covered with uniquely patterned calcite plates. For many years, scientists thought the two were different organisms, classifying the active form as a haptomonad and the resting stage as a coccolithophorid. Like some other protists, *Emiliana huxleyi* can multiply very quickly in favourable conditions, when it accounts for up to 90 per cent of the phytoplankton. These blooms cover areas of up to 100,000 square km (38,600 square miles) and are visible from space because they turn the water from deep blue to milky turquoise, as seen here in a satellite image taken off the coast off Cornwall, UK. The coccolithophorids cause the change in colour of the water as the calcite plates act like mirrors reflecting the incoming sunlight. They have been found worldwide in chalk deposits dating from 65 million years ago. Chemicals called alkenones are present within the fossils and are used to gather information about sea surface temperatures in the past.

RED AND BROWN SEAWEEDS

DOMAIN	Eucarya
KINGDOMS	Rhodophyta
	Phaeophyta
SPECIES	ca 5,500 and ca 2,000

IN SHALLOW SEAS, RED AND BROWN seaweeds are highly successful primary producers, providing shelter and a vital food source for marine animals. Both are kinds of algae and are not true plants, but they have many plant-like features. Like plants, they obtain food by photosynthesis using solar energy, but unlike plants, they use pigments other than chlorophyll to trap the sunlight, hence their red and brown colours. The classification of seaweeds is not agreed. Red and brown seaweeds are treated here as two kingdoms, while green seaweeds are considered to belong to the plant kingdom. Some researchers regard red and brown seaweeds as protists and green seaweeds as plants, or all seaweeds as protists, or all seaweeds as plants.

ANATOMY

Red and brown seaweeds live in, and are supported by water, so they have no need for strong skeletal supports or moisture-retaining tissues. Some seaweeds, including the large brown kelps and seashore wracks, have stiff stalks (stipes) or gas-filled bladders (pneumatophores) that hold up their fronds towards the light, and away from grazing sea-bed animals. Unlike most plants, but like green seaweeds, red and brown seaweeds do not need roots to absorb water and nutrients, or complex vascular systems to transport them; they absorb these directly from the water and photosynthesize over their entire surface. In place of roots, seaweeds have a holdfast, which acts as an anchor. Red and brown seaweeds vary enormously in form from tiny single cells and delicate filaments to giant kelp, more than 100m (330ft) long. Depending on the species, red and brown seaweeds can reproduce asexually by fragmenting or division – where parts of the plant break off and grow into new individuals – and sexually by producing spores.

midrib

stipe

holdfast

frond

PERENNIAL SPECIES
This beautiful red seaweed is called Sea Beech. It grows new fronds each year from a perennial stipe, and reproduces from spores in winter.

FARMING THE OCEANS

Seaweeds are harvested wild, but are also increasingly grown or enhanced artificially, especially in Asia. They are used for food, and seaweed extracts are used in a wide range of products – for example, in gels as a stabilizer, in cosmetics and pharmaceuticals, in beer-making, and as a fertilizer.

SEAWEED HARVEST IN ZANZIBAR
Many seaweeds grow readily on floating rafts, as seen here. They flourish in strong light, away from grazing invertebrates, and provide vital income for coastal communities.

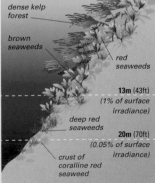

Depth 0
(100% irradiance)

dense kelp forest

brown seaweeds

red seaweeds

13m (43ft)
(1% of surface irradiance)

deep red seaweeds

20m (70ft)
(0.05% of surface irradiance)

crust of coralline red seaweed

LIGHT AND DEPTH
Light is absorbed rapidly by temperate coastal water rich in sediment and plankton. This diagram shows the depths at which kelps, smaller red and brown seaweeds, and encrusting red seaweeds grow in such water.

HABITATS AND DISTRIBUTION

Red and brown seaweeds thrive in fast-moving water on exposed coasts or on current-swept sea beds. In cooler climates, they often dominate rocky seashores. Great underwater forests of brown kelps grow in colder waters. Because of light limitations (see left), seaweeds do not usually grow below 30m (100ft) deep. However, in the clearest waters, such as in the Mediterranean, they can grow below 100m (330ft). Red seaweeds have pigments that enable them to grow in deeper water and shade than brown ones. Red and brown seaweeds are less abundant in tropical waters, an exception being the red coralline encrusting seaweeds, which have an important role in cementing coral reefs, and in building other carbonate reefs in warm, shallow water. Some seaweeds grow unattached, in sheltered lagoons, and a few grow in salt marshes, anchored in mud. The brown sargassum or Gulfweed, *Sargassum natans*, is unusual in that it floats at the surface of the open ocean, forming the basis for a unique ecosystem (see p.444). Drifting seaweeds may have an important role in the long-distance transport of marine life to isolated islands.

EVOLVING IN ISOLATION
Many of the seaweeds of Tristan da Cunha, in the south Atlantic, are found nowhere else, having evolved in isolation on an extremely remote, geologically young island group. The klipfish seen here is also endemic.

LIFE STRATEGIES

Red and brown seaweeds must cope with a disturbed environment. Seashore and rock pool species in particular experience daily and seasonal extremes of salinity or moisture and temperature, and the attentions of animal grazers. Because of regular exposure to air when the tide is out, the most dessication-resistant brown seaweeds usually live at the top of the shore. Although they may dry out at low tide, they can rapidly absorb seawater and resume photosynthesis as soon as the tide returns. Many seashore brown seaweeds produce mucus, both to protect from drying out and to deter grazers and colonizing animals. Some seaweeds are annuals, so they grow, reproduce, and die within a year. Others are perennial, or have parts from which new fronds grow each year. Many red seaweeds that colonize disturbed habitats have a two-phase life history, with conspicuous, erect filaments or fronds, present only during the calm season, and a perennial crust or creeping filament that helps it withstand abrasion during storms. These phases look so different that they were first described as separate species. The distinctive red seaweeds called coralline algae have a heavily calcified, pink frond, too hard for most grazers to eat. In some species, the frond is jointed, while others form crusts on rock. An unattached version, maerl, forms hard, free-living nodules on the sea bed. Some seaweeds are parasites of other seaweeds, obtaining at least part of their nutrition from the host.

NEW KELP GROWS FROM OLD
A new, yellow frond is growing from the top of this kelp stipe. The old frond, which will drop off, is covered in white animals called bryozoans, which block vital photosynthesis.

LIMPET GARDENS
On South African seashores, the Pear Limpet encourages fast-growing and nutritious red algae to grow on the surrounding rock, which it then grazes, fertilizes with its own waste, and defends from other limpets.

BARBED COLONIZER
This red seaweed has specialized barbed branches, enabling detached fragments to hook onto other marine growth and travel to new areas, either carried by currents or on ships' hulls.

SEAWEED CLASSIFICATION

Red and brown seaweeds have been classed as divisions, phyla, or classes of plants or protists, but here they are treated as independent kingdoms. Their defining features are their photosynthetic pigments.

BROWN SEAWEEDS
Kingdom Phaeophyta

About 1,500–2,000 species
Of the 14 or so orders of brown seaweeds, the most conspicuous marine orders are the kelps and wracks. Their brown colour is due to the pigment fucoxanthin; other yellow pigments (xanthophylls) may also be present. They also have beta-carotene, and chlorophylls a, c1, and c2.

RED SEAWEEDS
Kingdom Rhodophyta

About 5,000–5,500 species
There are two classes and 18 orders of red seaweeds. The majority are in the order Gigartinales, which contains a variety of frond-bearing (frondose) and crust-forming (crustose) species. The red colour comes from the pigment phycoerythrin. Rhodophyta also have blue pigments, carotenoids, and chlorophyll a.

THRIVING IN SURF
To absorb nutrients, red and brown seaweeds rely on moving water. Many appear green, and they flourish on these wave-washed rocks in the Canary Islands.

Limey Petticoat
Padina gymnospora

HEIGHT	Up to 10cm (4in)
HABITAT	Rock pools and shallow subtidal rocks
WATER TEMPERATURE	20–30°C (68–86°F)

DISTRIBUTION Coasts in tropical and subtropical areas worldwide

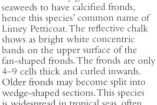

Padina is the only genus of brown seaweeds to have calcified fronds, hence this species' common name of Limey Petticoat. The reflective chalk shows as bright white concentric bands on the upper surface of the fan-shaped fronds. The fronds are only 4–9 cells thick and curled inwards. Older fronds may become split into wedge-shaped sections. This species is widespread in tropical seas, often growing in masses on shallow subtidal rocks, and on old coral and shells.

Giant Kelp
Macrocystis pyrifera

LENGTH	45m (150ft)
HABITAT	Rocky sea beds, occasionally sand
WATER TEMPERATURE	5–20°C (41–68°F)

DISTRIBUTION Temperate waters of southern hemisphere and northeastern Pacific

Giant Kelp (pictured on pp.242–43) is the largest seaweed on Earth. It can grow at the phenomenal rate of 60cm (24in) per day in ideal conditions, and reaches lengths of over 30m (100ft) in a year. Giant Kelp normally grows at a depth of 10–30m (30–100ft), but it can grow much deeper in very clear water. The huge branched holdfast, which is about 60cm (24in) high and wide after three years, is firmly attached to the sea bed. From it, a number of long, flexible stalks stretch towards the surface, bearing many strap-like fronds, each buoyed by a gas-filled bladder. The fronds continue to grow on reaching the surface, floating as a dense canopy.

Giant Kelp has a two-phase life cycle. Fronds (sporophylls) at the base of the kelp produce spores that develop into tiny creeping filaments. The filaments produce eggs and sperm, which combine to produce embryonic kelp plants.

Oyster Thief
Colpomenia peregrina

DIAMETER	Up to 10cm (4in)
HABITAT	Intertidal and subtidal rocks and shells
WATER TEMPERATURE	6–28°C (49–83°F)

DISTRIBUTION Coasts of western North America, Japan, and Australasia; introduced in Atlantic

The Oyster Thief gets its unusual name from its habit of growing on shells, including commercially grown oysters. The frond is initially spherical and solid, but as it grows, it becomes irregularly lobed and hollow and fills with gas. Sometimes, this can make it sufficiently buoyant to lift the oyster, which is not fixed to the sea bed, and they may both be carried away by the tide. This seaweed has a thin wall with only a few layers of cells. The outer layer is made of small, angular cells which contain the photosynthetic pigments that give the Oyster Thief its brown colour.

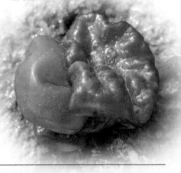

Landlady's Wig
Desmarestia aculeata

LENGTH	Up to 1.8m (6ft)
HABITAT	Subtidal rocks, and kelp forests
WATER TEMPERATURE	0–18°C (32–64°F)

DISTRIBUTION Near coasts in temperate, cold, and polar regions

This large seaweed has narrow brown fronds with many side-branches. Its bushy appearance is the reason for its common name of Landlady's Wig. The smallest branches are short and spine-like, hence the species name *aculeata*, which means "prickled". In summer, the whole plant is covered with delicate branched hairs. This species is particularly abundant on boulders and in kelp forests disturbed by waves.

Sea Palm
Postelsia palmaeformis

LENGTH	Up to 60cm (24in)
HABITAT	Wave-exposed shores
WATER TEMPERATURE	8–18°C (46–64°F)

DISTRIBUTION West Coast of North America

Sea Palms are kelps, which are large brown seaweeds that belong to the order Laminariales. Unusually for a kelp, Sea Palm grows on the midshore, where it forms dense stands on wave-exposed coasts. It has a branched holdfast, and a stout, hollow stalk, which stands erect when the tide is low. The top of the stalk is divided into many short, cylindrical branches, each of which bears a single frond up to 25cm (10in) long, with toothed margins and deep grooves running down both faces. Spores are released into the grooves and drip off the frond tips onto the holdfasts and nearby rocks at low tide, so that the developing seaweeds grow as dense clumps. Some Sea Palms attach to mussels and are later ripped off during storms, making more rock available for other Sea Palms to grow.

CROFTER'S WIG

In very sheltered bays and sea lochs, detached pieces of "normal" Knotted Wrack will continue to grow, lying loose on the sea bed. In situations where the fronds are alternately covered by salt and fresh water, they divide repeatedly to form a dense ball that has no bladders or reproductive structures. This attached form, which is known as Crofter's Wig, appears very different to the attached form, even though it is genetically identical.

KINGDOM PHAEOPHYTA

Knotted Wrack
Ascophyllum nodosum

LENGTH	Up to 3m (10ft)
HABITAT	Sheltered seashores
WATER TEMPERATURE	0–18°C (32–64°F)

DISTRIBUTION Coasts of northwestern Europe, eastern North America, and north Atlantic islands

Knotted Wrack belongs to a group of tough brown seaweeds that often dominate rocky seashores in cooler climates. It is firmly attached to the rocks by a disc-shaped holdfast, from which arise several narrow fronds that often grow to 1m (3ft) in length, and exceptionally to 3m (9ft) in very sheltered situations. Single oval bladders grow at intervals down the frond. The fronds produces about one bladder a year, so the seaweed's age can be roughly estimated by counting a series of bladders. The bladders hold the fronds up in the water so that they gain maximum light, which is an advantage in the often turbid waters where Knotted Wrack grows. This also makes it harder for grazing snails to reach the fronds when the tide is in.

The dark brown fronds may be bleached almost to yellow in summer. Reproductive structures that look like swollen sultanas are borne on short side-branches, and orange eggs can sometimes be seen oozing from them.

KINGDOM PHAEOPHYTA

Neptune's Necklace
Hormosira banksii

LENGTH	Up to 30cm (12in)
HABITAT	Lower shore and subtidal rocks
WATER TEMPERATURE	10–20°C (50–68°F)

DISTRIBUTION Coasts of southern and eastern Australia and New Zealand

Neptune's Necklace is one of the many brown seaweeds endemic (unique) to New Zealand and the cooler waters around Australia. Its distinctive fronds, which look like a string of brown beads, are made up of chains of ovoid, hollow segments joined by thin constrictions in the stalk. Small reproductive structures are scattered over each "bead".

Dense mats composed almost entirely of this one species can be found on seashore rocks. The fronds are attached to the rock by a thin, disc-shaped holdfast. Neptune's Necklace also lives unattached among mangrove roots. The shape of its segments varies according to habitat. They are spherical and about 2cm (¾ in) wide in fronds growing on sheltered rocks, mussel beds on tidal flats, or mangrove swamps. Fronds growing on subtidal rocks on moderately exposed coasts have smaller segments that are just 6mm (¼ in) long.

KINGDOM PHAEOPHYTA

Japweed
Sargassum muticum

LENGTH	2–10m (6–33ft)
HABITAT	Intertidal and subtidal rocks and stones
WATER TEMPERATURE	5–26°C (41–79°F)

DISTRIBUTION Coasts of Japan, introduced in western Europe and western North America

Japweed can reproduce all year and forms dense stands in quiet waters. Native to Japan (hence its common name), it was accidentally introduced to western North America and Europe, and is steadily expanding its range in these areas. It outcompetes other seaweeds and in these regions is regarded as an invasive species. This long, bushy seaweed has numerous side-branches, which have many leaflike fronds up to 10cm (4in) long. The fronds bear small, gas-filled bladders, either singly or in clusters.

GIANT KELP
This enormous seaweed can grow at a rate of 50cm (20in) per day in favourable conditions, such as the relatively cold water off California, USA (shown here). Air bladders help to keep the kelp's blades afloat as they grow upwards towards the surface, where there is an enhanced supply of light and nutrients.

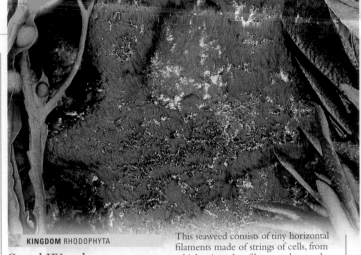

KINGDOM RHODOPHYTA

Laver
Porphyra dioica

LENGTH	Up to 50cm (20in)
HABITAT	Intertidal rocks
WATER TEMPERATURE	6–18°C (43–64°F)

DISTRIBUTION Coasts of northeastern and western Europe and Mediterranean around Italy

This species of red seaweed has only recently been separated from the very similar *P. purpurea* on the basis of how they reproduce. *P. dioica* is dioecious (male and female reproductive cells are on separate fronds), while *P. purpurea* is monoecious (male and female reproductive cells are on the same frond). *P. dioica* grows on intertidal, sandy rocks and is most abundant in the spring and early summer. The membranous frond is only one cell thick and is olive-green to purple-brown or blackish. This species appears to have a limited distribution in western Europe, but the genus is widespread throughout the world. All species of *Porphyra* are edible and are often harvested for food worldwide, especially in Japan where they are cultivated and known as *nori*. In the United Kingdom, wild Laver is collected and made into the Welsh delicacy laverbread.

KINGDOM RHODOPHYTA

Sand Weed
Rhodothamniella floridula

HEIGHT	Up to 3cm (1¼in)
HABITAT	Lower shore and subtidal rocks
WATER TEMPERATURE	4–18°C (40–64°F)

DISTRIBUTION Coasts of northeastern Europe and southern Africa

This seaweed consists of tiny horizontal filaments made of strings of cells, from which arise other filaments that reach a maximum height of 3cm (1¼in). The upright filaments have irregular side-branches, some of which interweave to form a complex network that traps sand grains. This network builds up to form a spongy mat about 1cm (½in) thick. The mat sometimes covers large areas of rocks that are next to sand and creates a habitat for small animals such as nematodes.

KINGDOM RHODOPHYTA

Small Jelly Weed
Gelidium foliaceum

LENGTH	5cm (2in)
HABITAT	Intertidal rocks
WATER TEMPERATURE	10–20°C (50–68°F)

DISTRIBUTION Coasts of southern Africa and southern Japan

There are many species of *Gelidium* worldwide, and they are difficult to identify because the plants can look very different depending on their habitat and whether they have been grazed by seashore animals such as limpets. Ongoing work on molecular sequencing is gradually resolving some of these problems, and *Gelidium foliaceum* is one species that has recently been reclassified. It has a flattened, much lobed and curled frond, which grows in dense clumps on rocky seashores. The fronds are tough and cartilaginous, and the seaweed is attached to the rock at frequent intervals by small hairlike structures, or rhizoids, from a creeping stem, or stolon. This creeping habit is probably the main method of spreading, but some species of *Gelidium* also reproduce sexually. Species of *Gelidium* are the main source of agar (see panel, right).

(see panel, right)

HUMAN IMPACT

SEAWEED JELLY

Red seaweeds have long been collected for making jellies, and agar (extracted from species of *Gelidium*, *Gracilaria*, and *Pterocladia*) is used for the preparation of gels on a large scale. Agar gel is used as a medium for growing fungi and bacteria in microbiological investigations, and in the food industry for producing jam and conserving meat and fish.

KINGDOM RHODOPHYTA

Coral Weed
Corallina officinalis

LENGTH	Up to 12cm (4¾in)
HABITAT	Rock pools and shallow subtidal rocks
WATER TEMPERATURE	0–25°C (32–77°F)

DISTRIBUTION Coasts worldwide except for far north and Antarctica

Coral Weed belongs to a group of red seaweeds known as coralline seaweeds, which have chalky deposits in the cell walls that give them a hard structure.

Coral Weed fronds have rigid sections that are separated by flexible joints. The branches usually lie in one plane, forming a flat, featherlike frond, but the shape is very variable. On the open shore, the fronds are often stunted, forming a short mat a few centimetres high in channels and rock pools and on wave-exposed rocks. These mats often harbour small animals, and other small red seaweeds attach to the hard fronds. Subtidally, the fronds grow much longer. The colour of Coral Weed varies from dark pink when it lives in the shade to light pink in sunny locations. When the seaweed dies, its hard skeleton becomes part of the sand.

KINGDOM RHODOPHYTA

Trottoir Coralline
Lithophyllum lichenoides

SIZE	Not recorded
HABITAT	Intertidal and subtidal rocks
WATER TEMPERATURE	15–25°C (59–77°F)

DISTRIBUTION Coasts of western Europe, Mediterranean, and southern Africa

Trottoir Coralline is one of many types of hard, encrusting coralline seaweeds that actively build chalky reefs in regions with warmer waters. The image below shows a complex coralline reef in the Mediterranean, where a low tide has exposed a horizontal platform, known as a "trottoir" (pathway), extending out from the coast. The seaweed grows best at the outer edge of the reef, where waves keep the crusts wet even during low tide. The crusts are riddled with animals that bore holes in it and live beneath it, including worms that live in hard tubes, which in turn become part of the reef structure.

KINGDOM RHODOPHYTA

Maerl
Phymatolithon calcareum

DIAMETER	Up to 7cm (2¾in)
HABITAT	Subtidal sea-bed sediments
WATER TEMPERATURE	0–25°C (32–77°F)

DISTRIBUTION Coasts of Atlantic islands, northern and western Europe, Mediterranean, and Philippines

The term "maerl" describes various species of unattached coralline seaweeds that live on sea beds. *Phymatolithon calcareum* forms brittle, purple-pink, branched structures that look more like small corals than seaweed. It grows as spherical nodules at sheltered sites, or as twigs or flattened medallions at more exposed sites. In places with some water movement from waves and tides, but not enough to break the maerl nodules, extensive beds can develop. Maerl is as much a habitat as a species, and both the living maerl and the maerl-derived gravel beneath it harbour many small animals. Maerl grows slowly and the beds are vulnerable to damage from bottom trawlers.

KINGDOM RHODOPHYTA

Cotton's Seaweed
Kappaphycus cottonii

LENGTH	50cm (20in)
HABITAT	Intertidal and shallow subtidal rocks
WATER TEMPERATURE	10–30°C (50–86°F)

DISTRIBUTION Coasts of Africa, southern and eastern Asia, and Pacific islands

Formerly called *Eucheuma cottonii*, this is a much-branched, cylindrical red seaweed that is farmed extensively in the Phillipines for extraction of carrageenan, a gelling agent similar to agar (see panel opposite). In the wild, it grows attached to rocks or lies loose in sheltered places. Like some other red seaweeds, its fronds are often shades of green and brown rather than red.

KINGDOM RHODOPHYTA

Irish Moss
Mastocarpus stellatus

LENGTH	17cm (7in)
HABITAT	Lower shore and subtidal rocks
WATER TEMPERATURE	0–25°C (32–77°F)

DISTRIBUTION Coasts of northeastern North America, northwestern Europe, and Mediterranean

This tough red seaweed is common on exposed shores, often forming a dense turf on the lower shore. Its frond is attached to rock by a disc-shaped holdfast, from which arises a narrow stipe (stalk) that gradually expands into a divided blade, which is slightly rolled to form a channel with a thickened edge. Reproductive structures housed in small nodules on the blade's surface produce a very different seaweed in the form of a thick black crust (it was originally named *Petrocelis cruenta* because it was thought to be an entirely different species). Spores from this crust grow back into the erect form, in a typical two-phase life history. *Mastocarpus stellatus* and the similar *Chondrus crispus* are both known as Irish Moss or Carrageen Moss and are collected on an industrial scale on both sides of the north Atlantic to produce the gelling agent carrageenan.

KINGDOM RHODOPHYTA

Spectacular Seaweed
Drachiella spectabilis

LENGTH	Up to 6cm (2½in)
HABITAT	Subtidal rocks at 2–30m (6–100ft)
WATER TEMPERATURE	8–18°C (46–64°F)

DISTRIBUTION Off western coasts of Scotland, UK, Ireland, France, and Spain

This colourful seaweed is rarely seen, except by divers, as it normally grows in relatively deep water, and is rarely washed ashore. It also grows in shallow water within kelp forests. It has a thin, fan-shaped frond, split into wedges, that spreads out over the rock and reattaches with small rootlike structures called rhizoids. Young plants have a purple-blue iridescence, which is lost as the seaweed ages. Sexual reproduction is unknown in this species and spores are produced asexually.

PLANT LIFE

DOMAIN	Eucarya
KINGDOM	Plantae
SPECIES	283,000

PLANTS FORM A GREAT kingdom of life-forms, all of which use the pigment chlorophyll to fix carbon dioxide from the atmosphere into organic molecules, using energy from sunlight. Most organisms in the plant kingdom are "higher" plants, which evolved on land and remain land-based. Of these, several unrelated families of flowering plants (see p.252) have since returned to the sea or taken up residence on the coast. The plant kingdom, as defined in this book, also encompasses more primitive organisms that first evolved in water – the microscopic green algae (microalgae) and the green seaweeds (also a type of algae). Red and brown seaweeds (see pp.238–39) may not be related to plants, and are classified in this book as separate, non-plant kingdoms.

BEACH PLANTS

Beach plants grow in places used by humans for recreation. We can coexist, especially when people use paths in coastal dune areas. In fact, paths maintain low-growing plants, such as mosses, which might otherwise become overgrown. However, fragile dunes are damaged by erosion, and plants that grow only in a limited strip of coastal habitat are highly vulnerable to human development.

SAND CROCUS
The Sand Crocus of the Canary Islands grows only in a few coastal locations on Lanzarote and Fuerteventura. It is protected, but threatened by tourism development.

MARINE PLANT DIVERSITY

Plants are united by their use of chlorophyll for photosynthesis. The higher plants include several major land-based groups, including ferns and conifers. Because higher plants evolved on land, they are adapted to life in air and to fresh water. They have tissues bearing vessels that transport water and food.

Of the higher plants, it is mainly the flowering plants (the largest group) that have invaded marine habitats. Along with mosses, they inhabit the coastal fringes, with only the seagrasses being fully marine. Green seaweeds and microscopic green algae (microalgae) lack stems and roots, have neither woody tissues nor transport vessels, and are mostly aquatic.

TROPICAL MARINE PLANTS
Seagrasses are abundant in tropical lagoons, and green seaweeds include calcified species. Mangroves line estuaries and creeks, and other flowering plants, including shrubs and trees, colonize the back of sandy beaches. Although not shown here, seaweeds may grow seasonally on rocky shores.

TEMPERATE MARINE PLANTS
Temperate seas are rich in phytoplankton, including green algae. Green seaweeds are commonly anchored to boulders or bedrock. Seagrasses have true roots and live in sediment in the shallow subtidal and intertidal zones, and in brackish lagoons. Above high water, cliffs, sand dunes, and salt marshes are home to flowering plants and mosses.

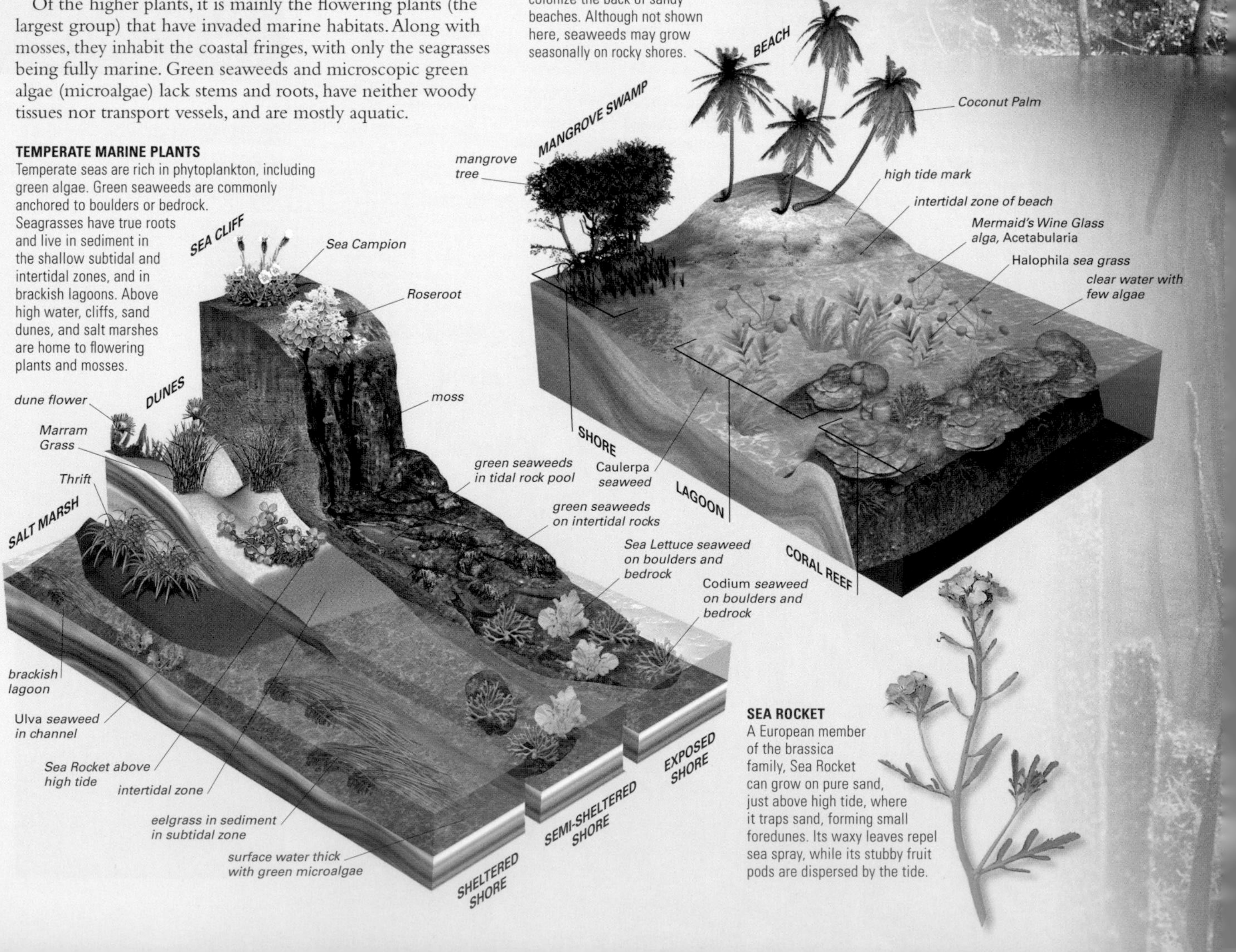

BEACH

MANGROVE SWAMP

Coconut Palm

mangrove tree

high tide mark

intertidal zone of beach

Mermaid's Wine Glass alga, Acetabularia

Halophila sea grass

clear water with few algae

SEA CLIFF

Sea Campion

Roseroot

dune flower

DUNES

moss

Marram Grass

Thrift

SALT MARSH

SHORE

green seaweeds in tidal rock pool

Caulerpa seaweed

LAGOON

green seaweeds on intertidal rocks

Sea Lettuce seaweed on boulders and bedrock

Codium seaweed on boulders and bedrock

CORAL REEF

brackish lagoon

Ulva seaweed in channel

Sea Rocket above high tide

intertidal zone

eelgrass in sediment in subtidal zone

surface water thick with green microalgae

SHELTERED SHORE

SEMI-SHELTERED SHORE

EXPOSED SHORE

SEA ROCKET
A European member of the brassica family, Sea Rocket can grow on pure sand, just above high tide, where it traps sand, forming small foredunes. Its waxy leaves repel sea spray, while its stubby fruit pods are dispersed by the tide.

PLANTS OF SHIFTING SHORES
Sea Mayweed, with abundant white flowers, and Oyster Plants, with their dark blue-green leaves, are salt-tolerant flowering plants that grow on semi-sheltered shores of shingle.

ABOVE HIGH WATER

Above the reach of the highest tides, the environment is essentially terrestrial, but its proximity to the sea makes life hard for all but a few specialized flowering plants and mosses. In places, the coast is covered with dunes of sand blown from the seashore. Dunes tend to be alkaline, because they are rich in calcium carbonate from the skeletons of marine organisms, and they are largely devoid of the humus that makes soil fertile. Dunes therefore support only hardy colonizers, tolerant of alkaline, infertile conditions. Marram, a grass with a fast-growing root network, and Golden Dune Moss stabilize the dunes and take the first steps towards soil formation. Plants that can fix nitrogen in root nodules, such as the Casuarina Tree, have an advantage here.

Farther inland, a wider variety of plants grow, but in more acidic soil, the plants characteristic of the shore become replaced by inland species. On rocky cliffs, plants are safe from large grazing animals, and more leafy plants often grow here, their roots reaching deep into rock crevices for moisture draining from the land.

ABOVE AND BELOW THE TIDE LINE
Forms of marine plant range from coastal plants above the high-tide mark, such as the Coconut Palm, to wholly aquatic, marine flowering plants, such as seagrass, below it.

BETWEEN THE TIDES

Plants between the tides have to live both in and out of water in conditions sometimes hot, dry, and salty, and at other times drenched with cold, fresh rainwater. Such intertidal zones support a small number of green seaweeds, seagrasses, and mangroves. Green seaweeds dry out quickly, so they are often confined to rock pools and freshwater seeps and streams on the upper shore. Lower down, they grow among or beneath tough brown seaweeds. Many are ephemeral, growing quickly in fair conditions and dispersing many spores. They may quickly cover tropical coasts where monsoons bring humid conditions, but then dry up and blow away when the sun returns. Seagrasses grow on lower intertidal flats; here the sediments retain moisture until the tide returns. In the tropics, mangroves colonize intertidal sediments, but only their roots are regularly submerged, while the rest of the plant remains in the air. In colder climates, salt-marsh vegetation develops on mudflats.

SUBMERGED SALT MARSH
The salt-tolerant Sea Pink grows in salt marshes, which develop on sheltered coasts in temperate regions. Like other salt-marsh plants, the Sea Pink is submerged only in the highest tides.

AQUATIC PLANTS

Only green microalgae, seaweeds, and seagrasses live permanently submerged in seawater. Green seaweeds can absorb nutrients and gases over their whole body surface, so they do not need the transport systems of land plants, and their holdfast simply attaches them to the sea bed. Seagrasses have a land-plant anatomy, so they need extra structures such as air spaces (lacunae) to aid gas exchange. Plants living in seawater can only thrive in the top few metres, because the light necessary for growth cannot penetrate beyond this. Marine plants must also deter grazing marine invertebrates, sometimes by producing toxic or distasteful chemicals.

SEAWEED BED
Green seaweeds in the English Channel include *Codium* (in the foreground) and Sea Lettuce (lower right). They grow together with an unrelated brown seaweed called Serrated Wrack.

GREEN SEAWEEDS

DOMAIN	Eucarya
KINGDOM	Plantae
DIVISION	Chlorophyta
CLASSES	4
SPECIES	1,200

GREEN ALGAE LARGE ENOUGH to be seen with the naked eye are known as green seaweeds. They are classified with the microscopic green algae, or microalgae (see p.250). True plants of the sea, they have pigments and other features in common with higher plants. They can be abundant in tropical lagoons, and proliferate seasonally on many temperate seashores. *Ulva* (sea lettuce) is grown for food.

frond

large, fibrous holdfast

SEAWEED BODY PARTS
Green seaweeds have a simple structure, with an erect frond and a disc-shaped or fibrous holdfast. This tropical *Udotea* species has calcified fronds with many branched siphons.

HABITATS

Green seaweeds often attach to rocks on rocky coasts, particularly in temperate and cold waters, and are ephemeral colonizers in seasonally disturbed tidal and shallow subtidal habitats. *Ulva* species, such as Sea Lettuce, dominate in high-level rock pools, or where fresh water seeps over the shore, since they can withstand changes in saltiness and temperature. The more delicate *Cladophora* and *Bryopsis* species live in rock pools or among red and brown seaweeds in the shallow subtidal zone. Green seaweeds also thrive in shallow, tropical lagoons, where species of *Caulerpa*, *Udotea*, and *Halimeda* are often abundant. *Caulerpa* species have runners (stolons), which creep through sand or cling to rock, while the bases of *Udotea* and *Halimeda* are a bulbous mass of fibres that anchor in sand. *Halimeda* (cactus seaweed) is heavily encrusted with calcium carbonate, which breaks up when the plant dies, contributing to the lagoon sand.

ANATOMY

The body structure of green seaweeds lacks stems and roots. Green seaweed shapes range from thread-like (filamentous) to tubes, flat sheets, and more complex forms. Their bright green colour is due to the fact that their chlorophyll is not masked by additional pigments, unlike red and brown seaweeds. Many of the features of green algae, including their types of chlorophyll, are shared by higher plants (mosses, liverworts, and vascular plants), so green seaweeds appear to be more closely related to higher plants than to red and brown seaweeds.

FRAGILE FRONDS
This delicate *Bryopsis plumosa* has coenocytic fronds, meaning its fronds do not have the crosswalls common in other green seaweeds.

FLEXIBLE SEAWEED
Able to withstand fluctuations in salinity and temperature, *Ulva* species thrive in this freshwater stream as it flows across the seashore.

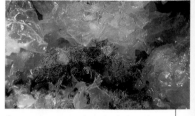

CODIUM FOREST
This mini-forest of *Codium fragile* is growing on shallow rocks in a sheltered bay in Scotland. The fronds are buoyant, holding the plants up to the light.

CLASS ULVOPHYCEAE

Flaccid Green Seaweed

Ulothrix flacca

SIZE	Up to 10cm (4in)
HABITAT	Intertidal on various shore types
HABITAT	0–20°C (32–68°F)

DISTRIBUTION Northern Atlantic, Mediterranean, waters off South Africa, Pacific

This seaweed is made up of many unbranched green filaments, which themselves consist of strings of cells. The filaments form soft, woolly masses or flat green layers that stick to intertidal rocks. Each filament is

attached to its rock by a single cell called a basal cell, which may be given additional anchorage by outgrowths called rhizoids. This seaweed reproduces by releasing up to a hundred gametes, each with two flagellae, from some of the cells. In another phase of its life cycle it is a single globular cell.

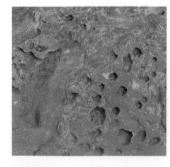

CLASS ULVOPHYCEAE

Sea Lettuce

Ulva lactuca

SIZE	Up to 100cm (40in)
HABITAT	Intertidal and shallow subtidal
WATER TEMPERATURE	0–30°C (32–86°F)

DISTRIBUTION Coastal waters worldwide

Sea Lettuce is common worldwide on seashores and in shallow subtidal areas, growing in a wide range of conditions and habitats. Its frond is a bright green, flat sheet, which is often split or divided, and has a wavy edge. The plant is very variable in shape and size, ranging from short, tufted plants on exposed shores to

sheets over a metre long in sheltered, shallow bays, especially where extra nutrients are available in polluted harbours. Sea Lettuce reproduces by releasing gametes from some cells, and it can also spread vegetatively by regeneration of small fragments. Large fronds lying on the sea bed may be full of holes made by grazing animals. It is a also popular food for humans in many parts of the world.

Sailor's Eyeball
Valonia ventricosa

SIZE	Up to 4cm (1½ in)
HABITAT	Rock and coral to 30m (100ft)
WATER TEMPERATURE	10–30°C (50–86°F)

DISTRIBUTION Western Atlantic, Caribbean, Indian and Pacific oceans

This odd seaweed looks like a dark green marble, and consists of a single large cell attached to the substrate (which is often coral rubble) by a cluster of filaments called rhizoids. Younger plants have a bluish sheen, but older ones become overgrown with encrusting coralline red seaweeds. Sailor's Eyeball has an unusual way of reproducing vegetatively: daughter cells are formed within the parent, which then degenerates, releasing the young plants in the process.

Cactus Seaweed
Halimeda opuntia

SIZE	Up to 25cm (10in)
HABITAT	Rock and sand
WATER TEMPERATURE	20–30°C (68–86°F)

DISTRIBUTION Red Sea, Indian Ocean, and western Pacific

The heavily calcified skeletons of species of *Halimeda* contribute much of the calcareous sediment in the tropics. The plant consists of strings of flattened, kidney-shaped, calcified segments, linked by uncalcified, flexible joints. By day, its chloroplasts are in the outer parts of the frond; at night they withdraw deep into the plant's skeleton. This, along with sharp crystals of aragonite, and the presence of toxic substances in the frond, protects them from nocturnal grazing.

Velvet Horn
Codium tomentosum

SIZE	Up to 20cm (8in)
HABITAT	Intertidal pools, shallow subtidal rocks
WATER TEMPERATURE	8–30°C (46–86°F)

DISTRIBUTION Coastal waters worldwide

The spongy fronds of Velvet Horn are made up of interwoven tubes, arranged rather like a tightly-packed bottlebrush, with each tube ending in a swollen bulb. Many of these bulbs packed together make up the outside of the frond, which is usually repeatedly branched in two. Many short, fine hairs cover the seaweed, giving it a fuzzy appearance when in water. The plants are attached to rocks by a spongy holdfast.

Although this seaweed is present all year round, its maximum development is in winter, and it also reproduces during the winter months. Velvet Horn, like all *Codium* species, is often grazed by sacoglossans, small sea slugs that suck out the seaweed's contents, but can keep the photosynthetic chloroplasts alive and use them to make sugars inside their own tissues. The chloroplasts colour the sea slugs green, which helps to disguise them from predators. There are about 50 species of *Codium*.

Giant Cladophora
Cladophora mirabilis

LENGTH	Up to 100cm (40in)
HABITAT	Subtidal rocks and kelp
WATER TEMPERATURE	10–15°C (50—59°F)

DISTRIBUTION Southern Atlantic off southwest Africa

A giant amongst *Cladophora* species, *C. mirabilis* grows to 1m (40in) long. It is bluish green and filamentous, with many straggly side-branches. It is made up of strings of cells, but individual cells in the main axis may be 12mm (½ in) long. The plant attaches using a disc made of interwoven extensions of its basal cell, and often has red algae growing on it. It has a very limited distribution in South Africa, but other species of *Cladophora* are common worldwide.

Sea Grapes
Caulerpa racemosa

HEIGHT	Up to 30cm (12in)
HABITAT	Shallow sand and rock
WATER TEMPERATURE	15–30°C (59–86°F)

DISTRIBUTION Warm waters worldwide

This seaweed has creeping stolons (stems) that anchor it to rocks or in sand, and from which arise upright shoots covered with round sacs, or vesicles, hence the common name Sea Grapes. Each plant is a single huge cell. Old plants may become densely branched and entangled, growing to 2m (6ft) across. There are many varieties of Sea Grapes, and around 60 species of *Caulerpa* worldwide

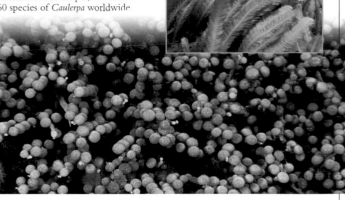

KILLER SEAWEED

A strain of *Caulerpa taxifolia* that is widely used in marine aquaria is an invasive species. It is toxic to grazers, grows rapidly, and forms a dense, smothering carpet on the sea bed. In 1984 it was discovered in the Mediterranean off Monaco, and has since spread rapidly along the coast, altering native marine communities.

Mermaid's Wineglass
Acetabularia acetabulum

SIZE	3cm (1¼ in)
HABITAT	Shallow subtidal rocks
WATER TEMPERATURE	10–25°C (50–77°F)

DISTRIBUTION Eastern Atlantic off North Africa, Mediterranean, Red Sea, Indian Ocean

This curious little green alga grows in clusters on rocks or shells covered with sand in sheltered parts of rocky coasts within its range. Although it grows to 3cm (1¼ in), it consists of just one cell. Its calcified frond appears white because it is encrusted with calcium carbonate, and it terminates in a small cup. The cup is made up of fused rays that produce reproductive cysts. The cysts are released after the remainder of the plant has decayed, and they then require a period of dormancy in the dark before they begin to germinate.

GREEN ALGAE

DOMAIN	Eucarya
KINGDOM	Plantae
DIVISION	Chlorophyta
CLASS	Prasinophyceae
SPECIES	200

THESE MICROSCOPIC, MOSTLY single-celled plants live in the surface layers of the ocean in immense numbers, and they form an important part of the phytoplankton (see p.212). Sometimes referred to as the "grasses of the sea", like most plants, they produce their own food through photosynthesis. Large green algae, visible to the naked eye, are called green seaweeds and are discussed elsewhere in this book (see p.248). Microscopic algae are often termed "microalgae". Green microalgae are frequently classified as protists. Numerous other groups of protists (see p.234) are also termed algae, and also live as phytoplankton.

HABITATS

With a few exceptions, marine microalgae swim and float, in countless millions, in the sunlit layers of the ocean so that photosynthesis can occur. They are more numerous in nutrient-rich waters, such as those benefitting from coastal run-off. In temperate coastal waters, green microalgae multiply rapidly each spring in response to rising nutrient and light levels, creating abundant food for zooplankton. Such population explosions, or blooms, can reduce the water clarity for weeks. Some green algae live inside the bodies of animals (see panel, right) and inside protist plankton – in the appendages (rhizopoda) of radiolarians (see p.237) and within compartments inside the dinoflagellate *Noctiluca* (see p.236).

GREEN TIDE
Green algae grow quickly and are the first to respond in spring when nutrients become available. While grazer levels are low, the algae are free to multiply until their density turns the ocean green.

HALOSPHAERA
These microalgae (shown greatly enlarged) are green with chlorophyll and bear hair-like swimming appendages called flagellae.

ANATOMY

Marine microscopic green algae mostly belong to a class of algae called the Prasinophyceae. Each consists of a single living cell that is generally too small to be visible to the naked human eye. Even the larger species, such as members of the genera *Halosphaera* and *Pterosperma,* measure just 0.1–0.8mm across, so appear as no more than a speck. Some green algae can swim, and beat two or more hair-like structures, called flagellae, to move through the water. Others lack flagellae and cannot propel themselves. Several groups of these plants have a two-stage life history, including both swimming and non-swimming forms. All green algae possess chloroplasts – structures that contain the green pigment chlorophyll that plants use in photosynthesis.

GREEN BEACHES

A few green algae and worms form symbiotic partnerships, in which both species gain. The beach-living worms ingest algae, giving them a green colour. At low tide, they move up through the sand to pools on the surface, where the algae photosynthesize. In return, the worms absorb food from the algae. Vast numbers of the worms tinge beaches green.

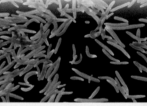

ANIMAL–ALGA PARTNERSHIP
When young, these marine flatworms ingest green algae, which may multiply until there are 25,000 algal cells living in each worm. The adult worms obtain all their nutrition from the algae.

CLASS PRASINOPHYCEAE

Halosphaera viridis

SIZE
20–30 micrometers (motile phase)

DISTRIBUTION Northeastern Atlantic, eastern Pacific

Halosphaera viridis is a small, pear-shaped cell with four swimming flagellae at one end. It reproduces by splitting in two, allowing it to reach high concentrations and from time to time some cells become small cysts whose contents divide into small discs. Each disc eventually becomes a flagellated cell that will be released into the sea. There can be hundreds of cysts per square metre in the open ocean, and they are probably a vital food source for larger zooplankton.

CLASS PRASINOPHYCEAE

Tetraselmis convolutae

SIZE
10 micrometers

DISTRIBUTION Northeastern Atlantic, off the western coasts of Britain and France

Although it can survive free-living, the tiny cells of *Tetraselmis convolutae* often live inside a worm host (see box, above) in a symbiotic relationship. The worm provides them with shelter and a constant environment inside its body. The worm's light-seeking behaviour gives the algae ideal conditions for photosynthesis, which in turn provides both algae and worm with nutrients and energy.

MOSSES

DOMAIN	Eucarya
KINGDOM	Plantae
DIVISION	Bryophyta
SPECIES	At least 10,000

MOSSES ARE LOW-GROWING plants that thrive in damp habitats on land, where they may carpet the ground or rocks. They dislike salty environments and only a few species manage to live in the intertidal zone of coasts, mainly in cooler climates. A much wider variety of mosses can be found slightly further inland, away from the direct effects of sea spray but within range of moisture-laden sea mists.

HABITATS

Mosses generally prefer moist, shady places and are most numerous in the cooler and damper climates of temperate regions. This is because they lack the thick cuticle that enables other types of plant to retain moisture. Without the protection of this skin-like surface, mosses soon shrivel up in dry conditions. However, some mosses have an amazing capacity to recover quickly when wetted after a long period of drought. A few species grow in salt marshes or among the lichens at the top of rocky shores; on sheltered coasts, where there is little salt spray, they may live only just above the high tide level. Sand-dune mosses grow rapidly to keep pace with accumulating sand, and blown fragments of moss can colonize new areas of dunes. Many more moss species grow on sea cliffs and in damp gulleys away from the intertidal zone.

SYNTRICHIA RURALIFORMIS
This moss grows in coastal sand dunes. Its leaves curl up when dry (left of picture) but unfurl a few minutes after wetting (on right).

ANATOMY

Most mosses have a recognizable structure of stems and leaves, which, as in other plants, gather sunlight and perform photosynthesis. However, unlike flowering plants (see p.252), they do not have woody tissues for support, and they also lack the conducting tissues that transport water and nutrients. Mosses have a very thin outer layer of cells, or cuticle, that can absorb (and lose) water, nutrients, and gases over their entire surface. Their "roots" are simple strands called rhizoids, which anchor the plant to its growing surface. Mosses reproduce sexually by means of wind-blown spores, or asexually by spreading across the ground.

SPORE PRODUCTION
Mosses have low-growing leaves, but sprout taller structures with bulbous tips called capsules, from which spores are released.

(see p.252)

CLASS BRYOPSIDA

Golden Dune Moss
Syntrichia ruraliformis

SIZE	1–4cm (½–1½in)
FORM	Yellow-green to orange-brown cushions and carpets
HABITAT	Mobile dunes

DISTRIBUTION Eastern Pacific, northwestern Atlantic, Mediterranean

This is one of the first mosses to colonize mobile dunes. It often forms extensive colonies that cover many square metres of sand, giving the sand a golden tinge. Its leaves are covered by hundreds of small papillae that enable swift absorption of water. The leaves gradually taper into long white hair points. This moss is able to establish new plants from fragments dispersed by the wind.

CLASS BRYOPSIDA

Salt marsh Moss
Hennediella heimii

SIZE	3mm (⅛in)
FORM	Single green plants
HABITAT	Salt marshes, other coastal areas

DISTRIBUTION Patchy distribution on temperate and cool waters worldwide

This tiny moss is a halophyte, meaning it is adapted to growing in highly saline conditions. It is rarely found growing inland. One of the few mosses that may be regularly found in salt marshes, it grows on patches of bare ground between the other vegetation in the upper parts of the salt marsh. It also grows in various other coastal habitats, including the banks of creeks, behind sea walls, and on footpaths. Although small, the plants may be abundant and may appear conspicuous from a distance, due to their prolific number of stout, dark, rusty-brown capsules, which are borne on short stalks less than 1cm (½in) tall. These have a little cap with a long point, which lifts to allow spores to escape, but remains attached to the capsule by a central stalk. The Salt marsh Moss has a wide distribution in colder climates.

CLASS BRYOPSIDA

Seaside Moss
Schistidium maritimum

SIZE	2cm (1in)
FORM	Dark blackish green, compact cushions
HABITAT	Hard, acid rocks, salt marshes

DISTRIBUTION Western and eastern coasts of North America, coasts of western Europe

This moss grows as small, dark green cushions on hard, acid rocks, with seashore lichens, just above high-tide mark. It also occurs in salt marshes. It is often soaked by salt spray and occasionally covered by

the highest tides. Seaside Moss appears to be a true halophyte, functioning normally even after immersion in sea water for a few days, and growing only in saline conditions; in Britain, it is found no further than 400m (1,300ft) from the sea. Its leaves curl when dry. In winter, it produces small brown capsules on short stalks.

CLASS BRYOPSIDA

Southern Beach Moss
Muelleriella crassifolia

SIZE	2–13cm (1–5in)
FORM	Black cushions and mats
HABITAT	Rocks

DISTRIBUTION Southern tip of South America, islands in the Southern Ocean

This moss is the southern version of Seaside Moss (see above), growing on coastal rocks in the usually lichen-dominated splash zone, where it is often inundated by the sea in stormy weather. It grows in southern Chile and on subantarctic islands, where it can become dominant. On Heard Island, for example, a saltspray community of plants found on exposed coastal lava rock, at elevations of less than 5m (16ft), is dominated by Southern Beach Moss, which has also colonized derelict buildings.

FLOWERING PLANTS

DOMAIN	Eucarya
KINGDOM	Plantae
DIVISION	Tracheophyta
CLASS	Angiospermae
SPECIES	235,000

PLANTS CONQUERED LAND, and then land-based flowering plants grew to be among the most abundant and diverse life-forms on Earth. However, relatively few have adapted to the poor soil, salt spray, and drying wind of coastal dunes and cliffs. These few include some fascinating plants found nowhere else. Few flowering plants have returned to the sea: salt-marsh plants and mangroves get wet at high tide, but only the seagrasses live fully submerged.

ANATOMY

Flowering plants, technically called angiosperms, uniquely possess fruit and flowers, unlike mosses, seaweeds, and other algae. They are adapted to life in air, absorbing fresh water through their roots. If they take salt water into their vascular system, water from their own cell sap is attracted to the more concentrated salts and sucked out by osmosis. This is fatal to cells, but mangroves cope by excreting the salt, while succulents partition it within their cells. Seagrasses have fully adapted by matching their cells' salt concentration to that of seawater. Most seagrasses have a similar form, with thin, grass-like blades that allow easy exchange of nutrients and gases. Mangroves grow in mud that lacks oxygen by growing aerial roots to assist gas exchange in their underground roots. Many angiosperm seeds are killed by seawater, but those of the coconut can stay viable at sea for long periods inside a waterproof case. Seagrasses are water-pollinated, and to increase the chances of a pollen grain catching onto a female stigma, pollen is released as a sticky string.

waterproof seed case

germinating plant

GERMINATING SEED
This coconut is a fruit – a defining characteristic of flowering plants. The coconut has the marine adaptations of buoyancy and a waterproof case.

SEAWATER PLANTS

Seagrasses are monocotyledons (the group of flowering plants with narrow, strap-like leaves), but are not true grasses, and they do not have a single evolutionary origin. There are 59 species in 5 families, although the Ruppiaceae, living mainly in brackish water, is not always accepted as a seagrass family. Salt-marsh plants are mainly small and herbaceous, with early colonizers including the salt-excreting Cord Grass and small succulents, such as Common Glasswort. Further salt-tolerant flowering plants grow further up the shore in established salt marshes, forming a dense, grassy turf. Salt marshes (see p.124) form in cooler climates, and are replaced in tropical seas by mangroves – trees with characteristic aerial roots. There are 16 families and 54 species of mangroves. Like seagrasses, they do not have a single origin, so the mangrove habit evolved separately, several times.

MANGROVES SUBMERGED
When the tide is in, mangroves form a mini-jungle of arching roots where small fish hide.

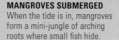

EELGRASS
Seagrasses, such as this eelgrass in a Scottish sea loch, can be found from the cold waters of Alaska to tropical seas.

COASTAL PLANTS

A greater variety of flowering plants can grow above the high-water mark. Salt-tolerant grasses are important constituents of the upper parts of salt marshes, and at the seaward edge of sand-dune systems, grasses are often the first to stabilize the shifting sand. In sand and sheltered shingle at the top of the shore, a few deep-rooted plants grow. A much wider variety of flowers and a few mosses colonize sand dunes and slacks just inland from the coast. Here they are subjected to salt spray but never inundated by tides. Nitrogen fixers thrive in these poor, sandy soils. In warmer climates, annuals bloom like desert flowers after seasonal rains. On cliff-tops, plants may be fertilized by sea-bird guano, stimulating lush growth.

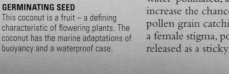

DUNE-SLACK FLORA
Here, in dune slacks behind a beach in the Canary Islands, annual plants bloom for a short period after rain. They flower and produce seeds quickly before drying up in the summer sun.

COASTAL FLOWERS
The beautiful pink flower heads of Sea Pink transform rocky seashores and salt marshes in late spring. The Sea Pink's compact cushions resist wind and cold.

ORDER NAJADALES

Neptune Grass
Posidonia oceanica

TYPE	Perennial
HEIGHT	30cm (12in)
HABITAT	Rocks and sand

DISTRIBUTION Mediterranean

Neptune Grass (also known as Mediterranean Tapeweed) forms meadows from shallow water to a depth of 45m (150ft) in the clearest waters. It grows on both rock and sand, has a tough, fibrous base, and persistent rhizomes (stems) that grow both horizontally and vertically. These build up into a structure known as "matte", which can be several metres high and thousands of years old. Around the island of Ischia, Italy, more than 800 species have been associated with Neptune Grass beds.

ORDER HYDROCHARITALES

Paddle Weed
Halophilia ovalis

TYPE	Perennial
HEIGHT	6cm (2½ in)
HABITAT	Sand

DISTRIBUTION Coasts of Florida, USA, East Africa, Southeast Asia, Australia, and Pacific islands

Members of the genus *Halophila* look quite unlike other seagrasses, having small, oval leaves that are borne on a thin leaf stalk. As its scientific name indicates, Paddle Weed is particularly tolerant of high salinities (*halophila* means "salt-loving"). Pollination takes place underwater, and the tiny, oval pollen grains are released in chains, which assemble into rafts like floating feathers. This is thought to increase the chances of pollination of a female flower. Despite its small size, Paddle Weed is an important food for the Dugong (see p.423). An adult can eat more than 40kg (88lb) of it a day.

ORDER POALES

Marram Grass
Ammophila arenaria

TYPE	Perennial
HEIGHT	0.5–1.2m (1½–4ft)
HABITAT	Coastal sand dunes

DISTRIBUTION Western Europe and Mediterranean (natural occurrence); introduced elsewhere

Marram Grass is a tall, spiky grass that plays a key role in binding coastal sand and building sand dunes. Its underground stems (rhizomes) spread

through loose sand, and upright shoots develop regularly along their length. When the tangle of stems and leaves impede onshore breezes, sand carried in the wind is deposited. Progressively, the sand builds up, the stems grow up through the sand, and a sand dune is formed. In dry weather, the leaves curl into a tube. The underside of the leaf then forms the outer surface and its waxy coating helps to reduce water loss from the plant. Marram Grass is widely planted to stabilize eroded dunes, and has been introduced for this purpose to North America (where it is known as European Beach Grass), Chile, South Africa, Australia, and New Zealand.

ORDER CARYOPHYLLALES

Common Glasswort
Salicornia europaea

TYPE	Annual
HEIGHT	10–30cm (4–12in)
HABITAT	Coastal mudflats and salt marshes

DISTRIBUTION Coasts of western and eastern North America, western Europe, and Mediterranean

Glasswort, also known as Marsh Samphire, is an early colonizer of the lower levels of salt marshes and mudflats, where plants are inundated twice a day by the tide. It is a small, cactus-like plant with bright green stems that later turn red. The tiny flowers and scale-like leaves are sunk into depressions in the fleshy stem. Glasswort is protected externally from salt water and moisture loss by a thick, waxy skin. It is able to prevent the salt absorbed through its roots from doing any damage by locking it away in vacuoles (small cavities) within its cells. The plant stores water inside its succulent stems, hence its cactus-like shape. For centuries, Glasswort was gathered and burnt to produce an ash rich in soda (impure sodium carbonate). The ash was then baked and fused with sand to make crude glass – hence its common name. Glasswort can also be eaten boiled or pickled in vinegar. It has a mild, salty flavour, and is also known as Poor Man's Asparagus.

ORDER PLUMBAGINALES

Common Sea Lavender
Limonium vulgare

TYPE	Perennial
HEIGHT	20–50cm (8–20in)
HABITAT	Muddy salt marshes

DISTRIBUTION Coasts of western Europe, Mediterranean, Black Sea, and Red Sea

This showy plant, which flowers in late summer, often forms dense colonies in salt marshes, particularly along the sides of muddy creeks. Several closely related species of sea lavender are highly localized in their distribution. For example, two species are confined to two rocky peninsulas in Wales, while others are only found in parts of Sicily or Corsica. Varieties of sea lavender, often called statice, are grown commercially as "everlasting" flowers. The coloured, papery "flowers" are actually what remains after the true flowers have fallen.

ORDER PAPAVERALES

Yellow Horned-poppy
Glaucium flavum

TYPE	Biennial or perennial
HEIGHT	50–90cm (20–36in)
HABITAT	Shingle, sometimes sand

DISTRIBUTION Coasts of western Europe, Mediterranean, and Black Sea

Also known as Yellowhorn Poppy, this plant has leaves covered in a waxy coating to protect it from salt spray and reduce water loss. Its taproot penetrates deep into shingle in search of water beneath. It blooms through most of the summer, producing flowers that are up to 9cm (3½in) across.

ORDER POLEMONIALES

Beach Morning-glory
Ipomoea imperati

TYPE	Perennial
LENGTH	Up to 5m (16ft)
HABITAT	Coastal beaches and grasslands

DISTRIBUTION Widespread on many coasts and islands with tropical or warm temperate climates

This pioneer plant helps to stabilize coastal sands, creating a habitat into which other species move. It can endure low nutrient levels, high soil temperatures, abrasion and burial by blown sand, and occasional frosts, but not hurricanes, according to studies in Texas. It also grows occasionally on disturbed ground inland. Beach Morning-glory is recorded on six continents and many isolated islands.

ORDER MYRTALES

Pacific Stilt-mangrove
Rhizophora stylosa

HABIT	Woody perennial
HEIGHT	Commonly 5–8m (16–26ft), but can be up to 40m (130ft)
HABITAT	Intertidal mudflats

DISTRIBUTION Coasts of northern Australia, Southeast Asia, and South Pacific islands

The aerial roots of Pacific Stilt-mangroves arch down from the main trunk, with secondary roots coming off the primary ones before they reach the ground to form a tangle of roots growing in all directions. When the tide is in, these form a sheltered refuge for many small fish. This species of mangrove can tolerate a wide range of soils, but thrives best in the fine, muddy sediments of river estuaries. Its roots absorb water selectively, so much of the damaging salt is not taken up, but it still has to excrete some salt through the leaves.

PROFILE CAPPARALES

Scurvy-grass
Cochlearia officinalis

HABIT	Biennial or perennial
HEIGHT	10–40cm (4–16in)
HABITAT	Coastal rocks and salt marshes

DISTRIBUTION Coasts of northern Europe and Asia and northern North America

The thick, fleshy leaves of this coastal plant help it to store water in an environment where fresh water soon drains away (Scurvy-grass plants found on mountains have thinner leaves and may belong to a different species). Scurvy-grass leaves are rich in vitamin C. They were once eaten, or pulped and drunk, to prevent scurvy – a disease caused by vitamin C deficiency to which sailors were prone ("grass" is Old English for any green plant).

ORDER CARYOPHYLLALES

Grand Devil's-claw
Pisonia grandis

TYPE	Woody perennial
HEIGHT	14–30m (46–98ft)
HABITAT	Coastal and island forests

DISTRIBUTION Coasts and islands in Indian Ocean, Southeast Asia, and South Pacific

The Grand Devil's-claw is typically found on small tropical islands and its distribution is associated with sea bird colonies. It can grow as tall as 30m (98ft), the trunk can be up to 2m (6½ft) in diameter, and it is often the dominant tree in coastal forests that are undisturbed by humans. The trees provide nesting and roosting sites for many species of sea bird, whose guano is an important fertilizer on isolated islands. The branches break easily, and can root in the ground.

BIRD-KILLING TREE

The seeds of Grand Devil's-claw are produced in clusters of 50–200 and exude a resin that makes them extremely sticky. They attach to the feathers of sea birds and may subsequently be flown to remote islands. This is an effective means of dispersal, but the seeds are so sticky that small birds often become completely entangled and die.

ORDER CASUARINALES

Casuarina
Casuarina equisetifolia

TYPE	Woody perennial
HEIGHT	20–30m (66–98ft)
HABITAT	Coastal and island forests

DISTRIBUTION Southeast Asia, eastern Australia, and islands in southeast Pacific

Casuarina has many common names, including Beach She-oak, Beefwood, Ironwood, and Australian Pine. It is typically found at sea level, but also grows inland to 800m (2,600ft). Casuarina is fast-growing, reaching a height of 20m (65ft) in 12 years. It is drought-tolerant, and can grow in poor soils because it can fix nitrogen in nodules on its roots. Its wood is very hard and is used as a building material and as firewood. The bark is widely used in traditional medicines.

ORDER ARECALES

Coconut Palm
Cocos nucifera

HABIT	Woody perennial
HEIGHT	20–22m (66–72ft)
HABITAT	Coastal rocky, sandy, and, coralline soils

DISTRIBUTION Tropical and subtropical coasts worldwide

The Coconut Palm was once the mainstay of life on Pacific islands. It provided food, drink, fuel, medicine, timber, mats, domestic utensils, and thatching for roofs. It remains an important subsistence crop on many Pacific islands today. Its original habitat was sandy coasts around the Indo-Malayan region, but it now is found over a much wider area, assisted by its natural dispersal mechanism, and deliberate planting by humans. The fibrous husk of the coconut fruit is a flotation aid that enables the seeds to be carried vast distances by ocean waves and currents. The Coconut Palm cannot develop viable fruits outside of the tropics and subtropics.

COCONUT FRUIT
The fruit of the Coconut Palm weighs 1–2kg (2¼–4½lb). It contains one seed, which is rich in food reserves and is part solid (flesh) and part liquid (coconut milk).

fibrous husk

edible flesh

COCONUT TREE
The Coconut Palm can live as long as 100 years, a mature tree producing 50–80 coconuts a year. The trunk is ringed with annual scars left by fallen leafbases.

FUNGI

DOMAIN	Eucarya
KINGDOM	Fungi
PHYLA	4
SPECIES	600,000

FUNGI FORM A GREAT KINGDOM OF SINGLE-CELLED and filamentous life-forms, including yeasts and moulds. Some organise their filaments into complex fruiting structures, such as mushrooms. Truly marine fungi are rare, but a few fungus-like organisms survive within a slime covering, avoiding contact with salt water. Fungi are abundant on shorelines, but only in close association with certain algae. Alga and fungus grow in partnership in a kind of symbiotic, compound organism called a lichen. Lichens proliferate in the hostile, wave-splashed zone of bare rock just above high tide.

LICHEN ENCRUSTATION
Fungi thrive on the coast if they grow in association with algae, in an intimate symbiosis called lichen. Here, encrusting and foliose lichens cover sandstone cliffs in the Shetland Isles, Scotland.

ANATOMY

A lichen's body (thallus) is composed mainly of fungal filaments called hyphae. The cells of the fungus's algal partner are restricted to a thin layer below the surface, where they cannot dry out. Lichens grow in one of four ways: bushy (fruticose); leaf-like (foliose); tightly clustered (squamulose); or lying flat (crustose).

Marine fungus-like organisms, such as slime nets (labyrinthulids) and thraustochytrids, are microscopic, usually transparent, and encased in a network of slimy threads. The cells move up and down within the threads and react positively towards food. They are increasingly recognized as protists, however, rather than fungi.

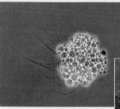

LICHEN COMPOSITION
This false-colour micrograph of a lichen (below) shows the smooth surface of the thallus, to the left, and fungal hyphae, to the right.

ENCASED IN SLIME
This thraustochytrid (above) is a fungus-like organism that lives as a parasite within certain bivalves. Its slime net forms a complete cover.

HABITATS

Most lichens require alternating dry and wet periods, but marine lichens can withstand continuous drought or dampness. On most rocky shores, yellow and grey lichens dominate surfaces splashed by waves at high tide (the splash zone). They endure both the drying Sun and wind, and the salt spray of the sea. Below, in the tidal zone, the brightly coloured lichens give way to black encrusting lichen, such as *Verrucaria maura*, which covers the bedrock and any large, stable boulders. *Verrucaria serpuloides* lives yet further down the shore and is the only lichen to survive permanent immersion in sea water.

Slime nets can live in the sea because they are protected from the dehydrating effects of salt water by slime, or because they live as parasites within seagrasses, green algae, or clams.

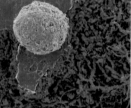

BELOW THE SPLASH ZONE
Some lichens, such as this crustose black *Verrucaria*, live below the splash zone, and may be surrounded by seaweeds.

PHYLUM ASCOMYCOTA

Sea Ivory
Ramalina siliquosa

LENGTH (BRANCHES)	2–10cm (1–4in)
HABITAT	Hard siliceous rocks above the splash zone

DISTRIBUTION Northeast and southwest Atlantic, coasts of Japan and New Zealand

Nutrient-poor siliceous rocks are the favourite habitat of grey lichens, such as Sea Ivory. This lichen is usually grey-green in colour, with a brittle, bush-like (fruticose) structure and disc-shaped fruiting bodies, called apothecia, at its branch tips. Sea Ivory cannot withstand being trampled or extensively grazed, and so it grows best on vertical rock faces, to which it fixes by a single basal attachment.

PHYLUM ASCOMYCOTA

Black Shields
Tephromela atra

WIDTH	Up to 10cm (4in)
HABITAT	In and above the splash zone

DISTRIBUTION Polar coasts, coast of California, USA, Gulf of Mexico, Mediterranean, Indian Ocean

Crustose lichens such as Black Shields, which form a crust over the rock, attach themselves so firmly using fungal filaments that they cannot be easily removed from it. Over time, these anchoring filaments break down the rock as they alternately shrink when dry and swell when moist. Black Shields is a thick, grey lichen with a rough, often cracked, surface from which project a number of characteristic black fruiting bodies.

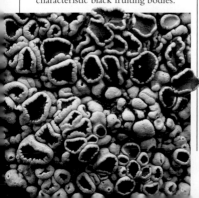

PHYLUM ASCOMYCOTA

Yellow Splash Lichen
Xanthoria parietina

WIDTH	Up to 10cm (4in)
HABITAT	Splash zone; favours surfaces high in nitrogenous compounds

DISTRIBUTION Temperate Atlantic, Gulf of Mexico, Indian and Pacific oceans

On most rocky shores, different species of lichen have a marked vertical territory related to their tolerance of salt exposure. The Yellow Splash Lichen is found in the splash zone and forms a bright orange band across the shore, with grey lichens above it and black lichens below. It has a leaf-like (foliose) form, with slow-growing, leafy lobes held more or less parallel to the rock on which it lives. Usually bright orange in colour, it tends to become greener if in shade. Lichens are widely used to monitor air pollution because they simply disappear when conditions deteriorate. The Yellow Splash Lichen is particularly sensitive to sulphur dioxide, a by-product of industrial processes and of burning fossil fuels.

PHYLUM ASCOMYCOTA

Black Tar Lichen
Verrucaria maura

THICKNESS	1mm (1/32 in)
HABITAT	Intertidal

DISTRIBUTION Temperate and polar coasts, Indian Ocean, Japan

This smooth, black, crustose lichen covers large areas of bedrock or stable boulders in a thin layer, making them appear as though they have been covered with dull black paint. Many types of lichen accumulate heavy metals and the Black Tar Lichen is no exception, having been found to have levels of iron that are about 2.5 million times more concentrated than the surrounding seawater. That may be an adaptation to deter grazers, such as gastropods, from eating it.

PHYLUM ASCOMYCOTA

Black Tufted Lichen
Lichina pygmaea

WIDTH (LOBES)	To 1.5cm (1/2 in)
HABITAT	Lower littoral fringe to middle shore, regularly covered by the tide

DISTRIBUTION Northeast Atlantic from Norway to northwest Africa

Typically found on exposed sunny rock faces, this lichen looks rather like a seaweed, being fruticose (bush-like) in form with branching, brownish black, flattened lobes. Its fruiting bodies form in small swellings at its branch tips. It is often seen growing in association with barnacles but does not tolerate algal (seaweed) growth. Its compact growth and rigid branches provide a refuge for several molluscs, particularly *Lasaea rubra*, a small, pink-shelled gastropod. All *Lichina* species are limited to coastal habitats.

PHYLUM ASCOMYCOTA

Grey Lichen
Pyrenocollema halodytes

SIZE	Not recorded
HABITAT	Upper shore on rocks and on shells of some sedentary invertebrates

DISTRIBUTION Temperate northeast and southwest Atlantic

Seen on hard, calcareous rocks, where it forms small, black-brown patches, Grey Lichen is unusual in being an association of three organisms – a fungus, a cyanobacteria, and an alga. The fungus anchors the lichen to the rock; the cyanobacteria and the alga contain chlorophyll and make food by photosynthesis. The cyanobacteria can also utilize nitrogen, a process that uses a lot of energy, and this comes from the sugar made during photosynthesis.

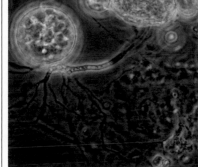

ANIMAL LIFE

DOMAIN	Eucarya
KINGDOM	Animalia
PHYLA	About 30
SPECIES	More than 1.5 million

ANIMAL LIFE FIRST APPEARED IN THE OCEAN over one billion years ago. It has since diversified into a vast array of different organisms. The range of scale among marine animals is immense: the smallest invertebrates are over half a million times smaller than the largest whales. Despite this huge disparity, animals all share two key features. First, they are heterotrophs, meaning they obtain energy from food. Second, they are multicellular, which distinguishes them from single-celled life forms.

INVERTEBRATE
This Yellow Tube Sponge, from the sea off Belize, is a typical sessile invertebrate. Instead of moving to find food, it filters out particles of food by pumping water through its pores.

MARINE ANIMAL DIVERSITY

Animals are classified into 30 or more major groups (phyla), all of which include at least some marine animals. Twenty-nine of these phyla are composed of animals without backbones (invertebrates), each phylum representing a completely different body plan. Only one phylum, the chordates, contains animals with backbones (vertebrates). In salt water, vertebrates include fish, reptiles, birds, and mammals – animals that are often described as the dominant forms of ocean life. However, in terms of abundance and diversity, invertebrates have a stronger claim to this title.

Invertebrates exist in all ocean habitats and outnumber marine vertebrates by a million to one. They include an array of fixed (sessile) animals, such as corals and sponges. They also form most of the zooplankton, a drifting community of animals and animal-like protists.

CHANGING SHAPE
Most invertebrates change shape as they develop. Feather stars start as drifting larvae, which eventually attach themselves to corals or rock before changing into swimming adults.

VERTEBRATE
Active predators, such as this Barracuda, need keen senses and rapid reactions to catch prey. Unlike invertebrates, they have fast-acting nerves and well-developed brains.

SUPPORT AND BUOYANCY

On land, most animals have hard skeletons to counteract gravity's pull. Life is different in the sea, because water is denser than air. It buoys up soft-bodied animals, such as jellyfish, enabling them to grow large. They use internal pressure to keep their shape, the same principle that works in balloons. Animals with hard body parts, such as fish and molluscs, are often denser than water, and would naturally sink. To combat this, many have a buoyancy device. Bony fish have an adjustable gas-filled swim bladder, while squid have an internal float made of chalky material, containing many gas-filled spaces. Some surface dwellers, such as the Violet Sea Snail, have gas-filled floats, which prevent them sinking.

BUBBLE RAFT
The Violet Sea Snail stays afloat by producing bubbles of mucus. The mucus slowly hardens, forming a permanent raft.

GROUPS AND INDIVIDUALS

Among marine animals, there is a social spectrum from species that live on their own to those that form permanent groups. The Whale Shark is a typical solitary species, spending its entire life on its own apart from when it breeds. It can do this because its huge size means it has few natural predators. Smaller fish often form shoals, which reduce each fish's chances of being singled out for attack. Many invertebrates, from corals to tunicates, live in permanent groups, known as colonies. In most coral colonies, the individual animals, or polyps, are anatomically identical and function as independent units, even though they are joined. Other animal colonies, such as the Portuguese Man-of-War, are made of individuals with distinct forms. Each form carries out a different task, like parts of a single animal.

COLONY ON THE MOVE
A diver films a pyrosome colony in the sea off Florida. It consists of thousands of tiny soft-bodied animals called tunicates, joined together to form a tube.

LONE GIANT
The Whale Shark (below) is a solitary species with a pantropical range. It only congregates in particular regions during the breeding season.

SAFETY IN NUMBERS
Crowded together in a ball, the gregarious Striped Catfish (right) make a confusing target for predators.

REPRODUCTION

Animals reproduce in two ways. In asexual reproduction, which occurs in many marine animals from flatworms to sea anemones, a single parent divides in two, or grows (buds off) parts that become independent. In sexual reproduction, the eggs of one parent are fertilized by the sperm of another. Sessile animals, such as corals and clams, usually breed sexually by shedding their eggs and sperm into the water, leaving them to meet by chance. In some fish, all mammals, and birds, fertilization is internal, which means that the two parents have to mate. Marine animals vary greatly in reproductive potential. Most whales have a single calf each time they breed, but an ocean sunfish can produce over 300 million eggs a year.

SINGLE PARENT
This sea anemone is budding off young that will eventually take up life on their own. Asexual reproduction is quick and simple, but it does not produce genetic variation, making it more difficult for a species to adapt to change.

SEA SYMPHONY
A wrasse feeds among coral in the Red Sea. Coral reefs contain the greatest diversity of animal life in the oceans, and are one of the few habitats that are actually created by animals.

COURTSHIP
Two Waved Albatrosses display to each other in the Galapagos Islands. Complex courtship rituals like this ensure that each parent finds a partner of the right species and the right sex, and they cement the bond once breeding begins.

OCEAN LIFE

SPONGES

DOMAIN	Eucarya
KINGDOM	Animalia
PHYLUM	Porifera
CLASSES	3
SPECIES	About 15,000

THIS ABUNDANT AND diverse group of often colourful invertebrates lives permanently attached to the sea floor. Naturalists once thought they were plants, but they are now known to be very simple animals with no close relatives. Sponges live by drawing water into their bodies through tiny holes called pores, filtering it for food and oxygen and pushing it out again. Many species are found on coral reefs or rocks, and a few live in fresh water.

HABITATS

Most sponges need a hard surface for attachment, but some can live in soft sediment; a few species are able to bore into rocks and shells. Sponges are common on rocky reefs, shipwrecks, and coral reefs in a wide range of temperatures and depths. The largest populations occur where there are strong tidal currents, which bring extra food. Animals such as crabs and worms sometimes live inside sponges, but little manages to settle and grow on their surface. This is because sponges produce chemicals to discourage predators.

CHANGING SHAPE
Many sponges grow different shapes in different habitats. This sponge develops fingers in strong currents (above), but has an encrusting form (right) when it grows in wave-exposed sites.

ANATOMY

The body plan of a sponge is based on a system of water canals lined with special cells known as collar cells. Collar cells are unique to sponges. They draw water into the sponge through pores, by each beating a long, whip-like flagellum. A ring of tiny tentacles around the base of the flagellum traps food particles, and the water and waste material then flows out of the sponge through larger openings. Rigidity is provided by a skeleton made up of tiny splinters (spicules) of silicon dioxide or calcium carbonate scattered throughout the body.

(diagram labels: osculum, central cavity, collar cell, flagellum, spicule, pore)

BODY SECTION
A sponge has specialized cells, but no organs. Water enters the sponge through hollow pore cells and exits via larger openings called osculae.

VARIETY OF FORM
Sponges come in many forms, including tubes, spheres, and thread-like shapes. Pictured are a brown tube sponge and an irregular deep red sponge.

CLASS HEXACTINELLIDA

Reef-forming Sponge
Heterochone calyx

HEIGHT	Up to 1.5m (5ft)
DEPTH	100–250m (300–800ft)
HABITAT	Deep hard sea bed

DISTRIBUTION Deep cold waters of north Pacific

The Reef-forming Sponge not only looks like a delicate glass vase, but its skeleton spicules are made from the same material as glass, silica. Each spicule has six rays, hence the Latin name of its class, *Hexactinellida*. Many glass sponges grow very large – off Canada's British Columbian coast, the Reef-forming Sponge forms huge mounds nearly 20m (65ft) high spread over several kilometres. Other members of their class also contribute to these reefs, which may have started forming nearly 9,000 years ago. Like coral reefs, sponge reefs provide a home for many other animals.

CLASS DEMOSPONGIAE

Barrel Sponge
Xestospongia testudinaria

HEIGHT	Up to 2m (6ft)
DEPTH	2–50m (6–165ft)
HABITAT	Coral reefs

DISTRIBUTION Tropical waters of western Pacific

These gigantic sponges grow large enough to fit a person inside. Their hard surface is deeply ridged, but their rim is thin and delicate. The Barrel Sponge belongs to the *Demospongiae*, the largest class of sponges, containing about 95 per cent of sponge species. The skeleton of sponges in this class is made from both scattered spicules of silica and organic collagen called spongin. An almost identical barrel sponge, *Xestospongia muta*, occurs in the Caribbean.

CLASS DEMOSPONGIAE

Blue Sponge
Adocia species

SIZE (LENGTH) Up to 1m (3¼ft)

DEPTH Shallow water

HABITAT Steep coral reefs

DISTRIBUTION Tropical waters off coast of northern Borneo

The exact identity of this beautiful, bright blue sponge has not yet been ascertained. It is soft and spongy in texture and grows in irregular branches with a row of large, round osculae running along each branch. It is especially common on the tops of prominent corals and rocks and on steep, shaded reef areas. This and many other tropical sponges are recognized as distinct species but have not yet been formally described and named. The Dorid Sea Slug (*Jorunna funebris*) feeds on this sponge.

CLASS DEMOSPONGIAE

Breadcrumb Sponge
Halichondria panicea

WIDTH To more than 30cm (12in)

DEPTH Shore to sublittoral zone

HABITAT Hard surfaces

DISTRIBUTION Temperate coastal waters of northeastern Atlantic and Mediterranean

The appearance of this soft encrusting sponge varies from thin sheets to thick crusts and large lumps. On wave-exposed shores, it usually grows under ledges as a thin, green crust, its osculae opening at the tops of small mounds. Its green colour is produced by photosynthetic pigments in symbiotic algae in the sponge's tissues. In deeper, shaded waters, the sponge is usually a creamy yellow. In waters with strong currents, this sponge may cover large rocky areas and kelp stems.

CLASS DEMOSPONGIAE

Tube Sponge
Kallypilidion fascigera

HEIGHT Up to 1m (3ft)

DEPTH Below 10m (33ft)

HABITAT Coral reefs

DISTRIBUTION Tropical reef waters of western Pacific; likely to be more widespread than shown

The elegant, tubular branches of this beautiful sponge are easily torn, and so it occurs only on deeper reef slopes, where wave action is minimal.

It sometimes grows as a single tube, but it is more often seen as bunches of tubes joined at the base. The tips of the tubes are translucent and slightly rolled in. The colour of this sponge is usually pinkish violet, although some specimens are pinkish blue. When this sponge releases sperm, it resembles smoking chimneys.

The taxonomic status of this species and its relationship to other species in the same family has not been fully determined, and it is listed under various names in different sources. Such uncertainties are not unusual in the study of sponges and mean that the exact distribution of this many other species is yet to be established.

CLASS DEMOSPONGIAE

Mediterranean Bath Sponge
Spongia officinalis adriatica

WIDTH Up to 35cm (14in)

DEPTH 1–50m (3–165ft)

HABITAT Rocks

DISTRIBUTION Mediterranean, especially the eastern part

The Mediterranean Bath Sponge, as its name suggests, is collected and processed for use as a bath sponge. It grows as rounded cushions and mounds, and is usually dull grey to black outside but yellowish white inside. It can be used as a sponge because it has no sharp skeletal spicules, just a network of tough fibres made from an elastic material called spongin. Huge numbers were once harvested, but today they are rare.

CLASS CALCAREA

Lemon Sponge
Leucetta chagosensis

WIDTH Up to 20cm (8in)

DEPTH Shallow

HABITAT Steep coral reef and rock slopes

DISTRIBUTION Tropical reef waters of western Pacific

The Lemon Sponge is a beautiful, bright yellow colour and is easy to spot underwater. It grows in the form of sacs, which may have an irregular, lobed shape. Each sac has a large opening – the osculum – through which used water flows out of the sponge. Through the osculum, entrances to the water-intake channels that run throughout the sponge can be seen. The Lemon Sponge belongs to a small class of sponges in which the mineral skeleton is composed entirely of calcium carbonate spicules, most of which have three or four rays. The densely packed spicules give the sponge a solid texture. Like all sponges, this sponge is hermaphroditic. It incubates its eggs inside and releases them as live larvae through the osculum. Each larva is a hollow ball of cells with flagellae for swimming.

CLASS SCLEROSPONGIAE

Coralline Sponge
Vaceletia ospreyensis

SIZE Not recorded

DEPTH At least 20m (65ft)

HABITAT Dark reef caves

DISTRIBUTION Not fully known, but includes tropical waters of western Pacific

Vaceletia ospreyensis is a living member of the coralline sponges group, most of which are known only from fossils. Coralline Sponges have a massive skeleton made of calcium carbonate, as well as silica spicules and organic fibres. They were the dominant reef-building organisms before the stony corals of modern reefs evolved. Although given a separate class here, the Sclerospongiae, it is quite possible that Coralline Sponges belong to the class Demospongiae.

CNIDARIANS

DOMAIN	Eucarya
KINGDOM	Animalia
PHYLUM	Cnidaria
CLASSES	4
SPECIES	8,000–9,000

THIS ANCIENT GROUP OF AQUATIC ANIMALS emerged in Precambrian times, about 600 million years ago. It includes reef-building corals, anemones, jellyfish, and hydroids, most of which are marine. Cnidarians have a radially symmetrical body shaped like a simple sac, with stinging tentacles around a single opening that serves as both mouth and anus. There are two body forms: the polyp form, typified by sea anemones, which is fixed to a solid surface and has an upward-facing mouth and tentacles; and the medusa, shown by adult jellyfish, which can swim and has a downward-facing mouth and tentacles.

HUMAN IMPACT

CORAL TRINKETS

Many corals are harvested for sale as souvenirs, and the most valued species are being over-collected. Particularly desirable are certain soft corals, in which the calcareous supporting column is so strong and dense it can be carved and polished. They include the red or precious coral, *Corallium rubrum*, which is made into trinkets and beaded necklaces (below). Black corals (order Antipatharia) also have strong skeletons that can be carved, and fortune-seeking divers sometimes take great risks to gather these deep-water species.

ANATOMY

Corals and anemones exist only as polyps, whereas other cnidarians can be either polyps or medusae at different stages of their life cycle. The body wall of both polyps and medusae consists of two types of tissue. On the outside is the epidermis, which acts like a skin to protect the animal. The inner tissue layer, lining the body cavity, is the gastrodermis, which carries out digestion and produces reproductive cells. Separating and connecting these two layers is a jelly-like substance called the mesoglea. The tentacles have stinging cells called cnidocytes, which are unique to this phylum and give it its name. A simple nervous system responds to touch, chemicals, and temperature.

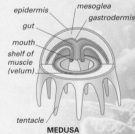

TENTACLE ARRANGEMENT
The number of tentacles on coral polyps varies from one group to another. The polyps of all soft corals (above) have eight tentacles, hence their alternative name of octocorals. Hexacorals (right) have tentacles arranged in multiples of six.

POLYP
- tentacle
- cnidocyte
- epidermis
- mesoglea
- gastrodermis
- budding juvenile
- gut
- basal disc

POLYP AND MEDUSA
Polyps are essentially a tube, closed at one end, that attaches to a hard surface by a basal disc. They live singly or in colonies. Medusae are bell-shaped and usually have a thicker mesoglea; some also have a shelf of muscle for locomotion.

MEDUSA
- epidermis
- gut
- mouth
- shelf of muscle (velum)
- mesoglea
- gastrodermis
- tentacle

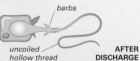

- epidermal cell
- coiled thread
- nematocyst
- **BEFORE DISCHARGE**
- barbs
- uncoiled hollow thread
- **AFTER DISCHARGE**

STINGING CELLS
Each cnidocyte contains a bulb-like structure, called a nematocyst, which houses a coiled, barbed thread. When triggered by touch or chemicals, the thread explodes outwards and pierces the prey's skin. The animal's tentacles are then used to haul the victim in.

SCLERITES
Small slivers of calcium carbonate called sclerites are scattered through the tissues of soft corals and sea fans. Here, they are visible as white shards under the colony's skin.

BUILDING REEFS
Coral reefs are built by colonies of coral polyps that secrete a hard exoskeleton of calcium carbonate. As the tiny polyps divide and grow, the reef expands.

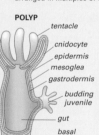

LOCOMOTION

The most mobile cnidarians are free-living jellyfish and medusae, which mainly drift in water currents but also swim actively using a form of jet propulsion. Most colonial cnidarians, such as corals and sea fans, cannot move from place to place. However, they can expand and contract their polyps to feed or escape danger, and some sea pens can withdraw the whole colony below the surface of the sediment in which they live. Unattached mushroom corals may move slowly, or even right themselves if overturned. Anemones can creep slowly over the sea bed on their muscular basal disc, and a few species swim if attacked.

JELLYFISH SWIMMING
A jellyfish swims by using muscles to contract its bell, forcing water out and pushing it along. The muscles then relax and the bell opens again.

bell relaxed and flattened, ready to propel forwards

bell begins to contract and force water out

bell fully contracted, with little water remaining inside

REPRODUCTION

Members of the class Anthozoa, such as corals and anemones, reproduce by asexual budding. A genetically identical copy of the adult grows on the polyp's body wall. This budding juvenile drops off or stays attached to form a colony. Anthozoans also reproduce sexually, producing eggs and sperm within the polyps. Fertilized eggs develop into hairy, oval larvae (planulae), which either swim free or are brooded internally and then released. Hydrozoans have a two-stage life cycle. Their polyps release tiny free-swimming medusae into the water which, when mature, shed eggs and sperm. The resulting fertilized eggs develop into planulae that settle in a new area to grow into polyps. In contrast, the medusa form of jellyfish is usually much larger than the fixed polyp form and the polyps bud asexually.

BUDDING JELLYFISH POLYPS
Jellyfish polyps are minuscule, and their sole function is to reproduce asexually by budding off baby jellyfish.

ZOOXANTHELLAE

The massive skeletons secreted by reef-building corals require energy for their construction. Corals cannot catch enough plankton in clear tropical waters to provide this energy. Instead, they rely on tiny, symbiotic single-celled algae, called zooxanthellae, living in their cells. These algae manufacture organic matter by photosynthesis, and make more food than they need, so the excess is used by the coral. The algae benefit from a safe place to live and obtain "fertilizer" from the coral by using its nitrogenous waste products. If stressed by disease or high temperatures, corals expel their zooxanthellae, in a process called coral bleaching, and may die of starvation.

ZOOXANTHELLAE
Reef corals have mainly colourless polyps and white skeletons – it is the tiny zooxanthellae living in their tissues that gives the corals their colour.

CNIDARIAN CLASSIFICATION

Cnidarians are divided into four classes and a large number of orders and families. This phylum used to be called the Coelenterata, a name still used by some authorities. Many species remain undescribed.

ANTHOZOANS
Anthozoa

About 6,000 species
These colonial or solitary polyps are diverse in shape and have no medusa phase. Octocorals (soft corals, sea fans, and sea pens) have polyps with eight feathery tentacles; hexacorals (including hard corals and anemones) have polyps with multiples of six simple tentacles; ceriantipatharians have polyps with unbranched tentacles.

RED SEA FAN *Paracis species*

JELLYFISH
Scyphozoa

About 200 species
These mostly free-swimming medusae are shaped like a bell or saucer with a fringe of stinging tentacles. The edges of the mouth, located on the underside, are drawn out to form trailing mouth tentacles or oral arms.

BOX JELLYFISH
Cubozoa

About 16 species
These jellyfish have a cube-shaped bell with four flattened sides and a domed top. There are four tentacles or clusters of tentacles, one at each corner. Most are virulent stingers.

HYDROIDS
Hydrozoa

About 2,700 species
These colonial cnidarians mostly resemble plant growths attached to the sea bed. A few have hard skeletons and resemble corals, and some colonies float at the surface like jellyfish. Most species have a free-living medusa stage.

Blue Buttons
Porpita porpita

DIAMETER	2cm (³/₄ in)
DEPTH	Surface
HABITAT	Surface waters

DISTRIBUTION Worldwide in warm waters

At first sight, Blue Buttons could be mistaken for a small jellyfish or even a piece of blue plastic. In fact, it is a hydrozoan colony that is modified for a free-floating existence. Swarms of these unusual creatures can be seen drifting on the water's surface or can sometimes be found washed up on the shore. The animal is kept afloat by a buoyant circular disc. Around the edge hang protective stinging polyps modified as knobbed tentacles. In the centre underneath hangs a large feeding polyp that acts as the mouth for the whole colony. In between this and the tentacles are circlets of reproductive polyps. Unlike the Portuguese Man-of-War (see p.214) to which it is related, Blue Buttons do not have a powerful sting.

CLASS HYDROZOA

Stinging Hydroid
Aglaophenia cupressina

HEIGHT	Up to 40cm (16in)
DEPTH	3–30m (10–100ft)
HABITAT	Coral reefs

DISTRIBUTION Tropical reefs in Indian Ocean and southwestern Pacific

While most hydroids are harmless to touch, the Stinging Hydroid has a powerful sting. The colonies look like clumps of feathers or ferns dotted around among the corals on a reef. Individual polyps are arranged along one side of the smallest branches and extend their stinging tentacles to catch small planktonic animals. The sting is not usually dangerous to humans, but it results in an itchy rash that can irritate for up to a week.

CLASS SCYPHOZOA

Deep-sea Jellyfish
Periphylla periphylla

HEIGHT	20–35cm (8–14in)
DEPTH	900–7,000m (3,000–23,000ft)
HABITAT	Open water

DISTRIBUTION Deep water worldwide, except Arctic Ocean

This jellyfish belongs to a group called coronate jellyfish, which are shaped like a ballet tutu. The upper part of the bell is a tall, stiff cone and the lower part a wider, soft, crown-shaped base with a scalloped edge. The 12 thin tentacles are often held in an upright position. The insides of the Deep-sea Jellyfish are a deep red colour and this may hide the bioluminescent light given out by its ingested prey. The jellyfish itself can squirt out a bioluminescent secretion that may help to confuse any predators. Unlike many jellyfish, the Deep-sea Jellyfish does not develop from a fixed bottom-living stage.

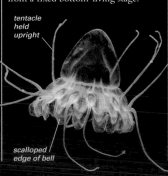

tentacle held upright

scalloped edge of bell

CLASS SCYPHOZOA

Moon Jellyfish
Aurelia aurita

DIAMETER	Up to 30cm (12in)
DEPTH	Near surface
HABITAT	Open water

DISTRIBUTION Worldwide; polar distribution unknown

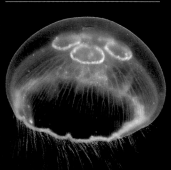

The Moon Jellyfish is possibly the most widespread of all jellyfish and can be found in almost every part of the ocean except for very cold waters. It exists mainly in coastal waters and is sometimes cast ashore in large numbers because it is not a strong swimmer and lives near the surface. The body is shaped like a saucer with a fringe of fine, short tentacles, which it uses to catch plankton. It can also trap plankton in sticky mucus on its bell and slide this down into its mouth on the underside. The gonads show through the translucent bell as four opaque horseshoe shapes.

CLASS SCYPHOZOA

Stalked Jellyfish
Haliclystus auricula

HEIGHT	Up to 5cm (2in)
DEPTH	0–15m (0–49ft)
HABITAT	On seaweed or seagrass

DISTRIBUTION Coastal waters of north Atlantic and north Pacific

Most jellyfish drift and swim freely in the water but Stalked Jellyfish spend their lives attached by a stalk to vegetation. The body of the jellyfish is shaped like a tiny funnel made up of eight equally spaced arms joined together by a membrane. Each arm ends in a cluster of tentacles on the funnel rim, and between each of these clusters is an extra anchor-shaped tentacle. This animal cannot swim, but it can move by bending over on its stalk and turning "head-over-heels", using the anchor tentacles to fix itself temporarily to the sea bed as it flips over and then reattaches its adhesive disc.

Stalked Jellyfish can be found attached to seaweed or seagrass in the intertidal zone and shallow water, where they feed by catching prey, such as small shrimps and fish fry, in their tentacles and passing it to the mouth inside the funnel. Undigested remains are expelled from the mouth.

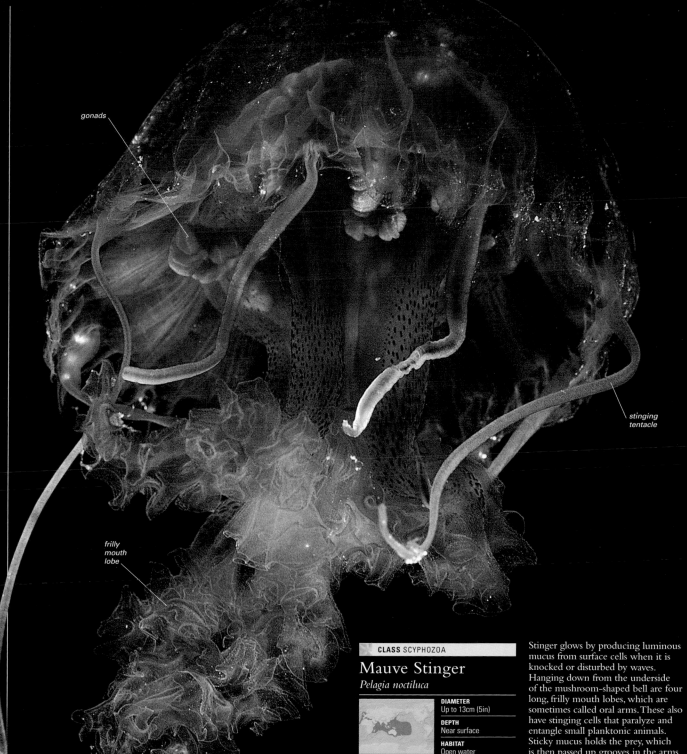

gonads

stinging
tentacle

frilly
mouth
lobe

CLASS SCYPHOZOA

Mauve Stinger

Pelagia noctiluca

DIAMETER	Up to 13cm (5in)
DEPTH	Near surface
HABITAT	Open water

DISTRIBUTION Northeastern Atlantic, Mediterranean, Indian Ocean, and western and central Pacific

This jellyfish produces bioluminescent lightshows, which are often admired from passing boats, but it also has a reputation as a ferocious stinger. As well as having eight stinging tentacles, it is covered in tiny red spots that are bundles of stinging cells. The sting is painful but not dangerous. The Mauve Stinger glows by producing luminous mucus from surface cells when it is knocked or disturbed by waves. Hanging down from the underside of the mushroom-shaped bell are four long, frilly mouth lobes, which are sometimes called oral arms. These also have stinging cells that paralyze and entangle small planktonic animals. Sticky mucus holds the prey, which is then passed up grooves in the arms and into the mouth.

Unlike most jellyfish, the life cycle of the Mauve Stinger does not involve a fixed stage. Eggs and sperm are shed into the water where the eggs are fertilized and develop into tiny, oval planula larvae covered in hair-like cilia. The planula larva changes directly into a tiny, lobed, saucer-shaped medusa called an ephyra, which gradually develops into an adult.

CLASS SCYPHOZOA

Upside-down Jellyfish
Cassiopeia xamachana

DIAMETER	Up to 30cm (12in)
DEPTH	0–10m (0–33ft)
HABITAT	Coastal mangroves

DISTRIBUTION Tropical waters of Gulf of Mexico and Caribbean

Divers who find this jellyfish upside-down on the sea bed often think they have found a dying specimen. However, the Upside-down Jellyfish lives like this, floating with its bell pointing downwards and its eight large,

branching mouth arms held upwards. The mouth arms have elaborate fringes consisting of tiny bladders filled with minute single-celled algae called zooxanthellae. The algae need light to photosynthesize, and the jellyfish behaves as it does in order to ensure its passengers can thrive. Excess food manufactured by the algae is used by the jellyfish, but it can also catch planktonic animals with stinging cells on the mouth arms. Its bell pulsates to create water currents that bring food and oxygen. When it wants to move, the Upside-down Jellyfish turns the right way up with the bell uppermost. A very similar jellyfish, *Cassiopeia andromeda*, is found in the tropical Indian and Pacific Oceans and may actually be the same species.

CLASS CUBOZOA

Box Jellyfish
Chironex fleckeri

DIAMETER	Up to 25cm (10in)
DEPTH	Near surface
HABITAT	Open water

DISTRIBUTION Tropical waters of southwest Pacific and eastern Indian Ocean

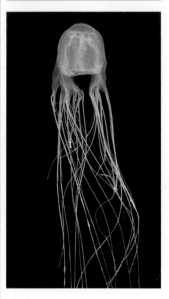

A sting from the Box Jellyfish can kill a person in only a few minutes, and this small animal is considered one of the most venomous in the ocean. At each corner of its box-shaped, transparent body is a bunch of 15 tentacles. When it is hunting prey such as prawns and small fish in shallow water, the tentacles extend up to 3m (10ft), and swimmers can be stung without ever seeing the jellyfish. In the middle of each flattened side is a collection of sense organs including some remarkably complex eyes. The exact range of this jellyfish in the Indo-Pacific region north of Australia is not known, but other smaller, less dangerous box jellyfish also occur in the Indian and Pacific oceans. Some sea turtles can eat the Box Jellyfish without being affected by its sting.

HUMAN IMPACT

LETHAL VENOM

The sting of a Box Jellyfish causes excruciating pain and skin damage and can leave permanent scars. In severe cases, death may occur from heart failure or drowning following loss of consciousness. A Box Jellyfish antivenom is available in Australia. In northern parts of the country, some beaches are closed to the public for periods between November and April when the jellyfish are most abundant.

CLASS ANTHOZOA

Organ Pipe Coral
Tubipora musica

DIAMETER	Up to 50cm (20in)
DEPTH	5–20m (15–65ft)
HABITAT	Tropical reefs

DISTRIBUTION Tropical reefs of Indian Ocean and western Pacific

Although the Organ Pipe Coral has a hard skeleton, it is not a true stony coral. Instead, it belongs in the group of cnidarians that includes soft corals and sea fans. Its beautiful red skeleton is made up of parallel tubes joined by horizontal links, and bits of this animal's skeleton are often found washed up on tropical shores. A single polyp extends from the end of each tube, and when the polyps expand their eight branched tentacles to feed, the skeleton cannot be seen.

CLASS ANTHOZOA

Mushroom Leather Coral
Sarcophyton species

DIAMETER	Up to 1.5m (5ft)
DEPTH	0–50m (0–165ft)
HABITAT	Rocks and reefs

DISTRIBUTION Tropical waters of Red Sea, Indian Ocean, and western and central Pacific

This distinctive soft coral has a conspicuous bare stalk topped by a wide, fleshy cap covered in polyps. When the colony is touched or is resting, the polyps are withdrawn into the fleshy body, and it looks and feels like leather. Within this genus there are many similar species.

CLASS ANTHOZOA

Dead Man's Fingers
Alcyonium digitatum

HEIGHT	Up to 20cm (8in)
DEPTH	0–50m (0–165ft)
HABITAT	Rocks and wrecks

DISTRIBUTION Temperate and cold waters of northeastern Atlantic

This soft coral's strange name comes from its appearance when thrown ashore by storms. It is shaped like a thick lump with stubby fingers, which can, with a little imagination, resemble a corpse's hand. When alive, it grows attached to rocks in shallow water and often covers large areas, especially where strong currents bring plenty of planktonic

food. With the polyps extended, the colonies have a soft, furry look. Most Dead Man's Fingers colonies are white but some, like those shown below, are orange with white polyps. Over the autumn and winter, the colony retracts its polyps and becomes dormant. In the spring, the outer skin is shed, along with any algae and other organisms that have settled on it.

Carnation Coral
Dendronephthya species

HEIGHT	Up to 30cm (12in)
DEPTH	10–50m (33–165ft)
HABITAT	Coral reefs

DISTRIBUTION Tropical reefs of Red Sea, Indian Ocean, and western Pacific

Carnation Corals are among the most colourful of all reef animals. They grow as branched and bushy colonies and often cover steep reef walls with pink, red, orange, yellow, and white patches. They prefer to live where there are fast currents. When the current is running, they expand to full size and the polyps, which are on the branch ends, extend out to feed. With little or no current, they often hang down as flaccid lumps. In some species, such as the one shown here, small slivers of coloured calcium carbonate show through the body tissue. These are called sclerites and help to give the soft branches some strength. Individual species of *Dendronephthya* are difficult to identify visually and many species have not yet been described.

Pulse Coral
Xenia species

HEIGHT	Up to 5cm (2in)
DEPTH	5–50m (15–165ft)
HABITAT	Coral reefs

DISTRIBUTION Tropical reefs of the Red Sea, Indian Ocean, and western Pacific

The most notable feature of this soft coral is the way the feathery tentacles of the polyps rapidly and continually open and close. A reef covered in Fast-Pulse Coral is alive with movement. The colonies have a stout trunk with a dome-shaped top covered with long polyps. Unlike the Mushroom Leather Coral (see opposite), Pulse Coral polyps cannot retract and disappear. The pulsating movements of the polyps may help to oxygenate the colony as well as bring food within range of their tentacles.

Common Sea Fan
Gorgonia ventalina

HEIGHT	Up to 2m (6ft)
DEPTH	5–20m (15–65ft)
HABITAT	Coral reefs

DISTRIBUTION Caribbean Sea

Sea fans grow attached to the sea bed and look like exotic plants. Unlike soft corals, they have a supporting skeleton that provides a framework and allows them to grow quite large. It is made mainly of a flexible, horny material called gorgonin and consists of a rod that extends down the inside of all except the smallest branches. In the Common Sea Fan, the branches are mostly in one plane and form a mesh that is aligned at right angles to the prevailing current. This increases the amount of planktonic food brought within reach of the polyps, which are arranged all around the branches. Fishing nets dragged over the reef can damage Common Sea Fans and, as they grow quite slowly, they take a long time to recolonize. They are also collected, dried, and sold as souvenirs.

White Sea Whip
Junceella fragilis

HEIGHT	Up to 2m (6ft)
DEPTH	5–50m (15–165ft)
HABITAT	Coral reefs

DISTRIBUTION Southwestern Pacific

Sea whips have a very similar structure to sea fans but grow up as a single tall stem. They have a very strong central supporting rod containing a lot of calcareous material as well as a flexible, horny material called gorgonin. The small polyps have eight tentacles and are placed all around the stem. White Sea Whips are often found in groups because they can reproduce asexually. As the whip enlarges, the fragile tip breaks off and drops onto the sea bed, where it attaches and grows.

CLASS ANTHOZOA

Mediterranean Red Coral
Corallium rubrum

HEIGHT	Up to 50cm (20in)
DEPTH	50–200m (165–650ft)
HABITAT	Shaded rocks and caves

DISTRIBUTION Mediterranean and warm waters of eastern Atlantic

Often called Precious Coral, Mediterranean Red Coral has been collected and its skeleton made into jewellery for centuries. In spite of its name, it is not a true stony coral but instead is in the same group as sea fans (see p.267). Like them, its branches are covered in small polyps, each of which has eight branched tentacles. However, the supporting skeleton is made mainly from hard calcium carbonate coloured a deep red or pink. This coral is now scarce in places that are easily accessible to collectors.

CLASS ANTHOZOA

Orange Sea Pen
Ptilosarcus gurneyi

HEIGHT	Up to 50cm (20in)
DEPTH	10–300m (33–1,000ft)
HABITAT	Sediment

DISTRIBUTION Temperate waters of northeastern Pacific

Unlike the majority of anthozoans, sea pens live in areas of sand and mud. They get their name from their resemblance to an old-fashioned quill pen. The Orange Sea Pen consists of a central stem with branches on either side. The basal part of the stem is bulbous and anchors the colony in the sediment. Single rows of polyps extend their eight tentacles into the water from each leaf-like branch, giving the front of the sea pen a downy appearance. The colony faces towards the prevailing current to maximize the flow of plankton over the feeding polyps. When no current is flowing, the colony can retract down into the sediment. Although they tend to stay in one place, colonies can relocate and re-anchor themselves if necessary. Predators of sea pens include sea slugs and starfish.

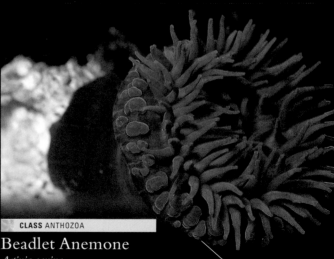

CLASS ANTHOZOA

Slender Sea Pen
Virgularia mirabilis

HEIGHT	Up to 60cm (24in)
DEPTH	10–400m (33–1,300ft)
HABITAT	Sediment

DISTRIBUTION Temperate waters of northeastern Atlantic and Mediterranean

The muddy bottoms of sheltered sea lochs in Scotland and Norway are often carpeted in dense beds of Slender Sea Pens. This species has a similar structure to the Orange Sea Pen (see below, left) but has a much thinner central stalk and thin branches. Almost half the stalk is buried in the sediment and the colony can withdraw into the sediment if disturbed.

CLASS ANTHOZOA

Beadlet Anemone
Actinia equina

DIAMETER	Up to 7cm (2¾ in)
DEPTH	0–20m (0–65ft)
HABITAT	Hard surfaces

DISTRIBUTION Coastal waters of Mediterranean, northeastern and eastern Atlantic

Most anemones cannot survive out of water but the Beadlet Anemone can do so provided it stays damp. At low tide, this anemone can be found on rocky shores with its tentacles retracted, looking like a blob of red or green jelly. The top of the anemone's body is ringed with blue beads called acrorhagi. These contain numerous stinging cells, which the anemone uses to repel any close neighbours. Leaning over, it will sting any anemone within reach and the defeated anemone will move slowly out of the victor's territory. The Beadlet Anemone broods its eggs and young inside the body and ejects them through its mouth.

CLASS ANTHOZOA

Giant Anemone
Condylactis gigantea

DIAMETER	Up to 30cm (12in)
DEPTH	3–50m (10–165ft)
HABITAT	Coral reefs and rocks

DISTRIBUTION Tropical waters of Caribbean Sea and western Atlantic

The long, purple-tipped tentacles of this large anemone bring a splash of colour to Caribbean reefs. Its columnar body is usually tucked away between rocks or corals leaving only the stinging tentacles exposed. Several small reef fish (mainly blennies) can live unharmed among the tentacles, where they gain protection from predators. The Giant Anemone can move slowly along on its basal disc if it wants to find a better position on the reef.

acrorhagi containing stinging cells

Plumose Anemone
Metridium senile

HEIGHT	Up to 30cm (12in)
DEPTH	0–100m (0–330ft)
HABITAT	Any hard surface

DISTRIBUTION Temperate waters of north Atlantic and north Pacific

This tall anemone resembles an ornate piece of architecture. It has a long column, topped by a collar-like ring and a wavy disc with thousands of fine tentacles. The commonest colours are white or orange but it can also be brown, grey, red, or yellow. Fragments from the base of large anemones can grow into tiny new anemones. The Plumose Anemone is often found on pier pilings and wrecks projecting out into the current.

Cloak Anemone
Adamsia carciniopados

DIAMETER	5cm (2in)
DEPTH	0–200m (0–650ft)
HABITAT	Hermit crab shells

DISTRIBUTION Temperate waters of northeastern Atlantic and Mediterranean

The Cloak Anemone lives with its wide base wrapped around the shell of a hermit crab and its tentacles trailing beneath the crab's head. In this position, the tentacles are ideally placed to pick up food scraps. The enveloping column of the anemone is off-white with distinct pink spots. Neither partner thrives without the other, though young Cloak Anemones can be found on rocks and shells between the tidemarks waiting to find a host.

ARMOURED VEHICLE

The Hermit Crab *Pagurus prideaux* is always seen with its protective anemone cloak. It does not have to find a bigger shell as it grows because the Cloak Anemone secretes a horny extension. The anemone on the crab on the left has thrown out pink stinging threads, called acontia, to repel another Hermit Crab.

Antarctic Anemone
Urticinopsis antarctica

SIZE	Not recorded
DEPTH	5–225m (15–740ft)
HABITAT	Rocky sea beds

DISTRIBUTION Southern Ocean around Antarctica and South Shetland Islands

Like many other Antarctic marine animals, the Antarctic Anemone grows to a large size but rather slowly. It has long tentacles with powerful stinging cells and is capable of catching and eating starfish, sea urchins, and jellyfish much larger than itself. As there are often many anemones living close together, two or more may hold a large jellyfish. As in most anemones, stinging cells on the tentacles fire barbed threads into the prey to hold it and to paralyse or kill it.

Jewel Anemone
Corynactis viridis

DIAMETER	1cm (½ in)
DEPTH	0–80m (0–260ft)
HABITAT	Steep rocky areas

DISTRIBUTION Temperate waters of northeastern Atlantic and Mediterranean

Jewel Anemones often cover large areas of underwater cliff faces creating a spectacular display. Individuals can be almost any colour and they reproduce by splitting in half, making two new identical anemones. This results in dense patches of different coloured anemones. Each anemone has a small saucer-shaped disc circled by stubby translucent tentacles. The tentacles have knobbed tips that are often a contrasting colour to the tentacle shafts, disc, and column of the anemone. The colour combination shown here is one of the most common. Jewel Anemones are not true anemones but belong to a group of anthozoans called coralliomorphs. These closely resemble the polyps of hard corals but have no skeleton. Coralliomorphs are found in all oceans but are most common in the tropics.

CLASS ANTHOZOA

Table Coral
Acropora hyacinthus

DIAMETER	
	Up to 3m (10ft)
DEPTH	
	0–10m (0–33ft)
HABITAT	
	Coral reefs

DISTRIBUTION Tropical waters of Red Sea, Indian Ocean, and western and central Pacific

The magnificent flat plates of Table Coral are ideally shaped to expose as much of their surface as possible to sunlight. Like most hard corals, the cells of Table Coral contain zooxanthellae that need light to photosynthesize and manufacture food for themselves and their host. Table Coral is supported on a short, stout stem that is attached to the sea bed by a spreading base. The horizontal plates have numerous branches that mostly project upwards from the surface, so each plate,

or table, resembles a bed of nails. Each of these branches is lined by cup-shaped extensions of the skeleton called corallites, from which the polyps extend their tentacles in order to feed, mainly at night.

The usual colour of Table Coral is a dull brown or green, but it is brightened up by the numerous reef fish that shelter under and around its plates. However, the shade the plates cast means that few other corals can live underneath a Table Coral. There are many other similar species that are also called table coral, but *Acropora hyacinthus* is one of the most abundant and widespread.

PEOPLE

CHARLIE VERON

Born in Sydney, Australia, in 1945, Charlie Veron has been dubbed the "King of Coral" for his lifelong work on coral reefs. He has formally named and described over 100 new coral species, including many from the genus *Acropora*. His three-volume book *Corals of the World* is a classic text.

CLASS ANTHOZOA

Hump Coral
Porites lobata

DIAMETER	
	Up to 6m (20ft)
DEPTH	
	0–50m (0–165ft)
HABITAT	
	Coral reefs

DISTRIBUTION Tropical waters of Red Sea, Persian Gulf, and Indian and Pacific oceans

It can be difficult to tell that Hump Coral is a living coral colony because it looks just like a large, lumpy rock. Closer inspection will show that the coral grows as a series of large lobes formed into a dome. The living polyps are tiny, with tentacles that are only about 1mm (¹⁄₃₂ in) long, and during the day, they are hidden in their shallow skeleton cups. At night, they extend their tentacles to feed and the colony takes on a softer appearance. Hump Coral is an important reef-building species.

CLASS ANTHOZOA

Daisy Coral
Goniopora djiboutiensis

DIAMETER	
	Up to 1m (3ft)
DEPTH	
	5–30m (15–100ft)
HABITAT	
	Turbid reef waters

DISTRIBUTION Tropical waters of Indian Ocean and western Pacific

In most corals it is difficult to see the tiny polyps, but the Daisy Coral has polyps that are several centimetres long. The head of each polyp is dome-shaped with the mouth in the middle, surrounded by a ring of about 24 tentacles. These are arranged rather like the petals of a daisy. Unlike the majority of corals, the polyps extend to feed during the day, though they will quickly withdraw if touched. Daisy Coral grows as a rounded lump, but the shape is difficult to see when the polyps are extended. While most corals need clear water to survive, this species often covers large areas where the water is made turbid by disturbed sediment.

CLASS ANTHOZOA

Mushroom Coral
Fungia scruposa

DIAMETER	Up to 2.5cm (1in)
DEPTH	0–25m (0–80ft)
HABITAT	Sediment and rubble

DISTRIBUTION Tropical waters of Red Sea, Indian Ocean, and western Pacific

Mushroom Coral is unusual in that it lives as a single individual rather than a colony. Juveniles start life as a small disc attached to dead coral or rock. By the time they reach about 4cm (1½in) in diameter, they become detached. The animal feeds at night and the tentacles are withdrawn during the day, leaving the skeleton clearly visible, with the mouth at the centre of the disc. The skeleton resembles the gills of a mushroom. Mushroom Coral uses its tentacles to turn itself the right way up if it is overturned by waves.

CLASS ANTHOZOA

Giant Brain Coral
Colpophyllia natans

DIAMETER	Up to 2m (6½ft)
DEPTH	1–55m (3–180ft)
HABITAT	Seaward side of coral reefs

DISTRIBUTION Tropical waters of Gulf of Mexico and Caribbean

This huge coral grows as giant domes or extensive thick crusts and can live for more than 100 years. The surface of the colony is a convoluted series of ridges and long valleys, as in other species of brain coral, and this is what gives it its name. The valleys and ridges are often differently coloured and the ridges have a distinct groove running along the top. Typically, the valleys are green or brown and the ridges are brown. The polyp mouths are hidden in the valleys and the tentacles are only extended at night. In recent years, Giant Brain Corals in the Tortugas Islands (south of the Florida Keys) have been attacked by a disease and some have died. Particularly large colonies are popular tourist attractions in islands such as Tobago. As well as attracting divers, the coral heads attract fish and some gobies live permanently on the coral.

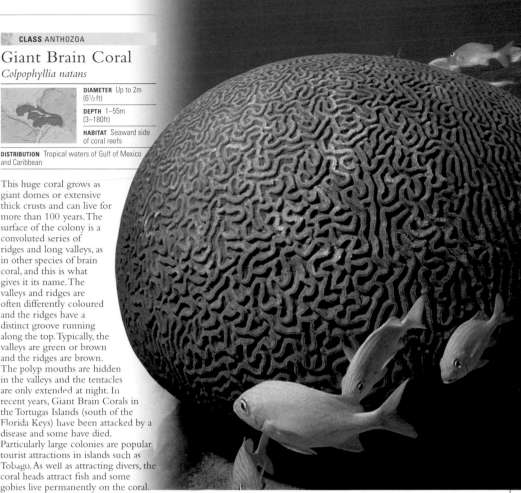

CLASS ANTHOZOA

Dendrophyllid Coral
Dendrophyllia species

HEIGHT	Up to 5cm (2in)
DEPTH	3–50m (10–165ft)
HABITAT	Steep rock faces

DISTRIBUTION Tropical waters in Indian Ocean and from western Pacific to Polynesia

With their large, flamboyant polyps, corals of the genus *Dendrophyllia* look more like an anemone than a coral. Dendrophyllids belong to a group called cup corals. They grow as a low-branching colony with each tubular individual distinct, and they do not develop the massive skeleton of reef-building corals. They have no zooxanthellae and grow in shaded parts of reefs such as below overhangs and especially on steep cliff faces. During the day, the polyps are entirely withdrawn and the coral looks like a dull reddish lump. As darkness falls, the polyps expand their orange tentacles to feed on plankton and make a spectacular display that often covers large areas. This genus of coral is very difficult to identify to species level and can also be confused with cup corals belonging to the genus *Tubastrea*.

CLASS ANTHOZOA

Devonshire Cup Coral
Caryophyllia smithii

DIAMETER	3cm (1¼in)
DEPTH	0–100m (0–330ft)
HABITAT	Rocks and wrecks

DISTRIBUTION Northeastern Atlantic and Mediterranean

While most corals grow as colonies in tropical waters, the Devonshire Cup Coral is solitary and lives in temperate parts of the ocean. It grows with its cup-shaped skeleton attached to a rock or even a shipwreck. When the tentacles are expanded, these tiny corals look just like anemones, with each tapering, transparent tentacle ending in a small knob. Devonshire Cup Coral occurs in a variety of colours from white to orange.

CLASS ANTHOZOA

Lophelia Coral
Lophelia pertusa

DIAMETER	At least 10m (33ft)
DEPTH	50–3,000m (165–10,000ft)
HABITAT	Deep-sea reefs

DISTRIBUTION Atlantic, eastern Pacific, and western Indian Ocean; distribution not fully known

Lophelia reefs more than 13km (8 miles) long and 30m (98ft) high have been recorded off the coast of Norway. Because it lives in deep, dark water, this cold-water coral has no zooxanthellae to help build its white, branching skeleton. It therefore grows very slowly, and such large reefs are many hundreds of years old. Each polyp has 16 tentacles, which it uses to capture prey such as zooplankton and even krill from the passing current. Stinging cells render the prey immobile and it is then transferred to the mouth.

CLASS ANTHOZOA

White Zoanthid
Parazoanthus anguicomus

HEIGHT	2.5cm (1in)
DEPTH	20–400m (65–1,300ft)
HABITAT	Shaded rocks, wrecks, and shells

DISTRIBUTION Temperate waters of northeastern Atlantic

Most zoanthids are found in tropical waters, but the White Zoanthid is common in the north Atlantic. Its white polyps arise from an encrusting base and it has two circles of tentacles around the mouth. One circle is usually held upwards while the other lies flat. As well as covering rocks and wrecks, this species also encrusts worm tubes and *Lophelia* reefs (see p.179).

CLASS ANTHOZOA

Whip Coral
Cirrhipathes species

LENGTH	Up to 1m (3 ft)
DEPTH	3–50m (10–165ft)
HABITAT	Coral reefs

DISTRIBUTION Tropical waters of eastern Indian Ocean and western Pacific

Whip corals, or wire corals, belong to a group of anthozoans called antipatharians to which the black corals (see right) also belong. Whip coral grows as a single unbranched colony that can either be straight

or coiled as in the species belonging to the genus *Cirripathes* shown here (whip corals are difficult to identify and many species remain undescribed). The feeding polyps of whip corals and black corals can be seen easily because, unlike sea fans, they cannot retract their short, pointed tentacles. Gobies live among the tentacles, hanging onto the coral with sucker-like pelvic fins.

CLASS ANTHOZOA

Bushy Black Coral
Antipathes pennacea

HEIGHT	Up to 1.5m (5ft)
DEPTH	5–330m (15–1,100ft)
HABITAT	Coral reefs

DISTRIBUTION Tropical waters of Gulf of Mexico, Caribbean Sea, and western Atlantic

Bushy Black Coral grows as a plant-like colony with branches shaped like large bird feathers. There are many different species of black corals and they get their name from the strong black skeleton that strengthens their branches. Made of a tough, horny material, the skeleton is valuable as it can be cut and polished to make jewellery, although this species is not widely used for this purpose.

CLASS ANTHOZOA

Tube Anemone
Cerianthus membranaceus

HEIGHT	35cm (14in)
DEPTH	10–100m (33–330ft)
HABITAT	Muddy sand

DISTRIBUTION Mediterranean and northeast Atlantic

The long, pale tentacles of the Tube Anemone make a spectacular display but at the slightest disturbance, the animal will disappear down its tube in an instant. Tube Anemones look superficially like true anemones but are more closely related to black corals (see above). They live in tubes made of sediment-encrusted mucus that can be up to 1m (3ft) long even though the animals are only about a third of this length. The slippery lining of the tube allows the animal to retreat rapidly. As well as about 100 long, slender outer tentacles, the animal has an inner ring of very short tentacles surrounding the mouth. The outer tentacles may look dangerous, but the Tube Anemone feeds only on plankton and suspended organic debris.

FLATWORMS

DOMAIN	Eucarya
KINGDOM	Animalia
PHYLUM	Platyhelminthes
CLASSES	4
SPECIES	20,000

POSSESSING EXTREMELY thin bodies, some of which are even transparent, flatworms are among the simplest of animals. Colourful marine species mostly belong to a group called polyclad flatworms – conspicuous, leaf-shaped animals, common on coral reefs and easily mistaken for sea slugs. Some flatworms are found in freshwater and many are parasitic. In the oceans, parasitic tapeworms and flukes are common in fish, mammals, and birds.

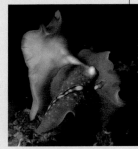

ONE-WAY RELATIONSHIP
This flatworm is living within the mantle of a bivalve mollusc, and scooping food from its host.

ANATOMY

The flatworm has a simple, solid structure with no internal cavity. It is so thin that oxygen can diffuse in from the water, and there are no blood or circulatory systems. The head end contains sense organs; advanced species have primitive eyes. The gut opens to the outside at one end, the opening serving as both mouth and anus. In polyclad flatworms, this opening is in the middle underside of the body. When feeding, they extend a muscular tube (pharynx) out of the mouth to grasp their food. Polyclad flatworms are covered in tiny hairs (or cilia) which, together with simple muscles, help them to glide over almost any surface. The anatomy of tapeworms and flukes is adapted to suit their parasitic lifestyle.

BODY SECTION
In flatworms, the space between the internal organs is filled with soft connective tissue criss-crossed by muscles.

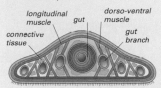

longitudinal muscle
gut
dorso-ventral muscle
connective tissue
gut branch

REPRODUCTION

Most flatworms are hermaphrodites so every individual has both ovaries and testes. The reproductive system is complex for such a primitive animal and includes special chambers and tubules where the ripe eggs are fertilized. When two polyclad flatworms meet, they may briefly touch heads and bodies in a short ritual before mating. After mating, the eggs are released into the water, laid in sand, or stuck to rocks. In some flatworms, the eggs develop directly into juvenile worms but in others they develop initially into an eight-lobed planktonic larva. Called Müller's larva, it swims for a few days and then settles onto the sea bed and flattens out into a young flatworm.

COMPLEX APPARATUS
Polyclad flatworms have a complex copulatory apparatus that includes a penis, muscular sac, and various glands.

CLASS TURBELLARIA

Acoel Flatworm
Waminoa species

LENGTH	Less than 5mm (1/4 in)
DEPTH	Not recorded
HABITAT	On Bubble Coral (*Pleurogyra sinuosa*)

DISTRIBUTION Tropical Indian and Pacific oceans

These diminutive flatworms look like coloured spots on the Bubble Coral on which they live. Their ultra-thin bodies glide over the coral surface as they graze, probably eating organic debris trapped by coral mucus. Acoel Flatworms have no eyes and instead of a gut, they have a network of digestive cells. They are able to reproduce by fragmentation, each piece forming a new individual. The genus is difficult to identify to species level and the distribution is uncertain.

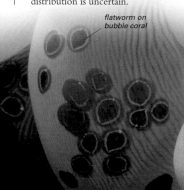

flatworm on bubble coral

CLASS TURBELLARIA

Green Acoel Flatworm
Convoluta roscoffensis

LENGTH	Up to 1.5cm (1/2 in)
DEPTH	Intertidal
HABITAT	Sheltered sandy shores

DISTRIBUTION Northeastern Atlantic; probably more widespread than shown

Although difficult to see individually, these flatworms show up when they collect together in puddles of water on sandy shores at low tide. Their bodies harbour tiny, single-celled algae that colour them bright green. In warm, sunlit pools the algae can photosynthesize and pass some of the food they make to their host. These flatworms are very sensitive to vibrations and quickly disappear down into the sand if footsteps approach.

CLASS TURBELLARIA

Candy Stripe Flatworm
Prosthecenaeus vittatus

LENGTH	Up to 5cm (2in)
DEPTH	0–30m (0–100ft)
HABITAT	Muddy rocks

DISTRIBUTION Temperate waters of northeastern Atlantic and Mediterranean

Most brightly coloured flatworms are found on tropical reefs, but the Candy Stripe Flatworm is an exception and can be found as far north as Norway. Generally a cream colour, it is marked with reddish-brown, lengthwise stripes. The head end of its flattened, leaf-shaped body has a pair of distinct tentacles and groups of primitive eyes. As it crawls along, the flatworm pushes the edges of its body up into folds; it is also able to swim using sinuous movements of the body. Usually found in rocky areas, it has also been seen on sand.

CLASS TURBELLARIA

Exquisite Lined Flatworm

Pseudobiceros bedfordi

LENGTH	Up to 8cm (3in)
DEPTH	Not recorded
HABITAT	Coral reefs

DISTRIBUTION Tropical waters of Indian and western Pacific oceans

Divers frequently come across this beautiful flatworm on coral reefs. Its striking pattern of pinkish transverse stripes and white dots against a black background make it easily recognizable. It is usually seen crawling over rocks in search of tunicates and crustaceans, but it is also a fairly good swimmer. Sometimes, the head end is reared up and a pair of flap-like tentacles can be seen.

CLASS TURBELLARIA

Divided Flatworm

Pseudoceros dimidiatus

LENGTH	Up to 8cm (3in)
DEPTH	Not recorded
HABITAT	Coral reefs

DISTRIBUTION Tropical waters of Indian and western Pacific oceans

Most species of flatworm display a distinctive pattern of colours that is more or less the same in every individual. However, the colour patterns of the Divided Flatworm vary greatly between individuals. The body is always black with an orange margin but the width and arrangement of the yellow or white lateral stripes, zebra-like bars, or narrow and wide longitudinal stripes is highly variable. These highly contrasting colours act as a warning to predators that Divided Flatworms are not good to eat. Like other flatworms, this species has numerous photo- and chemosensitive cells in its head region, which help the worm to find food and avoid danger.

CLASS TURBELLARIA

Imitating Flatworm

Pseudoceros imitatus

LENGTH	Up to 2cm (1in)
DEPTH	Not recorded
HABITAT	Coral reefs

DISTRIBUTION Waters around New Guinea and northern Australia, perhaps more extensive

Unlike the majority of polyclad flatworms, which have a relatively smooth skin, the Imitating Flatworm has a bumpy surface covered in small pustules. This appearance is an

SOURCE OF INSPIRATION
Phylidiella pustulosa is one of the most common and widespread sea slugs on Indo-Pacific reefs about 5–40m (15–130ft) deep.

GOOD IMITATION
The Imitating Flatworm has a creamy grey background colour and black reticulations surrounding pale pustules.

imitation of the skin of the sea slug *Phylidiella pustulosa*, and the flatworm's colour pattern is also almost identical to that of the sea slug. The sea slug secretes a noxious chemical to deter potential predators and it may be that the Imitating Flatworm gains protection by looking and feeling to the touch like the distasteful sea slug.

CLASS TURBELLARIA

Thysanozoon Flatworm

Thysanozoon nigropapillosum

LENGTH	Up to 8cm (3in)
DEPTH	1–30m (3–100ft)
HABITAT	Coral reef slopes

DISTRIBUTION Tropical waters of Indian and western Pacific oceans

The highly convoluted edge of the very thin Thysanozoon Flatworm is prominently displayed with a white outline. The rest of the upper side

of the body is black and covered in short papillae, or protuberances, each of which ends in a yellow tip. This gives the flatworm the appearance of being peppered with yellow spots. As is the case with most tropical reef flatworms, little is known of the biology of this species, but the Thysanozoon Flatworm has been found in association with colonial tunicates and is thought to feed on these and other colonial animals. It has been observed to swim well, rhythmically undulating its wide body. Much of what is known about this and other tropical reef flatworms has come from observations made by recreational divers and photographers. A similar species, *Thyanozoon flavomaculatum,* is found on Red Sea coral reefs.

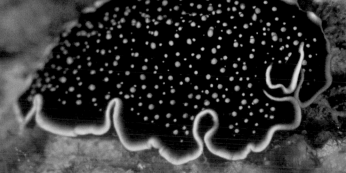

CLASS TURBELLARIA

Giant Leaf Worm

Kaburakia excelsa

LENGTH	Up to 10cm (4in)
DEPTH	Intertidal
HABITAT	Under coastal rocks

DISTRIBUTION Temperate waters of northeastern Pacific

This large, oval flatworm crawls around rocks, stones, and undergrowth on the Pacific shores of North America. Its colour is reddish-brown to tan, marked with darker spots, and when it is fully spread out, the branches of its digestive system may

be seen through the skin. It feeds in the same way as most polyclad flatworms, by everting its pharynx over its prey. Most intertidal flatworms in this region are only about 2cm (1in) long, making this species easy to identify. It is common on floating docks and in mussel beds.

CLASS CESTODA

Broad Fish Tapeworm

Diphyllobothrium latum

LENGTH	Up to 10m (33ft)
DEPTH	Dependent on host
HABITAT	Parasitic

DISTRIBUTION Probably worldwide, dependent on host species

Some flatworms, including tapeworms, have become highly modified and live as parasites. The Broad Fish Tapeworm has a complex life history. It begins life as a fertilized egg that is eaten by tiny freshwater crustaceans, inside which the larvae hatch. Freshwater, estuarine, and migratory marine fish, (such as salmon) become infected by the larvae when they eat either the crustaceans or other infected fish. The adult tapeworm lives in fish-eating mammals and may infect humans who eat raw fish. Other tapeworm species live as adults in the guts of marine fish.

RIBBON WORMS

DOMAIN	Eucarya
KINGDOM	Animalia
PHYLUM	Nemertea
CLASSES	2
SPECIES	900

ALSO CALLED NEMERTEAN worms, ribbon worms can reach great lengths of at least 50m (160ft), although many are small and inconspicuous. While they are commonly slightly flattened, the longest are cylindrical and are often called bootlace worms. The majority of ribbon worms live in the sea under rocks, amongst undergrowth or in sediment, and some are parasitic. A few species live inside the shells of molluscs and crabs.

ANATOMY

Nemertean worms have a long, unsegmented body with strong muscles in the body wall that can shorten the worm to a fraction of its full length. Unlike flatworms, ribbon worms have blood vessels and a complete gut with mouth and anus. It is often difficult to distinguish between the front and rear end of the worm, but most species have many simple eyes at the front. The most characteristic feature of these worms is a strong, tubular structure called a proboscis that lies in a sheath above the gut. It can be thrust out by hydrostatic pressure, either through the mouth or a separate opening, and is used to capture prey. In some species, the proboscis is armed with a sharp stylet.

Labels: stylet, proboscis, nerve ganglion, nerve, excretory organs, proboscis sheath, blood vessel, ovary, gut

BODY SECTION
Ribbon worms have no body cavity or gills; a simple circulatory system carries oxygen around the body.

REPRODUCTION

Most marine ribbon worms have separate sexes and their numerous, simple gonads produce either eggs or sperm. These are usually shed into the sea through pores along the sides of the body. Some species cocoon themselves together in a mucous net where the eggs are duly fertilized. In some types of ribbon worm the eggs develop directly into juvenile worms, while others initially hatch into various types of larvae. The long, fragile bodies of ribbon worms tend to break easily but they have the useful ability to regenerate any lost parts. Some species even use regeneration as a method of asexual reproduction, where the body breaks up into several pieces and each piece develops a new head and tail.

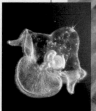

SWIMMING LARVA
Some ribbon worms develop from a planktonic larva called a pilidium. It is able to swim by beating hair-like structures, called cilia.

WARNING PATTERN
Some ribbon worms have bright patterns that may serve as a warning to predators that they are toxic. Drab-coloured species only emerge at night to hunt.

CLASS ANOPLA

Football Jersey Worm
Tubulanus annulatus

LENGTH	Up to 75cm (30in)
DEPTH	0–40m (0–130ft)
HABITAT	Gravel, stones, and sediment

DISTRIBUTION Cold and temperate waters of north Atlantic and north Pacific

One of the most strikingly coloured ribbon worms, the Football Jersey Worm has a patterning of longitudinal white lines and regularly spaced white rings. It may be found lying in an untidy pile beneath stones on the lower shore and may also be seen scavenging when the tide is out. More usually it lives below the shore on almost any type of sea bed, including mud, sand, and shell gravel. To camouflage itself, it secretes a mucous tube that becomes covered in surrounding sediment.

CLASS ANOPLA

Bootlace Worm
Lineus longissimus

LENGTH	Up to 10m (33ft)
DEPTH	Not recorded
HABITAT	Sediments and stones

DISTRIBUTION Temperate waters of northeast Atlantic

The Bootlace Worm makes up for its rather drab brown colour by its incredible length. Only a few millimetres in diameter, it reaches at least 10m (33ft) in length, and is one of the longest animals known. On the shore it appears as a writhing mass of knots lying on muddy sediment beneath boulders. Like all Anoplan worms, it has its mouth behind the brain. This worm is difficult to pick up, because it exudes large amounts of mucus when handled.

CLASS ENOPLA

Ribbon Worm
Nipponnemertes pulcher

LENGTH	Up to 9cm (3½in)
DEPTH	0–570m (0–1,900ft)
HABITAT	Coarse sediments

DISTRIBUTION Temperate and cold waters of Arctic, Atlantic, Pacific and Southern oceans

This worm belongs to a class of nemertean worms called Enoplan ribbon worms, whose mouth is located in front of the brain. *Nipponnemertes pulcher* has a short, stout body with a width of up to 5mm (¼in) that tapers to a pointed tail. The coloration varies from pink to orange or deep red and is paler beneath. This species has a distinctive, shield-shaped head with numerous eyes along its edges. The number of eyes increases with age. It is usually seen when dredged up by scientists from the coarse sediments in which it lives, but is sometimes found beneath stones on the lower shore. Its full distribution is unknown.

OCEAN LIFE

SEGMENTED WORMS

DOMAIN	Eucarya
KINGDOM	Animalia
PHYLUM	Annelidia
CLASSES	3
SPECIES	About 15,000

SEGMENTED WORMS include two familiar, predominantly land-based and freshwater groups, the earthworms and the leeches. In the oceans, a third group, the bristleworms or polychaetes, are numerous and diverse. These include burrowing lugworms, free-living predatory ragworms, and tube-dwelling worms. All segmented worms share one main characteristic – the long, soft body is divided into a series of almost identical, linked segments.

BRISTLEWORM
Fire worms have long, sharp bristles on each body section. These break off if the worm is attacked and can cause severe skin irritation.

ANATOMY

Each body segment is called a metamere and, except for the head and tail tip, all are virtually indistinguishable from each other. In bristleworms, flattened lobes (parapods) project from the sides of each segment, and are reinforced by strong rods made of chitin. The worm uses parapods for locomotion, and projecting bundles of bristles help it to grip. Internally, the segments are separated by partitions and filled with fluid. The gut, nerve cord, and large blood vessels run all along the body.

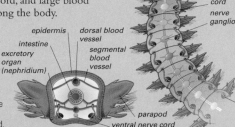

parapod
ventral nerve cord
nerve ganglion
epidermis
dorsal blood vessel
intestine
excretory organ (nephridium)
segmental blood vessel
parapod
ventral nerve cord
excretory organ (nephridium)

BODY SECTIONS
Most segments contain their own organs, including excretory and reproductive organs, and branches from the main blood vessels and ventral nerve cord.

JAWS OF A PREDATOR
This Bobbit Worm seizes prey using a proboscis tipped with sharp mandibles, which it shoots out from the mouth.

REPRODUCTION

In most polychaete worms, the sexes are separate and the eggs and sperm are shed into the water. Spawning is usually seasonal, especially at temperate latitudes. In many species, the fertilized egg develops into a larva (trochophore) that resembles a tiny spinning top. It floats and swims in the plankton, propelled by the beating of hair-like cilia around its middle. Eventually, the larva elongates and constricts into segments as it turns into an adult. Some species brood their eggs until the larvae are well developed. Many polychaete worms change shape as they become sexually mature, becoming little more than swimming bags of eggs or sperm. Known as epitokes, they swarm, burst open to release the eggs or sperm, then die.

epitoke

READY TO BURST
The egg- or sperm-laden epitoke of a palolo worm separates from the front segments, and bursts open.

CLASS POLYCHAETA

Lugworm
Arenicola marina

LENGTH	Up to 20cm (8in)
DEPTH	Shore and just below
HABITAT	Muddy sand

DISTRIBUTION Temperate shores of northeastern Atlantic, Mediterranean, and western Baltic

One of the most familiar sights on western European beaches is the neat, coiled casts of undigested sand deposited by Lugworms. The worm itself is rarely seen, remaining hidden in its U-shaped tube beneath the surface of the sand. The entrance to the tube is marked by a shallow, saucer-shaped depression in the sand. The worm may be pink, red, brown, black, or green. The first six segments of its front section are thick with bristles, while the next thirteen segments have red, feathery gills. The rear third of the body is thin, with no gills or bristles.

Lugworms feed by eating sand, extracting organic matter from it, and expelling the waste. These fleshy worms are a favourite food of many wading birds and are also used by fishermen as bait. They are most abundant at mid-shore level in sediments containing reasonable amounts of organic matter.

CLASS POLYCHAETA

Green Paddle Worm
Eulalia viridis

LENGTH	Up to 15cm (6in)
DEPTH	Shore and shallows
HABITAT	Rocky areas under stones, in crevices

DISTRIBUTION Temperate coastal waters of northeastern Atlantic

Although this beautiful green worm is usually found crawling over rocks, it can also swim well. The name Paddle Worm comes from the large, leaf-shaped appendages called parapodia that are attached to the side of each body segment and which help it to swim. The head has two pairs of stout tentacles on each side, a single tentacle on top, and four short, forward-pointing tentacles at the front. These tentacles and two simple black eyes help the worm in its hunt for food. The Green Paddle Worm is attracted to dead animals, especially mussels and barnacles, but will also hunt for live prey. However, unlike the King Ragworm (opposite), it does not have jaws to tackle large prey. Instead, carrion and debris sticks to its proboscis and is wiped off inside the mouth.

During spring, the Green Paddle Worm lays gelatinous green egg masses about the size of a marble on the shore and in shallow water, attaching them to seaweeds and rocks.

CLASS POLYCHAETA

Sea Mouse
Aphrodita aculeata

LENGTH	Up to 20cm (8in)
DEPTH	Shallow to moderate
HABITAT	Sand, muddy sand

DISTRIBUTION Temperate coastal waters of northeastern Atlantic and Mediterranean

The segmented structure of this pretty worm can be seen only if it is turned over, because its back is disguised by a thick felt of hairs that mask its segments. Running along each side of its body are numerous stiff, black bristles and a fringe of beautiful, iridescent hairs that glow green, blue, or yellow. The bristles can cause severe irritation if they puncture the skin. The Sea Mouse is so called because it looks like a bedraggled mouse when washed up dead on the seashore.

CLASS POLYCHAETA

King Ragworm

Neanthes virens

LENGTH
Up to 50cm (20in)

DEPTH
Shore and shallows

HABITAT
Muddy sand

DISTRIBUTION Temperate coastal waters of northeastern and northwestern Atlantic

This large worm has strong jaws that are easily capable of delivering a painful bite to a human. The jaws are pushed out on an eversible proboscis and are used for pulling food into its mouth as well as for defending itself. The King Ragworm lives in a mucus-lined burrow in the sand, and waits for the tide to come in before coming out to feed. It swims well by bending its long body into a series of S-shaped curves. Fishermen collect it for bait.

crown of spines in three concentric rings

finger-like gills on each body segment

CLASS POLYCHAETA

Honeycomb Worm

Sabellaria alveolata

LENGTH Up to 4cm (1½in)

DEPTH Shore and shallows

HABITAT Mixed rock and sand areas

DISTRIBUTION Intertidal areas of northeastern Atlantic and Mediterranean

Although Honeycomb Worms are tiny, the sand tubes they build may cover many metres of rock in rounded hummocks up to 50cm (20in) thick. The worms build their tubes close together, and the tube openings give the colony a honeycomb appearance. This worm's head is crowned by spines and it has numerous feathery feeding tentacles around the mouth, which it uses to trap plankton. The body ends in a thin, tube-like tail with no appendages.

WORM REEFS

Honeycomb Worms build their tubes by gluing together sand grains stirred up by waves. The glue is a mucus secreted by the worm, which uses a lobed lip around its mouth to fashion the tube. As new worms settle out from the plankton to build their own tubes, a reef develops and expands sideways and upwards, provided there is a good supply of sand. These structures provide a home to many other species.

LIVE REEF
Live reefs will survive for many years provided new larvae settle and grow to replace wave-damaged areas and dead worms.

CLASS POLYCHAETA

Magnificent Feather Duster

Sabellastarte magnifica

LENGTH
Up to 15cm (6in)

DEPTH
1–20m (3–65ft)

HABITAT
Coral reefs

DISTRIBUTION Shallow waters of the western Atlantic and Caribbean

The only part of this worm that is normally visible is a beautiful fan of feathery tentacles. The worm's segmented body is hidden inside a soft, flexible tube that it builds tucked beneath rocks or in a coral crevice or buried in sand. The tentacles are in two whorls and are usually banded brown and white. They are normally extended into the water to filter out plankton but at the slightest vibration or disturbance, such as the exhalation of a scuba diver, the worm instantly retracts the tentacles down into the safety of the tube.

CLASS POLYCHAETA

Christmas Tree Worm

Spirobranchus giganteus

LENGTH Up to 3cm (1¼in)

DEPTH 0–30m (0–100ft) or more

HABITAT Living coral heads

DISTRIBUTION Shallow reef waters throughout the tropics

Many large coral heads in tropical waters are decorated with Christmas Tree Worms, which occur in a huge variety of colours. The worm lives in a calcareous tube buried in the coral and extends neat, twin spirals of feeding tentacles above the coral surface. If disturbed, the worm pulls back into its tube in a fraction of a second. For added safety, the worm can also plug its tube with a small plate called an operculum.

CLASS POLYCHAETA

Pompeii Worm

Alvinella pompejana

LENGTH Up to 10cm (4in)

DEPTH 2,000–3,000m (6,500–10,000ft)

HABITAT Hydrothermal vent chimneys

DISTRIBUTION Eastern Pacific

This extraordinary worm lives in thin tubes massed together on the sides of chimneys of deep-sea hydrothermal vents. The tubes are close to the chimneys' openings, where water from deep inside the Earth pours out at temperatures of up to 350°C (662°F). The temperature within the worm tubes reaches 70°C (158°F). At its head end, the Pompeii Worm has a group of large gills and a mouth surrounded by tentacles. Each of the worm's body segments has appendages on the side known as parapodia. The posterior parapodia have many hair-like outgrowths that carry a mass of chemosynthetic bacteria. The bacteria manufacture food that the worm absorbs, and the worm also eats some of the bacteria.

OCEAN LIFE

MOLLUSCS

DOMAIN	Eucarya
KINGDOM	Animalia
PHYLUM	Mollusca
CLASSES	8
SPECIES	50,000

AMONG THE MOST SUCCESSFUL of all marine animals, molluscs display great diversity and a remarkable range of body forms, allowing them to live almost everywhere from the ocean depths to the splash zone. They include oysters, sea slugs, and octopuses. Some species lack eyes and shells and live passively in sediment or on the sea bed. Others are intelligent, active hunters with complex nervous systems and large eyes. Filter-feeding molluscs, such as clams, are crucial to coastal ecosystems, as they provide food for other animals and improve water quality and clarity. Many molluscs are commercially important for food, pearls, and their shells.

ANATOMY

Most molluscs have a head, a soft body mass, and a muscular foot. The foot is formed from the lower body surface and helps it to move. Molluscs have what is called a hydrostatic skeleton – their bodies are supported by internal fluid pressure rather than a hard skeleton. All molluscs have a mantle, a body layer that covers the upper body and which may or may not secrete a shell. The shell of bivalves (clams and relatives) has two halves joined by a hinge; these can be held closed by powerful muscles while the tide is out, or if danger threatens. Molluscs other than bivalves have a rasping mouthpart, or radula, which is unique to molluscs. Cephalopods (octopuses, squid, and cuttlefish) also have beak-like jaws as well as tentacles, but most lack a shell, while most gastropods (slugs and snails) have a single shell. This is usually a spiral in snails, but can be cone-shaped in other forms, such as limpets.

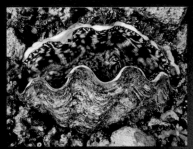

REEF-DWELLING GOLIATH
The tropical Giant Clam is the largest bivalve and may measure more than 1m (3ft) across and weighs over 220kg (440lb).

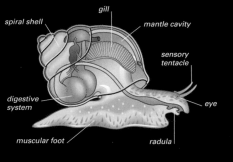

gill
spiral shell
mantle cavity
sensory tentacle
digestive system
eye
muscular foot
radula

SPIRAL SNAIL SHELL

GASTROPOD ANATOMY
The body plan (far left) of gastropods (slugs and snails) features a head, large foot, and usually a spiral shell (left). In shelled forms, all the soft body parts can be withdrawn into the shell for protection, or to conserve moisture while uncovered by the outgoing tide.

BIVALVE ANATOMY
Bivalves are housed within a shell of two halves (right) from which the siphons and muscular foot can be extended. The shell is opened and closed by the adductor muscles, labelled in the body plan (far right).

BIVALVE SHELL

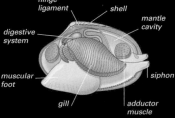

hinge ligament
shell
digestive system
mantle cavity
muscular foot
siphon
gill
adductor muscle

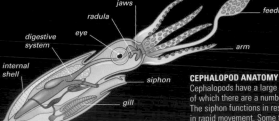

jaws
radula
eye
digestive system
internal shell
feeding arm
arm
siphon
gill
mantle cavity

CEPHALOPOD ANATOMY
Cephalopods have a large eye, in front of which there are a number of tentacles. The siphon functions in respiration and in rapid movement. Some forms have a small internal shell (cuttle).

SENSE ORGANS

Touch, smell, taste, and vision are well developed in many molluscs. The nervous system has several paired bundles of nervous tissue (ganglia), some of which operate the foot, and interpret sensory information such as light intensity. Photoreceptors range from the simple eyes (ocelli) seen along the edges of the mantle or on bivalve siphons, to the sophisticated image-forming eyes of cephalopods. Cephalopods are also capable of rapidly changing their colour according to their mood or surroundings.

PIGMENTED SKIN CELLS HELP CUTTLEFISH TO CHANGE COLOUR

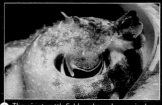 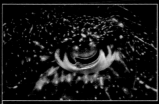

1 The giant cuttlefish's colour change is due to skin cells called chromatophores. It is pale when pigment is confined to a small area of each cell.

2 When the cuttlefish passes over a darker background, it disperses the coloured pigments throughout each of its chromatophores, and the animal darkens.

MOLLUSCAN BEAUTY
Displaying fabulous warning colours, this nudibranch is a shell-less example of the many thousands of marine species of gastropods (slugs and snails).

GRAFTING OYSTERS

Pearls form in oysters when a grain of sand or other irritant lodges in their shells. The oyster coats the grain with a substance called nacre, forming a pearl. Today many pearls are cultured artificially: the shell is opened just enough to introduce an an irritant into the mantle cavity.

SEEDING AN OYSTER
The best-shaped artificial pearls are produced by "seeding" oysters with a tiny pearl bead and a piece of mantle tissue from another mollusc.

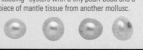

MOVEMENT

Molluscs move in many different ways. Most gastropods glide across surfaces using their mucus-lubricated foot. Exceptions include the sea butterfly, which has a modified foot with fin-like extensions for swimming. Some bivalves, such as scallops, also swim, producing jerky movements by clapping the two halves of their shell together. Other bivalves burrow by probing with their foot and then pulling themselves downwards by muscular action. Cephalopods are efficient swimmers; some have fins on the sides of their bodies that let them hover in the water, and they can accelerate rapidly by squirting water out through their siphons.

siphon

REDUCING DRAG
Swimming backwards reduces drag from the tentacles. The siphon, used for jet propulsion, is clearly visible in this Humboldt Squid.

AIDED BY MUCUS
Muscular contractions ripple through the fleshy foot of this marine snail. It secretes a lubricating mucus that helps it to move on rough surfaces.

RESPIRATION

Most molluscs "breathe" using gills, called ctenidia, which are situated in the mantle cavity. They are delicate structures with an extensive capillary network and a large surface area for gaseous exchange. In species that are always submerged, water can continually be drawn in and over the gills. Those living in the intertidal zone are exposed to the air for short periods so must keep their gills moist. At low tide, bivalves close their shells but retain a little water inside. Some gastropods also clamp down against rocks to retain moisture, but pulmonate snails have lost their ctenidia and instead have a lung formed from the mantle cavity. They take in air while exposed and respire through their skin while immersed. The respiratory pigment in most molluscan blood is a copper compound called haemocyanin. It is not as efficient at taking up oxygen as haemoglobin and gives molluscs' blood a blue colour.

external gills (ctenidia)

COLOUR CODING
Nudibranchs (sea slugs) have feathery external gills towards the rear of their bodies. The warning coloration of this species includes the bright orange gills.

FEEDING

The ways in which molluscs feed are almost as varied as their anatomy. Sedentary molluscs, such as many bivalves including clams and oysters, create water currents through tubular outgrowths of their mantle (siphons). They filter food from the moving water with their mucus-covered gills. Suitably sized particles are then selected and passed to the mouth by bristly flaps called palps. Sea slugs, chitons, and many sea snails graze algae from hard surfaces using their rasp-like radula. Radulae have tooth-like structures called denticles, many of which are reinforced with an iron deposit for durability. Larger molluscs feed on crustaceans, worms, fish, and other molluscs, which they locate either by scent or, in the case of some cephalopods such as octopuses, by sight. Cephalopods use their suckered arms to capture prey and their parrot-like beak to crush and dismember it. Some squid even appear to hunt in packs and swim in formation over reefs looking for prey.

SPECIES-SPECIFIC DENTICLES
The denticles on a mollusc's radula are often species-specific. This electron micrograph shows the distinctive radula of the gastropod *Sinezona rimuloides*.

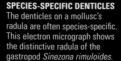

FEEDING TRAIL
Limpets continually graze the same area as the algae on which they feed regrow rapidly. The abrasive radula of the limpet wears a trail on the rock surface, as shown above.

REPRODUCTION

In many molluscs, reproduction simply involves releasing sperm or eggs (gametes) into the water. Fertilization is external and there is no parental care. Individuals may be of separate sexes or hermaphrodites (having both male and female reproductive organs). Hermaphrodites may function as either male or female or, as in nudibranchs, produce both eggs and sperm, although eggs can be fertilized only by cross-fertilization. Some species, such as slipper limpets, change sex with age, while oysters can change sex several times in a breeding season. Among cephalopods, males court females, fertilization is internal, and in some species, the eggs are protected by the females until they hatch.

DEVELOPING EMBRYOS
In 4 months, Australian Giant Cuttlefish eggs develop into mini-replicas of the adults.

LIMPET CHAIN
Slipper Limpets change from male to female as they grow. This chain of four such limpets has a female at the bottom and smaller males above her.

OYSTER DEMAND

Oysters have long been harvested as a food source. Their high market value and increasing demand has led to overexploitation of wild stocks. In the North Sea, the Common Oyster has vanished from much of its former range, and today most oysters are commercially farmed.

SLOW RECOVERY PERIOD
Relatively long-lived and reproducing only sporadically, the Common Oyster (right) takes a long time to recover from overexploitation.

READY AND WAITING FOR PREY
This cuttlefish hovers with its arms outstretched. When prey comes within reach, the two feeding arms, currently contracted and set above the two lower arms, will shoot forward to grab the prey.

LIFECYCLES

Most molluscs produce eggs that either float or are deposited in clusters, anchored to the substrate. Most forms have eggs that hatch into shell-less larvae, which live in the plankton. The larvae are called ciliated trochophores due to their bands of hair-like cilia, used in swimming. In gastropods, bivalves, and scaphopods, the trochophore larvae change into veliger larvae, which have larger ciliated bands, and sometimes adult features such as a mantle or a rudimentary shell or both. As they approach maturity, the larvae float down from the surface and, on reaching the sea bed, change into adults. Only those that land in a suitable environment survive to reach sexual maturity. Cephalopod eggs hatch into active predators. Some resemble mini-adults; others live in the plankton and initially look and behave differently from the adults.

PLANKTONIC LARVA
The visible bands of this veliger larva of the Common Limpet beat with tiny hair-like cilia, which are used in locomotion and feeding.

SECURING EGG CLUSTERS
This female Bigfin Reef Squid produces up to 400 egg capsules containing about 2,500 eggs. Here, she is securing egg capsules to a solid substrate.

each finger-shaped egg capsule holds up to seven eggs

MOLLUSC CLASSIFICATION

The phylum Mollusca is the second largest animal phylum, comprising over 50,000 species, and their diverse form has led to the identification of eight different classes. The majority of species live in marine habitats, but freshwater and terrestrial species are also numerous.

CAUDOFOVEATES
Class Caudofoveata

About 70 species
These are marine, shell-less, worm-like organisms of deep-water sediments. Their horny outer layer is covered with spines.

SOLENOGASTERS
Class Solenogaster

About 180 species
Another marine class of shell-less, worm-like organisms, solenogasters live in or on the ocean floor. Some lack a radula.

MONOPLACOPHORANS
Class Monoplacophora

8 species
These deep-sea molluscs lack eyes but have a radula and a cone-like shell. They are more abundant as fossils than as living species.

TUSK SHELLS
Class Scaphopoda

About 350 species
These animals have a tubular, tapering shell, open at both ends. The head and foot project from the wider end and dig in soft sediments.

BIVALVES
Class Bivalvia

More than 14,000 species
Bivalves, or clams and their relatives, have a hinged shell of two halves, but no radula. Most are sedentary and marine. Siphons create a water current through the shell, aiding feeding and respiration. Sexes are usually separate.

GASTROPODS
Class Gastropoda

More than 35,000 species
Familiar as slugs and snails, these molluscs are marine, freshwater, and terrestrial. They have a spiral shell and a large, muscular foot. The body is twisted 180° so the mantle cavity lies over the head. Many species can retract into their shell; hermaphrodite species are common.

CEPHALOPODS
Class Cephalopoda

About 650 species
Squid, octopuses, and cuttlefish are all cephalopods – fast-moving and intelligent, with a complex nervous system and large eyes. The shell is internal or absent, the head surrounded by arms, with or without suckers. The central mouth has a parrot-like beak and a radula. The sexes are separate.

CHITONS
Class Polyplacophora

About 500 species
Chitons have a uniform structure with a series of plates (usually 8) on their backs enclosed by an extension of the mantle. The underside is dominated by the foot.

CLASS BIVALVIA

Common Mussel
Mytilus edulis

LENGTH	10–15cm (4–6in)
HABITAT	Intertidal zones, coasts, estuaries

DISTRIBUTION North and southeastern Atlantic, northeastern and southwestern Pacific

Also called the Blue Mussel, this edible, black-shelled bivalve attaches itself in large numbers to various substrates using tough fibres called byssal threads. These fibres are extremely strong – five times tougher than a human tendon – and prevent the mussels from being washed away. When the mussel opens its shell, water is drawn in over the gills, or ctenidia, which absorb oxygen into the tissues and also filter food particles out of the water.

Common Mussels are very efficient filter feeders – they process about 45–70 litres (10–15 gallons) of water per day and consume almost everything they trap. The sexes are separate and so grouping together in "beds" helps to ensure that their eggs are fertilized. After hatching, the planktonic larvae are dispersed by the ocean currents. After about three months, they settle and mature further before moving once again to join the adult population.

CLASS BIVALVIA

Great Scallop
Pecten maximus

WIDTH	Up to 17cm (7½ in)
HABITAT	Sandy sea beds, at 5–150m (16–500ft), commonly 10m (33ft)

DISTRIBUTION Northeastern Atlantic

Also known as the King Scallop, the Great Scallop is usually found partly buried in sand. It is one of the few bivalves capable of rapid movement through water, which it achieves using a form of jet propulsion. It claps the two halves of its shell together, which pushes water out of the mantle cavity close to the hinge. It moves forwards with its shell gape first, producing jerky movements as it takes successive "bites" of water. These odd movements may be a useful strategy to escape from predators. These edible bivalves are now farmed to meet growing demand.

CLASS BIVALVIA

Black-lip Pearl Oyster
Pinctada margaritifera

LENGTH	Up to 30cm (12in) diameter
HABITAT	Hard substrata of inter- and subtidal zones; reefs

DISTRIBUTION Gulf of Mexico, western and eastern Indian Ocean, western Pacific

CLASS BIVALVIA

Atlantic Thorny Oyster
Spondylus americanus

LENGTH	Up to 11cm (4½ in)
HABITAT	Rocks to a depth of 140m (460ft)

DISTRIBUTION Southeast coast of USA, Bahamas, Gulf of Mexico, Caribbean

The Atlantic Thorny Oyster's spiny shell protects it from predators. The oyster pictured here is covered with an encrusting red sponge, which provides camouflage. This species is unusual in having a ball-and-socket type hinge joining the two halves of its shell,

rather than the more common toothed hinge seen in many other bivalves. The Atlantic Thorny Oyster cements itself directly to rocks rather than using byssal threads.

Black-lip Pearl Oysters begin life as a male before changing into a female two or three years later. Females produce millions of eggs, which are fertilized randomly and externally by the males' sperm, before hatching into free-swimming larvae. The mobile larvae pass through various larval stages for about a month before eventually settling on the sea floor, after metamorphosing into the sessile (immobile) adult form. This species is famous and much sought-after because it occasionally produces prized black pearls.

ORDER BIVALVIA

Shipworm
Teredo navalis

LENGTH	60cm (24in)
HABITAT	Wood burrows in high-salinity seas and estuaries

DISTRIBUTION Coastal waters off North, Central, and South America, and Europe

Despite its worm-like appearance, the shipworm is a type of clam that has become elongated as an adaptation to its burrowing lifestyle. Its bivalve shell, situated at the anterior end, is very small and ridged. The Shipworm uses it with a rocking motion to bore into wooden objects. Outside the shell its body is unprotected, except for a calcareous tube it secretes to line the burrow. These worms damage wooden structures, such as piers, irreparably and in the past caused many ships to sink. The burrow entrance is only about the size of a pinhead, but the burrow itself may be over 1cm (½ in) wide, so the extent of an infestation is often underestimated until it is too late.

Shipworms change from male to female during their lifetime, and the female form produces many eggs, from which free-swimming larvae hatch. When they mature and settle on a suitable piece of wood, the larvae quickly metamorphose into the adult form and start burrowing.

ORDER BIVALVIA

Common Piddock
Pholas dactylus

LENGTH	Up to 15cm (6in) across
HABITAT	Lower shore to shallow sublittoral

DISTRIBUTION South and east coasts of UK, Severn estuary in UK, west coast of France, Mediterranean

This mollusc has a pronounced "beak" covered in tooth-like projections at the front end of its shell. It uses this feature for boring holes into relatively soft substrates, such as mud, chalk, peat, and shale. Like the Shipworm (opposite), this piddock relies on its burrows for protection from predation, because the shell does not encase all of its body – its two fused siphons (tubes for eating, breathing, and excretion) trail out behind it. The shell is fragile, elliptical, and covered in a pattern of concentric ridges and radiating lines. If disturbed, the Common Piddock has an unusual defence strategy: it squirts a luminous blue secretion from its outgoing, or exhalant, siphon. Such bioluminescence is very rare in bivalve molluscs and is seen in only one other of the 14,000 species.

anterior beak of elliptical shell

fused siphons

ORDER BIVALVIA

Common Edible Cockle
Cerastoderma edule

LENGTH	Up to 5cm (2in)
HABITAT	Middle and lower shore, 5cm (2in) below surface of sand or mud

DISTRIBUTION Barents Sea, eastern north Atlantic from Norway to Senegal, West Africa

This edible bivalve has a robust, ribbed shell and burrows in dense populations just below the surface of sand or mud, filtering organic matter such as plankton from the water. Free-swimming larvae emerge from its fertilized eggs. Adult cockles sink to the sea floor, where they assume a more sedentary lifestyle.

ORDER BIVALVIA

Razor Shell
Ensis americanus

LENGTH	16cm (6in)
HABITAT	Sandy and muddy shores

DISTRIBUTION Atlantic coast of North America, introduced to North Sea

So called because their shells resemble a cut-throat razor, Razor Shells live in deep, vertical burrows on muddy and soft sandy shores. They are native to the northeast coast of North America, and the free-swimming larval stage is thought to have been introduced to the North Sea in 1978 when a ship emptied its ballast tanks outside the German port of Hamburg. This clam has subsequently spread along the continental coast. In places, it affects local polychaete worm populations, but it is not considered a pest.

ORDER BIVALVIA

Giant Clam
Tridacna gigas

LENGTH	Up to 1.5m (5ft)
HABITAT	Sandy beds of reef flats and shallow lagoons to 20m (65ft)

DISTRIBUTION Tropical Indo-Pacific from south China seas to northern coasts of Australia, and Nicobar Islands in the west to Fiji in the east

The largest and heaviest of all molluscs is the Giant Clam. Like other bivalves, it feeds by filtering small food particles from the water using its ingoing, or inhalant, siphon, which is fringed with small tentacles. However, it differs in obtaining most of its nourishment from zooxanthellae (unicellular algae that live within its tissues) – a type of relationship also associated with coral polyps. The algae have a constant and safe environment in which to live; in return they provide the clam with essential nutrients, the carbon-based products of photosynthesis. In fact, so dependent is the Giant Clam on these algae that it will die without them.

The adult is sessile (immmobile) and its inhalant and exhalant (outgoing) siphons are the only openings in its mantle. Although the scalloped edges of their shell halves are mirror images of one another, larger individuals may be unable to close their shells fully, so their brightly coloured mantle and siphons remain constantly exposed.

Many Giant Clams appear irridescent due to an almost continuous covering of purple and blue spots on their mantles, while others look more green or gold, but all have a number of clear spots, or "windows", that allow sunlight to filter into the mantle cavity. Fertilization is external and the eggs hatch into free-swimming larvae before settling onto the sea bed. The exhalant siphon expels water and at spawning time provides an exit point for the eggs or sperm.

SPAWNING

Reproduction in Giant Clams is triggered by chemical signals that synchronize the release of sperm and eggs into the water. Giant Clams start life as males and later become hermaphroditic, but during any one spawning event, they release either sperm or eggs in order to avoid self-fertilization. A large clam can release as many as 50 million eggs in 20 minutes.

GIANT CLAM
The largest living mollusc, the Giant Clam obtains energy by filter feeding and from photosynthetic algae living in its colourful mantle tissues. During the daytime, it spreads these tissues out to expose the algae to sunlight. Although not carnivorous, the clam does sometimes trap other animals when closing its shell.

CLASS GASTROPODA

Common Limpet

Patella vulgata

DIAMETER
6cm (2½ in)

HABITAT
Rocks on high shore to sublittoral zone

DISTRIBUTION Northeastern Atlantic from Arctic Circle to Portugal

Abundant on rocks from the high to the low water mark, the Common Limpet is superbly adapted to shore life. A conical shell protects it from predators and the elements. Limpets living at the low water mark are buffeted by the waves and so require smaller, flatter shells than those living at the high water mark, where wider,

orange foot with greenish tint

conical shell

MUSCULAR FOOT
The Common Limpet's muscular foot, seen here from below, holds it firmly to its rock, regardless of the strength of the waves.

taller shells allow for better water retention during periods of exposure. Limpets travel slowly during low tide, covering up to 60cm (24in) using contractions of their single foot. They graze on algae from rocks using a radula (a rasp-like structure), which has teeth reinforced with iron minerals.

RETURNING HOME
Limpets gradually grind a "scar" into their anchor spot on the rock, to aid their grip and help retain water. A mucus trail leads them back to the spot.

CLASS GASTROPODA

Zebra Nerite

Puperita pupa

LENGTH
Up to 1cm (½ in)

HABITAT
Rocky tide pools

DISTRIBUTION Caribbean, Bahamas, Florida

The small, rounded, smooth, black-and-white striped shell of the Zebra Nerite is typical of the species, but in examples from Florida the shell is sometimes more mottled or speckled with black. These gastropods are most active during the day, when they feed on micro-organisms such as diatoms and cyanobacteria, but if they become too hot or they are exposed at low tide, they cluster together, withdraw into their shells, and become inactive. This may be a mechanism for preventing excessive water loss.

Unusually for gastropods, there are separate males and females of Zebra Nerite and fertilization of the eggs occurs internally. The males use their penis to deposit sperm into a special storage organ inside the female. Later, she lays a series of small white eggs that hatch into planktonic larvae.

CLASS GASTROPODA

Dog Whelk

Nucella lapillus

LENGTH
Up to 6cm (2½ in)

HABITAT
Middle and lower rocky shores

DISTRIBUTION Northwestern and northeastern Atlantic

One of the most common rocky shore gastropods, the Dog Whelk has a thick, heavy, sharply pointed spiral shell. The shell's exact shape depends on its exposure to wave action and its colour depends on diet. Dog Whelks are voracious predators, feeding mainly on barnacles and mussels. Once the prey has been located, the whelk uses its radula to bore a hole in the shell of its prey before sucking out the flesh.

CLASS GASTROPODA

Top Shell

Trochus niloticus

LENGTH
16cm (6in)

HABITAT
Intertidal and shallow subtidal areas, reef flats to 7m (23ft)

DISTRIBUTION Eastern Indian Ocean, western and southern Pacific

Easily distinguished from most other gastropods by the conical shape of its spiral shell, the Top Shell moves slowly over reef flats and coral rubble, feeding on algae. Demand for its flesh and pretty shell has led to declining numbers, especially in the Philippines, due to unregulated harvesting. It has, however, been successfully introduced elsewhere in the Indo-Pacific, such as French Polynesia and the Cook Islands, from where some original sites are being restocked.

CLASS GASTROPODA

Red Abalone

Haliotis rufescens

LENGTH
15–20cm (6–8in)

HABITAT
Rocks from low tide mark to 30m (100ft)

DISTRIBUTION East Pacific coasts from southern Oregon, USA to Baja California, Mexico

The largest of the abalone species, the Red Abalone is so called because of the brick-red colour of its thick, roughly oval shell. There is an arc of

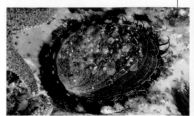

three to five clearly visible holes in the shell, through which it respires and expels waste products. These are filled and replaced with new holes as the abalone increases in size. Sea Otters are one of the Red Abalone's main predators, along with human divers.

CLASS GASTROPODA

Venus Comb

Murex pecten

LENGTH
Up to 8cm (3in)

HABITAT
Tropical warm waters to 200m (650ft)

DISTRIBUTION Eastern Indian Ocean and western Pacific

The tropical carnivorous snail known as the Venus Comb has a unique and spectacular shell. There are rows of long, thin spines along its longitudinal ridges, which continue onto the narrow, rod-like, and very elongated siphon canal. The exact function of these spines is unknown, but they are thought to be either for protection or to prevent the snail sinking into the soft substrate on which it lives. Its body is tall and columnar so that it can lift its cumbersome shell above the ground to move in search of food.

HIDING FROM VIEW

There are times when the Venus Comb buries itself just below the surface of the sea floor, displacing the sand with movements of its muscular foot. However, it leaves the opening of its tubular inhalant siphon above the sand's surface so that it can draw water into its mantle cavity to obtain oxygen and to "taste" the water for the presence of prey.

HALF BURIED
The spines of this Venus Comb can be seen sticking out of the sand. The siphon is visible to the right of the picture.

CLASS GASTROPODA

Tiger Cowrie
Cypraea tigris

LENGTH
Up to 15cm (6in)

HABITAT
Low tide to 30m (100ft)
on coral reefs

DISTRIBUTION Indian Ocean, western Pacific

One of the largest cowrie species, the Tiger Cowrie has a shiny, smooth, domed shell with a long, narrow aperture, and is variously mottled in black, brown, cream, and orange. Extensions of the cowrie's mantle (its body's outer, enclosing layer) cover parts of the exterior of the shell. These extensions have numerous projections, or papillae, whose exact function is unknown, but which may increase the surface area for oxygen absorption or provide camouflage of some sort. Tiger Cowries are nocturnal creatures, hiding in crevices among the coral during the day and emerging at night to graze on algae. The sexes are separate and fertilization occurs internally. Females exhibit some parental care in that they protect their egg capsules by covering them with their muscular foot until they hatch into larvae, which then enter the plankton to mature.

CLASS GASTROPODA

Giant Triton
Charonia tritonis

LENGTH
Up to 40cm (16in)

HABITAT
Coral reef, sandy substrate, intertidal and subtidal zones

DISTRIBUTION Indian Ocean, western and central Pacific

This gastropod is one of the very few animals that eats the Crown-of-thorns Starfish, itself a voracious predator and destroyer of coral reefs. The Giant Triton is an active hunter that will chase prey, such as starfish, molluscs, and sea stars, once it has been detected. It uses its muscular single foot to hold its victim down while it cuts through any protective covering using its serrated, tongue-like radula; it then releases paralysing saliva into the body before eating the subdued prey.

CLASS GASTROPODA

Common Periwinkle
Littorina littorea

LENGTH
Up to 3cm (1in)

HABITAT
Upper shore to sublittoral rocky shores, mud flats, estuaries

DISTRIBUTION Coastal waters of northwest Europe; introduced to North America

The Common Periwinkle has a black to dark grey, sharply conical shell and slightly flattened tentacles, which in juveniles also have conspicuous black banding. The sexes are separate and fertilization occurs internally. Females release egg capsules, containing two or three eggs, directly into the water during the spring tides. The eggs hatch into free-swimming larvae that float in the plankton for up to six weeks. After settling and metamorphosing into the adult form, it takes a further two to three years for the adult to fully mature. It feeds mainly on algae, which it rasps from the rocks. Recently, the Common Periwinkle was accidentally introduced to North America, where its selective grazing of fast-growing algal species has considerably affected the ecology of some rocky shores.

CLASS GASTROPODA

Flamingo Tongue
Cyphoma gibbosum

LENGTH
3–4cm (1–1½ in)

HABITAT
Coral reefs at about 15m (50ft)

DISTRIBUTION Western Atlantic, from North Carolina to Brazil; Gulf of Mexico, Caribbean Sea

The off-white shell of the Flamingo Tongue cowrie is usually almost completely hidden by the two fleshy, leopard-spotted extensions of its body's outer casing, or mantle. When threatened, however, its distinctive coloration quickly disappears as it withdraws all its soft body parts into its shell for protection. This snail feeds almost exclusively on gorgonian corals, which dominate Caribbean reef communities. Although these corals release chemical defences to repulse predators, the Flamingo Tongue cowrie is apparently able to degrade these bioactive compounds and eat the corals without coming to any harm. After mating, the female strips part of a soft coral branch and deposits the egg capsules on it. Each capsule contains a single egg that will hatch into a free-swimming planktonic larva.

CLASS GASTROPODA
Three-tooth Cavoline
Cavolinia tridentata

LENGTH	1cm (½ in)
HABITAT	100–2,000m (330–6,500ft); carried in ocean currents

DISTRIBUTION Warm oceanic waters worldwide

This species of sea butterfly has a small, almost transparent, spherical shell with three distinctive, posterior projections. The shell also has two slits through which large extensions of the mantle pass. These brownish "wings" are ciliated and so can create weak water currents as well as aid buoyancy. Sea butterflies are unusual among shelled molluscs in that they can live in open water. Like other members of this group, the Three-tooth Cavoline produces a mucous web very much larger than itself, which traps planktonic organisms, such as diatoms and the larvae of other species. It eats the web and the trapped food at intervals, then produces a new one. During their lifetime, sea butterflies change first from males into hermaphrodites and then into females.

CLASS GASTROPODA
Sea Hare
Aplysia punctata

LENGTH	Up to 20cm (8in)
HABITAT	Shallow water

DISTRIBUTION Northeast Atlantic and parts of the Mediterranean

The Sea Hare, a type of sea slug, has tentacles reminiscent of a hare's ears. It has an internal shell about 4cm (1½ in) long that is visible only through a dorsal opening in the mantle. If disturbed, it releases purple or white ink. It is not known if this response is a defence mechanism.

CLASS GASTROPODA
Bubble Shell
Bullina lineata

LENGTH	2.5cm (1in)
HABITAT	Sand, reefs to 20m (65ft), mainly intertidal; subtidal at range limits

DISTRIBUTION Tropical and subtropical waters of Indian Ocean and west Pacific

The pale spiral shell of the Bubble Shell (also known as the Red-lined Bubble Shell) has a distinctive pattern of pinkish red lines by which it can be identified. Its soft body parts are delicate and translucent with a fluorescent blue margin and, in form, reminiscent of the Spanish Dancer (opposite), which is a close relative that has lost its shell. If threatened, the Bubble Shell quickly withdraws into its shell and at the same time regurgitates food, possibly as a defence mechanism to distract predators. The Bubble Shell is itself a voracious predator, feeding on sedentary polychaete worms. This mollusc is hermaphroditic and produces characteristic spiral white egg masses.

CLASS GASTROPODA
Polybranchid
Cyerce nigricans

LENGTH	Up to 4cm (1½ in)
HABITAT	Reefs

DISTRIBUTION Western Indian Ocean, western and central Pacific

This colourful sea slug is a herbivore that browses on algae. It has no need of camouflage or a protective shell as it has two excellent alternative defence strategies. First, it can secrete distasteful mucous, by utilizing substances in the algae it feeds upon and secreting them from small microscopic glands over the body. Second, its body is covered with petal-like outgrowths called cerata, spotted and striped above and spotted below, that can be shed if it is attacked by a predator, in the same way as a lizard sheds its tail. This ability to cast off body parts to distract predators is called autonomy.

The cerata are also used in respiration, their large collective surface area allowing efficient gas exchange with the surrounding water. The head carries two pairs of sensory organs – the oral tentacles near the mouth and, further back, the olfactory organs (rhinophores). These are retractile and subdivide as the Polybranchid matures. They are used to assist in finding food and mates. There is some debate as to whether this sea slug is a separate species or is simply a colour variation of a similar mollusc, *Cyerce nigra*.

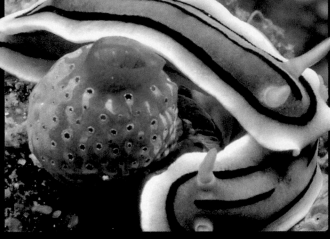

CLASS GASTROPODA
Chromodorid Sea Slug
Chromodoris lochi

LENGTH	4cm (1½ in)
HABITAT	Reefs

DISTRIBUTION Tropical and subtropical western and central Pacific

Protected from predators by its bright warning coloration and unpleasant taste, the Chromodorid Sea Slug forages in the open, rather than hiding away in cracks and crevices. Since it cannot swim, it glides over the tropical reefs on which it lives on its muscular foot, secreting a mucous trail much as terrestrial slugs do. The different species of the genus *Chromodoris* are distinguished by the pattern of black lines on their backs and the plain colour of their gills and rhinophores (a pair of olfactory organs at the head end).

The two Chromodorid Sea Slugs pictured here are possibly about to mate. To do so, they must face in opposite directions so that their sexual openings are aligned. As they are hermaphrodites, they both produce sperm, which they exchange during mating, and both later produce fertilized eggs.

CLASS GASTROPODA
Hermissenda Sea Slug
Hermissenda crassicornis

LENGTH	Up to 8cm (3in)
HABITAT	Mud flats, rocky shores

DISTRIBUTION Northwest and northeast Pacific

This sea slug, usually known simply as Hermissenda, has an unusual way of deterring predators. It separates the stinging cells from any organism it eats and stores them in the orange-red tips of petal-like tentacles, or cerata, that cover its back. Any creature that touches the cerata is stung. Unlikely though it seems, Hermissenda is used extensively by scientists conducting memory experiments. The animal has an excellent sense of smell that enables it to find its way around mazes to locate food, and it can be "taught" to respond to simple stimuli.

Spanish Dancer
Hexabranchus sanguineus

LENGTH
Up to 60cm (24in)

HABITAT
Shallow water on coasts
and reefs

DISTRIBUTION Parts of tropical Indian Ocean,
west Pacific

The largest of the nudibranchs is the Spanish Dancer – so called because when it swims, the undulating movements of its flattened body are reminiscent of a flamenco dancer. Adults are brightly but variably coloured, generally in shades of red, pink, or orange, sometimes mixed with white or yellow. While resting, crawling, or feeding, the lateral edges of its mantle are folded up over its back, displaying the less colourful underside. If disturbed, it will escape by swimming away, exposing its bright colours and possibly startling potential predators. Spanish Dancers are specialist predators that feed only on sponges, particularly encrusting species, from which they modify and concentrate certain distasteful compounds in their skin to use as another defence against predation. They have external gills for respiration, which are extensively branched and attached to the body wall in distinct pockets and which cannot be retracted. Like all nudibranchs, the Spanish Dancer is hermaphroditic, but it requires a partner in order to reproduce.

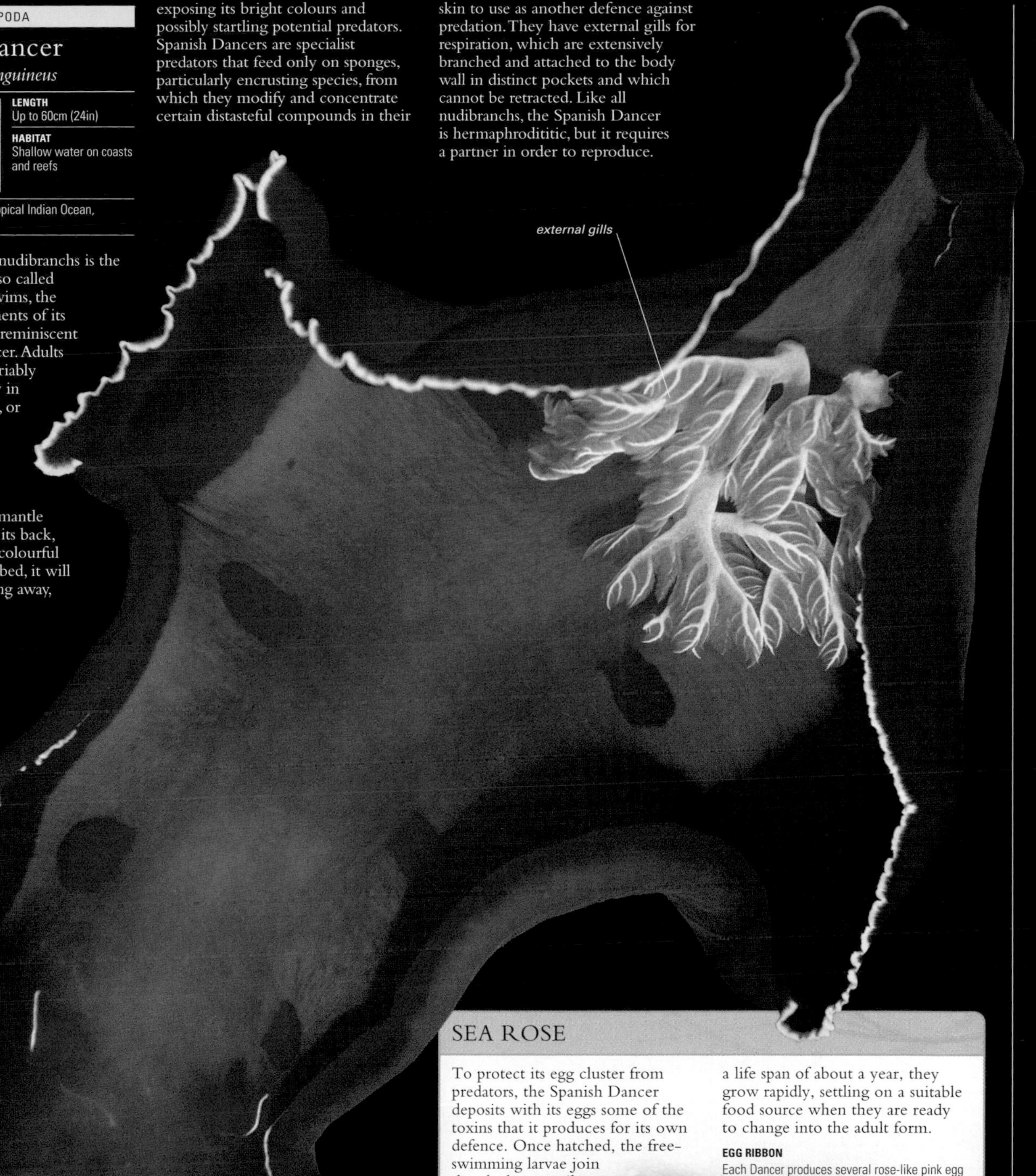

external gills

bright coloration

SEA ROSE

To protect its egg cluster from predators, the Spanish Dancer deposits with its eggs some of the toxins that it produces for its own defence. Once hatched, the free-swimming larvae join the plankton until they mature. With a life span of about a year, they grow rapidly, settling on a suitable food source when they are ready to change into the adult form.

EGG RIBBON
Each Dancer produces several rose-like pink egg ribbons about 4cm (1½in) across; together these may contain over one million eggs.

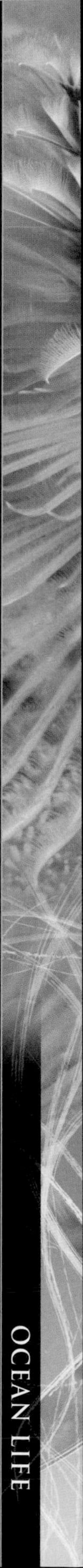

CLASS CEPHALOPODA

Dumbo Octopus
Grimpoteuthis plena

LENGTH	Up to 20cm (8in)
HABITAT	Deep water, to 2,000m (6,500ft)

DISTRIBUTION Northwest Atlantic

Little is known about the Dumbo Octopus as only a few have been recorded. Its common name derives from a pair of unusual, ear-like flaps extending from the mantle above its eyes. It has a soft body, an adaptation to its deep-water habitat, and eight arms connected to each other almost to their tips by "webbing". Its diet includes worms and snails.

CLASS CEPHALOPODA

Blue-ringed Octopus
Hapalochlaena maculosa

LENGTH	10–20cm (4–8in)
HABITAT	Shallow water, rock pools

DISTRIBUTION Tropical west Pacific and Indian Ocean (all species of Hapalochlaena)

The most dangerous cephalopod is the small Blue-ringed Octopus, which produces highly toxic saliva powerful enough to kill a human. To catch prey, it either releases saliva into the water and waits for the poison to take effect, or catches, bites, and injects prey directly. Its bright colouring is unusual for an octopus, and the numerous blue rings covering its body become more iridescent if it is disturbed.

CLASS CEPHALOPODA

Nautilus
Nautilus pompilius

WIDTH	Shell up to 20cm (8in)
HABITAT	Tropical open waters to 500m (1,600ft)

DISTRIBUTION Eastern Indian Ocean, western Pacific, and Australia to New Caledonia

The six remaining species of *Nautilus* belong to a once numerous group of shelled cephalopods that existed from 400 to 65 million years ago. They are often referred to as "living fossils" because they are so little changed from their ammonoid ancestors. Their shell protects them from predation, while gas trapped in its inner chambers provides buoyancy. The head protrudes from the shell and has up to 90 suckerless tentacles, which are used to capture prey such as shrimp and other crustaceans; the head also features a pair of rudimentary eyes that lack a lens and work on a similar principle to a pin-hole camera. The nautilus swims using jet propulsion, drawing water into its mantle cavity and expelling it forcefully through a tubular siphon, which can be directed to propel the nautilus forwards, backwards, or sideways. Unlike most other cephalopods, nautiluses mature late, at about ten years of age, and produce only about twelve eggs per year.

CLASS CEPHALOPODA

Giant Octopus
Enteroctopus dofleini

LENGTH	Up to 9m (30ft)
HABITAT	Bottom dwellers, to 750m (2,500ft)

DISTRIBUTION Temperate northwest and northeast Pacific

The Giant Octopus is one of the largest invertebrates as well as one of the most intelligent. It can solve problems, such as negotiating a maze by trial and error and remember the solution for a long time. It has large, complex eyes with colour vision and sensitive suckers that can distinguish between objects by touch alone. It changes colour rapidly by contracting or expanding pigmented areas in cells called chromatophores, enabling it to remain camouflaged regardless of background. It also uses its colour to convey mood, becoming red if annoyed and pale if stressed. Most cephalopods show little parental care, but female Giant Octopuses guard their eggs for up to eight months until they hatch. They do not eat during that time, and siphon water over the eggs to keep them clean and aerated.

DEFENCE MECHANISM

When threatened, a Giant Octopus squirts a cloud of purple ink out through its siphon into the water and at the same time moves backwards rapidly using jet propulsion. Potential predators are left confused and disorientated in a cloud of ink. The octopus can repeat this process several times in quick succession.

A QUICK GETAWAY
This Giant Octopus is making a rapid retreat, expelling an ink jet as a defence mechanism. The jet also propels the octopus backwards forcefully.

Australian Giant Cuttlefish

Sepia apama

LENGTH	Up to 1.5m (5ft)
HABITAT	Shallow water over reefs

DISTRIBUTION Coastal Australian waters

Of about 100 cuttlefish species, the Australian Giant Cuttlefish is the largest. Like all cuttlefish, it has a flattened body and an internal shell, known as the cuttle and familiar to many as budgerigar food. This species lives for up to three years and gathers in huge numbers to breed. Males have elaborate courtship displays, which involve hovering in the water while making rapid, kaleidoscopic changes of colour, as the male shown here is doing. When a female is receptive, the male deposits a sperm package in a pouch under her mouth. This later bursts, releasing sperm and fertilizing her 200 or more golfball-sized eggs, which she then deposits on a hard substrate. The eggs hatch into miniature adults after several months.

Common Squid

Loligo vulgaris

LENGTH	Up to 30cm (12in)
HABITAT	20–250m (60–800ft)

DISTRIBUTION Eastern Atlantic, Mediterranean

A tubular body and a small, rod-like internal skeleton are characteristic features of all species of squid. They also have very large eyes relative to body size. The Common Squid is an inshore, commercially important species that has been harvested for centuries and is probably the best known of all cephalopods. It is a fast swimmer that actively hunts its prey, such as crustaceans and small fish. Once caught, the squid passes the prey to its mouth, where it is dismembered by powerful, beak-like jaws.

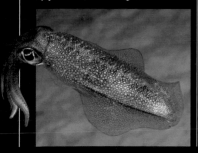

Glass Squid

Teuthowenia pellucida

LENGTH	1.4–3.8cm (½–1½ in)
HABITAT	Midwater

DISTRIBUTION Circumglobal in southern temperate waters

Like many molluscs, juvenile Glass Squid live in the plankton, then descend to deeper, darker levels as they mature. The presence of light organs, called photophores, in the tips of their arms and in the eye may help in locating a mate. Sexually mature females are also thought to produce a chemical attractant, or pheromone.

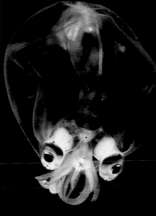

Vampire Squid

Vampyroteuthis infernalis

LENGTH	Up to 38cm (15in)
HABITAT	500–1,500m (1,600–5,000ft), oxygen-poor water

DISTRIBUTION Tropical and temperate oceans worldwide

This is the only squid that spends its entire life in deep, oxygen-poor water. Like many creatures that live at depth, the Vampire Squid is bioluminescent and has light organs, or photophores, on the tips of its arms and at the base of its fins. If threatened, it flashes these lights and writhes about in the water, finally ejecting mucus that sparkles with blue luminescent light. When the lights go out, the Vampire Squid will have vanished. Its predators include sea lions and deep-diving whales.

Lined Chiton

Tonicella lineata

LENGTH	3.5cm (1½ in)
HABITAT	Intertidal and subtidal zones, common on rocky surfaces

DISTRIBUTION Temperate waters of northeast and northwest Pacific

Chitons are molluscs with shells made up of eight arching and overlapping plates. The Lined Chiton is so called because of a series of zigzagging blue or red lines on its shell. The shell is usually pinkish in colour, which provides good camouflage as this chiton grazes from rocks that are covered with encrusting pink coralline algae. The Lined Chiton's mantle extends around the shell on all sides, forming an unusually smooth, leathery "girdle" that helps to hold its eight shell-plates together. It has a large, muscular foot, which it uses to move over rocks and, when still, to grip on to them in much the same way as limpets do. At low tide, it remains stationary to avoid water loss. Its head is small and eyeless. The sexes are separate and it reproduces by releasing its gametes into the water.

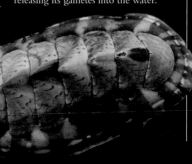

ARTHROPODS

DOMAIN	Eucarya
KINGDOM	Animalia
PHYLUM	Arthropoda
SUBPHYLA	5
SPECIES	More than 1.1 million

THE ANIMALS THAT HAVE ACHIEVED the greatest diversity on Earth are the arthropods, although relatively few can live in salt water. They include insects, spiders, and scorpions, but most marine arthropods are crustaceans, such as crabs, shrimps, and barnacles. Marine arthropods, particularly larval stages, form most of the ocean's zooplankton – the community of tiny, floating life forms that support all oceanic food chains. Like land arthropods, all marine forms have an external skeleton, segmented body, and jointed appendages, permitting some, such as robber crabs, to live on land as well as in water. Spiders and insects, although rarely fully marine, may live inshore and on coasts.

APPENDAGES
This Spotted Cleaner Shrimp has jointed walking appendages. Two furthers pairs of jointed appendages, which are located on its head, are modified into sensory antennae.

ANATOMY

Although arthropods may look very different from one another, they all have an external skeleton (exoskeleton), which is either thin and flexible or rigid and toughened by deposits of calcium carbonate. The body is segmented and has a variable number of jointed appendages – some are used for walking and swimming, while others are modified into claws and antennae or adapted for feeding. Muscles are attached across the joints to facilitate movement. Most of the body cavity is hollow; this space, called the haemocoel, contains the internal organs and a fluid – haemolymph – that is the equivalent to vertebrate blood, which is pumped around the body by the heart in an open circulatory system. Most marine forms use gills for respiration and have well-developed sense organs.

SPIDER FEATURES
Although it is called a crab and has a hinged carapace, this horseshoe crab is a close relative of spiders, ticks, and mites. Like them, it has four pairs of jointed legs.

SEASHORE INSECTS
Unlike many insects, the Springtail is not divided into a head, thorax, and abdomen. Instead, its thorax has three segments bearing three pairs of legs, and the abdomen has six segments.

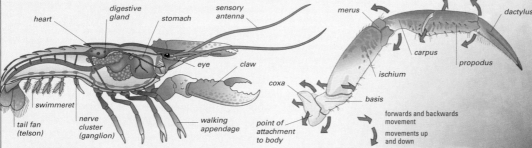

ARTHROPOD ANATOMY
Lobsters have a protective shell, called a carapace, large pincers, and well-developed walking appendages. The haemocoel contains the internal organs.

heart · digestive gland · stomach · sensory antenna · eye · claw · swimmeret · tail fan (telson) · nerve cluster (ganglion) · walking appendage

ARTHROPOD LIMB
Walking appendages, such as this crab's limb, comprise rigid sections linked with moveable joints. The joints move in different planes, allowing extensive movement.

merus · dactylus · carpus · propodus · ischium · coxa · basis · point of attachment to body · forwards and backwards movement · movements up and down

FILTER-FEEDING LIMBS
At high tide, barnacles feed by extending their long, feathery appendages from their "shell" and sweeping the water for plankton and detritus.

FEEDING

Among crustaceans, feeding is extremely varied. Many crabs are scavengers that feed on dead and decaying organic matter. They are therefore vital in helping to recycle nutrients. Others are hunters and have robust claws to stun (mantis shrimps) or crush (lobsters) their prey before tearing it apart and consuming it. Many small planktonic crustaceans, such as cyclopoids and other copepods, are filter feeders that make effective use of various appendages, including long antennae, to create water currents that waft food particles towards their mouths. Most barnacles, although they are attached to rocks, feed in a similar way, using their limbs to collect food. A few crustaceans are parasitic (some isopods, copepods, and the barnacle *Sacculina*) and obtain all their nourishment from their host. The shoreline is an ideal place for insects, such as kelp flies, that feed by releasing enzymes onto rotting seaweed and then taking in the resultant digested material. Further inland, among the dunes, there is more vegetation, so spiders and pollen- and nectar-feeding insects start to appear.

SCAVENGING IN THE SAND
As the tide retreats, this Sand Bubbler Crab emerges to feed on the microscopic material trapped between sand grains.

HARD SKELETON AND JOINTED LIMBS
These tiny porcelain crabs, less than 2.5cm (1in) wide, are filter feeders. They have typical arthropod features, such as a hard exoskeleton covering a segmented body, and jointed limbs.

GROWTH

Once crustaceans take on their adult form, they can only grow by moulting and replacing their exoskeleton with a larger one. This process, called ecdysis, is controlled by hormones and occurs repeatedly during adult life. The exoskeleton is produced from the layer of cells situated immediately below it. Before a moult starts, the exoskeleton detaches from this cell layer and the space in between fills with moulting fluid. Enzymes within this fluid weaken the exoskeleton so that it eventually splits at the weakest point, often somewhere along the back. The new exoskeleton is soft and wrinkled, so it needs to harden and expand. Marine arthropods absorb water rapidly after moulting to expand their new protective covering. They then remain hidden for a few days, as they are more vulnerable to predation until their exoskeleton hardens.

HUMAN IMPACT

KRILL DECLINE

Antarctic Krill are only 6cm (2½in) long but are among the most abundant arthropods. Numbers in the Southern Ocean have fallen by 80 per cent since the 1970s due to a water temperature rise of 2.5°C (4.5°F), causing a significant melt of ice. Krill feed on algae that form beneath the sea-ice, and their larvae shelter under the ice to avoid predation. With less ice, there are fewer krill and therefore fewer of the animals that feed on them.

THE MOULTING SEQUENCE

This sequence shows a Harlequin Shrimp moulting. The exoskeleton has split just behind the neck joint, allowing the shrimp to pull out its head. The rest of its body quickly follows as the split enlarges. It only takes a few minutes for the shrimp to free itself completely, after which it rests for a few seconds.

The new exoskeleton is soft, since it must be flexible to buckle up to fit inside the older, smaller skeleton. It stretches to accommodate the increased size of the shrimp. Complete hardening of the new exoskeleton will take about two days.

❶ The old exoskeleton splits along the back behind the Harlequin Shrimp's head. It eases out backwards.

❷ The shrimp emerges further and struggles to free itself from the old exoskeleton.

❸ Moulting is complete, and the old exoskeleton lies beside the shrimp, as the animal rests.

LIFESTYLES

All marine arthropods are free-living for at least part of their lives. Some, such as crabs, have planktonic larvae that sink to the sea floor and become bottom-living, or benthic, as they mature. They tend to live alone unless seeking a partner to breed with, and may defend territory. Others, such as krill and copepods, live in vast swarms, travelling hundreds of metres up and down the water column each day to feed (see p.221). Adult barnacles remain anchored to one spot, that is, they are sessile, but they also aggregate in large numbers on rocky shores where living conditions are most favourable. Deep-sea arthropod species are not well known but many have cryptic red or black coloration to make themselves invisible as opposed to krill, which have light organs and exhibit bioluminescence. A few arthropods live in close association with other species. Sometimes both partners benefit from a relationship (mutualism), sometimes only one has an advantage (commensalism), and sometimes one gains at some cost to the other (parasitism).

COMMENSALISM
This crab is camouflaged by a sea squirt in a commensal relationship. The crab benefits but the sea squirt neither gains nor loses.

PARASITISM
Some arthropods live closely with other species. This fish is being parasitized by an isopod, which is related to woodlice. There are two isopods, one under each eye, feeding on tissue fluid to the detriment of the fish.

SEA SPIDERS

As a group, sea spiders are typical of many problematic organisms whose classification is changing as information becomes available. They appear in different taxonomic groups depending on current scientific opinion. In this book, they form a group separate from the spiders and relatives (chelicerates). However, some recent research places them within the chelicerates and closely related to ticks and mites.

DECEPTIVE APPEARANCE
Sea spiders are so called because of their resemblance to land spiders, but their exact relationship to spiders is still not clear.

REPRODUCTION AND LIFE CYCLES

In most crustaceans, the sexes are separate, fertilization is predominantly internal, and the eggs must be laid in water. Some females store sperm and then let it flow over their eggs as they release them. Others protect their eggs by carrying them around, and keep them healthy by continually wafting water over them. On hatching, the larvae join the zooplankton, and pass through various stages before maturing into adults. Barnacles are both male and female (hermaphroditic) but only function as one sex at a time. The male has a long, extendable penis and mates with all neighbouring females within reach. In horseshoe crabs, fertilization occurs externally. Males and females pair up, the males fertilize the eggs as the females lay them in the sand, and then both sexes abandon them.

egg mass on underside of female crab

CRAB MOTHER AND LARVA
A velvet crab (above) carries eggs beneath her body until they hatch. The hatchlings enter a planktonic larval stage called a zoea (left). This moults four to seven times before it becomes a megalops larva, then once again to become an adult crab.

ARTHROPOD CLASSIFICATION

Here, the arthropods are split into five subphyla – Crustacea, Pycnogonida, Chelicerata, Hexapoda, and the non-marine Myriapoda (centipedes and millipedes – not described below). All contain land and freshwater species except the exclusively marine Pycnogonida.

CRUSTACEANS
Subphylum Crustacea

50,000 species
This group includes the familiar crabs, lobsters, shrimps, prawns, and barnacles as well as the smaller copepods, isopods, and krill. Most crustaceans have two pairs of antennae and only two body segments: the cephalothorax, fused together from the head and the thorax; and the abdomen. Paired appendages vary greatly – some are sensory, while others are adapted for walking or swimming; sometimes there is also a large pair of claws.

SEA SPIDERS
Subphylum Pycnogonida

1,000 species
All members of this group are small and spider-like with a leg span of between 1mm (¹/₃₂in) and 50cm (19½in). Many species have a unique pair of legs, called ovigers, which are situated in front of the first pair of walking legs. The females use them for grooming, courtship, and also to transfer eggs to the ovigers of the male, where they remain until they hatch. Sea-spiders are common in intertidal areas but they are rarely seen due to their excellent camouflage.

CHELICERATES
Subphylum Chelicerata

70,000 species
Spiders, scorpions, ticks, mites, and horseshoe crabs belong to this group. A few species of spiders live in the intertidal zone and some types of ticks and mites are either free living or parasitic in marine habitats.

The horseshoe crabs (class Merostomata) are completely marine. Like spiders, they have four pairs of walking legs and their body comprises two parts, called the prosoma and opisthosoma. In horseshoe crabs, the prosoma contains most of the body organs, and the opisthosoma has most of the musculature and the gill books, which are used for respiration and locomotion. What makes this species unique is the hinged carapace that protects the body and the long, tail-like telson which is used to right itself if the crab is accidentally inverted.

INSECTS
Subphylum Hexapoda

950,000 species
By far the largest group within the *Hexapoda* is the insects – the largest of all animal groups. It includes beetles, flies, ants, and bees. Most insects have compound eyes and three distinct body segments – the head, the thorax with its three pairs of walking appendages, and the abdomen. Many species also have wings.

Many insects live in coastal areas, but only a few live on the shore. Only one type of insect is truly marine – the marine skater, *Halobates*, a type of "true bug" (order Hemiptera). There are five species of this insect and they spend their entire life on the ocean. However, they require a solid object, such as a floating feather or lump of tar, on which to lay their eggs.

SUBPHYLUM PYCNOGONIDA

Giant Sea Spider
Colossendeis australis

LENGTH	25cm (10in) (leg-span)
WEIGHT	Not recorded
HABITAT	Bottom dweller

DISTRIBUTION Antarctic shelf and slope

Unlike most sea spiders, which have a leg-span of less than 2.5cm (1in), the Giant Sea Spider has a huge leg-span of about 25cm (10in). It has a large proboscis through which it sucks its food, but its tiny body is so small that the sex organs and parts of its digestive system are situated in the tops of the legs. Sea spiders are somewhat unusual among arthropods in that they exhibit parental care, the males having a modified pair of legs to carry the eggs until they hatch.

SUBPHYLUM CRUSTACEA

Water Flea
Evadne nordmanni

LENGTH	1mm (1/32 in)
WEIGHT	Not recorded
HABITAT	Open waters, to depths of 2,000m (6,500ft)

DISTRIBUTION Temperate and cool waters worldwide

The Water Flea feeds on planktonic organisms and is itself food for larger animals. It has a conspicuous eye and feathery swimming appendages, which are modified antennae. In spring, unfertilized eggs that have been nourished in a brood chamber by the female hatch as females. In autumn, males and females are produced in the same way. A large egg is also produced sexually, which overwinters and then invariably develops into a female.

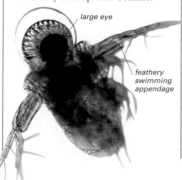

large eye

feathery swimming appendage

SUBPHYLUM CHELICERATA

American Horseshoe Crab
Limulus polyphemus

LENGTH	Up to 60cm (24in)
WEIGHT	Up to 5kg (11lb)
HABITAT	Sandy or muddy bays to 30m (100ft)

DISTRIBUTION Western Atlantic and Gulf Coast from southern Maine to the Yucatan Peninsula

Despite its name, the American Horseshoe Crab is more closely related to spiders than to crabs. It is mainly active at night and scavenges anything it can find, including small worms, bivalves, and algae. Its horseshoe-shaped, greenish-brown outer shell, or carapace, is for protection, and adults have few predators. It has six pairs of appendages: the first pair is used for feeding; the other five are for walking.

The crab's five plate-like "book" gills each contain many membranes, like the pages of a book, and are situated towards the tail. As well as being used for respiration, the gills are also used for propulsion when swimming. They can also absorb water, which helps the crab to fill its new shell after it has shed the old one. Its long, rigid tail is used for steering and for righting itself.

The reproductive cycle is closely linked to the spring and autumn high tides (especially the spring tides in the northern part of the range) and the lunar cycle. At full Moon, the adults gather in large numbers on sandy beaches to breed. Females lay up to 20,000 eggs in a nest near the high-tide mark, providing a vital food source for birds and other marine creatures.

After hatching, the young remain hidden in the sand for safety. They emerge some weeks later at high tide and take to the water until they moult for the first time, after which they look like small adults and start to live on the sea floor.

HUMAN IMPACT

MEDICAL RESEARCH

If the American Horseshoe Crab is injured, some of its blood cells form a clot, which kills harmful bacteria that are also dangerous to humans. In order to exploit this property for human benefit, crabs are collected from shallow waters on the Atlantic coast of North America during the summer months. Researchers then remove about 20 per cent of the blood from each crab. From this they extract a protein that is used to detect bacterial contamination in drugs, vaccines, and other medical products that are given intravenously. Bleeding the crabs is not fatal to them, and afterwards they are returned to the sea to recover.

OCEAN LIFE

SUBPHYLUM CRUSTACEA

Cyclopoid Copepod
Oithona similis

LENGTH
0.5–2.5mm (1/54–3/32 in)

HABITAT
Surface waters to a
depth of 150m (500ft)

DISTRIBUTION Atlantic, Mediterranean, Southern
Ocean, southern Indian and Pacific oceans

Copepods make up over 70 per cent
of zooplankton, and this one of the
most abundant, widespread species.
As the name suggests, cyclopoid
copepods have a single, central eye,
which is light sensitive. These
arthropods also have a T-shaped,
shrimp-like body that tapers towards
the rear and bears six pairs of
swimming limbs. Jerky movements of
its limbs sweep food particles
efficiently towards the mouth. Females
can be recognized when carrying egg
sacs attached to their abdomens.

As part of the zooplankton,
copepods of this genus are a vital
element of oceanic food chains. They
feed on marine algae and bacteria and
in turn are an important source of
protein for many ocean-dwelling
animals. Every night cyclopoid
copepods migrate from a depth of
about 150m (500ft) to the surface
layers of the ocean to feed. This daily
journey, which is undertaken by many
marine creatures, is one of the largest
mass movements of animals on Earth.

SUBPHYLUM CRUSTACEA

Gooseneck Barnacle
Pollicipes polymerus

LENGTH
Up to 8cm (3in)

HABITAT
Intertidal zone of rocky
shores

DISTRIBUTION Eastern Pacific coast of North
America, from Canada to Baja California, Mexico

So-called because of its resemblance
to a goose neck and head, the
Gooseneck Barnacle forms dense
colonies in crevices on rocky shores
with strong waves. Barnacles anchor
themselves to rocks by a tough,
flexible stalk (peduncle), which also
contains the ovaries. This is actually
their 'head' end. Once the barnacle has
attached itself to an object it secretes
a series of pale plates at the end of
its stalk, forming a shell around its
feather-like legs, which comb through
the water for food. The legs face away
from the sea, enabling the barnacle to
feed by filtering out particles of detritus
from returning tidal water as it funnels
past them through cracks in the rocks.
These barnacles become sexually
mature at about five years of age and
may live for up to 20 years. The larval
stage is free-living but depends on sea
currents for its transport and survival.
Colonies of Gooseneck Barnacle are
susceptible to the damaging effects of
oil pollution and they recover only
slowly from disturbance.

SUBPHYLUM CRUSTACEA

Acorn Barnacle
Semibalanus balanoides

LENGTH
Up to 1.5cm (1/2 in)
diameter

HABITAT
Intertidal zone of rocky
shores

DISTRIBUTION Northwest and northeast Atlantic,
Pacific coast of North America

Like all adult barnacles, the adult
Acorn Barnacle remains fixed in one
place once it has anchored itself to a
site. The free-swimming juveniles pass
through several larval stages before
moulting into a form that can detect
both other Acorn Barnacles and
suitable anchoring sites. Once a larva
fixes itself to a rock, using cement
produced by glands in its antennae, it
moults again. It then secretes six grey
calcareous plates, forming a protective
cone that looks rather like a miniature
volcano. Four smaller, moveable plates

at the top of the cone open, allowing
the Acorn Barnacle to feed. It does
this when the tide is in by waving its
modified legs, called cirri, in the water
to filter out food. When the tide is
out, the plates are closed to prevent
the barnacle drying out. Acorn
Barnacles are hermaphrodites that
possess both male and female sexual
organs, but they function as either a
male or a female. They do not shed
their eggs and sperm into the water;
instead they use extendable penises, to
transfer sperm to receptive neighbours.

CHARLES DARWIN

Before the British naturalist
Charles Darwin (1809–1882)
proposed his revolutionary theory
of evolution in *The Origin of
Species* (1859), he spent eight
years studying barnacles.
Realizing the impact his ideas on
evolution would have on existing
scientific and religious thinking,
he delayed writing and instead
produced four monographs on
the classification and biology
of barnacles. This work earned
him the Royal
Society's Royal
Medal in 1853,
validating his
reputation as
a biologist.

OCEAN LIFE

SUBPHYLUM CRUSTACEA

Giant Mussel Shrimp

Gigantocypris muelleri

LENGTH
1.4–1.8cm (½–¾in)

HABITAT
Planktonic, intermediate
to deep sea

DISTRIBUTION Atlantic, Southern Ocean, western
Indian Ocean

The body of the Giant Mussel Shrimp
is totally enclosed within its carapace, so
that its seven pairs of limbs are almost
hidden from view. It has large, mirror
eyes with parabolic-shaped reflectors
that focus light on to a flat plate in its
centre. It is planktonic, but lives at
greater depth than many forms of
plankton, usually below 200m (650ft),
where it feeds on detritus sinking
down from above. The picture shows
a large female carrying embryos, which
are clearly visible through the carapace.

SUBPHYLUM CRUSTACEA

Peacock Mantis Shrimp

Odontodactylus scyallarus

LENGTH
Up to 15cm (6in)

HABITAT
Warm water near reefs
with sandy, gravelly, or
shelly bottoms

DISTRIBUTION Indian and Pacific oceans

A relative of shrimps and lobsters,
the brightly coloured Peacock Mantis
Shrimp is a voracious predator. Its
large, mobile, compound eyes have
sophisticated stereoscopic and colour
vision that includes some ultraviolet
shades. It uses sight when hunting,
waiting quietly, like the praying
mantis, for its unsuspecting prey to
come within reach, then striking using
its powerful, club-like second pair of
legs with immense speed – about
120kph (75mph) – and force (up
100 times its own weight).
Such power is enabled by
a special, saddle-like hinge-joint in
these legs, which acts like a spring.
The Peacock Mantis Shrimp can
smash the shells of gastropods and
crabs and tackles prey larger than itself.
It excavates U-shaped burrows or
makes crevices in rocks or coral to live
in. After hatching,
its larvae enter the
plankton, where
they develop over
a few weeks before
drifting down towards
the sea floor to make
their own burrows.

SUBPHYLUM CRUSTACEA

Sea Slater

Ligia oceanica

LENGTH
Up to 3cm (1¼in)

HABITAT
Coasts with rocky
substrata

DISTRIBUTION Atlantic coasts of northwestern
Europe

Commonly found under stones and
in rock crevices, the Sea Slater is a
seashore-dwelling relative of the
woodlouse. It lives in the splash
zone, but can survive periods of
immersion in salt water. Its
head, which has a pair of well-
developed compound eyes and
very long antennae, is not
markedly separated from its
body, which is flattened, about twice
as long as it is broad, and ends in two
forked projections called uropods.
As adults, Sea Slaters have six pairs
of walking legs until their final moult,
after which they have seven. The Sea
Slater is not generally seen during
the day unless it is disturbed, and it
emerges from its hiding place only at
night to feed on detritus and brown
seaweed. Sea Slaters mature at about
two years of age and usually breed
only once before dying.

uropod

antenna

SUBPHYLUM CRUSTACEA

Sand Hopper

Orchestia gammarella

LENGTH
2–10mm (1/16–3/8 in)

HABITAT
Splash zone of sandy
shores

DISTRIBUTION Atlantic coasts of northeastern Canada
and northwestern Europe

Amphipod crustaceans, such as the
Sand Hopper, live in large numbers
in the splash zone of any shore where
there is rotting seaweed. Their life
cycle takes about 12 months and the
female usually produces only one
clutch of eggs, which she keeps in a
brood pouch, where they hatch after
one to three weeks. The young leave
the pouch about a week later when
their mother moults. Sand hoppers
are also known as sand fleas because
they move in the same way and have
similar laterally compressed bodies.

SUBPHYLUM CRUSTACEA

Antarctic Krill

Euphausia superba

LENGTH
Up to 5cm (2in)

HABITAT
Planktonic

DISTRIBUTION Southern Ocean

All oceans contain vast numbers of
krill – small, shrimp-like, planktonic
crustaceans that live in open waters.
This species lives in the subantarctic
waters of the Southern Ocean, where
it forms a vital link in the food chain,
being taken in vast quantities by
baleen whales, seals, and various fish.
Krill rise to the surface at night to
feed on phytoplankton, algae, and
diatoms. For safety they sink to greater
depths during the day. The feathery
appearance of this species is due to
its gills, which, unusually, are carried
outside the carapace. Their filamentous
structure increases the surface area
available for gaseous exchange.
Antarctic krill also have large light
organs, called photophores. The
light is thought to help them group
together. They spawn in spring, during
which females may release several
broods of up to 8,000 eggs.

SUBPHYLUM CRUSTACEA

Deep Sea Red Prawn
Acanthephyra pelagica

LENGTH	Not recorded
HABITAT	Deep water

DISTRIBUTION Atlantic

In the low light levels of the deep ocean, red appears black, making the Deep Sea Red Prawn invisible to potential predators. Its hard outer casing, or exoskeleton, is thinner and more flexible than that of shallow-water crustaceans, which prevents collapse under the immense water pressure of the ocean depths. The flesh of this prawn is oily to aid buoyancy. It uses its first three pairs of limbs to feed on small copepods. The remaining five pairs of limbs, the pereiopods, are used for locomotion. Gills attached to the tops of the legs are used for respiration.

SUBPHYLUM CRUSTACEA

Common Prawn
Palaemon serratus

LENGTH	Up to 11cm (4¼in)
HABITAT	Rock pools among seaweeds and lower parts of estuaries

DISTRIBUTION Eastern Atlantic from Denmark to Mauritania, Mediterranean, Black Sea

The Common Prawn has a semi-transparent body, making its internal organs visible, and is marked with darker bands and spots of brownish red. As with many other species of prawn, its shell extends forwards between its stalked eyes to form a stiff, slightly upturned projection called a rostrum. This feature has a unique structure by which the Common Prawn can be distinguished from all other members of the same genus. The rostrum curves upwards, splitting in two at the tip, where it has several tooth-like projections on the lower and upper surfaces.

To either side of the rostrum there is a very long antenna that can sense any danger close by and is also used to detect food. Of the prawn's five pairs of legs, the rear three pairs are used for walking, while the front two pairs are pincered and used for eating. Attached to the abdomen is a series of smaller limbs called swimmerets that the

HUMAN IMPACT

PRAWN FARMING

Nearly all the world's farmed prawns come from developing countries such as Thailand, China, Brazil, Bangladesh, and Ecuador, which use chemical-intensive farming to meet demand. More environmentally friendly techniques are now being encouraged.

SHARED RESOURCES
Fishermen in Honduras fish for wild prawns in a lagoon shared with prawn farmers.

prawn uses to swim. For a sudden, backwards movement, the prawn flicks its tail. Females produce and look after about 4,000 eggs until they hatch into larvae. The larvae float among the plankton until they mature.

tail fan, or telson

pincered leg

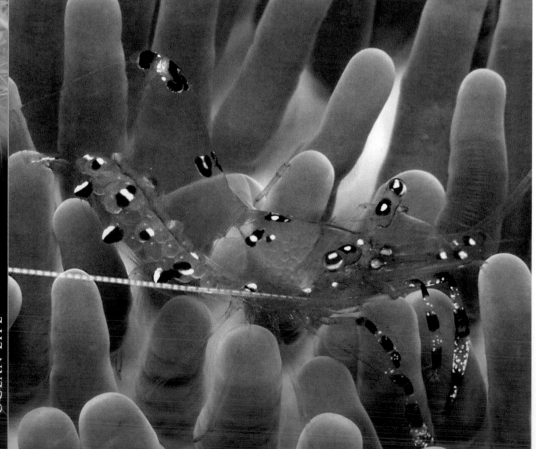

SUBPHYLUM CRUSTACEA

Anemone Shrimp
Periclimenes brevicarpalis

LENGTH	2.5cm (1in)
HABITAT	Shallow water reefs

DISTRIBUTION Indian Ocean, western Pacific

Nestling among the tentacles of an anemone, the Anemone Shrimp is safe from attack by predators. It rarely wanders far from its host, surviving by scavenging scraps that the anemone cannot eat. The shrimp may benefit the anemone by removing excess food particles as well as any waste it produces. This type of relationship is called commensalism: one individual in the partnership profits from the liaison and the other comes to no harm. Removed from its host, this shrimp is defenceless. The Anemone Shrimp belongs to the same family (Palaeomonidae) as the Common Prawn (above) and so they have several features in common. These include a pair of long, sensory antennae used to sense danger and detect food and a rostrum (the elongated projection of the shell from between the eyes). The Anemone Shrimp is almost completely transparent, with a few purple and white spots.

SUBPHYLUM CRUSTACEA
Spiny Lobster
Panulirus argus

LENGTH	60cm (24in)
HABITAT	Coral reefs in deep water

DISTRIBUTION Western Atlantic, Gulf of Mexico, Caribbean Sea

Being both nocturnal and migratory, the Spiny Lobster has excellent navigational skills. It can establish its position in relation to the Earth's magnetic field and then follow a particular route as well as any homing pigeon. This lobster prefers warm water and so remains in the shallows in summer before migrating in groups to deeper water in winter by walking in single file across the sea floor. It lacks the large claws of the Common Lobster (right) but is well protected from most predators by the sharp spines that cover its carapace.

SUBPHYLUM CRUSTACEA
Common Lobster
Homarus gammarus

LENGTH	Up to 1m (3ft), typically 60cm (24in)
HABITAT	Rocky coasts

DISTRIBUTION Eastern Atlantic, North Sea, Mediterranean

In life, the upper surface of the Common Lobster is blue mottled with yellow, while its underside is more yellowish. Individuals weigh up to 5kg (11lb). It has large, differently sized claws: the smaller one has sharper edges and is used for cutting prey, while the larger one is used for crushing. It lives in holes and crevices on the sea bed. The Common Lobster is commercially important and in danger of overexploitation because it matures slowly, not beginning to breed until it is about six years old.

crushing claw

cutting claw

SUBPHYLUM CRUSTACEA
Reef Hermit Crab
Dardanus megistos

WIDTH (LEGSPAN)	Up to 30cm (12in)
HABITAT	Near-shore tropical reefs

DISTRIBUTION Indian and Pacific oceans

Like other hermit crabs, the Reef Hermit Crab uses a "borrowed" shell to protect its hindquarters, being unable to make its own carapace. When it grows too big for its current shell, it simply looks for an unused larger one. It is while switching from one shell to the next that the Reef Hermit Crab is most vulnerable, as it risks exposing its soft, rather asymmetrical abdomen to predators. There are about 500 species of hermit crab worldwide – the Reef Hermit Crab lives in shallow-water tropical reef habitats, but some species live on land. The Reef Hermit Crab is a scavenger rather than a hunter, and drags itself over the sea floor looking for bits of animal matter and algae, tearing apart any carcasses that it finds with its dextrous mouthparts. It may attach stinging anemones to its shell as protection from predators.

eye stalk

single large claw

SUBPHYLUM CRUSTACEA
Porcelain Crab
Petrolisthes lamarckii

WIDTH (SHELL)	Up to 2cm (¾in)
HABITAT	Pools on rocky beaches and shorelines

DISTRIBUTION Indian Ocean, Pacific coast of Australia, western Pacific

The flat, rounded body of the Porcelain Crab allows it to slip easily into small rock crevices to hide. However, if it becomes trapped by a predator or stuck beneath a rock, it can shed one of its claws in order to escape, and a new one will grow over time. This crab's abdomen is folded under its body, but it can be unfolded and moved like a paddle when swimming.

SUBPHYLUM CRUSTACEA
Robber Crab
Birgus latro

LENGTH	Up to 60cm (24in) across
HABITAT	Rock crevices and sandy burrows

DISTRIBUTION Tropical waters of Indian and Pacific oceans

The Robber Crab, or Coconut Crab, is the largest terrestrial arthropod. An anomuran, like the hermit and porcelain crabs, it has evolved to look similar to brachyurans, the group to which most other crabs belong, but differs in having long antennae and only three pairs of walking legs instead of four. It lives on oceanic islands and offshore inlets. It scavenges but will also eat fruit and can even smash into coconuts using its powerful claws. Adults live and mate on land, but females release their eggs into water.

SUBPHYLUM CRUSTACEA
Nodose Box Crab
Calappa angusta

LENGTH	Not recorded
HABITAT	Offshore to depths of 15–200m (50–650ft)

DISTRIBUTION Western Atlantic, Gulf of Mexico, Caribbean Sea

Found in the warm waters around the Caribbean, the Nodose Box Crab is a true crab with a small abdomen that is tucked away underneath the body and four pairs of legs. This species may be recognized by the rows of nodules that radiate from behind its eyes across the upper surface of its yellowish shell, or carapace.

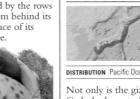

SUBPHYLUM CRUSTACEA
Japanese Spider Crab
Macrocheira kaempferi

WIDTH (SHELL)	Up to 37cm (14½in)
HABITAT	Deep-water vents and holes to depths of 50–300m (160–1,000ft)

DISTRIBUTION Pacific Ocean near Japan

Not only is the giant Japanese Spider Crab the largest of all crabs, with a legspan of up to 4m (13ft) and weighing 16–20kg (35–44lb), it may also be the longest living, surviving for up to 100 years. Living in the deep, cold waters around Japan, it moves slowly across the ocean floor on its spider-like legs, scavenging for food.

PORCELAIN CRAB
The Porcelain Crab uses its flat body to crawl out of reach of predators. Here, the tentacles of an anemone provide a secure retreat for a Porcelain Crab in the Andaman Sea in the northern Indian Ocean. The mouthparts of the crab fan out and trap plankton, which it then brushes into its mouth.

SUBPHYLUM CRUSTACEA

Long-legged Spider Crab

Macropodia rostrata

LENGTH
Up to 2.5cm (1in)

HABITAT
Lower shore, usually not beyond 50m (165ft)

DISTRIBUTION Northeastern Atlantic from southern Norway to Morocco, Mediterranean

Also called the Decorator Crab because it camouflages itself using fragments of seaweed and sponges, the Long-legged Spider Crab is covered in hook-shaped hairs that hold its disguise in place, enabling it to blend in with the seaweed among which it lives. This crab has a triangular-shaped carapace that extends forwards between the eyes into an eight-toothed projection called a rostrum. Its spider-like legs are at least twice as long as its body and can be used, somewhat ineffectively, for swimming. The Long-legged Spider Crab feeds on small shellfish, algae, small worms, and detritus. Breeding occurs all year round on Atlantic coasts, but takes place between March and September in the Mediterranean. The male transfers sperm to the female using its first pair of abdominal legs. The female carries the eggs until they hatch into larvae that live in the plankton.

SUBPHYLUM CRUSTACEA

Spotted Reef Crab

Carpilius maculatus

LENGTH
About 9cm (3½in)

HABITAT
Shoreline to 10m (33ft), inshore reefs

DISTRIBUTION Indian Ocean and western Pacific

The conspicuous colouring of the Spotted Reef or Coral Crab is highly distinctive. Its smooth, light brown carapace has two large red spots behind each eye, three across the middle, and either two or four at the rear. Between the eyes the carapace has four small, rounded projections, which are also characteristic of the species. It is a nocturnal, slow-moving crab that uses its disproportionally large claws to feed on corals, snails, and other small marine creatures.

SUBPHYLUM CRUSTACEA

Edible Crab

Cancer pagurus

LENGTH
Up to 16cm (6in)

HABITAT
Intertidal zone to 100m (330ft), in rock pools and muddy sand offshore

DISTRIBUTION Northeastern Atlantic and North Sea; introduced to parts of the Mediterranean

The oval carapace of the edible crab has a characteristically "scalloped" or "piecrust" edge around the front and sides. Its huge pincers are distinctively black-tipped, while the body is purple-brown in small individuals and reddish brown in larger ones. Edible crabs breed all year round, the females incubating their eggs for seven to eight months. This crab is caught in large numbers and is highly valued as a luxury food.

black eye
on short
eye stalk

mouth

SUBPHYLUM CRUSTACEA

Pea Crab

Pinnotheres pisum

LENGTH
Males 8mm (⅓in);
females 14mm (½ in)

HABITAT
Intertidal zone to 150m (500ft)

DISTRIBUTION Eastern Atlantic from northwestern Europe to West Africa, Mediterranean

Typically about the size of a pea, the tiny Pea Crab is usually found inside the shells of the Common Mussel. Protected from predators in the mantle cavity of its host, it feeds on any plankton that become trapped on the mussel's gills as water passes over them. Whether the presence of this guest is harmful to the mussel is unclear. Female Pea Crabs are substantially larger than males and have an almost translucent carapace through which their pink reproductive organs are visible. Males have harder, yellowish brown carapaces that protect them during the breeding season, which runs from April to October. During this time, males leave the safety of their host's shell and swim about looking for females with which to mate. In regions where shellfish are harvested commercially, the Pea Crab is considered a pest.

SUBPHYLUM CRUSTACEA

Common Shore Crab

Carcinus maenas

LENGTH
Up to 6cm (2½in)

HABITAT
Intertidal zone to 60m (200ft), all substrates; estuaries

DISTRIBUTION Northeastern Atlantic from Norway to West Africa; introduced elsewhere

The Common Shore Crab tolerates a wide range of salt concentrations and temperatures and so can live in salt marshes and estuaries as well as along the shoreline. Its dark green carapace has five marked serrations on the edge behind the eyes. This opportunistic hunter preys voraciously on many types of animals, including bivalve molluscs, polychaetes, jellyfish, and small crustaceans. Where introduced, it may be detrimental to local marine life. On the west coast of the USA, for example, it has had a considerable impact on the shellfish industry.

SUBPHYLUM CRUSTACEA

Blue Swimming Crab

Portunus pelagicus

LENGTH
Up to 7cm (2¾in)

HABITAT
Intertidal sandy or muddy sea beds to 55m (180ft)

DISTRIBUTION Coastal waters of the Indian and Pacific oceans, eastern Mediterranean

Unlike most crabs, the Blue Swimming Crab is an excellent swimmer and uses its fourth pair of flattened, paddle-like legs to propel itself through the water. Despite its common name, only the males are blue, and the females are a rather dingy greenish brown. Males also differ in having very long claws, more than twice as long as the width of their carapace. The claws are armed with sharp teeth that are used to snag small fish and other food items.

When a Blue Swimming Crab feels threatened, it usually buries itself in the sand. If this measure fails to deter the threat, the crab adopts its own threat stance, extending its claws sideways in an attempt to look as large as possible. The natural range of this crab has been extended to a small part of the eastern Mediterranean by the opening of the Suez Canal. It is a popular food in Australia.

SUBPHYLUM CRUSTACEA

Ghost Crab

Ocypode saratan

LENGTH
About 3.5cm (1½in)

HABITAT
Sandy shores in deep burrows above the water line

DISTRIBUTION Coastal waters of western Indian Ocean, Red Sea

Resting in the cool of its burrow during the day, the fast-moving Ghost Crab emerges at twilight to hunt. It will eat anything it can find, including other crabs, and also scavenges whatever was brought in by the last tide. During the mating season, males defend their burrows, but they rarely fight, any disputes being settled by ritualistic displays. Their burrows can be over 100m (330ft) from the sea and over 1m (3ft) deep.

SUBPHYLUM CRUSTACEA

Orange Fiddler Crab

Uca vocans

LENGTH
About 2.5cm (1in)

HABITAT
Near water on mud or sand

DISTRIBUTION Indian Ocean and western Pacific

Like its close relative the Ghost Crab (see above, right), the male Orange Fiddler Crab also exhibits ritualistic displays to deter rivals. Males are easily recognized because one of their claws is greatly enlarged. In a mature adult this claw makes up more than half the crab's body weight and is used both to attract potential mates and to ward off rival males. Observing the distinctive "courtship wave" of fiddler crabs is helpful in identifying different species.

Orange Fiddler Crabs are active during the day. As well as digging a main burrow up to 30cm (12in) deep, they create a number of bolt holes into which they can retreat if danger threatens. At high tide, they seal themselves into their burrows with a small pocket of air. The presence of air is essential for their survival because fiddler crabs obtain oxygen from air not water, despite having gills.

RITUAL DISPLAY

Each species of fiddler crab waves its claw in a slightly different way. If this ritual movement does not deter a rival male, then two crabs may "arm-wrestle" each other to resolve their dispute. The weaker individual usually retreats before any serious damage is done.

LEFT- OR RIGHT-HANDED?
Both crabs in this picture are right-handed, but in some males it is the left claw that is enlarged (see below).

RESTORING THE BALANCE
On arriving at the shore, the crabs head to the ocean, where they replace water and body salts lost during the arduous journey down from the forest plateau.

RED CRAB MIGRATION

The remarkable annual migration of the Red Crab (*Gecarcoidea natalis*) from the forests of Christmas Island (southwest of Indonesia) down to the sea to spawn is one of the wonders of the natural world. Until recently, about 120 million of the crabs made the journey each year, spending the rest of the time on a forested plateau about 360m (1,200ft) above sea level. Red Crabs are mainly herbivorous, feeding on fallen leaves, fruit, and flowers gathered from the forest floor, but they also eat other dead crabs and birds when the opportunity arises. They conserve water by restricting their activity to times of high humidity (over 70 per cent), retreating to their burrows during drier periods. Like their marine counterparts, Red Crabs have gills for respiration, but the gill chamber of this species is lined with tissue that acts as a lung and maximizes gaseous exchange.

Red Crabs on Christmas Island have undergone a noticeable decline in numbers, largely due to the accidental introduction of the Yellow Crazy Ant in the 1930s. Since the mid 1990s, about 20 million Red Crabs are thought to have been killed by the ants, which squirt formic acid on the crabs as a defence mechanism when they are disturbed. There is also pressure to increase the number of phosphate mines on the island, which would involve deforestation, depriving the crab of its habitat in the affected areas.

THE SEQUENCE OF MIGRATION

The onset of the wet season on Christmas Island in early November signals the start of the Red Crabs' migration, which takes place over three lunar cycles. The males set off first, followed by females. It takes about a week for the crabs to reach the shore. After dipping in the ocean, the males compete fiercely for space to dig their burrows, where it is thought mating takes place. The males then return inland, leaving the females in the burrows while their eggs develop. Egg release is synchronized with the lunar cycle and occurs over six or seven nights during the last quarter of the Moon.

STAGES OF MIGRATION

TRAVELLING ACROSS LAND

MAN-MADE OBSTACLES *Although the distance to the shore is only about 500m (1,600ft), Red Crabs have to negotiate a number of obstacles, including roads and hot railway tracks. In the past, it was not unusual for one million crabs to perish each year, yet this had little impact on the population. Today, various measures such as road closures and concrete underpasses offer some security to the crabs, but they still run the risk of dying from dehydration during their journey.*

RELEASING EGGS

EGG RELEASE *After incubating up to 100,000 eggs for 12–13 days, the females leave their burrows and gather on the shore to disperse them directly into the sea. They do so at night, as the high tide turns, by raising their claws and shaking their bodies vigorously to free the eggs from their pouches. Crabs on the cliffs may be 8m (26ft) above the water.*

MEGALOPAE LARVAE *Red Crab eggs hatch as soon as they hit the water. The young remain in the ocean for up to 30 days, and pass through several larval stages before returning to the shallows as shrimp-like megalopae larvae. They metamorphose into tiny crabs after 3–5 days and leave the water to start life on land.*

FROM SEA TO LAND

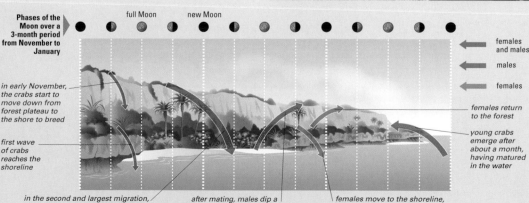

Phases of the Moon over a 3-month period from November to January

full Moon   new Moon

females and males
males
females

in early November, the crabs start to move down from forest plateau to the shore to breed

first wave of crabs reaches the shoreline

females return to the forest

young crabs emerge after about a month, having matured in the water

in the second and largest migration, the crabs dip in the sea to replenish body salts, then the males fight to establish burrows

after mating, males dip a second time, then either return inland immediately or remain and feed

females move to the shoreline, releasing eggs into the sea at the turn of the high tide during the last quarter of the third lunar cycle

OCEAN LIFE

Shore Bristletail

Petrobius maritimus

LENGTH	1cm (²⁄₅ in)
HABITAT	Rocky shores in the splash zone

DISTRIBUTION British Isles excluding Ireland

The Shore Bristletail, also known as the Shore Springtail, derives its common name from the three long filaments extending from the tip of its abdomen. Its long body is well-camouflaged by drab-coloured scales. It has long antennae and compound eyes that meet at the top of its head. The Shore Bristletail lives in rock crevices and feeds on detritus. It can move swiftly about the rocks using small spikes on its underside, called styles, to help it grip. When disturbed, it can leap small distances through the air using its abdomen to catapult it away from the rock.

Rock Springtail

Anurida maritima

LENGTH	Up to 3mm (¹⁄₈ in)
HABITAT	Upper intertidal zone of rocky shores

DISTRIBUTION Coasts of the British Isles

At low tide hundreds of Rock Springtails wander down the beach searching for food, returning to the shelter of their rock crevices an hour before the tide turns. Vast numbers

Dune Snail Bee

Osmia aurulenta

LENGTH	1cm (²⁄₅ in)
HABITAT	Sand dune systems

DISTRIBUTION Coasts of northeastern Atlantic, North Sea, Baltic, and Mediterranean

Important in the pollination of sand-dune plants, the Dune Snail Bee has a compact, brownish black body with a dense covering of golden red hairs that later fade to grey. Unlike the Honey Bee, which carries any pollen it collects in pouches on its legs, the Dune Snail Bee carries its pollen in a brush of hairs under its abdomen.

of them squeeze together in the fissures to avoid being immersed at high tide. It is here that they moult and lay their eggs safe from submersion and many of their predators.

Rock Springtails are blue-grey in colour with segmented bodies that are wider at the posterior end. They have three pairs of appendages used for locomotion, which also allow them to swarm over the surface of calm rock pools without sinking – they cannot swim. Springtails are so-named for their jumping organ, called the furcula, which acts like a spring, propelling the animal upwards if threatened. Unlike other springtails, however, the Rock Springtail does not have this feature.

Male bees of this species emerge between April and July, a little earlier in the year than the females, and seek out territories that contain a snail shell. They then leave scent marks (pheromones) on the stems of plants to attract passing females. Once a female has mated with her chosen partner, she will adjust the position of the shell so that the entrance is orientated in the most sheltered direction and lays her eggs inside it.

Intertidal Rove Beetle

Bledius spectabilis

LENGTH	Up to 2cm (³⁄₄ in)
HABITAT	Intertidal sandy and muddy shores

DISTRIBUTION Coasts of the British Isles and northern Europe

Unusual in that it lives in the intertidal zone after which it is named, this small arthropod has an elongated, smooth black body. Short reddish brown wing cases, or elytra, protect the wings but leave most of the flexible abdomen exposed. A mobile abdomen allows the Intertidal Rove Beetle to squeeze into narrow crevices and also to push its wings up under the elytra.

Most rove beetles are active either by day (diurnal) or by night (nocturnal), but the life of the Intertidal Rove Beetle is dictated by the tides. It builds a vertical, wine-bottle shaped burrow in the sand with a living chamber about 5mm (¹⁄₅ in) diameter and retreats into it whenever the tide comes in. The burrow entrance is so narrow – about 2mm (¹⁄₁₀ in) in diameter – that the air pressure within prevents any water from entering. The female lays her eggs in side chambers within the burrow and remains on guard, until her offspring have hatched and are mature enough to leave and construct their own burrows.

Marine Skater

Halobates sericeus

LENGTH	Females: 5mm (¹⁄₅ in)
HABITAT	Ocean surface

DISTRIBUTION Pacific Ocean between 40° and 5° north and south of the equator

This is a member of the only truly marine genus of insects. The Marine Skater spends its entire life on the surface of tropical and subtropical oceans where winter temperatures rarely fall below 20°C (68°F). Little is known about these insects due to the difficulty in studying them. Females are larger than males and after mating they lay 10–20 cream-coloured, oval eggs on a piece of flotsam, such as a piece of floating wood. The eggs hatch into nymphs that moult through five stages before becoming adults.

Because this insect never dives below the surface, its diet is restricted to small organisms such as floating fish eggs, zooplankton, and dead jellyfish. It feeds by releasing enzymes onto the surface of its food and then drawing in the pre-digested material through its modified mouthparts.

Kelp Fly

Coelopa frigida

LENGTH	3–10mm (¹⁄₈–²⁄₅ in)
HABITAT	Temperate shores with rotting seaweed

DISTRIBUTION North Atlantic and north Pacific shorelines

The most widely distributed of the seaweed flies, the Kelp Fly is found almost everywhere there is rotting seaweed along a strand line. They have flattened, lustrous black bodies, tinged with grey, and bristly, brownish yellow legs. Of the two pairs of wings, only the front pair is functional, the hind pair being modified to small club-shaped halteres that act as stabilizers when in flight. Kelp Flies can crawl through vast layers of slimy seaweed without getting stuck, and if immersed in seawater they simply float up to the surface and fly off. Their larvae are equally waterproof. Strongly attracted to rotting seaweed by its smell, the female Kelp Flies seek out warm spots in which to lay their eggs. The larvae hatch and feed on the seaweed around them. After three moults they pupate; the adults emerge and complete the life cycle about 11 days after the eggs were laid. Kelp Flies are an important food source for several coastal birds including Kelp Gulls and sandpipers.

BRYOZOANS

DOMAIN	Eucarya
KINGDOM	Animalia
PHYLUM	**Bryozoa**
SPECIES	4,000–5,000

THESE COLONIAL ANIMALS live attached to the sea bed and although numerous, they are often overlooked. The individuals making up the colony are usually less than 1mm (1/32 in) long, but the colonies may span over 1m (3ft). Bryozoans are also called ectoprocts or sea mats, the latter name referring to their tendency to encrust the surfaces of stones and seaweeds. Other colonial forms of bryozoan include coral-like growths, branched plant-like tufts, and fleshy lobes. Most species are marine, but a few live in fresh water.

ATTACHED TO SEAWEED
Bryozoans, such as this hard species encrusting a seaweed, often live in areas with strong currents.

ANATOMY

A bryozoan colony is made up of individuals called zooids, and may contain several or up to many millions. Each zooid is encased in a box-shaped body wall of calcium carbonate or a gelatinous or horn-like material, and a small hole links it to other zooids. To feed, the animal pushes a circular or horseshoe-shaped structure (a lophophore) out of an opening. This is crowned by tentacles covered in tiny, beating hairs that draw in planktonic food. In most species, fertilized eggs are stored in specialized zooids that form a brood chamber for developing larvae.

MAT OF ZOOIDS
This encrusting species of bryozoan has rectangular zooids joined in a single layer. The resulting mat spreads over seaweeds.

HABITATS

With their great variety of body form, bryozoans can be found in almost any habitat from the seashore to the deep ocean, and from Arctic waters to tropical coral reefs. Colonies are most often found firmly attached to submerged rocks, seagrasses, seaweeds, mangrove roots, and dead shells, but some encrusting species even hitch a ride on the shells of living crustaceans and molluscs. A few unusual species do not need a surface for support and can live in the sand; these bryozoan colonies can move slowly over or through the sand by co-ordinated rowing movements of a long projection found on specialized zooids.

Bryozoan colonies originate from a single larva that settles on the sea bed and becomes a zooid. More zooids are added to the colony by budding, a process in which a new zooid grows out from the side of the body wall. Most bryozoan larvae are short-lived and settle near to the parent.

BRYOZOANS UNDER ATTACK
Sea slugs often make a meal of encrusting bryozoans, breaking into each zooid and eating the insides.

Gelatinous Bryozoan
Alcyonidium diaphanum

SIZE (HEIGHT)	Up to 30cm (12in)
DEPTH	From shore down
HABITAT	Rocks and shelly sand

DISTRIBUTION Temperate waters of northeastern Atlantic

Colonies of this bryozoan have a firm, rubbery consistency and grow as irregular, lobed, or finger-like growths that attach to their substrate with a small, encrusting base. This species may cause an allergic dermatitis when handled, and North Sea fishermen are often affected when their trawl nets have gone through areas of dense bryozoan undergrowth.

Hornwrack
Flustra foliacea

SIZE (HEIGHT)	Up to 20cm (8in)
DEPTH	0–100m (0–330ft)
HABITAT	Stones, shells, rock

DISTRIBUTION Temperate and Arctic waters of northeastern Atlantic

This species is often mistaken for a brown seaweed. The colony grows up from a narrow base as thin, flat, fan-like lobes. These usually form dense clumps and cover the sea bed like a crop of tiny brown lettuces. They litter the strandline on many shores in dried clumps and, by using a magnifying glass, an observer can easily see the individual, oblong colony members.

Pink Lace Bryozoan
Iodictyum phoeniceum

SIZE (WIDTH)	Up to 20cm (8in)
DEPTH	15–40m (50–130ft)
HABITAT	Rocky reefs

DISTRIBUTION Temperate and tropical waters around Australia

Pink Lace Bryozoan colonies feel hard and brittle to the touch because the walls of the individual zooids are reinforced with calcareous material. The colony is shaped like curly-edged potato crisps with a lacework of small holes. Its beautiful dark pink to purple colour remains even after the colony is dead and dried. This species prefers to live in areas with some current, and its holes may help reduce the force of the water against it. Similar species are found on coral reefs throughout the Indo-Pacific region.

ECHINODERMS

DOMAIN	Eucarya
KINGDOM	Animalia
PHYLUM	Echinodermata
CLASSES	6
SPECIES	About 7,000

THE NAME OF THIS PURELY MARINE group of invertebrates is derived from the Greek for "hedgehog skin". The group includes starfish, sea urchins, brittlestars, feather stars, and sea cucumbers. Echinoderms have radiating body parts, so most appear star-shaped, disc-shaped or spherical, and all have a skeleton of calcium-carbonate plates under the skin. Inside is a unique system of water-filled canals, called the water-vascular system, that enables them to move, as well as to feed and breathe. Typically bottom-dwellers, they live on reefs, shores, and the sea bed.

RADIAL SYMMETRY
This tropical starfish has the five-rayed structure of echinoderms. Its arms are protected by hard plates, and its bright colours warn predators of its toxins.

ANATOMY

The echinoderm body is based on a five-rayed symmetry similar to the petals of a flower. This is apparent in starfish, brittlestars, and urchin shells (tests). Sea urchins are like starfish, with their arms joined to form a ball. Sea cucumbers resemble elongated urchins – their five-rayed symmetry can be seen end-on. The echinoderm skeleton is made of hard calcium-carbonate plates, which are fused to form a rigid shell (as in urchins) or remain separate (as in starfish). Usually, it also features spiny or knob-like extensions that project from the body. Sea cucumbers have minimal skeletons reduced to a series of small, isolated knobs.
The water-vascular system consists of a network of canals and reservoirs, as well as tentacles that extend through pores in the skin to form hundreds of tiny tube feet.

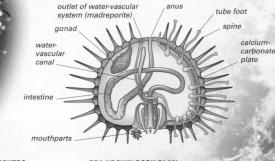

outlet of water-vascular system (madreporite)   anus   tube foot

gonad   spine

water-vascular canal   calcium-carbonate plate

intestine

mouthparts

MINI SUCKERS
Tube feet act like hydraulic suckers. They are operated by water squeezed in and out from a small reservoir similar to the bulb on the end of an eye dropper.

SEA URCHIN BODY PLAN
The body consists of a fluid-filled cavity inside the shell (test), which houses the organs. The mouth is in the centre of the underside, and the anus is on top of the upper side.

ECHINODERM CLASSIFICATION

The echinoderms were until recently split into five classes based on their shape, skeleton, and the position of their mouth, anus, and madreporite. The sixth class, comprising the newly discovered sea daisies, is regarded as part of the Asteroidea by some scientists.

FEATHER STARS, SEA LILIES
Class Crinoidea

About 625 species
Also known as crinoids, these animals have a saucer-shaped body extending into five repeatedly branching, feathery arms used

as filter-feeding appendages. Mouth and anus face upwards. Sea lilies attach to the sea bed by a jointed stalk, but feather stars break free when young to become swimming adults.

STARFISH OR SEA STARS
Class Asteroidea

About 1,500 species
The body of these mostly sea-bed scavengers is star-shaped, with five or more stout arms merging into a central body disc. On the underside of the arms are rows of numerous tube feet and a groove, along which they pass food to the central mouth. The mouth is on the underside, and the anus and madreporite are on the upper surface. The skeleton is a layer of plates (ossicles) embedded in the body wall.

BRITTLESTARS, BASKET STARS
Class Ophiuroidea

About 2,000 species
These echinoderms have a disc-shaped body with five narrow, flexible arms. Basket stars' arms are branched and finely divided. The skeleton is a series of overlapping plates. The mouth, on the underside, doubles as an anus.

SEA URCHINS, SAND DOLLARS
Class Echinoidea

About 940 species
Body shape ranges from a disc (sand dollars) to a sphere (urchins) with five double rows of tube feet. The skeletal plates join to form a rigid shell (test) with moveable spines.

SEA CUCUMBERS
Class Holothuroidea

About 1,250 species
These echinoderms have a sausage-shaped body with five double rows of tube feet, with those encircling the mouth modified into feeding tentacles. The skeleton comprises small, multi-shaped plates.

SEA DAISIES
Class Concentricycloidea

2 species
The body is disc-shaped, edged with spines, and has no arms. The upper surface is scaly. The underside has a membrane (vellum) through which food is absorbed.

REPRODUCTION

Most echinoderms have separate males and females, which reproduce by releasing sperm and eggs, respectively, into the water. Individuals often gather to spawn at the same time, thereby increasing their chance of success. This synchronized spawning is initiated by factors such as daylight length and water temperature. Each echinoderm group has its own type of larva with its own way of swimming, floating, and feeding. Some starfish, for example, keep their fertilized eggs and developing larvae in a pouch under their mouth, and nourishment comes in the form of yolk. In some brittlestars, the larvae are brooded in sacs inside the body, and the young are released after metamorphosis. In most species, however, the fertilized eggs drift in the plankton and develop into free-floating larvae. The larvae eventually transform into their adult form and settle on the sea bed.

FLOATING AIDS
Long, paired arms help sea-urchin larvae, such as this one from a Sea Potato or Heart Urchin, to float in the plankton. Brittlestars have similar larvae.

RELEASING SPERM AND EGGS
By rearing up to spawn, sea cucumbers ensure that their eggs have a chance of mixing with sperm released by another individual.

FEEDING

Echinoderms range from peaceful grazers and filter feeders to voracious predators. Carnivorous species of starfish extend their stomach over their prey and digest it externally. In contrast, most sea urchins are grazers, scraping rock surfaces using teeth that resemble the chuck of an electric drill. Combined with muscles and skeletal plates, they form a complex, powerful feeding apparatus called an Aristotle's lantern. Sea cucumbers have an important role as sea-bed cleaners, vacuuming up organic debris and mud.

FILTER FEEDING BY TUBE FEET
Feather stars raise their arms to trap plankton using finger-like tube feet. The food is coated in mucus and passed down the arms into the mouth.

DEFENCE

If they can be broken open, sea urchins make a good meal for fish, sea birds, and sea otters. So, along with many other echinoderms, they protect themselves from predators with their long, sharp spines. These spines are mounted on ball-and-socket joints and can move in all directions, which turns them into fearsome weapons. If an echinoderm is attacked, spines may break off and embed themselves in the predator, creating a wound. Some spines are also venomous, such as those belonging to the Crown of Thorns starfish. Fire Urchins and Flower Urchins also have enlarged and venomous, pincer-like pedicellariae (see below), which are strong enough to sting humans.

The cumbersome-looking sea cucumbers have no spines or protective plates, but they are far from defenceless. If attacked, many eviscerate their gut (and sometimes other internal organs) as a decoy and regrow them later. Similarly, brittlestars can break free from attack by discarding an arm. Some tropical sea cucumbers eject sticky white threads, called Cuverian tubules, which are strong enough to entangle and restrain an attacking crab.

PEDICELLARIAE
Slow-moving urchins and starfish can become overgrown by planktonic larvae looking for a place to settle. They defend themselves by using spines modified into tiny pincers, called pedicellariae, to catch and crush the larvae.

HUMAN IMPACT

FIGHTING OVER OYSTERS

In European waters, the Common Starfish has a voracious appetite for oysters and mussels. So, fishermen dredging the shellfish beds used to cut up the starfish and throw the pieces overboard. Unfortunately for the fishermen, this tactic proved ineffective, because starfish can not only regenerate their limbs, but if a lost limb retains part of the central body disc, it is able to completely regenerate the body.

REGENERATING LIMBS
This Common Starfish is regrowing its two lost arms. Sometimes, regrowth produces one or more extra limbs.

CLASS ASTEROIDEA

Seven Arm Starfish

Luidia ciliaris

DIAMETER	Up to 60cm (24in)
DEPTH	0–400m (0–1,300ft)
HABITAT	Sediment, gravel, rock

DISTRIBUTION Temperate waters of northeastern Atlantic and Mediterranean

While the majority of starfish species have five arms, this large species has seven arms and very occasionally eight. Its body and arms have a velvety texture and are coloured brick-red to orange-brown. Each arm is fringed with a conspicuous band of multidirectional, stiff white spines, which help the Seven Arm Starfish to bury

itself in the sediment and delve after its prey. Using its long tube feet it can move very quickly over rocks and gravel to latch onto its victims. This starfish is a voracious predator that feeds mainly on other echinoderms, including burrowing sea urchins, sea cucumbers, and brittlestars.

It breeds during the summer in southern Britain but much earlier, between November and January, in the Mediterranean.

multidirectional, stiff white spines on arm

velvety red or orange skin

CLASS ASTEROIDEA

Cushion Star

Culcita novaeguineae

DIAMETER	Up to 30cm (12in)
DEPTH	0–30m (0–100ft)
HABITAT	Coral reefs

DISTRIBUTION Adaman Sea and tropical waters of western Pacific

The Cushion Star looks more like a spineless sea urchin than a starfish; it gets its name from its plump, rounded body. Its arms are so short that they merge with its body and only their tips can be seen. Juveniles are much flatter than adults and have a clear pentagonal star shape, with obvious arms. They hide under rocks to escape predators, whereas the tougher adults are relatively safe in the open.

Cushion Stars occur in a wide range of colours, from predominantly red to green and brown. The underside has five radiating grooves that represent the arms and are filled with tube feet. If the starfish is turned over, it can right itself by stretching out the tube feet on one side, anchoring them to the sea bed, and pulling. It feeds mainly on detritus and fixed invertebrates, including live coral. Two other similar species are found in the Indo-Pacific tropics but this one is the most common and widespread.

thick, soft body

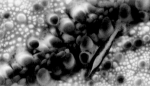

LIVE-IN GUESTS

The surface of the Cushion Star provides a home for a tiny shrimp, the Sea-star Shrimp *Periclimenes soror*. The shrimp does no harm to its host and is also found on other starfish. It often hides beneath the starfish and also matches its colour to that of its host.

COMMENSAL SHRIMP
The Sea-star Shrimp seen here is on the underside of the Cushion Star but will venture out onto the top to feed.

CLASS ASTEROIDEA

Icon Star

Iconaster longimanus

DIAMETER	Up to 12cm (5in)
DEPTH	30–85m (100–280ft)
HABITAT	Deep reefs and slopes

DISTRIBUTION Tropical waters of Indian Ocean and western Pacific

This strikingly patterned species has long, thin arms and a flat disc. The arms and disc are edged by rows of skeletal plates that protect the starfish

and give it a rigid feel. These plates may be pale or dark and they form unique patterns on each individual, which enable researchers to recognize, track, and monitor individuals in the field. Data from such studies indicates that Icon Stars grow very slowly and suggests that the largest individuals may live as long as humans. Although Icon Stars usually live in deeper, dark waters, they are common in waters at depths of 5–20m (15–65ft) around Singapore, probably because the water there is turbid and light levels are low. Females produce large orange eggs that develop into tiny orange larvae; the eggs contain chemicals that deter fish predators.

CLASS ASTEROIDEA

Goosefoot Starfish

Anseropoda placenta

DIAMETER	Up to 20cm (8in)
DEPTH	10–500m (30–1,600ft)
HABITAT	Gravel, sand, mud

DISTRIBUTION Temperate and warm waters of northeastern Atlantic and Mediterranean

The Goose Foot Starfish gets its name from the appearance of the flattened, web-like disc that joins each of its five short arms together and produces an almost pentagonal shape. The central

portion of its disc is marked with a dark red patch, and conspicuous red lines radiate outwards from this patch, along its arms. Its underside is coloured yellow. This starfish glides slowly over the sea bed searching out small crustaceans, molluscs, and other echinoderms to eat.

CLASS ASTEROIDEA

Mosaic Sea Star

Plectaster decanus

DIAMETER	Up to 16cm (6in)
DEPTH	10–180m (30–600ft)
HABITAT	Rocky reefs

DISTRIBUTION Temperate waters of South Australia

The incredibly bright colours of this starfish may be a warning that it contains toxic chemicals. If carried with bare hands for any length of time, it causes numbness. A mosaic of raised yellow ridges covers its red upper surface and it has a soft texture. The Mosaic Sea Star feeds mostly on sponges, and these may be the source of its toxins. When the females spawn, the fertilized eggs are retained and brooded on the underside of the body.

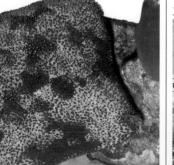

Crown-of-thorns Starfish

Acanthaster planci

DIAMETER	Up to 50cm (20in)
DEPTH	1–20m (3–65ft)
HABITAT	Coral reefs

DISTRIBUTION Tropical waters of Indian and Pacific oceans

Occasional plagues of this large and destructive starfish have killed extensive areas of coral on the Great Barrier Reef of Australia and the western Pacific reefs. There has been much debate on whether such plagues are natural or are caused by over-fishing of the few molluscs and fish that can eat this starfish, such as the Giant Triton, *Charonia*

tritonis (see p.287). With up to 20 arms and a formidable covering of long spines, this species has few predators. The spines are mildly venomous and may inflict a painful wound if the starfish is picked up with bare hands. Crown-of-thorns Starfish feed on

corals by turning their stomach out through their mouth and digesting the coral's living tissue. Pure white coral skeletons indicate that this starfish has been feeding recently in the area. In popular diving tourism areas, attempts are sometimes made to kill the starfish by injecting them with poison or removing them by hand, but with only limited success.

arm coiled around black coral branch

Serpent Star

Astrobrachion adhaerens

DIAMETER	Up to 30cm (12in)
DEPTH	15–180m (50–600ft)
HABITAT	Antipatharian black corals

DISTRIBUTION Tropical and temperate waters of Australia and southwestern Pacific

Serpent Stars are a type of brittlestar with long, flexible arms. During the day they wind their arms tightly round the branches of the deep-water black corals in which they live. At night they uncoil their arms and move around, feeding on the living polyps of their host. A black coral bush one or two metres high may be host for up to forty or so Serpent Stars.

Common Brittlestar

Ophiothrix fragilis

DIAMETER	Up to 12cm (5in)
DEPTH	0–150m (0–500ft)
HABITAT	Rocks, rough and gravely ground

DISTRIBUTION Temperate and warm waters of eastern Atlantic

This large brittlestar species gathers in dense groups that may cover several square kilometres of sea bed in areas where there are strong tidal currents. They have been recorded in densities of 2,000 individuals per square metre. Each brittlestar holds up one or two arms into the current to feed on plankton, while linking its remaining arms with surrounding individuals to form a strong mat and prevent itself being swept away. They vary greatly in colour, ranging from red, yellow,

and orange to brown and grey, and they often have alternate light and dark bands on the arms. The fragile arms of this species are covered in long, untidy spines, while its small disc, which is only 2cm (1in) across, has a covering of shorter spines. In the intertidal zone, Common Brittlestars are not usually found in groups, but occur as individuals hiding in crevices and beneath stones.

CLASS ECHINOIDEA

Sand Dollar
Echinodiscus auritus

DIAMETER	Up to 11cm (4in)
DEPTH	0–50m (0–165ft)
HABITAT	Clean sand

DISTRIBUTION Tropical and warm waters of Indian Ocean, Red Sea, western Pacific

Sand Dollars are sea urchins that have become extremely flattened as an adaptation for burrowing through sand. A mat of very fine spines covers the shell, or test, and the pattern of the animal's skeleton plates can often be seen through the skin. The mouth is on the underside. At the rear are two notches that open at the margins of the test, and water currents passing through these slits are thought to help to push the urchin down and prevent it being swept away.

CLASS ECHINOIDEA

Long-spined Sea Urchin
Diadema savignyi

DIAMETER	Up to 23cm (9in)
DEPTH	0–70m (0–230ft)
HABITAT	Coral and rocky reefs

DISTRIBUTION Tropical waters of Indian and western Pacific oceans

Many divers on coral reefs have learnt to avoid these sea urchins. They bristle with long, sharp spines that can easily wound, even through a wetsuit. The spines are mildly venomous and so brittle that they may break off in the wound. If a diver or predator comes near to it, this sea urchin waves its spines about vigorously. Only a few tough fish, such as the Titan Triggerfish, can successfully attack and eat such prickly prey. This species often has striped spines, while the other common Indo-Pacific long-spined species, *Diadema setosum*, has black spines.

CLASS ECHINOIDEA

Flower Urchin
Toxopneustes pileolus

DIAMETER	Up to 15cm (6in)
DEPTH	0–90m (0–300ft)
HABITAT	Sand, rubble, rocky reef

DISTRIBUTION Tropical waters of Indian Ocean, central and western Pacific Ocean

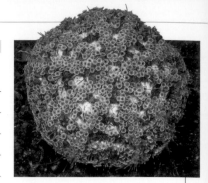

This species is extremely venomous and has caused rare fatalities. It has short, inconspicuous spines through which emerge an array of flower-like appendages called pedicellariae. These help to keep the urchin's surface clean but will sting animals that touch it. The pedicellariae also hold pieces of shell, rubble, and seaweed that shade the urchin from sunlight. Flower Urchins may partially bury themselves, despite having few predators.

CLASS ECHINOIDEA

Edible Sea Urchin
Echinus esculentus

DIAMETER	Up to 16cm (6in)
DEPTH	0–50m (0–160ft)
HABITAT	Rocky areas

DISTRIBUTION Temperate waters of northeastern Atlantic

This large, spherical urchin is covered with uniform short spines that give it the appearance of a fat hedgehog. It is generally a pinkish colour, with pairs of darker, radiating lines where its numerous tube feet emerge. These urchins are important grazers and can have much the same effect underwater as rabbits do on land, leaving the rocks covered only in hard pink encrusting algae. As their name suggests, the roe of this species can be eaten.

CLASS ECHINOIDEA

Purple Sea Urchin
Strongylocentrotus purpuratus

DIAMETER	Up to 10cm (4in)
DEPTH	0–40m (0–130ft)
HABITAT	Rocky reefs

DISTRIBUTION Temperate coastline of North America from Alaska to Mexico

This small sea urchin has been responsible for the demise of large areas of giant kelp forest off the North American coastline. Like most sea urchins, it feeds by scraping away at seaweeds and fixed animals and its favourite food is the giant kelp *Macrocystis pyrifera*. Its numbers reach densities of up to several hundred animals per square metre, and it can chew through kelp holdfasts, setting the plants adrift. Populations are normally kept in check by Sea Otters and by large fish such as sheepheads. In the past, when Sea Otters were hunted, urchin numbers increased explosively in some areas.

CLASS ECHINOIDEA

Sea Potato

Echninocardium cordatum

LENGTH Up to 9cm (3½in)

DEPTH 0–200m (0–650ft)

HABITAT Sand, muddy sand

DISTRIBUTION Temperate waters of northeastern Atlantic

Most sea urchins live in rocky areas, but the Sea Potato or Heart Urchin burrows in the sand. Unlike regular urchins it has a distinct front end and its spines are thin and flattened. Special spoon-shaped spines on the urchin's underside help it to dig, while longer spines on its back allow water to funnel down into its burrow to be used for respiration. The dried shell, or test, of this urchin resembles a potato, hence the common name.

CLASS CRINOIDEA

Sea Lily

Neocrinus decorus

HEIGHT Up to 60cm (24in)

DEPTH 150–1,200m (500–4,000ft)

HABITAT Deep-sea sediments

DISTRIBUTION Tropical waters of western Atlantic Ocean

Sea lilies are stalked relatives of feather stars and usually remain fixed in the same place after developing from a settled planktonic larva. However, while *Neocrinus decorus* and other, similar sea lilies cannot swim, like shallow-water feather stars, they have been filmed dragging themselves over the sea bed by their arms. To do this, they appear to break off the end of the stalk, then re-attach to the substrate using flexible, finger-like appendages on the stalk. In this way they can escape from predatory sea urchins. The stalk consists of a stack of disc-shaped skeleton pieces called ossicles, and looks like a simple vertebrate spinal column. Sea lilies feed by spreading out their numerous, feathery arms against the current and trapping plankton. Food particles are passed down the arms and into the mouth.

CLASS CRINOIDEA

Tropical Feather Star

Oxycomanthus bennetti

DIAMETER Up to 15cm (6in)

DEPTH 10–50m (33–165ft)

HABITAT Coral reefs

DISTRIBUTION Tropical waters of western Pacific

All that can usually be seen of the Tropical Feather Star is its numerous feathery arms held up into the water to trap food. This species has about a hundred arms, compared to the ten that most temperate water feather stars have. The arms are attached to a small, disc-like body and the mouth is on the upper side of the body, between the arms. The Tropical Feather Star clings onto corals using numerous articulated, finger-like appendages called cirri. It prefers elevated positions where it is exposed to food-bearing currents, and is active by both day and night. Like all feather stars, this species starts its early life by becoming attached to the sea bed by a stalk – at this stage, it closely resembles a small sea lily. As it matures, the feather star breaks away and becomes free-living, leaving the stalk behind.

CLASS CRINOIDEA

Passion Flower Feather Star

Ptilometra australis

DIAMETER Up to 12cm (5in)

DEPTH To at least 60m (200ft)

HABITAT Rocky reefs, rubble

DISTRIBUTION Endemic to temperate waters of southern Australia

This stout feather star has 18–20 arms with long, stiff side branches called pinnules; the arms are different lengths, giving it a flower-like appearance when viewed from above. They are called passion flowers by fishermen because they are brought up in large numbers by commercial trawlers, clinging tightly to their nets. These feather stars are found in reefs and also in very shallow, sheltered bays and estuaries. Like most feather stars, the Passion Feather Star is a filter feeder that grips onto the tops of rocks, sponges, and sea fans, where it spreads its arms wide to trap plankton and suspended detritus. It remains expanded both day and night but, like other feather stars, it can curl up its arms if disturbed or while resting. Its usual colour is a burgundy red.

CLASS HOLOTHUROIDEA

Prickly Redfish
Thelenota ananas

LENGTH	Up to 70cm (28in)
DEPTH	5–30m (16–100ft)
HABITAT	Sandy areas of coral reefs

DISTRIBUTION Tropical waters of Indian Ocean and western Pacific

This massive sea cucumber looks like an animated hearthrug as it crawls slowly over the sea bed. The large, star-shaped papillae, called caruncles, that cover its body make it an unattractive proposition to potential predators. However, its appearance does not deter humans, and Prickly Redfish fetches high prices in parts of eastern Asia, where it is considered a delicacy. Large specimens reach up to 5kg (11lb) in weight and are traditionally collected by reef walking at low tide or by breath-hold diving. Other predators include various fish and crustaceans. The flat underside of the Prickly Redfish is covered with orange tube feet, which it uses to crawl over the sea bed in search of its main food, calcareous *Halimeda* seaweeds. When spawning, Prickly Redfish gather together and rear up, then release eggs or sperm into the water from small pores near the head end of the body.

CLASS HOLOTHUROIDEA

Sea Cucumber
Bohadschia graeffei

LENGTH	Up to 30cm (12in)
DEPTH	5–50m (15–165ft)
HABITAT	Coral reefs

DISTRIBUTION Tropical waters of Indian Ocean, Red Sea, western Pacific

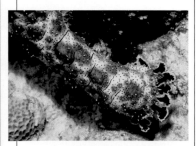

The juveniles of this sea cucumber look completely different from the adult (shown here). They are white with black lines and protruding yellow papillae, closely resembling sea slugs from the genus *Phyllidia*. These slugs are distasteful to fish, so this mimicry protects the young sea cucumbers. *Bohadschia graeffei* feeds by scooping sand and mud into its mouth using large black tentacles, which are modified tube feet. Organic material in the sediment is digested, while the remainder passes through the gut and is deposited outside, looking like a string of sausages. In areas where sea cucumbers are common, much of the surface sediment is hoovered up and cleaned several times a year in this way.

CLASS HOLOTHUROIDEA

Edible Sea Cucumber
Holothuria edulis

LENGTH	Up to 35cm (14in)
DEPTH	5–30m (15–100ft)
HABITAT	Sand, rock, coral reefs

DISTRIBUTION Tropical waters of Indian and Pacific oceans

As its name suggests, this is an edible species of sea cucumber, although it is not considered as good eating as others, such as the Prickly Redfish (above). It has a soft body, peppered with tiny warts, and is coloured black on its back and pinkish red to beige underneath. Like many other large sea cucumbers, it is sometimes host to small pearlfish (family Carapidae) that live inside its body cavity.

CLASS HOLOTHUROIDEA

Sea Apple
Pseudocolochirus tricolor

LENGTH	Up to 10cm (4in)
DEPTH	0–30m (0–100ft)
HABITAT	Rocks and reefs

DISTRIBUTION Tropical waters of Indo-Pacific region

The Sea Apple is one of the most colourful of all sea cucumbers and is widely collected for use in marine aquariums. It has a red and purple body and yellow tentacles. The species in the genus *Pseudocolochirus* are often hard to distinguish and for this reason the distribution map may include

several species. The Sea Apple uses its tube feet to attach itself to rocks then extends its branched tentacles into the water to trap organic particles. From time to time, each tentacle is pushed into the mouth and the food it has trapped is wiped off.

CLASS HOLOTHUROIDEA

Deep-sea Cucumber
Laetmogone violacea

LENGTH	Not recorded
DEPTH	To at least 2,500m (8,000ft)
HABITAT	Soft sediments

DISTRIBUTION Deep, cold waters of Atlantic, Indian, and Pacific oceans

In the deep ocean, sea cucumbers are one of the dominant sea-floor groups all over the world. *Laetmogone violacea* is one of a large number of species that crawl over the soft, muddy ocean bottom eating organic detritus. Its peg-like "legs" may help to keep it from sinking too far into the mud. Ingested mud that the cucumber is unable to digest leaves its body as faecal casts, which may then be eaten again by other sea cucumbers. Like many deep-sea animals, this sea cucumber is almost colourless but glows all over with bioluminescent light; exactly how the animal uses this light is not yet known. Some deep-sea starfish are known to light up when approached by a predator, which may scare it away. It may be that the Deep-sea Cucumber uses its bioluminescence in the same way.

SMALL, BOTTOM-LIVING PHYLA

DOMAIN	Eucarya
KINGDOM	Animalia
PHYLA	14
SPECIES	Many

MANY DISPARATE INVERTEBRATES play important parts in marine ecosystems but are seldom seen, because they are small or their habitats are difficult to study. Like all animals, they are grouped into phyla, each phylum representing a body plan as distinct as invertebrates are from vertebrates. Most of these small, bottom-living phyla live in sea-bed sediment and are loosely called "worms" due to their shape and burrowing lifestyle. However, the superabundant nematodes or roundworms (20,000 species) live in a wide range of environments including the sea bed. A selection of 14 bottom-living phyla are represented below.

SAND-GRAIN ANIMALS

A community of tiny animals, referred to collectively as meiofauna, lives in the surface water film between the sand grains on beaches and in shallow water. They range in size from 0.05mm to 1mm ($^1/_{1000}$ in to $^1/_{25}$ in) and so can only be seen properly with a microscope. Many have intricate and beautiful shapes. Almost every marine invertebrate phylum has representative species that live in this habitat, and several phyla, such as tardigrades (water bears) and gastrotrichs, occur virtually nowhere else. A diverse meiofauna is a good indication of a healthy environment since these minuscule organisms are the basis for many marine food chains.

SAND COMMUNITY
Many invertebrates exist in the watery spaces between grains of coastal sand. The worm-like gastrotrich (phylum Gastrotricha) shown in this photo-micrograph is typical of species found in this habitat.

FEEDING APPARATUS
Female spoonworms (phylum Echiura) sweep up organic material and sediment with a scoop like proboscis, seen here extending from a burrow.

MUD-SWALLOWERS

The consistency and structure of seashore and sea-bed sediments depends largely on the many worm-like phyla that live there. Millions of these burrowing animals continually mix and rework the sediment, a process called bioturbation. Lugworms (see p.275) are famous for their ability to eat sand, depositing the inedible material on the sand surface in the form of coiled heaps, or casts, but many of the less well-known groups, including peanut worms, acorn worms, and kinorhynchs, are just as important. Organic material washed onshore with each tide or carried by currents is quickly incorporated into the sediment as the animals move about, or is processed as the surface mud is eaten.

NEW PHYLA

Scientists have so far described only a fraction of the species that live in oceans. New species are being discovered all the time, mostly in groups such as sponges and soft corals that traditionally have been neglected. Occasionally a species is found that is fundamentally different from all other known organisms and so it is classified as a new phylum. Most of these exciting discoveries are from inaccessible areas such as deep-sea mud, and the animals are usually small. But when the abundant life around deep-sea hydrothermal vents (see p.188–89) was sampled in the 1970s, gigantic tube worms like no others were found.

DEEP-SEA WORM
The relatively recent discovery of giant tube worms living around hydrothermal vents led scientists to form a new phylum called Vestimentifera. They obtain food from symbiotic bacteria.

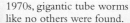

HORSESHOE WORMS
These sedentary worms (phylum Phoronida) live in small tubes buried in sand or mud or (as here) attached to sea-bed rocks. To feed, the worms extend a horseshoe-shaped net of tentacles.

PHYLUM PHORONIDA

Horseshoe Worm

Phoronis hippocrepia

LENGTH	Up to 10cm (4in)
DEPTH	0–50m (0–165ft)
HABITAT	Rocks and empty shells
DISTRIBUTION	Shallow coastal waters of Atlantic Ocean, northeastern and western Pacific

Horseshoe Worms are easily overlooked but they sometimes cover large areas of rock with their narrow, membranous tubes. The animal lives inside its tube, which encrusts rock surfaces or can bore into shells or limestone rock so that only the top part of the tube shows. The end of the worm-like body is thickened and anchors the animal in its tube. The feeding head with its horseshoe of delicate ciliated tentacles is extended to catch tiny planktonic animals while the body remains hidden in the tube. The feeding head is called a lophophore and is found in all members of the phylum. Horseshoe Worms brood their egg masses within the lophophore, and larvae are continually released to drift and develop in the water.

PHYLUM NEMATODA

Roundworm

Dolicholaimus marioni

LENGTH	Up to 5mm (½in)
DEPTH	Intertidal
HABITAT	Among algae in rock pools
DISTRIBUTION	Shores of the northeastern Atlantic

It is hard to see which end is which of a Roundworm as both ends of its thin body are pointed. The body is round in cross-section and has longitudinal muscles but no circular ones. This results in a characteristic way of moving in which the body is thrashed in a single plane forming C- or S-shapes in the process. This is a marine species, but Roundworms also occur in vast numbers in the soil and fresh water.

PHYLUM ECHIURA

Spoonworm

Bonellia viridis

LENGTH	Up to 15cm (6in)
DEPTH	1–100m (3¼–330ft)
HABITAT	Muddy rocks
DISTRIBUTION	Coastal temperate waters of northeastern Atlantic and Mediterranean

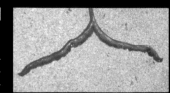

Female Spoonworms have a proboscis that stretches out like an elastic band and can reach at least a metre away in search of food. The worm's green, pear-shaped trunk remains hidden between rocks, safe from predators. In this species, the tip of the proboscis is forked and usually this is all that can be seen of the worm. The proboscis collects food particles with the help of sticky mucus, and the food is moved along the proboscis and into the mouth by the whipping movements of hair-like cilia. Male Spoonworms are tiny and parasitic on the females, their only function being to fertilize the female's eggs.

PHYLUM ENTEROPNEUSTA

Acorn Worm

Glossobalanus sarniensis

SIZE	Not recorded
DEPTH	Shallow water
HABITAT	Soft sediments
DISTRIBUTION	Coastal temperate waters of northeastern Atlantic

The soft, slimy body of this worm-like animal is divided into three regions. At the front end is a pointed proboscis, separated from the long, thin trunk by a tube-like collar. It lives in a U-shaped burrow and feeds by trapping small organisms in sticky mucus and eating sediment. The sexes are separate, and reproduction can be either asexual, by fragmentation or budding, or sexual. Some biologists group acorn worms with pterobranch worms (see below) in one phylum, the Hemichordata. Unusually for invertebrates, they have some vertebrate characteristics, which include a nerve cord that runs along the back.

PHYLUM PRIAPULIDA

Priapula Worm

Priapulus caudatus

LENGTH	Up to 10cm (4in)
DEPTH	Not recorded
HABITAT	Buried in sediment
DISTRIBUTION	Cold waters of north Atlantic and Arctic Ocean

PHYLUM SIPUNCULA

Peanut Worm

Golfingia vulgaris

LENGTH	Up to 20cm (8in)
DEPTH	0–2,000m (0–6,560ft)
HABITAT	Muddy sand and gravel
DISTRIBUTION	Northeastern Atlantic and eastern Mediterranean; possibly Indo-Pacific, Southern Ocean

The Peanut Worm is shaped like a half-inflated sausage balloon. It's body is stout and has a long, thin region at the front called the introvert, which can be stretched right out or withdrawn completely inside the body. The animal has a crown of short tentacles around the mouth at the end of the introvert. It lives buried in sediment, which it eats as it burrows and digests any organic matter.

plump body

extended introvert

PHYLUM PTEROBRANCHIA

Pterobranch Worm

Rhabdopleura compacta

LENGTH	Up to 5mm (½in)
DEPTH	Not recorded
HABITAT	Attached to sessile animals
DISTRIBUTION	Cold waters of northern hemisphere

Like the Acorn Worms to which they are related (above, left), Pterobranch Worms live in a thin tube and their bodies are divided into a proboscis, collar, and trunk. They also have a pair of arms covered in tentacles arising from the collar region. The tubes of many individuals are connected together with strands of soft tissue that join the trunks of the animals, enabling them to form a colony.

The stout, cylindrical body of this animal is divided into a short barrel-shaped proboscis at the head end, a longer trunk region, and a tail that consists of small bladders attached to a hollow stalk. The proboscis can be withdrawn into the trunk. The mouth on the end of the proboscis is edged with spines, which help the animal to seize other small marine worms for food.

Pogonophoran Worm
Siboglinum ekmani

LENGTH Up to 10cm (4in)

DEPTH At least 100m (330ft)

HABITAT Mud

DISTRIBUTION Temperate waters of north Atlantic

Pogonophoran Worms live inside tightly fitting tubes made of chitin, a substance that is also found in insect skin. The animal has an extremely long, thin body, like a piece of string, divided into different regions. The head end has a single tentacle, which can coil up when the animal contracts back into its tube. Behind the head end is a bridle, a raised ridge of tissue that runs obliquely around the body. The main length of the body is covered in small projections (papillae) and appears red under a microscope due to its blood, which contains haemoglobin. There is no gut and the animal feeds by absorbing dissolved organic matter from the water through the skin. The very end of the body has a segmented region with hair-like spines called setae that may help the animal to grip the sides of its tube.

Giant Tube Worm
Riftia pachyptila

HEIGHT Up to 2m (6½ft)

DEPTH 2,000–4,000m (6,560–13,120ft)

HABITAT Hydrothermal vents

DISTRIBUTION Pacific Ocean

When it was first discovered in the 1970s, living in great clusters in the deep sea around hydrothermal vents, the Giant Tube Worm caused a sensation. It has many structural similarities to Pogonophoran Worms (see left), but is huge. Like them, it lives in a permanent tube and has no mouth or gut. Most of the worm's body remains hidden but a brilliant red plume of gills sticks up out of the tube and absorbs chemicals and oxygen from the water. Living within the worm's tissues are bacteria that can make up over half the weight of the body.

LIVING WITHOUT SUNLIGHT

Giant Tube Worms obtain all their energy from chemicals in the hot water that pours out of hydrothermal vents. Chemosynthetic bacteria in their tissues oxidize sulphur to provide energy and fix carbon from vent chemicals. The worms thus gain their food with no direct or indirect reliance on sunlight.

VENT TUBE WORMS
Giant Tube Worms have their tubes attached deep within crevices to help keep them upright.

gill plume is red from haemoglobin in blood

rigid tube

Lamp Shell
Terebratulina septentrionalis

LENGTH Up to 3cm (1¼ in)

DEPTH 0–1,200m (4,000ft)

HABITAT Rocks and stones

DISTRIBUTION Temperate and cold waters of north Atlantic

It would be easy to mistake a lamp shell for a small bivalve mollusc as both have a hinged shell in two parts and live attached to the sea floor. Lamp Shells, however, have a very thin, light shell and the two parts are different sizes, with the smaller one fitting into the larger. The shell valves cover the dorsal and ventral surfaces of the animal whereas in bivalve molluscs they are on the left and right side of the body. Lamp Shells attach their pear-shaped shell to hard surfaces by means of a fleshy stalk that emerges from a hole in the ventral shell valve. With the shell valves gaping open, the animal draws in a current of water that brings plankton with it. Taking up most of the space inside the shell is a feeding structure called the lophophore, which consists of two lateral lobes and a central coiled lobe covered in long ciliated tentacles. The beating of the cilia creates the water current. Lamp Shells are found worldwide, but they are especially abundant in colder waters. In the northeastern Atlantic, *Terebratulina septentrionalis* is mostly found in deep water, while along the east coast of North America, it commonly occurs in shallow water. This species is very similar to *Terebratulina retusa*.

PHYLUM KINORHYNCHA

Mud Dragon
Echinoderes aquilonius

LENGTH	Less than 1mm
DEPTH	Shallow water
HABITAT	Muddy sediments
DISTRIBUTION	Northwestern Atlantic

Mud dragons look rather like miniature insect pupae. The body appears segmented on the outside but this is only superficial. It is covered with a thick, articulated cuticle and there are sharp spines on each body section. The tail end has a bunch of longer spines, and the head region has several rings of spines. The mouth is situated on the end of a cone-shaped structure and the animal can withdraw the entire head region into the rest of the body for protection rather as a tortoise does, but it can also close the resulting hole with special plates, which are called placids. The head spines are used to help the animal push its way through the sediment, feeding on organic debris, bacteria, protists, and diatoms.

There are about 100–150 species of mud dragons, all of which are marine. The sexes are separate but look similar. The eggs develop into free-living larvae that moult several times before attaining the adult form.

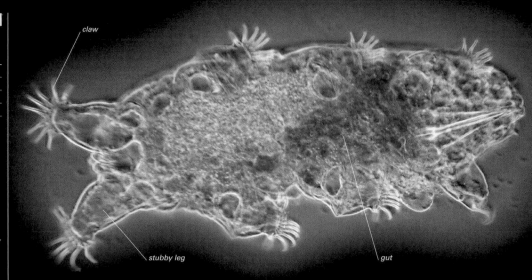

claw

stubby leg

gut

PHYLUM TARDIGRADA

Water Bear
Echiniscoides sigismundi

LENGTH	Less than 1mm
DEPTH	Not recorded
HABITAT	Marine sands
DISTRIBUTION	Worldwide

This species of water bear lives in the spaces between sand grains in marine sediments, as do most of the other 25 or so marine species. The rest of the 400 or so other species live in fresh water, especially in the thin layer of water around damp-loving plants such as mosses. Water bears have a short, plump body without a well-defined head but with eyespots and sensory appendages at one end. There are four pairs of short stubby legs each ending in a bunch of tiny claws on which the animal lumbers slowly along. The relatively thick skin protects against abrasion from sand grains. The sexes are separate, but there are few males and the eggs can probably develop without being fertilized. The nearest relatives of these tiny animals are thought to be the land velvet worms of the phylum Onchophora.

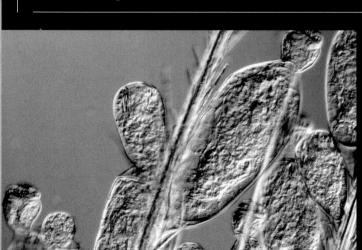

PHYLUM CYCLIOPHORA

Cycliophoran
Symbion pandora

LENGTH	0.3mm
DEPTH	Not recorded
HABITAT	Mouthparts of the Norway Lobster
DISTRIBUTION	North Sea

Symbion pandora is the only known representative of the phylum Cycliophora. This minute animal was first described in 1995 by two Danish biologists. It was found clinging to the mouthparts of a Norway Lobster (*Nephrops norvegicus*) that was dredged up from the North Sea, and the biologists must have looked very closely to have seen it at all. The female has a rounded body and is attached to the substratum by a short stalk and an adhesive disc. Attached to her are tiny dwarf males that never release their grip. It feeds by means of a mouth funnel surrounded by cilia and excretes via an anus next to the mouth. The reproductive cycle is complex and involves both sexually and asexually produced free-swimming larvae.

The structure and habits of this species are so unlike any other known organism that it was classified in a new phylum of its own. Molecular studies indicate that it may be related to rotifers (see p.319) and worm-like creatures called acanthocephalans. The asexual reproduction is similar to that seen in some bryozoans.

PHYLUM GASTROTRICHA

Gastrotrich
Turbanella species

LENGTH	Less than 1mm
DEPTH	Not recorded
HABITAT	Well-oxygenated sediments
DISTRIBUTION	Not recorded

Gastrotrichs are found in both fresh water and the sea, but *Turbanella* is a marine genus that lives in the spaces between sand grains in sea-floor sediments. It looks similar to a ciliated protist, but is a true multicellular animal with a mouth, gut, kidney cells, and other structures. It has several adhesive tubes, structures that secrete a sticky substance and help the animal attach to the substratum. By attaching and detaching the adhesive tubes at the front and rear of its body, it can loop around rather like a leech. Alternatively it can glide using its cilia, searching for bacteria and protists to eat.

gut

cilia

PLANKTONIC PHYLA

DOMAIN	Eucarya
KINGOM	Animalia
PHYLA	Ctenophora
	Chaetognatha
	Rotifera
SPECIES	2,170

OF THE MANY MAJOR GROUPS (phyla) of invertebrate animals, a few are entirely composed of planktonic animals. Like all animals, they are classified into different phyla based on their anatomy, which can be complex. All are ecologically important because the plankton community underpins all the ocean food chains. Three minor phyla (Ctenophora, Chaetognatha and Rotifera) are represented here.

PREDATORS

Carnivorous zooplankton have many methods of catching prey. Comb jellies are voracious predators – some trap their prey with a sticky secretion released from special cells (colloblasts) lining their tentacles. Others draw in prey using negative pressure created by rapidly opening their mouths. Species of *Haeckelia* even recycle the stinging cells from their cnidarian prey. Arrow-worms have vibration sensors to detect prey, which is caught and held by moveable hooks. The prey is then paralyzed by neurotoxins released from pores adjacent to the mouth.

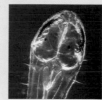

HOOKING PREY
This arrow-worm has brown-coloured hooks on either side of its circular mouth for holding prey.

GRAZERS

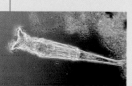

ROTIFER FEEDING METHOD
The rotifer's ciliated crown, used in locomotion and filter feeding, is visible on the left.

Some of the zooplankton are herbivorous, grazing on phytoplankton or filter feeding. Some planktonic rotifers feed on organic particles suspended in the water. The cilia on the crown that surrounds the oral cavity waft water into a food groove leading to the mouth. Here, food particles are sifted and returned to the pharynx where jaw-like structures, called trophi, grind the food before it passes into the stomach. Trophi are unique to rotifers.

GLEAMING CILIA
Comb jellies, or sea gooseberries, swim by beating eight vertical rows of cilia combs, which shimmer with iridescent colours.

Creeping Comb Jelly
Coeloplana astericola

DIAMETER	1cm (½in)
DEPTH	Not recorded
HABITAT	On the Orange Sea Star

DISTRIBUTION Tropical waters of western Pacific

While most comb jellies live in the plankton, Creeping Comb Jellies have taken up a bottom-living existence. Instead of the more usual rounded shape, they are flattened and look like a tiny squashed ball. The mouth is in the centre of the underside with the statocyst, a balancing organ found in all comb jellies, opposite it on the upper side. Comb rows are absent as they have no need to swim, and they move by muscular undulations of the body rather like a small flatworm. This species lives on the Orange Sea Star (*Echinaster luzonicus*), lying still and

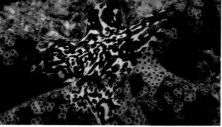

Creeping Comb Jelly

almost invisible during the day, its colour and mottled pattern matching its echinoderm host. At night, it extends its two long feeding tentacles to ensnare planktonic prey.

Predatory Comb Jelly
Mnemiopsis leidyi

LENGTH	Up to 7cm (3in)
DEPTH	0–30m (0–98ft)
HABITAT	Open water

DISTRIBUTION Temperate and subtropical waters of western Atlantic, Mediterranean, and Black Sea

This comb jelly is a slightly flattened pear shape and has two rounded lobes on each side of the mouth that help it to surround and enclose larger prey. As well as two main feeding tentacles, there are smaller secondary tentacles in grooves surrounding the mouth. The long tentacles are armed with lasso cells that secrete a sticky material to ensnare prey.

The Predatory Comb Jelly, which is native to the western Atlantic, was accidentally introduced to the Black Sea in the 1980s by the release of ship ballast water, and it has since spread to adjacent bodies of water, including parts of the eastern Mediterranean. In the Black Sea, it multiplied rapidly because of the ideal water conditions and the absence of its natural predators. This has had very serious effects on commercial fish catches because the Predatory Comb Jelly is a planktonic predator and consumes fish larvae and fry.

Venus's Girdle
Cestum veneris

LENGTH	Up to 2m (6½ft)
DEPTH	Near surface
HABITAT	Open water

DISTRIBUTION Tropical and subtropical waters of north Atlantic, Mediterranean, and western Pacific

The unusual name of this animal comes from the ribbon shape of its transparent, pale violet body. Eight rows of comb cilia are modified and run in two lines along one edge of the ribbon. The two main tentacles are short, and numerous other short tentacles occur along the lower edge of the body. As an escape response, Venus's Girdle can swim rapidly by undulating its body. However, more usually it moves slowly by beating its comb cilia.

TUNICATES AND LANCELETS

DOMAIN	Eucarya
KINGOM	Animalia
PHYLUM	Chordata
SUBPHYLA	Urochordata
	Cephalochordata
CLASSES	4 or 5
SPECIES	About 2,025

TUNICATES HAVE A LONG, bag-like body often attached to the sea floor; lancelets resemble small, stiff worms and live buried in sediment. Despite their simple appearance, these animals are included not with the world's other invertebrates but in the same group as backboned animals such as fish and mammals. This is because, uniquely among invertebrates, tunicates and lancelets possess an internal skeletal rod, or notochord. The best known tunicates are sea squirts, some of which form colonies, whereas lancelets are solitary.

ANATOMY

When tunicate larvae become adults, they lose the supporting notochord, but lancelets keep it during the adult stage. Tunicates are covered by a tough protective bag made out of cellulose called a tunic, which sticks to the sea floor by means of root-like projections. Inside is a big sieve-like structure, the pharynx, which connects the mouth and gut. This has a sticky mucus coating to trap plankton from the seawater passing through it. Lancelets also filter water through a pharynx, expelling it through an opening near the anus. A ring of stiff hairs (cirri) surrounding their mouth prevents sand getting in.

LIFESTYLE

Sea squirts live attached to hard surfaces such as rocks, reefs, and shipwrecks. They spend their time filtering seawater, drawing in food-rich water through one siphon (inhalent) and releasing waste water through another (exhalent). Most sea squirts occur in shallow coastal waters where there is plenty of plankton, but there are also a few deep-water species. In sheltered sea lochs, they can cover hundreds of square metres of sea bed. Some tunicates, including salps and pyrosomes, drift along on ocean currents with the plankton, often forming giant swarms. Lancelets are strong swimmers due to their flexible, muscular bodies, but they usually just burrow in sediment with only their head sticking out.

BOTTOM-LIVING SQUIRTS
Sea squirts sometimes grow together in clumps with cnidarians, such as sea sponges, and they can be very colourful.

LANCELET ADULT
A lancelet's muscular body is flattened from side to side and is supported by a stiff notochord.

pharynx dorsal nerve cord notochord
cirri surrounding mouth swimming muscles anus

TUNICATE LARVA
The tadpole-shaped tunicate larva's nerve cord and notochord are reabsorbed when it changes into the adult form.

siphon nerve cord notochord
pharynx
attachment sucker
heart

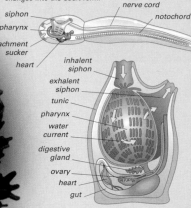

inhalent siphon
exhalent siphon
tunic
pharynx
water current
digestive gland
ovary
heart
gut

TUNICATE ADULT
Most of the space inside a tunicate is taken up by the huge pharynx, through which there is a nonstop current of water.

SWIMMING SQUIRT
Floating salps swim by jet propulsion, taking in water at one end and squirting it out of the other.

SUBPHYLUM UROCHORDATA

Common Sea Squirt
Ciona intestinalis

HEIGHT	Up to 15cm (6in)
DEPTH	0–500m (0–1,600ft)
HABITAT	Any hard substrate

DISTRIBUTION Atlantic, Pacific, Indian, Arctic oceans; possibly Southern Ocean

The Common Sea Squirt has no supporting structures in its adult form, so when it is seen out of water, it resembles a blob of jelly that may squirt out a jet of water when prodded. It is a typical solitary tunicate, whose internal structures are visible through its pale, greenish yellow, gelatinous outer covering, called a test or tunic, which is smooth and translucent. It has two yellow-edged

siphons and uses the larger of these, the inhalent siphon, to draw in water; the smaller, exhalent siphon is used to expel water, and its opening has six lobes, while that of the exhalent siphon has eight. The Common

Seasquirt lives up to its name and is found attached to a wide variety of rocks, reefs, seaweeds, and, in particular, man-made structures. The legs of oil platforms and jetties, for example, are often festooned with this sea squirt.

CLEANING UP

In sheltered sea lochs and harbours, the Common Sea Squirt often covers large areas of rock or wall. In spite of its small size, it is able to filter several litres of water per hour, filtering out plankton and other organic particles and leaving the water much clearer than it might otherwise be.

Colonial Sea Squirt

SUBPHYLUM UROCHORDATA

Didemnum molle

HEIGHT	Up to 3cm (1¼ in)
DEPTH	Shallow water
HABITAT	Coral reefs and rocks

DISTRIBUTION Widespread in tropical reef waters of Indian and western Pacific Oceans

Although it appears to be a solitary sea squirt, this species lives in urn-shaped colonies that share a single exhalent siphon (an opening through which water exits). The colony is dotted all over with the inhalent, or ingoing, siphons of the tiny zooids – the individuals that make up the colony. The colony's green colour results from the presence of the symbiotic cyanobacterium *Prochloron*.

Star Sea Squirt

SUBPHYLUM UROCHORDATA

Botryllus schlosseri

WIDTH (CLUSTER)	Up to 15cm (6in)
DEPTH	Shore and shallows
HABITAT	Rock, stones, seaweeds

DISTRIBUTION Coastal Arctic and temperate waters of north Atlantic

Individual Star Sea Squirts are only about 2mm (³⁄₃₂ in) long and cannot live on their own. Instead, they arrange themselves in star-shaped clusters, or colonies, embedded in a shared gelatinous casing, called a tunic or test, that encrusts the substrate. At the centre of each star is a shared outgoing (exhalent) opening through which used water is voided. The colonies vary greatly in colour and may be green, violet, brown, or yellow, with the individuals having a contrasting colour to the test.

Sea Tulip

SUBPHYLUM UROCHORDATA

Pyura spinifera

HEIGHT	Up to 30cm (12in)
DEPTH	5–60m (15–200ft)
HABITAT	Rocky reefs

DISTRIBUTION Temperate waters of Australia

This giant sea squirt is held up into the water on the end of a long, thin stalk. This means that its large inhalent siphon is in a better position to pull in plankton-rich water. The Sea Tulip's body is covered in warty outgrowths and is naturally a bright yellow. However, the growth of an encrusting commensal sponge on many of these sea squirts gives them a pink appearance. During rough weather, Sea Tulips are often battered down onto the sea bed, but they soon spring back on their flexible stalks.

Salp

SUBPHYLUM UROCHORDATA

Pegea confoederata

LENGTH	Up to 15cm (6in)
DEPTH	Near surface
HABITAT	Open water

DISTRIBUTION Warm waters worldwide

Salps are tunicates that resemble floating sea squirts. They swim by jet propulsion, taking in water through a siphon at one end of their bodies and expelling it at the other. Their transparent casing is loose and flabby and is encircled by four main muscles that form two distinct cross-bands. Individual salps are joined together in chains up to 30cm (12in) long, produced by the asexual reproduction (budding) of a young individual. The chains break up and disperse as they mature. Salps also reproduce sexually. Eggs are kept inside the body on the wall of the exhalent siphon, through which the developed larvae are expelled after being fertilized by sperm drawn in through the inhalent siphon.

Giant Pyrosome

SUBPHYLUM UROCHORDATA

Pyrosoma spinosum

LENGTH	Up to 10m (33ft) long
DEPTH	Near surface
HABITAT	Open water

DISTRIBUTION Warm waters between about 40° north and 40° south

The individuals that make up this giant, floating, colonial tunicate are only about 2cm (1in) long, but the colony, which resembles a gigantic hollow tube, can be large enough for a person to fit inside. Each individual lies embedded in the wall of the tube, with one end drawing in nutrient-laden water from outside and the other end expelling water and waste inside. The expelled water is used to propel the colony as a whole. A wave of bioluminescent light travels along the community if it is touched.

Appendicularian

SUBPHYLUM UROCHORDATA

Oikopleura labradoriensis

LENGTH	About 5mm (¼ in)
DEPTH	Near surface
HABITAT	Open water

DISTRIBUTION Cold waters of north Atlantic, north Pacific, and Arctic

Appendicularians are shaped like tiny tadpoles and live inside flimsy mucus dwellings that they build to trap plankton. Water enters the dwelling via two inlets covered by protective grids, passes through fine nets that trap any plankton in mucus, and passes out through an aperture. The animal eats the plankton-loaded mucus and beats its tail to create water currents.

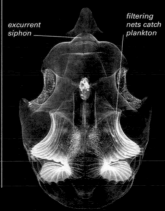

excurrent siphon

filtering nets catch plankton

Lancelet

SUBPHYLUM CEPHALOCHORDATA

Branchiostoma lanceolatum

LENGTH	Up to 6cm (2½ in)
DEPTH	Shore and shallows
HABITAT	Coarse sand

DISTRIBUTION Coastal temperate waters of northeastern Atlantic, and Mediterranean

Looking like a thin, semi-transparent, elongate leaf, the lancelet is difficult to spot in the coarse sediments in which it lives. It usually lies half-buried in the sand with its head end sticking out. Muscle blocks that run along both sides of its body show through the skin as a pattern of V-shaped stripes. At the head end, a delicate hood ringed by stiff tentacles overhangs the mouth. This feature filters out large sediment particles but allows smaller, organic particles to pass so that they can be ingested.

JAWLESS FISH

DOMAIN	Eucarya
KINGDOM	Animalia
PHYLUM	Chordata
CLASSES	Myxini
	Petromyzontida
SPECIES	About 88

JAWLESS FISH FORM AN ANCIENT group of vertebrates encompassing a diverse range of extinct groups. Today, there are only two small groups: the lampreys and the hagfish. They are considered to be the most primitive living vertebrates, although many scientists do not regard hagfish as true vertebrates. Hagfish and lampreys look similar, with elongated bodies and jawless mouths, but the two groups evolved along separate lines. Lampreys live in temperate coastal waters throughout the world and swim up rivers to breed, although some remain in fresh water. Hagfish are exclusively marine.

ANATOMY

At first glance, lampreys and hagfish could easily be mistaken for eels due to their long, thin bodies and slimy, scaleless skin. However, they lack a bony skeleton, and have only a simple flexible rod called a notochord running along the length of the body. In lampreys, the mouth is in the centre of a round oral disc armed with small, rasping teeth. Hagfish have a slit-like mouth surrounded by fleshy barbels on the outside and by tooth plates on the inside. The gills in both groups open to the outside through small, round pores behind the head, and there is a single nostril on top of the head.

LAMPREY MOUTH
The oral disc, or sucker, of lampreys is studded with horny teeth arranged in roughly concentric rows. Larger teeth surround the central mouth opening.

HAGFISH
Hagfish find their way and detect carrion using fleshy barbels around the mouth. Their eyes are undeveloped and hidden beneath the skin, so they are nearly blind.

dorsal fin   gill openings

notochord   spinal cord   round, fleshy mouth

BODY SECTION
The bodies of lampreys (left) and hagfish are supported by a simple notochord flexed by a series of muscle blocks along the back. With no true bony vertebrae, this makes their bodies very flexible. They have a tail and dorsal fin, but lack paired fins.

REPRODUCTION

Lampreys migrate from the sea into rivers and move upstream to spawn. In gravel, females lay thousands of tiny white eggs, which hatch into worm-like larvae called ammocoetes. These simple creatures have a horseshoe-shaped mouth without teeth. They live in muddy tunnels for about three years, feeding on debris, then transform into adults and swim out to sea. Hagfish lay a few large eggs on the sea bed, and these hatch into miniature adults.

HAGFISH EGGS
The eggs of hagfish are armed with tiny anchor-like hooks at both ends. When laid, they stick together like a string of sausages.

FEEDING

With the exception of a few freshwater species, lampreys are parasitic, feeding on both bony and cartilaginous fish. They attach to their living host using the teeth and lips of their oral disc to suck onto their victim. Teeth in the mouth are then used to rasp a hole in the fish, and its flesh, blood, and body fluids are all consumed. Sometimes, lampreys cause the death of their host through blood loss or tissue damage. In contrast, hagfish are mostly scavengers that feed on dead fish and whale carcasses, as well as live invertebrates. They can gain leverage to tear off chunks of flesh by literally tying themselves in a knot and using the knot to brace themselves against the carcass.

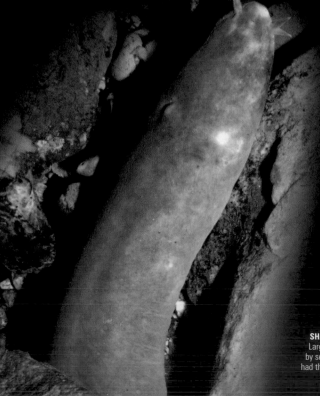

SHARK HOST
Large, slow-moving Basking Sharks are often parasitized by sea lampreys. The lampreys drop off when they have had their fill, leaving wounds that may get infected.

Sea Lamprey

Petromyzon marinus

LENGTH	Up to 1.2m (4ft)
WEIGHT	Up to 2.5kg (5½ lb)
DEPTH	1–650m (3–2,100ft)

DISTRIBUTION Coastal temperate waters of, and rivers adjacent to, north Atlantic

With its long, cylindrical body, the Sea Lamprey might at first be mistaken for an eel, but closer inspection reveals differences. Unlike eels, the Sea Lamprey has no jaws. Its body is flattened toward the tail and it has two dorsal fins. Its circular mouth lies beneath the head, and is surrounded by a frill of tiny skin extensions. Inside the mouth, the teeth are arranged in numerous concentric arcs, which helps to distinguish it from the similar, but smaller, Lampern (right). As a mature adult, the Sea Lamprey has dark mottling on its back. Adults live at sea and feed on dead or netted fish as well as attacking a wide variety of live ones. It uses a "sucker" to attach to its host, scrapes a hole through the skin, and sucks out flesh and fluids. It spawns in rivers and the larvae remain in fresh water for about five years before they mature and move out to sea. This species is now rare as a result of trapping, intentional poisoning, and the degradation of its river habitat.

Lampern

Lampetra fluviatilis

LENGTH	Up to 50cm (20in)
WEIGHT	Up to 150g (5oz)
DEPTH	0–10m (0–30ft)

DISTRIBUTION Coastal waters and rivers of northeastern Atlantic, northwestern Mediterranean

The Lampern is also known as the River Lamprey because the adults never stray far from the coast and often remain in estuaries. It may be distinguished from the Sea Lamprey (left) by its smaller size, uniform colour, and the smaller number and different arrangement of its teeth. Larvae that hatch in rivers migrate to estuaries, where they spend a year or so feeding on herring, sprat, and flounder.

Hagfish

Myxine glutinosa

LENGTH	Up to 80cm (30in)
WEIGHT	Up to 750g (1¾ lb)
DEPTH	40–1,200m (130–4,000ft)

DISTRIBUTION Coastal and shelf waters, below 13°C (55°F) in north Atlantic and western Mediterranean

This extraordinary fish can literally tie itself in knots and it does so regularly as a means of ridding itself of excess slime. Special slime-exuding pores run along both sides of the eel-like body, enabling it to produce sufficient slime to fill a bucket in a matter of minutes. The glutinous slime is usually more than adequate to deter most predators. Like all jawless fish, the Hagfish has no bony skeleton but simply a supporting flexible rod of cells, called a notochord, allowing it great flexibility. Fleshy barbels surround its slit-like, jawless mouth, and it has only rudimentary eyes. There is a single pair of ventral gill openings about a third of the way along the body.

The Hagfish spends most of its time buried in mud with only the tip of the head showing. It mainly eats crustaceans but will scavenge on whale and fish carcasses. Once the Hagfish has latched onto a carcass with its mouth, it forms a knot near its tail, then slides the knot forwards in order to provide itself with sufficient leverage to tear its mouth away along with a chunk of food.

Pacific Hagfish

Eptatretus stoutii

LENGTH	Up to 50cm (20in)
WEIGHT	Up to 1.4kg (3lb)
DEPTH	20–650m (65–2,100ft)

DISTRIBUTION Coastal and shelf waters of northeastern Pacific

The Pacific Hagfish is similar to the Hagfish found in the Atlantic. It is usually a brownish red colour and may have a blue or purple sheen. It has no true fins, only a dorsal finfold that continues around the tail but that has little function in swimming. The Pacific Hagfish lives in soft mud and feeds mainly on carrion. It causes great damage to fish caught in static nets and will enter large fish through either the mouth or the anus and proceed to eat them from the inside out, consuming their guts and muscles.

dorsal finfold

Japanese Hagfish

Eptatretus burgeri

LENGTH	Up to 60cm (24in)
WEIGHT	Insufficient information
DEPTH	10–270m (30–900ft)

DISTRIBUTION Inshore temperate waters of northwestern Pacific

This species is also known as the Inshore Hagfish because it lives in relatively shallow water compared to other species of hagfish. It is similar in shape and size to the Pacific Hagfish, with six gill apertures and a white line along its back. It lives in the sublittoral zone, and usually buries itself in the bottom mud, but migrates to deeper water to breed. Unlike other species of hagfish, it reproduces seasonally; this is thought to be a response to changing temperatures in the shallow waters in which it lives.

CARTILAGINOUS FISH

DOMAIN	Eucarya
KINGDOM	Animalia
PHYLUM	Chordata
CLASS	Chondrichthyes
SUBCLASSES	2
SPECIES	About 1,114

THIS GROUP INCLUDES SHARKS, skates, rays, and deep-water fish called chimaeras. Within the group are some of the Earth's most efficient predators, such as the White Shark, as well as filter feeders, such as the Manta Ray. Some have features unusual for fish, such as large brains, live birth, and warm blood. Fossils show that cartilaginous fish have changed little in form in hundreds of millions of years. All have a skeleton of cartilage, teeth that are replaced by new ones when necessary, and tooth-like scales covering their skin.

ANATOMY

The internal skeleton of all the fish in this group is made from flexible cartilage. In some species, parts of the skull and skeleton are strengthened by mineral deposits. The teeth are covered by very hard enamel and are formidable weapons. Sharks have several rows of teeth lying flat behind the active ones. These gradually move forwards, and individual teeth may be replaced as often as every 8–15 days. Cartilaginous fish have extremely tough skin. It is extra-thick in female sharks because males use their teeth to hold onto them when mating. A shark's skin is covered in tiny, backward-pointing, tooth-like structures called dermal denticles, which feel like sandpaper. Rays have scattered denticles, some enlarged to form spines, while chimaeras have no denticles. Unlike bony fish (see pp.338–41), cartilaginous fish do not have a gas-filled swim bladder. Sharks living in the open ocean, however, often have a very large, oil-filled liver, which aids buoyancy.

snout
nostril
mouth
gill slits
pectoral fin
pelvic fin
cloaca
tail

RAY BODY SHAPE
Skates and rays have flat bodies and large pectoral fins. The mouth is on the underside, so water for breathing is sucked in through a pair of holes, called spiracles, on the upper side, then passed over the gills.

first dorsal fin
second dorsal fin
gill slit
eye
anal fin
pelvic fin
pectoral fin
heterocercal tail
underslung mouth

SHARK BODY SHAPE
A typical shark has a sleek, streamlined body. The tail is asymmetrical (heterocercal) to give lift, and the paired pelvic fins are set far back. The mouth is underslung, and there are five gill slits on each side.

dagger-like point grips flesh

serrated, blade-like edge cuts like a knife

SANDTIGER SHARK

TIGER SHARK

palate, covered with flat teeth, crushes food

RAY

TEETH ADAPTATIONS
Sharks' teeth are shaped to suit their diet. Pointed ones are used for holding, while serrated teeth slice chunks from prey. Rays and chimaeras have teeth like grindstones to crush hard crustaceans and molluscs.

SHARK FISHING

Sharks are heavily fished all over the world for their meat, fins, liver oil, and skin. They reproduce very slowly, producing only a few young every one to two years, and many species take a decade or more to reach full maturity. This slow rate of both reproduction and growth means that shark populations cannot sustain heavy fishing pressure, and take many years to recover. Worldwide, most stocks are currently being fished at rates above safe biological limits.

SHARK FINS
Thousands of sharks are killed every year for their valuable fins, which are dried and then made into shark fin soup. The body is often discarded while the shark is still alive.

REPRODUCTION

In all cartilaginous fish, the eggs are fertilized inside the female's body. Adult males have organs on the belly called claspers – rod-like appendages that are a part of the pelvic fins. During mating, one or both claspers are inserted into the female's cloaca (the shared opening of the digestive and reproductive tracts) to introduce the sperm. In sharks, mating can be a rough affair, although there may be some courtship.

Chimaeras, as well as some sharks and rays, lay eggs – that is, they are oviparous. The eggs are protected by individual leathery egg capsules, often known as mermaid's purses. The young then hatch out several months later. In contrast, most sharks and rays give birth to live young after a long period of gestation. Of these, some are ovoviviparous, while the rest are viviparous. In ovoviviparous species, the eggs simply remain inside the mother until they hatch, prior to birth. In viviparous species, representing about 10 per cent of sharks, the young develop attached to a placenta-like structure and are directly nourished by the female's body. In all cases, the young are born fully formed, and they are able to hunt and feed. Immediately after birth, the female swims away and the young are left to fend for themselves.

CATSHARK EGGS
These egg capsules contain Catshark embryos, which will hatch after about a year. The tendrils anchor the capsules to seaweeds.

LEMON SHARK BIRTH
Lemon Sharks move to shallow, sheltered bays or lagoons to give birth. The young emerge tail first and swim away. The mothers then leave the nursery grounds.

HAMMERHEAD SHARK
The broad head of this shark provides space for extra electrical sense organs on the snout and gives a wide field of view, making it a formidable hunter.

Cartilaginous fish are divided into two subclasses: sharks and rays (the Elasmobranchii); and chimaeras (the Holocephali), which include ratfish and rabbitfish. Sharks and rays comprise nine orders, one of which is rays. Rays are divided into four suborders – considered by some to be orders. Chimaeras have one order, the Chimaeriformes.

CHIMAERAS
Order Chimaeriformes

About 34 species
The only member of subclass Holocephali, this order is distinct from sharks and rays and includes rabbitfish and ratfish. Chimaeras have a long, flabby body without scales, a large head with sensory canals, plate-like teeth, and one gill opening. The first of two dorsal fins is erectile with a venomous spine. Reproduction is oviparous.

FRILL AND COW SHARKS
Order Hexanchiformes

6 species
These sharks have a long, thin body, six or seven pairs of gill slits, small spiracles, and a single dorsal fin near the tail. Frill sharks have three-pointed teeth; cow sharks' teeth are saw-like. Reproduction is ovoviviparous.

BRAMBLE, SLEEPER, AND DOGFISH SHARKS
Order Squaliformes

130 species
Seven families form this large, varied order: bramble, dogfish, rough, lantern, sleeper, gulper, and kitefin sharks. All have spiracles, five gill slits, and two dorsal fins, but no anal fin. They are ovoviviparous.

SAWSHARKS
Order Pristiophoriformes

9 species
Small, slender sharks, these species have a flattened head and saw-like snout with barbels. They have two spineless dorsal fins, no anal fin, and are ovoviviparous.

ANGELSHARKS
Order Squatiniformes

20 species
These flattened, ray-like sharks have a rounded head with gill slits on the side, and spiracles. They have large pectoral and pelvic fins, two small dorsal fins, but no anal fin. Reproduction is ovoviviparous.

BULLHEAD AND HORN SHARKS
Order Heterodontiformes

9 species
Small, bottom-living sharks, these species have pointed front teeth and molar-like back teeth, a blunt, sloping head, nostrils connected to the mouth by a groove, paddle-like pectoral fins, an anal fin, and two spined dorsal fins. They are oviparous.

CARPETSHARKS
Order Orectolobiformes

About 33 species
These mainly bottom-living sharks include wobbegongs and nurse sharks. They have a broad, flattened head, barbels, and nostrils joined to the mouth by a deep groove. They have an anal fin and two spineless dorsal fins. Reproductive strategies vary.

MACKEREL SHARKS
Order Lamniformes

15 species
This order is made up of large, ovoviviparous sharks with a cylindrical body, conical head, two dorsal fins, an anal fin, and a long upper tail lobe. The White, Basking, and Megamouth sharks are all members. Many can maintain a high body temperature.

GROUND SHARKS
Order Carcharhiniformes

About 225 species
This is the largest and most diverse shark group. Body shapes vary, but most species are large. All species have two spineless dorsal fins and an anal fin. Reproductive strategies vary.

RAYS AND SKATES
Order Rajiformes

Over 600 species
These are mostly bottom-living fish with a flat, disc-shaped body, wing-like pectoral fins joined to the head, and a long, thin tail. Reproduction is mostly viviparous with many live young, but some are oviparous.

HUNTING SENSES

Cartilaginous fish have acute senses that help them to find prey, even if it is distant or buried in sediment. Predatory sharks smell or taste tiny amounts of blood as water passes over highly sensitive membranes in their nostrils, while catsharks also use smell to recognize each other. All cartilaginous fish have a system of pores called ampullae of Lorenzini that allows them to detect weak electrical signals given off by other animals. Most also have a lateral-line system, similar to that of bony fish, which detects water movements. Cartilaginous fish have eyes similar to those of mammals, and most have acute vision. They have no eyelids, but some sharks have a transparent "nictitating membrane," which protects their eyes when they are attacking prey.

AMPULLAE OF LORENZINI
The black spots on a shark's snout are tiny electrical sense organs that help it find prey even in complete darkness.

BARBELS
Active at night, nurse sharks can find buried prey by touch and smell, using their sensory barbels.

LENGTH	About 1.3m (4¼ft), plus tail filament
WEIGHT	Not recorded
DEPTH	330–1,500m (1,100–4,900ft)

DISTRIBUTION Parts of Pacific and eastern Indian Ocean

When scientists first hauled a Pacific Spookfish up from the ocean depths, they were astonished by the sight of its enormously long, conical snout. The long, brownish body of this strange-looking fish tapers to a thin tail, so that the fish gives the impression of being pointed at both ends. The snout is whitish, flexible, and covered in sensory pores and canals.

Living in the dark depths of the ocean where its small eyes are of at best limited use, the Spookfish uses its snout to find food and sense objects around it. The beak-shaped mouth under the base of the snout contains pairs of black, plate-like teeth. Its tail has only a small lower lobe, while the upper lobe consists of a row of fleshy tubercles. Like other chimaeras, the Spookfish relies mainly on its pectoral fins for propulsion rather than using its tail as most species of fish do. A very similar species of spookfish is found in the Atlantic Ocean.

LENGTH	Up to about 1.3m (4¼ft)
WEIGHT	Not recorded
DEPTH	At least 230m (750ft)

DISTRIBUTION Temperate waters in the southwest Pacific, off southern Australia and along the east coast of South Island, New Zealand

The Ploughnose Chimaera is also known as the Elephant Fish due to its most distinctive feature, a long, fleshy snout. The Ploughnose uses this bizarre appendage to snuffle through mud of the ocean floor in search of shellfish, which it crunches up using its plate-like teeth. A network of prominent sensory canals criss-crosses its head. In spring, these fish come inshore into estuaries and bays to breed, and lay their eggs in horny, yellow-brown capsules. This chimaera is fished commercially for food.

REPRODUCTIVE EMBRACE

Mating underwater is a slippery business, so the male Ploughnose Chimaera has a retractable, club-like, spiny clasper on its head that helps it hang onto the female. The male transfers his sperm when he inserts his pelvic clasper into the female's cloaca.

retractable clasper

LENGTH	Up to 1m (3¼ft)
WEIGHT	Not recorded
DEPTH	Close inshore to at least 900m (2,950ft)

DISTRIBUTION Northeastern Pacific

The scientific name of the Spotted Ratfish means "water rabbit", and it is also called the Blunt-nosed Chimaera. It belongs to the same family as the Rabbit Fish (see right) and is similar in shape, but unlike its relative it does not have an anal fin on the underside next to the tail. Its pattern of white spots on a dark background may provide camouflage in the same way as the spots help to camouflage deer in a forest. The Spotted Ratfish uses its large pectoral fins to glide and flap its way over the sea bed in search of its prey, which consists mainly of molluscs and crustaceans. Like other chimaeras, the female lays eggs, each one encased in a tadpole-shaped, protective capsule. The eggs are laid in the summer, two at a time, and are dropped onto the sea bed.

The Spotted Ratfish is not fished commercially as it is not very palatable, although it is sometimes unintentionally caught in nets along with other fish. It is not popular with fishermen because it has the ability to inflict a nasty wound with its sharp dorsal spine and can also deliver a painful bite. The Spotted Ratfish is frequently encountered at night by scuba divers, its large eyes glowing green in torchlight.

LENGTH	Up to 1.5m (5ft)
WEIGHT	Up to 2.5 kg (5½lb)
DEPTH	Typically 300–400m (1,000–1,300ft)

DISTRIBUTION Eastern Atlantic and Mediterranean

Beautifully patterned with wavy brown and white lines, the Rabbit Fish belongs to a family called Chimaeridae, whose members have rounded snouts, long, tapering bodies, and tails ending in a long, thin filament, giving rise to their alternative name of ratfishes. The long, sharp spine in front of the first dorsal fin of the Rabbit Fish is venomous and can inflict a serious wound. Unlike sharks but in common with all other chimaeras, the Rabbit Fish can raise and lower this fin. The second dorsal fin is low and long and almost reaches the tail fin. Rabbit Fish swim sluggishly in small groups and feed mainly on sea-bed invertebrates using their paired, rabbit-like teeth. These fish are often caught by accident in shrimp nets in the North Sea.

ORDER HEXANCHIFORMES

Frilled Shark

Chlamydoselachus anguineus

LENGTH	Up to 2m (6½ft)
WEIGHT	Not recorded
DEPTH	Mostly 120–1,570m (400–5,150ft)

DISTRIBUTION Worldwide but discontinuous

With its elongated, eel-like body and flattened head, the Frilled Shark bears little resemblance to other sharks. The most noticeable difference is that its mouth is at the front of its head instead of on the underside. In addition, while most modern sharks have five pairs of gill slits, the Frilled Shark has six, each with a frilled edge. Its small teeth are also unusual, each having three sharp points.

Frilled Sharks have been observed swimming with their mouths open, displaying their conspicuous white teeth, leading to the suspicion that the teeth act as a lure for prey. This shark lives near the sea bed in deep water but occasionally comes to the surface. It feeds on deep-water fish and squid. The male has two long claspers on the belly, which are used to transfer sperm to the female when mating. This species is ovoviviparous, meaning that the eggs hatch inside the mother, which then gives birth to live young. Up to 12 young are born as long as two years after fertilization.

Trawlers fishing for other deep-sea species often catch Frilled Sharks as by-catch. Because this species reproduces so infrequently, it is especially vulnerable, and is listed as Near Threatened on the IUCN Red List of endangered species.

ORDER HEXANCHIFORMES

Bluntnose Sixgill Shark

Hexanchus griseus

LENGTH	Up to 5.5m (18ft)
WEIGHT	Up to at least 600kg (1,300lb)
DEPTH	Up to 2,300m (7,550ft)

DISTRIBUTION Tropical and temperate waters worldwide

This enormous deep-water shark is sometimes spotted by divers in shallow water at night, but its more usual haunt is rocky seamounts and mid-ocean ridges. Its thick-set, powerful body has one dorsal fin, a large mouth lined with comb-like teeth, and six gill slits. Its fins are soft and flexible, not rigid like those of most sharks. Fish, rays, squid, and bottom-living invertebrates are this shark's typical prey, although larger adults sometimes also hunt for seals and cetaceans.

ORDER HEXANCHIFORMES

Sharpnose Sevengill Shark

Heptranchias perlo

LENGTH	Up to 1.4m (4½ft)
WEIGHT	Not recorded
DEPTH	Up to 1,000m (3,300ft), typically 27–720m (90–2,360ft)

DISTRIBUTION Tropical and temperate waters worldwide, except northeastern Pacific

The Sharpnose Sevengill Shark, as its name suggests, has a sharply pointed snout and is one of only two shark species that have seven gill slits – more than any other living shark species. It lives in deep water and hunts squid, crustaceans, and fish near the sea bed. Like the sixgill sharks (see above), it has comb-like teeth. Young fish have black markings, which fade with age, on the tip of the single dorsal fin and on the upper part of the tail. Females are ovoviviparous and give birth to 6–20 young at one time. The Sharpnose Sevengill Shark is rarely seen alive and little is known of its feeding and breeding behaviour, but it is lively and aggressive on the rare occasions when it is captured. It is occasionally caught up as by-catch in trawl nets, and this may be contributing to a reduction in its numbers. It is listed as Near Threatened on the IUCN Red List of endangered species.

ORDER SQUALIFORMES

Piked Dogfish

Squalus acanthias

LENGTH	Up to 1.5m (5ft)
WEIGHT	Up to 9kg (20lb)
DEPTH	Up to 1,460m (4,800ft), typically 0–600m (0–2,000ft)

DISTRIBUTION Worldwide, except tropics and polar waters

Sharks are not normally shoaling fish but Piked Dogfish aggregate into huge groups numbering thousands of individuals, often all of one sex and size. Also known as the Spurdog or Spiny Dogfish, this sleek, dark grey shark has two dorsal fins, each with a sharp spine in front of it. Irregular white spots decorate its sides, especially in young fish, and it has a pointed snout and large oval eyes. It was once very common and was possibly the most abundant species of shark but it is now threatened globally as a result of overfishing. These sharks do not begin to breed until they are 10–25 years old and may live to be 70–100 years old. They grow very slowly, and the young take up to two years to develop inside the mother. Some populations migrate thousands of kilometres seasonally in order to avoid very cold water.

ORDER SQUALIFORMES

Velvet Belly Lanternshark

Etmopterus spinax

LENGTH	Up to 45cm (18in), rarely 60cm (24in)
WEIGHT	Up to 850g (2lb)
DEPTH	Typically 200–500m (650–1,650ft), up to 2,500m (8,200ft)

DISTRIBUTION Eastern Atlantic and Mediterranean

As this small shark searches for fish and squid in the darkness, its black belly is illuminated by tiny, bright light organs called photophores. It has large eyes and two dorsal fins, each with a strong, grooved spine in front. It is one of about 30 similar species that include the smallest known sharks.

ORDER SQUALIFORMES

Greenland Shark

Somniosus microcephalus

LENGTH	2.4–4.3m (8–14ft)
WEIGHT	Up to 775 kg (1,710lb)
DEPTH	0–1,200m (0–4,000ft)

DISTRIBUTION North Atlantic and Arctic waters

The sluggish Greenland Shark has a heavy, cylindrical body that is usually brown or grey. It has a short, rounded snout and two equal-sized dorsal fins. As well as feeding on a variety of live prey, including fish, sea birds, and seals, this shark is also a scavenger, often eating dead cetaceans and drowned land animals, such as reindeer. It is often caught by hook-and-line at ice holes while it hunts seals, but its flesh is poisonous and must be boiled several times before it becomes edible.

broad, interlocking teeth on lower jaw

ORDER SQUATINIFORMES

Pacific Angel Shark

Squatina californica

LENGTH	Up to 1.5m (5ft)
WEIGHT	Up to 27kg (60lb)
DEPTH	Typically 0–300m (0–1,000ft), up to 185m (610ft)

DISTRIBUTION Continental shelf of the eastern Pacific

Resembling something between a squashed shark and a ray, the Pacific Angel Shark spends most of its time lying quietly on the sea bed. Its sandy or grey back, peppered with dark spots and scattered dark rings, provides good camouflage. Although superficially similar to a ray, this fish is marked out as a true shark by the gill slits on the side of its head, while rays have their gills underneath. It draws water in through large, paired holes called spiracles behind its eyes and pumps it over the gills. Rearing up like a cobra, the Pacific Angel Shark ambushes passing fish including halibut, croakers, and other bottom-dwellers. It has also been known to snap at divers and fishermen who have provoked it. At night, it swims for short distances above the sea bed, sculling along with its tail. Females give birth to litters of six to ten pups after a gestation of nine to ten months. Young fish do not mature until they are at least ten years old and can live until they are 35 years old. This fish used to be abundant in the waters off California until intense fishing caused a population collapse in the 1990s. A gill net ban ended the fishery. This shark is categorized as Near Threatened on the IUCN Red List of endangered species.

ORDER SQUALIFORMES

Cookiecutter Shark

Isistius brasiliensis

LENGTH	Up to 56cm (22in)
WEIGHT	Not recorded
DEPTH	0–3,500m (0–11,500ft)

DISTRIBUTION Atlantic, Pacific, and southern Indian Ocean

Many cetaceans and large fish, including other sharks, suffer when Cookiecutter Sharks are around. This small, cigar-shaped shark is a parasite that bites chunks out of its prey. Using its unique thick, flexible lips to hold onto its victim by suction, it then twists itself around so that its razor-sharp lower teeth bite out a cookie-shaped piece of flesh. It is active at night, luring its victims with glowing green bioluminescent lights on its belly. It also preys on squid and crustaceans.

ORDER PRISTIOPHORIFORMES

Longnose Sawshark

Pristiophorus cirratus

LENGTH	Up to 1.4m (4½ft)
WEIGHT	Not recorded
DEPTH	40–310m (130–1,200ft)

DISTRIBUTION Temperate and subtropical waters of southern Australia

Like all sawsharks, this species has a head that is flattened and extended to form a long, saw-like projection, or rostrum. This is edged with rows of large, sharp teeth. Two long sensory barbels hang down from the underside of the rostrum, which is studded with further sense organs, and the shark uses these to detect vibrations and electrical fields. It seeks out and kills prey, such as fish and crustaceans, by poking around on the sea bed and slashing out sideways with its rostrum.

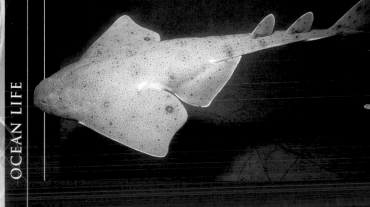

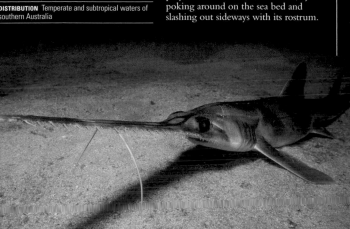

Tasselled Wobbegong

Orectolobus dasypogon

LENGTH	At least 1.3m (4¼ft)
WEIGHT	Not recorded
DEPTH	At least 40m (130ft)

DISTRIBUTION Southwestern Pacific off northern Australia and Papua New Guinea

While it lies still, the Tasselled Wobbegong looks like a seaweed-covered rock, which is exactly its objective. It is one of a group of flattened, bottom-living sharks that are masters of camouflage. The squashed shape and broad, paired fins are further adaptations to an existence on the ocean floor. This species has a beautiful reticulated pattern of narrow, dark lines against a paler background. Around its mouth is a fringe of skin flaps that resemble weeds. During the day, it rests unseen under overhangs and ledges on coral reefs. At night, it emerges onto the reef to find a good vantage point from which to snap up passing fish. There is no escape from the gape of its huge jaws and its needle-like teeth for any fish straying near, as the Tasselled Wobbegong lunges up and grabs its prey. This species has been reported to bite divers who disturb it. Reef destruction and overfishing have reduced its numbers.

MISLEADING SIMILARITY

The Tasselled Wobbegong looks remarkably similar to the Angler (see p.353), which is an unrelated species of bony fish. Both of these predators, which specialize in ambushing their prey, are flattened, have broad heads, wide mouths disguised by skin flaps, and sharp, pointed teeth. Following a similar lifestyle, these two species have come up with similar answers, an example of convergent evolution.

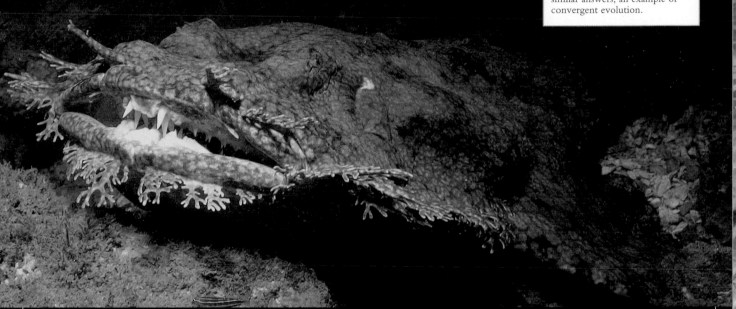

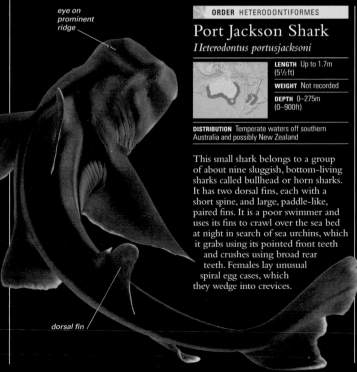

eye on prominent ridge

dorsal fin

Port Jackson Shark

Heterodontus portusjacksoni

LENGTH	Up to 1.7m (5½ft)
WEIGHT	Not recorded
DEPTH	0–275m (0–900ft)

DISTRIBUTION Temperate waters off southern Australia and possibly New Zealand

This small shark belongs to a group of about nine sluggish, bottom-living sharks called bullhead or horn sharks. It has two dorsal fins, each with a short spine, and large, paddle-like, paired fins. It is a poor swimmer and uses its fins to crawl over the sea bed at night in search of sea urchins, which it grabs using its pointed front teeth and crushes using broad rear teeth. Females lay unusual spiral egg cases, which they wedge into crevices.

Zebra Shark

Stegostoma fasciatum

LENGTH	Up to 2.4m (8ft)
WEIGHT	Not recorded
DEPTH	0–63m (0–210ft)

DISTRIBUTION Indian Ocean and southwestern Pacific

The Zebra Shark is often seen by divers around coral reefs. Its long, ridged body and densely spotted skin make it unmistakeable. Juveniles have stripes instead of spots and no ridges. This shark spends most of the day lying on the reef, usually facing into the current. At night-time it squirms its flexible body into cracks and crevices on the reef, searching for molluscs, crustaceans, and small fish.

Tawny Nurse Shark

Nebrius ferrugineus

LENGTH	Up to 3.2m (10½ft)
WEIGHT	Not recorded
DEPTH	1–70m (3–230ft), typically 30m (100ft)

DISTRIBUTION Indian Ocean, western and southwestern Pacific

The docile, bottom-living Tawny Nurse Shark is a favourite with underwater photographers because, although it may bite if harassed, it can be approached closely. During the day, it rests quietly in caves and channels in coral reefs, emerging at night to hunt for invertebrates. A pair of long sensory barbels on either side of the mouth helps the shark to find its prey, which it crushes using wide teeth.

ORDER ORECTOLOBIFORMES

Whale Shark
Rhincodon typus

LENGTH	12–20m (40–65ft)
WEIGHT	Over 12 tonnes (11¾ tons)
DEPTH	Surface, deep water in winter

DISTRIBUTION Tropical and temperate waters worldwide

The Whale Shark is a graceful, slow-moving giant and the largest fish in the world. At 1.5m (5ft) wide, its mouth is large enough to fit a human inside, but it is a harmless filter feeder that eats only plankton and small fish. To obtain the huge amount of food it needs, it sucks water into its mouth and pumps it out over its gills, where particles of food become trapped by bony projections called gill rakers and are later swallowed. This shark has the thickest skin of any animal, at up to 10cm (4in) thick. Prominent ridges run the length of its body and it has a large, sickle-shaped tail. The pattern of white spots on its back is unique to each fish, enabling scientists, through analysis of photographs, to identify individuals. While little is known of their ocean travels, satellite tagging has shown that some Whale Sharks migrate across entire oceans. Whale Shark eggs hatch inside the mother, and she gives birth to live young. Whale Sharks are killed for their meat and fins (used in soup), although they are legally protected in some countries.

FEASTING ON PLANKTON

Every year, around April, Whale Sharks migrate to Ningaloo Reef off northwestern Australia for a plankton feast. The plankton explosion results from a simultaneous mass spawning of the reef's corals, possibly triggered by the full Moon.

ORDER LAMNIFORMES

Megamouth Shark
Megachasma pelagios

LENGTH	At least 5.5m (18ft)
WEIGHT	Not recorded
DEPTH	0–165m (0–540ft)

DISTRIBUTION Little known, but probably worldwide in the tropics

This gigantic shark was discovered as recently as 1976, when one became entangled in the folds of a ship's sea anchor. Like the Whale Shark and

Basking Shark, it is a filter feeder, gulping down huge mouthfuls of shrimp, which it probably compresses with its huge tongue. At night it follows the shrimp towards the surface and is thought to attract them with bioluminescent tissue inside its mouth.

ORDER LAMNIFORMES

Basking Shark
Cetorhinus maximus

LENGTH	6–11m (20–36ft)
WEIGHT	Up to 7 tonnes (7 tons)
DEPTH	0–2,000m (0–6,500ft)

DISTRIBUTION Cold- to warm-temperate coastal waters worldwide

The world's second largest fish, the Basking Shark is protected in several countries. In summer, it swims open-mouthed at the surface, filtering out plankton. Every hour, the shark passes up to 1.5 million litres (330,000 gallons) of seawater through the huge gills that almost encircle its head. Its liver runs the length of the abdominal cavity and is filled with oil to aid buoyancy.

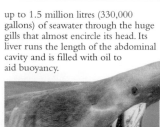

ORDER LAMNIFORMES

Sandtiger Shark
Carcharias taurus

LENGTH	Up to 3.2m (10½ ft)
WEIGHT	Up to 160kg (350lb)
DEPTH	0–190m (0–625ft)

DISTRIBUTION Warm-temperate and tropical coastal waters, except eastern Pacific

Also known as the Ragged-Tooth Shark and the Grey Nurse Shark, the Sandtiger Shark is fearsome to look at. It is heavily built and its dagger-like, menacing teeth protrude, even when its mouth is closed. Many people will have seen these sharks in aquariums, and they are quite docile in captivity despite their chilling appearance. The Sandtiger Shark has a flattened, conical snout, is light brown in colour, and its body is often speckled with darker spots. It lives in shallow coastal waters, especially on reefs and in rough, rocky areas with gullies and caves. Although it spends most of its time near the sea floor, it can hover in mid-water by filling its stomach with air gulped in at the surface.

The mother gives birth to two live young at a time, one from each of a pair of uteruses. Within each uterus there are many other embryos, and the strongest embryo in each uterus kills and eats its siblings along with any unfertilized eggs before it is born. Sandtiger Sharks are widely hunted for both sport and food.

ORDER LAMNIFORMES

White Shark
Carcharodon carcharias

LENGTH	Up to 7.2m (24ft)
WEIGHT	Over 3.4 tonnes (3½ tons)
DEPTH	0–1,300m (0–4,300ft)

DISTRIBUTION Wide range through most oceans except polar waters

The White Shark, or Great White, is one of the most powerful predators in the ocean and has a reputation as a killing machine. In fact, this shark is intelligent and capable of complex social interactions. It is, however, first and foremost a predator, feeding on prey that ranges from small fish to tuna, marine mammals (such as porpoises, seals, and sea lions), and birds (such as gannets and penguins).

Its powerful, tapered body and crescent-shaped tail are designed for sudden, swift attack, which may occur with such momentum that the shark leaves the water. It can sustain high speeds even in cold waters because it can maintain a body temperature well above that of the surrounding water due to adaptations in its circulatory system. This means that the shark's metabolism is more efficient than that of other sharks, allowing it to swim faster and with greater endurance.

Large numbers of these sharks are attracted to areas where there are sea mammal colonies, such as off South Africa. Satellite tags have shown they can migrate huge distances. Their numbers are declining due to sport fishing, netting, and commercial bycatch.

serrated edge

FEARSOME TEETH
This shark's teeth can be up to 7.5cm (3in) long. They are as hard as steel with razor-sharp, serrated edges developed to slice through the toughest flesh.

HUMAN IMPACT

SHARK ATTACK

The White Shark has made more unprovoked attacks on humans than any other shark. However, humans are not its natural prey and many such attacks can be put down to the shark mistaking a diver for a seal or turtle. When stimulated by bait in the water, White Sharks will bite anything, even a metal diving cage.

COUNTER-SHADED COLORATION
From above, the shark's dark back merges with the sea bed; from below, its white belly blends with the down-welling light.

ORDER LAMNIFORMES

Goblin Shark
Mitsukurina owstoni

LENGTH	Up to 3.9m (12¾ft)
WEIGHT	Up to 210kg (460lb)
DEPTH	300m–1,300m (1,000–4,300ft)

DISTRIBUTION Not fully known, but thought to be in temperate and tropical waters

One of the strangest-looking of all deep-water sharks, the Goblin Shark is pale pink, with a flabby body, tiny eyes, and a long, flattened, bill-like snout. This strange projection is covered in electro-receptors and is probably used to detect prey in the inky depths. Beneath the snout, the Goblin Shark has specialized jaws that can be shot forwards to grab fish and octopuses using long, pointed teeth. Not very much else is known about

this shark except that it gives birth to live young and after death changes from pinkish to a dirty, brownish grey colour. Only a few dozen Goblin Sharks have officially been caught, and this species is thought to be rare. Most data has come from sharks caught by boats fishing for deep-water fish using long lines. Fossils of sharks very similar to this species have been found in rocks over 100 million years old.

ORDER CARCHARHINIFORMES

Chain Catshark
Scyliorhinus retifer

LENGTH	0.5m (1½ft)
WEIGHT	Not recorded
DEPTH	35–550m (115–1,800ft)

DISTRIBUTION North and western Atlantic, Caribbean

With its chain-link pattern, this shark (also known as the Chain Catfish) is unmistakable. It is one of about 90 catsharks that make up the largest shark family, Scyliorhinidae. Living on the sea bed, it feeds on worms, crustaceans, and small fish. Deep furrows connect the nostrils to the mouth, and the eyes are cat-like. Catsharks lay 40–50 eggs per year in horny capsules with long tendrils at each corner. The empty cases may be washed ashore and are known as mermaid's purses.

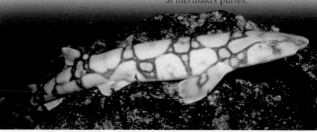

OCEAN LIFE

WHITE SHARK
One of the world's most formidable predators, the White Shark catches its prey with a short, fast attack, typically from below. Sometimes its speed takes it and its prey (in this case, a seal) clear of the water. The shark then often withdraws to wait for its victim to weaken before returning to finish it off.

Blue Shark
Prionace glauca

LENGTH	Up to 4m (13ft)
WEIGHT	Up to 200kg (440lb)
DEPTH	0–350m (0–1,150ft)

DISTRIBUTION Temperate and tropical waters worldwide

A true ocean wanderer, the Blue Shark makes seasonal trans-ocean crossings in search of food. It is streamlined and elegant, with a long, pointed snout, and characteristic white-rimmed black eyes. On long journeys, it may use its wing-like pectoral fins to help it glide on ocean currents. On the way, it makes frequent, deep dives, possibly to help it get its magnetic bearings. When chasing fish this shark may reach speeds of 70kph (43mph). It has been known to harass swimmers and has caused a few human fatalities. Although one of the commonest sharks, it is also the most exploited and its populations are declining.

Tiger Shark
Galeocerdo cuvier

LENGTH	Up to 7.4m (24ft)
WEIGHT	Up to 800kg (1,750lb)
DEPTH	To 350m (1,150ft)

DISTRIBUTION Tropical and warm temperate waters worldwide

The Tiger Shark is the second most dangerous shark to humans, after the White Shark (see p.331). It is huge and has a heavy head and a mouth filled with serrated teeth that have the characteristic shape of a cockscomb. One reason it is so dangerous is that it prefers coastal waters and is also found in river estuaries and harbours, and so it frequently comes into contact with humans. It is reputed to eat almost anything – as well as eating smaller sharks, including its own young, other fish, marine mammals, turtles, and birds, it is an inveterate scavenger, and a huge variety of rubbish has been found in Tiger Shark stomachs. The young, born live after hatching from eggs inside the mother, begin life marked with blotches, which become "tiger stripes" in juveniles and fade by adulthood.

Whitetip Reef Shark
Triaenodon obesus

LENGTH	Up to 2m (6½ft)
WEIGHT	Up to 18kg (37lb)
DEPTH	Typically 8–40m (26–130ft), recorded at 330m (1,080ft)

DISTRIBUTION Tropical waters of the Indian Ocean and Pacific

One of the sharks most often seen by divers is the Whitetip Reef Shark, which during the day may be found around coral reefs resting in caves and gullies, often in groups. The tip of its first dorsal fin and the upper tip of its tail are white, in contrast to its greyish brown back. At night, the Whitetip comes out to hunt reef fish, octopus, lobsters, and crabs hidden among the coral. Packs sometimes hunt together, sniffing out the prey and barging and banging the coral to get at them.

Scalloped Hammerhead Shark
Sphyrna lewini

LENGTH	Up to 4.3m (14ft)
WEIGHT	Up to 150kg (330lb)
DEPTH	0–500m (0–1,640ft)

DISTRIBUTION Tropical and warm temperate waters worldwide

Along with the seven other known species of hammerhead sharks, the Scalloped Hammerhead has a strange, flattened, T-shaped head. In this species, the front of the head has three notches, which produces the scalloped shape from which it takes its name. The eyes are located at the sides of the head. Hunting near the sea bed, the shark swings its head from side to side, looking for prey such as fish, other sharks, octopus, and crustaceans, and using sensory pits on its head to detect the electrical fields of buried prey such as rays. The head may also function as an aerofoil, giving the shark lift and helping it to twist and turn as it chases its prey.

Scalloped Hammerheads may be seen in large shoals of over a hundred individuals. They give birth to live young in shallow bays and estuaries, where the skin of the young darkens to give protection against sunlight.

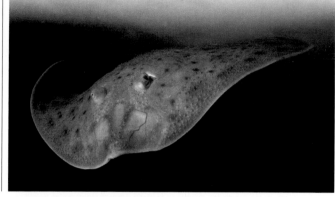

ORDER RAJIFORMES

Atlantic Guitarfish

Rhinobatos lentiginosus

LENGTH	Up to 75cm (30in)
WEIGHT	Not recorded
DEPTH	0–30m (0–100ft)

DISTRIBUTION Coastal waters of Gulf of Mexico, Caribbean, and western Atlantic

Guitarfish are elongated rays with a triangular snout and narrow pectoral fins. Like sharks, these rays use their spineless tails for swimming, while other rays swim by flapping only their pectoral fins. The Atlantic Guitarfish has two small dorsal fins set far back near the tip of its tail. Greyish brown with small white spots, it blends in with the sandy sea bed, but this ray can be seen in shallow water searching for molluscs and crabs, probing the sand with its snout. Females are ovoviviparous, giving birth to about six live young.

ORDER RAJIFORMES

Common Skate

Dipturus batis

LENGTH	Up to 2.9m (9½ft)
WEIGHT	Up to 100kg (220lb)
DEPTH	100–1,000m (330–3,300ft)

DISTRIBUTION Eastern Atlantic from northern Europe to southern Africa, Mediterranean

The Common Skate is the largest and heaviest of the European rays. Due to overexploitation, it is now a rare species throughout most of its range and has become extinct in some regions. Its snout is long and pointed, and the front margin of the wings is strongly concave, giving this skate an overall angular shape. Its tail has a row of spines along its length but, unlike the large stinging spine of stingrays, these are not venomous.

This species is sometimes called the Blue Skate because its underside is bluish grey. It can swim strongly and feeds on fish in mid-water as well as hunting over the sea bed for crabs, lobsters, bottom-dwelling fish, and other rays. Its oblong egg cases are up to 25cm (10in) long. They are laid in autumn or winter and hatch two to five months later. While mature specimens prefer deeper water, the young will spend time in shallow waters.

ORDER RAJIFORMES

Painted Ray

Raja undulata

LENGTH	Up to 1.2m (4ft)
WEIGHT	Up to 7kg (15lb)
DEPTH	45–200m (150–650ft)

DISTRIBUTION Eastern Atlantic and Mediterranean

Also known as the Undulate Ray, the Painted Ray is one of the most distinctive northern European rays. This species is patterned with long, wavy, dark lines edged with white spots that run parallel to the wing margins. Its ornate appearance makes it an attractive species for aquariums, but in its natural habitat, this pattern has the practical advantage of helping the ray to blend in with the gravel and sand on the sea bed, where it feeds on flatfish, crabs, and other bottom-living invertebrates.

The biology of this beautiful ray has not been fully studied, but during the breeding season males use paired claspers to transfer sperm during mating, and females are known to lay up to 15 eggs in muddy or sandy flats. Each egg is encased in a reddish-brown, oblong egg capsule up to 9cm (3 1/2 in) long, with a curved horn at each corner. Additional information on the distribution and status of this and other rays around Britain is currently being collected through an egg-case identification project. Members of the public are encouraged to collect empty egg cases that have been washed ashore, re-hydrate them and identify them. The number and location of egg cases found are collated each year.

pelvic fin

dorsal fin

ORDER RAJIFORMES

Smalltooth Sawfish

Pristis pectinata

LENGTH	Up to 7.6m (25ft)
WEIGHT	Up to 350kg (770lb)
DEPTH	0–10m (0–33ft)

DISTRIBUTION Subtropical waters in all oceans

Sawfish are elongated rays with a long, flat, saw-like snout, or rostrum, which they use to slash through shoals of fish and dig for shellfish and invertebrates. Like all rays, they have gill slits on the underside of the body rather than the sides. Females give birth to live young, which are about 60cm (2ft) long in the Smalltooth species. The saws of the pups are sheathed and flexible at birth, in order to prevent injury to the mother.

The Smalltooth Sawfish lives in coastal waters but also swims up river estuaries. Numbers of this species are severely depleted and it is listed as Critically Endangered on the IUCN Red List of endangered species.

ORDER RAJIFORMES

Reticulate Whipray

Himantura uarnak

LENGTH	About 4.5m (15ft) including tail
WEIGHT	About 120kg (265lb)
DEPTH	20–50m (65–165ft)

DISTRIBUTION Coastal waters of Arabian Gulf, Red Sea, Indian Ocean, and western Pacific

This beautifully patterned stingray belongs to a group called whiprays, which have long, thin, flexible tails. Its upper surface is densely covered with wavy brown lines or reticulations. Its disc, or body, is about 1.5m (5ft) long and the tail can be nearly three times this length. Its sting is a single, large spine located a short distance from the tail base; some individuals have two spines. The body is almost diamond-shaped and the snout is broadly triangular with a pointed tip. Found in warm waters, mainly near the coast, Reticulate Whiprays are sometimes seen by divers, lying quietly in sandy patches between rocks.

OCEAN LIFE

OCEAN LIFE

ORDER RAJIFORMES

Southern Stingray
Dasyatis americana

WIDTH (WINGSPAN)	2m (6½ft)
WEIGHT	Up to 135kg (300lb)
DEPTH	0–55m (0–180ft)

DISTRIBUTION Western Atlantic, Gulf of Mexico, Caribbean Sea

Stingrays are feared because their long tails are equipped with one or more dagger-like, venomous spines. The Southern Stingray has a single, serrated spine about midway along the tail and a flap of skin, also known as a finfold, on the underside of the tail. Its thick disc is dark grey on top and white underneath. This stingray spends most of the day lying buried in the sand at the bottom of shallow lagoons and off beaches. At night, the ray feeds by excavating holes in the sand and crunching up bivalve molluscs, crustaceans, and worms. Because its eyes are on top of its head, it cannot see its prey but uses smell and electro-receptors to detect it. While it is buried, its spiracles – through which it draws in water for breathing – are visible as a pair of holes in the sand. People are often stung when they inadvertently tread on Southern Stingrays; the stinging spine is sharp enough to cause a serious wound and the venom causes severe pain. The pain can be reduced by immersing the wound in hot water.

ORDER RAJIFORMES

Atlantic Torpedo
Torpedo nobiliana

LENGTH	Nearly 2m (6½ft) including tail
WEIGHT	Up to 90kg (200lb)
DEPTH	To 800m (2,600ft)

DISTRIBUTION Atlantic, Mediterranean

Electric rays use special organs to produce electricity, which they discharge to stun their prey or attack predators. The Atlantic Torpedo is the largest electric ray and can produce a shock of up to 220 volts – enough to stun a person. It can easily be recognized by its circular, disc-like body and short, thick tail ending in a large, paddle-shaped fin. It is a uniform dark brown or black on the back and white underneath. The electric organs are in the ray's wings, or pectoral fins, and like a battery they can store electricity. When hunting, the Atlantic Torpedo wraps its wings around its prey before stunning it.

ORDER RAJIFORMES

Round Stingray
Urolophus halleri

LENGTH	58cm (23in) including tail
WEIGHT	1.4kg (3lb)
DEPTH	0–90m (0–300ft)

DISTRIBUTION Eastern Pacific

As its name suggests, the Round Stingray has an almost circular disc. This species and its relatives the "stingarees" have shorter tails than other stingrays and the tail ends in a leaf-shaped fin. The Round Stingray varies in colour from pale to dark brown and can be either plain or mottled with darker spots and reticulations. These rays are most often seen in summer, when they move inshore into inlets and bays to forage for invertebrates among seagrass and bask in the warm-water shallows. Females arrive in the shallows around June ready to breed, and the males, who are already there, swim along the shoreline looking for suitable mates. Sexually mature females are reported to give off an electrical field that the males can sense. Females give birth about three months after mating to about six live young. The juveniles stay inshore, where there are fewer predators, until they mature. When out foraging they do not stray too far, remaining within an area of about 2.5 square km (1 square mile).

Predators of the Round Stingray include the Northern Elephant Seal and the Black Sea Bass. They are also likely to be hunted by large carnivorous fish such as sharks. The Round Stingray's sting is painful and can cause minor injuries.

ORDER RAJIFORMES

Blue-spotted Stingray
Taeniura lymma

LENGTH	Up to 2m (6½ft) including tail
WEIGHT	Up to 30kg (65lb)
DEPTH	Shallow water to about 20m (65ft)

DISTRIBUTION Indian Ocean, western Pacific, Red Sea

As it is active in the daytime, divers often see this beautifully coloured ray on coral reefs. It is most often spotted lying on sandy patches under coral heads and rocks. Often, its blue-striped tail sticks out and gives away its hiding place. Large, bright blue spots cover the disc, which is greenish brown. Like all stingrays, it has a venomous spine on its tail. As the tide rises, these rays move in groups into shallow water to hunt for invertebrates such as molluscs, crabs, shrimp, and worms.

ORDER RAJIFORMES

Manta Ray
Manta birostris

WIDTH (WINGSPAN) Up to 8m (26ft)	
WEIGHT Up to 1.8 tonnes (4,000lb)	
DEPTH 0–24m (0–80ft); usually near surface	

DISTRIBUTION Surface tropical waters worldwide, sometimes warm temperate areas

Divers often describe the experience
of swimming beneath a Manta Ray as
like being overtaken by a huge flying
saucer. This ray is the biggest in the
world, but like the biggest shark,
the Whale Shark, it is a harmless
consumer of plankton and small fish.
When feeding, it swims along with its
cavernous mouth wide open, beating
its huge triangular wings slowly up
and down. On either side of the
mouth, which is at the front of the
head instead of on the underside as
in other rays, are two long lobes, called
cephalic horns, that funnel plankton
into the mouth. These are the origin
of its other name of Devil Ray.
A short, stick-like tail trails behind.

On coral reefs Manta Rays tend
to congregate over high points where
currents bring plankton up to them.
Small fish called remoras often travel
attached to these giants. Despite their
huge size, these ovoviviparous rays can
leap clear of the water, occasionally
giving birth to their young as they
do so. The Manta Ray is also sociable
with divers in some sites, and has
been known to
"dance" with
them.

ORDER RAJIFORMES

Spotted Eagle Ray
Aetobatus narinari

WIDTH (WINGSPAN) Up to 3m (10ft)	
WEIGHT Up to 230kg (500lb)	
DEPTH 1–80m (3–260ft)	

DISTRIBUTION Tropical waters worldwide

Often solitary, Spotted Eagle Rays also
move around in huge shoals of at least
a hundred individuals in open waters
– a truly spectacular site when
silhouetted against a sunlit surface.
Unlike most other rays, the Spotted
Eagle Ray is a very active swimmer.
Most of its swimming time is spent
in open water, although it is also
commonly seen inshore. It appears to
"fly" through the water as it moves its
pointed "wings" – enlarged pectoral
fins – gracefully up and down. Besides
the beautiful patterning of spots on its
dorsal surface, another distinctive
feature of the Spotted Eagle Ray is its
head, which ends in a flattened, slightly
upturned snout that resembles a duck's
bill. It has a long, thin whip-like tail
with a venomous spine near the base.

These rays are very agile and can
twist and turn to escape predatory
sharks. Sometimes, small groups splash
around at the surface, making
spectacular leaps out of the water.
Why they do this is not clear but it
may be to help to dislodge parasites.

BONY FISH

DOMAIN	Eucarya
KINGDOM	Animalia
PHYLUM	Chordata
SUPERCLASS	Osteichthyes
ORDERS	47
SPECIES	More than 28,000

BONY FISH EXCEED ALL OTHER VERTEBRATE classes both in number of living species and in their abundance. They have evolved into myriad shapes and sizes, suiting every aquatic lifestyle and habitat and range from the shore to the deepest depths and from polar seas to hot deep-sea vents. Bony fish have an internal skeleton of bone, although that of a few primitive groups is part cartilage. The bony skeleton supports flexible fins that allow the fish to move with far greater precision than do the stiff fins of cartilaginous fish. About one-third of bony fish live only in fresh water, while the remainder lives in the oceans or migrates between the two habitats.

ANATOMY

Like other vertebrate animals, bony fish have a skull, backbone, and ribs, but the skeleton also extends out into the fins as a series of flexible rays. Bony fish, unlike sharks, can use their paired pectoral and pelvic fins for manoeuvring, braking, and even swimming backwards. Spiny-rayed fish, a group that includes most bony fish, also have sharp spines in the front portion of their dorsal, anal, and pelvic fins. A bony flap called the operculum covers the gills of bony fish. It can be opened to regulate the flow of water in through the mouth and out over the gills. A covering of overlapping, flexible scales made of thin bone protects most bony fish. Some primitive bony fish, such as sturgeon, are heavily armoured with thick, inflexible scales or plates.

SWIMMING
The sideways force and backwards force exerted when a fish moves its tail from side to side results in a thrust at an angle between the two. The resultant thrusts on left and right produce a net backwards thrust and so the fish is propelled forwards.

forward movement

movement of tail

sideways force

resultant thrust

backwards force

FISH FARMING

Most of the world's fish stocks are currently fished at unsustainable levels. Cod, turbot, and salmon are now often farmed instead of wild-caught. On a large scale, this might relieve pressure on wild stocks. However, most farmed species consume feed prepared from other wild-caught fish.

SALMON FARM
Salmon farms such as this one in Tasmania are a common sight in temperate seas. However, pesticides applied to kill sea-lice are causing concerns, as is the breeding of escaped salmon with wild fish, polluting the wild gene pool.

BONY SKELETON
Flexible rays and hard spines support all the fins in bony fish, such as this cod. The fins connect to spines extending from the vertebrae. The fish can precisely adjust the position of each fin.

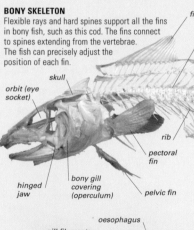
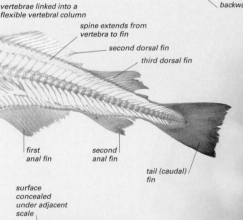

first dorsal fin

vertebrae linked into a flexible vertebral column

spine extends from vertebra to fin

second dorsal fin

third dorsal fin

skull

orbit (eye socket)

rib

pectoral fin

first anal fin

second anal fin

tail (caudal) fin

hinged jaw

bony gill covering (operculum)

pelvic fin

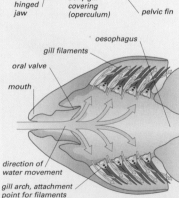

oesophagus

gill filaments

oral valve

mouth

direction of water movement

gill arch, attachment point for filaments

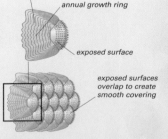

surface concealed under adjacent scale

annual growth ring

exposed surface

exposed surfaces overlap to create smooth covering

GILLS, VIEWED FROM ABOVE
As water passes over the gill filaments, gases are exchanged. Oxygen passes into the blood and carbon dioxide passes out into the water. Within the filaments, the blood flows in reverse relative to the water outside, so the concentration of gases in the fluids is opposed, which speeds the gases' transfer.

SCALES
Bony fish can be aged by their scales. Slow winter growth produces dark rings on the scales, so each dark ring indicates one year of life. The system works best for temperate-water fish such as cod.

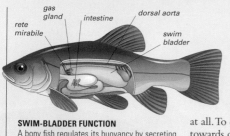

SWIM-BLADDER FUNCTION
A bony fish regulates its buoyancy by secreting gas, usually oxygen, from a gas gland into its swim bladder. The gland is supplied with blood (the source of the gas) by a network of capillaries called a rete mirabile.

BUOYANCY

Most bony fish have a gas-filled swim bladder that allows them to adjust their buoyancy, enabling them to hover in midwater and prevent themselves sinking. This is especially useful to fish that spend their lives in mid-water. Many bottom-living fish, such as flatfish, have a poorly developed swim bladder or none at all. To compensate for pressure changes as a fish swims towards or away from the surface, it regulates the amount of gas in the swim bladder, usually by secreting gas into it through a gland. In some primitive fish, such as the herring, the swim bladder is connected to the gut and is filled when the fish gulps air at the surface.

Many bony fish can vibrate the swim bladder with special muscles to produce sounds. Cartilaginous fish do not have a swim bladder. They gain buoyancy to some extent with their large, oil-filled livers and lightweight bones. However, cartilaginous fish must also use their large pectoral fins and tail to give them lift. Bony fish with a swim bladder have been freed from this necessity and, in many species, the fins have developed into versatile appendages used for courtship, feeding, attack, or defence.

MULTI-PURPOSE FINS
Triggerfish swim by undulating their second dorsal and anal fins, maintaining buoyancy with their swim bladder. Bright fins may also function as visual signals in communication, including courtship.

FLEXIBLE APPENDAGES
A frogfish displays one of the many functions of bony fish fins. Since they are not required to give the fish lift, the Warty Frogfish's paired fins have evolved into flexible appendages with which it clambers over the sea bed.

SENSES

Bony fish use vision, hearing, touch, taste, and smell. Vision is most important in well-lit habitats. Coral reef fish have good colour vision, and they use colours and patterns for recognition, warning, deception, and courtship. Colour receptors in the eyes do not operate well in dim light. Nocturnal fish and fish living in the twilight zone (see p.170) have large, sensitive eyes, but little sensitivity to different colours. Dark-zone fish often have only tiny eyes, but have a keen sense of smell and use pheromones for long-distance communication. Sound also carries well underwater (see p.39) and some fish produce intense sounds with their swim bladder. Bony fish move in unison in shoals with the help of their lateral-line sensory system, for which there is no equivalent in other vertebrates. Sense organs arranged in a canal along the head and sides of each fish pick up water movements created by the other fish. The wide field of view, due to having eyes set on the sides of the head, also helps precision shoaling.

SIGNALLING COLOURS
Colour is effective in communication on well-lit coral reefs. The gaudiness of the Mandarin Fish may warn predators that it is unpalatable.

LATERAL LINE
The lateral-line system can be seen in many bony fish, such as the Pollack shown here, as a white line along the sides of the fish. The shape of the line is a useful identification feature.

REPRODUCTION

The majority of bony fish, when mature, simply shed their eggs and sperm directly into the sea, where fertilization takes place. The eggs develop and the larvae hatch while drifting on ocean currents. Death rates of eggs and larvae are high, so many eggs are laid – up to 100 million by the giant Ocean Sunfish. Once the larvae have grown to juvenile fish, they often congregate in nursery grounds in sheltered estuaries and bays. In contrast to most oceanic fish, many coastal bottom-living species are able to protect their offspring, so they lay fewer, larger eggs, often hiding them or caring for them until they hatch. Some have evolved elaborate forms of care, such as mouth brooding.

MOUTH BROODING
When a female jawfish has laid her eggs, the male collects them up into his mouth to keep them safe. He will not feed until the eggs hatch and the fry disperse.

SEX CHANGE IN THE CUCKOO WRASSE

❶ Like most wrasse, Cuckoo Wrasse have a complex reproductive pattern featuring sex change. The majority of eggs develop first into pink females.

❷ Some older females develop the blue-and-orange pattern of males and change sex after about seven years of age. Others remain female.

❸ At the next spawning season, a sex-changed male acquires vibrant colours and courts all females in his territory, fertilizing their eggs (see p.369).

HUNTING AND PROTECTION

All fish must eat and in doing so may expose themselves to the risk of being eaten if they come out into the open to forage. Their ultimate aim is to survive long enough to reproduce successfully and so pass on their genes to the next generation. Bony fish have evolved many ingenious methods for catching prey and defending themselves against predators. Camouflage is an effective strategy and can serve to hide a fish, both from its predators and from its prey. Colour patterns can also deceive, and butterflyfish use false eye patterns to fool predators into lunging for their tail end. In the crowded environment of a coral reef, many small fish protect themselves with spines. Filefish erect a dorsal spine and lock it into position, thereby preventing larger fish from swallowing them. Out in the surface waters of the open ocean, there is nowhere to hide, and many small fish live in shoals for safety. Predators find it difficult to pick out a target as the shoal moves and swirls. Although the shoal is conspicuous, it is safer for each individual to join than to swim alone.

CAMOUFLAGE
Scorpionfish employ colour, shape, and behaviour in a combined camouflage strategy. Experts at keeping still, they can strike with lightening speed if a small fish strays within reach.

SHOALING
Even predatory fish need protection from larger predators, especially when young. Barracuda juveniles live in shoals during the day, while most adults hunt alone.

SHADOW-HUNTING
Trumpetfish often shadow predatory fish when they hunt, since the larger fish will often flush out suitable prey. This trumpetfish has chosen to swim with a Nassau Grouper similar in colour to itself.

BONY FISH CLASSIFICATION

Bony fish comprise the ray-finned fish (subclass Actinopterygii), and the lobe-finned fish (subclass Sarcopterygii), including the lungfishes (freshwater) and the coelacanths, but also giving rise to tetrapods (see cladogram, p.206). Below are 30 marine orders of both subclasses.

COELACANTHS
Order Coelacanthiformes

2 species
One family – the only marine lobe-finned fish. Fins arise from fleshy, limb-like lobes, vertebral column not fully formed.

STURGEONS AND PADDLEFISH
Order Acipenseriformes

25 species
Only sturgeon family is marine. Skeleton part bone, part cartilage. Sturgeons have asymmetrical tail and underslung mouth.

TARPONS AND TENPOUNDERS
Order Elopiformes

7 species
Two families, mostly marine. Spindle-shaped, silvery fish, one dorsal fin, forked tail. Unique bones in throat (gular plates). Swim bladder can be used as lung. Transparent larvae.

BONEFISH
Order Albuliformes

5 species
One family, mostly marine. Similar to tarpons, but smaller and very bony, with complex structural differences.

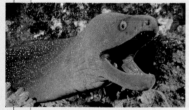

MORAY EEL, ORDER ANGUILLIFORMES

EELS
Order Anguilliformes

737 species
Marine and freshwater, 15 families. Body long and thin, no scales or pelvic fins, one long fin along back, tail, and belly.

SWALLOWERS AND GULPERS
Order Saccopharyngiformes

28 species
Four families of highly aberrant deep-sea, eel-like fish with huge, loose jaws; no tail fin, pelvic fins, scales, ribs, or swim bladder.

HERRINGS
Order Clupeiformes

397 species
Six families, mostly marine. Silvery body with keeled belly and forked tail. Anchovies and herrings comprise two biggest families.

MILKFISH
Order Gonorynchiformes

27 species
Four families, only milkfish and beaked salmon marine. Pelvic fins set far back.

CATFISH
Order Siluriformes

2,867 species
Only two marine families out of 33. Long body, up to four pairs of barbels around mouth. Sharp, sometimes venomous spine in front of dorsal and pectoral fins. Most with adipose fin.

SMELTS
Order Osmeriformes

227 species
Thirteen families, mostly marine or anadromous. Small, slim relations of salmon.

SALMONS
Order Salmoniformes

66 species
One family with marine, anadromous, and freshwater members. Powerful, spindle-shaped fish, with large mouth and eyes. One fin plus adipose fin on back. Small, rounded scales. Pelvic fins abdominal.

LIGHTFISH AND DRAGONFISH
Order Stomiiformes

391 species
Four abundant, deep-ocean families. Mostly elongate predators with large teeth and photophores. Significant part of ocean's fishes.

GRINNERS
Order Aulopiformes

228 species
Thirteen families, all marine. Diverse, slim coastal and deep-sea fish. Large mouth with many small teeth. Pelvic fins abdominal, one fin plus adipose fin on back, no fin spines.

LANTERNFISH
Order Myctophiformes

241 species
Two deep-ocean, widely distributed, abundant families. Small, slim fish, large eyes and mouth. One fin plus adipose fin on back. Many photophores. Daily vertical migration.

VELIFERS, TUBE-EYES, AND RIBBONFISH
Order Lampriformes

18 species
Seven families, all marine. Colourful, bright, often huge, open-water fish with crimson fins. Many have long rays from dorsal fin.

HERRING, ORDER CLUPEIFORMES

COD FISH
Order Gadiformes

475 species
Ten families, mostly marine and benthic. Most with two or three spineless dorsal fins and a chin barbel. Grenadiers have long, thin tails.

TOADFISH
Order Batrachoidiformes

69 species
One family, mostly marine, coastal, and benthic. Broad, flat head, wide mouth, eyes on top; one short, spiny and one long, soft dorsal fin.

CUSK EELS
Order Ophidiiformes

354 species
Five families, mostly marine. Eel-like fish with long dorsal and anal fin that may join with tail fin. Thin pelvic fins.

ANGLERFISH, ORDER LOPHIIFORMES

ANGLERFISH
Order Lophiiformes

300 species
Eighteen families, all marine. Large, flattened or rounded head with cavernous mouth and fishing lure on top. Shallow-water species benthic; deep-water species pelagic.

CLINGFISH
Order Gobiesociformes

120 species
One family, mostly marine. Small, shallow-water, benthic fish; pelvic fins forming sucker-disc. Eyes set high; single dorsal fin.

NEEDLEFISH
Order Beloniformes

186 species
Five families, marine and freshwater. Mostly long, thin fish with jaws extended as beaks. Flying fish have large pectoral and pelvic fins.

SILVERSIDES
Order Antheriniformes

312 species
Eight families, marine and freshwater. Small, slim, silvery fish; most with two dorsal fins, often in large shoals.

SQUIRRELFISH AND RELATIVES
Order Beryciformes

121 species
Seven families, all marine. Deep-bodied, big eyes (except deep-water), dorsal fin spiny at front, forked tail, large scales. Most nocturnal.

DORIES AND ALLIES
Order Zeiformes

42 species
Six families, all marine. Deep-bodied but thin fish; large, spiny head and protrusile jaws. Long dorsal and anal fins with spines at front.

STICKLEBACKS AND SEAMOTHS
Order Gasterosteiformes

16 species
Five families, mostly freshwater. Long, thin, stiff with bony scutes along sides, separate spines on back. Seamoths aberrant, flattened with enlarged pectoral fins.

PIPEFISH AND SEAHORSES
Order Syngnathiformes

240 species
Five families, marine and freshwater. Long body encased in armour of bony plates. Small mouth at end of tubular snout.

SCORPIONFISH AND FLATHEADS
Order Scorpaeniformes

1,326 species
Thirty-five families, mostly marine. Mainly shallow water, benthic. Large, spiny head, most with spiny dorsal fins, often venomous. Unique bony strut across cheek.

PERCH-LIKE FISH
Order Perciformes

About 9,500 species
156 families, marine and freshwater. Largest and most diverse vertebrate order. Most have both spines and soft rays in dorsal and anal fins. Pelvic fins close to pectorals and with one spine. Perciform classification subject to change.

TURBOT, ORDER PLEURONECTIFORMES

FLATFISH
Order Pleuronectiformes

572 species
Eleven families, mostly marine. Lie on sea bed. Body flattened from side to side, both eyes on upper side. Start life as normal, symmetrical fish larvae in plankton.

PUFFERS AND FILEFISH
Order Tetraodontiformes

353 species
Nine families, marine and freshwater. Very diverse group ranges from triggerfish to Ocean Sunfish. Small mouth with few large teeth or tooth plates. Scales usually modified as plates, spines, or shields.

ORDER COELACANTHIFORMES

Coelacanth
Latimeria chalumnae

LENGTH	Up to 2m (6ft)
WEIGHT	Up to 95kg (210lb)
DEPTH	150–700m (490–2,300ft)

DISTRIBUTION Western Indian Ocean

When it was first discovered in 1938, the Coelacanth was nicknamed "old four legs" because its pectoral fins had strange, fleshy, limb-like bases. The only other primitive group to have a similar arrangement are the freshwater lungfish. It is from fish like these that the first four-legged land animals are thought to have developed. This Coelacanth's tail has an extra small lobe in the middle, and its body is covered in heavy scales, which are made up of four layers of bone and a hard mineral material. In life, these shimmer an iridescent blue with white flecks. Coelacanths live in deep water on steep, rocky reefs and, so far, have been found at only a few sites off the south and east coasts of Africa and the west coast of Madagascar. By using small submersibles to study these fish, scientists have discovered that they retreat into caves at night. When out searching for food, they drift along in ocean currents or scull slowly with their fins. Having located a fish or squid, the Coelacanth then uses its powerful tail to propel itself forward so that it can seize its prey. The Coelacanth is listed as Critically Endangered on the IUCN Red List of Endangered Species. International trade in this species is banned.

FOSSIL EVIDENCE

Coelacanths were thought to have become extinct about 65 million years ago. When a live Coelacanth was caught in 1938, comparing it with fossil Coelacanths enabled scientists to confirm its identity. A living specimen of a fossil species had been found.

FOSSIL COELACANTH
Fossil specimens of Coelacanths have almost identical features to present-day Coelacanths, including the unique three-lobed tail.

ORDER COELACANTHIFORMES

Indonesian Coelacanth
Latimeria menadoensis

LENGTH	Up to 1.4m (4½ ft)
WEIGHT	Up to 90kg (200lb)
DEPTH	150–200m (490–655ft)

DISTRIBUTION Celebes Sea, north of Sulawesi, in the western Pacific

When the Indonesian Coelacanth was discovered in 1998, it was at first thought to be the same species as the one found in African waters (see left). The two are indeed very similar, but molecular studies suggest that they are different species. An entire ocean separates them, and because coelacanths are slow swimmers the populations are not thought to mix. The Indonesian Coelacanth has the same white markings and distinctive gold flecks as the African species, but it is brown rather than bluish. As yet, little is known of its life history, but because it is so physically similar to the African species it probably has the same behaviour and could therefore be endangered by fishing.

ORDER ACIPENSERIFORMES

European Sturgeon
Acipenser sturio

LENGTH	3.5m (11ft)
WEIGHT	Up to 400kg (880lb)
DEPTH	4–90m (13–295ft)

DISTRIBUTION Coastal waters of northeastern Atlantic, Mediterranean, and Black Sea

Like most sturgeon, this species swims from the sea into large rivers to spawn in gravelly areas. These prehistoric-looking fish belong to a primitive group in which only the skull and some fin supports are made of bone. The rest of the skeleton consists mainly of cartilage. Instead of scales, five rows of distinctive bony plates, or scutes, run along the body. Two pairs of barbels hang down from

CAVIAR BAN

Once common, the European Sturgeon is now extremely rare due to overfishing and poaching, and because locks and polluted estuaries have made many rivers unsuitable for spawning. Few active spawning sites remain. This sturgeon is critically endangered, and international trade in the fish itself and any products from it, including caviar (salted roe), has been banned.

the pointed snout and are used to search out bottom-living invertebrates. The European Sturgeon can live for up to 100 years.

ORDER ACIPENSERIFORMES

Beluga Sturgeon
Huso huso

LENGTH	5m (16ft)
WEIGHT	Up to 2,000kg (4,400lb)
DEPTH	70–180m (230–590ft)

DISTRIBUTION Northern Mediterranean, Black Sea, Caspian Sea, and associated rivers

The Beluga is both the largest species of sturgeon and the largest European fish to enter fresh water. Stouter and heavier than the European Sturgeon (see left), it has a more triangular snout with a very wide mouth. Four long barbels hang from the underside of the snout, reaching almost to the mouth. Like all sturgeon, the Beluga has an asymmetrical, shark-like tail with the backbone extending into the large upper lobe. Reputed to be the most expensive fish in the world, it also produces the most prized caviar, with large fish containing 100–200kg (220–440lb). The Beluga Sturgeon is endangered owing to poaching and damming of its spawning rivers.

flattened, bony head

barbels

bony scute

ORDER ELOPIFORMES

Tarpon
Megalops atlanticus

LENGTH	Up to 2.5m (8ft)
WEIGHT	160kg (350lb)
DEPTH	0–30m (0–100ft)

DISTRIBUTION Coastal waters of western and eastern Atlantic

With its large scales and intensely silvery body, the Tarpon resembles an oversized herring but, in fact, is closely related to the eels. It has an upturned mouth, and the base of the single dorsal fin is drawn out into a long filament, although this is not always easy to see. Living close inshore, this fish often enters estuaries, lagoons, and rivers. If it enters stagnant water, it surfaces and gulps air, which passes from the oesophagus into its swim bladder; this then acts like a lung.

Tarpon spawn mostly in open water at sea. A large female can produce over 12 million eggs, but larval and juvenile mortality is high. The larvae, which are thin and transparent and very like eel larvae except that they have forked tails, drift inshore into estuarine nursery grounds. Tarpon larvae are also found in pools and lakes that become temporarily cut off from the sea. Fishermen get to know the areas where Tarpon shoals can regularly be seen from year to year, hunting for

other shoaling fish such as sardines, anchovies, and mullet. They will also eat some bottom-living invertebrates such as crabs. Considered an excellent game fish in US and Caribbean waters, Tarpon make spectacular leaps when hooked. They are also fished commercially and, in spite of being rather bony, are considered delicious. Tarpon can live for 55 years and are often displayed in public aquariums. Their large scales are sometimes used in ornamental work.

ORDER ELOPIFORMES

Ladyfish
Elops saurus

LENGTH	Up to 1m (3ft)
WEIGHT	10kg (22lb)
DEPTH	0–50m (0–165ft)

DISTRIBUTION Coastal waters of western Atlantic and Caribbean Sea

The Ladyfish has a single dorsal fin in the middle of its back and a tail that is deeply forked. Shoals of this slim, silvery blue fish can be found close to the shore and will skip along the surface if alarmed by a boat's engine noise. The adult fish move offshore to spawn in open water, and the young larvae, which resemble eel larvae, eventually drift back into sheltered bays and lagoons. Also known as the Ten-pounder, the Ladyfish is considered a good game fish and will leap out of the water when hooked. It is fished commercially but it is not a very high-quality food fish and so is often used for bait.

ORDER ALBULIFORMES

Bonefish
Albula vulpes

LENGTH	Up to 1m (3ft)
WEIGHT	10kg (22lb)
DEPTH	0–85m (0–280ft)

DISTRIBUTION Tropical and subtropical coastal waters of western and eastern Atlantic

As its common name suggests, this fish is extremely bony. It is streamlined and silvery, with dark markings on its back, a single dorsal fin, and a blunt snout extending over the mouth. Bonefish have been found in tropical and subtropical waters of the Pacific Ocean, but it is not yet known if these populations are a different species to those found in the eastern and western Atlantic. Although Bonefish do not make good eating, they are one of the world's most important game fish. Fishermen enjoy stalking them through the shallows in bays and estuaries as they often swim at the surface with the dorsal fin showing.

ORDER ANGUILLIFORMES

European Eel
Anguilla anguilla

LENGTH	Up to 1.3m (4¼ft)
WEIGHT	6.6kg (14lb)
DEPTH	0–700m (0–2,300ft)

DISTRIBUTION Temperate waters of northeastern Atlantic, fresh water inland

Living most of its life in fresh water, the European Eel swims thousands of kilometres down to the sea and across the Atlantic Ocean to the Sargasso Sea to spawn. After spawning in deep water, the eels die, leaving the eggs to hatch into transparent, leaf-like (leptocephalus) larvae. Over the next year or so, the larvae drift back to the coasts of Europe. Nearing the coast, they change shape and become tiny transparent eels, or elvers, that swim and wriggle their way up rivers into fresh water. The European Eel is becoming increasingly scarce due to pressure on stocks caused by fishing and the damming of rivers.

ORDER ANGUILLIFORMES

Chain Moray Eel
Echidna catenata

LENGTH	Up to 1.7m (5½ft)
WEIGHT	Not recorded
DEPTH	0–12m (0–39ft)

DISTRIBUTION Tropical reefs of western and central Atlantic

The Chain Moray Eel is one of very few marine eels that can survive for some time out of water, and it will forage over wet rocks for up to 30 minutes at a time during low tide. As long as it remains wet, it can absorb some oxygen through its skin. The Chain Moray Eel is easily recognized by its short, blunt snout and chain-like yellow markings. Some of its teeth are broad and molar-like and help it to cope with heavily armoured prey such as crabs. It can swallow small crabs whole, but breaks up bigger ones first by twisting, tugging, and thrashing around. The Chain Moray Eel is a member of a large family of moray eels (Muraenidae) that live on reefs throughout the tropics. Most species of moray eel are nocturnal but the Chain Moray Eel is usually active during the day.

VERSATILE HUNTER

Most species of moray eel spend the day in holes in a reef with just their heads sticking out, emerging at dusk to hunt. They rely on their excellent sense of smell to find fish resting between corals and rocks. Unusually, the Chain Moray Eel also hunts over rocky shores and reefs at low tide during the day. It uses its keen eyesight to search for fish and crustaceans in crevices and holes and, when it has located its prey, it strikes, rather like a snake. Other moray eels will sometimes strike at passing prey from their holes during the day.

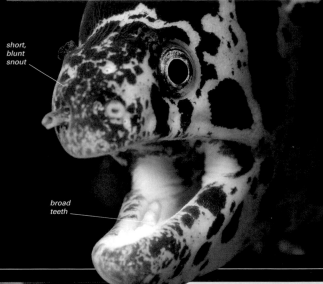

short, blunt snout

broad teeth

ORDER ANGUILLIFORMES

Ribbon Eel
Rhinomuraena quaesita

LENGTH	Up to 1.3m (4¼ft)
WEIGHT	Not recorded
DEPTH	1–60m (3–200ft)

DISTRIBUTION Tropical reefs of Indian and Pacific oceans

Unlike most other eels, Ribbon Eels change colour and sex during their life. Juveniles are nearly black with a yellow dorsal fin. As they mature, the black becomes bright blue and the snout and lower jaw turn yellow. This is the male colour stage. When they reach a body length of about 1.3m (4¼ft), the males turn yellow and become fully functional females, which lay eggs. Ribbon Eels live on coral reefs, mostly hiding in crevices. They have leaf-like nostril flaps, which sense vibrations in the water.

ORDER ANGUILLIFORMES

Conger Eel
Conger conger

LENGTH	Up to 3m (10ft)
WEIGHT	Up to 110kg (240lb)
DEPTH	0–500m (1,600ft)

DISTRIBUTION Temperate waters of northeastern Atlantic and Mediterranean

The large, grey head of a Conger Eel sticking out of a hole in a shipwreck is a familiar sight to many divers. Like their relatives the moray eels, Conger Eels hide in holes and crevices in rocky reefs during the day, only emerging at night to hunt for fish, crustaceans, and cuttlefish. This snake-like fish has a powerful body with smooth skin, no scales, and a pointed tail. A single dorsal fin runs along the back, starting a short distance behind the head, continuing around the tail, and ending halfway along the belly.

In the summer, adult Conger Eels migrate into deep water in the Mediterranean and Atlantic to spawn and then die. The female lays 3–8 million eggs, which hatch into long, thin larvae that slowly drift back inshore, where they grow into juvenile eels. They take 5–15 years to reach sexual maturity.

The Conger Eel is a good food fish and is caught in large numbers by anglers, but it sometimes manages to use its strength to escape with the bait.

ORDER ANGUILLIFORMES

Slender Snipe Eel
Nemichthys scolopaceus

LENGTH	Up to 1.3m (4¼ft)
WEIGHT	Not recorded
DEPTH	90–2,000m (300–6,600ft)

DISTRIBUTION Temperate and tropical seas worldwide

This long, slender, deep-sea eel has remarkable jaws, shaped like a bird's bill, with the ends turned out so that they can never fully close. It spends its life drifting in midwater, catching small crustaceans to eat. When the males mature and are ready to spawn, their jaws shorten, they lose all their teeth, and their front nostrils grow into large tubes. This probably enhances their sense of smell, helping them to find mature females. Little else is known of the Slender Snipe Eel's lifestyle as it is rarely caught.

Spotted Garden Eel
Heteroconger hassi

LENGTH Up to 40cm (16in)
WEIGHT Not recorded
DEPTH 7–45m (23–150ft)

DISTRIBUTION Red Sea and tropical waters of Indian Ocean and western Pacific

These eels spend their lives swaying gracefully to and fro with their heads up in the water and their tails in their sandy burrows. Several hundred fish live together in a colony, or "garden", looking like evenly spaced plants blowing in the breeze. Garden eels are much slimmer than their close relatives, the conger eels. They are only about 14mm (½in) in diameter and have very small pectoral fins. The Spotted Garden Eel usually has two large dark spots behind the head as well as many tiny ones all over the

body. It has an upturned mouth that is designed to pick tiny planktonic animals from the water as the current flows by. Colonies of these eels occur only on sandy slopes that are exposed to currents but sheltered from waves. When danger threatens, the eels sink back down into their burrows, using their tails as an anchor until only their small heads and eyes are visible. They are very difficult to photograph

underwater because they are able to detect the vibrations from a scuba diver's air bubbles and will disappear when they are approached.

Spotted Garden Eels stay in their burrows even when spawning. Neighbouring males and females reach across and entwine their bodies before releasing eggs and sperm. Mixed colonies of Spotted and Whitespotted garden eels sometimes occur.

Banded Snake Eel
Myrichthys colubrinus

LENGTH Up to 97cm (38in)
WEIGHT Not recorded
DEPTH Shallow water

DISTRIBUTION Tropical waters of Indian Ocean and western Pacific

Cleverly disguised to look like the venomous Yellow-lipped Sea Krait, the Banded Snake Eel is avoided by most predators. This allows it to hunt safely over sand flats and seagrass beds near coral reefs for small fish and crustaceans. Most individuals of this species are banded with broad black and white bands, but in some areas these eels have dark blotches between the bands. This colour variant may

eventually be identified as a different species. The Banded Snake Eel has a pointed head with a pair of large tubular nostrils on the upper jaw that point downwards. This arrangement gives the fish an excellent sense of smell that allows it to seek out prey hidden beneath the sand surface.

With no fins except for very small pectoral fins, the Banded Snake Eel swims by undulating its long body. When not hunting, it buries itself in the sand using the hard, pointed tip of its tail to burrow in tail-first. These fish are most active by night. They tend to remain in their burrows during the day and so are not often seen by divers.

The Banded Snake Eel belongs to a large family (Ophichthidae) which includes around 250 snake and worm eels, most of which burrow into sand and mud. All the members of this family have flattened transparent, leaf-like (leptocephalus) larvae.

Gulper Eel
Saccopharynx lavenbergi

LENGTH Up to 1.5m (5ft)
WEIGHT Not recorded
DEPTH 2,000–3,000m (6,600–9,800ft)

DISTRIBUTION Deep waters of eastern Pacific, from California to Peru

The Gulper Eel is best known for its ability to swallow prey as large as itself. This fish has a small head and tiny eyes but enormous jaws. Its mouth and throat can be hugely distended to engulf its prey and the teeth can be depressed backwards. Its stomach can be similarly extended to accommodate its gargantuan meals. The body ends in a luminous organ on a long, whip-like tail. This feature may be used as a lure or a decoy this has yet to be confirmed as no-one has been able to observe these deep-sea fish in the wild. The Gulper Eel has planktonic eggs that develop into long, thin larvae, like those of its shallow-water relatives, and it probably dies after spawning.

OCEAN LIFE

ORDER CLUPEIFORMES

Atlantic Herring

Clupea harengus

LENGTH	Up to 45cm (18in)
WEIGHT	Up to 1kg (2¼lb)
DEPTH	0–200m (0–650ft)

DISTRIBUTION North Atlantic, North Sea, and Baltic Sea

Until the middle of the 20th century, the Atlantic Herring was the mainstay of many fishing communities bordering the North Sea and north Atlantic. Along the East Anglian coast of Great Britain, the fish were known as Silver Darlings. In the 20th century, excessive fishing using new techniques led to a steep decline in stocks. Today, the stocks are managed, but they are still under pressure.

The Atlantic Herring feeds on plankton, coming to the surface at night after spending the day in deeper water. It lives in large shoals, and across its range the species is divided into distinct local races, which differ from each other in size and behaviour. Each race has several traditional spawning grounds. The females produce up to 40,000 eggs each, which form a thick mat on the sea bed.

ORDER CLUPEIFORMES

Peruvian Anchoveta

Engraulis ringens

LENGTH	Up to 20cm (8in)
WEIGHT	Up to 25g (1oz)
DEPTH	3–80m (10–260ft)

DISTRIBUTION West coast of South America and southeastern Pacific

The distribution of this tiny, silvery relative of the herring depends on the yearly extent of the Peruvian Current. This cold, deep current comes to the surface along the west coast of South America, bringing rich supplies of nutrients with it. Enormous shoals of Anchoveta feed on the plankton blooms triggered by the increase in nutrients. The fish shoal within about 80km (50 miles) of the coast, and many local people depend on them, as do many birds, including pelicans.

ORDER CLUPEIFORMES

South American Pilchard

Sardinops sagax

LENGTH	Up to 40cm (16in)
WEIGHT	Up to 485g (17½oz)
DEPTH	0–200m (0–650ft)

DISTRIBUTION West coast of South America and southeastern Pacific

Enormous shoals of these fish, made up of millions of individuals, were once found, but excessive fishing has reduced their numbers greatly.

Pilchards are an important food fish and are also used to produce oil and fish meal. The South American Pilchard may, in fact, be the same species as the California Pilchard, and, worldwide, all pilchard species are very similar. These silvery, medium-sized fish are blue-green on the back and have a series of black marks along the sides.

ORDER CLUPEIFORMES

Allis Shad

Alosa alosa

LENGTH	Up to 83cm (33in)
WEIGHT	Up to 4kg (9lb)
DEPTH	0–5m (16ft)

DISTRIBUTION Temperate waters of northeastern Atlantic and Mediterranean Sea

The Allis Shad is a silvery fish belonging to the herring family (Clupeidae) and is one of the few that enter fresh water. During April and May, mature adults migrate into rivers to spawn, swimming up to 800km (500 miles) upstream. In some parts of its range, the species is known as the May Fish. Its streamlined body is covered by large circular scales that form a keel under the belly, and it has a single dorsal fin. It is now very rare over much of its range.

ORDER GONORYNCHIFORMES

Milkfish

Chanos chanos

LENGTH	Up to 1.8m (6ft)
WEIGHT	Up to 14kg (31lb)
DEPTH	0–30m (0–100ft)

DISTRIBUTION Tropical and subtropical waters of Indian and Pacific oceans

The Milkfish is an elegant silvery fish with a streamlined body and a large, deeply forked tail. It is an important food fish in much of Southeast Asia and is extensively farmed. It feeds on plankton, soft algae, cyanobacteria, and small invertebrates. It is easy to keep in captivity as it is able to tolerate a wide range of salinity. Mature fish spawn in the sea, and the eggs and larvae drift inshore. Juveniles swim into estuaries and mangroves, where there are fewer predators, returning to the sea as they mature.

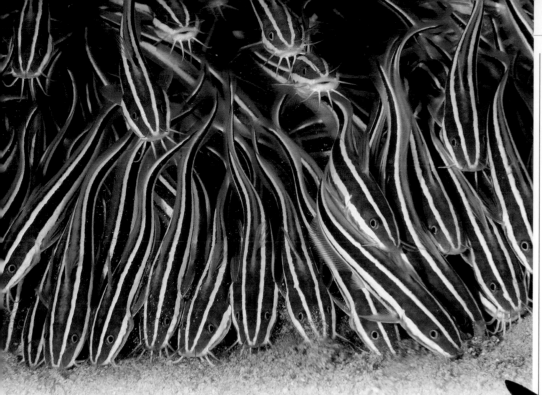

Gafftopsail Sea Catfish

Bagre marinus

LENGTH	Up to 70cm (28in)
WEIGHT	Up to 4.5kg (10lb)
DEPTH	To 50m (160ft)

DISTRIBUTION Gulf of Mexico, Caribbean Sea, and subtropical waters of western Atlantic

The most conspicuous feature of this silvery catfish is the pair of very long mouth barbels that extend back almost to the end of the pectoral fins. It has another pair of short barbels under the chin. The first rays of the large dorsal fin and the pectoral fins are drawn out as long flat filaments and these fins also have a venomous serrated spine. When threatened, this catfish erects its dorsal fin and spreads out its pectoral fins like the sails of a yacht.

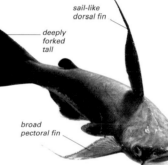

sail-like dorsal fin

deeply forked tail

broad pectoral fin

Striped Catfish

Plotosus lineatus

LENGTH	Up to 32cm (13in)
WEIGHT	Not recorded
DEPTH	1–60m (3–200ft)

DISTRIBUTION Red Sea and tropical waters in Indian and Pacific oceans

The juveniles of this distinctive black-and-white striped catfish of the family Plotosidae stay together in dense, ball-shaped shoals and are often seen by divers over coral reefs. Adults live on their own or in small groups, but are well protected by a venomous, serrated spine in front of the first dorsal fin and each of the pectoral fins. A sting from an adult Striped Catfish can be dangerous to humans and is very occasionally fatal. These fish hunt at night, using four pairs of sensory barbels around the mouth to find worms, crustaceans, and molluscs hidden in the sand. During the day, they hide among rocks. Plotosids are the only catfish found in coral reefs. This species also ventures along open coasts and into estuaries. It spawns in the summer months. Male Striped Catfish build nests in shallow, rocky areas and guard the eggs for about ten days. The larvae are planktonic.

European Smelt

Osmerus eperlanus

LENGTH	45cm (18in)
WEIGHT	Not recorded
DEPTH	To 50m (160ft)

DISTRIBUTION Temperate waters of northeastern Atlantic and Baltic Sea

The European Smelt is a small relation of salmon and trout and, like them, has a dorsal fin and an adipose, the latter of which is small and fatty. The name derives from the fact that, when fresh, this fish has a strong smell, which is reminiscent of cucumber. Adults swim in shoals in inshore waters, hunting small crustaceans and fish. They migrate up rivers to spawn, and the young fish are common in sheltered estuaries such as the Wash in southeast England (see p.134).

Barrel-eye

Opisthoproctus soleatus

LENGTH	Up to 10cm (4in)
WEIGHT	Not recorded
DEPTH	300–800m (1,000–2,600ft)

DISTRIBUTION Tropical and subtropical waters worldwide

Many fish that live in the twilight zone (see p.174), including the Barrel-eye, have large eyes to make full use of what little light is available. As well as being large, the eyes of this species are tubular and point upwards. This arrangement probably helps the Barrel-eye to stalk other fish from below. Looking up, it is likely that it can pick out the silhouette of its prey or spot fish with bioluminescent patches on their undersides.

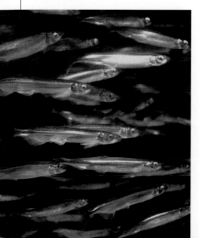

Capelin

Mallotus villosus

LENGTH	Up to 25cm (10in)
WEIGHT	Up to 52g (1⁴⁄₅oz)
DEPTH	0–300m (0–1,000ft)

DISTRIBUTION North Pacific, north Atlantic, and Arctic Ocean

This small, silvery relative of salmon forms large shoals in cold and Arctic waters and is a vital food source for sea birds and marine mammals. The breeding success of some sea bird colonies has been linked to the abundance of Capelin, and this in turn depends on environmental factors and exploitation by fishing. It is a major food source for Inuit peoples. Capelin are slim fish, with an olive-green back fading into silvery white on the sides. Shoals of this fish swim along with their mouths open, sieving out plankton, which is caught on their modified gills. While this is their main source of food, they also eat worms and small fish. In spring, the schools move inshore, the males arriving first and waiting for the females. The males develop a band of modified scales along their sides and use these to massage the female, stimulating her to lay her eggs in the sand.

TIDAL BREEDING

Capelin eggs make a good meal for many invertebrates and fish. To protect their eggs, large numbers of adult Capelin swim into very shallow water at high tide and spawn on sandy beaches just below the tideline. Each female produces about 60,000 reddish, sticky eggs, which lie in the sand. When the eggs hatch after about 15 days, the larvae are washed out of the sand by the incoming tide and then swept out to sea on the outgoing tide.

OCEAN LIFE

ORDER SALMONIFORMES

Atlantic Salmon
Salmo salar

LENGTH	Up to 1.5m (5ft)
WEIGHT	Up to 46kg (100lb)
DEPTH	Mostly surface waters

DISTRIBUTION Temperate and cold waters of north Atlantic and adjacent rivers

While most marine fish would quickly die in fresh water, the Atlantic Salmon can move easily between river and sea. Fish with this ability are called anadromous. Designed for long-distance swimming, this fish has a powerful, streamlined body and a large tail. During their spawning runs, Atlantic Salmon swim against strong river currents and leap up waterfalls to reach their spawning grounds. Before spawning, the salmon roam the north Atlantic for several years feeding on other fish. The Atlantic Salmon is highly prized as a game fish but wild salmon are becoming increasingly rare.

HUMAN IMPACT

SALMON FARMING

Floating fish farms rearing Atlantic Salmon are common in Scottish sea lochs and Norwegian fiords. Young salmon (fry) are hatched in hatcheries and reared in submerged nets. Environmental concerns about fish farms focus on the chemicals used to kill fish lice, and the effects of high levels of nutrients in the water.

ORDER SALMONIFORMES

Coho Salmon
Oncorhynchus kisutch

LENGTH	Up to 1m (3ft)
WEIGHT	Up to 15kg (33lb)
DEPTH	0–250m (0–820ft)

DISTRIBUTION Temperate and cold waters of north Pacific and adjacent rivers

Like most other salmon, the Coho Salmon is a fast, streamlined predator with excellent eyesight for spotting its prey. This makes it a challenging game fish for anglers. When ready to breed, mature fish find their way from the ocean back to the same river in which they were born. While swimming upstream, they develop bright red sides and a green head and back. When they reach the shallow waters at the river's head, the females dig a nest, which is called a redd, in the gravel of the riverbed and lay their sticky eggs while the male fertilizes them. After spawning, the adults die and their bodies provide a feast for scavenging bears and other animals.

ORDER SALMONIFORMES

Arctic Char
Salvelinus alpinus

LENGTH	Up to 1m (3ft)
WEIGHT	Up to 15kg (33lb)
DEPTH	0–70m (0–230ft)

DISTRIBUTION Arctic Ocean and northern freshwater rivers and lakes

Arctic Char are adapted for life in cold, oxygen-rich water and cannot tolerate warm or polluted water. There are two physiological races: a migratory form that lives in the sea but spawns in rivers, and a land-locked lake form. Migratory char grow to at least a metre (3ft) long and are regarded as excellent game fish. Shortly after the last ice age, they ranged much further south but are now restricted to Arctic waters. The char that live in mountain lakes are relics of this period.

ORDER STOMIIFORMES

Sloane's Viperfish
Chauliodus sloani

LENGTH	Up to 35cm (14in)
WEIGHT	Up to 30g (1oz)
DEPTH	475–1,800m (1,600–6,000ft)

DISTRIBUTION Tropical and temperate waters worldwide

Deep-water fish are some of the most bizarre of all fish and Sloane's Viperfish is no exception. At one end of its slender body it has a large head with huge, barbed teeth, while at the other it has a tiny forked tail. Rows of photophores run along the sides and belly and light the fish up like a night-flying aeroplane. During the day, it stays in deep water but at night it migrates upwards to feed where prey is more abundant. The single dorsal fin, just behind the head, has a very long first ray that can be arched over the head and may help entice prey within reach. Sloane's Viperfish spawns throughout the year. It is one of nine species of viperfish, all living at depth.

ORDER STOMIIFORMES

Pacific Blackdragon
Idiacanthus antrostomus

LENGTH	Up to 38cm (15in)
WEIGHT	Up to 55g (2oz)
DEPTH	200–1,000m (650–3,300ft)

DISTRIBUTION Deep, tropical and temperate waters of eastern Pacific

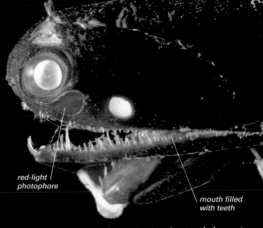

red-light photophore

mouth filled with teeth

The Pacific Blackdragon haunts the depths of the ocean, its black snake-like body lit up by photophores along its belly. When it opens its mouth, it reveals a set of long, dagger-sharp teeth. Hanging off the lower jaw is a barbel tipped by a glowing lure that can be moved to entice prey to venture within reach.

The Pacific Blackdragon is black on the inside as well as the outside, its black stomach preventing light from swallowed bioluminescent prey shining out. Male Pacific Blackdragons are only about a quarter the size of the females. In the closely related species *Idiacanthus fascicola*, the young fish are similar in shape to the adults, but their eyes stick out on very long stalks. The stalks are absorbed as the fish grows and the eyes eventually come to lie in their sockets.

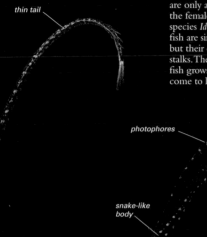

thin tail

photophores

snake-like body

barbel with lure

ORDER STOMIIFORMES

Stoplight Loosejaw
Malacosteus niger

LENGTH	Up to 24cm (9½in)
WEIGHT	Not recorded
DEPTH	1,000–4,000m (3,300–13,000ft)

DISTRIBUTION Deep tropical and temperate waters worldwide

Like many other deep-sea fish, the Stoplight Loosejaw is black, relatively small, and has a large mouth. However, it is unique in that it has no floor to its mouth, hence its name. Instead, a ribbon of muscle that joins the gill basket and the lower jaw contracts to shut the mouth. This arrangement may allow the fish a wider gape and a faster strike at prey. This fish is also a specialist in light production. It has two large photophores under each eye, one that produces normal blue-green bioluminescence and the other red. No natural red light reaches these depths so most deep-sea creatures cannot see it. The red bioluminescence reflects well off a red animal, such as a shrimp, but the shrimp will be unaware that it has been spotlighted.

ORDER STOMIIFORMES

Lovely Hatchetfish
Argyropelecus aculeatus

LENGTH	Up to 8cm (3in)
WEIGHT	Not recorded
DEPTH	100–600m (330–2,000ft)

DISTRIBUTION Tropical and temperate waters worldwide

An expert at hiding from predators, the Hatchetfish's silvery coloration and use of bioluminescence conceals it against the downwelling light. They are also so thin that they are difficult to see head-on. This fish lives at medium depths and has large bulging eyes to make best use of what little light there is. At dusk, it rises up to 100–300m (330–1,000ft) to feed on small planktonic animals.

ORDER AULOPIFORMES

Tripodfish
Bathypterois grallator

LENGTH	Up to 37cm (15in)
WEIGHT	Not recorded
DEPTH	875–3,500m (2,900–11,500ft)

DISTRIBUTION Deep waters of Atlantic, Pacific, and Indian oceans

The deep ocean floor where the Tripodfish lives consists largely of soft mud. So, to prevent itself from sinking into the ooze while lying in wait for its prey, this fish perches on a tripod made from elongated rays of its pelvic and caudal fins. Facing into the current, it waits for small crustaceans to drift within reach, catching them in its mouth, which has a large gape. The Tripodfish has very small eyes and is thought to detect its prey by feeling for tiny vibrations in the water.

ORDER AULOPIFORMES

Reef Lizardfish
Synodus variegatus

LENGTH	20–40cm (8–16in)
WEIGHT	Not recorded
DEPTH	5–90m (16–295ft)

DISTRIBUTION Tropical reefs in Red Sea, Indian Ocean, and western Pacific

The Reef Lizardfish habitually perches on the tops of rocks and corals, propped up on its long pelvic fins. From such vantage points, it keeps a lookout for passing shoals of fish, darting out and seizing one with its rows of sharp teeth. Its large mouth allows it to swallow quite big fish (as shown in the photograph). A variable blotchy brown and red coloration camouflages the Reef Lizardfish, hiding it from larger predators. It can also bury itself in patches of sand, leaving only its head and eyes showing. Confident of its disguise, this fish will remain completely still and allow divers to approach to within a few centimetres before darting away to a new perch. It is caught and eaten by reef fishermen.

ORDER MYCTOPHIFORMES

Spotted Lanternfish
Myctophum punctatum

LENGTH	Up to 11cm (4¼in)
WEIGHT	Not recorded
DEPTH	0–1,000m (0–3,300ft)

DISTRIBUTION Deep waters of north Atlantic and Mediterranean

The Spotted Lanternfish is one of over 250 species of lanternfish found in the world's oceans. Lanternfish are rather unprepossessing, small spindle-shaped fish with large eyes. However, in spite of their drab appearance they can put on an unrivalled display of light from an array of photophores along their sides and belly. In some species, males and females have different patterns of photophores and this helps them to find each other in the dark depths. Photophore patterns also differ between species.

Large shoals of Spotted Lanternfish are common in the north Atlantic. Along with other lanternfish, it is an important food source for larger fish, sea birds, and marine mammals. During the day it stays in deep water between about 250–750m (800–2,500ft), but at night it swims up to within about 100m (330ft) or even right to the surface, where it feeds on planktonic crustaceans and fish fry.

ORDER LAMPRIFORMES

Opah
Lampris guttatus

LENGTH	Up to 2m (6ft)
WEIGHT	50–275kg (110–600lb)
DEPTH	100–400m (330–1,300ft)

DISTRIBUTION Tropical, subtropical, and temperate waters worldwide

Roaming the oceans worldwide, the Opah leads a nomadic existence. Shaped like a gigantic oval dinner plate, this colourful fish is a steely blue and green with silvery spots and red fins. Although it is toothless, the Opah is an efficient hunter, catching squid and small fish. Rather than using its tail to swim, like most fish, it flies through the water by beating its long, narrow pectoral fins like a pair of wings.

Opah regularly reach a weight of 50kg (110lb), although specimens as heavy as 270kg (600lb) have been reported. They spawn in the spring, laying eggs midwater, which hatch into larvae after 21 days. Also known as the Moonfish, the Opah is a valuable food fish in the Hawaiian Islands and on the west coast of mainland USA. It is caught on long lines and with gill nets.

ORDER LAMPRIFORMES

Oarfish
Regalecus glesne

LENGTH	Up to 11m (36ft)
WEIGHT	Up to 270kg (600lb)
DEPTH	0–1,000m (0–3,300ft)

DISTRIBUTION Tropical, subtropical, and temperate waters worldwide

At up to 11m (36ft) in length, the Oarfish is the longest bony fish known to science and is thought to be responsible for many sea serpent legends. Its bizarre appearance is enhanced by a crest of long red rays on its short, bluish head, which are followed by a bright red dorsal fin that runs the length of its silvery body. Its name comes from the pelvic fins, both of which extend as a single, long ray ending in an expanded tip, which looks like the blade of an oar. In the open ocean, the Oarfish drifts in the currents, feeding on other fish and squid, its great length protecting it from most predators. Although it lives in tropical and temperate waters worldwide, the Oarfish is rarely caught or seen alive, so little is known about its behaviour. It was first photographed underwater in 1997 in the Bahamas.

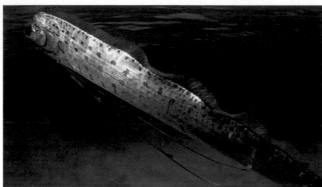

ORDER GADIFORMES

Bib
Trisopterus luscus

LENGTH	Up to 46cm (18in)
WEIGHT	Up to 2.5kg (5½lb)
DEPTH	3–100m (10–330ft)

DISTRIBUTION Temperate waters of northeastern Atlantic and western Mediterranean

Divers often see shoals of striped Bib around rocky reefs and shipwrecks. These are usually younger fish or adults that have moved inshore to spawn. Large old fish often lose their banded pattern and become very dark. The Bib has a much deeper body than most of its relatives in the family Gadidae. A long chin barbel and long pelvic fins help it to find crustaceans, molluscs, and worms to eat.

ORDER GADIFORMES

Atlantic Cod
Gadus morhua

LENGTH	Up to 2m (6ft)
WEIGHT	Up to 90kg (200lb)
DEPTH	0–600m (0–2,000ft)

DISTRIBUTION Temperate and cold waters of north Atlantic

The Atlantic Cod is a powerful, heavily built fish with a large head, an overhanging upper jaw, and a single long chin barbel. It has small, elongated scales. The coloration varies from reddish, especially in young fish, to a mottled brown with a conspicuous white lateral line. The Atlantic Cod is a shoal-forming fish, living in water over the continental shelf, and usually feeding at 30–80m (100–250ft) above areas of flat mud or sand. Adults migrate to established breeding grounds to spawn, usually in the early spring, with each female releasing several million eggs into the water.

Atlantic Cod can live for 60 years and mature fish can reach a weight of over 90kg (200lb), but sophisticated modern fishing techniques mean that most cod today are caught long before they reach this age and weight. The average weight is now 11kg (24lb) and specimens over 15kg (33lb) are rare. It is still one of the world's most commercially important species, however.

square-ended tail

FINS AND SCALES
The Atlantic Cod has three dorsal fins and two anal fins. Its small scales have growth rings, which can be counted to give its age.

HUMAN IMPACT

OVERFISHING

Stocks of Atlantic Cod were once thought to be inexhaustible, but numbers of this once-abundant fish have declined drastically over most of its range. Cod exist as a number of discrete populations, or stocks, that spawn in specific areas in water about 200m (660ft) deep. Important stocks include those in the North Sea, northeast Arctic Ocean, Labrador Sea, and the Grand Banks. The latter stock collapsed due to overfishing some years ago and has still not fully recovered.

FORAGING FOR FOOD
Atlantic Cod feed both in midwater and on the sea bed. They eat shoaling fish, such as herring, and also crustaceans, worms, and molluscs.

ORDER GADIFORMES

Shore Rockling

Gaidropsarus mediterraneus

LENGTH	Up to 50cm (20in)
WEIGHT	Up to 1kg (2¼ lb)
DEPTH	0–450m (0–1,500ft)

DISTRIBUTION Temperate waters of northeastern Atlantic and Mediterranean

Rockling are eel-like in appearance, with two dorsal fins. The first of these is a fringe of short rays that ripple constantly. The Shore Rockling can be found in rock pools, where it uses its mouth barbels to find food. Most are dark brown, but some are paler.

ORDER GADIFORMES

Torsk

Brosme brosme

LENGTH	Up to 1.2m (4ft)
WEIGHT	12–30kg (26–66lb)
DEPTH	20–1,000m (65–3,300ft)

DISTRIBUTION Temperate and cold waters of north Atlantic

This heavily built member of the order Gadiformes lurks among rocks and pebbles in deep water offshore, where it searches for crustaceans and molluscs. It has thick lips, a long chin barbel, and a long dorsal and anal fin, each edged in white. In summer, two to three million eggs are laid, which float and develop near the surface. This species can live for 20 years. Torsk is fished commercially, especially off Norway, using trawls and lines, and it is also caught by anglers.

ORDER GADIFORMES

Pacific Grenadier

Coryphaenoides acrolepis

LENGTH	Up to 1m (3ft)
WEIGHT	Up to 3kg (6½ lb)
DEPTH	300–3,700m (1,000–12,000ft)

DISTRIBUTION Deep, temperate waters of north Pacific

The Pacific Grenadier is one of about 300 different species of grenadiers that are found just off continental shelves and are abundant in every ocean. Grenadiers are also known as rat-tails because they have a large, bulbous head with big eyes, a sharp snout, and a long, scaly tail. The Pacific Grenadier is dark brown with a tall dorsal fin. Another low fin runs along the back and all the way round the tail.

This species spends most of its time near the sea bed searching for food but it sometimes swims up into midwater, where it can catch squid, prawns, and small fish.

ORDER OPHIDIIFORMES

Pearlfish

Carapus acus

LENGTH	Up to 21cm (8in)
WEIGHT	Not recorded
DEPTH	To 100m (330ft)

DISTRIBUTION Mediterranean; occasionally found in subtropical waters of eastern Atlantic

The adult Pearlfish has a most unusual home as it lives inside the body cavity of sea cucumbers. To allow it to slip in and out of its host easily, it has an eel-like body, no pelvic fins, and no scales. It is a silvery-white colour with reddish markings. At night, the Pearlfish may swim out of the sea cucumber's anus to go hunting for invertebrates to eat, returning to the body cavity tail first. However, the Pearlfish may also eat the gonads and other organs of its host.

ORDER OPHIDIIFORMES

Spotted Cusk-eel

Chilara taylori

LENGTH	Up to 37cm (15in)
WEIGHT	Not recorded
DEPTH	0–280m (0–900ft)

DISTRIBUTION Temperate and subtropical waters of eastern Pacific

As it is a favourite food of sea lions, cormorants, and other diving birds, the Spotted Cusk-eel is most active at night or on gloomy, sunless days. If danger threatens, this eel-shaped fish can quickly slip between rocky rubble or bury itself tail-first in sand or mud. Unlike true eels, it has scales and pelvic fins. The latter are reduced to one split ray set very far forward under the head. Its eggs are laid in open water and hatch into larvae that live close to the surface. These develop into juveniles that drift for an extended period before settling down to a sea-bed existence.

While the Spotted Cusk-eel lives in shallow water, one of its close relatives, the Basketweave Cusk-eel, has been found over 8,000m (26,000ft) deep in the abyssal zone (see p.182), the greatest depth for any fish.

ORDER BATRACHOIDIFORMES

Oyster Toadfish

Opsanus tau

LENGTH	Up to 43cm (17in)
WEIGHT	Up to 2.2kg (4¼ lb)
DEPTH	Not recorded

DISTRIBUTION Temperate and subtropical waters of northwestern Atlantic

Some people would consider the Oyster Toadfish an ugly animal, with its flat head, wide, toad-like mouth, and thick lips. It also has tassels around its chin, prominent eyes, and two dorsal fins, the first of which is spiny. Its shape and coloration provide camouflage in its home under rocks and debris. This hardy fish tolerates dirty and rubbish-strewn water and is often found under jetties. It has been reared in captivity for use in experiments. It also does well in aquariums and is a popular game fish.

COURTSHIP CALLS

People living in houseboats along the east coast of the USA are sometimes kept awake at night during April to October by loud grunting noises. The culprits are male Oyster Toadfish calling to attract females to lay their eggs in nests dug under rocks. The male makes these noises by vibrating the walls of his swim bladder using special muscles. The swim-bladder wall acts like the skin over a drum. The male guards the eggs until they hatch after about four weeks.

sensory
hairs

large
mouth

ORDER LOPHIIFORMES

Hairy Angler
Caulophryne jordani

LENGTH Females up to 20cm (8in); males not recorded, but tiny	
WEIGHT Not recorded	
DEPTH 100–1,500m (330–5,000ft)	

DISTRIBUTION Deep water worldwide

Anglerfish include some of the most bizarrely shaped fish in the ocean and the Hairy Angler certainly fits into this category. It has a huge mouth, tiny eyes, and large dorsal and anal fins

with very long projecting fin rays. It is also covered in sensory hairs, giving it a dishevelled appearance. Like most anglerfish, it has a moveable lure on top of the head that is formed from the first spine of the dorsal fin. The biology of the Hairy Angler is poorly known as only a few specimens have ever been captured. However, in other deep-sea anglerfish, this lure is is used to attract prey within reach. The fish then opens its mouth and creates a sudden, strong inward suction current. The prey is engulfed within a fraction of a second. Food is scarce in the deep sea and anglerfish living here usually have extra-large mouths and expandable stomachs that allow them

to swallow prey as big or bigger than themselves. The Hairy Angler belongs to the family Caulophrynidae, also known as fanfins. The males of fish in this family are tiny and do not have lures. They live as parasites on the females when they are adult. This

species is often difficult to identify because so few are caught and they are often damaged from contact with nets and from changes in pressure as they are brought to the surface.

ORDER LOPHIIFORMES

Polka-dot Batfish
Ogcocephalus radiatus

LENGTH Up to 38cm (15in)	
WEIGHT Not recorded	
DEPTH 0–70m (0–230ft)	

DIS... Subtropical waters of western Atlantic and Gulf of Mexico

Fish of the family Ogcocephalidae, to which the Polka-dot Batfish belongs, are among the most oddly shaped of the anglerfish. They prop themselves up on paired pectoral and pelvic fins that enable them to walk over the sea bed in search of worms, crustaceans, and fish. Although the Polka-dot Batfish has a fishing lure, this is very short and evidence suggests it may secrete an odour that attracts potential prey. A hard, spiny skin protects the fish from predators, but they are so sluggish that divers can pick them

ORDER LOPHIIFORMES

Deep-sea Angler
Bufoceratias wedli

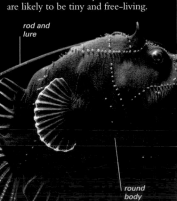

LENGTH Females up to 25cm (10in); males not recorded	
WEIGHT Not recorded	
DEPTH 300–1,750m (1,000–5,700ft)	

DISTRIBUTION Gulf of Mexico, Caribbean Sea, and Atlantic

Living in the deep sea, this small, dark-coloured anglerfish has a round body, delicate fins, and a luminescent lure at the end of a long rod called an illicium. A second, much smaller rod on the head is often hidden from view. It has a weak skeleton and small muscles that make it relatively light

and able to float more easily. It has no need to swim much as it lures its prey within reach. Female Deep-Sea Anglers have been caught undamaged by research submarines using a piece of equipment called a slurp gun that sucks animals into a container. The fish have then been photographed alive. Males have not yet been seen but are likely to be tiny and free-living.

rod and
lure

round
body

Coffinfish
Chaunax endeavouri

LENGTH	Up to 22cm (9in)
WEIGHT	Not recorded
DEPTH	50–300m (160–1,000ft)

DISTRIBUTION Temperate waters of southwestern Pacific, off east coast of Australia

The Coffinfish resembles a pink balloon covered in tiny spines and can make itself look bigger by inflating its body. It belongs to a family of anglerfish called Chaunacidae, or seatoads, that have large, flabby bodies and loose skin. Like other anglerfish, it has a lure but this is very small and can be hidden in a depression on the snout. Little is known of its life history, but it spends most of its time lying quietly on the bottom in muddy areas. It is usually found in deep water on the continental shelf and slope, but has also been found in water as shallow as 50m (165ft).

Sargassumfish
Histrio histrio

LENGTH	Up to 20cm (8in)
WEIGHT	Not recorded
DEPTH	About 0–11m (0–36ft)

DISTRIBUTION Tropical and subtropical seas worldwide; not recorded in eastern Pacific

This unusual frogfish (family Anternnariidae) lives in floating rafts of sargassum seaweed. It uses its prehensile, leg-like pectoral fins to

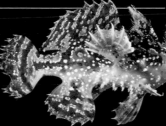

clasp clumps of weed and scramble around the rafts. With its skin tassels, mottled pattern, and variable colour, the Sargassumfish is well camouflaged and able to lure small fish and shrimps within striking range. If threatened, it can scramble onto the top of the seaweed raft. These fish are sometimes washed ashore with their rafts.

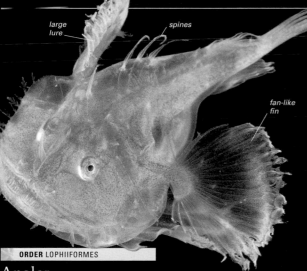

large lure / spines / fan-like fin

Angler
Lophius piscatorius

LENGTH	Up to 2m (6ft)
WEIGHT	Up to 57kg (125lb)
DEPTH	20–1,000m (65–3,300ft)

DISTRIBUTION Northeastern Atlantic south to West Africa, Mediterranean, and Black Sea

The Angler has a head like a flattened football fringed by a camouflage of seaweed-shaped flaps of skin, and a wide, flattened body that tapers towards the tail. Its darkly marbled greenish-brown skin also helps the Angler to blend into the sediment of the sea floor. It lies patiently on the sea bed, ready to suck in any fish it can entice within range by flicks of the fleshy fishing lure on its dorsal fin. Large Anglers have even been known to lunge up and catch diving birds. The species has well-developed pectoral fins, set on arm-like bases, with sharp "elbows" that allow it to shuffle along over the sea floor. Anglerfish of the genus *Lophius* are also known as Goosefishes or Fishing Frogs. This species is commercially exploited and sold as "monkfish".

Common Blackdevil
Melanocetus johnsonii

LENGTH	Females 18cm (7in); males 3cm (1¼ in)
WEIGHT	Not recorded
DEPTH	To 2,000m (6,600ft)

DISTRIBUTION Deep waters of Atlantic, Pacific, and Indian oceans

This deep-sea anglerfish is also known as the Humpback Angler. The female Common Blackdevil has a huge head and large jaws with very long, dagger-like teeth, which are used to catch prey that may be larger than herself. Her stretchy stomach and loosely attached skin help her to accommodate these huge meals. Although the female is not completely blind, her eyes are tiny and she probably cannot see her prey until she has enticed it within range using her glowing lure. By contrast, the male is tiny and uses his acute sense of smell to find a mate. He has no teeth but hangs on to the female with special hooks on his snout. When she has laid her eggs and he has fertilized them, he swims away, but how long he then lives for is not known. Both male and female juveniles live near the surface, where they feed on small planktonic animals.

glowing lure / long, sharp teeth

Regan's Angler
Haplophryne mollis

LENGTH	Females 8cm (3in); males 2cm (¾ in)
WEIGHT	Not recorded
DEPTH	200–2,000m (650–6,600ft)

DISTRIBUTION Tropical and subtropical deep waters worldwide

This unusual deep-sea anglerfish has unpigmented skin. The female of the species has an almost round body when mature, numerous very small teeth, spines above the eyes and behind the mouth, and a minimal fishing lure that consists of just a small flap on the snout. Like many other deep-sea anglerfish, the males of this species remain very small all their lives and their sole aim in life is to track down a female using their excellent sense of smell and latch onto her using special hooks. Finding a mate in the depths of the ocean is difficult and by keeping the male attached, the female is assured that her eggs will be fertilized. The males eventually turn into parasites, biting into the female's skin. In time their blood supplies fuse and the male then becomes nourished by the female. Up to three males have been found on a single female.

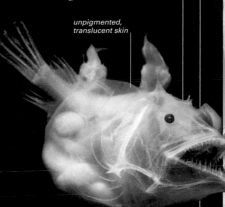

unpigmented, translucent skin

ORDER GOBIESOCIFORMES

Shore Clingfish
Lepadogaster lepadogaster

LENGTH	7cm (3in)
WEIGHT	Not recorded
DEPTH	0–2m (0–6ft)

DISTRIBUTION Temperate waters of northeastern Atlantic, Mediterranean, and Black Sea

Strong waves are no problem to this little fish as it can cling onto rocks with a powerful sucker formed from its pelvic fins. It also has a low-profile body and a flattened, triangular head with a long snout that resembles a duck's bill. This shape allows the fish to slip easily between the rocks and, because it is only a few centimetres long, it may be difficult to spot, but it can be found by turning over rocks and seaweeds and searching in rock pools. The colour of the Shore Clingfish is variable, but it always has two blue spots outlined in brown, red, or black behind its head, and it has a small tentacle in front of each eye. In the spring or summer, females lay clusters of golden yellow eggs on the undersides of rocks on the shore. The eggs are guarded by the parent fish until they hatch.

ORDER BELONIFORMES

Atlantic Flyingfish
Cheilopogon heterurus

LENGTH	Up to 40cm (16in)
WEIGHT	Not recorded
DEPTH	Surface waters

DISTRIBUTION Tropical and warm temperate waters worldwide

Also known as the Mediterranean Flyingfish, this species is distinguished by its very large, wing-like pectoral and pelvic fins. If a predator, such as a tuna, attacks from below, the fish will beat its powerful forked tail rapidly, spread its "wings" at the last moment, and lift clear of the surface away from danger. The fish continues to beat its tail even in mid-flight and it can remain airborne for over 100m (330ft). The Atlantic Flyingfish is edible, but is not commercially exploited.

ORDER BELONIFORMES

Hound Needlefish
Tylosurus crocodiles

LENGTH	Up to 1.5m (5ft)
WEIGHT	Up to 6.5kg (14lb)
DEPTH	0–13m (0–43ft)

DISTRIBUTION Tropical waters over coral reefs worldwide

Rendered almost invisible by its silvery colour and needle-like shape, the Hound Needlefish swims along just beneath the surface, hunting for other fish that also live over coral reefs. Its long, thin snout is shaped like a spear, and it has been known to puncture small boats and cause severe injury to people by shooting up into the air when frightened. Although edible, the Hound Needlefish is not popular as a food fish as it has green-coloured flesh.

ORDER BELONIFORMES

Atlantic Saury
Scomberesox saurus

LENGTH	Up to 50cm (20in)
WEIGHT	Not recorded
DEPTH	0–30m (98ft)

DISTRIBUTION North, northwestern, and eastern Atlantic and Mediterranean

Although not as thin as its needlefish relatives, the Atlantic Saury has a similar narrow body and a long, beak-like snout lined with tiny teeth. The body is clear green above and bright silver on the sides. It has a single dorsal and anal fin, each followed by a series of small finlets. This fish lives in large schools that chase and capture smaller fish and shrimp-like crustaceans while skimming along at the surface. It is fished commercially and is caught by being attracted to bright lights at night.

ORDER BERYCIFORMES

Whitetip Soldierfish
Myripristis vittata

LENGTH	Up to 25cm (10in)
WEIGHT	Not recorded
DEPTH	3–80m (9¾–260ft)

DISTRIBUTION Tropical waters of Indian and Pacific oceans

Soldierfish are nocturnal coral-reef residents that hide in groups in caves and beneath overhangs on steep reefs during the daytime. The Whitetip Soldierfish is red, as are most members of its family (Holocentridae). The leading edges of its median fins are white. At depth, where natural red light does not penetrate, the fish's red colour appears black or grey, providing it with camouflage, especially on the deeper parts of the reef. Like many nocturnal fish, the Whitetip Soldierfish has large eyes, which help it to spot planktonic animals by dim moonlight and then snap them up. It has a short, blunt snout, large scales, and a deeply forked tail. Divers have observed that some individuals in a group of Whitetip Soldierfish often swim along upside down.

ORDER BERYCIFORMES

Pineapplefish
Cleidopus gloriamaris

LENGTH	Up to 22cm (9in)
WEIGHT	Up to 500g (8oz)
DEPTH	3–200m (10–650ft)

DISTRIBUTION Temperate waters of eastern Indian Ocean and southwestern Pacific around Australia

The Pineapplefish is completely encased in armour consisting of large, thick, modified scales studded with spines. Each yellow scale is outlined in black, resembling a segment of pineapple skin. This fish, which lives in dark caves and under ledges on rocky reefs, has a pair of bioluminescent organs on its lower jaw that are hidden by the upper jaw when the mouth is closed. Orange during the day, the organs glow blue-green at night, when they are used to help find prey, such as crustaceans and small fish.

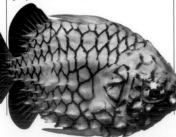

ORDER BERYCIFORMES

Eyelight Fish
Photoblepharon palpebratum

LENGTH	12cm (5in)
WEIGHT	Not recorded
DEPTH	7–25m (23–82ft)

DISTRIBUTION Tropical waters of western and central Pacific

The most characteristic feature of this small fish is the large light organ under each eye. The blue-green light can be turned on and off using a black membrane like an eyelid. These fish are active at night, often feeding in large groups, and use the light to signal to other individuals, startle predators, and find small planktonic animals to feed on. Eyelight Fish are sometimes seen at night by divers on steep reef faces. Daytime sightings are rare, as these fish usually hide in caves during the day.

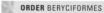

Common Fangtooth

Anoplogaster cornuta

LENGTH	15–18cm (6–7in)
WEIGHT	Not recorded
DEPTH	500–5,000m (1,600–16,000ft)

DISTRIBUTION Deep waters in temperate and tropical waters worldwide

The huge, sabre-like teeth of this deep-water predator are designed to grab and hold onto other fish that may be as big as itself. The teeth are no good for cutting or chewing and so the Common Fangtooth swallows its prey whole, rather like a snake does. Adults are uniformly black or dark brown in colour and can live as deep as 5,000m (16,000ft), but they are most common between 500 and 2,000m (1,600–6,500ft). They hunt by themselves or in small shoals, searching for other fish to eat. Juvenile Common Fangtooths look very different from the adults and were classified as a separate species until 1955. They are light grey in colour and have long spines on the head. They live in water as shallow as 50m (160ft) and feed mainly on crustaceans.

Adult females shed their eggs directly into the sea, where they develop into planktonic larvae. The juveniles take on the adult shape when they are about 8cm (3in) long.

Orange Roughy

Hoplostethus atlanticus

LENGTH	50–75cm (20–30in)
WEIGHT	Up to 7kg (15lb)
DEPTH	900–1,800m (3,000–6,000ft)

DISTRIBUTION North and south Atlantic, Indian Ocean, southwestern and eastern Pacific

This is one of the longest-lived fish species, with individuals having been recorded to reach at least 149 years old. It is a bright, brick-red colour, but appears black in the dark waters in which it lives and this helps to hide it from predators.

The Orange Roughy is a deep-bodied fish with a large head, armed with spines behind the eyes and on the gill covers. The scales on the belly also have sharp spines. Orange Roughy live in deep water in rugged, rocky areas and over steep, rough ground, and have a relatively limited home range.

soft rays

John Dory

Zeus faber

LENGTH	Up to 90cm (3ft)
WEIGHT	Up to 8kg (18lb)
DEPTH	5–400m (15–1,300ft)

DISTRIBUTION Eastern Atlantic, Mediterranean, Black Sea, Indian Ocean, western and southwestern Pacific.

The John Dory has one of the most distinctive appearances of all fish, with a rounded but very thin body, a heavy mouth, and tall fins. It is an expert hunter, stealthily approaching its prey head-on. In this attitude, its thin body is almost invisible and it can approach other fish closely. When it comes within striking range, it shoots out its protrusible jaws and engulfs its victim.

soft second dorsal fin

spiny first dorsal fin

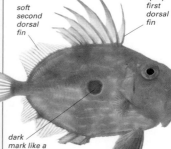

dark mark like a thumbprint

OCEAN LIFE

DEEP-SEA FISHING
Trawling for fish in small boats is an arduous and often hazardous way of earning a living. Many fishermen have perished at sea over the years.

FISHING

Exploitation of the sea's bounty has given humans high-quality food, many useful by-products, and sustains coastal communities around the world. Fish have long been seen as a vast resource that could never run out. However, there are signs that modern, industrial-scale fishing methods are taking their toll. Some fish stocks have already collapsed, and others are thought to be beyond recovery. The total global recorded catch of fish and shellfish rose steadily from 16 million tons in 1950 to 86 million tons in 1995, with a few dips associated with poor anchovy catches in El Niño years (see p.68–69). However, since then catches have levelled off and actually decreased between 2001 and 2004.

The problems of ensuring a sustainable harvest from the sea are many. One fundamental difficulty is the "ownership" of stocks. There is little incentive for some to stop fishing in order to conserve fish if others continue, legally or illegally. It is difficult to police fisheries on the high seas, and illegal fishing is rife in some areas. It is notoriously problematic to obtain an accurate assessment of mobile fish stocks, and illegal fishing and trading distort catch statistics.

Most large-scale fishing methods are indiscriminate and there is enormous waste, as unwanted fish and invertebrate species are discarded, and vast numbers of cetaceans, turtles, and sea birds are also caught. Huge amounts of fish such as Sand Eels are caught in industrial fisheries and turned into fishmeal to feed pigs, poultry, and other livestock, which is an inefficient conversion of protein. However, not all fishing is bad, and there is increasing guidance for consumers wishing to support well-managed fisheries and non-damaging fishing methods.

TRADITIONAL FISHING

Traditional fishing using small-scale fishing gear is rarely a threat to fish stocks. Fish is an important food source, particularly in countries in the developing world, where it provides up to 80 per cent of total protein needs. Fishing is also a vital part of the economy in these countries. And yet, such localized, traditional fisheries take only about 10 per cent of the global total catch.

STILT FISHING This method of fishing is still practised in parts of Sri Lanka and Thailand. The fishermen cast their lines while perching on poles in shallow water.

FISHING AND THE ENVIRONMENT

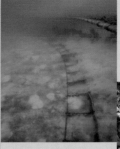

BOTTOM TRAWLING Fishing gear dragged across the sea bed damages marine life and stirs up sediment, smothering wildlife on nearby rocks. Heavy metal scallop dredges are particularly harmful.

PRAWNS AND BYCATCH In every catch of prawns, up to ten times their weight of other species is also caught in the net and subsequently discarded.

DAMAGE AND WASTE

FISHING GEAR Thousands of animals die needlessly each year entangled in fishing tackle. This Hawaiian Monk Seal is one of a total population of just 1,500.

GREEN TURTLE Drifting longlines for tuna, often tens of kilometres long with thousands of hooks, also kill turtles, sharks, and marine birds.

HAZARDS TO WILDLIFE

SCALLOP FARMING Farming scallops is an environmentally sound practice that avoids the adverse effects of trawling on the sea bed and other species.

PEN-RAISED TUNA Fattening of wild tuna in cages falls between fishing and aquaculture legislation, and there are fears that this practice is further depleting overfished stocks.

FISH FARMING

OCEAN LIFE

ORDER SYNGNATHIFORMES

Leafy Seadragon
Phycodorus eques

LENGTH	35cm (14in)
WEIGHT	Not recorded
DEPTH	4–30m (13–100ft)

DISTRIBUTION Eastern Indian Ocean, along the southern coast of Australia

It is hard to imagine anything less fish-like than the Leafy Seadragon. The bizarre tassels and frills that adorn its head and body form a spectacular camouflage that fools both predators and prey. Even its body and tail are bent and twisted to resemble seaweed stems. Closely related to seahorses, the Leafy Seadragon has a similar, but much longer, tubular snout. This is an effective feeding tool – the fish aims its snout at a small shrimp and then sucks hard, rather like a person would on a drinking straw. The Leafy Seadragon lives on rocky, seaweed-covered reefs and in seagrass beds. Unlike seahorses, it cannot coil its tail around an object. It moves very slowly and sways with the waves, mimicking the seaweed. Like seahorses and pipefish, the female deposits her eggs in a brood pouch under the male's tail and he carries them until they hatch.

ORDER SYNGNATHIFORMES

Trumpetfish
Aulostomus maculatus

LENGTH	Up to 1m (3ft)
WEIGHT	Not recorded
DEPTH	2–25m (7–80ft)

DISTRIBUTION Gulf of Mexico, Caribbean Sea, and subtropical waters of western Atlantic

The Trumpetfish looks like a piece of drifting wood, hiding itself among sea fans and other corals. It has a long, slender, straight body and when it flares open its mouth, its long snout resembles a thin trumpet. The Trumpetfish is usually brown, but some individuals have a yellow body.

It hunts by lying in wait to ambush passing shoals of fish, but it is also known to follow predatory fish such as moray eels and steal some of the fish that these flush from their hiding places.

ORDER SYNGNATHIFORMES

Snake Pipefish
Entelurus aequoreus

LENGTH	Up to 60cm (24in)
WEIGHT	Not recorded
DEPTH	10–100m (33–330ft)

DISTRIBUTION Temperate waters of northeastern Atlantic

At first sight, the Snake Pipefish could easily be mistaken for a small sea snake. It has a long, smooth, rounded body tapering to a thin tail with a minute tail fin. However, like all pipefish and seahorses, its head is drawn out into a distinctive tubular snout for sucking up small floating crustaceans and fish fry. Pipefish have no scales but, instead, the body is encased in segmented bony armour lying beneath the skin.

The Snake Pipefish has an orange-brown body with pale blue bands. It lives among seaweed, where it is well camouflaged. The female lays several hundred eggs into a shallow pouch along the male's belly during the summer. The eggs develop in the pouch and the young are released when they are about 1cm (½in) long, but before they are fully developed.

ORDER SYNGNATHIFORMES

Harlequin Ghost Pipefish
Solenostomus paradoxus

LENGTH	12cm (5in)
WEIGHT	Not recorded
DEPTH	Not recorded

DISTRIBUTION Tropical reefs in Indian Ocean and western and southwestern Pacific

The Harlequin Ghost Pipefish looks as though it has wings attached to the sides of its long, thin body. In reality these are greatly enlarged pelvic fins in which the female broods her eggs. The fins are modified to form a pouch where the eggs remain until they hatch. This uncommon species occurs in a wide variety of bright colours and patterns that mimic the reef feather stars and black corals among which it lives. It also often swims head down and so gains further camouflage by aligning its body with the branches among which it swims.

ORDER SYNGNATHIFORMES

Short-snouted Seahorse
Hippocampus hippocampus

LENGTH	15cm (6in)
WEIGHT	Not recorded
DEPTH	5–60m (16–200ft)

DISTRIBUTION Temperate and subtropical waters of northeastern Atlantic and Mediterranean

In the seahorse world, it is the males that give birth to the young. After an elaborate courtship dance, the female lays her eggs in a special pouch on the male's belly. The pouch seals over until the eggs hatch and the tiny baby seahorses emerge. This species is distinguished by its short snout, which is less than a third of the head length.

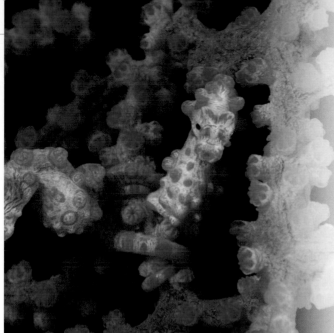

ORDER SYNGNATHIFORMES

Pygmy Seahorse
Hippocampus bargibanti

LENGTH
2.5cm (1in)
WEIGHT
Not recorded
DEPTH
15–50m (50–165ft)

DISTRIBUTION Tropical waters of southwestern Pacific

This miniature seahorse lives on *Muricella* sea fans and was originally discovered when a sea fan was collected for an aquarium. It is very difficult to spot as its body is covered in tubercles that exactly match the polyps of its host. Clinging on tightly with its prehensile tail, it reaches out into the water to suck in planktonic animals. Like other seahorses, it has a rigid body made up of bony plates and a head that is tucked in like a tightly reined carriage horse.

ORDER GASTEROSTEIFORMES

Three-spined Stickleback
Gasterosteus aculeatus

LENGTH
11cm (4in)
WEIGHT
Not recorded
DEPTH
0–100m (0–330ft)

DISTRIBUTION Temperate waters of north Atlantic and north Pacific

The Three-spined Stickleback is equally at home in fresh water and sea water. It has three sharp spines on its back and a series of bony plates along its sides. This species is best known for its breeding behaviour, which involves the male building a tunnel-like nest of plant material into which he entices one or more females to lay their eggs. He fans oxygenated water over the eggs as they develop.

ORDER SYNGNATHIFORMES

Razorfish
Aeoliscus strigatus

LENGTH
15cm (6in)
WEIGHT
Not recorded
DEPTH
1–20m (3–65ft)

DISTRIBUTION Tropical reefs in Indian Ocean and western Pacific

While some reef fish habitually swim upside down, Razorfish swim in synchronized groups in a vertical position, with their long, tubular snouts pointing down. These strange fish are encased in transparent bony plates that meet in a sharp ridge along the belly, like the edge of a razor, and also form a sharp point at the tail. A dark stripe along the body provides camouflage for the Razorfish when hiding among sea urchins and branched corals.

ORDER SCORPAENIFORMES

Stonefish
Synanceia verrucosa

LENGTH
Up to 40cm (16in)
WEIGHT
Up to 2.5kg (5½ lb)
DEPTH
1–30m (3–100ft)

DISTRIBUTION Tropical waters of Indian Ocean and western Pacific

The Stonefish is the world's most venomous fish and its sting is capable of killing a human. Each sharply tipped spine of the dorsal fin has a venom gland at the base from which a duct runs in a groove to the spine tip. Lying quietly on rocks or sediment in the shallows, the Stonefish matches its colour to its background and is easily trodden on. Its camouflage helps it to ambush passing fish, which are sucked into its cavernous mouth with lightening speed.

ORDER SCORPAENIFORMES

Lionfish
Pterois volitans

LENGTH
Up to 38cm (15in)
WEIGHT
Not recorded
DEPTH
2–55m (7–180ft)

DISTRIBUTION Tropical waters of eastern Indian Ocean and western Pacific

Although the Lionfish can inflict a painful sting, it is not dangerous to humans. Its flamboyant coloration of red stripes serves as a warning both to divers and to would-be predators. Also known as the Turkeyfish, it hunts at night using its wing-like pectoral fins to trap its prey of fish, shrimps, and crabs against the reef.

ORDER SCORPAENIFORMES

Long-spined Bullhead
Taurulus bubalis

LENGTH
Up to 25cm (10in)
WEIGHT
Not recorded
DEPTH
0–100m (0–330ft)

DISTRIBUTION Temperate waters of northeastern Atlantic and western Mediterranean

Bullheads are small, cold-water relatives of scorpionfish and the Stonefish (see above). Like them, they are stout, bottom-living fish with a broad head, large mouth, and spiny fins. The Long-spined Bullhead also has a long, sharp spine on each cheek. None of its spines is venomous. These small fish can be found in rock pools, but are difficult to spot as they match their colour to their background. In the winter, the female lays clumps of eggs between rocks. These are then guarded by the male until they hatch between five and 12 weeks later.

LIONFISH
During the daytime, Lionfish hide in rocky parts of the sea bed, where they remain motionless. They hunt at night, first cornering their prey and then, in one swift movement, stunning it with one of their venomous spines and swallowing it. There are some reports of Lionfish threatening humans.

ORDER SCORPAENIFORMES

Spotted Scorpionfish
Scorpaena plumieri

LENGTH	Up to 45cm (18in)
WEIGHT	Up to 1.5kg (3¼ lb)
DEPTH	1–60m (3–200ft)

DISTRIBUTION Western Atlantic and eastern Atlantic around Ascension Island and St. Helena

Resting quietly on the sea bed, the Spotted Scorpionfish is almost invisible thanks to its mottled colour and weed-like skin flaps that cover its head. However, if this fish is disturbed it can open its large pectoral fins to display dramatic black and white-spotted patches. Flashing these "false eyes" is often enough to frighten off a potential predator, but if this does not work, the spines on its dorsal fin can inflict a poisonous sting.

ORDER SCORPAENIFORMES

East Atlantic Red Gurnard
Aspitrigla cuculus

LENGTH	Up to 50cm (20in)
WEIGHT	Not recorded
DEPTH	15–400m (50–1,300ft)

DISTRIBUTION Temperate waters of northeastern Atlantic and Mediterranean

The East Atlantic Red Gurnard could be said to be a "walking-talking" fish. The first three rays of the pectoral fins are shaped as separate, thick, finger-like feelers, which are covered with sensory organs. These feelers are used to "walk" over the sea bed and probe for shrimps and crabs. The fish has a large head protected by hard, bony plates and spines and two separate dorsal fins.

These gurnards sometimes form shoals, and as the fish move around, they make short, sharp grunting noises by vibrating their swim bladder with special muscles and so stay in contact with other gurnards nearby. They spawn in spring and summer and the eggs and larvae float freely near the surface. Adults live for at least 20 years. Although caught commercially, this species is not a main target for fishing.

ORDER SCORPAENIFORMES

Lumpsucker
Cyclopterus lumpus

LENGTH	Up to 60cm (24 in)
WEIGHT	Up to 9.5kg (21lb)
DEPTH	2–400m (7–1,300ft)

DISTRIBUTION Temperate and cold waters of north Atlantic

The adult Lumpsucker (or Lumpfish) has a slightly grotesque appearance because the first dorsal fin becomes overgrown with thick, lumpy skin. Other bony lumps and bumps stick out in irregular rows along its large, rounded body. The pelvic fins form a strong sucker disc on its belly, which the Lumpsucker uses to cling to wave-battered rocks near the shore where it spawns. The male guards the eggs from crabs and also fans them. Lumpsucker eggs are marketed as substitute caviar.

ORDER PERCIFORMES

Wreckfish
Polyprion americanus

LENGTH	Up to 2m (7ft)
WEIGHT	Up to 100kg (220lb)
DEPTH	40–600m (130–2,000ft)

DISTRIBUTION Atlantic, Mediterranean, Indian Ocean, and Pacific

The name of this fish comes from the juveniles' habit of accompanying drifting wreckage. This is a large, solid fish with a pointed head and a protruding lower jaw. It has a spiny dorsal fin and a bony ridge running across the gill cover. Adult Wreckfish live close to the bottom of the sea floor and often lurk inside shipwrecks and caves. Juvenile Wreckfish prefer surface waters and can often be approached by snorkellers, presumably because the fish consider the swimmers to be floating wreckage. As the young fish grow, they give up their nomadic life to live near the sea bed. Adults are fished using lines and make good eating.

ORDER PERCIFORMES

Fairy Basslet
Pseudanthias squamipinnis

LENGTH	Up to 15cm (6in)
WEIGHT	Not recorded
DEPTH	0–55m (0–180m)

DISTRIBUTION Red Sea and tropical waters of Indian Ocean and western Pacific

Fairy Basslets live around coral outcrops and drop-offs. The larger, more colourful males have a long filament at the front of the dorsal fin, and they defend a harem of females. As they grow larger, the females change sex and turn into males.

ORDER PERCIFORMES

Potato Grouper
Epinephelus tukula

LENGTH	Up to 2m (7ft)
WEIGHT	Up to 110kg (240lb)
DISTRIBUTION	10–150m (33–500ft)

DISTRIBUTION Tropical waters of Red Sea, Gulf of Aden, and western Pacific

Groupers are large and important predators on coral reefs. They help to maintain the health of a reef by picking off weak fish. They also eat crabs and Spiny Lobsters. The Potato Grouper inhabits deeper reef channels and seamounts. It has a large head and heavy body with a single long, spiny dorsal fin. Irregular dark blotches cover the body and dark streaks radiate from the eyes. These fish are territorial, and in some areas, individuals are hand-fed by divers. However, one diver drowned after being rammed in the chest by a large Potato Grouper. The large size of these fish makes them an easy target for spearfishermen.

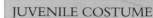

Harlequin Sweetlips
Plectorhinchus chaetodontoides

LENGTH	
Up to 72cm (28in)	
WEIGHT	
Up to 7kg (15lb)	
DEPTH	
1–30m (3–100ft)	

DISTRIBUTION Tropical waters of Indian Ocean and western Pacific

Small groups of Harlequin Sweetlips can often be seen gathered at dusk around large coral heads, waiting to be cleared of parasites by a Cleaner Wrasse (see p.365). These deep-bodied fish are patterned with small, brownish black spots that break up their outline as they swim among the ever-changing shadows on the reef. Their name comes from their thickened lips, which they use to dig out invertebrates from sand.

JUVENILE COSTUME

Juvenile Harlequin Sweetlips have a different patterning to the adults. They have brown bodies and white spots edged in black. By swimming in a weaving, undulating fashion the smallest juveniles mimic a toxic flatworm with a similar coloration and so escape predation. Their colour may also warn that they themselves are unpalatable to predators.

ORDER PERCIFORMES

Common Bluestripe Snapper
Lutjanus kasmira

LENGTH	
Up to 40cm (16in)	
WEIGHT	
Not recorded	
DEPTH	
3–265m (10–870ft)	

DISTRIBUTION Tropical reefs of Red Sea, Indian Ocean, and Pacific

Divers often see large shoals of Common Bluestripe Snapper around coral and rock outcrops during the day. Their streamlined bodies mean that they can swim fast when they disperse at night to feed on smaller fish and bottom-dwelling crustaceans. They have a single long dorsal fin, which, like all their fins, is bright yellow. The Common Bluestripe Snapper and many other similar species are important commercial fish. Their beautiful colours also make them popular specimens among aquarium-fish enthusiasts.

ORDER PERCIFORMES

Red Bandfish
Cepola macrophthalma

LENGTH	
Up to 80cm (30in)	
WEIGHT	
Not recorded	
DEPTH	
15–400m (50–1,300ft)	

DISTRIBUTION Temperate and subtropical waters of northeastern Atlantic and Mediterranean

Very little was known about this strange fish until the 1970s, when divers discovered a population in shallow water around Lundy Island off the west coast of Britain. The Red Bandfish is shaped like an eel but flattened from side to side, with a long, golden-yellow fin running the length of the body on both sides. In mature males, the fin has a bright blue edge. These fish live in deep mud burrows, emerging just far enough to feed on passing arrow worms and other plankton in the manner of tropical garden eels (see p.345). They also swim free of their burrows at times. In addition to single burrows, colonies of many thousands of individuals have been discovered. The burrows sometimes connect with those of burrowing crabs and this may be a deliberate association.

ORDER PERCIFORMES

Bluecheek Butterflyfish
Chaetodon semilarvatus

LENGTH	
Up to 23cm (9in)	
WEIGHT	
Not recorded	
DEPTH	
3–20m (10–65ft)	

DISTRIBUTION Coral reefs in Red Sea and Gulf of Aden

Butterflyfish provide testimony to the health of a coral reef. A wide variety and plentiful numbers of these brightly coloured, disc-shaped fish indicate that a reef is flourishing. Bluecheek Butterflyfish are usually seen in pairs and often hide under table corals. The blue eye-patch hides the eye and confuses predators.

ORDER PERCIFORMES

Ring-tailed Cardinalfish
Apogon aureus

LENGTH	
Up to 15cm (6in)	
WEIGHT	
Not recorded	
DEPTH	
1–40m (3–130ft)	

DISTRIBTION Red Sea and tropical waters of Indian Ocean and western Pacific

Cardinalfish are small nocturnal reef fish. The Ring-tailed Cardinalfish hides under corals and in crevices during the day and emerges at night to feed on plankton. It has a distinctive black band around the tail base and two blue and white lines running from the snout through the eyes. Like all of the 200 or so species of cardinalfish, it has two separate dorsal fins. The male does not feed during the breeding season. Instead, after the female has laid her eggs the male broods them in his mouth, protecting them until they hatch.

ORDER PERCIFORMES

Queen Angelfish
Holacanthus ciliaris

LENGTH	
Up to 45cm (18in)	
WEIGHT	
Up to 1.5kg (3¼ lb)	
DEPTH	
1–70m (3–230ft)	

DISTRIBUTION Gulf of Mexico, Caribbean Sea, and subtropical waters of western Atlantic

One of the most colourful Caribbean reef fish, the blue and yellow Queen Angelfish slips its slim body effortlessly between corals and sea fans. It uses its small mouth and brush-like teeth to nibble sponges, which are its main food. Like all angelfish, it has a sharp spine at the corner of the gill cover. Juveniles are brown and yellow with curved blue bars and feed on parasites that they pick from other fish.

ORDER PERCIFORMES

Bigeye Trevally
Caranx sexfasciatus

LENGTH
Up to 1.2m (4ft)

WEIGHT
Up to 18kg (40lb)

DEPTH
1–100m (3–330ft)

DISTRIBUTION Tropical waters of Indian Ocean and Pacific

During the day, shoals of Bigeye Trevally spiral lazily in coral reef channels and next to steep reef slopes, but at night, these fast-swimming predators split up and scour the reef for prey. Built for speed, these silvery fish have a narrow tail base, which is reinforced with bony plates called scutes, and a forked caudal fin. The first dorsal fin folds down into a groove to improve the streamlining of the fish and the pectoral fins are narrow and curved.

There are many different species of trevally, which are difficult to tell apart. The Bigeye Trevally has a relatively large eye and the second dorsal fin usually has a white tip. These fish make good eating and are common in local markets in Southeast Asia. Juvenile Bigeye Trevally live close inshore and may enter estuaries and rivers.

ORDER PERCIFORMES

Pilotfish
Naucrates ductor

LENGTH
Up to 70cm (28in)

WEIGHT
Not recorded

DEPTH
0–30m (0–100ft)

DISTRIBUTION Tropical, subtropical, and temperate waters worldwide

Although a member of the predatory trevally family Carangidae (see left), the Pilotfish has taken up a scavenging, nomadic existence, travelling with large, ocean-dwelling bony fish, sharks, rays, and turtles. Its slim, silvery to pale bluish body is marked with six or seven bold black bands. These may help the host fish recognize it and so leave it alone. Darting in when its host has made a kill, the Pilotfish eats any scraps it can find and also removes parasites. Young fish associate with jellyfish.

ORDER PERCIFORMES

Sharksucker
Echeneis naucrates

LENGTH
Up to 1m (3ft)

WEIGHT
Up to 2.3kg (5lb)

DEPTH
20–50m (65–165ft)

DISTRIBUTION Tropical, subtropical, and temperate waters worldwide

The distinctive feature of the Sharksucker is the powerful sucker disc on the top of its head, which enables it to attach itself securely to another fish. It feeds on its host's scraps and parasites, and also on small fish. Usually found attached to sharks or other large fish, cetaceans, and turtles, the Sharksucker also swims freely over coral reefs. Its body is long and thin and ends in a fan-like tail.

ridged oval sucker disc replaces first dorsal fin

lower jaw juts out beyond upper jaw

ORDER PERCIFORMES

Dolphinfish
Coryphaena hippurus

LENGTH
Up to 2.1m (7ft)

WEIGHT
Up to 40kg (88lb)

DEPTH
0–85m (0–280ft)

DISTRIBUTION Tropical, subtropical, and temperate waters worldwide

With its shimmering colours, a Dolphinfish leaping clear of the water is a spectacular sight. Metallic blues and greens cover its back and sides, grading into white and yellow on the underside. A fast ocean-dwelling fish, it is powered by a long, forked tail, with a single elongated dorsal fin providing stability. Also known as the Dorado, it is a valuable market fish.

ORDER PERCIFORMES

False Clown Anemonefish
Amphiprion ocellaris

LENGTH
Up to 11cm (4in)

WEIGHT
Not recorded

DEPTH
1–15m (3–50ft)

DISTRIBUTION Tropical waters of eastern Indian Ocean and western Pacific

The most surprising thing about the False Clown Anemonefish is its home. It lives inside a giant stinging anemone. This small orange and white fish spends its whole life with its chosen anemone, which can be one of three species. At night, it sleeps among the bases of the tentacles on the anemone's disc. The fish is not stung and eaten because the anemone does not know it is there: a special slime covers the fish's body and prevents the anemone from recognizing it as food. Each anemone usually supports a large female, her smaller male partner, and several immature fish.

If the female dies, the male changes sex and becomes female and the largest immature fish takes on the male role. Both the False Clown and the Clown Anemonefish are among the most popular aquarium fish, and numbers have been reduced in some areas by overcollecting.

ORDER PERCIFORMES

Sergeant Major
Abudefduf saxatilis

LENGTH
Up to 23cm (9in)

WEIGHT
Up to 200g (7oz)

DEPTH
1–15m (3–50ft)

DISTRIBUTION Tropical and subtropical waters of Atlantic Ocean

This small fish is a familiar sight on most coral reefs in the Atlantic. It is one of the most common members of the damselfish family (Pomacentridae). It feeds on zooplankton in large groups, gathering above the reef to pick tiny animals and fish eggs from the water. In tourist areas, the fish are attracted to divers and boats, and will eat almost anything that is offered. Male Sergeant Majors prepare a nesting area and guard the eggs laid by the females. A similar species, *Abudefduf vaigiensis*, is found on reefs in the Indo-Pacific region.

Cuckoo Wrasse

Labrus mixtus

LENGTH	Up to 40cm (16in)
WEIGHT	Not recorded
DEPTH	2–200m (7–650ft)

DISTRIBUTION Temperate and subtropical waters of northeastern Atlantic and Mediterranean

The Cuckoo Wrasse is one of the most colourful fish in northern European waters. Large mature males (shown here) are a beautiful blue and orange, while females are pink with alternate black and white patches along the back. When they are 7–13 years old, some females change colour and sex and become fully functional males. These males are known as secondary males and spawn in pairs with females. The male excavates a nest and attracts the female with an elaborate swimming display. To further complicate matters it has been found that a very few fish are born male but have the female colouring. These males are known as primary males and their role in reproduction has not been fully ascertained.

Cleaner Wrasse

Labroides dimidiatus

LENGTH	14cm (5½ in)
WEIGHT	Not recorded
DEPTH	1–40m (3–130ft)

DISTRIBUTION Tropical reefs in Indian Ocean and southwestern Pacific

The Cleaner Wrasse spends its life grooming other fish, turtles, and occasionally even divers. This little fish is silvery blue with a black band running from snout to tail. The "client" recognizes it from its markings and does not try to eat it. Groups of Cleaner Wrasse usually consist of an adult male and a harem of females. If the male dies, the largest female changes sex and takes on the male role, becoming fully functional within a few days.

distinctive black band

small mouth with strong teeth

MUTUAL BENEFIT

Skin parasites are irritating and fish can be debilitated by a heavy infestation. On coral reefs, large fish queue up at known "cleaning stations" such as a prominent coral head, spread their fins and open their mouths. The resident Cleaner Wrasse picks off parasites and dead tissue and gets a good meal in return.

Green Humphead Parrotfish

Bolbometopon muricatum

LENGTH	Up to 1.3m (4¼ ft)
WEIGHT	Up to 46kg (100lb)
DEPTH	1–30m (3–100ft)

DISTRIBUTION Tropical reefs in Red Sea, Indian Ocean, and southwestern Pacific

Parrotfish are aptly named because not only are they brightly coloured but also their teeth are fused together to form a parrot-like beak. The Green Humphead Parrotfish is much larger than most of its relatives. It has a huge crest-shaped hump on its head, a greenish body, large scales, and a single long dorsal fin. This destructive fish feeds by crunching up live coral and it often breaks up the coral with its head. However, on the positive side, the coral sand it defecates after a meal helps to consolidate the reef and build up patches of sand.

Blackfin Icefish

Chaenocephalus aceratus

LENGTH	Up to 72cm (28in)
WEIGHT	Up to 3.5kg (7½lb)
DEPTH	5–770m (16–2,500ft)

DISTRIBUTION Polar waters of Southern Ocean around northern Antarctica

In the freezing waters around Antarctica, the temperature can fall to nearly -2°C (28°F). This is below the temperature at which the blood of most fish would freeze. The Blackfin Icefish has a natural antifreeze in its blood that helps it survive in these conditions. It has no red blood cells and so appears a ghostly white. This makes its blood thinner so that it can flow freely in the cold temperatures. It is a sluggish hunter of small fish and krill and needs little oxygen.

Wolf-fish

Anarhichas lupus

LENGTH	Up to 1.5m (5ft)
WEIGHT	Up to 24kg (53lb)
DEPTH	1–500m (3–1,650ft)

DISTRIBUTION North Atlantic and Arctic Ocean

This large and ferocious-looking fish is normally found on rocky reefs in deep water. However, north of the British Isles, divers regularly see them in shallow water. They are not aggressive to divers unless provoked.

The Wolf-fish has a long body and a huge head with strong canine-like teeth at the front and molar-like teeth at the sides. These are used to break open hard-shelled invertebrates such as mussels, crabs, and sea urchins. Worn teeth are replaced each year. The skin is tough, leathery, and wrinkled and is usually greyish with darker vertical bands extending down the sides.

Spawning takes place during the winter. The female lays thousands of yellowish eggs in round clumps among rocks and seaweeds and the male guards them until they hatch. In spite of their unattractive appearance, Wolf-fish are good to eat and are caught by anglers. They are also sometimes caught in trawl nets.

BIGEYE TREVALLIES
These Bigeye Trevallies, also known as Bigeye Jacks, are shoaling in shallow water near the Solomon Islands in the western Pacific. They are usually slow-moving by day, but at night the shoals disperse and they hunt singly, moving quickly in search of the fish and crustaceans on which they feed.

ORDER PERCIFORMES

Common Stargazer

Kathetostoma laeve

LENGTH	Up to 75cm (30in)
WEIGHT	Not recorded
DEPTH	0–60m (200ft), possibly 150m (550ft)

DISTRIBUTION Temperate waters of Indian Ocean around southern Australia

Looking like a cross between a bulldog and a seal, the Common Stargazer normally lies buried in shelly sand. It has its eyes set right on top of its large, square head and its mouth slants obliquely upwards. This allows it to breathe and to see while remaining almost completely buried and is probably the reason behind its unusual name. Its large, white-edged pectoral fins help it to lunge out of the sand and engulf passing fish and crustaceans. Common Stargazers have also occasionally bitten divers who have inadvertently disturbed them while on night dives, when they are particularly difficult to spot. Anglers face a greater threat if they catch a Common Stargazer. Careless handling can result in a painful sting from a tough, venomous spine that lies behind each gill cover.

ORDER PERCIFORMES

Greater Weever

Trachinus draco

LENGTH	Up to 50cm (20in)
WEIGHT	Up to 2kg (4½ lb)
DEPTH	1–150m (3–500ft)

DISTRIBUTION Temperate waters of northeastern Atlantic and Mediterranean

The Greater Weever is one of very few venomous fish found in European waters. It has a long body, large eyes, and two dorsal fins, the first of which has venomous spines. During the day, the fish lies buried in the sand with just its eyes and fin-tip exposed. A painful wound can result from treading on the fish in shallow water.

ORDER PERCIFORMES

Sand Eel

Ammodytes tobianus

LENGTH	Up to 20cm (8in)
WEIGHT	Not recorded
DEPTH	0–30m (0–100ft)

DISTRIBUTION Temperate waters of northeastern Atlantic and Baltic Sea

Shimmering shoals of Sand Eels are a familiar sight in shallow sandy bays around northern Europe. These small silvery fish have long, thin bodies with a pointed jaw and a single long dorsal fin. Large shoals patrol the waters just above the sea bed, feeding on planktonic crustaceans, tiny fish, and worms. If threatened, they dive down and disappear into the sand. In winter, they spend most of the time buried. Sand Eels form a very important part of the diet of larger fish such as cod, herring, and mackerel, and of sea birds, especially Atlantic puffins. When Sand Eels are scarce, local puffin colonies produce very few young. In some areas, overexploitation of Sand Eels for processing into fishmeal has been linked to sea-bird declines (see p.403).

ORDER PERCIFORMES

Tompot Blenny

Parablennius gattorugine

LENGTH	Up to 30cm (12in)
WEIGHT	Not recorded
DEPTH	1–30m (3–100ft)

DISTRIBUTION Temperate and subtropical waters of northeastern Atlantic and Mediterranean

With its thick lips, bulging eyes, and a pair of tufted head tentacles, the Tompot Blenny is a comical-looking fish. Like all blennies, it has a long body, a single long dorsal fin, and peg-like pelvic fins, which it uses to prop itself up. Inquisitive by nature, the Tompot Blenny will peer out at approaching divers from the safety of a rock crevice.

broad caudal fin

large eye

large pelvic fins

ORDER PERCIFORMES

Mandarinfish

Synchiropus splendidus

LENGTH	Up to 6cm (2½ in)
WEIGHT	Not recorded
DEPTH	1–18m (3–60ft)

DISTRIBUTION Tropical waters of southwestern Pacific

With its yellow and orange body and distinctive green and blue markings, the Mandarinfish is one of the most colourful of all reef fish. Its skin is covered with a distasteful slime and its bright colours warn predators not to touch it. Small groups live inshore on silt-covered sea beds among coral and rubble. Most members of the dragonet family (Callionymidae), to which it belongs, are coloured to match their surroundings. It is a popular aquarium fish but is very difficult to maintain.

ORDER PERCIFORMES

Yellow Shrimp Goby

Cryptocentrus cinctus

LENGTH	Up to 8cm (3in)
WEIGHT	Not recorded
DEPTH	1–15m (3–50ft)

DISTRIBUTION Tropical waters of northeastern Indian Ocean and southwestern Pacific

Shrimp gobies share their sandy burrows with snapping shrimps belonging to the genus *Alpheus*. The shrimps have strong claws and excavate and maintain the burrow, while the gobies have good eyesight and act as a lookout at its entrance. The Yellow Shrimp Goby has bulging, high-set eyes, thick lips, and two dorsal fins. Although the usual coloration is yellow with faint, dusky bands, it can also be greyish white. This species lives in sandy areas of shallow lagoons and bays.

ORDER PERCIFORMES

Bignose Unicornfish

Naso vlamingii

LENGTH	Up to 60cm (24in)
WEIGHT	Not recorded
DEPTH	1–50m (3–165ft)

DISTRIBUTION Tropical waters of Indian Ocean and southwestern Pacific

Unicornfish are so called because many have a horn-like projection on their forehead. However, the Bignose Unicornfish just has a rounded bulbous snout. At the base of the tail are two pairs of fixed, bony plates that stick out sideways like sharp knives, and the fish can inflict a serious wound on a potential predator. These blades are characteristic of surgeonfish (Acanthuridae), the family to which unicornfish belong.

Usually dark with blue streaks, the Bignose Unicornfish can pale instantly to a silvery grey. This often happens when the fish is being cleaned by a Cleaner Wrasse (see p.365). It favours steep reef slopes where it can feed on zooplankton in the open water.

Great Barracuda

Sphyraena barracuda

LENGTH	Up to 2m (6½ ft)
WEIGHT	Up to 50kg (110lb)
DEPTH	0–100m (0–330ft)

DISTRIBUTION Tropical and subtropical waters worldwide

Barracuda are fast-moving predators with needle-sharp teeth and an undeserved reputation for ferocity. The Great Barracuda has a long, streamlined body with the second dorsal fin set far back near the tail. This fin arrangement, along with a large, powerful tail, allows it to stalk its prey and then accelerate forward at great speed. Large individuals in frequently dived sites will often allow divers to approach closely. Very occasionally a lone fish may attack a diver if it mistakes a hand or shiny watch for a silvery fish. Eating even small amounts of barracuda can result in ciguatera poisoning, caused by toxins accumulated from its food.

GREAT BARRACUDA SKULL
Barracuda have flat-topped, elongated skulls with large, powerful jaws and knife-like teeth.

long front teeth

BARRACUDA SHOAL
While adults are normally solitary, juvenile Great Barracuda often swim together in large shoals in sheltered areas for protection.

Atlantic Mackerel

Scomber scombrus

LENGTH	Up to 60cm (24in)
WEIGHT	Up to 3.5kg (7½ lb)
DEPTH	0–200m (0–650ft)

DISTRIBUTION Temperate waters of north Atlantic, Mediterranean, and Black Sea

dark lines on back

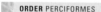

silvery belly

The Atlantic Mackerel is designed for fast swimming. It has a torpedo-shaped, streamlined body, small dorsal fins, close-fitting gill covers, and small, smooth scales. In the summer, large shoals feed close inshore, voraciously preying on small fish and sieving plankton through their gills. From March to June, they lay their floating eggs in habitual spawning areas, the eggs hatching after a few days. In winter, the fish move into deeper water offshore and hardly feed. Several separate stocks exist within the north Atlantic, all of which are commercially exploited.

Northern Bluefin Tuna

Thunnus thynnus

LENGTH	Up to 4.5m (15ft)
WEIGHT	Up to 680kg (2,230lb)
DEPTH	0–3,000m (0–9,900ft)

DISTRIBUTION Northern and central Atlantic and Mediterranean

The Northern Bluefin Tuna is one of the world's most valuable commercial fish and is heavily overexploited. Like mackerel, it is designed for high-speed swimming and is one of the fastest bony fish, attaining speeds of at least 70kph (43mph). The pectoral, pelvic, and first dorsal fins can be slotted into grooves to further streamline the torpedo-shaped body. To provide for long-distance, sustained swimming, the fish has large amounts of red muscle, which has a high fat content and can store oxygen. Other, similar species of bluefin tuna occur in the Pacific Ocean and southern parts of the Atlantic Ocean.

Atlantic Sailfish

Istiophorus albicans

LENGTH	Up to 3.2m (10ft)
WEIGHT	Up to 60kg (130lb)
DEPTH	0–200m (0–650ft)

DISTRIBUTION Temperate and tropical waters of Atlantic and Mediterranean

Like swordfish and marlin, the Atlantic Sailfish has its upper jaw extended into a long spear. This is used to slash through shoals of fish, stunning and maiming them. It has a huge sail-like dorsal fin, which is used in displays, but is folded away for fast swimming. A similar sailfish occurs in the Pacific and may be the same species.

OCEAN LIFE

ORDER PLEURONECTIFORMES

European Plaice

Pleuronectes platessa

LENGTH	Up to 1m (3ft)
WEIGHT	Up to 7kg (15lb)
DEPTH	0–200m (0–655ft)

DISTRIBUTION Arctic Ocean, northeastern Atlantic, Mediterranean, and Black Sea

This species is the most important commercial flatfish for European fisheries. Heavy fishing, however, has resulted in a progressive reduction in the size and age of fish landed. It is a typical oval-shaped flatfish with long fins extending along both edges of its thin body. Flatfish have both eyes on one side of their body and are either "right-eyed" or "left-eyed". Plaice are right-eyed: they lie on the sea bed with their left side down. Their upward-facing right side is brown with orange or red spots.

Plaice spend the day buried in the sand, emerging at night to feed on shellfish and crustaceans, which they crush using special teeth in the throat (pharyngeal teeth). Young Plaice are also expert at nipping off the breathing siphons of shellfish that they spot sticking up out of the sand.

ORDER PLEURONECTIFORMES

Common Sole

Solea solea

LENGTH	Up to 70cm (28in)
WEIGHT	Up to 3kg (6½ lb)
DEPTH	0–150m (0–500ft)

DISTRIBUTION Temperate waters of northeastern Atlantic, Baltic, Mediterranean, and Black seas

Common Sole are masters of camouflage and can subtly alter their colour to match the sea bed on which they lie. The basic greyish brown colour can be lightened or darkened and the pattern of darker splotches changed. Sole have a rounded snout and a semicircular mouth and their head is fringed with short filaments, giving them an unshaved appearance.

Like all flatfish, the Common Sole starts life as a tiny larval fish floating near the surface. As it grows, it gradually undergoes a radical metamorphosis. The eye on the left side moves around the head to join the eye on the right and the body starts to flatten. When it is about a month old, it settles on the sea floor with its eyeless side facing down. The skin on the underside stays white but the upper side develops pigment. Although Common Sole can live to be nearly 30 years old, they are a valuable food fish and most are caught when only a few years old.

ORDER TETRADONTIFORMES

Titan Triggerfish

Balistoides viridescens

LENGTH	Up to 75cm (30in)
WEIGHT	Not recorded
DEPTH	1–50m (3–160ft)

DISTRIBUTION Tropical reefs of Red Sea, Indian Ocean, and southwestern Pacific

The Titan Triggerfish is also known as the Moustache Triggerfish due to a dark line above its lips. It has large, strong front teeth and strong spines in its first dorsal fin. The first and longer spine can be locked in an upright position and released by depressing the second smaller "trigger" spine. This allows the fish to jam itself into a reef crevice where it can rest safely, away from potential predators. The Titan Triggerfish preys on shellfish and crustaceans, which it crunches up using its tough mouth and teeth. It can even make a meal of sea urchins by flipping them over and biting them on their vulnerable underside, where the spines are shorter.

PROTECTIVE PARENT

In the breeding season, Titan Triggerfish dig a nest in a sandy patch of coral rubble using their mouth as a water jet. The female lays her eggs in the nest and one or both of the parents remains nearby to guard it. Normally a wary fish, parent Titan Triggerfish will attack divers that come too close to the nest and can inflict severe bites that need medical attention.

ORDER TETRADONTIFORMES

Scrawled Filefish

Aluterus scriptus

LENGTH	Up to 1.1m (3½ ft)
WEIGHT	Up to 2.5kg (5½ lb)
DEPTH	2–120m (6–400ft)

DISTRIBUTION Tropical and subtropical waters of Atlantic, Pacific, and Indian oceans

Beautiful blue, irregular markings like a child's scribbles give this reef fish its name. Filefish are closely related to triggerfish (see below, left), but are thinner and, except for the Scrawled Filefish, usually smaller. This large species has one large and one tiny spine on its back over the eyes. The fish uses these spines to help wedge itself into crevices for safety.

ORDER TETRADONTIFORMES

Spotted Boxfish

Ostracion meleagris

LENGTH	Up to 25cm (10in)
WEIGHT	Not recorded
DEPTH	1–30m (3–100ft)

DISTRIBUTION Tropical reefs in Indian Ocean and South pacific, possibly extending to Mexico

Instead of a covering of scales, all boxfish are protected by a rigid box of fused bony plates under the skin. This means they cannot bend their body and must swim by beating their pectoral fins. A large tail gives some propulsion and is also used to help steer them like a rudder. Male Spotted Boxfish are more colourful than the females, which are brown with light spots. These fish secrete a poisonous slime from their skin that protects them from predators.

Ocean Sunfish
Mola mola

LENGTH	
Up to 4m (13ft)	
WEIGHT	
Up to 2,300kg (5,070lb)	
DEPTH	
0–480m (0–1,600ft)	

DISTRIBUTION Tropical, subtropical, and temperate waters worldwide

The Ocean Sunfish is the world's heaviest bony fish and has a distinctive disc-like shape. Instead of a caudal fin, it has a rudder-like structure (clavus) formed by extensions of the dorsal and anal fin rays, and it swims by flapping its tall dorsal and anal fins

from side to side. Its common name comes from the fish's habit of drifting in surface currents while lying on its side. It also swims upright with its dorsal fin sticking above the surface. The Ocean Sunfish has no scales, but its skin is very thick and stretchy. Like the Porcupinefish (see below, left), to which it is related, the Ocean Sunfish has a single fused tooth-plate in each jaw, but it feeds mainly on soft-bodied jellyfish and other slow-moving invertebrates and fish. Females produce the most eggs of any bony fish, laying up to 100 million in the open ocean. Lone fish make grating noises with pharyngeal (throat) teeth and this may help them to make contact with potential mates.

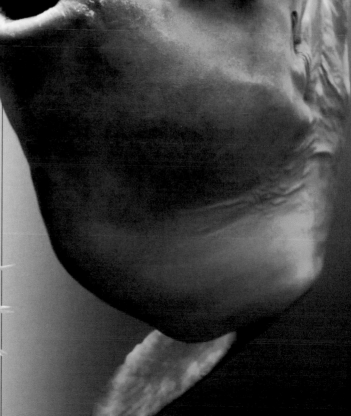

Star Pufferfish
Arothron stellatus

LENGTH	
Up to 1.2m (4ft)	
WEIGHT	
Not recorded	
DEPTH	
3–60m (10–200ft)	

DISTRIBUTION Tropical reefs in Indian Ocean and south Pacific

Compared with most other pufferfish, the Star Pufferfish is a relative giant. Its black-spotted skin is covered in small prickles and, if threatened, it will swallow water and swell up to an even larger size. At night, it searches out hard-shelled reef invertebrates and crushes them with powerful jaws that have fused, beak-like teeth.

HUMAN IMPACT

FUGU FISH

Pufferfish produce tetrodotoxin, a lethal poison that is stronger than cyanide and for which there is currently no antidote. In spite of this, these fish are eaten in Japan as a delicacy called "fugu". The poison is in the skin and some of the internal organs and only licensed chefs, who have been specially trained, are permitted to prepare this dish. Pufferfish from the genus *Takifugu* are considered to be the best eating. A few people die every year from eating fugu and the Emperor of Japan is officially banned from eating this delicacy for his own protection.

Porcupinefish
Diodon histrix

LENGTH	
Up to 90cm (35in)	
WEIGHT	
Up to 3kg (6½ lb)	
DEPTH	
2–50m (6–160ft)	

DISTRIBUTION Tropical and subtropical waters of Atlantic, Pacific, and Indian oceans

When a Porcupinefish is frightened, it pumps water into its body until it looks like a prickly football. Few predators are large enough or brave enough to swallow a fish in this state. Left to itself, the Porcupinefish deflates and its long spines lie flat against its body. During the day, it hides in caves and reef crevices, emerging at night to feed on hard-shelled invertebrates such as gastropod molluscs.

prominent eye

erect spine

REPTILES

DOMAIN	Eucarya
KINGDOM	Animalia
PHYLUM	Chordata
CLASS	Reptilia
ORDERS	4
SPECIES	About 7,500

DURING THE JURASSIC PERIOD, over 140 million years ago, reptiles were the largest animals in the oceans. Their place has since been taken by mammals, leaving few reptiles that are wholly marine. Of these, turtles are the most widespread, and sea snakes are the most diverse. Apart from the Leatherback Turtle, almost all are confined to warm-water regions, with the largest numbers around coasts and on coral reefs.

EXPLOITATION

All marine reptiles, apart from sea snakes, have a long history of exploitation by humans for food, skins, or shells. Turtles face the additional hazard of being accidentally caught in fishing nets, and numbers of all seven species have steeply declined. Marine turtles are now protected by international legislation.

ILLEGAL SOUVENIRS
Stuffed marine turtles – seen here on a beach in Peru – are still sold to tourists, despite being liable to seizure by customs officials.

ANATOMY

Marine reptiles have several adaptations for life in the sea. Turtles have a low, streamlined shell, or carapace, and broad, flattened forelimbs that beat up and down like wings. Marine lizards and crocodiles use their tails to provide most of the power when swimming, while most sea snakes have flattened tails that work like oars. Unlike land snakes, true sea snakes do not have enlarged belly scales, since they do not need good traction for crawling on land. All reptiles breathe air, and marine species have valves or flaps that prevent water entering their nostrils when they dive. Crocodiles also have a valve at the top of the throat, which enables them to open their mouths beneath the surface without flooding their lungs with water. Marine reptiles all need to expel excess salt. Sea snakes and crocodiles do this through salt glands in their mouths, while marine turtles lose salt in their tears. The marine iguana has salt glands located on its nose.

pointed scales (scutes)   streamlined shell (carapace)

head

short rear flippers

long front flippers

STREAMLINED SHELL
The Hawksbill Turtle has a tapering carapace with conspicuous scales, or scutes. Unlike most terrestrial tortoises, it cannot retract its head or legs inside its shell.

REPLACEMENT TEETH
A Saltwater Crocodile's teeth are constantly shed and replaced. During its lifetime, it may use over 40 sets.

HABITAT

Most marine reptiles live close to the shore, or return to it to breed. The only fully pelagic species are true sea snakes – those in the family Hydrophiidae. They remain in the open ocean for their entire lives. Sea snakes are also the deepest divers, feeding up to 100m (330ft) below the surface. Apart from the Leatherback Turtle, most marine reptiles depend on external warmth to remain active, which restricts them to tropical and subtropical waters. They also show striking variations in regional spread. This is particularly true of sea snakes: up to 25 species are found in some parts of the Indo-Pacific, but the Atlantic Ocean has none.

FOOD AND FEEDING

Most marine reptiles are carnivorous. Sea snakes typically feed on fish, although a few are specialized predators of fish eggs. They use their venom mainly in feeding, rather than for defence, killing their prey by biting it, and then swallowing it whole. Green Turtles feed on seagrass when they become adult, while other marine turtles are carnivorous throughout their lives. The Marine Iguana is the only marine reptile that is a fully herbivorous. When young, it feeds on algae close to the waterline, but as an adult, it grazes seaweed growing on submerged rocks. Reptiles are cold blooded (ectothermic), so they use less energy than mammals or birds. This means that they need less food, and can go for long periods between meals. Sea snakes, for example, can survive on just one or two meals a month.

GRAZING ON ALGAE
Marine Iguanas have blunter heads than most lizards, enabling them to tear seaweed from rocks. Sharp claws act as anchors.

KEY
Number of sea snake species

- 12–25 species
- 2–12 species
- 1 species

PACIFIC OCEAN

ATLANTIC OCEAN

INDIAN OCEAN

SOUTHERN OCEAN

SEA SNAKES WORLDWIDE
Although diverse in the Indo-Pacific, sea snakes are absent from the Atlantic. Cold waters off southern Africa prevent them from spreading west.

REEF SNAKE
A Yellow-Lipped Sea Krait searches for prey in a coral reef. Reefs are prime habitats for sea kraits, which generally live in shallow water.

REPRODUCTION

True sea snakes are the only reptiles that reproduce at sea. They give birth to live young (they are viviparous) after a gestation period of up to 11 months – much longer than most terrestrial species. All other marine reptiles, including sea kraits and marine turtles, lay their eggs on land. Many of these animals breed on remote beaches and islands, and the adults sometimes arrive simultaneously and in large numbers. The eggs

are incubated by ambient warmth, and in crocodiles and turtles, the nest temperature determines the sex ratio of the hatchlings. Once the eggs have hatched, growth is fast, but mortality can be high. Parental care is rare in marine reptiles; female crocodiles are an exception, guarding their nests and carrying their young to water after they have hatched.

NEST IN THE SAND
After excavating a nest, a female Leatherback lays her eggs. Turtle eggs are almost spherical, and have soft, leathery shells, which tear open when they hatch.

REPTILE CLASSIFICATION

Three orders of living reptiles contain marine species. The fourth order includes only the tuataras, which are terrestrial. Snakes make up the vast majority of marine reptiles. Others, such as wart snakes and terrapins, live in fresh water, occasionally entering the sea.

TURTLES AND TORTOISES
Order Chelonia

About 300 species
Seven turtle species are exclusively marine. Typical marine turtles (six species), have a hard carapace. The separately classified Leatherback Turtle has a rubbery carapace.

SNAKES AND LIZARDS
Order Squamata

About 7,400 species
About 70 species of snake live in salt water. True sea snakes, belonging to the family Hydrophiidae, spend all their lives at sea, while sea kraits (members of the Elapidae) breed on land. Sea-going lizards are all semi-terrestrial; only one species, the Marine Iguana, gets all its food offshore.

CROCODILES AND ALLIGATORS
Order Crocodilia

23 species
Only the American Crocodile and the Saltwater Crocodile live in both fresh water and the sea. Crocodilians usually feed at the surface, rarely diving more than a few metres when at sea.

MARINE ADAPTATIONS
Thanks to their low metabolic rate, marine reptiles can remain underwater for long periods. This young Saltwater Crocodile is lurking on the sea bed off New Guinea.

ORDER CHELONIA

Green Turtle

Chelonia mydas

LENGTH	0.8–1m (2½–3¼ft)
WEIGHT	65–130kg (140–290lb)
HABITAT	Open sea, coral reefs, coasts

DISTRIBUTION Tropical and temperate waters worldwide

Elegantly marked and very effectively streamlined, this species is the most common turtle in subtropical and tropical waters, where it is often seen in eelgrass beds and on coral reefs.

Its colour varies from green to dark brown, but its scales and shell plates (scutes) are lighter where they meet, giving it a distinctive, chequered pattern. Like all marine turtles, it has front flippers that are long and broad and beat up and down like wings. They provide the power for swimming, while the much shorter rear flippers act as stabilizers. Young Green Turtles are carnivorous, eating molluscs and other small animals, but the adults feed mainly on eelgrass and algae – a diet that keeps them close to the coast.

Green Turtles breed on isolated beaches, and they are remarkably faithful to their nesting sites. To reach them, some make journeys of more than 1,000km (600 miles), navigating their way to remote islands that may be just a few kilometres across. They mate in the shallows, and the females then crawl ashore after dark to dig their nests and lay eggs. Green Turtles lay up to 200 eggs, burying them about 75cm (30in) beneath the sand. The eggs take about 6–8 weeks to hatch. All the young emerge simultaneously and scuttle for the safety of the waves.

The Green Turtle has been hunted for centuries, mainly for food, and its numbers have declined significantly. Conservation measures include protection of the turtles' nest sites, so that the young have a better chance of reaching the sea.

EARLY LIFE

After hatching while buried in the sand, the young turtles use their front flippers to dig towards the surface. They then make a dash for the sea, trying to avoid becoming a meal for waiting predators, including birds, crabs, snakes, and ants. Very little is known about their early life, as young Green Turtles are rarely observed in the wild, but it is certain that they face many predators in the sea. Their growth rate is known to average more than 5kg (11lb) a year.

ORDER CHELONIA

Hawksbill Turtle

Eretmochelys imbricata

LENGTH	0.8–1m (2½–3¼ft)
WEIGHT	45–75kg (100–165lb)
HABITAT	Coral reefs and coastal shallows

DISTRIBUTION Tropical and warm-temperate waters worldwide

Named after its conspicuous beaked snout, the Hawksbill has a carapace with a raised, central keel and pointed shell plates (scutes) around its rear margin. It lives in warm-water regions, feeding on sponges, molluscs, and other sedentary animals, and rarely strays far from shallows and coral reefs. It is less migratory than other marine turtles, breeding at low densities all over the tropics instead of gathering at certain beaches. On land, it has a distinctive gait, moving its flippers in diagonally opposite pairs – other marine turtles move their front flippers together – the same action they use when swimming.

The Hawksbill is the chief source of tortoiseshell – detached, polished scutes. Despite being classified as Critically Endangered by the IUCN, Hawksbills are often killed and stuffed when young to be sold as curios, particularly in Southeast Asia. Attempts at farming these turtles have not been successful.

ORDER CHELONIA

Loggerhead Turtle

Caretta caretta

LENGTH	0.7–1m (2¼–3¼ft)
WEIGHT	75–160kg (165–350lb)
HABITAT	Open sea, coral reefs, coasts

DISTRIBUTION Tropical and warm temperate waters worldwide

After the Leatherback (opposite), the Loggerhead is the second largest marine turtle. It has a blunt head, powerful jaws, and a steeply domed carapace. It hunts and eats hard-bodied animals, such as crabs, lobsters, and clams. This species takes about 30 years to mature and breeds every other year.

ORDER CHELONIA

Leatherback Turtle
Dermochelys coriacea

LENGTH	1.3–1.8m (4¼–6ft)
WEIGHT	Up to 900kg (2,000lb)
HABITAT	Open sea

DISTRIBUTION Tropical, subtropical, and temperate waters worldwide

The Leatherback is the world's largest marine turtle. Its carapace has a rubbery texture, having no hard plates, and has a tapering, pear-like shape. Its head is not retractable, and the Leatherback is unique among turtles in having flippers without claws. It spends most of its life in the open sea, returning to the coast only when it breeds. It feeds on jellyfish and other planktonic animals, and while it gets most of its food near the surface, it can dive to depths of 1,000m (3,300ft).

Leatherbacks breed mainly in the tropics, on steeply sloping sandy beaches, laying up to nine clutches of eggs in each breeding season.

Unusually for a reptile, the Leatherback Turtle can keep its body warmer than its surroundings, thanks partly to the thick layer of insulating fat beneath its skin. This allows it to wander much more widely than other turtles, reaching as far north as Iceland and almost as far south as Cape Horn. Individuals may roam huge distances – one Leatherback tagged off the coast of South America was later found on the other side of the Atlantic, 6,800km (4,200 miles) away.

carapace with parallel ridges

large head on short neck

THROAT SPINES

The Leatherback's throat contains dozens of backward-pointing spines that prevent jellyfish escaping before they are completely swallowed. These endangered turtles often die after eating discarded plastic bags, which they mistake for jellyfish.

JELLYFISH TRAP
The Leatherback's throat spines can be over 1cm (½in) long. They are regularly replaced during the animal's life.

ORDER CHELONIA

Atlantic Ridley Turtle
Lepidochelys kempi

LENGTH	50–90cm (20–35in)
WEIGHT	25–40kg (55–90lb)
HABITAT	Coral reefs, coasts

DISTRIBUTION Caribbean, Gulf of Mexico, occasionally as far north as New England

Also known as the Kemp Ridley Turtle, this is the smallest marine turtle, and also the most threatened, largely as a result of its unusual breeding behaviour. Unlike most marine turtles, Atlantic Ridleys lay their eggs by day, and the females crawl out of the sea simultaneously, during mass nestings called *arribadas* (Spanish for "arrivals"). At one time, these nestings took place throughout the turtle's range, but because the eggs were laid in such large concentrations in daylight they were easy prey for human egg-harvesters and natural predators. Today, the vast majority of Atlantic Ridleys breed on a single beach in Mexico, where their nests are protected. These turtles were also often caught as bycatch in shrimp nets, but Turtle Excluding Devices (TEDs) fitted to nets have helped to reduce this threat. Several weeks after an *arribada*, young Atlantic Ridleys emerge from their eggs in their thousands to make the dangerous journey down the beach and into the relative safety of the sea.

The adults are carnivorous bottom-feeders that mainly hunt crabs. They have an unusually broad carapace, and their small size makes them agile swimmers. The carapace changes colour with age: yearlings are often almost black, while adults are light olive-grey. A closely related species, the Olive Ridley Turtle (*L. olivacea*), lives throughout the tropics. It is much less endangered than the Atlantic Ridley, thanks to its wider distribution.

ORDER CHELONIA

Flatback Turtle
Natator depressus

LENGTH	1–1.2m (3¼–4ft)
WEIGHT	Up to 85kg (190lb)
HABITAT	Coasts, shallows

DISTRIBUTION North and northeastern Australia, New Guinea, Arafura Sea

Named after its carapace, which is only slightly domed, the Flatback has the most restricted distribution of any marine turtle. It lives in shallow waters between northern Australia and New Guinea, reaching south along the Great Barrier Reef. When adult, it is largely carnivorous, feeding on fish and bottom-dwelling animals such as molluscs and sea squirts.

Despite their restricted range, adult Flatbacks may swim over 1,000km (600 miles) to reach nesting beaches. Females dig an average of three nests each time they breed and lay a total of about 150 eggs. The young feed at the surface on planktonic animals. Instead of dispersing into deep oceanic water, like the young of other turtle species, they remain in the shallows over the continental shelf.

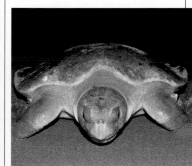

HAWKSBILL TURTLE
This turtle owes its common name to its sharp, powerful beak shaped like that of a bird of prey. The specimen photographed here, on a reef in the southern Red Sea, is holding a piece of soft coral, but its jaws are strong enough to detach even hard corals. It has two claws on each flipper.

ORDER SQUAMATA

Yellow-lipped Sea Krait
Laticauda colubrina

LENGTH	1–3m (3¼–10ft)
WEIGHT	Up to 5kg (11lb)
HABITAT	Coral reefs, mangrove swamps, estuaries

DISTRIBUTION Eastern Indian Ocean and southwestern Pacific

This species is the most widespread of the sea kraits – a group of four closely related species that lay eggs on land, instead of giving birth at sea.

It has a pale blue body, marked with eyecatching dark blue rings, and distinctive yellow lips, which give it its common name. The Yellow-lipped Sea Krait feeds on fish in shallow water, and although it has highly potent venom, it presents very little danger to humans because it is not aggressive and even when handled it rarely bites.

Unlike many other marine snakes, sea kraits have large ventral scales that give them good traction when they crawl, allowing them to move around comfortably on land. During the breeding season, they come ashore in large numbers to mate and lay clutches of up to 20 eggs. Once they have hatched, the young make their way to the shallows, before dispersing along coasts and out to sea.

ORDER SQUAMATA

Yellow-bellied Sea Snake
Pelamis platurus

LENGTH	1–1.5m (3¼–5ft)
WEIGHT	Up to 1.5kg (3lb)
HABITAT	Open water

DISTRIBUTION Tropical and subtropical waters in Indian Ocean and Pacific

This boldly striped yellow-and-black snake has venom that is more toxic than that of a cobra. It is also the world's most wide-ranging snake and one of the very few that lives in the surface waters of the open ocean. Its distinctive colours warn that it is poisonous, protecting it from many predators. It feeds on small fish trying to shelter in its shade, swimming forwards or backwards with equal ease to grab them with its jaws. Although its fangs are tiny, its potent venom occasionally causes human fatalities.

At sea, these snakes may form vast flotillas hundreds of thousands strong, and after storms, they may be washed up on beaches that lie far outside their normal range. However, the species has never managed to colonize the Atlantic Ocean, because cold currents stand in its way. Yellow-bellied Sea Snakes give birth to up to six young each time they breed.

ORDER SQUAMATA

Beaked Sea Snake
Enhydrina schistosa

LENGTH	1–1.5m (3¼–5ft)
WEIGHT	Up to 2kg (4½lb)
HABITAT	Shallow inshore waters

DISTRIBUTION Indian Ocean and western Pacific, from Persian Gulf to northern Australia

Notoriously aggressive and readily provoked, this widespread species is responsible for nine out of every ten deaths from sea-snake bite. Light grey with indistinct blue-grey bands, it has a sharply pointed head, slender body, and paddle-like tail. Its fangs are less than 4mm (⅛in) long, but its jaws can gape widely to accommodate large prey. It feeds mainly on catfish and prawns, swimming near the bottom in shallow, murky water, in coastal waters, mangrove swamps, estuaries, and rivers, locating its victims by smell and touch. Like all fish-eating snakes, it waits until its prey has stopped struggling, before turning it so that it can be consumed head-first.

Beaked Sea Snakes give birth to up to 30 young each time they breed, but their mortality is high, and only a small proportion of the young survive to become parents themselves. Despite their venom, these snakes are eaten by inshore predators, such as fish and estuarine crocodiles.

Turtle-headed Sea Snake

Emydocephalus annulatus

LENGTH	60–120cm (2–4ft)
WEIGHT	Up to 1.5kg (3lb)
HABITAT	Coral reefs and coral sand banks

DISTRIBUTION Indian Ocean and Pacific, from northern Australia to Fiji

This Australasian sea snake is highly notable for its colour variation, and also for its highly specialized lifestyle as a predator of fish eggs. The colour it most commonly takes is a plain blue-grey, which is found throughout its range. A striking ringed form lives in some parts of the Great Barrier Reef, while a rarer, dark or melanistic form is found on isolated reefs further east in the Coral Sea. The Turtle-headed Sea Snake moves slowly among living corals, methodically searching for egg masses either glued to the coral's branches or laid directly on the coral sand. When it finds an egg mass, it scrapes the eggs off with an enlarged scale on its upper jaw, which works like a blade. In most cases, parent fish leave the eggs unguarded, so the snakes can feed unhindered, but some species – such as damselfish – guard their eggs aggressively and try to keep the snakes away.

Little is known about this snake's reproductive habits, apart from the fact that the females give birth to live young. In keeping with their lifestyle, Turtle-headed Sea Snakes have tiny fangs (less than 1mm, 1/32 in, long) and they rarely try to bite. Their venom is one of the weakest of any sea snake, and instead of striking back at predators, they react to danger by disappearing into crevices in the reef.

Olive Sea Snake

Aipysurus laevis

LENGTH	1–2.2m (3–7ft)
WEIGHT	Up to 3kg (6½ lb)
HABITAT	Coral reefs, coastal shallows, estuaries

DISTRIBUTION Eastern Indian Ocean and western Pacific, from western Australia to New Caledonia

Plain brown or olive-brown above, with a paler underside, this common sea snake is one of six closely related species found in the reefs and shallow coastal waters of northern Australasia. Like its relatives, it has a cylindrical body, a flattened tail, and enlarged ventral scales – a feature normally found in snakes that spend some or all of their life on land. However, it is fully aquatic, hunting fish among the crevices and recesses of large corals. Instead of roaming throughout a reef, it often stays in the same small area of coral, rarely venturing into open water except after dark.

Olive Sea Snakes give birth to live young, producing up to five finger-sized offspring after a gestation period of nine months. Unlike the adults, the young are dark in colour, with a boldly contrasting pattern of lighter bands. This is gradually lost as they become mature. Olive Sea Snakes are naturally inquisitive and often approach divers. They have short-fangs and bite readily if provoked. Their venom is toxic and has been known to be fatal.

Leaf-scaled Sea Snake

Aipysurus foliosquama

LENGTH	Up to 60cm (2ft)
WEIGHT	Up to 0.5kg (1lb)
HABITAT	Coral reefs and coral sand banks

DISTRIBUTION Timor Sea (Ashmore and Hibernia reefs)

This fish-eating snake has one of the most restricted ranges of any sea snake, being confined to a group of remote coral reefs about 300km (185 miles) off the northwest coast of Australia. It is marked with contrasting bands or rings and gets its name from the characteristic shape of its dorsal scales. It lives in shallow water and rarely dives deeper than about 10m (33ft). Although venomous, it is rarely aggressive. Female Leaf-scaled Sea Snakes are larger than the males and give birth to live young.

EQUIPPED TO GRIP
An adult Marine Iguana sprawls on the sand, displaying the broad feet and long claws it uses to grip submerged rocks while it tears off mouthfuls of food.

ORDER SQUAMATA

Marine Iguana
Amblyrhynchus cristatus

LENGTH	Up to 1.5m (5ft), but often smaller
WEIGHT	Females 500g (1⅛lb); males 1.5kg (3⅓lb), sometimes larger
HABITAT	Rocky coasts

DISTRIBUTION Galapagos Islands

Restricted to the Galapagos Islands, this primeval-looking reptile is the only lizard that feeds exclusively at sea once it is an adult. The size and weight of this species varies between islands.

spiky crest

It has a blunt head with powerful jaws and a distinctive spiky crest that runs down its head, neck, and back. This lizard's powerful claws help it clamber over rocks, while its tail propels its through water. It feeds on seaweeds and other algae. The young feed above the water, but adults dive up to 10m (33ft), and can hold their breath for over an hour. During the day, they spend their time feeding and sunbathing to raise their body temperature.

During the breeding season, male Marine Iguanas engage in lengthy headbutting contests as they compete for mates. Females lay up to six eggs in the sand, and the young emerge after an incubation period of up to three months. Marine Iguanas have many natural predators, including sharks and birds of prey, and have been severely affected by introduced animals, such as rats and dogs.

blunt snout

PROFILE OF A GRAZER
Unlike predatory lizards, the Marine Iguana has blunt but powerful jaws. It secretes the surplus salt derived from its diet from glands near its nose.

SURVIVING THE COLD

Although the Galapagos Islands are on the Equator, they are bathed by the chilly Humboldt Current, which flows northward along the west coast of South America. Being a reptile, the Marine Iguana cannot generate its own body heat and needs special adaptations for feeding in these conditions. When it dives, its heart rate drops by about half, helping it conserve energy to keep its core temperature higher than the water around it. At night, the iguanas often huddle together to keep themselves warm.

BASKING IN THE SUN
When it returns to land, the Marine Iguana sprawls over rocks just above the surf to soak up warmth from the Sun through its skin.

ORDER SQUAMATA

Water Monitor
Varanus salvator

| **LENGTH** Up to 2.7m (9ft) |
| **WEIGHT** 15–35kg (35–75lb) |
| **HABITAT** Low-lying coasts, estuaries, rivers |

DISTRIBUTION Indian Ocean, western Pacific, from Sri Lanka to the Philippines and Indonesia

An opportunistic predator with a wide-ranging diet, this is one of the largest lizards that regularly ventures into salt water. It has a long neck, strong legs, and a flattened tail, which lashes from side to side when it swims. It feeds on anything it can overpower, diving to catch prey in the shallows or running it down onshore. Like other monitors, it also feeds on carrion. In some places, it can often be seen on the outskirts of coastal villages, where it scavenges on discarded remains. Water monitors breed by laying eggs, which the female places at the end of a burrow.

conspicuous markings when young

ORDER SQUAMATA

Mangrove Monitor
Varanus indicus

| **LENGTH** Up to 1.2m (4ft) |
| **WEIGHT** Up to 10kg (22lb) |
| **HABITAT** Mangrove swamps, coastal forests, estuaries, rivers |

DISTRIBUTION Western Pacific, from Micronesia to northern Australia

Similar in shape to the Water Monitor (left), this lizard has a comparable lifestyle, although it rarely swims far from the shore. Like all monitors, it has a long, supple neck and powerful clawed feet. Its tail is flattened laterally and is double the length of its body. Mangrove Monitors are very good swimmers and excellent climbers, hunting on the ground, in shallow water, and in trees. Fish make up a large part of their diet, although they eat a wide range of other food, including crabs, birds, other lizards, and even scavenged fishing bait. The

Water Monitor has had human help in expanding its range. In the past, it was introduced by humans throughout the western Pacific as a source of food, and more recently, the species has been introduced into some Pacific islands as as a way of controlling rats. Water Monitors lay up to a dozen eggs each time they breed and, like most lizards, their young hatch and develop without parental protection.

flexible neck longer than head

long, narrow head

OCEAN WANDERER
The Saltwater Crocodile is a strong swimmer. It has been seen 1,000km (600 miles) from the nearest coast.

ORDER CROCODILIA

Saltwater Crocodile
Crocodylus porosus

| **LENGTH** Up to 7m (23ft) |
| **WEIGHT** Up to 1 tonne (2,200lb) |
| **HABITAT** Open sea, rivers, estuaries, coasts |

DISTRIBUTION Indian Ocean, Pacific, from southern India to New Guinea and Australia

Also known as the Estuarine or Indo-Pacific Crocodile, this formidable predator is the world's largest reptile, and is also one of the few crocodilians

tail with vertical scutes

body slung between legs

that frequently swims far out to sea. Its power and ferocity are legendary, and it is thought to be responsible for more than 1,000 human deaths a year. The Saltwater Crocodile has powerful jaws housing teeth up to 13cm (5in) long. Its immensely tough skin is covered with thick scales. The scales on its back are armoured with bony deposits called osteoderms, while its tail has a double row of upright bony plates (scutes). Its nostrils close when it dives, but it cannot exclude water from its mouth. Instead, it has a valve at the entrance to its throat, which opens only when it swallows food.

pointed jaws

ARMOUR PLATING
The Saltwater Crocodile is protected by parallel rows of bony protruberances along its back.

It controls its body temperature by cooling down in water and warming up in the sun. Like other large crocodiles, the Saltwater Crocodile hunts by stealth, lurking close to the shore, hiding beneath the water with little more than its eyes and nose visible. When an animal comes within range, it bursts out of the water with explosive force, grabs its victim, and then drags it under until it drowns. Crocodiles cannot chew their food – instead, they tear it to pieces, digesting scales, skin, and even bones. Their natural prey includes birds, fish, turtles, and a wide variety of mammals, such as wild boar, monkeys, horses, and water buffalo. Females lay up to 90 eggs in a waterside mound, carrying their young to the water when they hatch. Saltwater Crocodiles are hunted in many parts of their range, making large specimens rarer than they once were.

ORDER CROCODILIA

American Crocodile
Crocodylus acutus

| **LENGTH** Up to 5m (16½ ft) |
| **WEIGHT** 180–450kg (400–1,000lb) |
| **HABITAT** Estuaries, open sea, coasts, lagoons |

DISTRIBUTION Caribbean Sea, adjoining areas of Atlantic, Pacific coast of Central and South America

Of the four species of crocodile found in the Americas, this is the only one that – as an adult – is equally at home in both fresh water and the sea. When fully grown, it is olive brown, with a narrow-tipped snout, broad back, and a powerful, tapering tail. Its bony deposits (osteoderms) are smaller than those of other crocodiles. When young, American Crocodiles feed on fish and small land animals, but adults often eat turtles, cracking them open in their jaws. Females bury their eggs in sand, laying about 40 every time they breed. Like all crocodiles, this species has been affected by being hunted for its skin, and by coastal development. Its stronghold is in Central America, but a few hundred individuals live in Florida, at the north of its range.

protruding eyes with vertical pupils

powerful jaws

BIRDS

DOMAIN	Eucarya
KINGDOM	Animalia
PHYLUM	Chordata
CLASS	Aves
ORDERS	29
SPECIES	About 9,500

BIRDS THAT HAVE ADAPTED to life at sea spend their lives in the air above the surface, in the upper layers of the open ocean, or along shorelines. Shore–(littoral) based birds rarely range far from land, and some visit the coast only at certain times of year. Others are pelagic, often remaining at sea for months on end and returning to land only to breed. Unlike land birds, many pelagic sea birds breed in large colonies on islands and cliffs, deserting them when the breeding season ends.

GUILLEMOT DIVING
Using their wings as hydrofoils, Common Guillemots speed through icy water in search of fish. Members of the auk family, they are common in northern seas.

ANATOMY

There is no such thing as a typical sea bird, although pelagic birds share many adaptations for life at sea. These include webbed feet, highly waterproof plumage, and glands that get rid of excess salt. Most terrestrial birds have hollow, air-filled bones (an adaptation that helps to save weight), but in diving species, such as penguins, the bones are denser and the air spaces reduced.

Some plunge–divers, including gannets and pelicans, have air sacs under their skin. These cushion the impact as they hit the water and help them to bob back to the surface with their prey. Compared to these marine species, shoreline birds show few specific adaptations for life in or near salt water but, like all birds, they have bills specialized for dealing with different kinds of food.

streamlined bill

narrow wings ideal for long-distance flight

FLYING DIVER
The Northern Gannet's streamlined shape is typical of a plunge diver. Its nostrils open inside its bill, enabling it to keep out water when it hits the surface.

webbed feet

BILL ADAPTATIONS
Apart from waterfowl, most birds of the sea and shore are carnivores, with bills that are adapted for different kinds of animal prey. A pelican's bill and pouch work like a scoop, while an albatross's hooked bill can grip slippery prey, such as jellyfish. Sea eagles catch their prey with their talons, but then use their bills to tear it into pieces. Curlews have long bills that can probe for animals buried in mud.

food pouch
PELICAN

tubular, external nostril
hooked tip
ALBATROSS

hooked bill
SEA EAGLE

long bill can probe deep into estuarine mud
CURLEW

HABITATS

Birds live throughout the world's oceans and shorelines, from the Equator to the poles. Less than 200 species are truly pelagic, meaning that they ply the oceans. These oceanic birds include albatrosses, which have wingspans of up to 3.5m (11ft), and much smaller species, such as shearwaters and terns. Although they feed on sea animals, their true habitat is the air: the Sooty Tern, for example, hardly ever rests on the water and may spend its first five years entirely on the wing. However, food is widely scattered in the open oceans, which is why the majority of sea birds live closer to land.

Most diving sea birds feed in the shallow waters over continental shelves, while rocky coasts and mudflats are key habitats for waders and gulls. Estuaries are important habitats for coastal birds. Their muddy silt often harbours numerous worms and molluscs, accessible at low tide. In the tropics, mangrove swamps attract birds for the same reason; they also have the added bonus of trees, in which birds nest and roost.

OCEAN WANDERER
The Black-Browed Albatross travels long distances in search of good feeding grounds. Its diet includes crustaceans, fish, squid, and carrion.

COASTAL WADERS
Eurasian Oystercatchers feed in a variety of coastal habitats, from rocky shores to mudflats. These birds are waiting for the tide to turn so that they can start to feed.

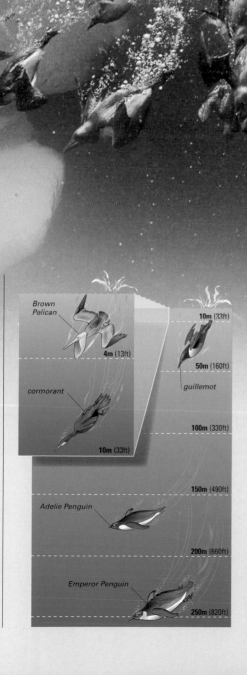

FEEDING METHODS

Marine birds have evolved several ways of hunting their food. Most spectacular are the plunge divers – birds such as gannets, boobies, and Brown Pelicans – which slam into shoals of fish from heights of up to 30m (100ft). Diving sea birds also include many that operate from the surface, such as cormorants and penguins. Emperor Penguins typically dive to 150m (500ft), but can descend to more than 265m (870ft) – the greatest depth for any bird.

Many oceanic birds, such as albatrosses and petrels, hunt on the wing, snatching animals or scraps from the surface. Kleptoparasitic birds, such as frigatebirds, which harass other birds into disgorging their catch, also hunt on the wing. Coastal birds often probe for food in the shallows or along the tideline, but skimmers slice through the water, holding their lower bill underwater while in flight – a remarkable technique that works only if the surface is flat and calm.

MAXIMUM FEEDING DEPTHS
Plunge divers (left), such as the Brown Pelican, rarely reach more than a few metres beneath the surface. Deeper divers, such as the penguins, use their wings or feet to propel themselves, often staying under for several minutes.

DAWN PATROL
Trailing its beak in the water, a skimmer searches for food in the calm waters of a lagoon.

SEA BIRDS UNDER THREAT

The inexorable increase in fishing and shipping has had a significant impact on many coastal and marine birds. Sea birds are often harmed directly, becoming entangled in nets or caught in oil spills. They can also be harmed indirectly, when fishing reduces their food supply. Global warming poses yet another threat: changing sea temperatures can trigger major changes in the fish stocks on which birds feed.

COLLATERAL DAMAGE
Caught in a fishing net, this cormorant is one of thousands of birds that drown every day. Diving birds have difficulty seeing plastic netting underwater and often become trapped.

DISPERSAL AND MIGRATION

Marine birds can range over a huge distance in their lifetime. Some, such as the Northern Gannet, disperse over wide areas of ocean, returning to isolated colonies to breed. The dispersal instinct of Northern Gannets is strongest in young birds and slowly declines during the four years that it takes them to mature sexually. From then onwards, adults congregate at their colonies in spring and summer, dispersing again when their chicks have left the nest.

Many other birds, such as the Grey Phalarope, migrate between distinct summer and winter ranges. During their migrations, they can be seen "on passage" between their two homes. In the species profiles on the following pages, distribution maps show all the places where a species occurs – its summer and winter ranges, as well as those regions it migrates through.

NORTHERN GANNET
A typical dispersing species, this bird nests in colonies scattered around the north Atlantic. When not breeding, it wanders as far south as the tropics, usually over continental shelves.

- summer distribution
- winter distribution

GREY PHALAROPE
This migrant nests in the high Arctic, and overwinters in the southeast Pacific and eastern Atlantic. An extensive network of migration routes means that it is seen in many parts of the world.

- summer distribution
- winter distribution

TREE NESTER
Frigatebirds are unusual among marine birds in that they nest in shrubs and tress.

BREEDING

Once they reach adulthood, all sea birds have to come to land to breed. Some species nest on their own, but many form large colonies – often because secure nesting sites are few and far between. Cliffs and islands are favourite locations, as they offer the best protection from predatory mammals. Fulmars and auks nest in burrows or fallen rocks, but most sea birds lay their eggs in the open, using little or no nesting material. Compared to terrestrial birds, they have small clutches. Cormorants often lay three or four eggs, but many other marine birds, such as albatrosses and puffins, lay a single egg each year. These birds are often long-lived, but their low reproductive rate makes them vulnerable to environmental problems, such as oil spills or climate change.

MIXED COLONY
Guanay Cormorants, boobies, and Brown Pelicans nest in dense colonies on the desert islands off the coast of Peru – an area rich in fish.

MARINE BIRD CLASSIFICATION

Of the world's 27–29 bird orders, only two are exclusively marine: the penguins, and the albatrosses and petrels. A further eight orders contain a mixture of terrestrial, coastal, and marine species.

WATERFOWL
Anseriformes

149 species
Most species of ducks, geese, and swans live on, or near, fresh water and often move to coasts for the winter. A few are totally marine, and live in inshore waters.

KING PENGUIN

PENGUINS
Sphenisciformes

17 species
These exclusively marine birds have lost the ability to fly. Most species are found in the Southern Ocean, but their range also extends northwards in cold-current regions, reaching as far as the Galapagos islands.

DIVERS AND LOONS
Gaviiformes

5 species
These sleek, fish-eating birds dive from the water's surface, propelling themselves with their feet. Divers are found mainly in the far north. They breed inland by fresh water, but often overwinter at sea.

ALBATROSSES AND PETRELS
Procellariiformes

108 species
Totally marine birds occuring throughout all oceans, albatrosses and petrels return to land only to breed. Their external nostrils lend a good sense of smell. Most remain airborne for days, snatching food from the sea's surface.

GREBES
Podicipediformes

22 species
These fish-eating birds have lobed feet set far back along their bodies. Most grebes live in freshwater habitats, but some migrate to coastal waters after the breeding season.

HERONS AND EGRETS
Ciconiiformes

119 species
These long-legged birds typically stalk their prey in shallow water or in marshy habitats. Most live inland, but several are found on coasts and coral reefs and in mangrove swamps. They often roost communally at night.

PELICANS AND RELATIVES
Pelecaniformes

65 species
This large group of sea birds includes pelicans, cormorants, tropicbirds, frigatebirds, and gannets. All are fish eaters, catching their food either by plunging into the water from the air or by diving from the surface. Found worldwide, they live on coasts and at sea and often feed in flocks. Some species, particularly cormorants, also frequent freshwater habitats.

BIRDS OF PREY
Falconiformes

307 species
Predatory birds, these species have hooked bills and sharp talons for snatching their prey. As a group, birds of prey are largely terrestrial, but some species specialize in catching fish, and can often be seen on coasts. They rarely venture far out to sea.

KELP GULL

WADERS, GULLS AND AUKS
Charadriiformes

343 species
This diverse order contains coastal and oceanic species, including many long-distance migrants. Diets are varied and feeding methods range from plunge-diving to shoreline scavenging. Many species are gregarious, feeding and nesting in colonies.

KINGFISHERS AND RELATIVES
Coraciiformes

191 species
These are primarily birds of forests or fresh water, although some species feed along coasts and inshore waters. They dive on prey from the air, taking off again directly after catching it, although they can swim.

ORDER ANSERIFORMES

Brent Goose
Branta bernicla

LENGTH	55–66cm (22–26in)
WEIGHT	1.3–1.6kg (3–3½lb)
HABITAT	Estuaries, tundra, coastal grassland

DISTRIBUTION Arctic (breeding); North America, northwest Europe, China, Japan (non-breeding)

A compact bird with a grey body, black head, and black neck, the Brent Goose breeds in the High Arctic but winters on coasts at temperate latitudes – a pattern followed by many other wildfowl. Its preferred food is eelgrass, a marine plant that grows in shallow water, but in its winter quarters it also grazes in coastal fields. Brent Geese nest in colonies in low-lying coastal tundra, laying up to five

eggs and raising a single brood each year. Like many birds in the High Arctic, their numbers undergo steep fluctuations. In mild summers, most of their goslings survive, but if conditions are unusually cold, very few young live long enough to migrate when summer comes to an end.

ORDER ANSERIFORMES

Common Eider
Somateria mollissima

LENGTH	50–71cm (20–28in)
WEIGHT	1.2–2.8kg (2¾–6¼lb)
HABITAT	Shallow coasts, estuaries

DISTRIBUTION Arctic Ocean, north Atlantic, north Pacific

This heavily built duck is a common sight on Arctic coasts, where it dives to catch molluscs and crabs, cracking them open with its powerful bill. The females are mottled brown, while the males (below) are mainly black and white, with a pink breast and greenish neck. Common Eiders breed in groups, building their nests close to the sea. After the breeding season, they move to more temperate zones in the south of their range for the winter months.

HUMAN IMPACT

EIDERDOWN

To keep the eggs and young warm, female eiders line their nest with down feathers plucked from their breast. Eiderdown is a superb insulator and has long been used as a filling for clothes and bedding. It is still collected in Iceland, although demand has dwindled following the introduction of synthetic fibres.

DOWN OF THE COMMON EIDER

ORDER ANSERIFORMES

Magellanic Flightless Steamer Duck
Tachyeres pteneres

LENGTH	61–76cm (24–30in)
WEIGHT	4.0–4.5kg (8¾–10lb)
HABITAT	Rocky coasts, inshore waters

DISTRIBUTION Southern South America

This heavily built duck is one of three closely related species, all from South America, that have lost the ability to fly. Like other steamer ducks, it has mottled grey plumage, yellow legs, and a robust, yellow-orange bill. It feeds on mussels, crabs and other small animals, diving among kelp beds to find its food. If threatened, it paddles noisily across the water with its wings – behaviour known as "steaming".

ORDER ANSERIFORMES

Red-breasted Merganser
Mergus serrator

LENGTH	52–58cm (20–23in)
WEIGHT	1–1.25kg (2¼–2¾lb)
HABITAT	Coasts, estuaries, lakes, rivers

DISTRIBUTION Arctic and subarctic (breeding); temperate coasts (non-breeding)

This is one the most widespread sawbill ducks – ducks that have narrow beaks with serrated edges, like the teeth of a saw. All these birds dive for fish, using their specially adapted bills to grip their slippery prey. Like other sawbills, the Red-breasted Merganser has an elongated body, a long neck, and orange-red legs. Males (right) have a metallic green head and shaggy crest, while the female's head is rust-coloured, with a less flamboyant crest. These birds breed near fresh water, but spend the winter on coasts, where the water is less likely to freeze. Females build a nest in dense cover, or in a tree-hole, lining it with down. They lay up to 11 eggs, raising a single brood a year. Red-breasted Mergansers are hunted in some parts of their range in order to protect fish stocks, although there is little evidence that they actually do much harm.

ORDER ANSERIFORMES

Common Shelduck
Tadorna tadorna

LENGTH	58–67cm (23–26in)
WEIGHT	0.85–1.45kg (2–3¼lb)
HABITAT	Coasts, estuaries, salt lakes

DISTRIBUTION Europe, North Africa, Asia

With its brightly coloured body and orange-red bill, the Common Shelduck is an eyecatching inhabitant of muddy shores. Normally seen in pairs, it feeds by dabbling in mud to collect small animals exposed by the falling tide. It nests in holes, and raises up to nine young each year. After breeding, Common Shelduck gather together to moult in flocks of up to 100,000 birds.

ORDER SPHENISCIFORMES

King Penguin

Aptenodytes patagonicus

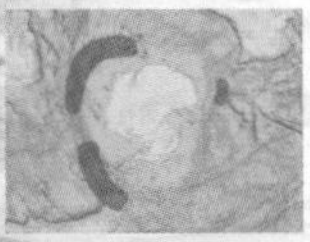

HEIGHT	85–95cm (33¹/₂–37¹/₂ in)
WEIGHT	12–14kg (26–31lb)
HABITAT	Rocky coasts, open ocean

DISTRIBUTION Southern Ocean, subantarctic islands including Falkland Islands

This is the largest penguin found on shores outside Antarctica. Like its close relative the Emperor Penguin, it has a blue-black body with a white chest and conspicuous, yellow-orange markings on its head. Males and females look identical, and they share the task of incubating the single egg. Instead of building a nest, they cradle the egg on their broad webbed feet, where it is kept warm by a flap of skin. Their bodies are protected from the cold by short, densely-packed feathers and a thick layer of blubber. King Penguins feed on fish and squid, diving to depths of over 200m (650ft) to hunt their prey. At one time, these birds were exploited commercially for their blubber, oil, and feathers, but today they are fully protected.

BREEDING OUT OF STEP

King Penguins have a breeding cycle found in no other sea bird. The cycle begins in November – the start of the southern summer – when the female lays her first egg. The chick takes 55 days to hatch, then stays with its parents for 11 months. Once the chick is independent, the female must complete her moult before laying again, this time in late autumn. As a result, the King Penguin's breeding cycle takes 18 months and moves in and out of phase with the calendar year.

KING PENGUIN CHICKS

ORDER SPHENISCIFORMES

Emperor Penguin
Aptenodytes forsteri

HEIGHT	110–115cm (43–45in)
WEIGHT	35–40kg (77–88lb)
HABITAT	Sea-ice, rocky coasts, open ocean

DISTRIBUTION Southern Ocean, Antarctica

The Emperor is the world's largest penguin and the only species that breeds in Antarctica during the southern winter. In shape and markings, it is very similar to the King Penguin, but it can be over twice its weight. Rarely found outside Antarctic waters, it feeds among broken sea-ice, diving to depths of up to 530m (1,750ft). It can remain underwater for as long as 20 minutes, and may travel distances of up to 1,000km (600 miles) in search of food. The Emperor Penguin breeds in scattered colonies on the ice itself. Adult females lay a single egg in early winter, and then transfer it to the male. During the winter darkness, while the females feed at sea, the males huddle together with their eggs balanced on their feet and protected within a fold of feathery skin. The incubation period lasts about two months. By the end of it, the males have lost about half their body weight. The females return when the chicks hatch, releasing the males, who head out to sea.

ORDER SPHENISCIFORMES

Chinstrap Penguin
Pygoscelis antarctica

HEIGHT	71–76cm (28–30in)
WEIGHT	3–4.5kg (6½–10lb)
HABITAT	Rocky coasts, open ocean

DISTRIBUTION Southern Ocean, Antarctic Peninsula, subantarctic islands

Easily identified by the black line around its chin, the Chinstrap Penguin is one of the most abundant penguin species. Males and females look identical, with blue-black bodies, white undersides, and straight black bills. They live at sea for most of the year, feeding in open water north of the polar ice. When swimming at speed, they often leap clear of the water, or "porpoise", which allows them to breathe and coats their bodies with a layer of air bubbles, reducing friction with the water.

In November, Chinstraps return to their breeding colonies on ice-free shores in Antarctica and on islands in the Southern Ocean. Here, they make their nests by scraping together small stones to form a shallow cup. Chinstraps tend to be more aggressive than other penguins, particularly when breeding. They steal stones from their neighbours and chase away any larger penguins that attempt to nest nearby. The female lays two eggs, and her chicks fledge and set off for the sea by February or March, when the southern autumn begins. Chinstraps feed almost entirely on krill, and their current population growth, like that of Antarctica's krill-eating seals, may be linked to the decline of krill-eating baleen whales.

chinstrap marking

ORDER SPHENISCIFORMES

Macaroni Penguin
Eudyptes chrysolophus

HEIGHT	70cm (27½in)
WEIGHT	4.2kg (9¼lb)
HABITAT	Rocky coasts, open ocean

DISTRIBUTION Southern Ocean, Antarctic Peninsula, subantarctic islands, southern South America

Macaroni Penguins are often seen together with a similar species of penguin, the Rockhopper. However, Macaronis are significantly larger and have distinctive flame-yellow crests that run above each eye and meet on the forehead. They are also found further south, breeding on ice-free coasts on the Antarctic Peninsula. Their breeding colonies are extremely noisy, some containing over a million pairs spaced out just beyond pecking distance of each other. Macaroni Penguins lay two eggs a year, and both parents help with incubation. Their reproductive rate is low, because only one nestling normally survives.

ORDER SPHENISCIFORMES

Little Penguin
Eudyptula minor

HEIGHT	40–45cm (16–18in)
WEIGHT	1kg (2¼lb)
HABITAT	Rocky and muddy coasts, open ocean

DISTRIBUTION Southern Australia, New Zealand, Tasman Sea and Southern Ocean

This is the smallest penguin, and it is also the only one that remains offshore during daylight, coming onto land after dark. It has a white underside, a grey-blue back and head, and no distinctive markings. During daylight, Little Penguins are often seen in small flotillas offshore, resting on the surface and periodically diving to catch fish.

SAFETY AFTER DARK

In some parts of their range – such as Phillip Island, near Melbourne – thousands of Little Penguins can be seen scrambling ashore as the light fades. This behaviour protects them from most predators, although not from introduced mammals such as foxes and domestic dogs.

When feeding, they circle around small fish to concentrate them into a close-knit group, before swimming through the shoal and snapping them up. Unlike other penguins, they do not leave the water when they travel at speed. Little Penguins usually nest in burrows or among fallen rocks, but may set up home in breakwaters and under houses and sheds. Each female lays a clutch of two eggs and raises up to two broods a year.

ORDER SPHENISCIFORMES

Magellanic Penguin
Spheniscus magellanicus

HEIGHT	71cm (28in)
WEIGHT	5.5kg (12lb)
HABITAT	Rocky coasts, open ocean

DISTRIBUTION Southern South America, Falkland Islands, south Atlantic and south Pacific

One of two species of black-and-white penguin from South America, the Magellanic Penguin is identified by the two black bands across its breast (see below). It feeds in the cold waters that flow northward from the Southern Ocean, eating small, shoal-forming fish such as sardines. Like its close relative, the Humboldt Penguin, it nests in burrows, raising up to two chicks each year.

ORDER SPHENISCIFORMES

Jackass Penguin
Spheniscus demersus

HEIGHT	60–70cm (24–28in)
WEIGHT	5kg (11lb)
HABITAT	Rocky coasts, open ocean

DISTRIBUTION Coastal waters of southern Africa, south Atlantic and southern Indian Ocean

Also known as the Cape Penguin, the Jackass is the only penguin that breeds in Africa. Physically, it bears a strong resemblance to the Magellanic Penguin (see left) from South America, although it has a single black breast band rather than two. It feeds on small fish such as pilchards, sardines, and anchovies, and gets its name from its braying call, which may be heard onshore when it breeds. Jackass Penguins nest in burrows, and in the past, many of their nesting sites were destroyed by farmers collecting their droppings, or guano, for use as fertilizer. Today, depletion of food stocks due to overfishing and oil spills are two major threats that they face, along with competition from fur seals for breeding sites. Their numbers are in sharp decline.

OCEAN LIFE

EMPEROR PENGUINS
The Emperor Penguin is one of the hardiest animals in the world, able to withstand exposure to blizzards on land and deep dives in the freezing waters of the Southern Ocean. The young and adult birds seen here are sheltering from a winter snowstorm in temperatures of -20 to -30°C (-4 to -22°F).

ORDER PROCELLARIIFORMES

Black-footed Albatross

Phoebastria nigripes

LENGTH	68–74cm (27–29in)
WEIGHT	3–3.5kg (6½–7¾lb)
HABITAT	Open ocean, atolls, isolated islands

DISTRIBUTION North Pacific, Johnston Island and Marshall Islands

This dark-coloured sea bird is often seen in summer off North America's west coast. One of three species of albatross found in the north Pacific, its dark underwings distinguish it from the other two. It is fond of scavenging, and often follows trawlers and shrimping boats to catch discarded offal. Black-footed Albatrosses breed in colonies on islands in the central and western Pacific. Like other albatrosses, they perform elaborate courtship displays. All albatrosses are monogamous, pairing up to breed with the same partner each autumn.

HUMAN IMPACT

LONG-LINE FISHING

The Black-footed Albatross is a frequent victim of long-line fishing, which involves trailing lines that carry thousands of baited hooks. Albatrosses swallow the bait and become caught. Long-line fishing is estimated to kill at least 300,000 sea birds of all kinds each year.

VICTIM OF DROWNING
Albatrosses usually swallow their food whole, so long-line hooks become lodged in their stomachs and drag the birds underwater to drown.

ORDER PROCELLARIIFORMES

Short-tailed Albatross

Phoebastria albatrus

LENGTH	84–94cm (33–37in)
WEIGHT	3–5kg (6½–11lb)
HABITAT	Remote islands (breeding); open ocean

DISTRIBUTION North Pacific, Tori Shima Island and Senkaku Islands

This north Pacific albatross almost became extinct during the early 1900s due to demand for its feathers. By 1950, only about 20 birds were left. Thanks to conservation measures, the population now stands at nearly 2,000, all breeding on remote islands in the far west of its range. Adults are mostly white, with black flight feathers, pink bills, and golden-yellow heads.

ORDER PROCELLARIIFORMES

Light-mantled Sooty Albatross

Phoebetria palpebrata

LENGTH	79–98cm (31–39in)
WEIGHT	2.5–4.5kg (5–10lb)
HABITAT	Remote islands (breeding); open ocean

DISTRIBUTION Southern Ocean, isolated islands in south Atlantic and southern Indian Ocean

Together with its close relative the Sooty Albatross, this is one of two southern albatrosses that have sooty brown plumage, as opposed to white and black. The Sooty Albatross is brown all over, but the Light-mantled species has a pale grey nape and back – the feature that gives it its name. A graceful glider, it feeds on fish, squid, and crustaceans. It is also highly inquisitive and often follows ships. After spending the winter at sea, it returns to its breeding sites by August, the start of the southern spring. Female birds lay a single egg in early summer, and the chicks become independent about four months after they hatch – a relatively rapid development compared with that of the larger albatrosses.

ORDER GAVIIFORMES

Great Northern Diver

Gavia immer

LENGTH	70–90cm (28–35in)
WEIGHT	3–4.5kg (6½–10lb)
HABITAT	Freshwater lakes (breeding); coasts

DISTRIBUTION Northern North America, Greenland, Iceland, Europe, north Pacific, north Atlantic

This striking bird is best known for its haunting cry, which echoes across freshwater lakes during the summer breeding season. During the winter, the same bird is a common visitor in coastal waters, although at this time of year its black-and-white breeding plumage is replaced by less eye-catching shades of brownish black and grey. Like other divers, it has a streamlined body, small wings, and webbed feet set far back – a feature that makes it clumsy on land. On water, it is far more graceful. It floats with its bill held at a characteristic upward slant and can dive to depths of over 75m (250ft) to catch fish – its principal food.

During the summer months, Great Northern Divers usually live in pairs, carrying out spectacular courtship displays. When the breeding season comes to an end, they migrate to sheltered coasts, where there is less risk of icing. In North America, flocks of several hundred often gather on the Great Lakes, before heading south as far as the coast of Florida. In Europe, they winter on Atlantic coasts, dispersing as far south as Portugal.

ORDER PROCELLARIIFORMES

Wandering Albatross

Diomedea exulans

LENGTH 1.1–1.35m
(3½–4½ft)

WEIGHT 8–11.5kg
(18–25lb)

HABITAT Remote islands
(breeding); open ocean

DISTRIBUTION Southern Ocean, south Atlantic,
southern Indian and Pacific oceans

This legendary sea bird has the largest
recorded wingspan of any bird, at
up to 3.5m (11½ft). It is restricted
to the windswept southern oceans,
where it feeds mainly on squid,
snatching its food from the surface
of the water. It is capable of remaining
airborne for weeks at a time and
frequently follows ships, soaring over
the waves on its stiff, outstretched
wings. The Wandering Albatross takes
up to 11 years to mature, and during
that time it gradually loses its juvenile
plumage, becoming all white except
for black markings on the tips and
trailing edges of its wings. These birds
nest on remote islands, typically
breeding in alternate years.

LONG INCUBATION

Wandering Albatrosses build
large, mound-like nests from
mud, grass, and moss. Their
single egg has one of the longest
incubation periods of any egg,
taking between 75 and 82 days
to hatch. The solitary chick then
remains in the nest for up to nine
months, where it is fed by both
its parents. During very severe
weather, the chick may be left
unattended for days at a time.

ORDER PROCELLARIIFORMES

Black-browed Albatross

Thalassarche melanophrys

LENGTH 83–93cm
(33–37in)

WEIGHT 3–5kg
(6½–11lb)

HABITAT Remote islands
(breeding); open ocean

DISTRIBUTION Southern Ocean, south Atlantic,
southern Indian and Pacific oceans

Also known as the Black-browed
Mollymawk, this is the most
numerous and widespread of the
albatrosses. It is found from Antarctica
to the edge of the tropics, and in
places even further north. Its wings,
back, and tail are greyish black, and
it has a distinctive black brow above
each eye. It feeds on fish, squid,
octopus, and crustaceans, and is also
a frequent ship-follower, congregating
in large numbers when waste is
thrown overboard. Black-browed
Albatrosses breed on remote islands
and take at least five years to become
mature. They are among the few
southern albatrosses that regularly
cross the Equator – isolated sightings
have been recorded as far north as
the British Isles.

ORDER PROCELLARIIFORMES

Southern Giant Petrel

Macronectes giganteus

LENGTH	86–99cm (34–39in)
WEIGHT	5kg (11lb)
HABITAT	Coasts, open sea; nests on ice-free coasts

DISTRIBUTION Southern hemisphere, from Antarctica as far north as the tropics

Part-scavenger and part-predator, this large petrel is often seen on the fringes of penguin colonies or near the carcasses of dead seals and whales. It uses its powerful bill to tear apart carrion and to kill young birds. Most adults have a pale head and a dark greyish brown back, but some are almost completely white with scattered black flecks.

tubular nostrils

ORDER PROCELLARIIFORMES

Northern Fulmar

Fulmarus glacialis

LENGTH	45–51cm (18–20in)
WEIGHT	700–900g (1½–2lb)
HABITAT	Rocky coasts, open sea

DISTRIBUTION North Pacific, north Atlantic, ice-free areas of Arctic Ocean

Often mistaken for a gull, this fulmar is actually a petrel and, like other petrels, has distinctive tubular nostrils. Common throughout northern waters, it is often seen flying over cliffs on its stiff, outstretched wings. Its weak feet make it clumsy on land, and its eyes are dark with a distinct brow ridge. Most Northern Fulmars in the Atlantic have white bodies and blue-grey upper wings, but in the Pacific many of the birds are much darker. Northern Fulmars feed on small animals at or near the sea's surface, and they gather in large flocks to scavenge around fishing boats. They breed on exposed cliff ledges, with each female laying a single egg directly onto the rock. The incubation period is 52 days – almost twice as long as that in gulls of similar size. Despite its low reproductive rate, the Northern Fulmar has increased both in range and in numbers in recent years. It is exceptionally long-lived for its size, with ages of over 50 years recorded.

ORDER PROCELLARIIFORMES

Snow Petrel

Pagodroma nivea

LENGTH	30–35in (12–14in)
WEIGHT	250–450g (9–16oz)
HABITAT	Rocky and ice-bound coasts

DISTRIBUTION Antarctica, subantarctic islands, Southern Ocean

Despite its dainty appearance, the Snow Petrel is one of the world's most southerly breeding birds. This entirely white petrel nests on ice-free cliffs in and near Antarctica, to within 1,100km (680 miles) of the South Pole. It picks food from the surface of the sea, rarely straying far from the polar ice. Flocks of Snow Petrels are often seen sitting on icebergs.

ORDER PROCELLARIIFORMES

Bonin Petrel

Pterodroma hypoleuca

LENGTH	30cm (12in)
WEIGHT	225g (8oz)
HABITAT	Oceanic islands (breeding); open ocean

DISTRIBUTION Northwestern Pacific

There are over two dozen species of *Pterodroma* petrels, mostly in tropical and subtropical regions, and these are often difficult to distinguish at sea. The Bonin Petrel is a typical example from the northwestern Pacific, where it nests on scattered islands westwards from Hawaii. It has a small, short, slightly hooked bill and sharply pointed wings, and it is fast and agile as it speeds through the air just above the waves. It eats small planktonic animals, usually landing on the surface to feed. This petrel nests in burrows but has difficulty moving on land. To reduce the risk of attack from predators, the Bonin Petrel generally returns to land at night, when it may deliver regurgitated food to its single chick. The parents share the task of egg incubation over about 49 days. On remote islands, petrel colonies can be decimated by introduced predators, such as rats and cats. This species is one that has been badly affected.

ORDER PROCELLARIIFORMES

Fairy Prion

Pachyptila turtur

LENGTH	25–28cm (10–11in)
WEIGHT	150–225g (5–8oz)
HABITAT	Islands (breeding); open ocean

DISTRIBUTION Southern Ocean and adjoining waters

A small, oceanic petrel, the Fairy Prion has a pale body and blue-grey upper wings with a distinct, M-shaped black band. It lives in flocks and feeds at night, using its bill to sieve planktonic animals from the water. It breeds on isolated coasts, laying a single egg either in a burrow or in a hollow deep among fallen rocks.

ORDER PROCELLARIIFORMES

Great Shearwater

Puffinus gravis

LENGTH 46–53cm (18–21in)

WEIGHT 800–900kg (1¾–2lb)

HABITAT Oceanic islands (breeding); open ocean

DISTRIBUTION Atlantic Ocean, except off west coast of Africa south of Sierra Leone

This ocean-living sea bird is a wide-ranging migrant and is found across most of the Atlantic Ocean during the course of the year. It breeds in the far south, on some of the world's remotest islands. One of these, Nightingale Island in the Tristan da Cunha group, is home to about four million birds. Great Shearwaters have pointed wings with dark brown upper surfaces. Their undersides are much paler, making the birds look alternately black then white during their tilting flight. Sometimes vocal at sea, they wail and scream noisily from their burrows when breeding. These calls are thought to help incoming birds to locate their mates after dark.

ORDER PROCELLARIIFORMES

Wilson's Storm Petrel

Oceanites oceanicus

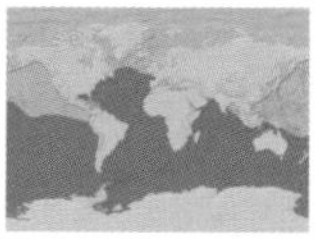

LENGTH 15–19cm (6–7½in)

WEIGHT 30–40g (1–1½oz)

HABITAT Coasts, islands (breeding); open ocean

DISTRIBUTION Worldwide except for north Pacific and extreme north Atlantic

Little bigger than a sparrow, Wilson's Storm Petrel is reputed to be the world's most numerous ocean-going sea bird. It breeds in widely scattered colonies, and its total population is unknown but may exceed 20 million. At sea this bird may be difficult to distinguish from its close relatives, but its plumage is uniformly sooty brown, apart from a band of white at the base of its tail. When feeding, storm petrels rarely settle on the water. Instead, they flutter their wings and patter the surface with their feet, pecking up planktonic animals. When food is abundant, they may suddenly appear in huge numbers then disappear with equal abruptness. Wilson's Storm Petrel breeds as far south as Antarctica, digging a burrow with its bill and feet. It migrates northward when the southern summer comes to an end.

ORDER PROCELLARIIFORMES

Short-tailed Shearwater

Puffinus tenuirostris

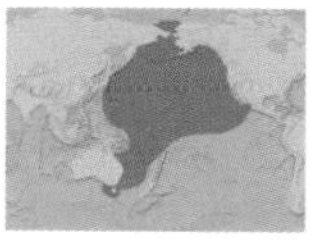

LENGTH 41–43cm (16–17in)

WEIGHT 500–700g (1–1½lb)

HABITAT Open ocean, offshore islands

DISTRIBUTION North Pacific, southwestern Pacific around southern coast of Australia

Awkward and ungainly on land, the Short-tailed Sheerwater is a tireless flier, skirting around most of the north Pacific during its annual migration. Like other shearwaters, it travels just centimetres above the waves in fast-moving flocks, interrupting its flight whenever it spots food. However, it has a narrower bill than other shearwaters, and its overall colour is a dark, smoky brown. It nests in vast island colonies, each pair producing a single chick. Fed on a rich diet of oily food, the chicks weigh more than their parents by the time they leave the nest. For several centuries, shearwater chicks have been harvested for their oil and meat. The practice continues today, although the numbers killed are now strictly controlled.

MIGRATION

The Short-tailed Shearwater has a unique figure-of-eight migration route, 33,500km (20,800 miles) long, that takes advantage of prevailing winds. After laying eggs in November and December, the birds head north in April and May, reaching the Bering Sea by August. They then move south along North America's west coast, before returning to their breeding colonies.

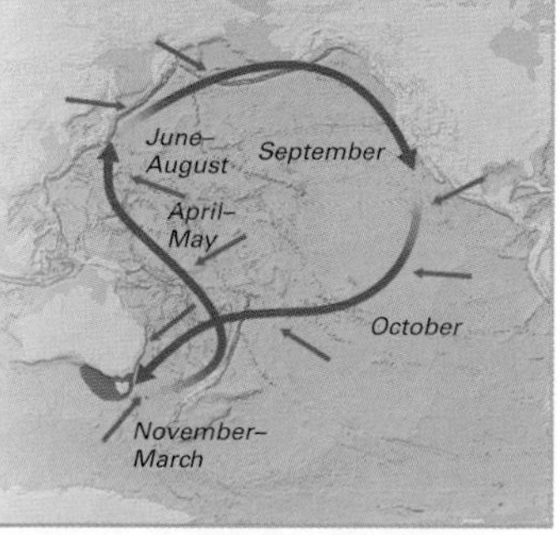
KEY
breeding area
migration route
wind direction
June–August
September
April–May
October
November–March

ORDER PROCELLARIIFORMES

Leach's Storm Petrel

Oceanodroma leucorhoa

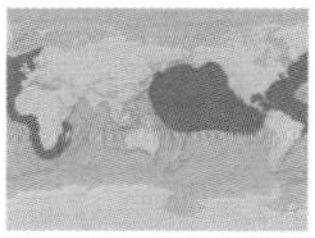

LENGTH 19–22cm (7½–8½in)

WEIGHT 40–50g (1½–1¾oz)

HABITAT Coasts, islands (breeding); open ocean

DISTRIBUTION North Pacific, north Atlantic, coastal North America and Aleutian Islands

Leach's Petrel is silent at sea, but it makes a high-pitched purring sound, interrupted by sharp whistles, in and near its nest. Unlike Wilson's Storm Petrel (see above), this species breeds in the northern hemisphere. It migrates southward in late summer, roaming throughout the north Pacific and much of the Atlantic. Small and brownish black, with a sharply forked tail, it flies rapidly, changing direction frequently as it scans the water's surface for food. It feeds on planktonic animals and small fish, pattering on the surface with its feet and occasionally settling on the water to rest. Leach's Storm Petrels breed in colonies, laying a single egg and returning to their burrows at night with food for their hatched young. In the far north, some birds delay nesting until August to avoid the 24-hour daylight of the Arctic summer, during which they would be more vulnerable to predators.

ORDER PROCELLARIIFORMES

Common Diving Petrel

Pelecanoides urinatrix

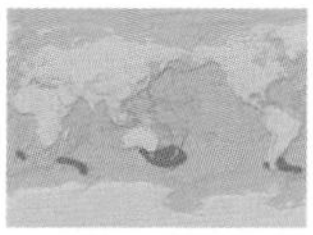

LENGTH 20–25cm (8–10in)

WEIGHT 110–130g (4–4½oz)

HABITAT Coasts, islands (breeding); open ocean

DISTRIBUTION Southern Ocean and adjoining waters and islands

This stubby bird with pointed wings and pale blue feet is the southern hemisphere's counterpart of the auklets (see p.403). Despite being unrelated, it shares the auklets' fast, low flight and their feeding technique. Instead of searching for food on the wing, like other petrels, it dives, using its wings to swim. It frequently flies straight through waves, emerging with rapidly whirring wings on the other side. This species nests in burrows, returning to its nests after dark. There are three other similar-looking species, all found in southern seas.

ORDER CICONIIFORMES

Grey Heron
Ardea cinerea

LENGTH	90–100cm (34–39in)
WEIGHT	1.6–2kg (3½–4½lb)
HABITAT	Estuaries, lagoons, coasts

DISTRIBUTION Europe, mainland Asia (except far north), Japan, Indonesia, Africa, Madagascar

Commonly seen in fresh water, the Grey Heron also frequently visits shores, especially in areas where lakes and ponds freeze in winter. Tall, grey-backed, and often immobile, it waits patiently for fish or other animals to come within range, then seizes them with a rapid jab of its dagger-like bill. On coasts, its feeding method restricts it to shallow water on rocky and low-lying shores, where it often follows the falling tide. Grey Herons fly with slow wingbeats, their heads hunched into their shoulders and their legs trailing behind. They nest in trees, typically inland near water.

ORDER CICONIIFORMES

Little Egret
Egretta garzetta

LENGTH	56–65cm (22–27in)
WEIGHT	300–450g (11–16oz)
HABITAT	Muddy coasts, mangrove swamps

DISTRIBUTION Southern Europe, Africa, southern Asia, Southeast Asia, Australasia

Pure white with black legs, a black bill, and bright yellow facial skin, the Little Egret is usually seen on its own or in scattered groups, wading quietly through shallow water on coasts. This bird feeds on fish and other shoreline animals that are disturbed by its approach. During the breeding season, both males and females grow long, lacy feathers on their heads and backs. They nest in trees, building flimsy nests out of sticks.

ORDER CICONIIFORMES

Pacific Reef Egret
Egretta sacra

LENGTH	60–70cm (24–27in)
WEIGHT	400–750g (14–26oz)
HABITAT	Coastal and freshwater wetlands

DISTRIBUTION Australasia, Pacific islands, western Pacific coast from Southeast Asia to Japan

This compact shoreline egret has two contrasting colour forms, so different that they look like separate species. One form (or morph) is completely white, with a pale yellow bill and yellow-grey legs. The other form has a similarly coloured bill and legs, but its plumage is dark grey. The balance between the two forms varies. In some islands in the tropical Pacific the white form predominates, but in New Zealand, the overwhelming majority are grey. Pacific Reef Egrets forage alone or in small groups, feeding on small fish, crabs, and molluscs. When hunting, they hold their heads and bodies almost horizontally and often shade the water with their half-spread wings. Unlike most egrets, they frequently nest on the ground, among fallen rocks or in coastal caves, as well as in low-growing trees.

ORDER PELECANIFORMES

Great Frigatebird
Fregata minor

LENGTH	86–100cm (34–39in)
WEIGHT	1.4–1.8kg (3–4lb)
HABITAT	Coasts, islands (breeding); open ocean

DISTRIBUTION Tropical regions in Indian Ocean and Pacific, sporadic in tropical Atlantic

With their extraordinarily long wings and slender bodies, frigatebirds are unrivalled experts at gliding flight. The five species all have glossy black plumage, strong, hooked bills, and small, webbed feet. The males also have a bright red throat pouch, which they inflate during courtship displays. Despite weighing less than a large gull, the Great Frigatebird has a wingspan of up to 2.3m (7½ft), allowing it to glide for hours while making only the merest flick of its wings. As it flies, it observes other sea birds as they feed, then pursues them to steal their catch. Frigatebirds also hunt their own food, snapping it up from the sea's surface. They nest in coastal bushes, where they make flimsy nests out of twigs.

ORDER PELECANIFORMES

Red-billed Tropicbird
Phaethon aethereus

LENGTH	Up to 50cm (19½in) excluding tail
WEIGHT	600–800g (1¼–1¾lb)
HABITAT	Coasts, islands (breeding); open ocean

DISTRIBUTION Eastern Pacific, Caribbean, tropical Atlantic, northeast Indian Ocean

The largest of the three species of tropicbird, this elegant sea bird spends most of its life flying over the open ocean, often hundreds of kilometres from land. From a distance the Red-billed Tropicbird resembles a dove, but for two highly distinctive tail streamers that flutter behind it as it flies. It feeds by plunge-diving, hovering to locate its prey before diving with half-folded wings into the sea. Despite being very buoyant, it seldom swims. Like other tropicbirds, it nests on remote coasts and oceanic islands and is rarely seen outside tropical waters.

ORDER PELECANIFORMES

Brown Pelican
Pelecanus occidentalis

LENGTH	1.2–1.6m (4–5¼ft)
WEIGHT	3.5–4.5kg (7¾–10lb)
HABITAT	Coastal waters, estuaries, islands

DISTRIBUTION Pacific and Atlantic coasts of North and South America, Galapagos Islands

Commonly seen inshore and in harbours, the Brown Pelican is the heaviest sea bird that fishes by plunge-diving. Groups of birds often fish together, skimming over the waves before rising into the air, folding back their wings, and hitting the water with a spectacular splash. The pelican's throat pouch balloons outwards underwater, scooping up prey, which it then swallows at the surface.

ORDER PELECANIFORMES

Northern Gannet
Morus bassanus

LENGTH	87–100cm (34–39in)
WEIGHT	2.8–3.2kg (6¼–7lb)
HABITAT	Open sea, rocky coasts, offshore islands

DISTRIBUTION Eastern and western coasts of north Atlantic

This highly streamlined bird with its gleaming white body and black-tipped wings is the most striking plunge-diver in the north Atlantic. Northern Gannets roam the seas with a distinctive pattern of flapping and gliding flight, attacking shoals of fish by diving from heights of up to 30m (100ft). They breed in crowded colonies on rocky islands and clifftops, laying a single egg

each year. Juveniles take five years to mature, gradually losing their brown plumage. During that time, they roam far over the ocean before returning to their native colony to breed.

ORDER PELECANIFORMES

Brown Booby
Sula leucogaster

LENGTH	64–76cm (25–30in)
WEIGHT	0.7–1.5kg (1½–3¼lb)
HABITAT	Inshore waters, rocky coasts, islands

DISTRIBUTION Tropical oceans worldwide, except southeastern Pacific

This booby is a superb diver. It is the most widespread booby and has distinct colour variations. Most Brown Boobies are brown all over, apart from a white underside. However, birds from the eastern Pacific have white heads and their bills are grey rather than the typical bright yellow. They all live in the same way, diving for fish and squid from heights of up to 30m (100ft). They also skim low over the surface, looking for flying fish, which they catch in mid-air. They often fly in front of ships, watching for fish caught up in the bow-waves, and they like to fish close to land, roosting on buoys or coastal trees. Despite their agility in the air, they are clumsy at take-off and landing.

PLUNGE-DIVING

Gannets and boobies all show adaptations for a plunge-diving lifestyle: forward-facing eyes, streamlined heads and bills; and nostrils with no external openings. Their wings fold back along the body just before the moment of impact, and the force is absorbed by air sacs under their skin.

AERIAL ATTACK
This sequence of photos shows how the wings fold during a dive.

ORDER PELECANIFORMES

Blue-footed Booby
Sula nebouxii

LENGTH	76–84cm (30–33in)
WEIGHT	1.5–2kg (2¼–4½lb)
HABITAT	Inshore waters, rocky coasts, islands

DISTRIBUTION Pacific coast of Central America, Galapagos Islands

This is one of six species of booby – a group of plunge-diving birds, closely related to gannets, that often have brightly coloured feet. The Blue-footed Booby is brown with white undersides. Its feet are greyish brown in juveniles but brilliant turquoise-blue in adults. Blue-footed Boobies often feed in flocks, hitting the water almost simultaneously when they locate a shoal of fish. Smaller than gannets, they are able to fish closer

inshore, sometimes diving into water less than 1m (3ft) deep. They nest in small colonies on offshore islands, laying their eggs on the ground.

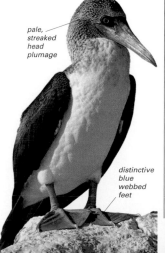

pale, streaked head plumage

distinctive blue webbed feet

ORDER PELECANIFORMES

Guanay Cormorant

Phalacrocorax bougainvillii

LENGTH	74–78cm (29–31in)
WEIGHT	1.75–2.25kg (4–5lb)
HABITAT	Desert coasts, islands, inshore waters

DISTRIBUTION Pacific coast of Peru and northern Chile

Boldly marked in black and white, with a conspicuous red patch around each eye, the Guanay Cormorant nests in huge colonies along the coast of the Atacama Desert, the most arid region on Earth. It feeds on anchovetas – small fish that abound in the cold waters of the Humboldt Current. Like other cormorants, it pursues fish underwater, holding its wings against its body and propelling itself with its legs. It floats low down in the water, periodically dipping its head beneath the surface to check for food. Guanay Cormorants have nested on the same offshore islands for millennia, depositing deep layers of desiccated droppings known as guano. During El Niño years, when the ocean temperature rises, shortage of food forces these cormorants to forage far afield, often as far north as Panama.

THE GUANO TRADE

Before the invention of synthetic fertilizers, nitrogen-rich guano was an extremely valuable commodity. Thousands of tonnes were exported from the South American coast to the northern hemisphere. Guano was also used in the manufacture of explosives.

GUANO MINING
Using picks and shovels, workers dig up compacted guano on an island off the coast of southern Peru.

ORDER PELECANIFORMES

Great Cormorant

Phalacrocorax carbo

LENGTH	80–101cm (32–40in)
WEIGHT	2–2.5kg (4¼–5½ lb)
HABITAT	Coasts, inshore waters, rivers, lakes

DISTRIBUTION Northeast North America, Europe, Africa, Asia, Australasia

Equally at home in fresh water and at sea, the Great Cormorant can be found across a vast swathe of the world, from Greenland to Australasia. From a distance, its plumage looks jet black, but close up it has a greenish metallic sheen, with white patches that vary between local races. Like its many relatives, it fishes by pursuit diving and its feathers are only partly waterproof. After feeding, it rests with its wings spread apart to dry. Great Cormorants have a strong, direct flight, with steady flapping interspersed with short glides. They can often be seen in small groups, skimming just above the surface of the sea or following rivers inland. They nest on rocky ledges and in trees, making a platform out of seaweed, flotsam, or twigs, and the females lay three or four greenish-white eggs. Great Cormorants are sometimes persecuted by anglers, particularly in trout-fishing regions, but they remain highly successful.

black flight feathers in adult

short, wedge-shaped tail

ORDER FALCONIFORMES

White-bellied Sea Eagle

Haliaeetus leucogaster

LENGTH	70–90cm (28–35in)
WEIGHT	2.5–4.2kg (5½–9¼ lb)
HABITAT	Inshore waters, rivers, lakes, reservoirs

DISTRIBUTION South and Southeast Asia, New Guinea, Australia

This black-and-white eagle makes an impressive sight as it soars over water with its wings, up to 2m (6½ft) wide, held in a shallow V shape. Its wide-ranging diet includes fish, water birds, turtles, and sea snakes, which it snatches from the surface, rarely entering the water. It also scavenges and forces smaller sea birds to drop their catch. It breeds close to water, building a large nest in a high tree.

ORDER FALCONIFORMES

Brahminy Kite

Haliastur indus

LENGTH	43–51cm (17–20in)
WEIGHT	400–700g (14–25oz)
HABITAT	Beaches, estuaries, rivers

DISTRIBUTION South and Southeast Asia, northern Australia, islands of western Pacific

A common scavenger in parts of its range, the Brahminy kite is also an effective hunter, criss-crossing the water from a height of a few metres, dropping to the surface to catch fish, or to pick up scraps of waste. It also feeds on beaches and mudflats, and is seen in the outskirts of coastal towns. Adults have deep chestnut plumage, and a distinctive white chest and head. Their breeding season varies according to location, but they often nest in mangroves, making a platform-shaped nest from seaweed and sticks. Both parents help to raise the one to two young.

ORDER FALCONIFORMES

Osprey

Pandion haliaetus

LENGTH	50–65cm (20–26in)
WEIGHT	1.2–2kg (2¾–4½ lb)
HABITAT	Coasts, reefs, lagoons, rivers, lakes

DISTRIBUTION Worldwide except polar regions, southern South America and New Zealand

This fish-eating hawk has one of the widest distributions of any bird of prey, breeding mainly in the northern hemisphere and migrating south for the winter. The Osprey is easy to distinguish from other birds of prey on coasts, thanks to its light build, its conspicuous, dark eye-stripe, and its narrow, slightly kinked wings. It feeds entirely on fish, plunging from heights of up to 50m (165ft) and entering the water feet-first. Its wings are strong, its legs are heavily muscled, and its toes have long, hooked talons and spiny soles – an adaptation that gives it a firm grip on its slippery prey. These birds have been known to take prey that approaches their own weight. They nest in the tops of high trees and hatch a single brood of two to three chicks each year. During the 20th century, Ospreys suffered severely as a result of pesticide pollution, particularly from DDT. Their population has now recovered, and in some regions – for example, northern Britain – they have resumed breeding after a gap of many years.

AIRBORNE ATTACK

The Osprey cruises high above water looking for food. Once it spots a fish, it hovers for a few seconds before half folding its wings and going into a steep dive. It hits the water at speed, sometimes partly submerging, before gripping its prey with one foot and climbing laboriously back into the air. Once airborne, it shakes the water off its plumage, before heading to a perching post or to its nest.

ORDER CHARADRIIFORMES

Snowy Sheathbill
Chionis alba

LENGTH	34–41cm (13½–16in)
WEIGHT	450–775g (1–1¾ lb)
HABITAT	Rocky coasts, inshore waters, sea-ice

DISTRIBUTION Antarctic Peninsula, subantarctic islands, southern South America, Falkland Islands

Sheathbills are the only birds with non-webbed feet that breed on the shores of Antarctica. Stocky and short-legged, they bear a superficial resemblance to chickens, particularly when they escape from danger by running away. Almost wholly carnivorous, they scavenge carrion along the shoreline, and also loiter around penguin colonies to steal eggs and food from adult birds.

ORDER CHARADRIIFORMES

Eurasian Oystercatcher
Haematopus ostralegus

LENGTH	40–48cm (15½–19in)
WEIGHT	400–800g (14–28oz)
HABITAT	Rocky shores, damp inland habitats

DISTRIBUTION Iceland, Europe, N. and E. Asia, (breeding); S. Europe, Africa, S. Asia (non-breeding)

With its bright orange bill and loud piping call, this is one of the most conspicuous waders on European shores. Often seen in small parties, it feeds on mussels, limpets, and other molluscs, using its bill to smash or prise apart their shells. To locate good feeding sites, it often flies along the tideline, calling loudly to other oystercatchers. On coasts, it nests on shingle and gravel, laying two to four camouflaged eggs. The Eurasian Oystercatcher is one of 11 species of oystercatcher (family Haematopodidae). All have the same overall shape and brightly coloured bills, but in some species, the plumage is totally black.

brightly coloured bill

ORDER CHARADRIIFORMES

Black-winged Stilt
Himantopus himantopus

LENGTH	35–40cm (14–16in)
WEIGHT	150–200g (5–7oz)
HABITAT	Shallow coasts, salt marshes, wetlands

DISTRIBUTION Worldwide except far north and northeast Asia; summer visitor only in north of range

The Black-winged Stilt's immensely long legs trail far behind its tail when it flies. It has several geographical races and breeds in a broad range of wetland habitats. It feeds in calm fresh or salt water, striding through the shallows, scything its bill through the water to catch small animals or picking them from the surface.

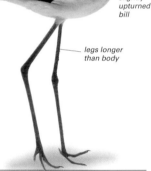

slender, slightly upturned bill

legs longer than body

ORDER CHARADRIIFORMES

Pied Avocet
Recurvirostra avosetta

LENGTH	42–45cm (16½–18in)
WEIGHT	225–400g (8–14oz)
HABITAT	Shallow coasts, salt marshes, wetlands

DISTRIBUTION Europe, temperate Asia (breeding); W. Europe, Africa, S. and S.E. Asia (non-breeding)

Instantly recognizable by their long upturned bills, avocets are elegant waders that feed in shallow water, both on coasts and inland. There are four species, all similar in shape and size. Of these, the Pied Avocet is by far the most widespread and is the only species that is found in Europe and Africa, as well as Asia. Pied Avocets feed by dipping their bill in water, and then sweeping it from side to side. The tip of the bill is highly sensitive to touch, so the bird can catch food even in the turbid water of estuaries and lagoons. Pied Avocets swim well and sometimes upend to find food in the same way as dabbling ducks. They nest in groups, making cup-shaped hollows on mudflats, where they lay a clutch of four eggs. Despite their dainty appearance, they can be aggressive if their nests are threatened. Parents charge at intruders with their heads lowered, and they are able to chase away much bulkier birds, such as geese and ducks.

ORDER CHARADRIIFORMES

Grey Plover
Pluvialis squatarola

LENGTH	26–28cm (10–11in)
WEIGHT	170–240g (6–8oz)
HABITAT	Arctic tundra, coasts, estuaries

DISTRIBUTION Arctic (breeding); temperate and tropical coasts worldwide (non-breeding)

This long-distance migrant, one of the most widespread waders, is found on coasts in every continent except Antarctica. In their breeding plumage, seen only in the Arctic tundra, males have a black underside and face, but by the time they head south to winter on coasts, both sexes are a speckled grey. Grey Plovers feed on insects in summer and on marine worms and crustaceans in winter.

Ruddy Turnstone
Arenaria interpres

LENGTH	21–25cm (8½–10in)
WEIGHT	80–110g (3–4oz)
HABITAT	Rocky/sandy coasts, coastal lowlands

DISTRIBUTION Arctic coasts (breeding); temperate and tropical coasts worldwide (non-breeding)

Found on coasts all over the world, the Ruddy Turnstone feeds in a distinctive way, scuttling along the tideline, flicking stones aside with a deft movement of its bill. This often reveals sandhoppers and other small animals, which it snaps up or chases. Ruddy Turnstones, like many waders, nest in the far north, but their feeding habits restrict them to coastal areas. After breeding, their southward migration takes them to coasts on every continent except Antarctica.

ORDER CHARADRIIFORMES

Grey Phalarope
Phalaropus fulicarius

LENGTH	20–22cm (8–9in)
WEIGHT	50–75g (2–3oz)
HABITAT	Marshy coastal tundra, plankton-rich open ocean

DISTRIBUTION Arctic coasts (breeding); South Atlantic and eastern South Pacific (non-breeding)

Also known as the Red Phalarope, this short-billed wader shows a remarkable reversal of roles when it breeds. Unlike most birds, the female – shown here – has a much brighter breeding plumage than the male. Once she has mated and laid her eggs, she takes no part in incubation or raising the young. By comparison with other waders, Grey Phalaropes are highly aquatic birds and spend much of their time afloat. They breed close to coasts, and once they have migrated south, they often overwinter far out at sea.

ORDER CHARADRIIFORMES

Whimbrel
Numenius phaeopus

LENGTH	40–46cm (15½–18in)
WEIGHT	270–450g (10–16oz)
HABITAT	Arctic tundra, coasts, reefs, wetlands

DISTRIBUTION N. Europe, Arctic (breeding); temperate and tropical coasts worldwide (non-breeding)

Using its long, downcurved bill, the Whimbrel feeds by probing into wet mud or by extracting animals from rocky crevices. It is one of eight similar species – collectively known as curlews – which have mottled brown plumage, sharply pointed wings, and bills up to 20cm (8in) long. The Whimbrel's bill is only half this length, but it is a precision instrument, with sensitive nerve-endings at its tip that enable the bird to feel for buried food. The Whimbrel is strongly migratory, nesting inland across much of the far north, in marshy open country. At this time of the year, the male sings from high in the air, gradually descending on widely spread wings. After breeding, Whimbrels head south along coastlines, reaching as far south as the tip of South America and New Zealand.

ORDER CHARADRIIFORMES

Dunlin
Calidris alpinus

LENGTH	16–22cm (6½–8½in)
WEIGHT	40–50g (1½–1¾oz)
HABITAT	Coasts, marshes, tundra

DISTRIBUTION Arctic, subarctic (breeding); temperate and tropical coasts in N. hemisphere (non-breeding)

In winter, flocks of Dunlins create a breathtaking spectacle, as they wheel in their thousands over coastal feeding grounds. Up close, the Dunlin is a typical calidrid wader, one of over two dozen similar species that feed on coasts worldwide. It has a compact body, narrow wings, a tapering tail, and a black, finely pointed bill. Its plumage is variable, but breeding males usually have a black patch on the underside, which fades when they moult. Dunlins mainly eat small crustaceans and molluscs that live just beneath the surface of the shore. When feeding, they usually stay close to the water's edge, alternately pecking into the mud or sand, and then running forwards at high speed. Dunlins breed in the Arctic and subarctic, where they nest in a range of habitats from moorland to tundra, often some distance inland. Both parents help to incubate the eggs and raise the young. After breeding, they gather in flocks to migrate to warmer coasts, but rarely travel into the southern hemisphere. Other members of this genus include many other flock-forming species, such as the Red Knot and Sanderling, most of which travel as far north as the Arctic Ocean to breed.

WINTER FLOCKS

Wintering waders form some of the largest bird flocks to be found on coasts. Flocking makes it harder for predators to approach unseen and helps young birds to locate good feeding sites by following adults. Some waders, such as the Purple Sandpiper and Ruddy Turnstone, frequently form mixed flocks.

AERIAL MANOEUVRES
Flocks of overwintering Dunlins show extraordinary coordination, with thousands of birds changing direction almost simultaneously.

ORDER CHARADRIIFORMES

Swallow-tailed Gull
Creagrus furcatus

LENGTH	55–60cm (21½–23½ in)
WEIGHT	600–900g (21–32oz)
HABITAT	Coasts, inshore waters, open sea

DISTRIBUTION Galapagos Islands and Malpelo Island (breeding); Pacific coast of South America

Distinguished by its sharply forked tail, this South American gull is atypical in feeding at night. It eats squid and fish, spotting them with its large eyes, which are surrounded by distinctive red rings and angled forwards to give a wide field of binocular vision. Swallow-tailed Gulls nest on islands and disperse far out to sea during the rest of the year.

ORDER CHARADRIIFORMES

Great Black-backed Gull
Larus marinus

LENGTH	71–79cm (28–31in)
WEIGHT	1.2–2.1kg (2¾–4¾ lb)
HABITAT	Rocky coasts, islands, inland in winter

DISTRIBUTION North Atlantic, breeding north to Svalbard

With a wingspan of up to 1.7m (5½ ft), the Great Black-backed is one of the world's largest gulls. Heavily built, with black upperwings and a powerful bill, it scavenges food, but it is also a highly predatory bird. It frequently preys on other sea birds and their young, and will attack mammals as large as rabbits. It breeds alone or in colonies, nesting on cliff ledges or on open ground.

ORDER CHARADRIIFORMES

Herring Gull
Larus argentatus

LENGTH	56–66cm (22–26in)
WEIGHT	750g–1.25kg (1¾–2¾ lb)
HABITAT	Coasts, reservoirs, urban areas

DISTRIBUTION Worldwide in northern hemisphere

Noisy, assertive, and always on the lookout for a meal, this is the most widespread gull in the northern hemisphere. It has grey upperparts and black wingtips, and a large yellow bill with a conspicuous red spot near the tip. Young Herring Gulls are mottled brown, and it takes them three years to develop the full adult plumage. Often seen in flocks, Herring Gulls are highly adaptable birds, feeding on anything edible that they can find. They rarely venture far out to sea, but their range extends a long way inland, where they are often associated with humans – following tractors to eat earthworms turned up by the plough, or wheeling noisily over rubbish heaps. Herring Gulls nest on the ground and on rooftops, typically laying three eggs. They can be highly aggressive if their nests are disturbed.

SCAVENGING

Scavenged food forms a large part of the Herring Gull's diet, both on land and at sea. This gull has benefited from urban expansion and the growth in fishing, both of which generate a large supply of edible waste. Herring Gulls may cause problems at inland rubbish tips by picking up waste and carrying it away.

FOOD OVERBOARD
Large numbers of Herring Gulls follow fishing ships operating close to coasts. Unlike pelagic birds, they usually return to land at night.

ORDER CHARADRIIFORMES
Black-legged Kittiwake
Rissa tridactyla

LENGTH	39–46cm (15½–18in)
WEIGHT	300–500g (11–18oz)
HABITAT	Rocky coasts, inshore waters, open sea

DISTRIBUTION Northern hemisphere; breeds north to Svalbard and Greenland

Kittiwakes get their name from their call – a loud, three-syllable shriek that echoes around their nesting colonies on northern coasts. A medium-sized, grey-backed gull, the Black-legged Kittiwake breeds on narrow cliff ledges but spends the rest of the year wandering far out to sea. It feeds mainly on small fish, and often follows fishing vessels. Unlike most gulls, however, it rarely shows any interest in scavenging food on land. Black-legged Kittiwakes have evolved several adaptations for breeding on bare rock. Their feet have longer claws than those of most other gulls, and they build cup-shaped nests out of seaweed and mud, which help to keep their eggs secure. Both parents help to incubate the eggs and feed the young, and the adults' recognition calls can make a deafening noise when several hundred pairs nest close together. After breeding, these birds disperse away from the coast, travelling as far south as tropics off West Africa. They are monogamous, with pairs meeting up again at the same nesting site after spending up to eight months apart.

ORDER CHARADRIIFORMES
Laughing Gull
Larus atricilla

LENGTH	38–43cm (15–17in)
WEIGHT	300–500g (11–18oz)
HABITAT	Coasts, inshore waters

DISTRIBUTION North America, Caribbean, Central America (breeding); N. South America (non-breeding)

A widespread summer visitor to North American coasts, the Laughing Gull rarely wanders far inland. It feeds mainly by scavenging and often follows ferries and fishing boats. Bold and self-confident, it is a familiar sight to picnickers on beaches, where it pushes larger gulls aside in the competition to get at food. It nests in large colonies on coasts. Like many dark-headed gulls, it loses its black cap during the non-breeding season, when its head turns a dull white.

ORDER CHARADRIIFORMES
Ivory Gull
Pagophila eburnea

LENGTH	40–46cm (16–18in)
WEIGHT	450–600g (1–1¼ lb)
HABITAT	Coasts, open sea, sea-ice

DISTRIBUTION Arctic Ocean, north Atlantic, wintering in south of range

Completely white, apart from its yellow-tipped bill, black eyes, and black feet, the Ivory Gull is the world's most northerly breeding bird. With its buoyant flight and pigeon-like walk, it ranges across open water and sea-ice, and it can be found almost anywhere over the Arctic Ocean. It feeds largely by scavenging and is quickly attracted to the carcasses of dead seals and whales. The Ivory Gull is currently undergoing a steep decline. The reasons for this are unclear.

ORDER CHARADRIIFORMES
Brown Noddy
Anous stolidus

LENGTH	40–45cm (16–18in)
WEIGHT	200–250g (7–9oz)
HABITAT	Open sea, inshore, oceanic islands

DISTRIBUTION Worldwide in tropical waters; present on some islands year-round

Noddies are dark, tropical terns that often feed far out to sea. There are three species of noddy, and the Brown Noddy is the largest and most widespread. Brownish black all over, apart from a paler crown, it has slender wings, a long, sharp bill, and small, jet-black legs. Brown Noddies feed mainly on fish and squid, hovering and then plunging in the same way as terns. They nest on islands throughout the tropics, making nests from twigs and seaweed in trees or on the ground.

ORDER CHARADRIIFORMES
Caspian Tern
Sterna caspia

LENGTH	48–59cm (19–23in)
WEIGHT	550–750g (1¼–1¾ lb)
HABITAT	Coasts, lakes, reservoirs, gravel pits

DISTRIBUTION North America, Eurasia, Africa, Australia (breeding); northern South America, Southeast Asia (non-breeding)

Despite its name, this large, black-crested tern has a global distribution. Grey-backed, with a large, dark red bill, it has a black cap that is darkest when it breeds. It plunge-dives for food in shallow water, and nests in colonies, laying its eggs directly on shingle or mud.

ORDER CHARADRIIFORMES
Inca Tern
Larosterna inca

LENGTH	40–42cm (16–17in)
WEIGHT	175–225g (6–8oz)
HABITAT	Coasts and inshore waters

DISTRIBUTION Pacific coast of South America from Ecuador to central Chile

With its curling white "moustache" plumes, this South American tern is easy to identify. It feeds in the cold, nutrient-rich waters of the Humboldt Current, dipping down to the surface to catch small fish. Inca Terns often follow sealions and whales, preying on shoals of fish as they try to escape the larger predators. They nest among rocks or in abandoned burrows.

ORDER CHARADRIIFORMES
White Tern
Gygis alba

LENGTH	28–33cm (11–13in)
WEIGHT	100–125g (3½–4½ oz)
HABITAT	Open sea, inshore, oceanic islands

DISTRIBUTION Tropical waters worldwide

Also known as the Fairy Tern, this delicate and graceful bird wanders far out over tropical oceans, where it is known for its habit of fluttering close to boats. Slim and lightly built, with black eyes and a straight black bill, it is the only tern whose plumage is entirely white. It spends most of its time flying a few metres above the surface, periodically dropping down in order to catch small fish and squid. Unlike most terns, it is a solitary breeder, nesting on widely scattered islands. It lays its single egg on a rocky ledge, or in a slight hollow in a sloping branch. The parents take turns to cradle the egg throughout its five-week incubation period – an unusually long time for an egg of its size. The chick emerges with strong feet and claws for clinging to its nesting site.

ORDER CHARADRIIFORMES

Black Skimmer
Rynchops niger

LENGTH	40–50cm (16–20in)
WEIGHT	250–400g (9–14oz)
HABITAT	Estuaries, lagoons, lakes, coasts

DISTRIBUTION Pacific and Atlantic coasts of North, Central, and South America, north to Massachusetts

Similar to terns in overall shape, skimmers have remarkable and highly distinctive bills. The lower part, or mandible, of the bill is at least a third longer than the upper part and is laterally compressed, giving it a shape like a scissor blade. When feeding, a skimmer flies low over calm water with its lower mandible slicing through the surface. If the mandible touches food, the skimmer snaps its bill shut, flicking its catch into its mouth. The Black Skimmer is one of three species of skimmer, all of which are dark above, with white underparts. Like its relatives, it often feeds at dawn and dusk, and it will also feed during the night if the moonlight is bright enough. It lives in small flocks and nests on beaches and sand spits, laying its eggs in an unlined hollow on the ground. It is migratory in the far north and south of its range.

ORDER CHARADRIIFORMES

Arctic Skua
Stercorarius parasiticus

LENGTH	46–65cm (18–26in)
WEIGHT	400–600g (14–21oz)
HABITAT	Coasts, tundra, moorland, open sea

DISTRIBUTION Northern waters (breeding); throughout southern hemisphere (non-breeding)

This slender-winged sea bird is exceptionally fast and manoeuvrable in the air – a skill that is central to the way it feeds. There are several colour forms, which differ in their proportion of brown and grey, but all Arctic Skuas have elongated streamers that give their tails a sharp central point. This species catches fish, but it is better known as a kleptoparasite, which steals food from other birds. It swoops down on gulls and terns as they return from the sea, chasing them at speed and often gripping their tail feathers with its bill. Its victims react by disgorging food, which this Skua deftly intercepts in mid-air. Arctic Skuas also hunt small land animals and steal eggs and chicks from nests. They nest on the ground and winter at sea.

ORDER CHARADRIIFORMES

Great Skua
Stercorarius skua

LENGTH	51–66cm (20–26in)
WEIGHT	1.2–1.6kg (2³⁄₄–3¹⁄₂ lb)
HABITAT	Coasts, inshore waters, open sea

DISTRIBUTION North Atlantic (breeding), dispersing south to Equator (non-breeding)

Powerfully built, with short, broad wings, the Great Skua is shaped like an unusually thickset gull, but it has mottled, dark brown plumage that changes only slightly as it matures. It is a rapacious predator, eating fish, small mammals, and also other birds, as well as raiding nests for eggs and chicks. Normally slow and ponderous in the air, it becomes swift and agile when it hunts, and chases birds as large as gannets to force them to regurgitate their food, which it then eats. The Great Skua nests on the ground and spends the rest of the year at sea.

ORDER CHARADRIIFORMES

Common Guillemot
Uria aalge

LENGTH	39–42cm (15¹⁄₂–16¹⁄₂ in)
WEIGHT	850–1100g (1³⁄₄–2¹⁄₂ lb)
HABITAT	Inshore waters, rocky coasts, open sea

DISTRIBUTION North Atlantic, north Pacific

Conspicuously marked in brownish black and gleaming white, the Common Guillemot spends most of the year at sea. It dives for fish from the surface, swimming underwater using its wings. In spring, Common Guillemots crowd together on narrow cliff ledges, where each female lays a single egg directly on to the rock. When the chick is fully grown, the male parent escorts it into the sea.

ADAPTED EGGS

Guillemot eggs are distinctly pointed at one end and will roll around in a circle if disturbed. This adaptation prevents them falling off the narrow cliff ledges where they are laid. Their colour varies greatly and their irregular surface markings of dark blotches and intricate scribbling may aid identification by the parents.

markings unique to each egg

ORDER CHARADRIIFORMES

Atlantic Puffin
Fratercula arctica

LENGTH	28–30cm (11–12in)
WEIGHT	400g (14oz)
HABITAT	Inshore waters, rocky coasts, open sea

DISTRIBUTION North Atlantic, breeding north to Greenland and Svalbard

With its vividly marked bill, bright red feet, and red-and-black eye patches, this is the most colourful sea bird in the north Atlantic. Like other members of the auk family, it feeds by pursuing fish underwater, using its strong, stubby wings to swim. In the air, it flies rapidly on fast-beating wings, skimming over the waves as it returns to its nest with food. Atlantic Puffins breed in large clifftop colonies, digging burrows in coastal turf. The parents take turns to incubate the single egg, and they both help to feed the developing nestling. Instead of regurgitating food, as most sea birds do, they return with small fish held in their bills, carrying about six fish simultaneously, arranged alternately head to tail. Each nestling is fed continuously for about six weeks, after which the parents abandon it and head out to sea. After going without food for several days, the young bird crawls out of the burrow and flutters down to the sea after dark. Puffins disperse out to sea in autumn, when they lose the bright bill colours that make them so conspicuous during the summer months.

COMPETING FOR FOOD

The puffin population has fallen sharply of late, especially in the eastern Atlantic. This may be due to the growing fishery for sand eels, a fish that puffins rely on, especially in breeding season. Sand eels are used in fertilizers, animal foods, and as a source of edible oil.

UNFAIR SHARES
A catch of sand eels is brought aboard a boat. These finger-shaped fish, unrelated to true eels, are an important food for some fish and sea birds.

ORDER CHARADRIIFORMES

Least Auklet
Aethia pusilla

LENGTH	15cm (6in)
WEIGHT	85g (3oz)
HABITAT	Inshore waters, rocky coasts, open sea

DISTRIBUTION North Pacific, breeding mainly in the Aleutian Islands and islands in Bering Sea

This tiny bird is probably the most abundant species of auk, a family that also includes guillemots and puffins. Short, plump, and grey-backed, with a stubby red-tipped bill, it nests in vast colonies off the Alaskan coast, some of which contain more than a million birds. Least Auklets also feed together, floating on the surface in large gatherings known as "rafts". They are pursuit divers that eat mainly zooplankton.

ORDER CHARADRIIFORMES

Crested Auklet
Aethia cristatella

LENGTH	24–27cm (9½–10½in)
WEIGHT	250g (9oz)
HABITAT	Inshore waters, rocky coasts, open sea

DISTRIBUTION North Pacific, breeding mainly in the Aleutian Islands and islands in Bering Sea

The north Pacific is home to more species of auk than anywhere else. The Crested Auklet is a typical example, with a compact body, sooty-grey plumage, and a feathery crest that curves forwards from its forehead over its orange-red bill. Like other auks, it flies low on rapidly whirring wings and feeds in flocks so dense that they resemble swarms of insects wheeling over the water. Crested Auklets breed among fallen rocks on island coasts, in colonies containing thousands of birds. Their courtship displays are energetic

and noisy, as they throw back their heads and make loud grunts and trumpeting sounds. When the breeding season is over, they disperse out to sea and spend the winter as far south as Japan.

ORDER CORACIIFORMES

Pied Kingfisher
Ceryle rudis

LENGTH	25cm (10in)
WEIGHT	90g (3¼oz)
HABITAT	Coasts, lagoons, estuaries, rivers, marshes

DISTRIBUTION Africa, Middle East, south Asia

This boldly patterned, black-and-white bird is the only kingfisher that regularly fishes offshore. Instead of watching for prey from a perch, as many other kingfisher species do, it flies rapidly above the surface with its head facing down as it scans the water below. If it spots food, it hovers on the spot, and then dives down to make a catch. It can also eat while in flight, another unique adaptation. Male and female Pied Kingfishers look similar, although the female has a double breast band compared to the male's single band. Pairs nest in burrows in sandy banks and are often helped by the previous year's young to collect food for the nestlings. The adults have a loud, high-pitched call, which may be heard as they speed past.

ORDER CORACIIFORMES

Collared Kingfisher
Todirhamphus chloris

LENGTH	28cm (11in)
WEIGHT	120g (4½oz)
HABITAT	Forests, coasts, beaches, mangrove swamps, estuaries

DISTRIBUTION Red Sea, Persian Gulf, Southeast Asia, Australasia

Also known as the Mangrove Kingfisher, this bird lives in a variety of habitats, although in Australia it is restricted to the coast. Greenish blue above, with a white belly and collar, it has a black eye-stripe and a sharply pointed bill. On coasts, it hunts crabs as well as fish and, like all kingfishers except the Pied (see above), beats its prey against a perch before swallowing it. It often nests in hollows in mangrove trees, and lays three or four eggs. In the far south of its range, this bird is a summer visitor only.

OCEAN LIFE

MAMMALS

DOMAIN	Eucarya
KINGDOM	Animalia
PHYLUM	Chordata
CLASS	Mammalia
ORDERS	27
SPECIES	5,000

ONLY A SMALL MINORITY OF THE WORLD'S mammals live in seawater, but taken together, they show an extraordinary range of shapes, sizes, and lifestyles. They include cetaceans (whales and dolphins), sirenians (manatees and Dugongs), and carnivores, particularly the pinniped carnivores (seals, sea lions, and Walruses). All marine mammals breathe air, like their terrestrial counterparts, and they give birth to live young, either in the sea or onshore. Many species are migratory, with a sophisticated navigational sense.

ANATOMY AND PHYSIOLOGY

Marine mammals have many adaptations for life at sea, not only in their anatomy, but also in their physiology, regulating how their bodies work. Cetaceans and sirenians have lost all visible traces of hind limbs; instead, they propel themselves with their tail flippers or flukes, which beat up and down. Fur seals and sea lions swim with their front flippers, while true seals use their rear flippers, bringing them together like a pair of hands. Despite needing to breathe air, many marine mammals are superb divers. Some, such as the Elephant Seal, can reach depths of over 1,000m (3,300ft) and stay underwater for up to two hours. When they dive, their heart rate drops, and blood flow is modified so that vital organs receive enough oxygen until they resurface. Instead of breathing in before they dive, the deepest divers often exhale. This helps them to avoid decompression sickness, or the "bends".

DIVING MAMMAL
When a Harbour Seal dives, its heart rate falls below 10 beats a minute. Blood diverted from its muscles and digestive system flows to its heart and brain.

heart beats rapidly after surfacing

heart rate drops as seal dives

rate remains low throughout dive

HEART RATE — 180 160 140 120 100 80 60 40 20

TIME (minutes) — 0 2 4 6

SHARED PATTERNS
A sea lion's front flipper has the same arrangement of bones as a human arm. The "arm" bones are short and sturdy, helping to bear the animal's bulk on land. Long finger bones make up the flipper's blade.

phalange — radius — humerus — metacarpal — ulna — scapula

FLIPPERS AND FLUKES
A Humpback Whale's flippers contain bones, and beat like a pair of wings. Its flukes, or tail fins, are made of rubbery tissue, and contain no bones at all.

VARIED DIET
Penguins are just one item on the Leopard Seal's menu. Despite its reputation for ferocity, at least half of its diet consists of krill, which it filters with its cheek teeth.

INSULATING BLUBBER
Compared to air, seawater drains much more heat from mammals' bodies. To keep warm, many polar species, such as this Walrus, have a thick layer of insulating fat, called blubber, under the skin.

blowhole — sonic lips (source of sound) — outgoing clicks (to prey) — melon — incoming (reflected) clicks — ear drum — sound channel in jaw

USING ECHOLOCATION
Dolphins and toothed whales use pulses of high-pitched sound to locate prey. The forehead contains an oil-filled organ called the melon, which is thought to function as an "acoustic lens" to focus outgoing sound.

FEEDING

Apart from plant-eating manatees and Dugongs, most marine mammals are exclusively carnivorous. In open water, many pursue individual prey, tracking it by sight or by echolocation. Some seals have a twin strategy. They catch prey individually, but they can also filter out planktonic animals in bulk, using complex cheek teeth that interlock to form a sieve. This efficient feeding method reaches extremes in the baleen whales, which cruise through shoals of fish or krill, often swallowing over 100kg (220lb) of food at a time. Not all marine mammals catch moving prey. Sea Otters dive to collect clams, mussels, and sea urchins, while Walruses and Grey Whales suck molluscs out of sea–bed sediment.

BREEDING

Marine mammals typically produce a single young each time they breed. Cetaceans and sirenians give birth in water, as do Sea Otters, but all other marine mammals have to return to land. In species with a harem system, such as fur seals and Elephant Seals, fighting between rival males for control of mates can be ferocious. After mating, the females of most marine mammals raise their young on their own. For their size, true seals develop fastest, some being weaned in as little as five days. At the other end of the spectrum, a dolphin calf may suckle for over 20 months – the start of a mother-calf bond that can last for six years.

SEA OTTER PUP
A young Sea Otter rides on its mother's chest, while she floats in calm water. The pup depends on her for at least five months.

THREATS AND CONSERVATION

Historically, marine mammals have been heavily exploited for their food, oil, and fur, bringing some species close to extinction. Whales and seals are the primary targets. In 1986, the International Whaling Commission agreed on a moratorium on all commercial whaling. Despite dissent, this ban remains in force. Seals continue to be hunted, or culled, to control populations, but the rarest species are protected by international agreements.

COLONY BREEDING
Many seals and sea lions, such as these South American Sea Lions, are highly sociable in the breeding season, forming large colonies on beaches to mate and have their pups.

ENGRAVED WHALE TOOTH
The art of scrimshaw, or engraving on whale teeth and Walrus tusks, was popular among whalers during the 17th and 18th centuries. Whalebone carving still takes place in areas where small-scale native whaling is permitted.

MARINE MAMMAL CLASSIFICATION

Two orders of mammals – the cetaceans and sirenians – are wholly marine. Seals and sea lions are also aquatic, but like other members of the carnivore order, they give birth on land. Several other carnivore species feed at sea, but of these only the Sea Otter is entirely marine.

CARNIVORES
Order Carnivora

293 species
Most carnivores are terrestrial, but a few spend some of their lives in the sea. The Polar Bear is equally at home on dry land, on sea-ice, and in salt water. Seven species of otter often enter salt water, but the Sea Otter is the only one to spend all of its time offshore. The most fully aquatic carnivores are the 34 species of pinnipeds, until recently classified in their own order, the Pinnipedia. They are split into three families. One family comprises the sea lions and fur seals, which have external ears, use their forelimbs for propulsion, and use all four flippers to move on land. The second family is composed of the true seals, which lack external ears, use hind limbs for propulsion and are less mobile on land. The final family contains only the Walrus, which has very wrinkled skin and long tusks.

CETACEANS
Order Cetacea

85 species
Cetaceans are divided into two suborders. The 13 baleen whales lack teeth, and filter food from the water using a fibrous material called baleen. The 72 toothed whales are predators that hunt individual prey. Cetaceans give birth at sea, and are helpless if stranded on land.

SIRENIANS
Order Sirenia

4 species
Living mainly in the tropics, sirenians, or sea cows, are barrel-shaped vegetarians that live in salt and fresh water. They include the Dugong and three species of manatees, such as the Caribbean Manatee, below. Slow-moving and thick-skinned, sirenians have broad muzzles, paddle-like front flippers, and a broad, horizontally flattened tail.

OCEAN LIFE

into a long, powerful neck. Its huge paws may be over 30cm (12in) wide and are furred on their undersides, providing grip while retaining body heat. Its hearing and sense of smell are acute: it can hear prey that is under one metre (3ft) or more of ice and can smell carrion 5km (3 miles) away. Polar Bears spend most of the year at sea, roaming the drifting pack ice and swimming across open areas. Naturally buoyant, they can swim for hours, although they hunt mainly on the ice.

The main prey of Polar Bears is seals, often caught at breathing holes. They also eat sea birds and fish, and the corpses of beached whales are a favourite food. During the summer, many of them live on land and eat a wider range of food, from reindeer to berries. Females give birth in winter, suckling their cubs in a den dug in the snow. For centuries, the Polar Bear has been hunted by native peoples of the Arctic, without its numbers declining. However, thinning of the Arctic's sea-ice by global warming could seriously reduce its access to food.

ORDER CARNIVORA

Polar Bear
Ursus maritimus

LENGTH	Up to 2.5m (8ft)
WEIGHT	Females up to 300kg (660lb); males up to 800kg (1,760lb)
HABITAT	Arctic tundra, pack ice, open sea

DISTRIBUTION Circumpolar in the Arctic, southwards as far as Newfoundland and the Pribilof Islands

Icon of the Arctic, the Polar Bear is the largest mammalian carnivore and has incomparable stamina, resilience, and power. Its body is streamlined, the head grading almost imperceptibly

thick fur over layer of blubber for insulation

large paws, furred on both sides

ORDER CARNIVORA

Sea Otter
Enhydra lutris

LENGTH	0.7–1.6m (2¼–5¼ ft) including tail
WEIGHT	15–45kg (33–100lb)
HABITAT	Inshore waters along rocky coasts

DISTRIBUTION North Pacific from Japan to Alaska and California

Unlike other otters, the Sea Otter is able to spend its whole life in the ocean. It has a blunt head, a stocky body, webbed rear feet, and small front paws with sharp claws. It uses these to gather food and pick up large stones. At the surface, it floats on its back, using a stone that rests on its chest as an anvil to smash open its prey. Sea Otters feed on molluscs, sea urchins, and crabs. While they can dive to 40m (130ft), they rarely venture more than about 1km (½ mile) from the shore.

SLEEPING SECURELY

Sea Otters often sleep in beds of Giant Kelp, using the seaweed to stop themselves drifting away. Their fur is the densest of any mammal. The hairs are packed so tightly that they prevent water penetration, ensuring that the otter's skin never gets wet. This is a vital adaptation, because Sea Otters live in cold water and do not have insulating fat.

ORDER CARNIVORA

Marine Otter
Lontra felina

LENGTH	Up to 95cm (3ft) including tail
WEIGHT	4–6kg (9–13lb)
HABITAT	Exposed rocky shores

DISTRIBUTION Pacific coast of South America from Peru to Cape Horn

This lithe predator lives on some of the world's stormiest coastlines, particularly in the remote southern part of its range. The Marine Otter's

closest relatives live mainly in fresh water, but it spends almost all its time in the sea. Like typical river otters, this coast-dwelling otter has short brownish yellow fur, webbed toes on all four feet, and sensitive whiskers that help it to find prey. It fishes along rocky coasts, in the rich waters of the Humboldt Current, and instead of making burrows, it shelters in sea caves just above the level of the highest tides.

Marine Otters have long been hunted for their pelts, and current estimates of the population are as low as 1,000 animals. The species is now protected, but preservation of its habitat may be equally important in guaranteeing its long-term survival.

ORDER CARNIVORA

European Otter
Lutra lutra

LENGTH	90–110cm (3–3½ ft) including tail
WEIGHT	7–10kg (15–22lb)
HABITAT	Rivers, lakes, estuaries, rocky coasts

DISTRIBUTION Temperate and tropical Eurasia, south to Indonesia

Once widespread throughout Europe and Asia, the European Otter has been badly affected by pollution and habitat change and by being hunted for its fur. It has a streamlined body, short but dense coat, and webbing on all four paws, and is extraordinarily agile underwater, twisting and turning to catch fish. Inland, European Otters are largely nocturnal, spending the daytime in their dens, or holts. Those that live on the coast, however, can often be seen during the day.

ORDER CARNIVORA

Antarctic Fur Seal
Arctocephalus gazella

LENGTH	1.5–2m (5–6½ ft)
WEIGHT	50–160kg (110–350lb)
HABITAT	Rocky coasts, open sea in polar waters

DISTRIBUTION Southern Ocean

Ranging further south than any other fur seal, this polar species feeds on fish, squid, and krill in the icy waters off Antarctica. In spring it comes ashore after spending winter at sea. Males are up to three times heavier than females, with an imposing mane and thickened neck that gives them a front-heavy appearance. This species breeds on islands, such as South Georgia and Kerguelen, and is rising in number. This may be a side-effect of the whaling industry, which has reduced competition for krill.

ORDER CARNIVORA

Northern Fur Seal
Callorhinus ursinus

LENGTH	1.4–2.1m (4½–7ft)
WEIGHT	50–270kg (110–600lb)
HABITAT	Coasts and sea in cold-water regions

DISTRIBUTION North Pacific, Bering Sea

Of the nine species of fur seal, most live in the southern hemisphere – this is the only northern species that exists in significant numbers. Like other fur seals, it has a thick, dark coat, external ears, and long front flippers that it uses for swimming and for moving about on land. Males can be five times heavier than females, but both sexes have short muzzles, giving them a characteristic snub-nosed look. Their large eyes allow them to see at night, which is when they do most of their feeding as their prey is closer to the surface. They feed mainly on fish, but also on squid and sea birds, and migrate far out into the Pacific after they breed. Most Northern Fur Seals breed on islands in the Bering Sea. Decimated by commercial hunters from the mid-1700s onwards, they are now protected by hunting controls.

ORDER CARNIVORA

South American Fur Seal
Arctocephalus australis

LENGTH	1.4–1.9m (4½–6¼ ft)
WEIGHT	60–200kg (130–440lb)
HABITAT	Coasts and sea in cold-water regions

DISTRIBUTION Pacific and Atlantic coasts of southern South America, Falkland Islands

Once found along the entire length of South America's southern coasts, this fur seal now breeds on offshore islands, where it faces less disturbance from humans. It is blackish grey, with paler undersides in females, and is agile on land, using its flippers to climb steep rocks. Males may be about three times the weight of females. This species feeds mainly at night, hunting fish, squid, lobsters, and crabs, and is itself hunted by sharks and killer whales.

ORDER CARNIVORA

Californian Sea Lion
Zalophus californianus

LENGTH	2–2.5m (6½–8¼ ft)
WEIGHT	110–400kg (240–880lb)
HABITAT	Rocky coasts and open sea

DISTRIBUTION Pacific coast of the USA, Galapagos Islands

Famed for its acrobatic antics in marine aquariums, the Californian Sea Lion is just as agile in the wild. Its sleek body is covered with short fur, which ranges in colour from brownish black in males to light brown in females and young; mature males may be more than three times as heavy as females and have a distinctive bony hump on their heads. They feed on fish and squid.

ORDER CARNIVORA

Walrus
Odobenus rosmarus

LENGTH	3.1–3.5m (10¼–11½ ft)
WEIGHT	1,250–1,700kg (2,750–3,750lb)
HABITAT	Coasts and shallow open water

DISTRIBUTION Arctic Ocean, Bering Sea, Hudson Bay

Instantly recognizable by its tusks, the walrus is the second largest pinniped after the elephant seals. Its skin is unlike any other mammal's, with deep creases and wrinkles, but very little hair. Its colour varies enormously: young walruses can be very dark, while old individuals are sometimes a mottled pink. Beneath the skin is a thick layer of fat, or blubber, which keeps their bodies warm. Walruses feed on shellfish, which they find in the sea-bed sediment at depths of up to 50m (165ft). They locate their food mainly by touch, using stiff whiskers that resemble a moustache. At one time, it was thought that they used their tusks to dredge up their food, but it is now known that they uncover it by squirting water with their mouths. Once their prey has been uncovered, they separate the soft parts from the shells. It is unclear how they do this, but their feeding technique probably involves suction rather than crushing, because intact shells are often found around their breathing holes.

Females give birth to a single calf after a 15-month gestation, and they breed only every other year. Walruses are highly gregarious, making them easy prey for hunters. They have been hunted by indigenous peoples for at least 15,000 years, both for food and for their hides.

ORDER CARNIVORA

Common Seal

Phoca vitulina

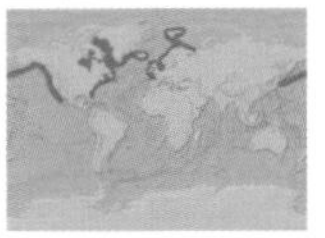

LENGTH	1.4–1.9m (4³⁄₅–6¼ ft)
WEIGHT	55–170kg (120–375lb)
HABITAT	Inshore waters, estuaries, rivers

DISTRIBUTION North Pacific and north Atlantic, reaching as far south as Baja California

Also known as the Harbour Seal, this species has the widest distribution of any seal and the widest variety of markings. Its background colour ranges from pale grey to brown, with dark spots and rings and sometimes a dark stripe along the back. It has a smoothly domed head and a dog-like muzzle. It feeds primarily on fish, often catching them in shallow water close to the shore. It dives

clawed front flipper

for up to five minutes, but rarely to any great depth. The Common Seal spends much of its time on rock flats and sandbanks, and it is here that the females give birth. The pups shed their soft natal coat before they are born, starting life with a dark version of the adult coat, unlike the pups of some other seals. Although they can swim almost immediately, they often use their front flippers to ride on their mother's back. They are weaned at about four weeks. True to its name, Common Seals are still abundant, but in the North Sea they have been adversely affected by pollution, and also by a highly infectious viral disease that broke out in the late 1980s.

ORDER CARNIVORA

Harp Seal

Pagophilus groenlandicus

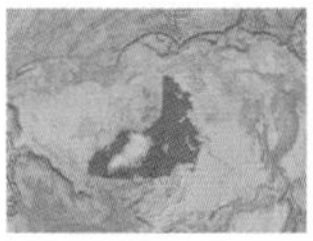

LENGTH	1.7–1.9m (5½–6¼ ft)
WEIGHT	120–140kg (265–310lb)
HABITAT	Polar waters

DISTRIBUTION North Atlantic and adjoining regions of the Arctic Ocean, extending eastwards to Siberia

One of the commonest seals in the far north, the Harp Seal is born with an exceptionally luxurious coat of long white fur, which camouflages the pups as they lie on sea-ice. Adult Harp Seals are silvery-grey with a mottled pattern of dark patches, which become more prominent as they age. They feed mainly on fish and shrimps, living on the southern edge of the Arctic pack ice, and resting on it when they moult. In early spring, adult females give birth to a single pup each, which they wean after just 12 days. At this point, the pup gradually sheds its white coat and takes up life in the sea. For many decades, the pups have been the subject of a controversial hunt, which supplies their pelts to the fur trade. Despite campaigns by conservationists, over 250,000 pups are still culled every year. Harp Seals are also hunted by sharks, Polar Bears, and Killer Whales.

ORDER CARNIVORA

Ringed Seal

Pusa hispida

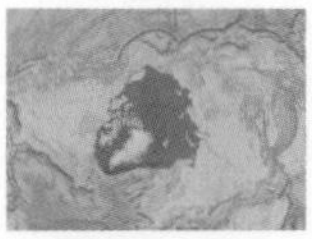

LENGTH	1.3–1.5m (4¼–5ft)
WEIGHT	45–95kg (100–210lb)
HABITAT	Polar waters around sea-ice

DISTRIBUTION Arctic Ocean, north Pacific, north Atlantic, Baltic Sea, Sea of Okhotsk

Named after its conspicuous circular markings, the Ringed Seal is found throughout the Arctic, in open water near sea-ice and also under the ice

itself, where it digs breathing holes. It can dive for over an hour, feeding on fish and zooplankton. Female Ringed Seals breed on the ice, where they dig dens in the snow. These seals are a favourite prey of Polar Bears, which hunt them in their dens and when they surface to breathe.

ORDER CARNIVORA

Grey Seal

Halichoerus grypus

LENGTH	1.8–2.3m (6–7½ ft)
WEIGHT	250–400kg (550–880lb)
HABITAT	Rocky coasts, offshore islands

DISTRIBUTION Discontinuous populations in northwest Atlantic, Iceland, British Isles, Baltic Sea

The Grey Seal has a distinctive convex muzzle, which gives it a "Roman-nosed" appearance. Adults vary in colour: males are usually grey overall, with pale patches on their undersides, while females often have a marbled pattern of dark patches over a much lighter background. Males may be two or

three times heavier than females – a difference exceeded by few other true seals. When not hunting for their usual diet of fish, Grey Seals spend their time either resting on rocks or "bottling" – sleeping in the water with their bodies vertical and their nostrils just above the surface. They breed onshore, hauling themselves out onto beaches or grass further inland. Their pups have a white natal coat, and they stay onshore for two to three months before venturing into the sea.

ORDER CARNIVORA

Mediterranean Monk Seal

Monachus monachus

LENGTH	2.5–2.7m (8¼–9ft)
WEIGHT	250–300kg (550–660lb)
HABITAT	Rocky coasts in warm-water regions

DISTRIBUTION Atlantic coast of North Africa, Mediterranean

Both of the two species of *Monachus* seals are endangered. The larger of the two, the Mediterranean Monk Seal, is listed by the IUCN as critically endangered. Its coat varies from dark brown to light tan. Females are larger than males, and the pups, unusually for seals, are born with black fur. This seal was once common, but centuries of hunting and disturbance have reduced its population to a few hundred. Most exist in the Mediterranean, but the largest colony is on the Atlantic coast of Morocco. Its closest living relative is the rare Hawaiian Monk Seal.

ORDER CARNIVORA

Northern Elephant Seal

Mirounga angustirostris

LENGTH	3–5m (10–16½ft)
WEIGHT	900–2,700kg (2,000–6,000lb)
HABITAT	Islands in deep water off rocky coasts

DISTRIBUTION Pacific coast of North America, from San Francisco to Baja California

Male elephant seals are the largest of all pinnipeds, and the colossal males dwarf the females. There are two species, one in each hemisphere. They are very similar in appearance and have similar life histories. The Northern Elephant Seal is grey or brown, with no obvious markings. The male has a huge, muscular neck, powerful jaws, and an inflatable proboscis resembling a shortened trunk. Both sexes have a layer of insulating blubber and a short, stiff coat, without any soft underfur. They are superb divers: the northern species has been tracked to depths of over 1.6km (1 mile). They eat squid and deep-water fish, although it is still not clear exactly how they find their prey.

FIGHTING MALES

Elephant seals have a winner-takes-all breeding system, in which rival males battle for the right to mate. During these contests, the two rivals face each other and then rear up, roaring noisily with their trunks inflated. They then lunge at each other with their teeth, often inflicting deep, scarring cuts. Winning males may mate with dozens of females during the course of the breeding season, while consistent losers do not mate at all.

ORDER CARNIVORA

Weddell Seal

Leptonychotes weddellii

LENGTH	2.5–2.9m (8¼–9½ft)
WEIGHT	400–600kg (880–1,300lb)
HABITAT	Polar waters around sea-ice

DISTRIBUTION Southern Ocean, extending northwards to South Georgia

The Weddell Seal is found around the entire coast of Antarctica and is the world's most southerly marine mammal. Its head looks small in proportion to its body, and it has a short, dense coat of bluish black fur, with light streaks on the sides. It feeds mainly on fish, diving to depths of 600m (2,000ft), and is able to stay underwater for up to an hour. Weddell Seals are so well adapted to life in cold water that they bask on ice in preference to bare ground. They breed on ice, and their winter survival depends on keeping open their breathing holes. They gouge these out with their canine teeth, starting when the ice is thin, and maintaining them as the ice thickens, to depths of up to 2m (6½ft).

ORDER CARNIVORA

Crabeater Seal

Lobodon carcinophagus

LENGTH	2–2.4m (6½–8ft)
WEIGHT	200–300kg (440–660lb)
HABITAT	Polar waters around sea-ice

DISTRIBUTION Southern Ocean and adjoining regions north of the Antarctic Convergence

Despite its name, this seal feeds only on krill and other planktonic animals. It filters water using its strange molar teeth, which have elongated cusps that look like a set of stubby fingers. When its jaws close, the cusps act like a sieve, letting water out but keeping food in. Crabeater Seals have slender bodies, with fur that may be light or dark brown and darker flippers. They live close to pack ice and breed on it, and they are extremely nimble on land. Their mummified remains have been found over 50km (30 miles) inland in Antarctica's Dry Valleys. Their total population is thought to be 10–20 million, making them more numerous than all other seal species combined.

ORDER CARNIVORA

Leopard Seal

Hydrurga leptonyx

LENGTH	2.5–3.2m (8½–10½ft)
WEIGHT	200–450kg (440–1,000lb)
HABITAT	Polar waters, rocky coasts

DISTRIBUTION Southern Ocean and adjoining regions north of the Antarctic Convergence

With its long muzzle and sharply constricted neck, this solitary predator looks very different from other seal species found off Antarctica. Unlike most true seals, it propels itself forward through the water with its front flippers rather than its rear ones – a characteristic that it shares with fur seals and sea lions. Its body is black or dark grey with a silvery underside, marked with darker flecks and spots. Its jaws are exceptionally powerful, with an unusually wide gape, and they are armed with long incisors and canine teeth, as well as elaborate cheek teeth that can strain food from the water. About half of the Leopard Seal's diet consists of krill, but the remainder is made up of much larger animals that it hunts individually. For example, Leopard Seals are adept at catching penguins as they enter the water, throwing them into the air to rip the skin and feathers from their bodies. They also prey on squid, fish, and other seals. Females give birth to a single pup each year, weaning it at the age of four weeks.

GREY SEAL
In the northeast Atlantic, Grey Seals give birth to their young in autumn (September to October), while the western Atlantic population give birth in winter (January to February). Having spent the first months of its life on land, this juvenile from the eastern population has shed its pup fur and gone to sea in search of fish.

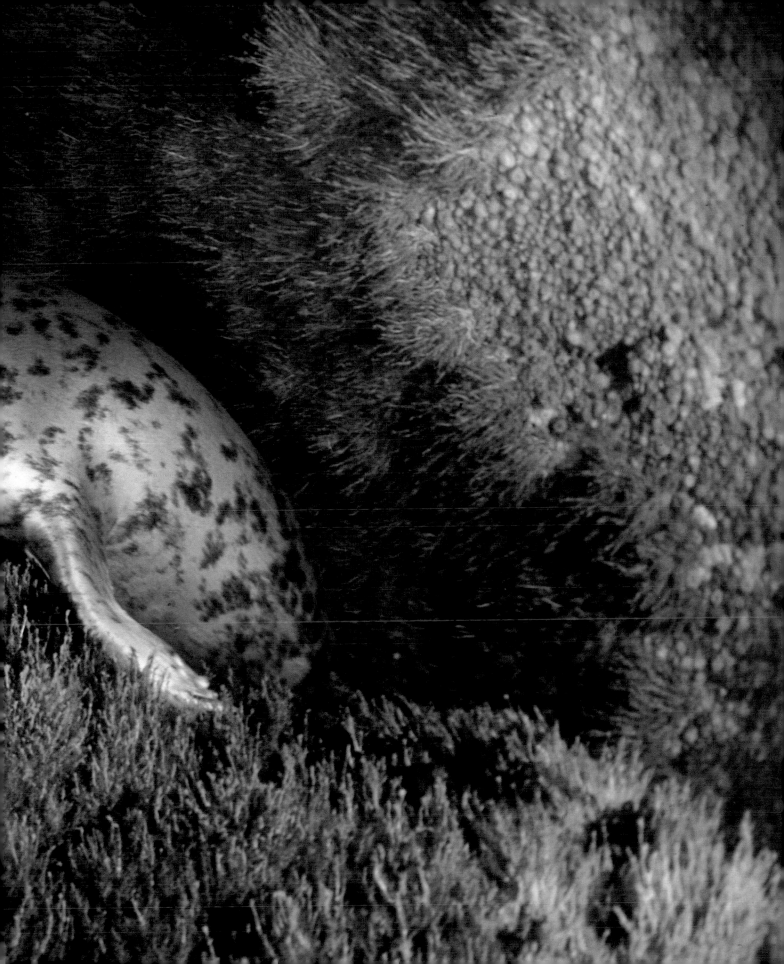

ORDER CETACEA

Northern Right Whale

Eubalaena glacialis

LENGTH	13–17m (43–56ft)
DEPTH	30–80 tonnes (30–79 tons)
HABITAT	Temperate and subpolar waters

DISTRIBUTION Northwestern Atlantic, vestigial populations in northeastern Atlantic and Pacific

The Northern Right Whale was one of the first whales to be hunted commercially and is now one of the most critically endangered species, with a total population of about 500 individuals. A deep bluish black, apart from white markings on its belly, it has a deeply arched mouth, with a lower jaw shaped like a gigantic scoop. Its head its covered with distinctive areas of hard pale skin – known as callosities – which scientists use to identify individuals. Like all baleen whales, it feeds by filtering food from seawater, using brush-like strips of baleen that hang from its upper jaw. Northern Right Whales feed at high latitudes, but they migrate to warmer waters to breed. An almost identical species, the Southern Right Whale, is found in the southern hemisphere. Unlike its northern counterpart, its numbers are gradually increasing and are currently estimated to be about 5,000.

HUMAN IMPACT

WHALING

Commercial whaling has exploited many species. The Northern Right Whale was one of the first to be seriously affected. This whale was decimated by Basque whalers who then expanded operations to Canada in the 1500s. The Sperm Whale was the quarry of American whalers in the Pacific from the 1780s. Modern whaling, targeting species such as the Blue Whale, expanded rapidly in the 20th century, using factory ships and explosive harpoons.

WHALING STATION
Hauled ashore in the Southern Ocean, a whale is flensed, or stripped of its blubber and flesh.

ORDER CETACEA

Bowhead Whale

Balaena mysticetus

LENGTH	14–18m (45–60ft)
WEIGHT	50–60 tonnes (49–59 tons)
HABITAT	Polar and subpolar waters

DISTRIBUTION Arctic Ocean, Bering Sea, adjoining regions of north Atlantic and north Pacific

Named after its arching lower jaw, the Bowhead has the longest baleen plates of any whale at up to 4.5m (15ft). Greyish black with a paler chin, it has a huge head in proportion to its body and remarkably thick blubber, which insulates it in near-freezing water. Bowheads can break upwards through ice over 30cm (12in) thick, allowing them to maintain open water holes throughout the Arctic winter.

ORDER CETACEA

Gray Whale

Eschrichtius robustus

LENGTH	12–15m (39–49ft)
WEIGHT	15–35 tonnes (15–34 tons)
HABITAT	Temperate and subpolar coastal waters

DISTRIBUTION North Pacific, Bering Sea, Arctic Ocean

Unlike other baleen whales, the Gray Whale feeds on the sea floor, filtering animals out of the sediment. Its body is grey with white mottling, and it has a narrowish head, with yellowish baleen plates up to 40cm (16in) long. Its entire body is often heavily encrusted with barnacles and whale lice. Although Gray Whales stay close to the coast, they carry out record-breaking migrations. On the west coast of North America, large numbers migrate between the Bering Sea and Baja California in Mexico, a round journey of up to 20,000km (12,400 miles). Unfortunately, their coast-hugging habits make them easy prey for whalers. By the mid-1900s, they had been almost wiped out, but legal protection has allowed their numbers to recover.

ORDER CETACEA

Humpback Whale

Megaptera novaeangliae

LENGTH	12–15m (40–50ft)
WEIGHT	25–30 tonnes (25–30 tons)
HABITAT	Open oceans, from subpolar to tropical

DISTRIBUTION Worldwide, except extreme north and south

The Humpback's lively behaviour makes it a favourite with whale-watchers. This whale has a blue-black body, deeply notched tail fins (flukes) and extremely long, wing-like flippers. Its flukes and flippers are often splashed with white markings – the pattern, unique as a fingerprint, is used to identify individuals. Unlike most baleen whales, Humpbacks often trap their prey by lunging upwards from below. To concentrate shoals of fish or krill, they often spiral around them while exhaling air. This "bubble-netting" may be carried out by several individuals working as a team. Humpbacks spend the summer in cold, food-rich waters, moving to lower latitudes to give birth in winter. They often feed near coasts. Although protected, current Humpback populations are about a fifth of those of pre-whaling days.

ORDER CETACEA

Minke Whale

Balaenoptera acutorostrata

LENGTH	7–10m (23–33ft)
WEIGHT	5–10 tonnes (5–10 tons)
HABITAT	Open ocean and coastal waters

DISTRIBUTION Worldwide, except extreme north and south

This is the smallest of the rorquals – a name given to baleen whales that have expandable, pleated throats. It is also the most numerous, with a global population as high as 1 million. Like its much larger relative, the Blue Whale, it has a torpedo-shaped body with a single dorsal fin set far back, towards its tail. It is grey or brown above, with a paler underside, and short, pointed flippers that may have a white band.

Minke Whales live alone or in small groups. They are naturally inquisitive and regularly approach boats. They eat small fish and planktonic animals and, like other rorquals, they feed mainly in cold-water regions, eating much less during the breeding season, when they migrate towards the tropics. The Minke is the only rorqual that is still hunted commercially, despite a moratorium observed by most member countries of the International Whaling Commission (IWC).

WHALE SONG

Like all whales, mature male Humpbacks use sound to communicate. They produce the longest, most complex sound sequences of any animal, with each "song" lasting up to 30 minutes. The song is heard miles away by other Humpbacks. Each regional population has its own song, sung only in the breeding season. To sing, the whale vibrates air inside itself, but exactly how is not known, because whales have no vocal cords.

HUMPBACK WHALE
In common with all baleen whales, the Humpback Whale has large jaws and a long head in relation to the rest of its body. It has widely spaced throat grooves and knob-like projections on the upper and lower jaws. Despite its great size, it is an energetic swimmer and often breaches spectacularly.

ORDER CETACEA

Blue Whale
Balaenoptera musculus

LENGTH	24–27m (80–90ft)
WEIGHT	Up to 120 tonnes (118 tons)
HABITAT	Open ocean

DISTRIBUTION Tropical, temperate, subpolar waters worldwide, except in regions with permanent sea-ice

The Blue Whale, one of the rorqual whales, is probably the largest animal that has ever lived. Its heart is the size of a small car and its call, at about 180 decibels, is louder than the sound of a jet aircraft taking off. This animal's future hangs in the balance after decades of whaling. Although it is no longer hunted, it remains seriously endangered.

The Blue Whale has a flattened head, a pointed snout, and a pleated, expandable throat. The rest of the body tapers to a pair of enormous tail fins (flukes). Blue Whales are a mottled blue mixed with grey on their backs, but their undersides vary from white to yellow. They feed by filtering small animals, mostly krill and other small crustaceans, from the water. Their baleen plates can collect over 3,000kg (6,600lb) of food a day. Females give birth to a single calf every 2–3 years.

BALEEN

Instead of teeth, baleen whales have flexible strips of baleen, or whalebone, which hang from the upper jaw. To feed, the whale takes in a mouthful of water, then sieves it through its baleen. The water is expelled, leaving small animals trapped, which the whale then swallows.

smooth outer face

fringed inner face

BALEEN STRIPS
Baleen strips are made of keratin, like human fingernails. The inner face of each strip is divided into hundreds of parallel fibres.

ORDER CETACEA

Sperm Whale
Physeter macrocephalus

LENGTH	Up to 20m (65ft)
WEIGHT	Up to 50 tonnes (49 tons)
HABITAT	Deep water, especially close to edges of continental shelves

DISTRIBUTION Worldwide, except extreme north and south

The largest toothed whale, the Sperm Whale is also the largest predator that hunts individual prey. Even in poor light, it is unmistakable, with a huge, square-ended head. Adult males are typically 4m (13ft) longer than females and twice as heavy. This species has wrinkled skin and a row of knobbly projections between its dorsal fin and its tail. It dives to over 3,000m (9,840ft) to hunt giant squid. Its head contains a store of a waxy oil called spermaceti, which is thought to act as a buoyancy regulator. The oil may also help to focus beams of high-pitched sound, which the whale uses to detect its prey.

ORDER CETACEA

Cuvier's Beaked Whale
Ziphius cavirostris

LENGTH	5.5–7m (18–23ft)
WEIGHT	Up to 3 tonnes (3 tons)
HABITAT	Deep water

DISTRIBUTION Tropical, subtropical, and temperate waters worldwide, except in far north and south

There are at least 20 species of beaked whale, but little is known about most of them. Cuvier's Beaked Whale is probably one of the most widespread, because stranded specimens have been found in many parts of the world.

Like its relatives, it has an almost cylindrical body, a small dorsal fin placed far back, and relatively short flippers for its size. Its jaws are short and beak-like, with an upturned mouthline. Females are toothless, but in males, the lower jaw has two peg-like teeth at its tip, which project when the mouth is closed. The overall colour varies from grey and dark brown to yellow, with a swirling pattern of darker markings.

Cuvier's Beaked Whale lives in deep water and can dive for more than half an hour. Its feeding behaviour is poorly known, apart from the fact that it preys on squid and fish. It has never been hunted commercially, but it is occasionally as accidental bycatch in fishing nets, an occurrence that has become more common with the spread of deep-water trawling.

ORDER CETACEA

Northern Bottlenose Whale
Hyperoodon ampullatus

LENGTH	8.5–10m (28–33ft)
WEIGHT	Up to 7.5 tonnes (7.5 tons)
HABITAT	Deep water

DISTRIBUTION Arctic Ocean, temperate and subpolar waters of the north Atlantic

One of the largest beaked whales, this species has a grey body and a bulbous forehead, which sometimes overhangs its jaws. Males have two to four teeth, at the tip of the lower jaw; females rarely have any. Its tail fins (flukes) are large and powerful, but the front flippers are unusually small and set far forwards, just behind the head. These whales are exceptionally good divers, capable of staying underwater for over two hours. Unlike other beaked whales, this species was commercially hunted for many years, but it is still locally abundant.

ORDER CETACEA

Narwhal
Monodon monoceros

LENGTH	4–5m (13–16½ft)
WEIGHT	Up to 1.5 tonnes (1.5 tons)
HABITAT	Polar waters, open leads in sea-ice

DISTRIBUTION Arctic Ocean, north as far as Svalbard and Franz Josef Land

The male Narwhal is instantly recognizable by its unicorn-like tusk, which is up to 3m (10ft) long. A highly modified upper tooth,

the tusk emerges through the animal's upper lip, developing spiral grooves as it grows. Apart from this outstanding feature, males and females are similar, with a long cylindrical body, a bulbous head, and very short, beak-like jaws. They are dappled grey above and pale or white beneath. The function of the Narwhal's tusk is unclear. It may be used for ritual combat during the breeding season, or even as a navigational aid as it is packed with nerves.

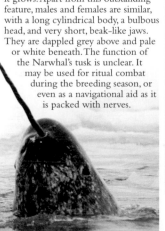

ORDER CETACEA

Beluga Whale
Delphinapterus leucas

LENGTH	4–5m (13–16½ft)
WEIGHT	Up to 1.5 tonnes (1.5 tons)
HABITAT	Coastal waters, sometimes rivers

DISTRIBUTION Arctic Ocean, Bering Sea, Sea of Okhotsk, Hudson Bay, Gulf of St. Lawrence

With its distinctive, yellowish white coloration, the Beluga or White Whale is easy to identify. In overall shape it is similar to its close relative the Narwhal (see left), although it has no tusk. Its colour changes with age: newborn Belugas are dark grey, and it can take them up to ten years to assume the adult colour, which comes with sexual maturity. Belugas are slow swimmers and feed on a wide variety of fish and other animals. They often live close inshore during summer months and may enter the lower reaches of large

rivers. They are remarkably sociable and vocal, making a range of different sounds, including trills, clicks, and chirps. In the days of wooden sailing-ships, these sounds were easily audible through hulls – earning Belugas the nickname "canary of the seas".

Formerly abundant throughout the Arctic, Belugas have been reduced to localized populations by centuries of hunting. They are still hunted today, although on a reduced scale, but they face growing threats from pollution and shipping traffic.

MASSED RANKS

During the breeding season, Belugas gather in herds that may be thousands strong. Within each herd, the whales are grouped according to age and sex, with pregnant and nursing mothers keeping close together with their young. Belugas communicate using sound, but they can also make facial expressions – a unique attribute among whales. They also often hunt in groups.

ORDER CETACEA

Indo-Pacific Humpback Dolphin
Sousa chinensis

LENGTH	2–2.8m (6½–9¼ft)
WEIGHT	Up to 200kg (440lb)
HABITAT	Coastal waters, lagoons, estuaries

DISTRIBUTION Red Sea, Persian Gulf, Indian Ocean, and southwestern Pacific

This warm-water dolphin gets its name from the conspicuous hump beneath its dorsal fin. Generally seen alone or in pairs, it shows a wide variation in colour – some specimens are bluish grey, while others are almost white, particularly when they age. It feeds on fish, octopus, and squid, and it rarely strays far from the shore. A similar species exists off the Atlantic coast of Africa, and both species roll their bodies when they breathe, instead of jumping clear of the water.

ORDER CETACEA

Bottlenose Dolphin
Tursiops truncatus

LENGTH	2–3m (6½–10ft)
WEIGHT	Up to 650kg (1,450lb)
HABITAT	Coastal waters, open oceans

DISTRIBUTION Temperate and tropical regions worldwide

A familiar sight worldwide in marine aquariums, the Bottlenose Dolphin is a playful and inquisitive mammal, with a habit of interacting with humans in the wild. Its colour varies from light blue to slate grey, with a paler underside. It has a pronounced beak, a slightly hooked dorsal fin, and up to 25 pairs of peg-like teeth in each jaw.

These dolphins are highly sociable and often travel in groups of several dozen. Like other dolphins, they find prey by echolocation, but they also use sound to communicate, using a complex repertoire of whistles, clicks, and squeaks. They frequently ride the bow-waves of ships and body-surf on breaking waves. Females give birth to a single calf once every 2–3 years. In some parts of the world, Bottlenose Dolphins play with human swimmers – a form of learned behaviour that can persist for many years.

ORDER CETACEA

Long-snouted Spinner Dolphin
Stenella longirostris

LENGTH	1.3–2.1m (4¼–7ft)
WEIGHT	Up to 75kg (165lb)
HABITAT	Open oceans

DISTRIBUTION Tropical and subtropical waters worldwide

Graceful, energetic, and highly acrobatic, this dolphin gets its name from its habit of leaping out of the water and then spinning around up to seven times before splashing back into the sea. Smaller than many other oceanic dolphins, it is dark grey with white on its underside – the white varies from a small patch to a wide zone extending from its head almost to its tail. It has up to 64 pairs of teeth in each jaw, and it feeds on fish, often far out to sea. Females give birth to a single calf, suckling it for up to two years.

Spinner Dolphins are sociable, swimming in groups that range in size from less than 50 to several thousand and often travelling with other species. These dolphins and their close relatives often swim in large groups above shoals of Yellowfin Tuna, and thousands are drowned every year in purse-seine nets, which are intended to catch tuna but trap other marine life indiscriminately.

ORDER CETACEA

Common Dolphin
Delphinus delphis

LENGTH	1.7–2.4m (5½–8ft)
WEIGHT	Up to 110kg (240lb)
HABITAT	Coastal waters, open oceans

DISTRIBUTION Temperate, subtropical, and tropical waters worldwide

The Common Dolphin is beautifully marked with a complex pattern of coloured bands and has inspired artists since classical times. Its markings are extremely variable – so much so that experts differ on whether it is a single species or several. Often seen in large groups, this dolphin is highly active and acrobatic, and is among the fastest swimmers of all cetaceans, with a top speed of about 40kph (25mph).

Common Dolphins usually feed far out to sea, where they prey on squid and small fish. Adult females give birth every 2–3 years, producing up to five calves during their lives. This dolphin is one of the commonest cetaceans and has a global population estimated at several million. However, like other oceanic dolphins, it is threatened by both the expansion of fishing and deliberate hunting.

ORDER CETACEA

Risso's Dolphin
Grampus griseus

LENGTH	3–4.3m (10–14ft)
WEIGHT	Up to 500kg (1,100lb)
HABITAT	Deep water

DISTRIBUTION Tropical and warm-temperate waters worldwide

Also known as the Grey Grampus, this large dolphin is typically blackish blue with a square-looking head that is quite different to the pointed heads of beaked dolphins. Close up, this dolphin's skin often appears scarred, especially in older individuals. Scarring is mainly due to fights between rivals, but some of it is due to encounters with squid, which make up a large proportion of its prey. When feeding, it can dive for up to half an hour. Risso's Dolphin is less sociable than many other dolphins, but it often swims alongside ships.

ORDER CETACEA

Killer Whale

Orcinus orca

LENGTH 5.5–9m
(18–30ft)

WEIGHT Up to 9 tonnes
(9 tons)

HABITAT Open waters,
areas of broken sea-ice

DISTRIBUTION Tropical, temperate, and polar waters
worldwide

With its conspicuous black-and-white markings, the Killer Whale, or Orca, is – despite its name – the largest and most striking member of the dolphin family (Delphinidae). Apart from its bold patterning, its most eye-catching feature is its huge dorsal fin, which is up to 1.8m (6ft) high in older males. It has large, paddle-shaped flippers and a massive, barrel-shaped body that tapers towards streamlined jaws, which are armed with interlocking teeth up to 5cm (2in) long. Killer Whales are the largest hunters of warm-blooded prey.

Their diet includes fish, squid, birds, seals, and other whales. Their hunting strategy is remarkably varied: they deliberately upend ice floes to tip seals into the sea, and they even lunge onto beaches to catch seals lying near the waterline. Intelligent, vocal, and highly sociable, they live in stable groups (pods), which develop their own cultural characteristics. Despite their ferocity towards prey animals, killer whales are easily tamed in captivity and have never been known to attack humans in the wild.

PODS AND CLANS

An average Killer Whale pod contains 20 animals, which stay together for life, often sharing care of the young. Pods within the same geographical range make up a clan – a regional group that is thought to have a distinctive "dialect" that is passed on from adults to their young.

WHALE MIGRATION

Many land mammals make long treks in search of seasonal feeding grounds, but the distances they cover are dwarfed by the vast annual journeys made by some whale species. Moreover, these journeys take them to places where there is nothing to eat. Humpback Whales, for example, spend the summer in rich feeding grounds in cold waters (see below), satiating themselves on krill, zooplankton, and fish. With the onset of winter, their food supply dwindles and the whales migrate towards the Equator. There is no food supply in these waters for the whales, and they fast for several months. It is another imperative that drives them to these warm and sheltered waters – they provide a suitable environment in which to give birth and begin to rear their calves.

Some scientists question whether these migrations are necessary, arguing that most migratory species of whale could breed perfectly well in colder waters as Bowhead Whales do. It is suggested that the whales evolved in areas at lower latitudes, which once had high concentrations of plankton and fish. As a result of continental drift and changes in ocean currents, the species on which they fed moved towards the polar regions. The whales followed them, but habitually return each year to traditional breeding grounds.

It is not precisely known how whales navigate on these long journeys. Although the mechanism is unclear, magnetite (an oxide of iron) has been found in tissues around the brains of some cetaceans, including Humpbacks. It is thought that this helps the whale sense gradients in the Earth's geomagnetic field, which in turn acts as a guide to navigation.

HUMPBACK WHALE MIGRATION

Humpback Whales from the northern hemisphere spend the summer in feeding grounds in the northern Pacific and Atlantic oceans. In winter, these whales migrate south to warmer waters to breed. Humpback Whales from the southern hemisphere feed in waters off Antarctica and breed in warmer waters off Australia, the Pacific islands, southern Africa, or South America. The northern Indian Ocean population may be resident all year.

← main migration   ←--- possible migration
major feeding areas (summer)   ▢ major breeding areas (winter)

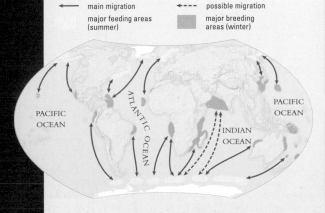

PACIFIC OCEAN   ATLANTIC OCEAN   INDIAN OCEAN   PACIFIC OCEAN

WHALE BEHAVIOUR

ALASKAN FEEDING GROUNDS
Humpback Whales feed by scooping water into their huge mouths, from which they mostly filter krill and other large zooplankton. In Alaska, however, they gather close inshore in summer to feed on shoals of fish.

FEEDING

BUBBLE-NETTING
A group of Humpback Whales swims slowly around a fish shoal, releasing bubbles, which form a net. Then, one whale bursts up through the bubble spiral to snatch a mouthful of fish.

SOLITARY BLUE WHALE
Although Blue Whales are known to make major migrations, they swim further offshore than other migratory whales and do not have defined breeding areas, so their migration is less well understood.

MIGRATING

SPY-HOPPING *Many whales, such as this Gray Whale, "spy-hop", rising vertically in the water with the head well above the surface. They may be checking for landmarks while migrating.*

BREEDING

WARM-WATER DISPLAY *From June to December each year, about 3,700 Southern Right Whales gather close to shore in bays east of the Cape Peninsula, South Africa. The females give birth to a single calf in these waters, then mate immediately afterwards. The area offers the best land-based whale-watching in the world, providing spectacular views, such as this whale breaching.*

WHALE-WATCHING

CLOSE ENCOUNTER *Whale-watching is a growth industry that is worth over US$1 billion a year, with over 11 million people taking part. These tourists are observing a Gray Whale in its summer breeding grounds in the waters off Baja California, Mexico.*

OCEAN LIFE

ORDER CETACEA

Long-finned Pilot Whale

Globicephala melas

LENGTH	3.5–7m (11½–23ft)
WEIGHT	Up to 3.5 tonnes (3.5 tons)
HABITAT	Cold coastal waters, open oceans

DISTRIBUTION Temperate and subpolar waters worldwide, except north Pacific

There are two species of pilot whale, distinguished primarily by the length of their flippers – a feature that is difficult to observe at sea. The Long-finned Pilot Whale lives mainly in cold-water regions. It has glossy, jet black coloration, with an anchor-shaped pale patch on the throat and chest. This species has a bulbous head and short jaws. Its long dorsal fin has a hooked shape in males. Its flippers have a sharp backward bend, or "elbow", and are up to a fifth of its body length. Long-finned Pilot Whales feed mainly on deep-water squid and octopus. They are highly gregarious, living in groups that can be hundreds strong, and often associate with other cetaceans. They easily become disoriented in shallow coastal waters, often becoming stranded in large numbers. This tendency to herd together has been exploited for centuries by whale hunters, who were able to drive them into shallow water for slaughter. In some locations – such as the Faroe Islands – pilot whales are still hunted today.

STRANDING

Pilot whales often become stranded on beaches. If one whale strands, others frequently follow, leading to a mass stranding. Theories to explain stranding involve factors that disrupt the whales' navigational systems, such as temporary anomalies in the Earth's magnetic field, ships' sonar, sickness, and storms.

ORDER CETACEA

Harbour Porpoise

Phocoena phocoena

LENGTH	1.4–2m (4½–6½ft)
WEIGHT	Up to 65kg (145lb)
HABITAT	Coastal waters, tidal regions of rivers

DISTRIBUTION Cold-temperate and subpolar waters in northern hemisphere

One of the commonest cetaceans in the northern hemisphere, the Harbour Porpoise, as its name suggests, rarely strays into deep water. It prefers shallow, coastal waters and sometimes swims into rivers. It has a short, barrel-like body, with small flippers and a blunt dorsal fin. Its overall colour is dark grey, while its underside is paler. Unlike most dolphins, this porpoise has a blunt snout, which houses 21–28 pairs of spade-shaped teeth in each jaw. Harbour Porpoises often live alone, or sometimes in pairs or small groups; they feed on fish and shellfish.

Females give birth after a gestation period of up to 11 months, and the single calf is tiny by cetacean standards, weighing as little as 6kg (13lb). In the past, Harbour Porpoises were often hunted for meat and as a source of oil. Today, a greater threat is posed by fishing nets – being small, it is easy for them to become accidentally trapped.

ORDER SIRENIA

West African Manatee

Trichechus senegalensis

LENGTH	3–4m (10–13ft)
WEIGHT	Up to 500kg (1,100lb)
HABITAT	Mangrove swamps, lagoons, inland waterways, estuaries

DISTRIBUTION West Africa, from Senegal to Angola

One of three species of manatee, this docile vegetarian lives mainly in fresh water but also feeds in the mangrove swamps on Africa's west coast. It has a barrel-shaped body covered in coarse grey skin and front flippers with tiny nails. Like all sirenians, it has no hind limbs and swims with its spoon-shaped tail, which slowly beats up and down as it cruises through the shallows.

Using its fleshy lips, it feeds on plants above and below the water line. Manatees lack the complex stomachs of terrestrial plant-eaters such as cattle and antelopes. Most digestion occurs in their intestines, which may be 45m (150ft) long. West African Manatees live in groups of up to six and give birth to young about 1m (3ft) long. Their slow reproductive rate makes them vulnerable to environmental change, and to hunters who target them for meat and skin.

ORDER SIRENIA

West Indian Manatee

Trichechus manatus

LENGTH	3.7–4.6m (12–15ft)
WEIGHT	Up to 1,600kg (3,500lb)
HABITAT	Coastal waters, inland waterways

DISTRIBUTION Western Atlantic from southeast USA to northeast South America, Caribbean Sea

This is the largest species of manatee, and also the best studied – something explained partly by its distribution, which extends northward as far as Florida. Unlike the West African Manatee, it often ventures into coastal waters, although it avoids regions where the winter temperature drops below 20°C (68°F). Its skin is grey, but its upper surface is often colonized by algae, which gives it a greenish tinge. Its vision and hearing, provided by small eyes and ears, are not very acute, but its mobile lips are covered with sensitive bristles, which it uses to find underwater plants in depths of up to about 4m (13ft). It needs to consume approximately one-quarter of its body weight in food each day. Although its diet is mainly vegetarian, it sometimes eats fish to obtain protein.

Manatees and Dugongs (see below) owe their blimp-like shapes partly to the large amounts of gas generated as they digest their food. To compensate for this, they have unusually dense bones, which help them to maintain neutral buoyancy. West Indian Manatees usually live in groups of up to 20 animals, and when food is plentiful, groups may increase to over a hundred individuals.

HUMAN IMPACT

COLLISION RISK

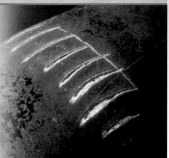

In the past, West Indian Manatees were hunted for their meat, skin, and oil, which was sometimes used in lamps. Today, the main threats facing them are pollution and collisions with boats. In Florida, where boat traffic is heavy, many manatees bear the scars of their encounters with boats.

PROPELLER INJURY
These parallel scars on a manatee's back were caused by a propeller. Fortunately, the cuts were not deep enough to be fatal.

ORDER SIRENIA

Dugong

Dugong dugon

LENGTH	2.5–4m (8–13ft)
WEIGHT	250–900kg (550–1,900lb)
HABITAT	Coastal shallows, lagoons, estuaries

DISTRIBUTION Indian Ocean and western Pacific, from East Africa to South Pacific islands

Unlike manatees, the Dugong is essentially a marine animal, grazing in seagrass beds in warm, shallow waters. Its body is blimp-shaped, like that of manatees, but it has a crescent-shaped tail and a broad head with a large, U-shaped upper lip. Part of its diet consists of buried stems or rhizomes, which it collects by nuzzling its way into the sediment, while steadying itself with its front flippers. Dugongs feed in scattered herds, which may contain more than a hundred animals.

Their main predators are sharks, but they are more threatened by hunting in many places. The species is already extinct in the Mediterranean, where it may have existed until classical times, and it is under threat in many parts of the Indian Ocean. However, it appears to be thriving around the coastline of Australia, which is home to over half the world's Dugongs.

STELLER'S SEA COW

A close relative of the Dugong, Steller's Sea Cow lived in the icy waters of the Bering Sea, feeding on kelp and other seaweeds. It was hunted to extinction in 1768, 27 years after it was first recorded by the German naturalist Georg Steller (1709–46).

ARTIST'S IMPRESSION
Steller's Sea Cow weighed up to 10 tonnes (10 tons) and was probably the largest marine mammal of its time, after whales.

ATLAS OF THE OCEANS

OCEANS OF THE WORLD

OCEANS COVER 71 PER CENT of the Earth's surface and contain 97 per cent of its water. The geography of the ocean basins tells us much about the Earth's past and the geological forces that continue to shape the world.

There are five oceans separating the world's major landmasses, and numerous marginal seas, gulfs, and connecting straits. The continents are surrounded by shallow shelves, which extend a variable distance from the shore before descending into the deep ocean basins. The ocean basins contain the flattest parts of the Earth's surface – the abyssal plains – but also the greatest extremes of elevation, from deep ocean trenches to the peaks of the world's largest volcanoes. The longest mountain chains on Earth are the mid-ocean ridges, which circle the planet along the boundaries of the major tectonic plates.

The maps in this chapter draw on the most detailed knowledge of the topography of the global sea floor yet assembled. They combine the latest measurements from satellite altimeters with more than 100 years of ship-borne hydrographic surveys to give the clearest possible portrayal of the shape of the sea bed.

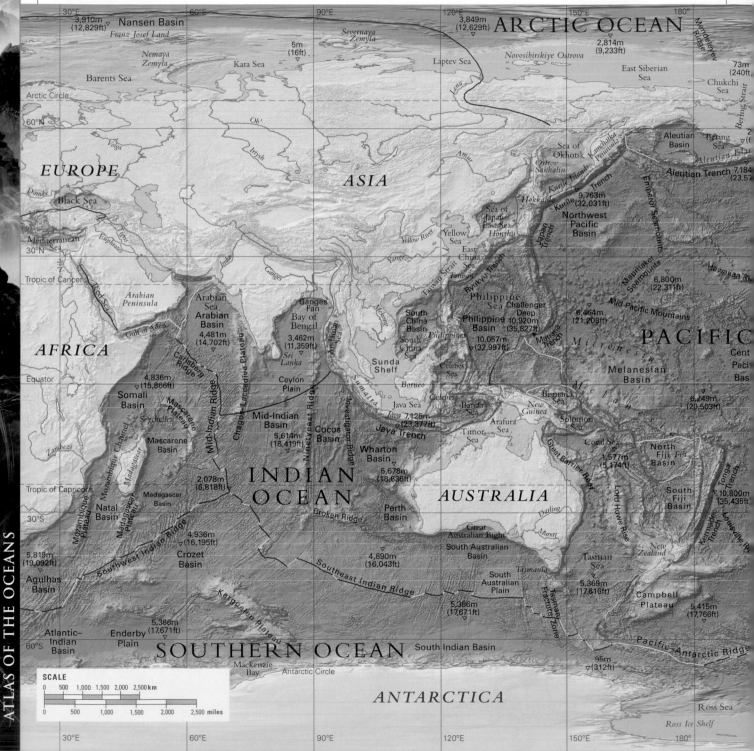

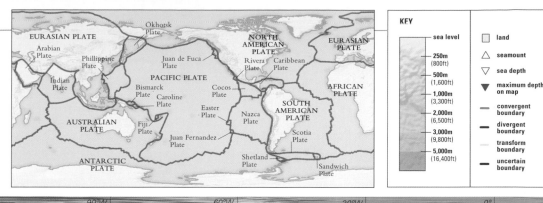

TECTONIC PLATES

The Earth's surface is split into 7 major plates and at least 15 minor ones. Their boundaries are most clearly expressed in the topography of the sea floor, where mid-ocean ridges, ocean trenches, and fracture zones represent divergent, convergent, and transform boundaries respectively. Although some plates have continental and oceanic parts, the largest, the Pacific Plate, is only oceanic crust.

KEY

sea level

250m (800ft)	
500m (1,600ft)	
1,000m (3,300ft)	
2,000m (6,500ft)	
3,000m (9,800ft)	
5,000m (16,400ft)	

□ land
△ seamount
▽ sea depth
▼ maximum depth on map
— convergent boundary
— divergent boundary
— transform boundary
— uncertain boundary

ATLAS OF THE OCEANS

THE ARCTIC OCEAN

THE SMALLEST OF THE OCEANS, the Arctic Ocean is nearly enclosed by Asia, Europe, Greenland, and North America. In winter it is almost entirely covered by pack ice, which halves in area during summer. Exploration of the Arctic in the 18th and 19th centuries was driven by the search for trade routes between the Atlantic and the Pacific oceans. The North Pole was first reached in 1909 by an American expedition using dogs and sleds, led by Robert Peary.

ARCTIC SEA-ICE
Sea-ice covering the Arctic expands from about 7 to 15 million square km (3 to 6 million square miles) from summer to winter.

OCEAN CIRCULATION

The Arctic receives a huge influx of fresh water from the great Siberian rivers – the Ob', Yenisey, and Lena. Together with the freezing and melting of sea-ice, this produces a layer of relatively fresh surface water. A clockwise gyre is established over the Canada Basin, while the Transpolar Current flows from the Chuckchi Sea to the Greenland Sea. Warm, salty water enters the Arctic from the Atlantic at moderate depth, while very cold, very salty "bottom water" flows out into the Atlantic. Eighty per cent of the Arctic's water exchange is with the North Atlantic and 20 per cent is with the Pacific. About two per cent of the water leaving the Arctic is in the form of icebergs calved from the Greenland Ice Sheet. Arctic sea-ice has reduced in area in recent decades and open water has even been found at the North Pole. The effect of global warming on the Arctic means that the ocean's sea-ice could disappear altogether by the end of the 21st century.

ICEBREAKER
Ships with strengthened bows and powerful engines – icebreakers – are needed to penetrate Arctic sea-ice.

OCEAN FLOOR

The floor of the Arctic Ocean consists of two main basins separated by the sharp Lomonsov Ridge. On the North American side lie the Canada and the Makarov basins, separated by the Alpha Cordillera. On the Eurasian side the Fram and Nansen basins are split by the Gakkel Ridge – an extension of the Mid-Atlantic Ridge. The young Arctic Basin started to open about 36 million years ago, completing the separation of North America from Europe, and connecting the Arctic to the Atlantic. There is an unusually broad continental shelf on the Asian side of the ocean, with shallow seas extending more than 1,600km (1,000 miles) from the coast in places, compared with the more typical 50–125km (30–75 miles) on the North American side.

SCALE

KEY

sea level

250m (800ft)
500m (1,600ft)
1,000m (3,300ft)
2,000m (6,500ft)
3,000m (9,800ft)
5,000m (16,400ft)

☐ land
△ seamount
▽ sea depth
▼ maximum depth on map

Map labels

D · E · F · G · H · I

70°N · 120°E · 70°N · 90°E · Arctic Circle

ASIA

Tiksi
Buorkhaya Guba
Dmitraya Lapteva
Ostrov Bol'shoy Lyakhovskiy
Ostrov Kotel'nyy
Novosibirskiye Ostrova
Ostrov Novaya Sibir'

6m (20ft)

Laptev Sea

Poluostrov Taymyr

Venisey

Gydanskiy Poluostrov

Dikson

Obskaya Guba
Poluostrov Yamal
Ostrov Belyy
Baydaratskaya Guba

5m (16ft)

Kara Sea

Proliv Vil'kitskogo
Ostrov Bol'shevik
Severnaya Zemlya
Ostrov Oktyabr'skoy Revolyutsii
Ostrov Komsomolets

3,849m (12,629ft)

80°N

ARCTIC OCEAN

2,814m (9,233ft)

Wrangel Plain

Makarov Basin

Pole Plain

Lomonosov Ridge

Gakkel Ridge

Nansen Basin

Voronin Trough

Central Kara Plateau

Syvataya Anna Trough

Franz Josef Land

East Novaya Zemlya Trough

328m (1,076ft)

Kara Strait

60°E

Novaya Zemlya

170m (558ft)

109m (358ft)

Barents Sea

limit of summer pack ice

102m (335ft)

Kvitøya

Stor Bank

Thor Iversen Bank

limit of winter pack ice

4,484m (14,712ft)

3,910m (12,829ft)

Barents Plain

Barents Trough

30°E

Alpha Cordillera

1,250m (4,101ft)

2,590m (8,498ft)

Fram Basin

North Pole

Nordaustlandet

Yermak Plateau

Like Trough

Edgeøya

Barentsøya

Storfjordrenna

Bjørnøya Bank

Bjørnøya

257m (843ft)

North Cape

Hammerfest

EUROPE

5

Kap Morris Jesup

Wandel Sea

Nord

Lena Trough

Spitsbergen Fracture Zone

5,601m (18,377ft)

Spitsbergen

Longyearbyen

Knipovich Ridge

Fugløya Bank

Tromsø

Alex Heiberg Island

Ellesmere Island

Nares Strait

Lincoln Sea

Cape Columbia

Alert

80°N

Hovgaard Fracture Zone

Boreas Plain

Greenland Fracture Zone

15m (49ft)

Belgica Bank

Mohns Ridge

2,580m (8,465ft)

Bodø

Vestfjorden

Røst Bank

6

Grise Fiord

Qaanaaq

77m (253ft)

Devon Shelf

732m (2,402ft)

Devon Slope

Cape Sherard

ncaster Trough

ncaster Sound

Bylot Island

Danebora

3,900m (12,796ft)

Greenland Sea

Greenland Plain

Greenland

210m (689ft)

Jan Mayen Fracture Zone

Jan Mayen

Kolbeinsey Ridge

Jan Mayen Ridge

Norwegian Sea

1,280m (4,200ft)

Dumshaf Plain

Vøring Plateau

Halten Bank

222m (728ft)

Norwegian Basin

Norwegian Trench

7

Kullorsuaq

Upernavik

Ittoqqortoormiit

Iceland Plateau

Denmark Strait

0°

Faroe-Shetland Trough

Shetland Islands

Faroe Islands

68m (223ft)

8

Baffin Basin

2,377m (7,799ft)

Baffin Bay

Uummannaq

Qeqertarsuaq

70°N

60°W

Arctic Circle

Iceland

Reykjavik

30°W

30°E

Britain

Baffin Island

Surface Currents inset

SURFACE CURRENTS

Transpolar Current
Beaufort Gyre
West Greenland Current
East Greenland Current
Norwegian Current

Surface Winds inset

SURFACE WINDS

Polar Easterlies
Polar Easterlies
Polar Easterlies
Westerlies

Beaufort Scale	Speed
0–3	0–16 kph (0–10 mph)
3–5.5	16–40 kph (10–25 mph)
over 5.5	over 40 kph (over 25 mph)

NORTHWEST PASSAGE

THE ATLANTIC AND PACIFIC OCEANS are linked by the Northwest Passage through the Arctic Ocean. A difficult route to navigate, it includes many narrow straits between islands, and its surface is often frozen, even in summer. The region contains broad sedimentary basins, which may harbour rich mineral deposits, but exploration is hampered by the extremely cold climate.

ARCTIC OCEAN E2

Baffin Bay

AREA 689,000 square km (266,000 square miles)

MAXIMUM DEPTH 2,100m (6,900ft)

INFLOWS Arctic Basin, Labrador Sea, glaciers of West Greenland

Baffin Bay lies between Greenland and Baffin Island, and is really the extreme northwest arm of the Atlantic. The surface ices over each winter, but warmer water from the Labrador Sea flows up its eastern shore, keeping parts of the adjacent Greenland coast ice-free. Water returns south along the western shore as the cold Labrador

GLACIER MEETING SEA

Current, often carrying icebergs into the North Atlantic. Seals have long been hunted in the area. Large numbers were killed each year as recently as the 1980s, but commercial hunting of marine mammals is now controlled.

DISCOVERY

EXPLORING THE PASSAGE

Much effort was expended in the 17th century in search of the Northwest Passage from the Atlantic to the Far East, but no viable route was found due to the year-round presence of sea-ice and the numerous islands. Interest revived in the 19th century when expeditions were undertaken by the Royal Navy. In 1820 an expedition from Baffin Bay got as far as Melville Island before being blocked by ice. A group of 129 men was lost off King William Island in 1848. British explorer Robert McClure crossed from the Beaufort Sea to Baffin Bay in 1854, but he had to walk part of the way. It was the Norwegian Roald Amundsen who finally sailed a ship via Lancaster Sound, south of Victoria Island, and out through the Bering Strait in 1906.

ARCTIC OCEAN A1

Beaufort Sea

AREA 476,000 square km (184,000 square miles)

MAXIMUM DEPTH 4,680m (15,350ft)

INFLOWS Chukchi Sea, Arctic Basin, rivers Mackenzie, Colville

The deep Beaufort Sea lies to the west of the Canadian Arctic Archipelago. Oil was discovered off the Alaskan shore in 1968, and is also extracted off the MacKenzie delta. Artificial islands have been built to protect some production wells from drifting sea-ice. Gas has also been found near Melville Island.

OIL EXPLORATION

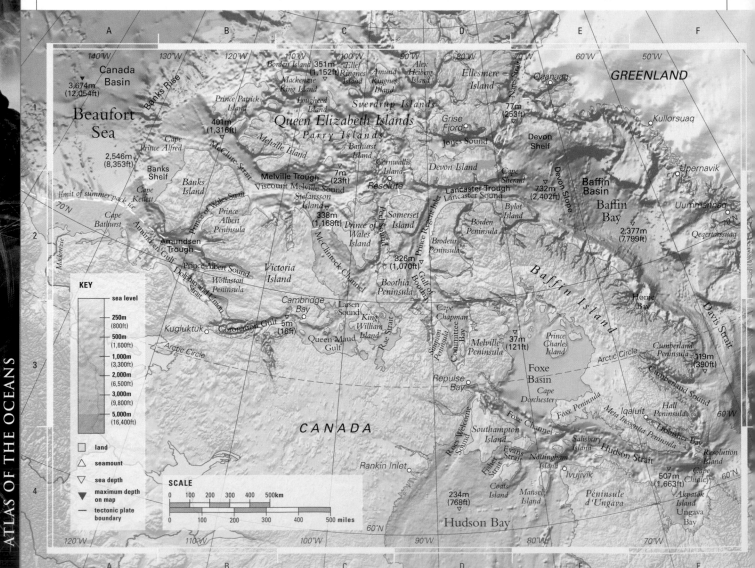

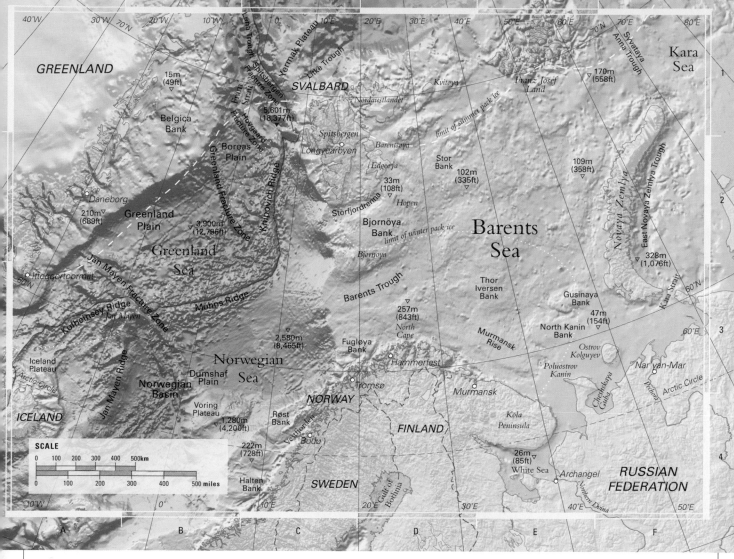

THE BARENTS SEA

THE BARENTS AND GREENLAND SEAS MARK the boundary between the Arctic Ocean and the Atlantic. Most of the Arctic Ocean's water exchange is with the north Atlantic. This exchange occurs at the Fram Strait, north of the Greenland Sea.

ARCTIC OCEAN E2

Barents Sea

AREA 1.4 million square km (540,000 square miles)

MAXIMUM DEPTH 600m (2,000ft)

INFLOWS Norwegian Sea, Arctic Basin

The Barents Sea is relatively shallow, lying north of Europe and south of the islands of Svalbard and Franz Joseph Land. To the east, Novaya Zemlya is an extension of the Ural Mountains, which mark the geographical boundary between Europe and Asia. Large areas along the mainland and around the islands are continental shelf of less than 200m (660ft) deep. Warm water from the North Atlantic Drift flows in from the southwest, keeping most of the sea ice-free in the summer. The Russian port of Murmansk

remains free of ice even in the winter. The warm, salty Atlantic water meets cold, less saline Arctic water, and warm, moderately salty coastal water, producing an area of high biological productivity. The spring bloom of phytoplankton starts near the ice edge, where fresh water from melting ice produces a stable surface layer. The phytoplankton form the basis of a food chain that supports a rich fishery, and cod is the most important catch. During the Cold War, Russia maintained a large northern fleet of warships and submarines. Many of these vessels now lie deteriorating in naval ports along the Kola Peninsula, raising fears of possible damage to the marine environment. Particular concerns have been raised about contamination from the nuclear reactors of abandoned submarines.

ARCTIC OCEAN B2

Greenland Sea

AREA 1.2 million square km (463,000 square miles)

MAXIMUM DEPTH 4,800m (16,000ft)

INFLOWS Arctic Basin, Norwegian Sea

The Greenland Sea, which stretches between Greenland, Svarlbad, and Jan Mayen Island, is a major area of sea-ice formation in the Arctic Ocean. The East Greenland Current carries surface water and ice south along the coast of Greenland, but the Jan Mayen Current takes some surface water to the east. This divergence leaves an area of open water where new sea-ice is

continually formed in the winter. An ice tongue, known as the Odden, develops eastwards from the main ice edge and dissolved salt is left behind in a layer of cold, briney water beneath the new ice. Being more dense, this very salty water sinks to the sea floor, where it pools before spilling over the ridges between Greenland and Jan Mayen to the south. This downwelling plays a major role in the global thermohaline circulation (see pp.60–61) of the oceans.

ICY PANCAKES

Pancake ice is formed in rough water as cakes of icy slush bump into each other, producing a raised rim.

THE ATLANTIC OCEAN

THE ATLANTIC SEPARATES the "old world" of Europe
and North Africa from the "new world" of the Americas.
The Portuguese pioneered the ocean's exploration, with
Bartholomeu Diaz reaching Africa's southern tip in 1488,
but a Spanish expedition
led by Christopher
Columbus was the first to
cross the Atlantic to reach
the West Indies in 1492.

**ATLANTIC SURF
BATTERS BERMUDA**

OCEAN CIRCULATION

A clockwise gyre controls surface currents in the
north Atlantic. The strong Gulf Stream current
brings in warm, salty water from the Caribbean.
It then continues northwestwards as the North
Atlantic Drift, giving western Europe a milder
climate than its latitude alone allows. Some
water moves north toward the Arctic Ocean,
but most returns south as the Canaries
Current. Warm water returns to the west
with the North Equatorial and Guiana
currents. The South Atlantic Gyre
turns anticlockwise, with the
Antarctic Circumpolar Current
forming its southern boundary.
The cold Benguela Current flows
north up the coast of Africa. In the
west, the Brazil Current flows rather
weakly south from the Equator.

TIDAL POWER
The Atlantic's high tidal range makes many
locations suitable for harnessing tidal power,
such as this installation off Devon, England.

OCEAN FLOOR

The Atlantic Ocean floor is dominated by the Mid-Atlantic Ridge,
marking where the continents to the east and west are splitting apart.
The ridge breaks through the sea surface at some points, most notably
as the island of Iceland. Transverse faults scar the flanks of the ridge for
many hundreds of kilometres east and west, until they are buried under
the marine sediments of the abyssal plains. These flat areas are quite
narrow in the Atlantic, occupying the edge of the deep ocean between
the broad ridge and the continental rise. The sedimentary margins of
the Atlantic bear some rich mineral deposits, including oil
and gas in the North Sea, the Gulf of Mexico, off
Venezuela, and West Africa. The Atlantic has
only two deep ocean trenches – the
Puerto Rico Trench and the
South Sandwich Trench.

ICEBERG SCOURING
This image shows a trench
scoured out by the keel of
a drifting iceberg – a fairly
common occurrence in shallow
waters around Greenland.

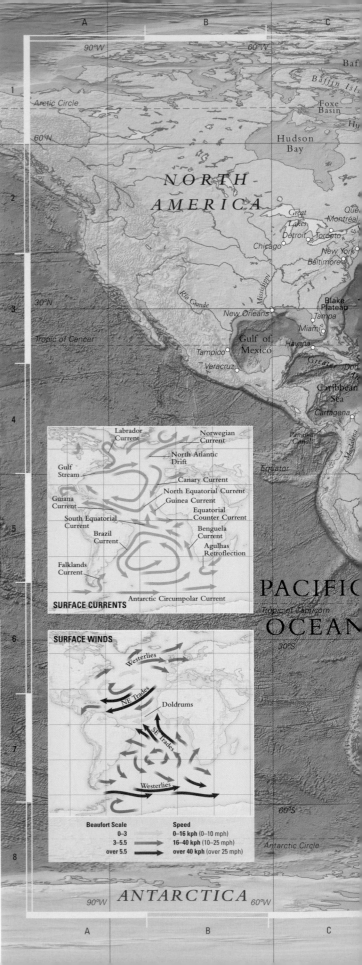

Column/Row coordinates
D · E · F · G · H · I

0°W · 0° · 30°E · 60°E · 90°E

Greenland

ffin · *sin*
218m (715ft) ▽

Nuuk

Labrador Sea
Labrador Basin

Reykjanes Basin
Eirik Ridge
Imarssuak Channel
Reykjanes Ridge
Reykjavik
Iceland · *Iceland Basin*
Hatton Ridge
Rockall Bank

Jan Mayen
Norwegian Sea · *Voring Plateau*
Norwegian Basin
Faeroe Islands

Scandinavia

North Cape
Arctic Circle

Helsinki
60°N

Oslo
Stockholm
St Petersburg
Riga
Baltic Sea
Gdansk

13m (43ft)

Newfoundland
Gulf of St Lawrence
69m (226ft) ▽

Northwest Atlantic Mid-Ocean Canyon
Charlie-Gibbs Fracture Zone
691m (2,267ft) ▽
4,139m ▽ (13,580ft)

Grand Banks of Newfoundland
Newfoundland Basin

Dublin Ireland
Britain
London
Celtic Shelf
Porcupine Plain

North Sea

Hamburg
Amsterdam
Elbe
Oder
Vistula
Dnieper
Dniester

St Petersburg
Odesa

EUROPE

New England Seamounts
muda Rise

Sohm Plain
5,464m (17,927ft)
Oceanographer Fracture Zone
Atlantis Fracture Zone

Azores-Biscay Rise
East Azores Fracture Zone
Azores
Great Meteor Tablemount
Madeira Plain
Madeira
Canary Islands

Iberian Plain
Horseshoe Seamounts
Strait of Gibraltar

Biscay Plain
Seine

Venice
Marseille
Barcelona
Lisbon
Naples
Danube
Istanbul
Black Sea

Athens
Algiers
Tunis
Mediterranean Sea
Tripoli
Beirut

Casablanca
Alexandria
Suez Canal
Nile
30°N

62m · 404ft

ATLANTIC
254m ▽ (833ft)

Sargasso Sea

Nares Plain
3,780m (12,402ft) ▽

Barracuda Fracture Zone

rto Rico
ch

Cape Verde Plain
Cape Verde Terrace
Cape Verde Basin
Cape Verde Islands
Dakar

Senegal
Niger

Tropic of Cancer

S a h a r a

AFRICA

Demerara Plain
Vema Fracture Zone
Gambia Plain
4,700m (15,421ft)
Doldrums Fracture Zone

Georgetown
Paramaribo
Cayenne
Amazon Fan

Conakry
Monrovia
Abidjan
Accra
Lagos
Douala

Sierra Leone Basin
Saint Paul Fracture Zone
Romanche Fracture Zone
Chain Fracture Zone

Gulf of Guinea
Niger Fan
Guinea Basin
Libreville

OCEAN

Ceara Plain
Amazon
Belém
Fortaleza

Fernando de Noronha
Pernambuco Basin
Ascension Fracture Zone
Ascension Island

Congo
Congo Fan
Luanda

Recife

Brazil Basin
5,706m (18,721ft)
Bode Verde Fracture Zone
Saint Helena
Saint Helena Fracture Zone

Angola Basin
5,042m (18,543ft)
Namibe

SOUTH MERICA

Tocantins
Sao Francisco
Salvador
Hotspur Seamount
Illas Martin Vaz
Isla Trindade

Rio de Janeiro

636m (2,087ft) ▽
Santos Plateau
Rio Grande Rise

Namib Desert
Walvis Ridge
Walvis Bay

Tropic of Capricorn

INDIAN OCEAN

Uruguay
Porto Alegre
Montevideo
uenos Aires

Rio Grande Fracture Zone

Namibia Plain

1,739m (5,706ft) ▽
Tristan da Cunha

Cape Basin
5,115m (16,782ft) ▽
Cape of Good Hope
Cape Town
Orange

30°S

Argentine Basin

Zapiola Ridge
Falkland Escarpment
Falkland Plateau
3,667m (12,031ft)
1,748m (5,735ft) ▽
South Georgia
Islas Orcadas Rise
South Sandwich Trench

Gough Fracture Zone

Discovery Tablemounts

Mid-Atlantic Ridge

alkland Islands
e Horn
Yaghan Basin
e Passage
th Shetland Islands
eninsula

Scotia Sea
East Scotia Basin
South Orkney Islands

Davis Seamounts
Atlantic-Indian Ridge
Southwest Indian Ridge

America-Antarctica Ridge

SOUTHERN OCEAN
Atlantic-Indian Basin
60°S

Weddell Plain
Weddell Sea
ne Ice Shelf

Antarctic Circle
Maud Rise
Lazarev Sea
Astrid Ridge

ANTARCTICA

0°W · 0° · 30°E · 60°E · 90°E

1 · 2 · 3 · 4 · 5 · 6 · 7 · 8

Equator

ATLAS OF THE OCEANS

D · E · F · G · H · I

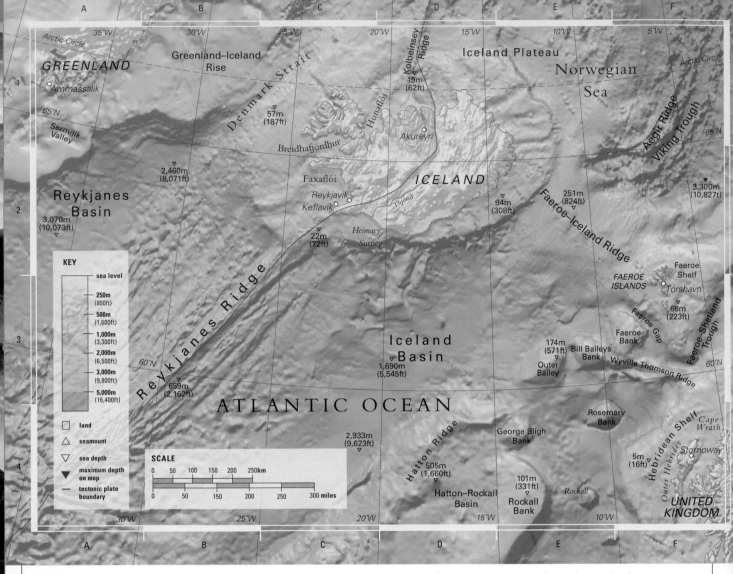

KEY

	sea level
	250m (800ft)
	500m (1,600ft)
	1,000m (3,300ft)
	2,000m (6,500ft)
	3,000m (9,800ft)
	5,000m (16,400ft)

□ land
△ seamount
▽ sea depth
▼ maximum depth on map
— tectonic plate boundary

SCALE

0 50 100 150 200 250km
0 50 150 200 250 300 miles

ICELAND

THE ISLAND OF ICELAND straddles the Mid-Atlantic Ridge and is one of the few places on Earth where it is possible to walk on newly created oceanic crust. It is the site of sea floor spreading that was responsible for linking the Atlantic to the Arctic Ocean around 36 million years ago. The surrounding seas are areas of water and heat exchange between the two oceans.

ATLANTIC OCEAN B1

Denmark Strait

LENGTH 480km (300 miles)

MINIMUM WIDTH 290km (180 miles)

Most of the water leaving the Arctic Ocean flows into the north Atlantic through the Denmark Strait, propelled by the East Greenland Current. Icebergs from the eastern side of the Greenland Ice Sheet are carried south by this cold current, while the warm North Atlantic Drift flows northeast on the eastern side of the island, between Iceland and the Faeroe Islands. At depth, cold, dense Arctic bottom water pools to the northeast

of Iceland until it overflows the Greenland–Iceland Rise and cascades 2,000m (6,500ft) down into the main Atlantic basin.

This is the start of a global journey as the dense water circulates around the deepest parts of the world's oceans – the deep-water leg of the "great ocean conveyor belt" (see p.61).

In winter, sea-ice builds up along the Greenland coast. Sometimes, cold winds blow east off the Greenland Ice Sheet, pushing sea-ice offshore. More sea-ice is created as the wind cools the exposed surface water, and a tongue of sea-ice can extend south from the Greenland Sea through the Denmark Strait.

ATLANTIC OCEAN B3

Reykjanes Ridge

LENGTH 1,500km (930 miles)

HEIGHT ABOVE SEA FLOOR 2,000m (6,500ft)

RATE OF SPREAD 1.8cm (³/₄in) per year

The Reykjanes Ridge is the part of the Mid-Atlantic Ridge that rises up to the ocean surface to the southwest of Iceland. The ridge very clearly displays the parallel ridges and valleys that are left behind either side of the central rift as the sea floor spreads at a divergent plate boundary. Here, the North American and Eurasian plates are moving apart at 1–2cm (½–1in) per year. The parallel

features become less distinct away from the ridge, as the older crust is draped in sediment in the Reykjanes Basin and Iceland Basin.

Before this rifting started, Greenland and Britain were almost adjacent, connected by a land bridge. The Hebrides and Faeroe Islands, Rockall, and the other banks on the eastern side of the Iceland Basin are the result of basalt floods associated with the early stages of the rift.

SURTSEY

Surtsey was born in 1963 when a volcano on the western flank of the Mid-Atlantic Ridge breached the surface of the sea off Iceland.

THE WESTERN NORTH ATLANTIC

IN THE WESTERN NORTH ATLANTIC, the warm, fast-flowing Gulf Stream pulls away from the American coast and continues northeastward as the North Atlantic Drift. The Labrador Current brings cold water south along the coast as far as the Gulf of Maine.

ATLANTIC OCEAN C2
Gulf of St. Lawrence

AREA 155,000 square km (60,000 square miles)

MAXIMUM DEPTH 2,300m (7,550ft)

INFLOWS Atlantic Ocean, St. Lawrence River

The gulf lies between the mouth of the St. Lawrence River and the islands of Newfoundland and Cape Breton. The Laurentian Trough, between the two islands, was scoured out by the Laurentide Ice Sheet during the last ice age. It channels sediment from the river over the edge of the continental shelf and onto the Laurentian Fan. The St. Lawrence River is the largest freshwater input to the Atlantic from the North American east coast, and its mouth is the largest estuary of its type in the world. The St. Lawrence Seaway, which was opened in 1959, gives vital shipping access to the Great Lakes.

SEA-ICE IN THE GULF OF ST. LAWRENCE

ATLANTIC OCEAN B3
Gulf of Maine

AREA 90,700 square km (35,000 square miles)

MAXIMUM DEPTH 377m (1,240ft)

INFLOWS Atlanic Ocean, rivers St. John, Penobscott

Like much of the continental shelf off the east coast of North America, the Gulf of Maine was above sea level during the last ice age. Georges Bank stands 100m (330ft) above the floor of the Gulf, and was an island until 6,000 years ago. Cape Cod and the islands of Nantucket and Martha's Vineyard are the highest standing of a series of moraines left behind as the glaciers retreated and the sea level rose.

Occasionally, the Gulf Stream lies not far offshore and the temperature of the sea off Nantucket beaches can be several degrees higher than it is off nearby Cape Cod.

North of the Gulf of Maine, the Bay of Fundy extends more than 200km (120 miles) inland. The bay acts like a funnel, producing a tidal range of 13m (43ft) at its northern end, which is the highest in the world.

ATLANTIC OCEAN E2
Grand Banks

AREA 280,000 square km (108,000 square miles)

AVERAGE DEPTH 100m (330ft)

The Grand Banks is a large area of continental shelf, extending up to 500km (311 miles) off Newfoundland. The area is renowned for dense sea fogs, which arise when warm, moist air from the south is chilled by the cold Labrador Current, causing condensation. The Labrador Current presents another shipping hazard by bringing icebergs to the area – the *Titanic* famously sank south of the Grand Banks in 1912.

Although turbidity currents have never been directly observed, their power was felt in 1929 when an earthquake triggered a huge sediment flow (submarine landslide) down the continental slope off the Grand Banks. Submarine telegraph and telephone cables were broken over a distance of 800km (500 miles) – from the timing of the breaks, the speed of the flow was estimated at 40–55kph (25–34mph).

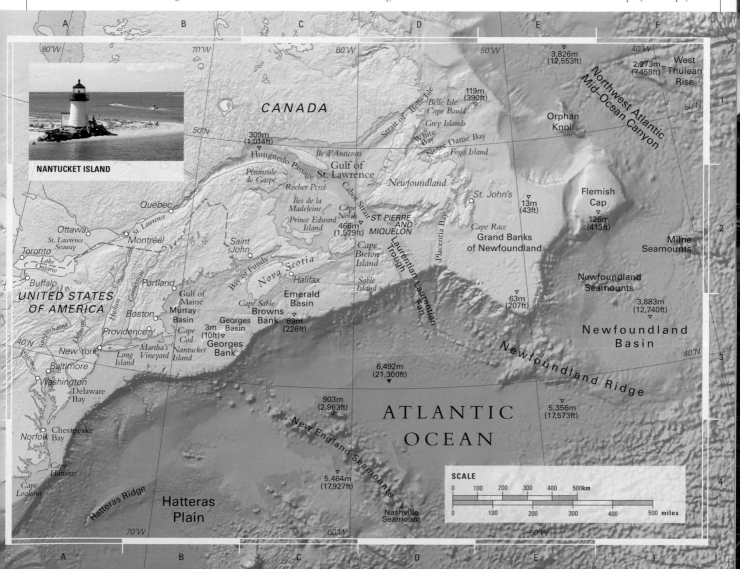

NANTUCKET ISLAND

THE NORTH SEA AND BALTIC SEA

A NUMBER OF SHALLOW SEAS COVER the continental shelf off the northwest coast of Europe, the largest of which is the North Sea. The North Atlantic Drift brings warmer water into the region, producing the mild climate enjoyed by adjacent coastal areas.

ATLANTIC OCEAN D1

Norwegian Sea

AREA 1.4 million square km (534,000 square miles)

MAXIMUM DEPTH 3,970m (13,020ft)

INFLOWS Central North Atlantic, numerous Norwegian fiords

The Norwegian Sea lies between Norway and Iceland, separated from the main part of the north Atlantic by the submarine Faeroe–Iceland Ridge. Although situated at high latitude, this sea is kept free of ice by the warm, salty North Atlantic Drift, which flows from the southwest between Scotland and Iceland and continues into the Barents Sea (see p.431) as the Norwegian Atlantic Current. This relatively warm water is the reason for the Norwegian port of Bergen's reputation as the wettest place in Europe, with rain expected at least 275 days a year.

PULPIT ROCK AT LYSEFJORD
Fiords cut into the Atlantic coast of Norway, showing where ice-age glaciers scoured deep valleys below today's sea level.

ATLANTIC OCEAN G5

Baltic Sea

AREA 386,000 square km (149,000 square miles)

MAXIMUM DEPTH 449m (1,473ft)

INFLOWS Rivers Vistula, Oder, Western Divina

The Baltic is a shallow, virtually enclosed inland sea with little tide. It does not benefit from the warmth of the North Atlantic Drift, and its northern branches, the Gulf of Bothnia and the Gulf of Finland, ice over in the winter. A large influx of river water gives the Baltic a low salinity – it is the largest area of brackish water in the world. Its only outflow is to the North Sea via the Danish Straits (three channels linking the Baltic to the Kattegat), the Kattegat Bay, and the Skagerrak Strait. There is a weak influx of dense salt water at depth that isolates the basin floor from the surface waters, producing an oxygen-depleted dead zone. Without significant out-flows, the Baltic Sea is vulnerable to pollution carried in by rivers and from large population centres on its coasts. Although there is a sea route to the North Sea, there is also a shorter, more sheltered route through the Kiel Canal.

ACROSS THE ORESUND STRAIT
The 16km- (10-mile-) long Oresund Bridge links the Danish capital Copenhagen with Malmo in Sweden.

ATLANTIC OCEAN D6

North Sea

AREA 570,000 square km (220,000 square miles)

MAXIMUM DEPTH 700m (2,300ft)

INFLOWS North Atlantic, rivers Elbe, Weser, Ems, Rhine, Scheldt, Thames, Humber

Water from the Atlantic enters the North Sea between the Shetland and Orkney Islands, flowing south down the Scottish and English coasts. Warmer Atlantic water also enters from the English Channel and flows east along the Dutch coast, resulting in an anticlockwise circulation. The largest sand banks on the North Sea floor, including the Dogger, Jutland, and Fisher banks, are terminal moraines marking the southern edge of the ice sheet during the last ice age, when the bottom of the sea was exposed by lower sea levels. A trough located to the west of the Norwegian Trench is buried under thick sediments which contain oil and gas deposits.

OIL RIG IN THE NORTH SEA

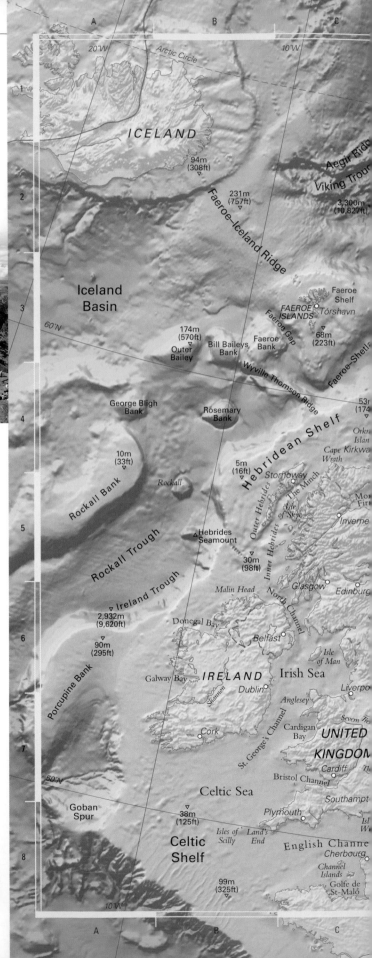

ICELAND

20°W   Arctic Circle   10°W

94m (308ft)

231m (757ft)

Aegir Ridge
Viking Trough

3,300m (10,827ft)

Faeroe–Iceland Ridge

Iceland Basin

Faeroe Shelf

FAEROE ISLANDS   Tórshavn

Faeroe Gap

60°N

174m (570ft)

Bill Baileys Bank   Faeroe Bank

68m (223ft)

Outer Bailey

Wyville Thomson Ridge

Faeroe-Shetland

53m (174ft)

George Bligh Bank

Rosemary Bank

Hebridean Shelf

Orkney Islands

Cape Kirkwall
Wrath

10m (33ft)

5m (16ft)   Stornoway

The Minch

Rockall Bank   Rockall

Outer Hebrides   Isle of Skye   Moray Firth

Rockall Trough

Hebrides Seamount

30m (98ft)

Inner Hebrides   Glasgow   Inverness

Edinburgh

Ireland Trough

Malin Head   North Channel

2,932m (9,620ft)   Donegal Bay   Belfast

Isle of Man

90m (295ft)

Galway Bay   IRELAND   Irish Sea

Shannon   Dublin   Liverpool

Porcupine Bank   Anglesey

Cardigan Bay   Severn

St. George's Channel   UNITED KINGDOM

50°N   Cork   Cardiff

Bristol Channel

Celtic Sea   Southampton

Goban Spur   38m (125ft)   Plymouth

Isle of Wight

Isles of Scilly   Land's End   English Channel   Cherbourg

Celtic Shelf   Channel Islands

99m (325ft)   Golfe de St-Malo

10°W

A   B   C

Norwegian
Basin

Voring
Plateau
1,280m
(4,200ft)

Røst
Bank

Vestfjorden

Traena Deep

Bodø

Norwegian Sea

Arctic Circle

Traena
Bank
222m
(728ft)

Sklinna
Bank

Halten
Bank

Luleå

Oulu

85m
(279ft)

SWEDEN

Umeå

FINLAND

Vaasa

Gulf of Bothnia

Trondheimsfjorden
Frøya
Bank
94m
(308ft)

Trondheim

1,320m
(4,331ft)

ugh

Hitra

Glomma

Shetland
Islands

Lerwick

Sognefjorden

NORWAY

Turku
(Åbo)

Helsinki

St Petersburg
RUSSIAN
FEDERATION

60°N

r Isle

Hardangerfjorden

Viking
Bank

Bergen

Gävle

Åland

Gulf of Finland

Tallinn

Lake
Peipus

Walker
Bank

Boknafjorden

Stavanger

Oslo

Stockholm

175m
(574ft)

Hinnmaa

ESTONIA

Lake
Pskov

uth
nk

67m
(220ft)

Kristiansand

Lindesnes

700m
(2,297ft)

Vänern

Norrköping

Vättern

Gotland
Basin

Saaremaa

Gulf of
Riga

LATVIA

75m
246ft)

Devil's
Hole

Skagerrak

Gothenburg

Kattegat

Gotland

Ventspils

Riga

Western Dvina

North Sea

Great
Fisher
Bank

Jutland
Bank

Läsø

Halmstad

Kalmar

Öland

Baltic
Sea

9m
(30ft)

Liepaja

LITHUANIA

13m
(43ft)

Aalborg

Karlskrona

Klaipeda

armade
Bank

Dogger
Bank

Weiss
Bank

DENMARK

Århus

Helsingborg

Copenhagen

Malmö

Courland
Lagoon

RUSSIAN
FEDERATION

15m
(49ft)

Odense

Fyn

Sjaelland

Bornholm

Gulf of
Danzig

Kaliningrad

ngston
on Hull

Jylland

Lolland

Falster

Bay

Rügen

Pomeranian
Bay

Gdansk

BELARUS

The Wash

Well
Bank

Vlieland
Bank

Friesian Islands

Kiel
Bay

Mecklenburg
Bay

Rugen

Szczecin

POLAND

Helgoland
Bay

Kiel

Kiel
Canal

ndon

NETHERLANDS

Amsterdam

Rotterdam

Hamburg

Bremen

Elbe

Weser

Ems

Mittelandkanal

Dover

Strait of Dover

Calais

Antwerp

BELGIUM

Maas

Rhine

GERMANY

Elbe

Meuse

Havre

FRANCE

Seine

Moss

LUXEMBOURG

Main

50°N

CZECH
REPUBLIC

10°E

20°E

SCALE
0 50 100 150 200 250km

0 50 100 150 200 250 miles

KEY

sea level

250m
(800ft)

500m
(1,600ft)

1,000m
(3,300ft)

2,000m
(6,500ft)

3,000m
(9,800ft)

5,000m
(16,400ft)

□ land

△ seamount

▽ sea depth

▼ maximum depth
on map

— tectonic plate
boundary

ATLAS OF THE OCEANS

MIDDELGRUNDEN WIND FARM
Arranged in an elegant curve 2km
(1.2 miles) east of Copenhagen, Denmark,
each of this wind farm's 20 turbines can
produce 2 megawatts of electricity.

WIND-FARMING IN THE BALTIC

Declining oil stocks, threats to fuel supplies, and the risks of climate change are increasingly focusing attention on alternative sources of energy that do not generate greenhouse gases. After hydroelectricity, wind-farming is the most advanced source of renewable energy. It is easiest to build wind farms on land, but many people do not want turbines built near their homes. Wind farms built at sea are less controversial, although costlier to build and maintain. However, it is usually windier at sea and there are no hills or trees to cause turbulence, so offshore turbines are more efficient than land-based ones.

The world's first commercial wind farm was built in the Baltic (see below). Other large wind farms are in place in British and Irish waters, and Japan also wants to develop such schemes. Canada has a proposal for a large wind farm off the coast of British Columbia, and there are plans for major wind farms off Cape Cod and Long Island in the US, although these are proving controversial and some members of Congress are attempting to stall them.

Because they burn no fossil fuels, wind farms can help reduce greenhouse gases. However, the overall benefits may take some time to appear. It can take 1,000 tonnes of concrete just to build the foundations of an offshore turbine, and concrete production is one of the biggest sources of greenhouse gases.

BALTIC WIND FARMS

The countries around the Baltic Sea have played a central role in the development of offshore wind-energy. The world's first commercial offshore wind farm was commissioned in 1991 near the Danish fishing port of Vindeby. It is easiest to build in shallow water, and the first of these wind farms were sited in water less than 10m (33ft) deep. The Baltic is ideal for wind-farming, with an average depth of just 55m (180ft).

OFFSHORE WIND FARMS

MASSIVE BLADES
Wind-turbine blades are assembled on the dockside, then carried to the tower on a barge. The rotors reach 95m (295ft) in diameter.

FOUNDATIONS
The concrete foundations of a turbine are cast onshore in a dry dock. They are then floated out to sea and sunk in shallow water on site.

UNDER CONSTRUCTION

SUBSTATION *The electricity generated by offshore turbines has to be transmitted to land. This substation collects power from 72 turbines at the Nysted Offshore Wind Farm off southern Denmark, and transforms it from 33,000 to 132,000 volts for transmission ashore via 48km (30 miles) of submarine cable.*

POWER DISTRIBUTION

HORNS REV *Sited in the North Sea off the coast of Jutland, Horns Rev is Denmark's largest wind farm and is one of the largest wind farms in the world. It began producing electricity in 2002 and has 80 turbines, positioned between 14 and 20km (9–13 miles) offshore. Horns Rev can produce enough power for about 150,000 Danish households.*

NORTH SEA WIND FARMS

WHITE-TAILED EAGLE *Although wind energy may bring about potentially huge benefits for wildlife by reducing climate change, turbines do sometimes harm birds. Four dead White-tailed Eagles were found at an offshore Norwegian wind farm in early 2006.*

CASUALTIES

A   B   C   D   E   F

50°N

40°W

30°W

20°W

Charlie Gibbs Fracture Zone

Hecate
Seamount

1

691m
(2,267ft)

East
Thulean
Rise

Porcupine Pla

Faraday Fracture Zone

2,193m
(7,195ft)

Flemish
Cap

126m
(413ft)

4,579m
(15,024ft)

719m
(2,339ft)

4,139m
(13,580ft)

2

Maxwell Fracture Zone

4,885
(16,028

Milne
Seamounts

691m
(2,267ft)

2,500m
(8,203ft)

Newfoundland
Seamounts

102m
(335ft)

Altair
Seamount

Olympus
Knoll

458m
(1,503ft)

Kings Trough

3

3,883m
(12,740ft)

Antialtair
Seamount

Newfoundland
Basin

A T L A N T I C

O C E A N

6,324m
(20,749ft)

Azores-Biscay Rise

40°N

4

Akademik Kurchatov Fracture Zone

Mid-Atlantic Ridge

Pico Fracture Zone

Corvo

Flores

Graciosa

Terceira

5

2,160m
(7,087ft)

São Jorge

Faial

Pico

Terceira Rift

Azores

5,143m
(16,874ft)

Azores Plateau

São Miguel

Ponta
Delgada

117m
(384ft)

Oceanographer Fracture Zone

Santa
Maria

East Azores Fracture Zone

772m
(2,533ft)

Hayes Fracture Zone

275m
(902ft)

6

2,194m
(7,199ft)

Madeira Ridge

Func

Ma

5,485m
(17,996ft)

Atlantis Fracture Zone

30°N

7

238m
(781ft)

Cruiser
Tablemount

**Madeira
Plain**

Great Meteor
Tablemount

La Palma

8

4,645m
(15,240ft)

Gome

**Canary
Basin**

Hierro

40°W

30°W

20°W

A   B   C   D   E   F

SCALE

0 50 100 150 250 300km

0 50 100 150 250 300 miles

THE EAST ATLANTIC

THE EASTERN NORTH ATLANTIC is renowned for its winter storms, which batter the western coasts of Europe. The energy for these storms is provided by the Gulf Stream feeding warm water into the North Atlantic Drift, which flows to the northeast. The North Atlantic Gyre pulls some of this water south along the African coast as the Canaries Current.

ATLANTIC OCEAN I3
Bay of Biscay

AREA 223,000 square km (86,000 square miles)

MAXIMUM DEPTH 4,735m (15,535ft)

INFLOWS Rivers Loire, Dordogne, Garonne, Adour

The Bay of Biscay lies between Brest, on the Brittany Peninsula, and the north coast of Spain. The northern half of the bay is quite shallow, overlying the continental shelf, but this steeply drops away to the Biscay Plain, which is a small, partially opened ocean basin. Ships crossing the bay experience heavy seas, as

LIGHTHOUSE ON THE BAY
The Old Lighthouse at La Raz Cap is one of several that mark treacherous rocks off the Brittany Peninsula in the Bay of Biscay.

full-size Atlantic rollers become amplified by the sudden shallow depth. There is a weak anticlockwise surface current within the bay. The Charcot Seamounts, Azores–Biscay Rise, and Kings Trough mark an inactive crustal fracture where the sea floor was once splitting apart.

ATLANTIC OCEAN D5
Azores

TYPE Volcanic islands

AREA 2,300 square km (890 square miles)

NUMBER OF ISLANDS 9

The Mid-Atlantic Ridge is the dominant sea-floor feature in the eastern Atlantic region, with a central trough and numerous transform fracture zones. Just east of the ridge, the Azores island group straddles the triple junction between the Eurasian, African, and North American plates.

The islands rise from the extensive Azores Plateau, an area of thickened ocean crust. Although volcanic in origin, the oldest islands also include substantial accumulations of limestone and clay sediments. A mantle hot spot (see p.51) underlies the plateau and seems to be slowly spreading it apart at the Terceira Rift, a fracture that links the East Azores Fracture Zone to the Mid-Atlantic Ridge. The last volcanic eruption in the Azores was in 1957, when the Capelinhos volcano produced a cinder island (an island composed of lava fractures called cinders) off Faial's coast.

ATLANTIC OCEAN G8
Canary Islands

TYPE Volcanic islands

AREA 7,400 square km (2,900 square miles)

NUMBER OF ISLANDS 7

The name of these islands derives not from the yellow bird of the same name, but from the Latin word for dogs, *Canaria*. The islands are volcanic, overlying a mantle hot spot. Pico del Teide on Tenerife is the third largest volcano on Earth, rising more than 3,700m (12,000ft) above sea level, or almost 7,000m (23,000ft) from the sea floor. It

last erupted in 1909. Teide's slopes are unstable and there is evidence of huge landslides having occurred in the past. There is also a risk that volcanic activity or earth tremors could cause part of La Palma island to slip into the sea, resulting in an enormous tsunami. Such an event would threaten the coasts of the north Atlantic, including heavily populated parts of North America, with inundation.

TENERIFE ISLAND

Map labels

10°W

IRELAND
Cork
Bristol Channel
Porcupine Bank
Porcupine Seabight
Celtic Sea
Plymouth
Isle of Wight
UNITED KINGDOM
English Channel
Cherbourg
Seine
50°N
1
38m (125ft)
Isles of Scilly
Land's End
Channel Islands
Golfe de St-Malo
Goban Spur
Celtic Shelf
Brest
Pointe du Raz
FRANCE
99m (325ft)
Loire
Belle Île
Nantes
Plateau de Rochebonne
Île de Ré
Île d'Oléron
2
Biscay Plain
Bay of Biscay
Bordeaux
4,870m (15,978ft)
Charcot Seamounts
Cabo de Ajo
Biarritz
Theta Gap
Cabo Ortegal
Costa Verde
Donostia–San Sebastián
Santander
Gijón
3
Galicia Bank
Cabo Fisterra
La Coruña
492m (1,614ft)
Vigo
Douro
Iberian Plain
Oporto
SPAIN
40°N
4
PORTUGAL
Tagus
Cabo Mondego
5,536m (18,164ft)
Guadiana
Lisbon
Guadalquivir
Cabo de Roca
Cabo Espichel
Cabo de Sines
5
Algarve
Tagus Plain
Gulf of Cadiz
Gorringe Ridge
Cabo de São Vicente
Gibraltar
Mediterranean Sea
Horseshoe Seamounts
Strait of Gibraltar
Ceuta
Gettysburg Seamount
Tangiers
Ampere Seamount
Rabat
6
20m (66ft)
Seine Plain
Casablanca
Seine Seamount
4,265m (13,993ft)
Safi
MOROCCO
7
Dacia Seamount
Agadir Canyon
Cap Rhir
Agadir
Conception Bank
30°N
Canary Islands
Lanzarote
Santa Cruz
Fuerteventura
Tenerife
Cap Juby
Las Palmas
WESTERN SAHARA
10°W
8

KEY

sea level

	250m (800ft)
	500m (1,600ft)
	1,000m (3,300ft)
	2,000m (6,500ft)
	3,000m (9,800ft)
	5,000m (16,400ft)

☐ land
△ seamount
▽ sea depth
▼ maximum depth on map
— tectonic plate boundary

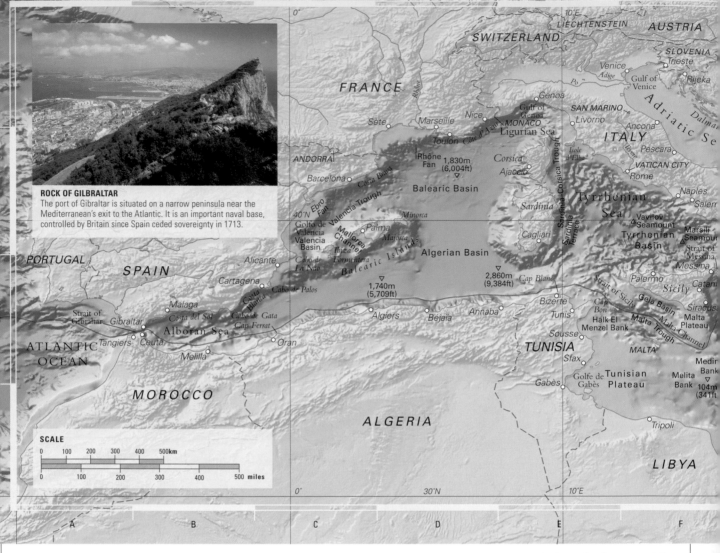

ROCK OF GIBRALTAR
The port of Gibraltar is situated on a narrow peninsula near the Mediterranean's exit to the Atlantic. It is an important naval base, controlled by Britain since Spain ceded sovereignty in 1713.

SCALE

0 100 200 300 400 500km

0 100 200 300 400 500 miles

THE MEDITERRANEAN SEA AND BLACK SEA

THE MEDITERRANEAN IS AN ALMOST enclosed sea, with high evaporation and salinity, a very small tidal range, and a complex floor. The adjacent Black Sea is the last remnants of the Tethys Ocean, which closed as Africa converged with Eurasia.

ATLANTIC OCEAN D2

Western Mediterranean

AREA 850,000 square km (328,000 square miles)

MAXIMUM DEPTH 3,600m (11,800ft)

INFLOWS Atlantic Ocean, rivers Ebro, Rhône

The entire Mediterranean loses three times more water by evaporation than it gains from rainfall and rivers combined. This loss is balanced by a surface inflow from the Atlantic through the Strait of Gibraltar. The inflow continues as an eastward current along the north African coast, giving rise to an anticlockwise circulation in the western Mediterranean. At depth there is a strong undercurrent of outflowing salty water. The flat floors of the Algerian and Balearic basins are underlain by deep sediments. In contrast, the Tyrrhenian Sea contains many seamounts and ridges. A chain of active volcanoes (including Etna, Stromboli, and Vesuvius) is found on the sea's eastern margin, where the African Plate is subducting beneath the Eurasian Plate. The eastward flow of surface water continues through the Strait of Sicily into the eastern Mediterranean. The narrower Strait of Messina, between Sicily and mainland Italy, is notorious for its whirlpool, possibly the inspiration for the Greek mythological sea monster Charybdis.

ATLANTIC OCEAN H3

Eastern Mediterranean

AREA 1.65 million square km (637,000 square miles)

MAXIMUM DEPTH 5,095m (16,720ft)

INFLOWS Black Sea, rivers Adige, Nile, Po

The eastern and western parts of the Mediterranean are separated by Sicily and the submerged Malta and Tunisian plateaus. The eastward flow from the western Mediterranean continues along the African coast, and an anticlockwise circulation prevails in the eastern Mediterranean, and in the Ionian, Aegean, and Adriatic seas. Surface water becomes more saline through evaporation as it travels east, and starts to sink after cooling by winter winds. It then returns westwards, exiting through the Strait of Gibraltar about 150 years after entering. The sea floor is dominated by the Mediterranean Ridge, a result of compression between the convergent African and Eurasian plates. These sediments are older – 70 million years compared with 25 million years in the western Mediterranean. The Adriatic Sea is a shallow branch of the eastern Mediterranean. Rising sea levels at the end of the last ice age flooded valleys parallel to its eastern shore, giving rise to the islands of the Dalmatian coastline.

VENICE LAGOON
Venice was built in the shallow waters of a lagoon in the Adriatic. Its merchants grew rich by controlling access to the Silk Route.

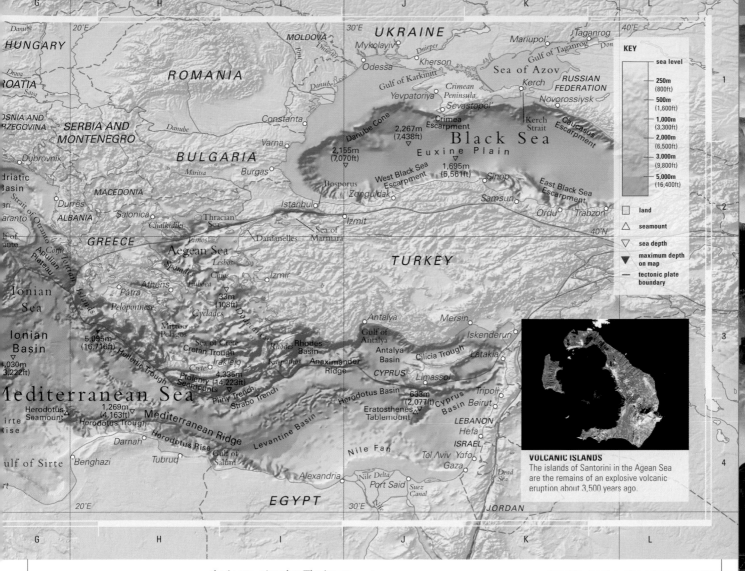

KEY

	sea level
	250m (800ft)
	500m (1,600ft)
	1,000m (3,300ft)
	2,000m (6,500ft)
	3,000m (9,800ft)
	5,000m (16,400ft)

□ land
△ seamount
▽ sea depth
▼ maximum depth on map
— tectonic plate boundary

VOLCANIC ISLANDS
The islands of Santorini in the Agean Sea are the remains of an explosive volcanic eruption about 3,500 years ago.

ATLANTIC OCEAN H2

Aegean Sea

AREA 214,000 square km (83,000 square miles)

MAXIMUM DEPTH 3,294m (10,800ft)

INFLOWS Black Sea, Mediterranean Sea

The Aegean Sea contains more than 1,000 islands and is the source of most of the Mediterranean's cold, saline deep water. Before 1990 this source was in the Adriatic, but climate changes have led to increased winter cooling in the Aegean. It is a geologically complex area as the Aegean microplate and the Anatolian Plate to the east are caught between the converging African and Eurasian plates. The Aegean crust is of continental thickness, but has been stretched and thinned, notably in the area of the Cretan Trough, so that much of it is now below sea level. The Hellenic Trough and Pliny Trench mark where the African Plate is subducting beneath the Aegean microplate. The Aegean Volcanic Arc stretches from Greece to Turkey through the southern Cyclades. These volcanoes are dormant or extinct, but earthquakes still occur at a depth of 150–170km (95–105 miles). The islands of Santorini, in the southern Cyclades, are the remains of an explosive volcanic eruption around 1640 BC.

This was the largest volcanic event of the last 10,000 years and may have caused the downfall of Crete's Minoan civilization. Behind the volcanic arc, the main Cyclades sit on top of a subsided plateau. At the northern end of the Aegean, a transform fault marks the contact with the Eurasian Plate, an area prone to strong, shallow earthquakes.

SKIATHOS ISLAND
Aegean islands consist mostly of hard metamorphic and volcanic rocks, so their coasts often show steep cliffs, headlands, and wave-cut features.

ATLANTIC OCEAN J2

Black Sea

AREA 422,000 square km (163,000 square miles)

MAXIMUM DEPTH 2,200m (7,200ft)

INFLOWS Mediterranean Sea, Sea of Azov, rivers Danube, Dniester, Dnieper, Kizil Irmak

The Black Sea is an enclosed inland sea, connected to the Mediterranean Sea via the Dardanelles, the Sea of Marmara, and the Bosporus. There is negligible exchange of water with the Mediterranean, and the surface waters of the Black Sea are about half as saline as the eastern Mediterranean. A previous small outflow through the Bosporus to the Aegean appears to have been reversed due to reduced inflow after the damming of some of the rivers feeding the Black Sea.

Although the surface waters are relatively fresh, below about 100–150m (330–490ft) lies a highly saline water body with very slow turnover. Decaying organic matter consumes all the oxygen in this water, making the Black Sea the world's largest oxygen-free marine system – the deep water is essentially dead. The basin is an isolated remnant of the north shore of the ancient Tethys Ocean. The southern part of the Black Sea is deep, but it is not as deep as the Mediterranean, and the underlying crust is thicker than most ocean crust.

The northern parts – the Sea of Azov and the Gulf of Odessa – overlie a shallow continental shelf. The delta of the Danube, Europe's longest river, extends from the western shore, and Danube waters have carried sediment across the edge of the shelf to build up a thick cone of sediment.

BLACK SEA SHIPPING
The Bosporus, the narrowest strait open to international navigation, connects the Black Sea with the Sea of Marmara.

THE GULF OF MEXICO AND CARIBBEAN SEA

THE CARIBBEAN SEA AND GULF OF MEXICO form a semi-enclosed extension of the north Atlantic. A low input of fresh water and high evaporation rates make the surface waters highly saline. The area is subject to violent storms and some volcanic activity.

ATLANTIC OCEAN B3

The Gulf of Mexico

AREA 1.6 million square km (618,000 square miles)

MAXIMUM DEPTH 5,203m (17,070ft)

INFLOWS Caribbean Sea, rivers Mississippi, Rio Grande, Apalachicola

The Gulf of Mexico is almost enclosed by parts of North America. The broad continental shelves to the north and south of the deep central basin are rich in oil deposits.

ATLANTIC OCEAN E6

The Caribbean Sea

AREA 2.75 million square km (1.06 million square miles)

MAXIMUM DEPTH 7,685m (25,215ft)

INFLOWS Atlantic Ocean, river Magdalena

The Caribbean Sea is a tropical body of water bounded to its south and west by South and Central America, and to its north and east by the Greater and Lesser Antilles. Most of the Antilles, and some parts of the mainland coast, are fringed with coral reefs and small, low-lying islets called cays (or keys).

The underlying Caribbean Plate was once part of the Pacific Ocean floor and it is still moving slowly eastwards between the North and South American plates. A subduction zone separates the Caribbean Plate from the Atlantic Plate to the east, giving rise to the volcanic island arc of the Lesser Antilles.

An east-to-west surface current permeates the whole of the Caribbean, with water from the Guiana Current flowing in via gaps between the small islands in the east, and flowing out in to the Gulf of Mexico via the Yucatan Channel in the northwest.

Circulation is weak, and the water becomes more salty as it is heated up. Inflows from rivers and the Caribbean are balanced by an outflow of warm, salty water – the beginnings of the Gulf Stream – via the Straits of Florida to the east. This channel runs between two limestone plateaus – the Florida peninsula, above sea level to the north, and the Bahamas, a submerged plateau topped by low-lying islands to the south. The coasts of the Gulf are affected by powerful hurricanes in the late summer and autumn.

SOUFRIERE VOLCANO ON MONTSERRAT

HUMAN IMPACT

PANAMA CANAL

Opened in 1914, the Panama Canal links the Atlantic Ocean with the Pacific, allowing ships to avoid the long journey around Cape Horn. Its construction was one of the most difficult engineering projects ever attempted, taking ten years and costing many lives. Each year, 14,000 ships use the canal.

PANAMA GATES IN 1913

ATLANTIC OCEAN H2

The Sargasso Sea

AREA 5.2 million square km (2 million square miles)

MAXIMUM DEPTH 7,000m (23,000ft)

INFLOWS None

The Sargasso Sea is a large area of the north Atlantic southeast of Bermuda. It is bounded by ocean currents: the Gulf Stream to its west and north, the Canary Current far to the east, and the North Equatorial Current to its south. The area between these currents rotates slowly in a clockwise direction and is often quite calm.

Large mats of yellow-brown Sargassum seaweed float on its surface, providing shelter and food for communities of small crustacea and fish, including freshwater eels. Adult eels migrate to the Sargasso every year to mate and spawn, and their young are carried back to the rivers of North America and Europe by the Gulf Stream. Deep water in this part of the Atlantic flows from north to south.

FLORIDA KEYS
The Florida Keys are small, low-lying islands composed mainly of ancient coral reefs that are underlain by limestone.

HURRICANE LILI
Hurricanes occur frequently in the Gulf of Mexico and Caribbean Sea, often causing much damage to coastal regions.

D   E   F   G   H   I

80°W   70°W   60°W

KEY

	sea level
	250m (800ft)
	500m (1,600ft)
	1,000m (3,300ft)
	2,000m (6,500ft)
	3,000m (9,800ft)
	5,000m (16,400ft)

☐ land
△ seamount
▽ sea depth
▼ maximum depth on map
— tectonic plate boundary

F AMERICA

Norfolk

Nashville Seamount

Cape Hatteras

Myrtle Beach
Onslow Bay
Long Bay

Charleston

Savannah

Richardson Hills

Florida-Hatteras Slope

Hatteras Ridge

BERMUDA

Bermuda Rise

Sohm Plain

5,464m (17,927ft)

80°N

Jacksonville

Hoyt Hills

Escarpment

Hatteras Plain

5,255m (17,242ft)

Sargasso Sea

Blake Spur

Blake Plateau

Blake-Bahama Ridge

Blake Abyssal Plain

Eastward Knoll

Researcher Seamount

Tampa

West Florida Shelf

Cape Canaveral

Great Abaco Canyon

Blake Basin

A T L A N T I C

Fort Lauderdale

Little Bahama Bank

Great Bahama Island

Great Abaco Canyon

Bahama Basin

O C E A N

Tropic of Cancer

Miami

BAHAMAS
Andros Island

Nassau

Eleuthera Island

Tongue of the Ocean

Exuma Sound

Bahama Escarpment

Florida Keys

Pourtalès Escarpment

Great Bahama Bank

Exuma Valley

Vema Gap

Nares Plain

6,081m (19,952ft)

Slope

Straits of Florida

Cay Sal Bank

Santaren Channel

Long Island

20°N

Havana

Archipiélago de Sabana

Archipiélago de Camagüey

Acklins Island

TURKS & CAICOS ISLANDS

5,868m (19,253ft)

Golfo de Batabanó

Cienfuegos

Great Inagua

Mouchoir Passage

Isla de la Juventud

CUBA

Puerto Rico Trench

8,962m (29,404ft)

Hispaniola Basin

647m (2,123ft)

BRITISH VIRGIN ISLANDS

Leeward Islands

6,691m (21,953ft)

Greater ucatan Basin

Golfo de Guacanayabo

Santiago de Cuba

Guantánamo Bay

Windward Passage

Mona Canyon

Mona Passage

PUERTO RICO
San Juan

Anegada Gap

Anegada Passage

Tintamarre Spur

ANGUILLA

ANTIGUA & BARBUDA
St John's

Barracuda Ridge

CAYMAN ISLANDS

Cayman Ridge

Cabo Cruz

Montego Bay

Jamaica Channel

NAVASSA ISLAND

Golfe de la Gonâve

HAITI

Hispaniola

Santo Domingo

Ponce

ST KITTS & NEVIS
VIRGIN ISLANDS

MONTSERRAT

Guadeloupe Passage

GUADELOUPE

Cayman Trench

7,680m (25,198ft)

Jérémie

Cap Haïtien

DOMINICAN REPUBLIC

Port-au-Prince

Muertos Trough

Gibbs Seamount

Basse-Terre

Dominica Passage

DOMINICA

5,830m (19,128ft)

JAMAICA

Kingston

Antilles

5,550m (18,210ft)

Martinique Passage

MARTINIQUE

Fort-de-France

Pedro Bank

4,536m (14,883ft)

960m (3,150ft)

Venezuelan Basin

St Lucia Channel

ST LUCIA

Caraguan Rise

Rosalind Bank

Beata Ridge

4,828m (15,841ft)

Aves Ridge

Grenada Basin

St Vincent Channel

BARBADOS
Bridgetown

Sue Ridge

Caribbean Sea

ST VINCENT

2,997m (9,833ft)

Tobago Basin

Barbados Trough

Mosquito Bank

San Andrés Trough

Pedro Escarpment

4,557m (14,952ft)

GRENADA
St George's

2,319m (7,609ft)

Calarca Bank

Colombian Basin

Aruba Gap

Los Roques Basin

Tobago

TRINIDAD & TOBAGO

Barbados Plain

Saury Seaumont

Punta Gallinas

ARUBA
Oranjestad

Curaçao
Willemstad
NETHERLANDS ANTILLES

Bonaire
Bonaire Basin

Islas los Roques

Isla de Margarita

Isla de Tortuga

Gulf of Paria

Port-of-Spain (Trinidad)

10°N

tibara Trough

Mono Rise

Ríohacha

Gulf of Venezuela

Coro

1,390m (4,561ft)

Cumaná

Clark Basin

3,531m (11,585ft)

Barranquilla

Santa Marta

Maracaibo

Caracas

Barcelona

3,493m (11,460ft)

Cartagena

Lesser Antilles

VENEZUELA

Volcán Bank

Colón
Panama Canal

Gulf of Darién

Magdalena

Lake Maracaibo

Orinoco

PANAMA

Panama City

COLOMBIA

SCALE

0 100 200 300 400 500km

0 100 200 300 400 500 miles

80°W   70°W   60°W

ATLAS OF THE OCEANS

THE CENTRAL ATLANTIC

THE MID–ATLANTIC RIDGE, THE WORLD'S LONGEST mountain
chain, is the main sea-floor feature in the central Atlantic. Either
side of the ridge are two flat abyssal plains, the Angola and Brazil
basins. The dominant Atlantic gyres meet in the central Atlantic.
Both are westward-flowing near the Equator, but are separated
by the Equatorial Countercurrent, a strong eastward surface flow,
and the Equatorial Undercurrent,
an even stronger flow 100m (330ft)
deep. The Canaries Current flows
south along the North African coast,
becoming the North Equatorial
Current. In the south, the cold
Benguela Current flows up the
African coast, then away from the
coast as the South Equatorial Current.
This current splits where it reaches
South America, becoming the Guiana
and rather weak Brazil currents.

PILLOW LAVA
Pillows of lava form at the Mid-Atlantic
Ridge when extruded lava rapidly
cools upon contact with cold water.

ATLANTIC OCEAN G3
Mid-Atlantic Ridge

LENGTH	10,000km (6,200 miles)
AVERAGE HEIGHT ABOVE SEA FLOOR	3,000m (9,800ft)
RATE OF SPREAD	2–5cm (1–2in) per year

The Mid-Atlantic Ridge dissects the
entire length of the Atlantic in a series
of rifts and fractures. In the central
Atlantic, with relatively narrow
continental shelves, it is easy to see
where the coasts on either side of the
ocean were once joined. The ridge
mostly lies 1,500–3,000m (4,900–
9,800ft) below sea level, although the
volcanoes of Ascension Island and
Saint Helena breach the surface. The

ASCENSION ISLAND
Rising just to the west of the Mid-Atlantic
Ridge, Ascension Island has 44 distinct
volcanic craters.

sea floor gets deeper and older away
from the centre, and smoother as its
features are covered in sediments.
The featureless abyssal plains of the
Angola and Brazil basins lie to the
east and west respectively. The ridge
is displaced east–west at numerous
points by transform faults, where the
African and South American plates
are moving past each other. These
fracture zones extend some distance
from the ridge, sometimes as active
faults where parts of the same plate
are moving at different speeds.

MOUTH OF THE AMAZON
The Amazon accounts for nearly 20 per cent of the water input
from rivers into the world's oceans, discharging 300,000 cubic
metres (10.6 million cubic ft) per second in the rainy season.

Gulf of Guinea

AREA 1.4 million square km (500,000 square miles)

MAXIMUM DEPTH 5,204m (17,070ft)

INFLOWS Atlantic Ocean, rivers Niger, Volta

Part of the north Atlantic's Canaries Current continues along the African coast and into the Gulf of Guinea as the eastward-flowing Guinea Current. The main freshwater input to the gulf is provided by the River Niger, which has an extensive depositional fan, up to 4km (2.5 miles) thick. An even greater source of fresh water for the south Atlantic is from the River Congo to the south. Large oil and gas reserves have accumulated in the sediments of the Niger Delta and Fan, and Nigeria is Africa's biggest oil producer. Smaller deposits lie in the Congo Fan and in the continental shelf off Gabon, and deeper water in the Gulf of Guinea is now being explored for oil. When the Atlantic Ocean basin started to open 180 million years ago, three rifts opened up in the crust, forming a tectonic triple-junction. Two of the rifts continued opening to the south and the west, forming today's south Atlantic Ocean. Activity in the third rift, to the northeast, ceased rather quickly. The site of this stalled spreading centre is marked by a chain of extinct volcanoes, including the islands of Annabon, São Tomé, Principe, and Bioco in the Gulf of Guinea, and Mount Cameroon inland. São Tomé rises 2,020m (6,640ft) above sea level.

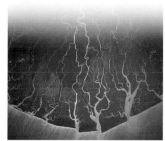

NIGER DELTA
This image, taken from a space shuttle, shows the delta coastline of the River Niger, and sediments being carried offshore.

Skeleton Coast

TYPE Secondary coast

LENGTH 1,400km (870 miles)

The cold Benguela Current hugs the west coast of southern Africa and dominates its climate. Although prevailing winds are from the sea, the air above the cold water carries little moisture, and the adjacent coast is a desert. When warm air from the land meets the cold sea air, dense fogs often form. This can be a hazard to navigation, as testified to by the numerous ship hulks along the notorious Skeleton Coast. Even without the fog, any vessel disabled by engine trouble is driven towards the shore by wind and current, and the nearest ports are quite distant. Many sailors who have survived being shipwrecked here have had little choice but to attempt the arduous journey out of the Namib Desert on foot. Namibia's coastal waters are dredged for diamonds, as the Benguela Current carries sediments from the Orange River north along the coast. These sediments include large quantities of gem-quality diamonds washed down from the South African interior. A rich fishery is another by-product of the Benguela Current, which causes upwelling of nutrient-rich waters.

SHIP'S SKELETON
Although there are doubtless human skeletons on the Skeleton Coast, the ones most likely to be seen are those of rusting ships.

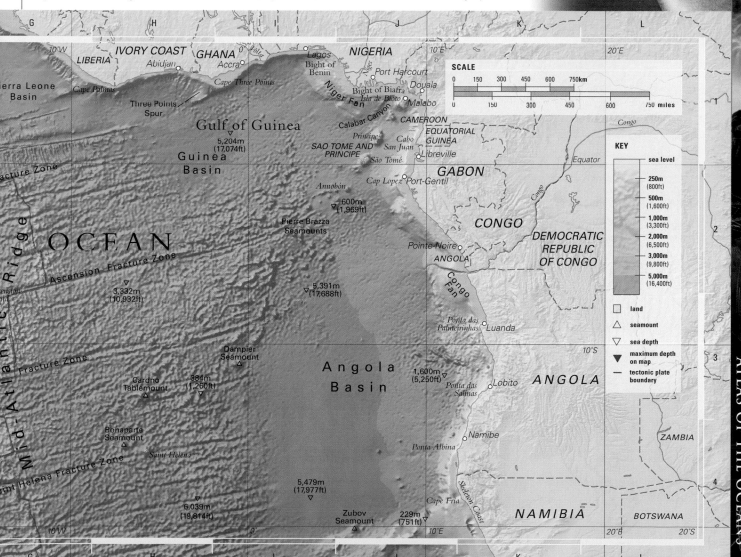

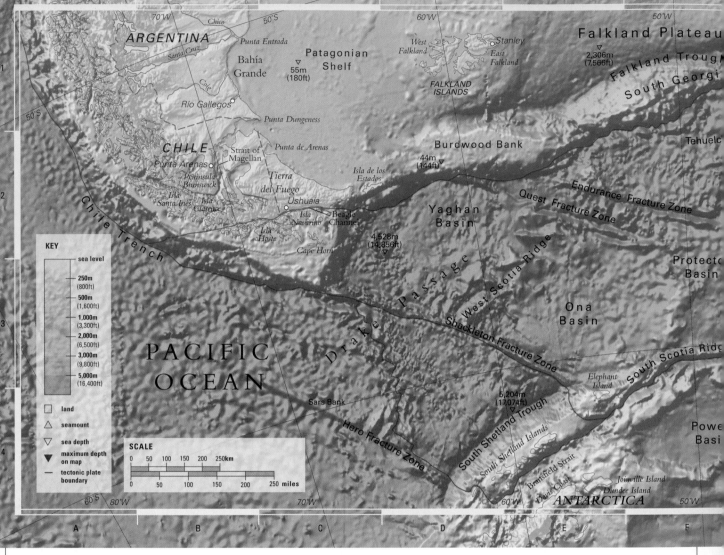

A B C D E F

70°W Chico 50°S 60°W 50°W

ARGENTINA Punta Entrada
West
Falkland Stanley Falkland Plateau
Santa Cruz Bahía Patagonian East 2,306m
Grande Shelf Falkland (7,566ft) Falkland Trough
55m FALKLAND South Georgia
(180ft) ISLANDS
Río Gallegos Punta Dungeness Tehuelc

CHILE Punta de Arenas Burdwood Bank
Strait of Endurance Fracture Zone
Punta Arenas Magellan 44m Quest Fracture Zone
Peninsula Punta de Arenas (144ft)
Brunswick Tierra Isla de los Yaghan Protecto
del Fuego Estados Basin Basin
Isla Ushuaia
Santa Inés Isla 4,528m West Scotia Ridge
Isla Navarino Beagle (14,856ft) Ona
Clarence Channel Shackleton Fracture Zone Basin
Isla Drake South Scotia Ridg
Hoste Cape Horn Passage
Elephant
Island South Scotia Ridge
KEY 5,204m
(17,074ft) Powe
PACIFIC Basi
sea level South Shetland Trough
250m **OCEAN** South Shetland Islands
(800ft)
500m Bransfield Strait
(1,600ft) Sars Bank Davis Coast Joinville Island
1,000m Hero Fracture Zone Dundee Island
(3,300ft) **ANTARCTICA**
2,000m
(6,500ft)
3,000m
(9,800ft)
5,000m
(16,400ft)

☐ land
△ seamount
▽ sea depth
▼ maximum depth on map
— tectonic plate boundary

SCALE
0 50 100 150 200 250km
0 50 100 150 200 250 miles

Chile Trench

60°S 80°W 70°W 60°W 50°W

A B C D E F

THE SCOTIA SEA

THE COLD SCOTIA SEA AND THE SUBPOLAR waters that adjoin
it lie between the south Atlantic and the Southern Ocean. Sea-ice
is present around the region's shorelines in winter, and icebergs
calved from the Antarctic ice sheets can be found all year round.

ATLANTIC OCEAN G2

Scotia Sea

AREA 900,000 square km (350,000 square miles)

MAXIMUM DEPTH 5,576m (18,300ft)

INFLOWS Southern Ocean

The Scotia Sea is bounded by Tierra
del Fuego and South Georgia to the
north, the South Shetland and South
Orkney islands to the south, and the
South Sandwich Islands to the east.

It is swept by the Antarctic
Circumpolar Current, which flows
from the Pacific into the Atlantic
through the Drake Passage. Part of this
flow turns north along the eastern
shore of South America as the cold
Falklands Current. Where it meets the
warm waters of the Brazil Current
north of the Falkland Islands,

upwelling of nutrients supports a rich
fishery. The Scotia Plate is moving
eastward relative to the South
American and Antarctic plates. The
separation of South America and
Antarctica began around 100 million
years ago, opening up a route for
Pacific Ocean currents to flow into
the young south Atlantic and Indian
Ocean basins – the first step in the
thermal isolation of Antarctica.

ROCKHOPPER PENGUINS

ATLANTIC OCEAN B2 AND C2

Strait of Magellan

LENGTH 530km (330 miles)

MINIMUM WIDTH 4km (2½ miles)

The first-known European to sail
from the Atlantic into the Pacific
was Portuguese explorer Ferdinand
Magellan, and the strait he used
between the South American mainland
and Tierra del Fuego is named after
him. The route is sheltered from the
full might of the Southern Ocean,
although it has some narrow passages
that can be hazardous to navigate.
It was the preferred route for Atlantic–
Pacific sea trade until the confirmation
of an open ocean route around Cape
Horn in 1616.

Another sheltered route through
the Tierra del Fuego archipelago is the
Beagle Channel, named after the survey
ship that carried British naturalist
Charles Darwin, on his scientific
voyage of 1831–1836. Cape Horn is
the southernmost point of South
America, situated on Hoorn Island,
one of the Hermite Islands to the
south of Tierra del Fuego. The most

southerly passage between the major
oceans, Cape Horn was discovered
and named Kaap Hoorn in 1616 by
a merchant navigator, in honour of his
sponsors in the Dutch town of Hoorn.
However, most commercial traffic
between the Atlantic and the Pacific
now travels via the Panama Canal.

CAPE HORN
Cape Horn is notorious for its atrocious
weather conditions. Sailing around it is
the peak of many sailors' ambitions.

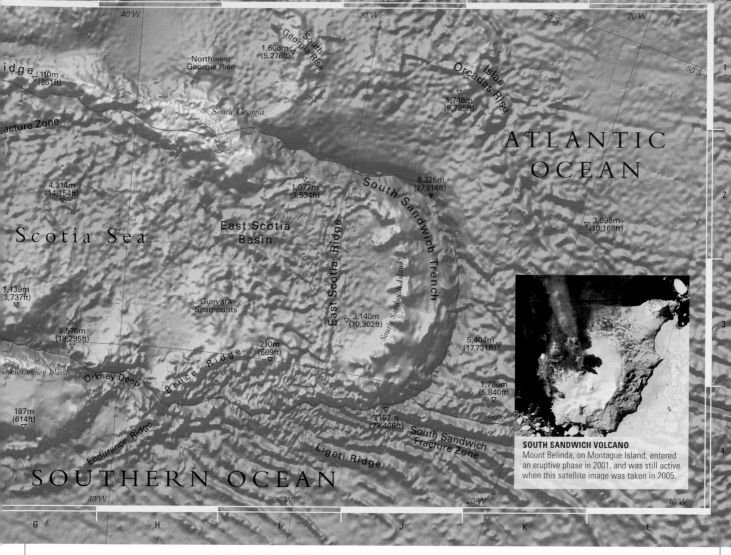

G   H   I   J   K   L

40°W   30°W   50°S   20°W

South Georgia Rise

Northwest Georgia Rise — 1,608m (5,276ft)

Islas Orcadas Rise — 1,748m (5,735ft)

ATLANTIC OCEAN

idge — 110m (361ft)

South Georgia

acture Zone

4,314m (14,154ft)

1,077m (3,534ft)

South Sandwich Trench — 8,325m (27,314ft)

3,099m (10,168ft)

Scotia Sea

East Scotia Basin

East Scotia Ridge

South Sandwich Islands

1,139m (3,737ft)

Guevara Seamounts

3,140m (10,302ft)

5,404m (17,731ft)

5,576m (18,295ft)

210m (689ft)

Bruce Ridge

South Orkney Islands

Orkney Deep

1,780m (5,840ft)

187m (614ft)

Endurance Ridge

7,152m (23,466ft)

South Sandwich Fracture Zone

Ligeti Ridge

SOUTHERN OCEAN

SOUTH SANDWICH VOLCANO
Mount Belinda, on Montague Island, entered an eruptive phase in 2001, and was still active when this satellite image was taken in 2005.

40°W   30°W   20°W   10°W

G   H   I   J   K   L

ATLANTIC OCEAN J2

South Sandwich Trench

LENGTH 965km (600 miles)

MAXIMUM DEPTH 8,325m (27,300ft)

RATE OF CLOSURE 7cm (2¾in) per year

Although discovered by James Cook in 1775, the South Sandwich Islands were not visited until 1818, when seal hunters landed. They were never permanently settled and remain uninhabited. With volcanic peaks rising up to 1,000m (3,300ft) above sea level, the islands are mostly composed of basaltic lava and covered by glaciers. North of the islands is the Protector Shoal – an undersea volcano that rises to within 30m (100ft) of the surface. The South Sandwich Islands mark the eastern boundary of the Scotia Sea, and the

South Sandwich Trench lies a little further to the east. Both features are caused by tectonic processes occurring where the Scotia and South Atlantic plates meet. The Scotia Plate is split and spreading at the East Scotia Ridge, forming a new plate at its eastern end – the South Sandwich microplate. This plate is geologically young, at about eight million years old, and buoyant. Moving eastward at about 7cm (2¾in) per year, it is converging with the South Atlantic Plate, resulting in the older South Atlantic Plate sinking beneath the South Sandwich Plate at a subduction zone. This zone is marked by the South Sandwich Trench and the volcanic island arc of the South Sandwich Islands (or the Scotia Arc).

SEA-ICE
Sea-ice clings to the shore of Bellinghausen Island, of the South Shetland group, named after the Russian explorer who discovered it in the 19th century.

ATLANTIC OCEAN F3

South Georgia Ridge

LENGTH 2,500km (1,600 miles)

HEIGHT ABOVE SEA FLOOR 3,000m (9,800ft)

RATE OF RELATIVE MOTION 0.7cm (¼in) per year

The South Georgia Ridge marks the northern edge of the Scotia Plate, a boundary that continues east through the Tierra del Fuego archipelago. This is a transform boundary (see p.50) with the South Atlantic Plate to the north. There is a similar transform boundary marked by the South Scotia Ridge, with the Antarctic Plate to the south. Fragments of continental crust, such as Burdwood Bank and South Georgia, seem to have been left behind as South America moved west. The island of South Georgia was named by James Cook in 1775, but may have been sighted as early as 1675. It was a base for seal hunters in the 19th century, and

in the 20th century seven whaling stations were established on the more sheltered northern shore. The last of these closed in 1965. North of the South Georgia Ridge lies the Falkland Plateau, an area of thickened ocean crust of moderate depth, and the broad continental shelf off the east coast of South America – the Patagonian Shelf. The Falkland Islands are a continental fragment left over from the breakup of Gondwana (see p.46) and the subsequent opening of the south Atlantic.

ABANDONED WHALING STATION
Old, rusting whaling ships lie in the harbour at Grytviken, a whaling station from 1904–65, on South Georgia.

THE INDIAN OCEAN

THE INDIAN OCEAN IS THE THIRD-LARGEST ocean on Earth, lying between Africa and Australia. Sea routes across the northern Indian Ocean were opened up by traders from the Persian Gulf, and by the Chinese Admiral Zheng He between 1405 and 1433. The Portuguese explorer Vasco de Gama was the first European to circumnavigate Africa, reaching India in 1498.

SUNRISE IN THE STRAIT OF MALACCA

OCEAN CIRCULATION

The southern Indian Ocean is dominated by the anticlockwise South Indian Gyre. This drives the South Equatorial Current, which in turn feeds the Agulhas Current. The circulation north of the Equator is complicated by the Indian subcontinent, and the annual wind reversal that characterizes the monsoon climate. High pressure over India from November to April pushes surface water in the Arabian Sea away from India, generating the North Equatorial Current and Equatorial Countercurrent. In the summer, low pressure over India gives rise to southwesterly winds, the Southwest Monsoon Current replaces the North Equatorial Current, and the Somali Current flows strongly northeast along the East African coast.

STILT FISHING
Poles are used as perches by some fishermen in Sri Lanka, so as not to scare away the fish.

OCEAN FLOOR

The Indian Ocean floor is dominated by three mid-ocean ridges – the Southwest Indian Ridge, the Mid-Indian Ridge, and the Southeast Indian Ridge – which meet at a triple junction. The Indian Ocean started opening when Africa separated from Antarctica and Australia, achieving its present form when India collided with Asia 36 million years ago. Two long, linear features record India's rapid movement northwards: Ninetyeast Ridge and the Chagos–Lacadive Plateau. The Indian Ocean has just one large oceanic trench – the Java–Sunda Trench, where the Australian and Indian plates are subducting beneath the Eurasian Plate. The Indian and Australian plates now appear to be moving independently, but the location of their boundary is uncertain.

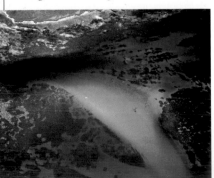

INDIAN WATERS
Clear water off the Kenyan coast reveals rock shoals, coral growth, and sand bars.

ATLAS OF THE OCEANS

KEY

sea level
- 250m (800ft)
- 500m (1,600ft)
- 1,000m (3,300ft)
- 2,000m (6,500ft)
- 3,000m (9,800ft)
- 5,000m (16,400ft)

□ land
△ seamount
▽ sea depth
▼ maximum depth on map

AFRICA
Suez, Tropic of Cancer, Arabian Peninsula, Jedda, Massawa, Aden, Djibouti, Berbera, Horn of Africa, Mogadishu, Kismaayo, Mombasa, Tanga, Pemba, Zanzibar, Dar es Salaam, Cabo Delgado, Comoros, Pemba, Zambezi, Quelimane, Beira, Maputo, Durban, Port Elizabeth, Transkei Basin, Africana Seamount, Agulhas Plateau, Agulhas Basin

Equator, Tropic of Capricorn, 30°S, 60°S, Antarctic Circle

Persian Gulf, Dubai, Abu Dhabi, Bandere Bushehr, Bande Abbas

3,039m (9,970ft), 1,128m (3,701ft), Gulf of Aden, Raas Caseyr, Owen, Andrew Seamount, Chain Ridge, Somali Basin, 4,886m (16,031ft), Seychelles Bank, Fred Seamount, Mascarene Basin, Comoro Basin, Mahajanga, Madagascar, Mascarene Plain, Toamasina, 4,976m (16,326ft), Mauritius, Réunion, Toliara, Madagascar Basin 7,023m (23,042), Davie Ridge, Mozambique Channel, 1,984m (6,510ft), Madagascar Plateau, 69m (226ft), Mozambique Plateau, Natal Basin, Mozambique Escarpment, Prince Edward Fracture Zone, Prince Edward Islands, Del Cano Rise, Indomed Fracture Zone, Southwest Indian Ridge, 4,936 (16,195), Crozet Plateau, Crozet Basin, Crozet Islands, 5,819m (19,092ft)

Atlantic-Indian Ridge, Lena Seamount, 5,386m (17,671ft)

Atlantic-Indian Basin, Enderby Plain, SOUTHERN OCEAN

ANTARCTICA

30°E

WINTER SURFACE CURRENTS

Somali
Current

North Equatorial
Current

Equatorial Counter Current

South Equatorial Current

Agulhas Current

Antarctic Circumpolar Current

SUMMER SURFACE CURRENTS

Somali
Current

SW Monsoon
Current

South Equatorial Current

Agulhas Current

Antarctic Circumpolar Current

SUMMER SURFACE WINDS

South West Monsoon

South East Trades

Westerlies

Beaufort Scale		Speed	
0–3		0–16 kph (0–10 mph)	
3–5.5		16–40 kph (10–25 mph)	
over 5.5		over 40 kph (over 25 mph)	

ASIA

Indus

Karachi

Ganges

Brahmaputra

Narmada

Tropic of Cancer

Mumbai

Godavari

Kolkata
(Calcutta)

Chittagong

3,427m
11,244ft)

20m
(66ft)

Goa

Krishna

Chennai
(Madras)

Ganges Fan

Rangoon

Irrawaddy

Salween

Arabian
Sea

Mangalore

Pondicherry

Bay of
Bengal

2,429m
(7,970ft)

Andaman
Islands

Andaman
Sea

Arabian
Basin

Cochin

Colombo

Sri Lanka

Andaman
Basin

Phuket

4,481m
(14,702ft)

Laccadive
Islands

3,462m
(11,359ft)

Nicobar
Islands

Maldives

4,846m
(15,900ft)

Ceylon Plain

Strait of Malacca

Malay Peninsula

Singapore

Equator

Mid-Indian Ridge

Chagos-Laccadive Plateau

Chagos
Archipelago

Chagos Trench

Mid-Indian
Basin

Cocos Basin

Sunda Trench

Sumatra

Java

Java Trench

Java Ridge

Timor

Nipety-east Ridge

Investigator Ridge

4,464m
(14,646ft)

7,125m
(23,377ft)

North
Australian
Basin

Sahul Shelf

Osborn
Plateau

Gascoyne
Plain

Exmouth
Plateau

Rowley Shelf

Broome

Wharton
Basin

5,678m
(18,636ft)

5,614m
(18,418ft)

Port Hedland

Egeria Fracture Zone

Wallaby
Plateau

Cuvier
Basin

Coolingua

Batavia
Seamount

Cuvier
Plateau

Camarvon

AUSTRALIA

Tropic of Capricorn

INDIAN
OCEAN

4,023m
(13,199ft)

East Indiaman Ridge

2,078m
(6,818ft)

Broken Ridge

Hartog Ridge

Perth
Basin

Perth

Amsterdam Fracture Zone

Ob' Trench

Naturaliste
Plateau

Great Australian
Bight

Port Augusta

Naturaliste Fracture Zone

Murray

Adelaide

Amsterdam Island
St Paul Island

Diamantina Fracture Zone

4,980m
(16,339ft)

5,852m
(19,200ft)

Melbourne

South Australian Basin

Bass
Strait

Kerguelen

South Australian
Plain

Tasmania
Hobart

Kerguelen Plateau

Heard and
McDonald Islands

4,285m
(14,059ft)

Southeast Indian Ridge

5,386m
(17,671ft)

Tasman
Plateau

184m
(604ft)

4,684m
(15,368ft)

Banzate Seamounts

South Indian Basin

SOUTHERN OCEAN

Prydz Bay

Antarctic Circle

ANTARCTICA

SCALE

0 200 400 600 800 1,000km

0 200 400 600 800 1,000 miles

90°E

120°E

30°N

60°E

120°E

Equator

Tropic of Capricorn

30°S

60°S

D E F G H I

THE RED SEA AND ARABIAN SEA

CIRCULATION IN THE NORTHEAST Indian Ocean uniquely reverses twice a year due to the monsoon winds. For thousands of years navigators used this to run trade routes in the region. Today, oil and the Suez Canal make the area strategically important.

INDIAN OCEAN B4

Red Sea

AREA	450,000 square km (175,000 square miles)
MAXIMUM DEPTH	3,040m (9,975ft)
INFLOWS	Arabian Sea

The Red Sea is an embryonic ocean, and it has been opening over the last 25 million years, ever since the Arabian Plate began its gradual rift away from Africa. A central trough is flanked by relatively shallow shelves, and its warm waters contain many fringing coral reefs. Since 1869 the Red Sea has been linked with the Mediterranean Sea via the 160km-(100-mile-) long Suez Canal.

THE SUEZ CANAL

INDIAN OCEAN D2

Persian Gulf

AREA	241,000 square km (93,000 square miles)
MAXIMUM DEPTH	110m (360ft)
INFLOWS	Rivers Tigris, Euphrates, Karun

The Persian Gulf (also known as the Gulf) is a warm, semi-enclosed sea, mostly less than 100m (330ft) deep. It is connected to the Arabian Sea via the Strait of Hormuz and the Gulf of Oman. The shallow waters are well-mixed and more productive than the Red Sea owing to the nutrient runoff from the land to the north and east. Corals have adapted to the very warm water temperature, which can reach 33°C (91°F).

The Arabian Plate, spreading from the Red Sea rift, is moving northeast and sliding under the Eurasian Plate, so the northeastern side of the Persian Gulf is deeper. This tectonic activity has folded and uplifted sediments up to 280 million years old and produced structural traps for oil, which has accumulated in large reservoirs beneath the Gulf and surrounding land. Oil now dominates the region's economy.

INDIAN OCEAN G5

Arabian Sea

AREA	3.9 million square km (1.5 million square miles)
MAXIMUM DEPTH	4,481m (19,038ft)
INFLOWS	Rivers Indus, Namada

The Arabian Sea lies between the Arabian Peninsula and India. It is underlain by the abyssal plain of the Arabian Basin. This oceanic part of the Indian Plate is bounded to the west by the Owen Fracture zone, a transform boundary with the Arabian Plate, and to the south by the Carlsberg Ridge, a mid-ocean spreading ridge where India and Africa are diverging.

To the west lies the Gulf of Aden, a precursor to the Red Sea rift, with a well-established spreading ridge. The continental shelf of the Indian subcontinent to the east is broad in places, and the depositional fan of the Indus River extends some distance across the deep floor of the basin. Mumbai, formerly known as Bombay, is the largest port on the Indian coast.

INDIAN OCEAN I8

Maldives

TYPE	Coral atoll islands
AREA:	298 square km (115 square miles)
NUMBER OF ISLANDS	1,192

The Maldives lie midway along the Chagos–Laccadive Plateau. The Laccadive Islands and a number of submerged banks mark the northern end of the ridge. There are more than 1,000 Maldive islands, grouped into 27 atolls, composed of coral and sandbars. The highest island is less than 3m (10ft) above sea level. With a warm climate, shallow lagoons, and refreshing sea breezes, the Maldives are an idyllic holiday destination. Although tourism plays an increasingly important role in the economy of the islands, fishing remains the main occupation of the islanders.

MALDIVE ISLAND

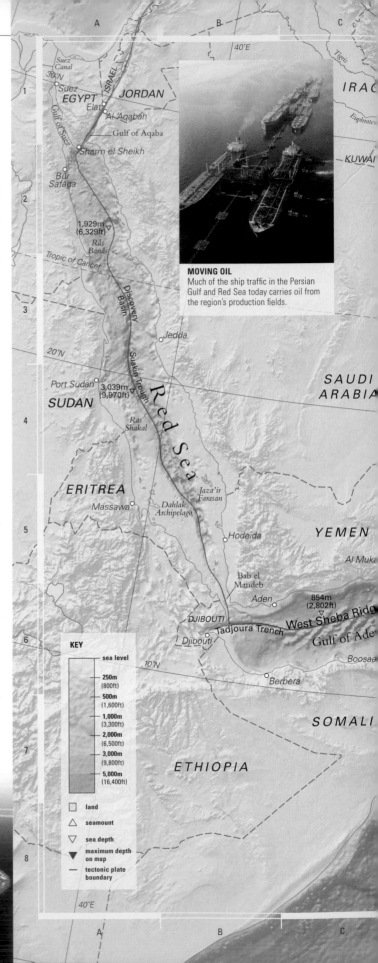

MOVING OIL
Much of the ship traffic in the Persian Gulf and Red Sea today carries oil from the region's production fields.

KEY

	sea level
	250m (800ft)
	500m (1,600ft)
	1,000m (3,300ft)
	2,000m (6,500ft)
	3,000m (9,800ft)
	5,000m (16,400ft)
□	land
△	seamount
▽	sea depth
▼	maximum depth on map
—	tectonic plate boundary

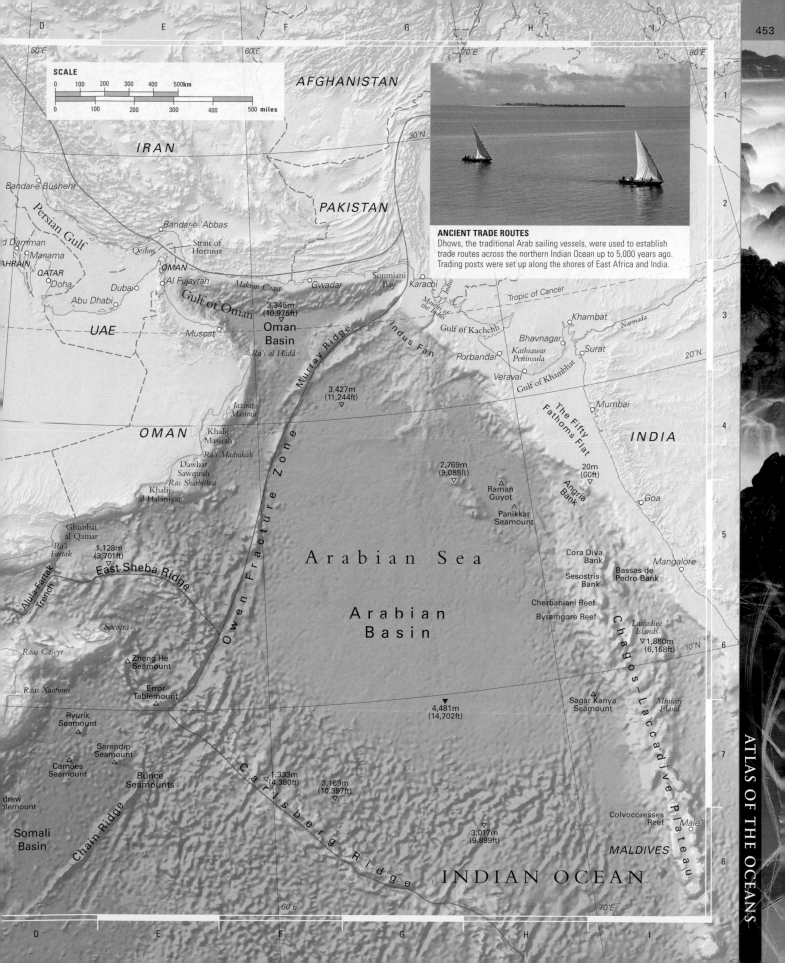

SCALE

| 0 | 100 | 200 | 300 | 400 | 500km |
| 0 | 100 | 200 | 300 | 400 | 500 miles |

AFGHANISTAN

IRAN

PAKISTAN

Bandar-e Bushehr

d Damman
Manama
HRAIN
QATAR
Doha

Persian Gulf

Bandar-e 'Abbas

Qeshm
Strait of Hormuz
OMAN

Dubai
Abu Dhabi

UAE

Makran Coast
Gwadar
Sonmiani Bay
Karachi

Mouths of the Indus

Tropic of Cancer

Gulf of Oman

3,345m
(10,975ft)
Oman Basin

Muscat

Ra's al Hadd

Al Fujayrah

Indus Fan

Khambat

Gulf of Kachchh

Bhavnagar
Porbandar
Kathiawar Peninsula
Surat
Veraval
Gulf of Khambhat

Narmada

20°N

Jazirat Masirah

3,427m
(11,244ft)

OMAN

Khalij Masirah

Ra's Madrakah

Dawhat Sawqirah
Ras Sharbithat

Khalij al Halaniyat

Mumbai

The Fifty Fathoms Flat

INDIA

2,769m
(9,085ft)

Raman Guyot
Panikkar Seamount

Angria Bank

20m
(66ft)

Goa

Ghubbat al Qamar
Ra's Fartak

1,128m
(3,701ft)

East Sheba Ridge

Alula-Fartak Trench

Owen Fracture Zone

Arabian Sea

Arabian Basin

Cora Diva Bank
Sesostris Bank

Bassas de Pedro Bank

Mangalore

Cherbaniani Reef
Byramgore Reef

Laccadive Islands

Chagos-Laccadive Plateau

1,880m
(6,168ft)

10°N

Socotra

Raas Caseyr

Zheng He Seamount

Error Tablemount

4,481m
(14,702ft)

Sagar Kanya Seamount

Minicoy Island

Raas Xaafuun

Ryurik Seamount

Serendip Seamount

Camões Seamount

Bunce Seamounts

Carlsberg Ridge

1,333m
(4,380ft)

3,169m
(10,397ft)

Colvocoresses Reef

Male'

drew
lemount

Somali Basin

Chain Ridge

3,017m
(9,899ft)

MALDIVES

INDIAN OCEAN

60°E

70°E

ANCIENT TRADE ROUTES
Dhows, the traditional Arab sailing vessels, were used to establish trade routes across the northern Indian Ocean up to 5,000 years ago. Trading posts were set up along the shores of East Africa and India.

50°E

60°E

70°E

80°E

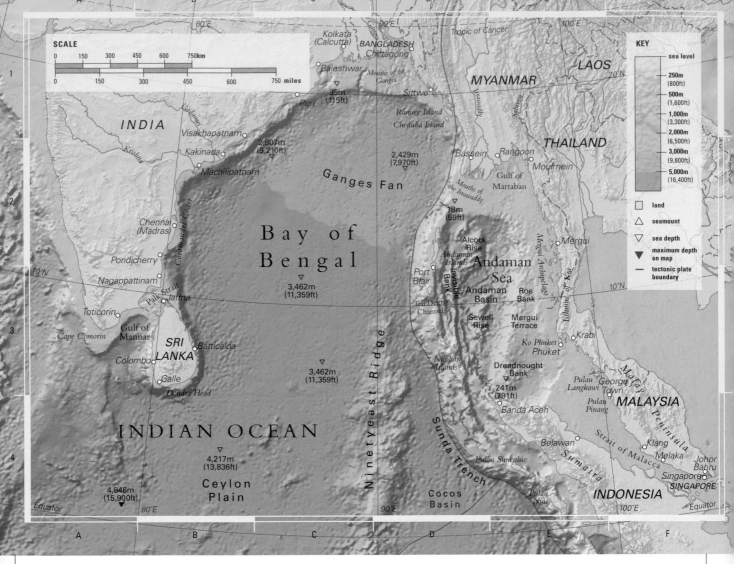

Map labels:

SCALE
0 150 300 450 600 750km
0 150 300 450 600 750 miles

KEY
sea level
250m (800ft)
500m (1,600ft)
1,000m (3,300ft)
2,000m (6,500ft)
3,000m (9,800ft)
5,000m (16,400ft)

□ land
△ seamount
▽ sea depth
▼ maximum depth on map
— tectonic plate boundary

INDIA
Kolkata (Calcutta)
BANGLADESH
Chittagong
Baleshwar
Mouths of the Ganges
Puri
35m (115ft)
Sittwe
MYANMAR
LAOS
Tropic of Cancer
20°N
Visakhapatnam
Ramree Island
Cheduba Island
Kakinada
2,807m (9,210ft)
Machilipatnam
Ganges Fan
2,429m (7,970ft)
Bassein
Rangoon
THAILAND
Moulmein
Gulf of Martaban
Mouths of the Irrawaddy
18m (59ft)
Alcock Rise
Andaman Islands
Andaman Sea
Bay of Bengal
Chennai (Madras)
Coromandel Coast
Port Blair
Invisible Bank
Andaman Basin
Roe Bank
Pondicherry
3,462m (11,359ft)
Nagappattinam
Ten Degree Channel
10°N
Sewell Rise
Mergui Terrace
Krabi
Palk Strait
Jaffna
Toticorin
Gulf of Mannar
Nicobar Islands
Ko Phuket
Phuket
Cape Comorin
SRI LANKA
Batticaloa
3,462m (11,359ft)
Dreadnought Bank
Colombo
241m (791ft)
Pulau Langkawi
George Town
Galle
Banda Aceh
Pulau Pinang
MALAYSIA
Dondra Head
Mergui
Mergui Archipelago
Isthmus of Kra
Malay Peninsula
INDIAN OCEAN
Ninetyeast Ridge
Sunda Trench
Sumatra
Belawan
Klang
Melaka
4,217m (13,836ft)
Pulau Simeulue
Johor Bahru
Singapore
SINGAPORE
Ceylon Plain
4,846m (15,900ft)
Cocos Basin
Pulau Nias
INDONESIA
Equator
80°E
90°E
100°E
Equator

THE BAY OF BENGAL

THE NORTHEAST CORNER of the Indian Ocean is enclosed on three sides by land. This area of tropical sea is subject to a monsoon climate, and vulnerable to cyclones between the months of June and November.

INDIAN OCEAN C2
Bay of Bengal

AREA 2.9 million square km (1.1 million square miles)

MAXIMUM DEPTH 4,695m (15,400ft)

INFLOWS Rivers Ganges, Brahmapurta, Mahanadi, Godavari, Krishna, Kaveri, Irrawaddy

Circulation in the Bay of Bengal is clockwise during the northwest monsoon, with a westward flow in the main ocean. This flow reverses during the southwest monsoon. The northern half of the bay is underlain by the Ganges Fan, a thick cone of sediment extending from the continental rise across the abyssal plain. This is the fastest-accumulating sediment in the world, originating high in the Himalayas, and supplied by the Brahmaputra River and the Ganges.

INDIAN OCEAN F4
Strait of Malacca

LENGTH 963km (600 miles)

MINUMUM WIDTH 15km (9 miles)

The Strait of Malacca links the Indian Ocean with the Pacific Ocean, via the South China Sea. It is one of the busiest waterways in the world, with about 140 ships passing through the strait each day. The cargo includes about one quarter of the world's oil, on its way from the Persian Gulf to markets including Japan and China. The high levels of traffic have lead to a dramatic increase in piracy in this area in recent years. Of the 251pirate attacks on ships worldwide in 2004, 70 (more than a quarter) were in the Malacca region.

INDIAN OCEAN E3
Andaman Sea

AREA 798,000 square km (308,000 square miles)

MAXIMUM DEPTH 3,777m (12,400ft)

INFLOWS Bay of Bengal, Strait of Malacca, rivers Irriwaddy, Salween

The Andaman Sea lies between the Andaman Islands, Sumatra, and the Malay Peninsula. There is a broad continental shelf in the east and the north, where the sediment is dredged for cassiterite, an ore of tin. Alcock Rise and Sewell Rise are separated by an area of deep ocean floor, where a spreading centre has been pushing the Burma and Sunda microplates apart for the last 3–4 million

years. This divergence created the Andaman Sea. The eastern half of the sea lies over the Sunda Plate, which includes most of Sumatra and the Malay Peninsula. The western half includes the Andaman and Nicobar Islands and sits on the Burma Plate, which forms a junction with the Indian Plate at the Sunda Trench. At this subduction zone, the Indian Plate is being overridden by the younger Burma Plate. The southern part of this zone was the source of the 2004 Indian Ocean tsunami (see pp.456–57).

HAVELOCK ISLAND
Mangroves line the eastern shore of Havelock Island, part of the Andaman Islands group. These volcanic islands are also fringed by coral reefs.

THE JAVA TRENCH

IN THE EASTERN INDIAN OCEAN, the South Equatorial Current carries water from east to west during the southwest monsoon, but shifts south during the northeast monsoon to be replaced by the eastward-flowing Equatorial Counter Current. The area includes the deepest part of the Indian Ocean, the Java Trench, where the Australian Plate meets the Eurasian Plate.

INDIAN OCEAN D2

Java Trench

LENGTH 2,600km (1,600 miles)

MAXIMUM DEPTH 7,125m (23,377ft)

RATE OF CLOSURE 6cm (2½in) per year

The Java Trench is a continuation of the Sunda Trench, where the oceanic part of the Australian Plate is being subducted beneath the continental Eurasian Plate. A string of volcanoes has resulted behind the trench, strung out across Sumatra, Java, and the Lesser Sunda Islands. The Australian Plate is moving north at a rate of 6cm (2½in) per year. The Indian Ocean floor south and west of the trench shows tectonic features aligned in this northerly direction. Between Investigator Ridge and Ninetyeast Ridge lie a series of north–south fractures formed to accommodate different rates of motion in the Australian and Indian plates. The Ninetyeast Ridge itself is the longest underwater mountain chain, 5,000km (3,100miles) in length. It followed in the wake of India's rapid motion north as the Indian Ocean opened up. The ridge represents piles of extruded volcanic material formed above the Kerguelen Hotspot, and were carried north as the sea floor spread between India and Antarctica.

INDIAN OCEAN F2

Timor Sea

AREA 610,000 square km (235,000 square miles)

MAXIMUM DEPTH 3,300m (10,800ft)

INFLOWS Indian Ocean, Arafura Sea

The Timor Sea marks the eastward boundary of the central Indian Ocean. Pacific water flows in from the Arafura Sea (see p.473) during the southwest monsoon, feeding the South Equatorial Current. This flow is reversed during the northeast monsoon. Australia's aboriginal people probably arrived from southeast Asia by island-hopping across the Timor Sea. It is mainly shallow, but with the deep Timor Trough lying along its northern edge. Significant reserves of oil and gas are thought to lie in the continental shelf sediments beneath the sea, and exploitation rights are disputed between Australia and East Timor. The warm shallow tropical waters make the Timor Sea a breeding ground for tropical storms and cyclones from January to March. Such storms proceed southwestwards into the Indian Ocean, sometimes turning inland to hit the coast of Western Australia. There are fishing grounds, including a shrimp fishery, in coastal waters on the Australian side of the Timor Sea.

TIMOR FISHERMEN
Traditional, shore-based fishing is practised by Timorese fishermen, seen here hauling in a net at Areia Branca Beach near Dili.

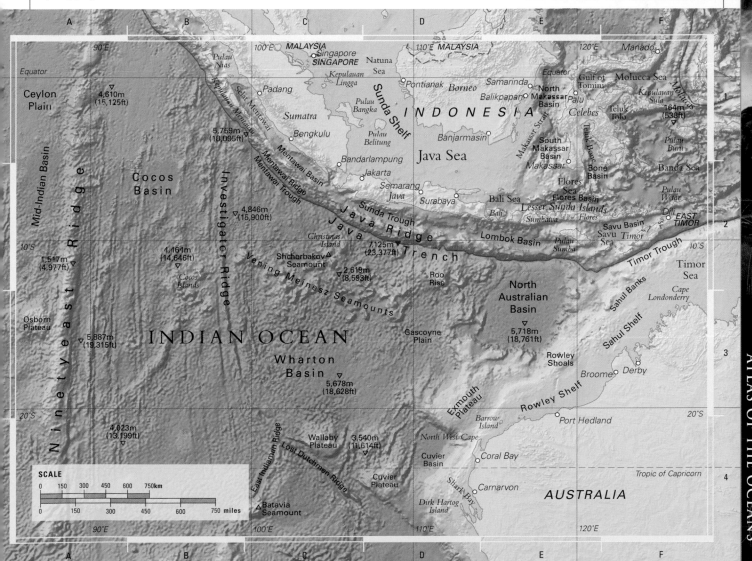

APPROACHING DISASTER
Tourists run for their lives in the path of the first of six giant tsunami waves that hit Hat Rai Lay beach on the western coast of southern Thailand.

THE 2004 INDIAN OCEAN TSUNAMI

At 7:58am on 26 December 2004, the Indonesian island of Sumatra was shaken by an earthquake that measured 9.1 on the Richter scale. It was the world's most powerful earthquake for 40 years, but because its epicentre was 160km (100 miles) offshore it did not itself produce the worst of the damage that day. Fifteen minutes later, a wall of water hit the western coast of Sumatra. By the end of the day, more than 200,000 people had been killed and a million left homeless in 11 countries, making the Indian Ocean tsunami the most destructive on record.

The earthquake, in a subduction zone between two of the Earth's tectonic plates, tore a rift 1,200km (745 miles) across the ocean floor and uplifted the sea bed by about 5m (16ft). A huge volume of water was displaced and broad waves spread out across the sea surface, reaching a height of 15m (50ft) in the shallow water of nearby shores. For coastal populations across the Indian Ocean, there was little time to escape to high ground. Even distant coasts received no warning of the approach of the tsunami, due to the absence of any planning for such an event. Natural warning signs helped some to survive. These included the earthquake itself and the retreat of the sea up to 2.5km (1½ miles) from the coast a few minutes before the first wave struck.

Dealing with the aftermath of shattered buildings, flooded land, polluted water supplies, blocked harbours, and displaced people has required a huge relief and reconstruction effort, which is set to continue for several years.

WAVE SPEED AND HEIGHT

A tsunami wave travels at the speed of a jet – 500–800kph (300–500mph), depending on the ocean depth, slowing to 150kph (90mph) as it approaches the shore. The 2004 tsunami crossed the Indian Ocean to Somalia in just eight hours. A tsunami does not lose much energy as it crosses an ocean and can cause much damage on distant coasts. Waves over 4m (13ft) high hit the coasts of the Seychelles and Somalia, a similar size to those that struck Thailand. Deaths were reported in South Africa, 8,500km (5,300 miles) from the earthquake's epicentre.

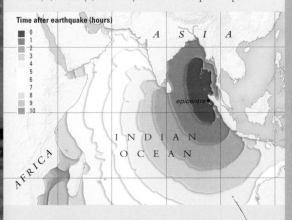

Time after earthquake (hours)

0
1
2
3
4
5
6
7
8
9
10

A S I A

INDIAN OCEAN

AFRICA

epicentre

EFFECTS OF THE TSUNAMI

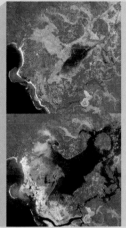

LOHKNGA *These satellite pictures show the Indonesian village of Lohknga before the disaster, on 10 January 2003 (top image), and afterwards, on 29 December 2004 (bottom). The village was flattened, as was the nearby town of Leupung, where only 1,800 survived out of a population of 10,000. All the sand on the beach and nearly all the vegetation around the village were stripped away, and 65 square km (25 square miles) of agricultural land was flooded behind the village. The wave is thought to have been 15m (50ft) high when it struck the shore and it ran 4km (2½ miles) inland up to 25m (80ft) above sea level.*

SUMATRA *The devastation stretched 225km (140 miles) along the Sumatran coast. Across Indonesia, 127,000 people were killed.*

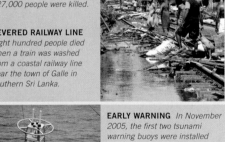

SEVERED RAILWAY LINE *Eight hundred people died when a train was washed from a coastal railway line near the town of Galle in southern Sri Lanka.*

EARLY WARNING *In November 2005, the first two tsunami warning buoys were installed off Indonesia. These are vital components of a planned Indian Ocean early warning system to match the one that is already operating in the Pacific.*

DAMAGE PREVENTION *Mangrove saplings are prepared for planting along the shore near Banda Aceh, Indonesia. Mangrove forests protected some coasts against the 2004 tsunami by absorbing most of its power before it reached settlements.*

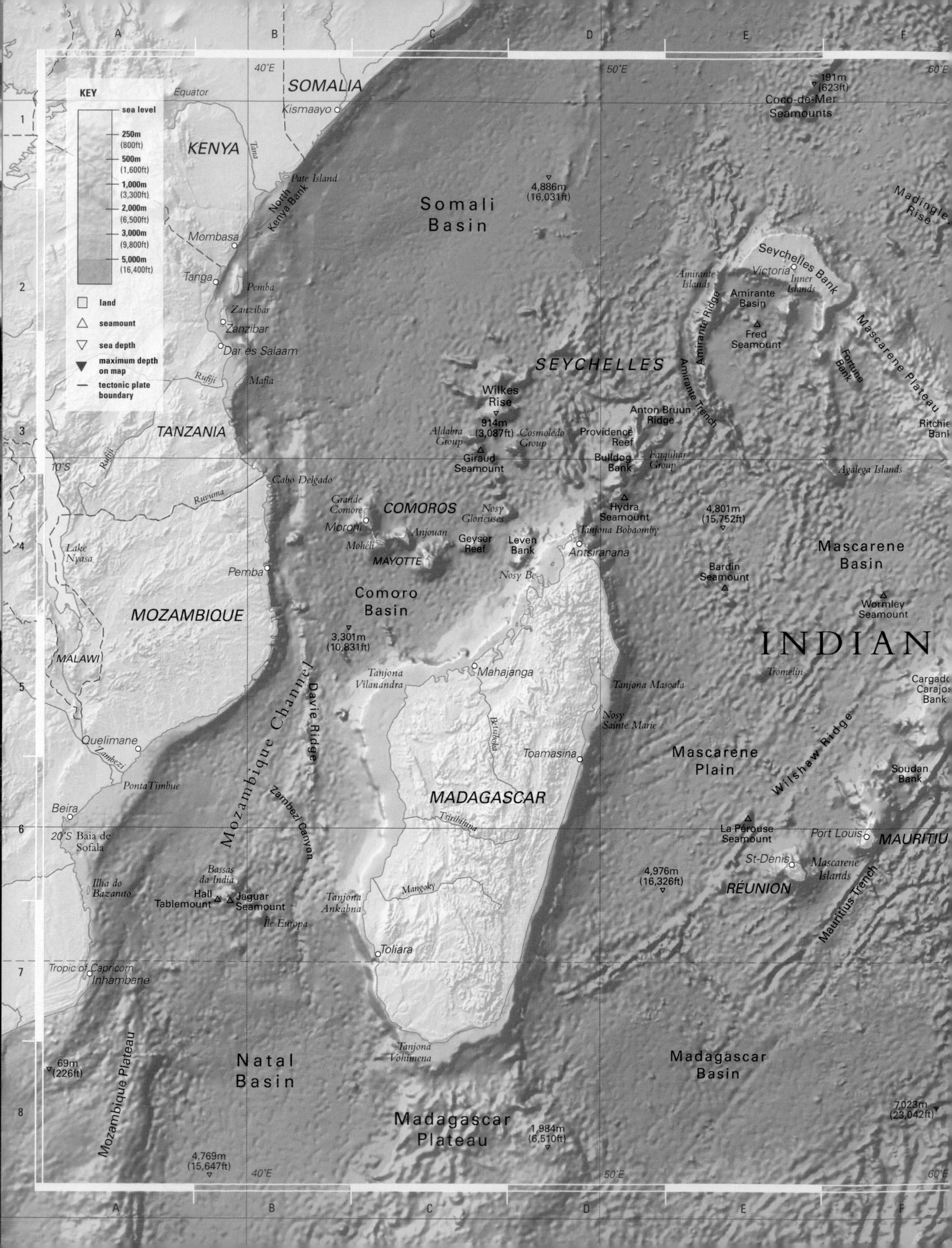

A   B   C   D   E   F

KEY

sea level
250m (800ft)
500m (1,600ft)
1,000m (3,300ft)
2,000m (6,500ft)
3,000m (9,800ft)
5,000m (16,400ft)

☐ land
△ seamount
▽ sea depth
▼ maximum depth on map
— tectonic plate boundary

Equator

SOMALIA

Kismaayo

KENYA

Tanu

North Kenya Bank

Pate Island

Mombasa

Tanga

Pemba

Zanzibar
Zanzibar

Dar es Salaam

Mafia

Rufiji

10°S

Rufiji

TANZANIA

Ruvuma

Cabo Delgado

Lake Nyasa

MOZAMBIQUE

MALAWI

Pemba

Quelimane
Zambezi

PontaTimbue

Beira

20°S Baia de Sofala

Ilha do Bazaruto

Inhambane

Tropic of Capricorn

Mozambique Plateau

69m (226ft) ▽

4,769m (15,647ft) ▽

40°E

Somali Basin

4,886m (16,031ft) ▽

191m (623ft) ▽
Coco-de-Mer Seamounts

50°E   60°E

Madingley Rise

Seychelles Bank

Amirante Islands
Victoria
Inner Islands

Amirante Ridge

Amirante Basin

Fred Seamount

Mascarene Plateau

Fortune Bank

SEYCHELLES

Wilkes Rise
914m (3,087ft) ▽

Aldabra Group

Cosmoledo Group

Giraud Seamount

Providence Reef

Anton Bruun Ridge

Amirante Trench

Ritchie Bank

Bulldog Bank

Farquhar Group

Agalega Islands

Grande Comore

COMOROS

Nosy Glorieuses

Hydra Seamount △

4,801m (15,752ft) ▽

Moroni

Anjouan

Mohéli

MAYOTTE

Geyser Reef

Leven Bank

Tanjona Bobaomby

Antsiranana

Bardin Seamount △

Mascarene Basin

Comoro Basin

3,301m (10,831ft) ▽

Nosy Be

Wormley Seamount △

INDIAN

Mahajanga

Tanjona Vilanandra

Betsiboka

Tanjona Masoala

Tromelin

Cargado Carajos Bank

Davie Ridge

Mozambique Channel

Zambezi Canyon

Nosy Sainte Marie

Toamasina

Mascarene Plain

Wilshaw Ridge

Soudan Bank

MADAGASCAR

Tsiribihina

Mangoky

La Pérouse Seamount △

Port Louis

MAURITIUS

Bassas da India

Hall Tablemount △ Jaguar Seamount

Île Europa

Tanjona Ankabna

St-Denis

Mascarene Islands

RÉUNION

4,976m (16,326ft) ▽

Mauritius Trench

Toliara

Natal Basin

Tanjona Vohimena

Madagascar Basin

Madagascar Plateau

1,984m (6,510ft) ▽

7,023m (23,042ft) ▽

40°E   50°E   60°E

A   B   C   D   E   F

Map labels (left page)

G H I

70°E
MALDIVES
Equator

Mascarene Fracture Zone

Sealark Fracture Zone

Chagos–Laccadive Plateau

4,301m
(14,112ft)

Mid-Indian Ridge

Chagos
Bank

Chagos Archipelago

Diego
Garcia

Chagos Trench

3,658m
(12,002ft)

6,102m
(20,021ft)

Vema Fracture Zone

...ya de
...alha Bank

10°S

Argo Fracture Zone

OCEAN

Mid-Indian
Basin

569m
(1,869ft)

Marie Celeste Fracture Zone

...lix
...eamount

...drigues Ridge
Rodrigues

Egeria Fracture Zone

Mid-Indian Ridge

20°S

Tropic of Capricorn

2,078m
(6,818ft)

Southwest Indian Ridge

SCALE

0 100 200 300 400 500km

0 100 200 300 400 500 miles

70°E

G H I

THE SEYCHELLES AND MADAGASCAR

THE EASTERN INDIAN OCEAN FLOOR is littered with scars documenting the break-up of Gondwanaland over the last 150 million years. The warm waters of the Indian Ocean have also proved to be an ideal environment for diverse marine life.

INDIAN OCEAN D3

Seychelles

VOLCANIC Continental islands

AREA 455 square km (176 square miles)

NUMBER OF ISLANDS 115

The main islands of the Seychelles – the Inner Islands – are made of granite, rising over 900m (3,000ft) above sea level on top of the Seychelles Bank. The other islands to the southwest – the Outer Islands – are coral islands (atolls) on top of seamounts. The Seychelles Bank is the most northerly part of the submarine Mascarene Plateau, which extends as far as the island of Réunion in the south. This continental fragment broke off from India around 65 million years ago as the current Mid-Indian Ridge started spreading.

GRANITE BOULDERS IN THE SEYCHELLES

INDIAN OCEAN B5

Mozambique Channel

AREA 1 million square km (386,000 square miles)

MAXIMUM DEPTH 110m (370ft)

INFLOWS Rivers Zambezi, Rio Lúrio

The Mozambique Channel separates Madagascar from the mainland of Africa. The area is home to the ancient coelacanth, found on both sides of the channel and off the Comoros. An anticlockwise gyre is found around the Comoros, and anticlockwise eddies dominate the water flow in the main part of the channel. The warm Aghulas Current arises over the Natal Basin, fed by the South Equatorial Current.

INDIAN OCEAN H3

Mid-Indian Ridge

LENGTH 3,400km (2,100 miles)

AVERAGE HEIGHT ABOVE SEA FLOOR 1,500m (5,000ft)

RATE OF SPREAD 3cm (1¹/₄in) per year

The Indian and African plates are moving apart due to spreading at the Mid-Indian Ridge, which is marked by a series of transform fracture zones. Rifting was triggered 65 million years ago when the Réunion Hot Spot erupted a vast amount of basalt through the Indian continental plate, forming a plateau called the Deccan Traps. An older spreading ridge, which first separated India from Africa, lies subsided between the Mascarene Basin and the Mascarene Plain.

INDIAN OCEAN F6

Mauritius and Réunion

TYPE Volcanic islands

AREA 4,550 square km (1,800 square miles)

NUMBER OF ISLANDS 2

The Mascarene Islands, Mauritius and Réunion, are the largest and youngest islands associated with the Mascarene Plateau, rising 6,500m (21,300ft) above the sea floor. Like the older banks of the plateau to the northeast and the Rodrigues Ridge to the east, they are volcanic in origin, having formed above a deep

FRINGING REEF
A reef fringes the lagoon on the north coast of the volcanic island of Mauritius.

mantle hot spot. After the Deccan Traps eruption (see above), the Réunion Hot Spot continued to punch through the crust as India moved north, leaving a trail of volcanic structures across the ocean floor, including the Laccadive and Maldive islands and the Chagos Bank on the other side of the Mid-Indian Ridge. Réunion's main peak, Piton de la Fournaise, is one of the most active volcanoes in the world.

THE PACIFIC OCEAN

THE PACIFIC IS THE LARGEST OCEAN. It is twice the size of the Atlantic and covers more than a third of the planet's surface. Many Pacific islands were colonized by Micronesians and Polynesians before Europeans arrived in the 16th century. The Portuguese explorer Ferdinand Magellan died after crossing the Pacific in 1521, leaving his crew to complete the first circumnavigation of the world.

BORA BORA ISLAND IN THE SOUTH PACIFIC

OCEAN CIRCULATION

The Pacific is cut off from the Arctic Ocean, but exchanges water with the Southern Ocean. The North Equatorial Current is the world's longest westward-flowing current, carrying water 14,500km (9,000 miles) across the ocean. The warm Kuroshio Current flows north as the North Pacific's western boundary current, and the Kuroshio Extension returns warm water to the western Pacific. The anticlockwise gyre in the South Pacific is formed by the South Equatorial Current, the warm East Australia Current, the Antarctic Circumpolar Current, and the Humboldt Current. A strong upwelling occurs where the cold Humboldt Current diverges from the coast, but this routinely fails as part of the El Niño Southern Oscillation (see pp.68–69).

ROUGH SEAS
The north Pacific is a breeding ground for storms. Here the bow of a ship ploughs through violent storm waves in the Bering Sea.

OCEAN BASIN

The Pacific Basin has been shrinking since the opening of the Atlantic and Indian oceans. It has more subduction zones, where oceanic crust is consumed, than any other ocean. Violent volcanic eruptions are associated with these zones, producing the Ring of Fire around the Pacific's shores (see p.184). The western Pacific is studded with chains of volcanic islands and marked by deep ocean trenches where the Pacific Plate meets the continental Eurasian Plate and smaller oceanic plates. The floor of the eastern Pacific is fairly smooth in comparison, sloping gently away from the coast of North America and the East Pacific Rise. Mid-ocean island chains and seamounts have arisen from intermittent eruptions above mantle hot spots.

EAST PACIFIC RISE
Pillow lavas are extruded at all mid-ocean ridges and form the top layer of the crust throughout the oceans.

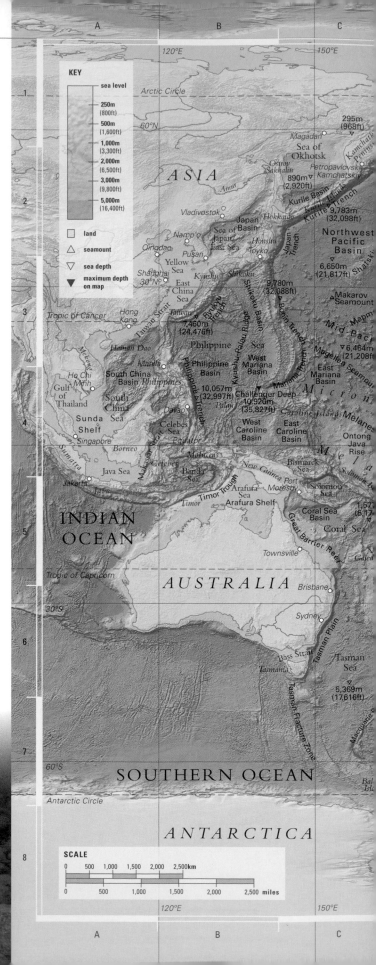

KEY

	sea level
	250m (800ft)
	500m (1,600ft)
	1,000m (3,300ft)
	2,000m (6,500ft)
	3,000m (9,800ft)
	5,000m (16,400ft)
☐	land
△	seamount
▽	sea depth
▼	maximum depth on map

ASIA

Magadan 295m (968ft)
Sea of Okhotsk
Ostrov Sakhalin Petropavlovsk-Kamchatskiy
890m (2,920ft) Kamchatka Peninsula
Vladivostok
Kurile Basin Kuril Islands Kurile Trench 9,783m (32,098ft)
Namp'o Japan Basin
Japan Hokkaido
Qingdao Sea of Japan/East Sea Honshu Tokyo Japan Trench
Pusan Shikoku Northwest Pacific Basin
Yellow Sea Kyushu 9,780m (32,088ft) 6,650m (21,817ft)
Shanghai 30°N Ryukyu Trench Shikoku Basin Makarov Seamount
East China Sea Shatsky
Hong Kong Taiwan Strait Bonin Trench Mid-Pac
Tropic of Cancer Taiwan 7,460m (24,476ft) Shikoku Basin 6,464m (21,208ft)
Hainan Dao Philippine Sea Mid-Pac
Manila West Mariana Basin Magellan Seamount
Mekong Kyushu-Palau Ridge Mariana Trench East Mariana Basin
Ho Chi Minh South China Basin Philippines Philippine Trench Micron
Gulf of Thailand 10,057m (32,997ft) Challenger Deep 10,920m (35,827ft)
South China Sea Davao Talaud Caroline Islands Melanes
Sunda Shelf Celebes Sea West Caroline Basin East Caroline Basin
Singapore Borneo Equator Ontong Java Rise
Sumatra Moluccas Mela
Java Sea Celebes Banda Sea New Guinea Trough Bismarck Sea Solomon Is
Jakarta Java Timor Trough Arafura Sea Port Moresby Solomon Sea
INDIAN OCEAN Timor Arafura Shelf Coral Sea Basin 1,57 (8,17
Tropic of Capricorn Great Barrier Reef Coral Sea
AUSTRALIA Townsville Caled
Brisbane
30°S Sydney Tasman Plain
Bass Strait Tasman Sea
Tasmania 5,369m (17,616ft)
Tasman Fracture Zone Macquarie
60°S
SOUTHERN OCEAN Bal Isl
Antarctic Circle
ANTARCTICA

120°E 150°E
Arctic Circle 60°N Amur

SCALE

0	500	1,000	1,500	2,000	2,500km
0	500	1,000	1,500	2,000	2,500 miles

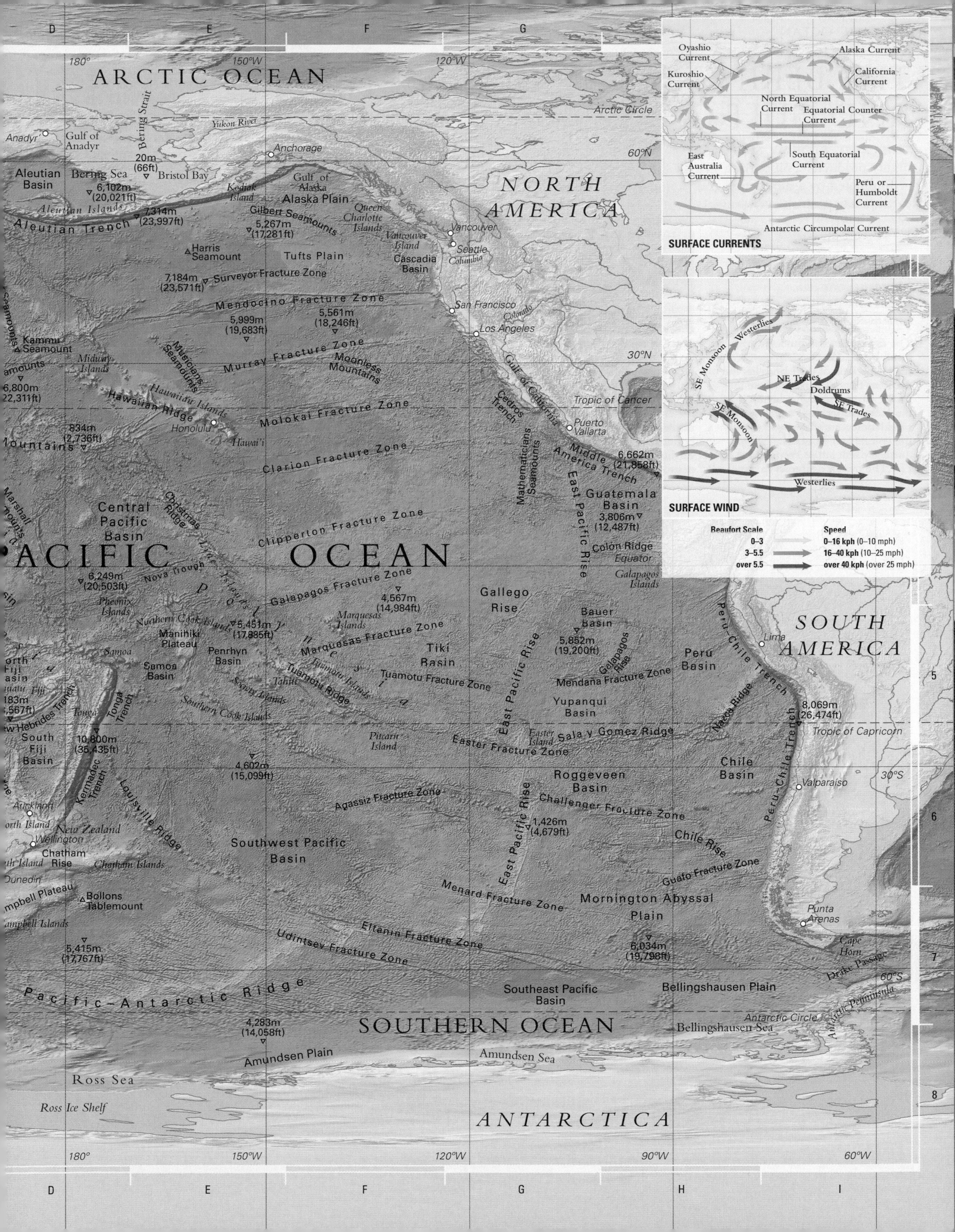

ARCTIC OCEAN

Anadyr • Gulf of Anadyr

Aleutian Basin

Bering Sea
6,102m (20,021ft)

Aleutian Islands

Aleutian Trench 7,314m (23,997ft)

NORTH AMERICA

Anchorage

20m (66ft) Bristol Bay

Kodiak Island

Gulf of Alaska Alaska Plain

Gilbert Seamounts 5,267m (17,281ft)

Queen Charlotte Islands

Vancouver Island

Vancouver
Seattle
Columbia

Cascadia Basin

Harris Seamount

Tufts Plain

7,184m (23,571ft) Surveyor Fracture Zone

Mendocino Fracture Zone 5,999m (19,683ft) 5,561m (18,246ft)

San Francisco

Los Angeles

Kammu Seamount

Murray Fracture Zone

Moonless Mountains

Colorado

Gulf of California

30°N

6,800m (22,311ft)

Midway Islands

Musicians Seamounts

Hawaiian Islands

Molokai Fracture Zone

Cedros Trench

Tropic of Cancer

834m (2,736ft) Honolulu

Hawaiian Ridge

Hawai'i

Clarion Fracture Zone

Puerto Vallarta

Mathematicians Seamounts

Middle America Trench

6,662m (21,858ft)

Mountains

Central Pacific Basin

Christmas Ridge

East Pacific Rise

Guatemala Basin 3,806m (12,487ft)

PACIFIC OCEAN

Clipperton Fracture Zone

Colón Ridge

Equator

Galápagos Islands

6,249m (20,503ft)

Nova Trough

Galapagos Fracture Zone 4,567m (14,984ft)

Gallego Rise

Phoenix Islands

Northern Cook Islands 5,451m (17,885ft)

Marquesas Islands

Bauer Basin

SOUTH AMERICA

Manihiki Plateau

Marquesas Fracture Zone

Tiki Basin

5,852m (19,200ft)

Galapagos Rise

Peru-Chile Trench

Lima

Penrhyn Basin

Tuamotu Islands

Tuamotu Fracture Zone

Mendana Fracture Zone

Peru Basin

Samoa

Samoa Basin

Tuamotu Ridge

Tahiti

Yupanqui Basin

North Fiji Basin

Tonga

Southern Cook Islands

Nazca Ridge

Tropic of Capricorn

New Hebrides Trench 183m (567ft)

Tonga Trench

Pitcairn Island

East Pacific Rise

Easter Island Sala y Gomez Ridge

8,069m (26,474ft)

South Fiji Basin 10,800m (35,435ft)

Easter Fracture Zone

Chile Basin

Kermadec Trench

4,602m (15,099ft)

Roggeveen Basin

Valparaiso 30°S

Auckland

Louisville Ridge

Agassiz Fracture Zone

Challenger Fracture Zone

Chile Rise

North Island New Zealand Wellington

1,426m (4,679ft)

Chatham Rise

Chatham Islands

Southwest Pacific Basin

Menard Fracture Zone

Mornington Abyssal Plain

Guafo Fracture Zone

Punta Arenas

Dunedin

Campbell Plateau

Bollons Tablemount

Eltanin Fracture Zone

6,034m (19,798ft)

Cape Horn

Campbell Islands 6,415m (17,767ft)

Udintsev Fracture Zone

Bellingshausen Plain

Drake Passage 60°S

Pacific–Antarctic Ridge

Southeast Pacific Basin

Bellingshausen Sea

Antarctic Peninsula

SOUTHERN OCEAN 4,283m (14,058ft)

Antarctic Circle

Amundsen Plain

Amundsen Sea

Ross Sea

Ross Ice Shelf

ANTARCTICA

180° 150°W 120°W 90°W 60°W

Yellow Sea

SOUTH KOREA

Sea of Japan/ East Sea

JAPAN

P'ohang

Pusan

Cheju Strait

Cheju-do

Goto-retto

Shanghai

CHINA

East China Sea

Great Yangtze Bank

Hiroshima
Kitakyushu
Fukuoka
Nagasaki
Kagoshima

Kyushu

Inland Sea
Shikoku
Tokushima

Honshu
Nagoya
Osaka

Tokyo

8,130m
(26,673ft)

Northwest Pacific Basin

6,650m
(21,819ft)

1,339m
(4,393ft)

Isakov Seamount

Makarov Seamount

5,66
(18,57

Choyo Seamount

Michelson Ridge

Marcus Island
1,082m
(3,550ft)

Tanega-shima
Yaku-shima

Amami-O-shima

Senkaku Islands

Sakishima-shoto

TAIWAN
Hualien

Okinawa Trough
Ryukyu Islands
Okinawa
Naha

24m
(79ft)

7,460m
(24,476ft)

Ryukyu Ridge

Ryukyu Trench

Daito Ridge

88m
(239ft)

Okidaito Ridge

Shikoku Basin

Nankai Trough

Izu Spur

Iwo-Jima Ridge

Izu Trench

9,780m
(32,088ft)

Kazan-retto

Ogasawara Trough

Ogasawara-shoto
Bonin Ridge

Bonin Trench

9,157m
(30,044ft)

PACIFIC

4,506m
(14,784ft)

6,326m
(20,756ft)

Basin Channel

11m
(36ft)

Luzon Ridge

Philippine Sea

Central Basin Trough

Philippine Basin

Kyushu-Palau Ridge

2,050m
(6,726ft)

265m
(869ft)

West Mariana Ridge

West Mariana Basin

NORTHERN MARIANA ISLANDS

Mariana Ridge

Mariana Trench

Saipan
Tinian
9,888m
(32,442ft)

GUAM

6,464m
(21,208ft)

Magellan Seamo

East Mariana Basin

M i

Luzon

Manila

Mindoro

Benham Plateau

20m
(66ft)

Samar

PHILIPPINES
Panay
Cebu
Cebu
Leyte
Negros
Bohol
Bohol Sea

Sulu Sea

Mindanao
Moro Gulf
Davao

Cape San Agustin

Tinaca Point
Tinaca Point

Philippine Trench

10,057m
(32,997ft)

5,513m
(18,088ft)

8,054m
(26,425ft)

8,510m
(27,820ft)

Yap
PALAU

Yap Trench

Palau Trench

Challenger Deep
10,920m
(35,827ft)

Ulithi

West Caroline Rise

Sorol Trough

Caroline Islands

Caroline Ridge

Chuuk Islands

MICRONESIA

c

Kepulauan Talaud
Kepulauan Sangir

Pulau Morota

West Caroline Trough

4,859m
(15,942ft)

Eauripik Rise

1,554m
(5,099ft)

West Caroline Basin

East Caroline Basin

4,262m
(13,984ft)

Lyra Basin

1,402m
(4,600ft)

Celebes Sea

Celebes Basin

5,499m
(18,042ft)

Manado

Equator

Molucca Sea
Kepulauan Sula

Ternate

Pulau Halmahera

Halmahera Sea

Ceram Sea

1,733m
(5,686ft)

Manokwari
Jazirah Doberai

22m
(72ft)

Jayapura

New Guinea Trench

7,249m
(23,784ft)

Manus Trench

Mussau Trench

Oli

Lyra Reef

164m
(538ft)

Celebes

Pulau Buru
Ambon

Pulau Seram

INDONESIA

Banda Sea

Moluccas

Wewak

Sepik

New Guinea

140°E

Bismarck Archipelago

Manus Island

Bismarck Sea

PAPUA NEW GUINEA

New Ireland

150°E Rabaul

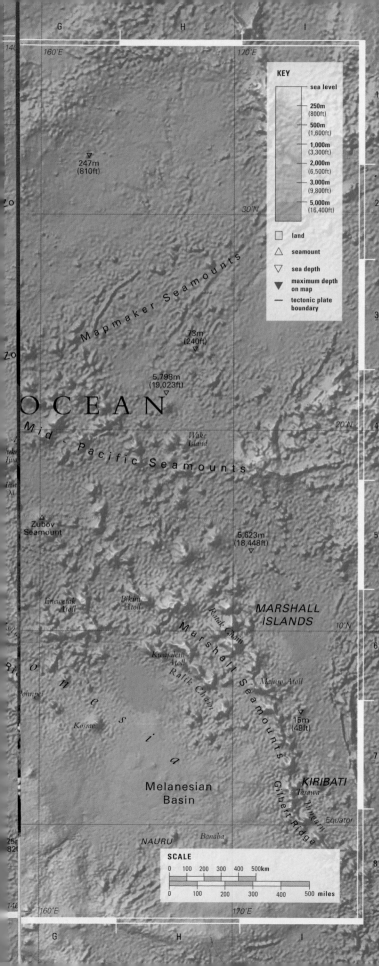

KEY

- sea level
- 250m (800ft)
- 500m (1,600ft)
- 1,000m (3,300ft)
- 2,000m (6,500ft)
- 3,000m (9,800ft)
- 5,000m (16,400ft)

□ land
△ seamount
▽ sea depth
▼ maximum depth on map
— tectonic plate boundary

247m (810ft) ▽

Mapmaker Seamounts

73m (240ft) ▽

5,798m (19,023ft) ▽

OCEAN

Mid-Pacific Seamounts

Wake Island

5,623m (18,448ft) ▽

Zubov Seamount

Enewetak Atoll   Bikini Atoll

MARSHALL ISLANDS

10°N

Radak Chain

Marshall Seamounts

Pohnpei

Kwajalein Atoll

Rilik Chain

Majuro Atoll

15m (49ft) ▽

Kosrae

M
i
c
r
o
n
e
s
i
a

Melanesian Basin

KIRIBATI

Tarawa

Gilbert Ridge

Equator

NAURU   Banaba

SCALE

0   100   200   300   400   500km

0   100   200   300   400   500 miles

MICRONESIA

THE NAME MICRONESIA APPLIES TO AN AREA in the western Pacific, north of the Equator. Its stretches to the Caroline and Mariana islands in the west, and Nauru, the Marshall Islands, and Kiribati (or Gilbert) Islands to the east.

PACIFIC OCEAN B4

Philippine Sea

AREA	5 million square km (1.9 million square miles)
MAXIMUM DEPTH	10,540m (35,580ft)
INFLOWS	Pacific Ocean, South China Sea

The Philippine Sea stretches east to west between the Marianas Islands and the Philippines, and from north to south between Japan and Palau. This warm sea is swept by the North Equatorial Current, which turns north to form the Kuroshio Current. The water becomes very warm in the summer, and the area is a breeding ground for typhoons. The Philippine Sea is underlain by the Philippine Plate, an oceanic plate that is subducting at the Philippine and Ryukyu trenches. The plate is split into two main basins. The westernmost

TYPHOON DESTRUCTION

Philippine Basin is the deepest and oldest, separated from the West Mariana Basin by the Kyushu–Palau Ridge. This ridge, and the Iwo-Jima and West Mariana ridges, are the remnants of island arcs associated with ancient subduction zones.

PACIFIC OCEAN E5

Mariana Trench

LENGTH	2,542km (1,580 miles)
MAXIMUM DEPTH	10,920m (35,827ft)
RATE OF CLOSURE	4cm (1½in) per year

At the eastern edge of the Philippine Plate lies the volcanic island arc of the Northern Mariana Islands. To the east lies the Mariana Trench, where the Pacific Plate is subducting beneath the Philippine Plate. The Mariana Trench includes Challenger Deep – at

10,920m (35,827ft) this is the deepest known part of the ocean. It was named after a British survey ship that measured its depth in 1951. It was explored for the first time in 1960 by the deep-sea submersible *Trieste*, which found flatfish and shrimps living at the bottom of the trench.

MARIANA ISLANDS

The southern members of the Mariana Islands are limestone platforms with fringing coral reefs.

PACIFIC OCEAN I6

Marshall Islands

TYPE	Coral atoll islands
AREA	180 square km (70 square miles)
NUMBERN OF ISLANDS	34

Most of the seamounts scattered across the floor of the western Pacific are far from any plate boundary. The seamounts are found in groups, often strung out in lines running southeast–northwest – the direction of motion of the Pacific Plate. They are caused by hot spots in the Earth's mantle, which periodically punch through the ocean crust to form volcanoes. Some may reach the surface as islands and in the Marshall Islands, coral atolls were formed as the plate moved away from the hot spot and the volcanic islands subsided. The atolls of Enewetak and

Bikini were chosen for their remote location as the site of American nuclear bomb tests in the 1940s and 1950s. Several ships were sunk in these tests, but their wrecks are now considered safe for recreational diving.

GARDEN EELS

These Garden Eels are among the sea life to be found at the bottom of Rongelap Atoll in the Marshall Islands.

ATLAS OF THE OCEANS

SOUTHEAST AUSTRALIA AND NEW ZEALAND

TO THE NORTHEAST OF AUSTRALIA, ocean currents flow from the east, feeding the warm East Australia Current, which sweeps south along the Australian coast, before turning east to flow north of New Zealand. South of New Zealand, the Antarctic Circumpolar Current flows from west to east. New Zealand straddles a major tectonic boundary between the Pacific and Australian plates.

PACIFIC OCEAN G4

Kermadec–Tonga Trench

LENGTH 2,500km (1,550 miles)

DEPTH 10,800m (35,430ft)

RATE OF CLOSURE 15–24cm (6–9in) per year

The Kermadec–Tonga Trench runs between the North Island of New Zealand and the island of Tonga. It is a subduction zone, where the Pacific Plate is converging with the Australian Plate. At its northern end, closure rates of 24cm (9in) per year have been measured – the fastest plate motion yet recorded. The older oceanic crust of the Pacific Plate is sinking below the more buoyant, young oceanic crust of the Australian Plate. The Tonga Ridge and the older Lau Ridge to the west formed as arcs of volcanoes above the subduction zone. The rapid speed of this subduction has caused extension of the over-riding Australian Plate and the opening of a back-arc basin (an isolated basin behind a subduction zone) between the two ridges, in the Lau Basin. Together with Fiji and Samoa, the 36 inhabited islands of Tonga are the cradle of the Polynesian seafaring culture, which had stretched across the South Pacific by the 12th century.

PACIFIC OCEAN B6

Bass Strait

LENGTH 400km (250 miles)

MINIMUM WIDTH 100km (62 miles)

The Bass Strait separates Tasmania from Australia, overlying a shallow shelf around 50m (160ft) deep. Strong winds and currents from the Southern Ocean combine with the shallow depth to make its waters notoriously rough. Hundreds of ships were wrecked on its shores during the 19th century, before the erection of lighthouses made navigation safer. Natural gas fields were discovered beneath the eastern Bass Strait in the 1960s and 1990s.

PHILLIP ISLAND

PACIFIC OCEAN C5

Tasman Sea

AREA 2.3 million square km (890,000 square miles)

MAXIMUM DEPTH 5,945m (19,500ft)

INFLOWS Southern Ocean, Coral Sea

This warm sea was discovered by Dutch explorer Abel Tasman in 1642, while looking for Terra Australis (the Southern Land). On this voyage he become the first European to reach the islands of Tasmania, New Zealand, Tonga, and Fiji. The area was not visited again until James Cook's voyage in 1768. On a later voyage, in 1644, Tasman succeeded in finding the continent of Australia. The Tasman Sea has a subtropical climate in the north, but the influence of cold sub-Antarctic water makes it temperate in the south.

LORD HOWE ISLAND
The warm waters of the East Australia Current allow Lord Howe Island to host the world's most southerly coral reef.

PACIFIC OCEAN H3

Southwest Pacific Basin

AREA 23 million square km (8.9 million square miles)

MAXIMUM DEPTH 5,655m (18,500ft)

INFLOWS Pacific Ocean, Southern Ocean

The Southwest Pacific Basin lies east of New Zealand and the Kermadec–Tonga Trench. It is bounded in the east by the East Pacific Rise (see p.478), in the south by the Pacific–Antarctic Rise, and in the north by the Polynesian island chains. The Louisville Ridge is the only significant chain of seamounts and much of the basin floor is an abyssal plain. There are extensive deposits of manganese in the northern and southern parts of the basin.

PACIFIC OCEAN E5

New Zealand

TYPE Micro-continental island group

AREA 268,680 square km (103,700 square miles)

NUMBER OF ISLANDS 2 main islands (700 smaller islands)

New Zealand separated from Australia and Antarctica 80 million years ago, and is now positioned at the boundary between the Pacific and Australian plates. The largely transverse Alpine Fault runs 700km (435 miles) across the South Island. Crustal compression and distortion across a 250km- (155-mile-) wide zone has raised the Southern Alps over 4,000m (13,000ft) above sea level. The plate boundary continues north as the Hikurangi Trench, a classic subduction zone producing volcanism on North Island, and south as the Macquarie Ridge, where shallow subduction has uplifted the Australian Plate. The main islands of New Zealand are the highest points of an extensive area of continental crust that includes the Challenger Plateau, Chatham Rise, and Campbell Plateau. To the southeast, Campbell Plateau is the largest area of submerged continental crust in the world. Lord Howe Rise was left isolated between Australia and New Zealand by sea-floor spreading in the Tasman Sea and by rifting in the New Caledonia Basin and Bellona Valley.

CAMPBELL ISLAND
The southernmost of New Zealand's subantarctic islands, Campbell Island is primarily volcanic in origin.

Map labels:

Coral Sea Islands • Mellish Rise • Coral Sea • 150°E • 1 • Great Barrier Reef • 20°S • Mackay • 2 • Tropic of Capricorn • Fraser Island • Brisbane • Gold Coast • 30°S • AUSTRALIA • Newcastle • Sydney • Wollongong • Tasman Plain • Melbourne • 5 • Cape Everard • 2,620m (8,596ft) • South East Point • Cape Otway • Furneaux Group • Bass Strait • 40°S • King Island • Tasmania • 6 • Hobart • East Tasman Plateau • 3,460m (11,352ft) • 730m (2,395ft) • Tasman Plateau • 140°E • 50°S • Tasman Fracture Zone • 140°E • A • B

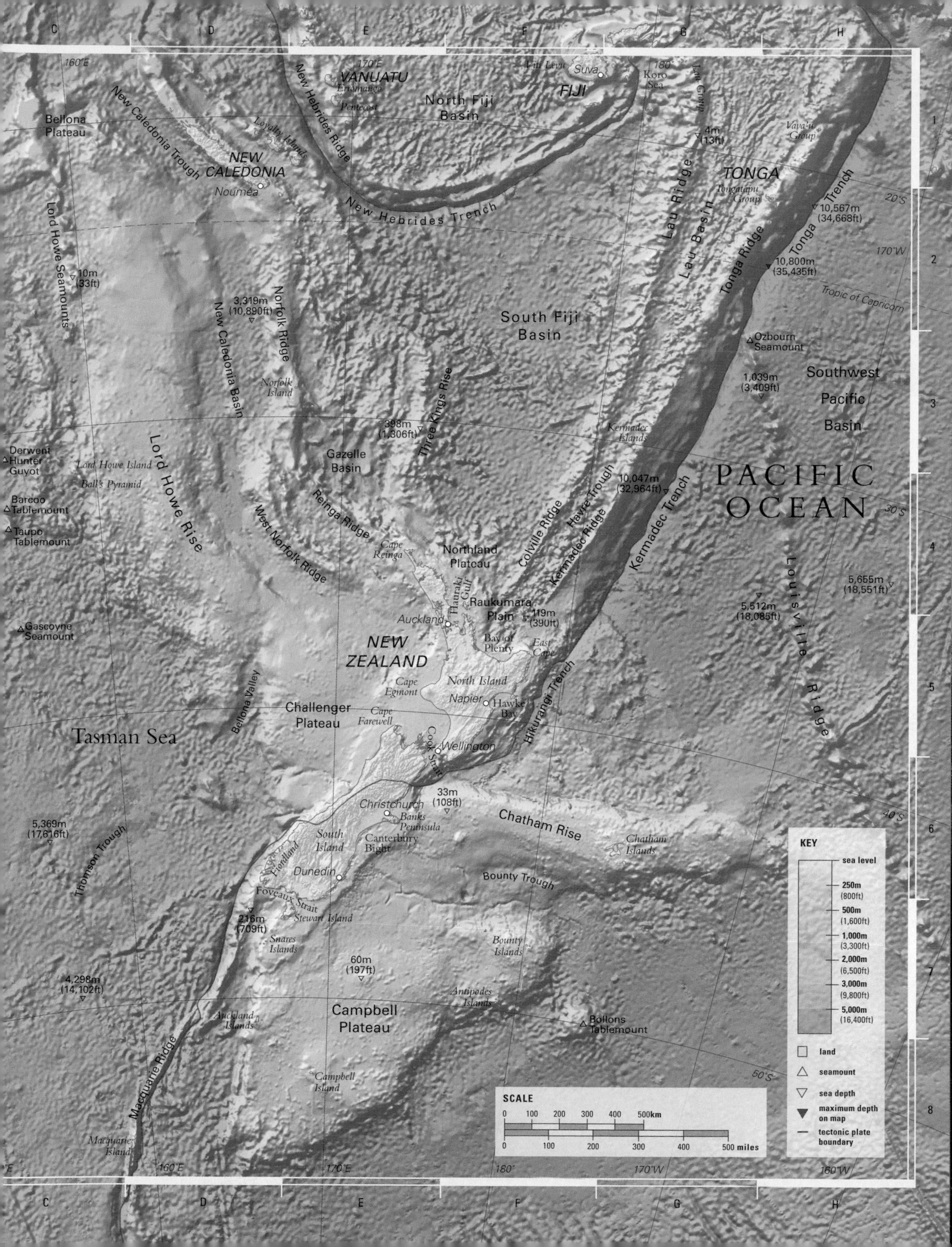

160°E

C D E F G H

VANUATU
Erromango
Pentecost

FIJI
Viti Levu · Suva
Koro Sea
Lau Group

North Fiji Basin

170°E

180°

NEW CALEDONIA
Nouméa

New Caledonia Trough

New Hebrides Ridge

Loyalty Islands

New Hebrides Trench

4m
(13ft)

Vava'u Group

TONGA

Lau Ridge

Tongatapu Group

10,567m
(34,668ft)

20°S

Bellona Plateau

10m
(33ft)

Lord Howe Seamounts

New Caledonia Basin

Norfolk Ridge

3,319m
(10,890ft)

South Fiji Basin

Lau Basin

Tonga Ridge

Tonga Trench

10,800m
(35,435ft)

170°W

2

Tropic of Capricorn

Ozbourn Seamount

1,039m
(3,409ft)

Southwest Pacific Basin

3

Derwent Hunter Guyot

Lord Howe Island
Ball's Pyramid

Norfolk Island

398m
(1,306ft)

Three Kings Rise

Gazelle Basin

Kermadec Islands

PACIFIC OCEAN

Barcoo Tablemount

Lord Howe Rise

Reinga Ridge

Colville Ridge

Havre Trough

Kermadec Ridge

10,047m
(32,964ft)

Taupo Tablemount

West Norfolk Ridge

Cape Reinga

Northland Plateau

Kermadec Trench

30°S

4

Gascoyne Seamount

Hauraki Gulf

Auckland

Raukumara Plain

Bay of Plenty

119m
(390ft)

East Cape

5,512m
(18,085ft)

5,655m
(18,551ft)

Louisville Ridge

NEW ZEALAND

Hikurangi Trench

5

Cape Egmont

North Island

Napier *Hawke Bay*

Tasman Sea

Bellona Valley

Challenger Plateau

Cape Farewell

Cook Strait

Wellington

5,369m
(17,616ft)

Thomson Trough

33m
(108ft)

Christchurch
Banks Peninsula

Canterbury Bight

Chatham Rise

Chatham Islands

40°S

6

Fiordland

Dunedin

Foveaux Strait
Stewart Island

Bounty Trough

Bounty Islands

KEY

sea level

250m
(800ft)

500m
(1,600ft)

4,298m
(14,102ft)

216m
(709ft)

Snares Islands

60m
(197ft)

Antipodes Islands

1,000m
(3,300ft)

2,000m
(6,500ft)

3,000m
(9,800ft)

5,000m
(16,400ft)

7

Auckland Islands

Campbell Plateau

Bollons Tablemount

land

seamount

sea depth

Macquarie Ridge

Campbell Island

50°S

maximum depth on map

tectonic plate boundary

8

Macquarie Island

160°E

170°E

180°

170°W

160°W

SCALE

0 100 200 300 400 500km

0 100 200 300 400 500 miles

THE SOUTHERN OCEAN

THE SOUTHERN OCEAN COMPLETELY surrounds
Antarctica and links the Indian, Atlantic, and Pacific
oceans. Antarctica's coast was not sighted until 1820,
and its shores were not fully explored until the 20th

century. The Southern Ocean is
generally described as being south
of 60° latitude, but is physically
better defined by the extent of the
Antarctic Circumpolar Current.

WEDELL SEAL DIVING BENEATH THE ICE

OCEAN CIRCULATION

The eastward-flowing Antarctic Circumpolar Current is the strongest in
the world. It flows up to 3,000m (9,800ft) deep and carries 135,000 cubic
metres (4.8 million cubic ft) of water per second through Drake Passage.
It diverts heat flowing from the Equator, isolating Antarctica and causing
the build up of the thick Antarctic ice cap. Cold currents branch off up the
eastern sides of the Indian, Atlantic, Pacific oceans. The Circumpolar

Current (also known as the West
Wind Drift) is driven by the
prevailing westerly winds, which
blow uninterrupted by any
landmass. Wind speeds in the
Southern Ocean are the highest
in the world: the "roaring
forties" give way to the "furious
fifties" and the "screaming
sixties" as one sails south.

SEA-ICE AND ICEBERGS
Tabular icebergs and sea-ice can be
found drifting in much of the Southern
Ocean throughout the year.

OCEAN FLOOR

The Southern Ocean is unusual in not having a well-defined basin
bounded by land masses: it surrounds the South Pole and extends for
360° of longitude. There are a series of deep basins lying between the
continental shelf and the ridges at the edge of the Antarctic Plate. The
continental shelf is narrow, and deeper than that of other continents due
to the depression of the crust under the weight of the 2.5km- (1½-mile-)
thick Antarctic Ice Sheet. The surrounding spreading ridges drove the
break up of the supercontinent of Gondwana, starting when Africa
began to move north 165
million years ago. An area
of thick ocean crust, the
Kerguelen Plateau, lies
between the Southern
and Indian oceans. This
is one of the largest
submarine plateaus, and
it is composed of flood
basalts that erupted about
97 million years ago.

FROZEN WATERFALLS
Fresh glacial water re-freezes on
contact with cold ocean water,
which is kept liquid by its salt.

AFRICA

Port
Elizabeth

Cape
Town
Cape of
Good Hope

Transkei
Basin

Mozambique Escarpment

Africana
Seamount

Agulhas
Plateau

5,819m
(19,092ft)

Prince Edward Fracture Zone

Prince
Edward
Islands

Crozet
Plateau

Del Cano
Rise

Lena
Seamount

Cape
Basin

Atlantic-Indian Ridge

Atlantic-
Indian
Basin

5,115m
(16,782ft)

Davis
Seamounts

SOUTHERN

Discovery
Seamounts

Astrid Ridge

ATLANTIC

OCEAN

Maud
Rise

Lazarev
Sea

Cape Norvegia

American-Antarctic Ridge

Gough Fracture Zone

Mid-Atlantic Ridge

South Sandwich Fracture Zone

5,012m
(16,444ft)

Weddell
Plain

Islas Orcadas Ridge

1,748m
(5,735ft)

South Sandwich Trench

7,152m
(23,466ft)

East
Scotia
Basin

187m
(614ft)

South
Orkney
Islands

South
Georgia

3,667m
(12,031ft)

Scotia Sea

Falkland Plateau

Bransfield Strait

South
Shetland
Islands

Zapiola
Ridge

Falkland Escarpment

Yaghan
Basin

Cape
Horn

Argentine
Basin

Falkland
Islands

SOUTH

AMERICA

SCALE

| 0 | 200 | 400 | 600 | 800 | 1,000km |

| 0 | 200 | 400 | 600 | 800 | 1,000 miles |

INDIAN OCEAN

4,285m
(14,959ft)

Kerguelen

*Heard and
McDonald Islands*

Kerguelen Plateau

Southeast Indian Ridge

5,386m
(17,671ft)

184m
(604ft)

4,684m
(15,368ft)

Banzare Seamounts

Enderby
Plain

OCEAN

*Cape
Batterbee*

*Cape
Darnley*

Antarctic Circle

*Davis
Sea*

*Mackenzie
Bay*

*Vincennes
Bay*

Cape Poinsett

South
Indian
Basin

*Lützow-
Holm
Bay*

ANTARCTICA

*Dumont
D'Urville
Sea*

*South
Pole*

95m
(312ft)

*Balleny
Islands*

Ross Ice Shelf

Cape Adare

*Ross
Sea*

Challenger
Plateau

Campbell
Plateau

Dunedin

*South
Island*

Berkner Island

*Roosevelt
Island*

New Zealand

Wellington

Weddell
Sea

*Ronne
Ice Shelf*

*North
Island*

5,415m
(17,767ft)

4,283m
(14,052ft)

Antarctic Peninsula

*Ellsworth
Land*

Amundsen Plain

*Bollons
tablemount*

Chatham Rise

*Alexander
Island*

Amundsen Sea

*Bellingshausen
Sea*

*Thurston
Island*

limit of summer pack ice

Pacific–Antarctic Ridge

SOUTHERN OCEAN

4,094m
(13,432ft)

Antarctic Circle

limit of winter pack ice

Bellingshausen
Plain

Southeast
Pacific
Basin

Udintsev Fracture Zone

PACIFIC
OCEAN

6,034m
(19,798ft)

Eltanin Fracture Zone

Mornington Abyssal
Plain

Menard Fracture Zone

East Pacific Rise

Southwest
Pacific
Basin

SURFACE CURRENTS

Benguela
Current

West Australia Current

Weddell
Gyre

Ross Sea
Gyre

Falklands
Current

Peru or Humboldt
Current

Antarctic Circumpolar Current

SURFACE WIND

Westerlies

Westerlies

Westerlies

Polar Easterlies

Westerlies

Beaufort Scale	Speed
0–3	0–16 kph (0–10 mph)
3–5.5	16–40 kph (10–25 mph)
over 5.5	over 25 (over 25 mph)

KEY

sea level
250m
500m (1,600ft)
1,000m (3,300ft)
2,000m (6,500ft)
3,000m (9,800ft)
5,000m (16,400ft)

☐ land
△ seamount
▽ sea depth
▼ maximum depth on map

ATLAS OF THE OCEANS

ICE-SHELF EDGE
Antarctic sea-ice is rarely more than 2m (6ft) thick. The Larsen Ice Shelf rises 20m (65ft) above sea level at its edge and extends 200m (650ft) below the water.

ICE-SHELF BREAKUP

Ice shelves cover 44 per cent of the Antarctic coastline. The floating extensions of Antarctica's continental ice sheet, they are continually pushed away from the land by the weight of accumulating snow. They typically advance over the ocean for years or decades, until the front of the ice shelf breaks off under its own weight to form a tabular iceberg. This advance and retreat is part of a natural cycle, but in the Antarctic Peninsula small ice shelves have recently suffered catastrophic collapses as a result of a regional warming of 2.5°C (4.5°F) over the last 50 years (see below and right).

Although the loss of floating ice does not affect global sea levels, it seems that the adjacent continental ice sheet may become unstable if it loses the "buffer zone" provided by an ice shelf. After the Larsen B Ice Shelf collapsed in 2002, scientists measured nearby glaciers flowing between two and eight times faster than they had before. It is not yet clear whether the larger Ronne and Ross ice shelves act as a similar brake on the West Antarctic Ice Sheet. If the regional warming continues and the West Antarctic Ice Sheet collapses as a result, global sea levels could rise more than 5m (16ft), threatening densely populated coastal areas worldwide.

LARSEN ICE SHELF COLLAPSE

The Larsen Ice Shelf occupies the eastern shore of the Antarctic Peninsula. In 1995, the northern part of the ice shelf, Larsen A, broke into tiny fragments during a storm. In 2002, most of the central part, Larsen B, disintegrated in a similar manner over a few weeks. At the moment, the largest part of the shelf, Larsen C to the south, seems to be stable, although it too lost a large area in 1986.

Weddell Sea

Ronne Ice Shelf

Larsen Ice Shelf

West Antarctica

Transantarctic Mountains

East Antarctica

Ross Ice Shelf

Ross Sea

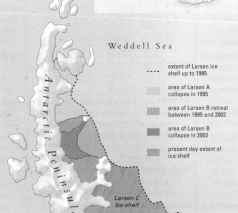

Weddell Sea

Antarctic Peninsula

- - - - extent of Larsen ice shelf up to 1995

area of Larsen A collapse in 1995

area of Larsen B retreat between 1995 and 2002

area of Larsen B collapse in 2002

present day extent of ice shelf

Larsen C Ice shelf

ICE SHELVES IN RETREAT

MELTWATER POOL *In the summer, meltwater collects in low-lying parts of the surface of an ice shelf, including crevasses and depressions. Melting also occurs on the underside.*

CRACK IN LARSEN A *The additional weight of meltwater may increase the pressure at the base of a crevasse, causing it to penetrate deeper into the ice shelf and to widen.*

LARSEN B COLLAPSE
Extensive meltwater pools are visible in a satellite image of the Larsen B Ice Shelf taken on 31 January 2002, before it broke up (top). The collapse itself, on 7 March 2002, is shown in the lower image. The ice shelf broke into a multitude of small fragments, and a few larger ones, which quickly dispersed into the Weddell Sea. It is possible that meltwater helped push surface crevasses through the entire 220m (720ft) thickness of the Larsen B Ice Shelf.

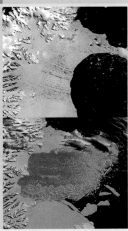

ICEBERG B-15 *One of the largest icebergs ever seen, at 300km (185 miles) long, 40km (25 miles) wide, and 60m (200ft) high, B-15 broke off from the Ross Ice Shelf in March 2000. It drifted around the Ross Sea for several years, disrupting navigation and penguin migration. By November 2005, it had broken into several smaller pieces and moved into the Southern Ocean.*

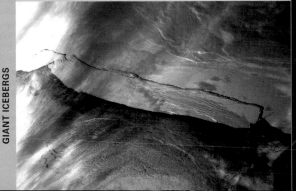

chlorophyll The green pigment of plants and seaweeds that allows them to make their own food by using the Sun's energy. See *photosynthesis*.

chromatophore A skin cell in which the distribution of coloured pigment can be altered, allowing an animal to change colour. Colour change may be fast, as in cephalopods, or slower, as in crustaceans and some fish.

cilia Tiny beating hair-like structures on the surfaces of some cells. Used to aid movement in small organisms, or to create water currents. Singular **cilium**.

cloaca The combined opening of the digestive, urinary, and reproductive systems of many vertebrates (e.g. fish, birds) and some invertebrates.

cnidarians A major group (phylum) of invertebrate animals with simple bodies bearing tentacles that surround a single opening (mouth). Cnidarians include corals, anemones, and jellyfish, and are often colonial. Their two typical body forms are the polyp and the medusa. In some cnidarians, both forms occur during the life cycle. See also *colonial, coral, medusa, nematocyst, polyp*.

coast See *concordant coast, depositional coast, discordant coast, drowned coast, emergent coast, erosional coast, primary coast, secondary coast*.

cold seep A natural seepage of oil or other energy-containing chemicals on the sea floor, often supporting dense concentrations of marine life.

colonial Of an animal: living in colonies. A colony can consist of separate individuals, as in the case of sponge shrimps, or animals joined by strands of living tissue, as in the case of many marine invertebrates, such as corals and bryozoans. Individuals may be specialized for different roles, such as feeding, reproduction, and defence, in which case the colony may behave like a single animal. See also *bryozoans, cnidarians, zooid*.

comb jellies see *ctenophores*.

commensal Living in close association with an organism of another species, for example, by sharing its burrow, without either helping or damaging it. See also *mutualism, symbiosis*.

concordant coast Coast on which hills and valleys are roughly parallel to the shore, resulting either in a straight coastline or one with rocky islands running parallel to the shoreline. See also *discordant coast*.

continental crust The material in the Earth's crust that forms the continents, including the continental margins. It is lighter and thicker than oceanic crust.

continental margin A continent's edge below sea level, including the continental shelf and continental slope.

continental rise The gently sloping sea bed around the edge of ocean basins that adjoins the bottom of the continental slope.

continental shelf The gently sloping sea bed around the edges of most continents, formed of continental crust and averaging around 130m (425ft) deep.

continental slope Sloping sea bed at the seaward edge of the continental shelf. It descends relatively steeply to the continental rise.

convection Circulating currents in a fluid, for example air, water, or hot rock, that result from heated portions rising because they are less dense, and sinking later as they cool.

copepods Small, swimming crustaceans, usually less than 2mm (1/16in) long, that make up a large part of the zooplankton. There are also many parasitic and burrowing species. See also *zooplankton*.

coral Any of various cnidarians that live fixed to the ocean bottom, secrete skeletons for support, and are usually colonial. The true corals lay down hard skeletons of calcium carbonate outside their bodies that eventually form coral reefs. Other coral groups include the sea fans. See also *cnidarians, sea fans, zooxanthellae*.

coral bleaching Phenomenon in which coral animals lose their tiny symbiotic algae (zooxanthellae), usually in response to a stress in the environment. Bleached corals may later die. See also *zooxanthellae*.

coral reef A rock-like, often ridge-shaped structure of calcium carbonate built in shallow tropical seas by generations of coral animals. See also *barrier reef, fringing reef*.

coralline Resembling coral; mainly applied to red seaweeds that form hard, calcareous crusts on rocks or in coral reefs.

Coriolis effect Phenomenon resulting from the rotation of the Earth, in which winds and currents travelling towards or away from the Equator are deflected to the right in the Northern Hemisphere and to the left in the Southern Hemisphere. The effect helps to explain the direction of prevailing winds and the existence of gyres.

crabs see *crustaceans*.

crinoids Stalked echinoderms, also called sea lilies, that filter-feed using their branching arms. Some species have no stalks and are known as **feather stars**. See also *echinoderms*.

crustaceans The most diverse and abundant group of arthropods in the oceans. It includes crabs, lobsters, shrimps, barnacles, krill, copepods, isopods, and amphipods. Their jointed appendages are variously modified as claws, legs, swimming organs, or filter-feeding devices, depending on the species. See also *arthropod*.

ctenophores Transparent jellyfish-like animals that hunt in the plankton. They swim using beating hair-like structures arranged in rows called comb plates. Also called **comb jellies**.

current Any sustained horizontal flow of water. See also *drift, surface current, thermohaline circulation, turbidity current, western boundary current*.

cusp Any shape formed by two concave lines meeting at a point. Cusp-shaped ridges of sand are often created on beaches by wave action.

cyanobacteria A group of minute, single-celled organisms, which can photosynthesize like plants. They are classified as bacteria, because they have a similar structure. Also called **blue-green algae**, although they are not closely related to other algae. See also *photosynthesis*.

cyclone (1) Also called a **depression**, a pattern of circulating air in the atmosphere with low pressure at the centre. Cyclones normally form over oceans outside the tropics and are associated with wet and windy weather. (2) See *tropical cyclone*.

D

dark zone Vertical zone of the sea bed and water column at around 1,000–4,000m (3,300–13,000ft), between the twilight zone and abyssal zone. Virtually no light penetrates this deep. See also *abyssal, twilight zone*.

delta An often fan-shaped structure of sediment built by the deposition of material by a river at its mouth.

demersal Of a fish: living mainly near the sea floor.

deposit feeding Feeding by extracting food particles from mud or other deposits. See also *filter feeding*.

depositional coast A coast that is growing seawards due to deposition of sand and other sediment supplied by rivers or ocean currents. See also *emergent coast, erosional coast*.

detritus Fragments of dead organisms and organic waste material, often mixed with sediment or suspended in ocean currents. A **detritivore** is an animal that feeds on detritus.

diatoms A group of plant-like protists that are part of the algae and major primary producers in the plankton. They are single-celled but often grow as chains or colonies. Diatoms secrete intricate cases of silica around themselves. See also *algae, primary producer, protists*.

dimorphism see *sexual dimorphism*.

dinoflagellates A group of protists that bear two flagella. They are common in ocean plankton. Some are animal-like (eating other organisms), while others are plant-like (photosynthesizing) and are therefore part of the alga. See also *algae, flagellum, protists*.

discordant coast Coast on which hills and valleys are roughly at right angles to the shore, resulting in an indented coastline of headlands and bays. See also *concordant coast*.

doldrums The region of very light winds close to the Equator.

dorsal Relating to the back or upper surface of an animal. See also *ventral*.

drift A broad, slow-moving flow of surface water; for example, the North Atlantic Drift.

drowned coast A coast where the land has sunk or the sea level has risen compared with the previous level. It may show features such as rias or fiords. See also *emergent coast, fiord, ria*.

dune A hill or ridge-shaped structure of sand formed by wind action along some coasts and in deserts. Coastal dunes are usually formed on low-lying land behind beaches.

E

echinoderms A major group (phylum) of marine invertebrates that includes starfish, brittle stars, sea urchins, sea lilies, sea cucumbers, and sea daisies. Echinoderms have bodies arranged in parts rather like the spokes of a wheel (so-called "radial symmetry"). They have chalky protective plates under their skin, and use a unique system of hydraulic "tube feet" for moving, or for capturing prey, or both.

echolocation Method of locating and characterizing nearby objects, used by dolphins, bats, and some other animals, by emitting high-pitched sounds and interpreting their echoes.

echo-sounding The use of sound equipment to measure the depth of objects or the ocean floor; also used as a synonym for echolocation. See also *sonar*.

eddy A circular motion of any size and speed in a fluid. **Mesoscale eddies** of more than 100km (60 miles) across are important features of ocean circulation. In tidal currents and whirlpools, an eddy is a circular motion slower than a whirlpool. See also *gyre, vortex, whirlpool*.

Ekman effect Tendency for a wind or current to cause air or water above or below it to move, but in a different direction to the original wind or current. The effect results from the rotation of the Earth. At the ocean surface, the net result is usually that a prevailing wind creates a water current at 90° to the wind direction. See also *Coriolis force*.

El Niño Phenomenon by which the waters of the eastern Pacific off South America become warmer than usual every 4–7 years. The opposite phenomenon, in which eastern Pacific waters are unusually cold, is called **La Niña**. The term El Niño is also used as shorthand for the larger phenomenon called the El Niño–Southern Oscillation. See *ENSO*.

emergent coast A coast where the land has risen or sea level has fallen compared with a former level. See also *drowned coast, isostasy*.

ENSO Used as an abbreviation for the El Niño–Southern Oscillation. A worldwide variation in the Earth's climate pattern and ocean circulation, including the El Niño phenomenon, associated with a change in the position of warm surface waters in the eastern Pacific.

erosional coast A coast that is being eroded by the action of the sea. Rocky coasts are typically erosional, but so are some low-lying, sandy coasts. See also *depositional coast*.

estuary The mouth of a large river. Used more broadly, the term includes any bay or inlet where sea water becomes diluted with fresh water.

eustatic Of sea-level changes: occurring worldwide simultaneously, for example, as a result of melting ice sheets. See also *isostasy*.

eutrophication The altering of an aquatic ecosystem by the addition of plant nutrients, such as nitrate and phosphate. Often caused by humans, it can greatly change the character of an ecosystem by, for example, causing algal blooms. See also *bloom*.

exoskeleton A skeleton on the outside of an animal's body, often also acting as a protective barrier. Arthropods, such as crustaceans and insects, have an exoskeleton. See also *arthropods*.

F

fast ice Sea-ice forming a continuous sheet. See also *sea-ice, pack ice*.

fathom The traditional unit of depth measurement at sea, equivalent to 1.83m (6ft).

fault A fracture in the Earth's crust where rocks have moved relative to one another either vertically or horizontally.

feather stars see *crinoids*.

Ferrel cell A large-scale circulation of air in temperate regions, involving air rising at around 60°N and S, flowing southwards at a high altitude, descending at around 30°N or S, and returning north as the westerlies (westerly winds). See also *Hadley cell*.

fertilization The union of a male and female sex cell (such as a sperm and an egg cell in animals) as the first step in the production of a new organism by sexual reproduction. Some marine animals release eggs and sperm into the sea to meet by chance (**external fertilization**), while in others, the male transfers sperm directly into the female's body (**internal fertilization**).

fetch The distance of open ocean across which a wind is able to blow, and across which waves generated by the wind are travelling. A longer fetch tends to result in larger swell waves. See also *swell wave*.

filter feeding Feeding by collecting and separating food particles from the environment. When the food particles are suspended in water it is also called **suspension feeding**. See also *deposit feeding*.

fiord A narrow, steep-sided, deep inlet of the sea, once occupied by a glacier. Fiords have a shallower **sill** where they meet the open sea. See also *ria*.

flagellum A flexible, microscopic, hair-like structure used for propulsion by some single-celled organisms and for creating a water current by sponges. It is longer than a cilium. Plural **flagella**. See also *cilia, sponges*.

flatworms A major group (phylum) of invertebrates with simple, usually flattened bodies. Free-living forms are carnivorous; there are also many parasitic species, including tapeworms.

fluke Either of the lobes forming a whale's, dolphin's, or dugong's tail.

foraminiferans A group of protists whose empty, chalky skeletons are a major part of some deep-sea sediments. They are animal-like (they feed on other organisms) and include both planktonic and bottom-living types. See also *protists*.

forced wave A water wave created by storm winds at sea. Forced waves are taller and have a shorter wavelength than swell waves. See also *swell wave*.

foreshore The part of a shoreline that lies between the average high- and low-water marks. See also *tides*.

frazil ice Ice in the form of tiny crystals floating on or near the sea surface. It is the first stage in the formation of sea-ice. See *sea-ice*.

fringing reef A coral reef just offshore, without an intervening lagoon or stretch of water. See also *barrier reef*.

front A vertical or oblique region at the boundary of two masses of air or water with different characteristics.

G

gabion A wire cage filled with stones. Gabions are used to protect coastlines artificially against erosion.

gastropods The group of molluscs that includes snails, slugs, and pteropods (sea-butterflies). See also *molluscs*.

gill rakers Projections on the insides of the gill supports of some fish that sieve particles entering their mouths.

glacier An elongated mass of compressed ice that flows slowly downhill. Glaciers that reach the sea give rise to icebergs.

grease ice Stage of formation of sea-ice in which frazil ice crystals congeal to form a soupy texture. See also *frazil ice, sea-ice*.

greenhouse gas A gas, such as water vapour, carbon dioxide, or methane, that prevents heat radiating from the Earth, causing the Earth's surface to warm (the **greenhouse effect**). Some greenhouse gas emissions are natural; others are caused by human activities.

groyne An artificial barrier built down a beach and into the sea to hinder transport of materials by longshore drift. See also *longshore drift*.

guyot A flat-topped submarine mountain, also called a **tablemount**. See also *seamount*.

gyre A large-scale circulation of surface ocean currents, typically spanning a whole ocean. See also *eddy*.

H

hadal Relating to the deepest oceanic regions below 6,000m (20,000ft), within ocean trenches; deeper than the abyssal zone. See also *abyssal*.

Hadley cell A large-scale circulation of air in warmer regions, caused by warmed air rising near the Equator, travelling to mid-latitudes, cooling and descending, and returning to the Equator as the *trade winds*.

halocline A boundary between waters of different salinities, across which salinity changes rapidly. See also *pycnocline, thermocline*.

headland A promontory on a shoreline, usually high and rocky and under strong forces of coastal erosion. See also *erosional coast*.

heat capacity The amount of heat energy that a given substance can absorb for a given rise in temperature. Water has a high heat capacity and so can act as a store of heat.

hermaphrodite An animal that is both male and female. Animals that are both sexes at once are called **simultaneous hermaphrodites**. Others start as males then become females, or vice versa. Some species change sex repeatedly.

holdfast A root-like structure that anchors a seaweed to rocks but does not absorb nutrients like a true root.

holoplankton Planktonic organisms that spend all of their life as plankton. See also *meroplankton, plankton*.

holothurians Soft-bodied, sausage-shaped echinoderms, also called **sea cucumbers**, that feed mainly by swallowing mud and detritus. Their radial symmetry is not obvious at first glance. See also *echinoderms*.

hot spot A localized region of the Earth that experiences large-scale upwelling of magma. As oceanic crust moves over a hot spot, a line of volcanic islands, such as the Hawaiian islands, may form over millions of years.

hurricane (1) A name for a tropical cyclone, especially one occurring in the Atlantic. See *tropical cyclone*. (2) A wind speed greater than 116kph (72mph).

hydrocarbon Any chemical compound made only of carbon and hydrogen atoms.

hydroids Cnidarians that grow as small, branching colonies of polyps attached to rocks or seaweed. Each polyp is specialized either for feeding, reproduction, or sometimes for defence. See also *cnidarians, polyp*.

hydrothermal vent A fissure in a volcanically active region of the ocean floor from which superheated, chemical-laden water emerges. The energy in the chemicals fuels rich biological communities via the activities of chemosynthetic bacteria and archaea. See also *chemosynthesis*.

I

ice age Any episode in which the Earth's temperatures were much lower than today and ice cover more extensive. The **Ice Age** (with capitals) refers to a series of such episodes within the last 2 million years, the last ending around 10,000 years ago.

iceberg A large fragment of ice calved from the end of a glacier or ice sheet that is in contact with the sea. See also *calve*.

ice cap A mass of permanent ice similar to an ice sheet but smaller in extent.

ice lead A channel of open water among sea-ice.

ice rafting Transport of rocky debris out to sea, frozen into icebergs. When the icebergs melt, the material is deposited as sediment.

ice sheet A very large mass of permanent ice covering land, such as the Antarctic Ice Sheet.

ice shelf An extension of an ice sheet into the ocean. Ice shelves are anchored to the sea floor at their landward end, but further from the coast, they float on water.

igneous rock Any rock that originates from the cooling of magma, such as basalt or granite.

intermediate coast A coast whose features are intermediate between a primary and secondary coast. See also *primary coast, secondary coast*.

internal wave A wave occurring at the boundary of two different layers of the same fluid rather than at the surface, for example at the boundary between two layers of ocean water.

intertropical convergence zone The region of air close to the Equator where the north and south trade winds converge.

invertebrate Any animal without a backbone, ranging from flatworms to spiders. Of a total of around 30 major groups (phyla) of animals, 29 are composed of invertebrates.

irradiance The amount of radiation falling on a given area.

island arc Chain of islands, usually including active volcanoes, created by the collision of the oceanic crust of two tectonic plates. One of the plates is subducted beneath the other, creating a trench on one side of the arc. See also *subduction, ocean trench*.

isopods A group of crustaceans that usually have flattened bodies. The group is mainly marine but also includes the land-living woodlice.

isostasy A state of equilibrium; applied especially to the relatively light rocks of the continental crust, which can be thought of as floating like icebergs among the heavier rocks of the ocean floor and mantle. **Isostatic rebound** is the tendency of land that was formerly ice-covered to rise slowly to its equilibrium level, often creating emergent coasts. See also *continental crust, emergent coast*.

IUCN The initials still used to designate the World Conservation Union (formerly the International Union for the Conservation of Nature). This organization carries out conservation-related activities, including gathering and publishing information on the current status of endangered species.

J

jawless fish Two groups of primitive fish called lampreys and hagfish, which branched off the line of fish evolution before jaws had evolved.

jellyfish Cnidarians that typically drift among the plankton and catch prey using stinging tentacles. The body form of true jellyfish is a medusa. Some apparently similar forms such as the Portuguese Man-of-War are not true jellyfish, but siphonophores. See also *cnidarians, medusa, siphonophores*.

K

katabatic wind A wind that blows downwards from an ice sheet, glacier, or cold valley, usually at night.

krill Swimming, shrimp-like crustaceans typically growing to 2–6cm (¾–2⅜in) long, which form a large part of the zooplankton and an important link in the Southern Ocean's food chain.

L

La Niña see *El Niño*.

lagoon A stretch of coastal water almost cut off from the sea by a spit or other barrier; also, the shallow water within the ring of an atoll.

larva A young stage of an animal, especially when completely different in structure from the adult. The larvae of many marine animals, such as starfish, live as part of the plankton. See also *metamorphosis*.

latent heat The heat absorbed or released when a substance changes its state – from gas to liquid, for example. The heat released when water vapour condenses is the main source of energy for hurricanes.

latitude A position on the Earth expressed in terms of its angle north or south of the plane of the Equator. **Low latitudes** are those close to the Equator, while **high latitudes** are nearer the poles.

levee A natural raised bank around some rivers, or an artificial bank built around a river or estuary.

littoral Relating to the area of shore between high- and low-water marks.

longitude A position on the Earth expressed in terms of its angle east or west of an agreed line called the prime meridian circling the Earth from pole to pole and passing through Greenwich, London, UK.

longshore drift Process by which sediment is transported along a coast as a result of waves breaking at an oblique angle to the shoreline.

M

magma Molten rock rising from deep in the Earth.

mangrove Any of various trees growing on muddy shores in the tropics and adapted to live with their roots and lower trunks immersed in salt water.

mangrove swamp Forest-like ecosystem formed by mangroves growing in muddy tidal areas and river mouths. Mangrove swamps only occur in the tropics and subtropics.

mantle All the rock lying between the Earth's crust and its core. The mantle extends to a depth of about 2,900km (1,800 miles).

medusa One of the two main body forms of cnidarians. Medusae are wide and saucer-shaped, as well as usually free-floating and able to swim. A jellyfish is an example of a medusa. See also *cnidarians, polyp*.

meroplankton Planktonic animals that are the larvae of animals that are not planktonic as adults, such as crabs.

metamorphosis The process of transforming body form from that of the young (larval) form to a radically different adult form. It is common in marine invertebrates such as starfish, whose larvae live in the plankton but whose adults live on the sea floor.

mid-ocean ridge A submerged range of mountains running along any part of the deep-ocean floor, marking the place where sea-floor spreading is taking place. Also called a **spreading ridge**. See also *sea-floor spreading*.

mimicry Phenomenon in which one species of animal has evolved to look similar to another, unrelated animal.

mixed layer The upper layer of the ocean that is kept mixed by winds and currents, so that its temperature and chemical characteristics are roughly uniform throughout.

molluscs A major group (phylum) of invertebrate animals that includes the gastropods (snails and slugs), bivalves (clams and relatives), and cephalopods (octopuses, squid, cuttlefish, and nautiluses). Molluscs are soft-bodied and typically have hard shells, though some subgroups have lost the shell during their evolution.

mucus A sticky or slimy substance secreted by animals for protection, trapping prey, helping with movement, or other purposes.

mutualism A close relationship between two different species in which both benefit.

N

nanoplankton Planktonic organisms of 0.002–0.2mm in diameter. Not as small as picoplankton. See also *picoplankton, plankton*.

neap tide The tide with the smallest range within an approximately two-week cycle, caused by the gravity of the Sun partly cancelling out the effect of the Moon. See also *spring tide, tides*.

nearshore The part of the shore affected by waves and tides under normal conditions. It includes the foreshore plus an area beyond whose bed is shallow enough to be stirred up by wave action. See also *foreshore*.

nekton Animals of the open ocean that can swim strongly enough not to be at the mercy of ocean currents. Nekton include squid, adult fish, and marine mammals. See also *plankton*.

nematocyst The coiled structure within the stinging cell of a jellyfish or other cnidarian that shoots out and injects toxin via a dart-like tip. See also *cnidarians*.

nudibranchs see *sea-slugs*.

O

ocean basin A region of low-lying oceanic crust within which a deep ocean (or part of one) is contained, and usually surrounded by land or shallower seas.

oceanic crust The type of Earth's crust that forms the deep ocean bed. Made mainly of basalt, it is thinner, denser, and heavier than continental crust.

ocean trench Elongated low-lying region of the ocean floor. Trenches are the deepest parts of the ocean. See also *subduction*.

ooze Sediment on the deep ocean floor containing a large proportion of the remains of the skeletons of planktonic organisms, such as foraminiferans or radiolarians.

overfall A stretch of rough water produced when a tidal current flows in the opposite direction to the wind.

ovoviviparous Producing live young by retaining eggs so that they hatch while still in the female's body.

P

pack ice A mosaic of floating ice formed when continuous sea-ice is broken up by storms or waves. See also *fast ice, sea-ice*.

pancake ice Stage of formation of sea-ice consisting of small flat areas of ice, curled at the edges where they bump into each other.

pectoral fin Either of the front pair of fins in most fish and marine mammals, mainly used for steering but sometimes for propulsion. See also *pelvic fin*.

pelagic Relating to or living in the waters of the open ocean, without immediate contact with the shore or the sea bottom. See also *demersal*.

pelvic fin Either of the pair of fins located further back than the pectoral fins in most fish. See also *pectoral fin*.

perennial Of plants: living for three or more years.

pheromone An odour produced by an animal to communicate with others of the same species, to attract the opposite sex, for example.

photic zone see *sunlit zone*.

photophore A light-producing organ.

photosynthesis Process in green plants, algae, and cyanobacteria whereby the Sun's energy is used to build energy-containing food molecules from carbon dioxide and water. See also *chemosynthesis, chlorophyll*.

phylum The highest-level grouping in the classification of the animal kingdom. Each phylum has a unique basic body plan. Molluscs, arthropods, and echinoderms are examples.

phytoplankton Planktonic organisms, for example, microscopic algae and cyanobacteria, which produce their own food by photosynthesis.

picoplankton The smallest planktonic organisms, typically bacteria, of 0.0002–0.002mm in diameter. See also *nanoplankton*.

plankton Marine or freshwater organisms, living in open water, that cannot swim strongly and so drift with the currents. Although small life forms dominate, larger creatures, such as jellyfish, are also planktonic. See also *nanoplankton, nekton, phytoplankton, zooplankton*.

plate boundary A border between two tectonic plates. The plates may be converging (**destructive boundary**), diverging (**constructive boundary**), or sliding past (**conservative or strike-slip boundary**). See also *transform fault*.

plate, tectonic see *tectonic plate*.

plate tectonics Phenomena linked to the relative movement of the Earth's tectonic plates, including continental drift, sea-floor spreading, earthquakes, and mountain-building; also, the theory explaining these occurrences.

polychaetes A large subgroup of segmented worms common in the oceans, often with bristles down the sides of the body. (Polychaete means "many bristles"). Some species can move around, while others anchor themselves within tubes or burrows and filter-feed. See also *segmented worm, tube worm*.

polynya An area of open water in an otherwise ice-covered sea, especially in the Arctic.

polyp One of the two main body-forms of cnidarians. An anemone or coral animal is a polyp. Polyps are typically tubular and attached to a surface at their base. See also *cnidarians, medusa*.

prevailing wind A wind that tends to blow from a particular direction. See *trade winds, westerlies*.

primary coast A coast whose features have not been significantly altered by marine erosion, the activity of animals such as corals, or human intervention. See also *secondary coast*.

primary producer Often called simply a producer, an organism that makes food, using energy either from the Sun or from naturally occurring inorganic chemicals. See also *autotroph, chemosynthesis, photosynthesis*.

productivity Rate at which living material is produced by organisms by growth and reproduction. See also *primary producer*.

prokaryotes Organisms such as bacteria and archaea, whose cells are smaller and simpler in structure than the cells of animals, plants, and protists. Cells of prokaryotes have no nucleus. See also *archaea, bacteria*.

protein A large molecule built by organisms from smaller molecules called amino acids. Proteins range from the enzymes that promote chemical reactions in body cells, to structural materials such as keratin – the tough protein that makes up hair, horn, and nails.

protists A wide grouping of often unrelated, microscopic organisms, traditionally classified as a single kingdom. It includes mostly single-celled forms, either animal-like (formerly called protozoa) or plant-like (many of which are termed algae). Some experts also include larger algae (seaweeds). Protist cells contain nuclei, like the cells of animals and plants, but unlike those of bacteria.

pteropods Swimming, planktonic gastropod molluscs, also called **sea butterflies**. The crawling foot of their snail-like ancestors has evolved into muscular "wings" that propel them along. See also *gastropod, plankton*.

pycnocline A boundary region in ocean waters within which density changes rapidly. It typically results from a combination of temperature and salinity levels, both of which affect density. See also *thermocline*.

R

radiation The emission of high-energy particles or waves. **Electromagnetic radiation** consists of electromagnetic waves: listed from long-wave to short-wave forms, these are radio waves, microwaves, infra-red (heat) rays, visible light, ultraviolet light, X-rays, and gamma rays. Short-wavelength electromagnetic radiation has the highest energy.

radiolarians Single-celled predatory organisms mainly living as plankton, often with a delicate, perforated, spherical skeleton. Radiolarian remains of are an important part of some oceanic sediments.

reclamation The artificial conversion of a former coastal sea or wetland area into dry land.

reef see *coral reef*.

refraction The change of direction of a wave when it passes into a different medium, for example, light waves passing from air into water. Ocean waves are also refracted when they reach shallow water.

respiration (1) Breathing. (2) Also called **cellular respiration**, the biochemical processes within cells that break down food molecules, usually by combining them with oxygen, to provide energy for an organism. See also *anaerobic*.

revetment A sloping structure of spaced wooden or concrete beams, constructed to protect a beach or low cliff against erosion.

ria A winding inlet of the sea, a drowned former river valley. Most present-day rias were created when sea levels rose at the end of the last ice age. Unlike a fiord, a ria was never occupied by a glacier.

ribbon worms A major group (phylum) of narrow-bodied, unsegmented marine worms, also called proboscis worms, some of which can reach 50m (160ft) in length.

rip current A current flowing away from a shoreline, carrying water that has been pushed shoreward by waves. See also *tide rip*.

rip-rap Boulders piled deliberately on a shoreline to prevent erosion.

S

salinity Degree of saltiness.

salps Barrel-shaped, delicate-bodied tunicates that live as filter-feeders in the plankton. See also *tunicates*.

salt marsh An ecosystem developing on sheltered, flat, muddy coastlines, where tidal flats are colonized by salt-tolerant land plants. See also *tidal flat*.

sand dune see *dune*.

scute Any of the horny plates that form the outer covering of the shells of turtles; also used to described a similar protective structure on some fish and other animals.

sea arch A natural arch on a rocky shoreline, usually created by two sea caves on either side of a headland eroding into each other.

sea butterflies see *pteropods*.

sea cave A cave created at the foot of a cliff by wave action.

sea cucumbers see *holothurians*.

sea fans Fan-shaped corals belonging to the gorgonian or horny coral group. Though often growing on coral reefs, they are not reef formers themselves. See also *coral*.

sea pens A group of soft-bodied, colonial cnidarians. Each colony resembles a single individual, with one large, burrowing polyp anchoring the colony in sea-floor mud, and smaller polyps feeding and reproducing. See also *cnidarians, polyps*.

sea slugs Shell-less marine gastropods, often with bright colours and tufty gills (**ctenidia**) on their backs. Sea slugs are carnivores and are not closely related to land slugs. Also called **nudibranchs**. See also *gastropods*.

sea stack An isolated pillar of rock left standing offshore on a rocky coastline after all the surrounding land has been eroded away.

sea urchins A group of echinoderms, usually with a rigid case called a **test**, a globular body, long spines, and a downward-facing mouth. Most graze algae from hard surfaces, though the heart urchins and sand dollars are burrowers. See also *echinoderms*.

sea-floor spreading The creation of new oceanic crust by the upwelling of magma at mid-ocean ridges and consequent spreading of the sea floor on either side. See also *plate tectonics*.

seagrasses Any of various plants able to grow and root in shallow, sandy sea bed along coastlines, especially in warmer seas. Although not actually grasses, they are true flowering plants, unlike seaweeds, which are algae.

sea-ice Ice that forms on the surface of the sea, as distinct from ice shelves and icebergs, which originate on land. Some sea-ice forms only in winter, while other sea-ice is semi-permanent. Sea-ice forms and evolves in several stages. See *frazil ice, grease ice, pack ice, pancake ice*.

seamount A submarine mountain, usually an extinct volcano.

sea spiders A group of eight-legged predatory marine arthropods. It is not agreed whether sea spiders are closely related to land spiders or not.

sea-squirts see *tunicates*.

seaweed A member of any of three main groups of large-bodied algae. Seaweeds can make their own food by photosynthesis, but they lack roots. Their classification is not agreed, but green seaweeds seem to be related to plants, while red and brown seaweeds may represent two unrelated lines of evolution. See also *algae*.

secondary coast A coast with features significantly altered by marine erosion, the activity of animals such as corals, human intervention, or all three. See also *primary coast*.

sedentary Of animals such as worms: habitually staying in one position. See also *sessile*.

sediment An accumulation of solid particles that have settled out from water; also used for deposits left by other agencies such as the wind.

sedimentary rock Any rock originating from sediment that has later become compacted and hardened, such as sandstone.

segmented worms A major group (phylum) of worms, also called **annelids**, whose body is built from repeating units (segments) each bearing copies of organs, such as kidneys. The phylum includes earthworms, plus many marine species, mostly within a subgroup called the polychaetes. See also *phylum, polychaetes, worm*.

sessile Of an animal: attached permanently to a surface, especially without a stalk, and not able to move around. See also *sedentary*.

sexual dimorphism Situation in which the males and females of a species differ in appearance, for example, in colour, shape, or size.

shrimp Any of various small, usually swimming crustaceans. True shrimps are relatives of crabs and lobsters.

siphon In molluscs: a fleshy tubular extension of the body that aids the flow of oxygenated seawater to the gills or sometimes transports food particles for filtering. Cephalopods use their siphons for jet propulsion. See also *cephalopods*.

siphonophores Floating, predatory, colonial cnidarians, such as the Portuguese Man-of-War. The colony members have specialized functions but act together so that the colony functions like a single animal. See also *cnidarians, colonial, polyp, zooid*.

sonar A method of echo-sounding; often used more broadly as a synonym for echolocation. See also *echolocation, echo-sounding*.

Southern Oscillation see *ENSO*.

spit A peninsula of sand or shingle or both created by longshore drift, usually at a point where the shoreline changes direction. See also *bar, barrier island, longshore drift, tombolo*.

sponges A large group (phylum) of marine animals with a very simple structure that feed by creating currents through their bodies and filtering small particles from the water. They have no muscles or nerve cells, and sometimes no symmetry.

spore (1) A tiny structure produced (usually in large quantities) by non-flowering plants, fungi, and some protists, from which a new individual can grow. Spores are much smaller than seeds and usually produced asexually, sometimes forming part of a complex life history. (2) The inactive, resistant form of some bacteria that helps them survive unfavourable conditions. See also *asexual reproduction*.

spreading ridge see *mid-ocean ridge*.

spring tide The highest high tide and lowest low tide within an approximately two-week cycle, caused by the Sun and the Moon being in positions in which their gravitational effects add together most strongly. See also *tides, neap tide*.

squid see *cephalopod*.

stack see *sea stack*.

standing wave A wave that stays in the same position rather than moving along, found in particular situations such as tidal races.

starfish A group of echinoderms, also called **sea stars**, having five or more "arms" (extensions to the body) and both mouth and anus on the underside. They swallow their prey whole, which can be very large for their size. See also *echinoderms*.

storm beach The topmost ridge of sediment on a beach, usually formed by the highest spring tides in combination with storm conditions. See also *berm, spring tide*.

storm surge A rapid rise in sea level caused by storm winds driving water towards a shoreline. It can cause disastrous coastal flooding, especially if occurring at the same time as a high spring tide.

subantarctic Relating to latitudes immediately north of the Antarctic Circle.

subarctic Relating to latitudes immediately south of the Arctic Circle.

subduction The forcing down of oceanic crust belonging to one tectonic plate beneath another plate when two plates are colliding. Ocean trenches are the location of such **subduction zones**.

sublittoral Relating to the coastal marine environment below the low-water mark.

submersible A vessel built to operate underwater. Some submersibles are designed to be able to withstand great pressures in order to explore the ocean depths.

sunlit zone The topmost layer of ocean water, where enough light penetrates for photosynthesis to occur. Also called the **photic zone**, it extends from the surface to up to 200m (660ft). See also *dark zone, twilight zone*.

surf zone The zone on a shore where waves break and create foaming, turbulent water.

surface current Any current flowing at the surface of the ocean, for example, the Gulf Stream. Surface currents are mainly caused by friction from prevailing winds. See also *current, thermohaline circulation*.

surface tension The attraction between water molecules at a water surface, which creates a thin film with the strength to resist small deflections, allowing some insects, for example, to walk on the water surface.

suspension feeding see *filter feeding*.

swash The movement of turbulent water up a shore after a wave breaks. The **swash zone** is the zone of a shore where swash typically occurs.

swell waves Regular, smoothly travelling waves on the open ocean, especially when at a distance from the winds or storms that originally caused them. See also *fetch*.

swim bladder A gas-filled organ in many fish, used to control buoyancy, and sometimes for other purposes such as sound production.

symbiosis A close living relationship between two species, especially one in which both benefit. See also *mutualism, commensalism*.

T

tablemount see *guyot*.

tabular Of an iceberg: very wide and flat-topped.

tectonic plate Any of the large rigid sections into which the Earth's crust and uppermost mantle are divided, whose relative movement is the subject of plate tectonics. The African Plate and the Pacific Plate are examples. See *plate tectonics*.

terrigenous Of marine sediments: originating on the land (for example, carried to the sea by rivers).

thermocline A region at a particular depth in the ocean or height in the air where average temperature changes rapidly. See also *pycnocline*.

thermohaline circulation The part of the ocean's water circulation powered by differences in the salinity and temperature of different water masses, rather than by the wind. Thermohaline circulation is the cause of most deep-water and some surface currents. See also *surface current*.

tidal bore A single large wave created when an incoming tide moves up a narrowing channel, such as an estuary.

tidal bulge or trough see *tides*.

tidal current see *tides*.

tidal flat A flat, muddy area covered at high tide; characteristic of sheltered areas such as estuaries.

tidal race A strong current created when a tide-generated water flow moves through a narrow channel.

tide rip A stretch of turbulent water that results from different tidal currents meeting.

tides Fluctuation in sea level resulting from the gravitational attraction of the Sun and the Moon on the Earth's oceans, combined with the Earth's own rotation. In the open oceans, each tidal cycle of just over 12 hours generates a small but measurable vertical rise (**tidal bulge**) and fall (**tidal trough**) in the water. Tidal effects are much more obvious near the coast, and lead to horizontal water movements (**tidal currents**) as well as vertical movements.

tombolo A spit linking an island to the mainland or another island. See *spit*.

trade winds Prevailing winds blowing from the east towards the Equator in subtropical and tropical latitudes.

transform fault A fault in which the rocks on either side are displaced horizontally. Numerous transform faults occur at right angles to mid-ocean ridges. See also *plate tectonics*.

trench see *ocean trench*.

tropical Relating to the warm regions of the Earth that lie between the Equator and the tropics of Cancer and Capricorn, at latitudes of 23.5° north and south, respectively. The term is sometimes used loosely for phenomena typical of these regions, even when occurring north or south of the two tropics.

tropical cyclone A large-scale, circulating weather system in warmer latitudes, called by different names, such as hurricane and typhoon, in different parts of the world. It generates intense winds and torrential rain. Its energy comes from the water vapour rising from warm seas and then condensing. A less powerful version of the phenomenon is called a **tropical storm**. See also *cyclone, hurricane, latent heat, typhoon*.

tsunami A sometimes huge water wave usually generated by displacement of water by an earthquake and capable of devastating shorelines thousands of miles from its origin. Sometimes inaccurately called a "tidal wave".

tube worms Worms that live anchored and protected in tubes, which are either secreted or built of material such as sand grains. Tube worms include the giant worms living around some hydrothermal vents, as well as many segmented worms. See also *polychaetes*.

tunicates A group of mainly filter-feeding marine invertebrates closely related to backboned animals (vertebrates). There are both solitary and colonial species. They include non-moving attached forms (**sea-squirts**) and others that drift in the plankton. See also *salps*.

turbidity current A phenomenon similar to an underwater avalanche or landslide, involving water laden with sediments slipping down a slope.

twilight zone The vertical zone of the water column and sea bed lying between approximately 200 and 1,000m (660ft and 3,300ft) deep, into which some light penetrates, but not enough to support photosynthesis.

typhoon see *tropical cyclone*.

U

upwelling The upward motion of deep-ocean water towards the surface. Some upwelling increases ocean fertility by recirculating nutrients from deeper layers.

V

ventral Relating to the lower surface or belly of an animal. See also *dorsal*.

vertical migration Behaviour of many zooplankton, fish, and squid of the open ocean, in which they rise nearer the surface by night and sink deeper by day, probably to escape predators.

vertical transport Any large-scale vertical flow of ocean water.

vortex A fast-rotating eddy in a fluid; sometimes used as a synonym for whirlpool. See also *eddy, whirlpool*.

W

water column The volume of water between the ocean surface and the bottom of the ocean.

wave A motion or disturbance that transfers energy. The water in a wave crossing the open ocean does not move significantly except up and down as the wave passes. The high point of a wave is its **crest** and the low point its **trough**. Water motion becomes more complex and turbulent in waves breaking on shores (**breakers**).

westerlies Prevailing winds that blow from the west. Westerlies are the most common winds in temperate regions.

western boundary current A relatively narrow, fast-moving surface current formed at the western boundary of an ocean basin, usually as part of a gyre. The Gulf Stream is an example. Deep-water western boundary currents also exist. See also *gyre*.

whirlpool A powerful eddy or vortex formed at the sea's surface, often caused when two separate tidal currents meet. See also *eddy, vortex*.

white smoker A deep-ocean hydrothermal vent in which the emerging hot water appears white because of light-coloured mineral particles suspended in it.

worm Any of a variety of usually non-swimming invertebrate animals that are long, slender and flexible, and lack legs and shells. See *flatworms, ribbon worms, segmented worms, tube worms*.

Z

zoea The planktonic larval stage of certain crustaceans, including crabs. They are different in structure from their adult forms, having long spines.

zooid An individual in a colony of interconnected animals, such as bryozoans. The term is not applied to colonial coral animals, which are termed polyps. See also *polyp*.

zooplankton Any animals or animal-like protists that are part of the plankton. See also *plankton*.

zooxanthellae Symbiotic, microscopic algae living in the tissues of many corals. See also *symbiosis*.

INDEX

Page numbers in **bold** indicate feature profiles or extended treatments of a topic. Page numbers in *italic* indicate pages on which the topic is illustrated. Features shown on the maps in the Atlas of the Oceans are listed in a separate index on pp.505–509.

ATLAS INDEX

This index lists features shown in the Atlas of the Oceans (pp.426–85). It includes the names of features on the sea floor and sea surface, such as the names of seas. It also includes physical features on land, such as islands and headlands.

ACKNOWLEDGMENTS

Dorling Kindersley would like to thank several people for their help in the preparation of this book. At the American Museum of Natural History, Udayan Chattopadhyay was unfailingly helpful and John Sparks provided many valuable comments on the text and images. Georgina Garner and Erin Richards worked on early versions of the contents list. Frances Dipper and Robert Dinwiddie drew up the original lists of species and physical features described in the catalogue sections. Additional design work was done by Janis Utton and Pankaj Sharma. Tamlyn Calitz and Amy Walters provided editorial assistance, and Klara Kayser contributed design assistance. Neil Fletcher did additional picture research for the Birds section.

PICTURE CREDITS

Dorling Kindersley would like to thank the following for their help in supplying images: Romaine Werblow in the DK Picture Library; All at SeaPics.com; All at FLPA; Jonathan Hamston at OSF; Teresa Riley at Getty Images; All at Alamy Images.

KEY:
(a-above; b-below/bottom; c-centre; f-far; l-left; r-right; t-top)

SIDEBAR IMAGES
Corbis: David Keaton (*Atlas of the Oceans*); Jeffrey L. Rotman (*Ocean Environments*). **Getty Images:** National Geographic/Raul Touzon (*Ocean Life*); Photonica/Anna Grossman (*Introduction*).

1 Getty Images: Taxi/Peter Scoones. **2—3 Getty Images:** Stone/Warren Bolster. **4 Corbis:** (tc); Lawson Wood (bc). **4–5 Getty Images:** National Geographic/Brian Skerry. **5 Getty Images:** Image Bank/Mike Kelly (tc). **NASA:** Jacques Descloitres, MODIS Rapid Response Team, NASA/GSFC (cra). **6–7 FLPA:** Minden Pictures/Norbert Wu (*Background*). **8–9 FLPA:** Minden Pictures/Chris Newbert, **Carrie Vonderhaar. 10–11 Still Pictures:** Kim Heacox. **12–13 Oceanwide Images:** Gary Bell. **14 SeaPics.com:** Kevin Schafer. **15 DeepSeaPhotography.Com:** Kim Westerskov. **16–17 David Hall (www.seaphotos.com). 18–19 Marine Wildlife:** Paul Kay. **20 Getty Images:** Iconica/John W. Banagan. **21 FLPA:** Minden Pictures/Frans Lanting. **22–23 Angel M. Fitor (www.seaframes.com). 24 Getty Images:** Image Bank/Jeff Hunter. **25 FLPA:** Minden Pictures/ZSSD. **26–27 naturepl.com:** Mark Carwardine. **28–29 Corbis. 30–31 Getty Images:** Taxi/Jason Childs. **32 Alamy Images:** Pictor International/ImageState (bc). **DK Images:** Frank Greenaway (bl). **32–33 Alamy Images:** Hawkeye (c). **33 Alamy Images:** Bryan & Cherry Alexander Photography (br). **DK Images:** (fbl); Brian Cosgrove (bc); Zena Holloway (bl). **NASA:** GSFC/MODIS Rapid Response Team, Jacques Descloitres (tl); Liam Gumley, MODIS Atmosphere Team, University of Wisconsin-Madison Cooperative Institute for Meteorological Satellite Studies (ca). **34 Alamy Images:** David Wall (br). **Science & Society Picture Library:** Science Museum, London (bl). **35 Alamy Images:** PHOTOTAKE Inc./Carolina Biological Supply Company (bc); Stephen Frink Collection/James D. Watt (ca); Visual&Written SL/Kike Calvo (cra). **NASA:** Provided by the SeaWiFS Project, Goddard Space Flight Center, and ORBIMAGE (br). **36 NASA:** MODIS Instrument Team, NASA Goddard Space Flight Center, (c); The U.S.-French TOPEX/Poseidon mission is managed by JPL for NASA's Earth Science Enterprise, Washington, D.C. JPL is a division of the California Institute of Technology in Pasadena (tr). **37 Alamy Images:** Roger Cracknell (cl); Chris A Crumley (tr). **DK Images:** Frank Greenaway/Courtesy of the University Marine Biological Station, Millport, Scotland (crb). **SeaPics.com:** Bob Cranston (ca). **38 Alamy Images:** Brandon Cole Marine Photography (cla); Reinhard Dirscherl (cb). **Dive Gallery/Jeffrey Jeffords (www.divegallery.com):** (bc). **Image Quest Marine:** Y. Kito (bl). **38–39 Alamy Images:** Visual&Written SL/Takaji Ochi (c). **39 AguaSonic Acoustics:** Mark Fisher (cl). **41** [illegible] Cunningham Photographic (cra); James Davis Photography (cr); Dinodia Images/Ashvin Mehta (tr). **Science Photo Library:** (cb). **40–41 Corbis:** Brenda Tharp. **43 Japan Agency for Marine-Earth Science and Technology (JAMSTEC):** (cra). **44 Alamy Images:** Danita Delimont (c). **Corbis:** Raymond Gehman (bc). **DK Images:** Harry Taylor (cra). **45 NASA:** JPL (br). **Science Photo Library:** Bill Bachman (tr). **47 Corbis:** Bettmann (tc); David Lawrence (cra). **48 Alamy Images:** Norman Price (tr). **Planetary Visions** (bl). **48–49 Alamy Images:** Nordicphotos/Sigurgeir Sigurjonsson (b). **49 Planetary Visions** (tr). **50 DK Images:** Colin Keates/Courtesy of the Natural History Museum, London (tr). **51 Alamy Images:** Douglas Peebles Photography (br). **Woods Hole Oceanographic Instittution:** Jayne Doucette (cra). **52–53 Corbis:** Image by Digital image © 1996 CORBIS; Original image courtesy of NASA. **54 NASA:** JPL (fcr); Image courtesy Quickscat Science Team (cr). **55 Alamy Images:** Kos Picture Source (t). **DK Images:** Peter Wilson (bl). **56 Action Images:** Reuters/Carlo Borlenghi. **57 Alamy Images:** Kos Picture Source (br). **Corbis:** Emmanuelle Thiercelin (cr). **Getty Images:** AFP/Marcel Mochet (tr); Clive Mason (cra). **Rex Features:** RAAF-AUSTRAL/Corbis Sygma (crb). **58–59 NASA:** Image courtesy the SeaWiFS Project, NASA/Goddard Space Flight Center, and ORBIMAGE (c). **59 Alamy Images:** Chris Linder (br). **Corbis:** Bettmann (tr). **NASA:** (cr); Image processed by Robert Simmon based on data from the SeaWiFS project and the Goddard DAAC (c). **60 SeaPics.com:** Doug Perrine (bl). **61 S.M.R.U:** Simon Moss (bl). Courtesy of **Andreas M. Thurnherr:** (br). **62 Getty Images:** Nordic Photos/Kristjan Fridriksson. **63 Alamy Images:** Bryan & Cherry Alexander Photography (tr); Apex News and Pictures Agency/Tim Cuff (br). **Corbis:** Lowell Georgia (crb). **FLPA:** Minden Pictures/Flip Nicklen (c). **64 Alamy Images:** Danita Delimont (bl). **65 Corbis:** Bettmann (cla); Sygma/Gyori Antoine (bl). **Getty Images:** Photographer's Choice/Kerrick James (tr). **NASA:** Image courtesy Jacques Descloitres, MODIS Land Rapid Response Team at NASA GSFC (clb). **NOAA:** Michael Van Woert, NOAA NESDIS, ORA (br). **66 Alamy Images:** Aflo Foto Agency (tr); Boating Images Photo Library/Keith Pritchard (cra); Michael J. Kronmal (br); Tribaleye Images/J Marshall (bl). **Corbis:** Image by Digital image © 1996 CORBIS; Original image courtesy of NASA (clb). **67 Alamy Images:** Mark Lewis (b); PHOTOTAKE Inc./Dennis Kunkel (tc). **Getty Images:** Photographer's Choice/Malcolm Fife (tr). Courtesy of **US Navy:** Photo courtesy of Ian R. MacDonald, Texas A&M Univ. Corpus Christi (cla). **68 Alamy Images:** Bill Brooks (cb); Images&Stories (br). Courtesy of **Chris Baisan, University of Arizona:** (bl). **NASA:** JPL (cl). **69 Corbis:** Jonathan Blair (ca); EPA/Josue Fernandez (b). **NOAA:** Lieutenant Mark Boland, NOAA Corps (cla). **70 NASA:** Image by Jesse Allen, NASA Earth Observatory; data provided by the MODIS Land Rapid Response Team, NASA GSFC (cl); Jacques Descloitres, MODIS Land Rapid Response Team, NASA/GSFC (cr); Jeff Schmaltz, MODIS Rapid Response Team, NASA/GSFC (c). **NOAA:** Aircraft Operations Center (br). **71 Corbis:** EPA/Alejandro Ernesto (bl). **OSF/photolibrary:** Warren Faidley (t). **SeaPics.com:** Doug Perrine (cr). **Still Pictures:** Michel Gunther (br). **72 Corbis:** Reuters/Vincent Laforet. **73 Corbis:** Dallas Morning News/Irwin Thompson (cr); Dallas Morning News/Smiley N. Pool (br); EPA/Vincent Laforet (cra); Jim Reed Photography/Mike Theiss (tr). **Getty Images:** AFP/Robert Sullivan (crb). **NASA:** Jeff Schmaltz, MODIS Land Rapid Response Team (clb). **74–75 Getty Images:** Lonely Planet Images/Karl Lehmann. **76 Alamy Images:** David Gregs (cr); ImagePix (bc). **iStockphoto.com:** Dan Brandenburg (cra). **NOAA:** Captain Andy Chase (bl). **OSF/photolibrary:** Pacific Stock (crb). **77 Alamy Images:** Michael Diggin (bl). **Getty Images:** Taxi/Helena Vallis (t). **iStockphoto.com:** Paul Topp (br). **78 Alamy Images:** Mooch Images (clb). **78–79 Getty Images:** Robert Harding World Imagery/Lee Frost (b). **80 Alamy Images:** Mary Evans Picture Library (br); Ian Simpson (bl). **Don Dunbar (www.easternmaineimages.com):** (tl) (c). **81 Alamy Images:** Malcolm Fife (cb); Peter L. Hardy (bc). **Still Pictures:** Markus Dlouhy (tr). **www.uwphoto.no:** Erling Svensen (cl). **82 Alamy Images:** Shaughn F. Clements (b); phototramp.com/Maciej Tomczak (cr). **83 Corbis:** Dave Bartruff (l); Christie's Images (br). **84–85 Corbis:** Lawson Wood. **86–87 Getty Images:** Photonica/Photolibrary.com. **88 Corbis:** Yann Arthus-Bertrand (br). **NASA:** (clb). **89 Corbis:** Michael Busselle (bc); Lloyd Cluff (t); Ecoscene/John Wilkinson (clb). **DK Images:** Colin [illegible] Courtesy of the Natural History Museum, London (crb). **90 Rex Features:** Sipa Press (SIPA). **91 Alamy Images:** Louise Murray (br). **Corbis:** Matthieu Paley (c); Sygma/Kapoor Baldev (crb). **Still Pictures:** Bryan Lynas (tc); Mark Lynas (tr) (cr). **92 Alamy Images:** Michael Howell (b). **Corbis:** Yann Arthus-Bertrand (cra); Jack Fields (c); Frans Lanting (tr). **93 Alamy Images:** FLPA (clb); geogphotos (fcla). **Corbis:** Jim Sugar (bc). **iStockphoto.com:** Andrew Dorey (ca); Gregor Erdmann (cla). **94 Alamy Images:** Jack Stephens (cl). **Rob Havemeyer Acadia National Park ME:** (tr). **94–95 Steven Russell (www.pbase.com/nodfather):** (b). **95 Alamy Images:** Eric Nathan (cr). **Corbis:** Kevin Fleming (tc). **DK Images:** Jon Spaull (tr). **96 Alamy Images:** Atmosphere Picture Library/Bob Croxford (tr). **Corbis:** Ric Ergenbright (bl). **DK Images:** Rough Guides/Ian Aitken (crb). **www.undiscoveredscotland.co.uk:** (cla). **97 Alamy Images:** Sean Burke (t); CuboImages srl/Marco Casiraghi (bl). **NASA:** Johnson Space Center - Earth Sciences and Image Analysis (br). **98 Corbis:** Peter Johnson (c); Richard T. Nowitz (br). **Wombat Pitts:** (cra). **99 Alamy Images:** Simon Reddy (cr). **Getty Images:** Robert Harding World Imagery/Neil Emmerson (b). Mark Kitching: (cl). **NASA:** Image courtesy Jacques Descloitres, MODIS Land Rapid Response Team at NASA GSFC (tc). **100–101 Corbis:** Digital image © 1996 CORBIS; Original image courtesy of NASA. **102 Alamy Images:** Danita Delimont (bc). **Corbis:** Bettmann (br). **Still Pictures:** Christoph Papsch (t). **Dr Sandy Tudhope, Institute of Geology and Geophysics, Edinburgh University:** (bl). **103 Alamy Images:** Danita Delimont (b). **Getty Images:** Stone/James Randklev (cl). **Marco Nero:** (cr). **104 WaterLand Neeltje Jans:** RWS MD afd. Multimedia. **105 Alamy Images:** Florida Images (cr); geogphotos (cra); Rodger Tamblyn (tc). **Corbis:** Lowell Georgia (bl). **Natural Visions:** Heather Angel (c). **NOAA:** NOAA Restoration Center, Chris Doley (cb). **Sky Pictures luchtfotografie (www.skypictures.nl):** (br). **106 Alamy Images:** Patrick Mallette (cl). **Getty Images:** Altrendo/altrendo nature (tr); Lonely Planet Images/Bethune Carmichael (clb). **106–107 Corbis:** Martin Harvey (b). **107 Alamy Images:** Guillen Photography (cra); Wildscape (c). **DK Images:** Shaen Adey (tr); James Stevenson (tl). **108 Alamy Images:** Danita Delimont (tc); Peter Lewis (b). Paul Yung: (cr). **109 Alamy Images:** Atmosphere Picture Library/Bob Croxford (bl); imagebroker/Harald Theissen (cla). **Corbis:** Jason Hawkes (r). **DK Images:** Geoff Dann (br). **110 Alamy Images:** Mark Boulton (tl). **Corbis:** Tony Arruza (bl); Yann Arthus-Bertrand (cr). **111 Alamy Images:** Simon Reddy (tc); Laurie Wilson (b). **Corbis:** (cr). **112 Alamy Images:** Ian Dagnall (tr); Danita Delimont (bc). **Corbis:** Douglas Peebles (br). **DK Images:** Lloyd Park (cl). **113 Alamy Images:** Jon Arnold Images (crb). **Corbis:** Neil Rabinowitz (bl). **Getty Images:** National Geographic/Skip Brown (t). **iStockphoto.com:** Judi Ashlock (tr). **114 Corbis:** Post-Houserstock/Dave G. Houser (cl). **114–115 Alamy Images:** Jon Arnold Images/Doug Pearson (c). **115 Alamy Images:** Tim Graham (ca). **Corbis:** (bc). **DK Images:** Dave King (tr). **OSF/photolibrary:** Richard Herrmann (cra). **116 Getty Images:** Stone/Paul Souders (bl); Stone/Tom Bean (cr). **US Geological Survey:** (bc). **116–117 Still Pictures:** Guy Boily (t). **117 Corbis:** Dale C. Spartas (bl). **NASA:** Jacques Descloitres, MODIS Rapid Response Team, NASA/GSFC (br). **118 Corbis:** Reuters/Sergio Moraes (cr). **NASA:** Jacques Descloitres, MODIS Rapid Response Team, NASA/GSFC (b). **Still Pictures:** Jacques Jangoux (t). **119 Alamy Images:** JL Images (tr). **Corbis:** Sygma/Annebicque Bernard (bc). Courtesy of **Clive Griffin (www.pbase.com/clivegriffin):** (clb). **Still Pictures:** Christiane Eisler (cla). **120 Alamy Images:** Jack Sullivan (cr). Pierre-Yves Lagrée, LMM CNRS Université Paris VI: (t). **NASA:** Earth Sciences and Image Analysis Laboratory at Johnson Space Center (b). **121 Alamy Images:** Karsten Wrobel (tr). **Corbis:** (clb); Yann Arthus-Bertrand (cla). **DK Images:** Rob Reichenfeld (crb). **122 Alamy Images:** Eddie Gerald (b). **Corbis:** Carl & Ann Purcell (cr). **NASA:** Provided by the SeaWiFS Project, NASA/Goddard Space Flight Center, and ORBIMAGE (tl). **123 Alamy Images:** Tibor Bognar (cl). **Corbis:** Peter Guttman (br); Galen Rowell (b); Michael S. Yamashita (tr). **124 Alamy Images:** Florida Images (crb); Renee Morris (bc). **FLPA:** Skylight (cl). **125 Corbis:** James L. Amos (bc); Carol Havens (t). **DK Images:** Mike Linley (cr). **Getty Images:** Photographer's Choice/Cameron Davidson (bl). **Natural Visions:** Heather Angel (br). **126 Alamy Images:** Jon Sparks (t). **Corbis:** Rob Howard (bl). **Still Pictures:** Cal Vornberger (cr). **127 Alamy Images:** David Poole (br); J. Schwanke (tr). **Corbis:** Annie Griffiths Belt (cla); Reuters/Darren Staples (crb). **128 Alamy Images:** Mark Boulton (cb); Rod Edwards (ca); Robert Harding Picture Library Ltd (br). **129 Corbis:** Natalie Fobes (b); Steve Kaufman (br). **Nial Moores/Birds Korea (www.birdskorea.org):** (tc); Mr. Jeon Shi-Jin (cla). **130 Alamy Images:** David Hosking (br). **Getty Images:** National Geographic/Tim Laman (tr). **SeaPics.com:** Jeremy Stafford-Deitsch (bc). **130–131 Oceanwide Images:** Bob Halstead (c). **131 Alamy Images:** Danita Delimont (cra); Reinhard Dirscherl (tr). **Corbis:** Michael S. Yamashita (br). SeaPics.com: D.R. Schrichte (ca). **132 DK Images:** Rough Guides/Demetrio Carrasco (br); Peter Wilson (ca). **132–133 SeaPics.com:** Masa Ushioda (t). **133 Alamy Images:** Mireille Vautier (clb). **Corbis:** Stephen Frink (br). **US Geological Survey:** (cra). **134 Alamy Images:** Tim Graham (bc). **Corbis:** (crb); Yann Arthus-Bertrand (ca). **135 Corbis:** Arne Hodalic (cl). **Getty Images:** National Geographic/Timothy Laman (tr). **SeaPics.com:** Jeremy Stafford-Dietsch (bc). **Still Pictures:** Alan Watson (br). **136–137 naturepl.com:** Jurgen Freund. **138–139 naturepl.com:** Aflo. **140 Alamy Images:** Aqua Image (cla). **DK Images:** Frank Greenaway/Courtesy of the Natural History Museum, London (crb). **iStockphoto.com:** Ingvald Kaldhussæter (cra). **Sue Scott:** (bl). **SeaPics.com:** Mark Conlin (c). **141 British Marine Aggregate Producers Association (www.bmapa.org):** (br). **Getty Images:** Image Bank/Astromujoff (t). **OSF/photolibrary:** Michael Brooke (bc). **142 DK Images:** Frank Greenaway/Courtesy of the Weymouth Sea Life Centre (bc); Jerry Young (cr). **Sue Scott:** (cl). **SeaPics.com:** Mark Conlin (bl). **142–143 David Hall (www.seaphotos.com):** (c). **143 Image Quest Marine:** Jim Greenfield (crb). **Marine Wildlife:** Paul Kay (tr). **NOAA:** Dr. James P. McVey, NOAA Sea Grant Program (bc). **Sue Scott:** (c). **144 DK Images:** Frank Greenaway (ca). **Sue Scott:** (cl) (bl) (br). **144–145 Getty Images:** National Geographic/Bill Curtsinger (t). **145 Alamy Images:** Guillen Photography (bl). **Sue Scott:** (crb) (cra). **146 Alamy Images:** Fabrice Bettex (bc); Gavin Parsons (cr). **DK Images:** Tim Ridley (tl). **Marine Wildlife:** Paul Kay (cla). **Sue Scott:** (clb). **147 Corbis:** Ralph A. Clevenger (b). **OSF/photolibrary:** Tobias Bernhard (tc). **Sue Scott:** (cla) (cra). **148 Corbis:** (tl). **Sue Scott:** (cra) (bl). **SeaPics.com:** Doug Perrine (crb). **149 Alamy Images:** Mark Lewis (b); PNR Photography (crb). **Dr. Alberto V. Borges/Chemical Oceanography Unit from the University of Liège, Belgium:** (tr). **150 Alamy Images:** Ross Armstrong (tr); Joel Day (clb); Andre Seale (cla). **Sue Scott:** (bc). **US Fish and Wildlife Service National Image Library:** Chris Dau (br). **151 SeaPics.com:** Phillip Colla. **152 Alamy Images:** Danita Delimont (c); Nick Hanna (cl). **Corbis:** Yann Arthus-Bertrand (cr). **152–153 OSF/photolibrary:** Pacific Stock (c). **153 Dive Gallery/Jeffrey Jeffords (www.divegallery.com):** (tc). **JM Roberts, Scottish Association for Marine Science:** (crb). **SeaPics.com:** Clay Bryce (br); James D. Watt (cra). **154 Alamy Images:** Michael Patrick O'Neill (bl); Sylvia Cordaiy Photo Library Ltd (tc). **DK Images:** Jerry Young (clb). **155 Alamy Images:** Stephen Frink Collection (cra); Karen & Ian Stewart (c). **Dive Gallery/Jeffrey Jeffords (www.divegallery.com):** (br). **SeaPics.com:** Andrew J. Martinez (tl); James D. Watt (bl). **156 Alamy Images:** Stephen Frink Collection (bl). **Corbis:** Stephen Frink (c); Lawson Wood (tc). **157 Corbis:** Bob Krist (b). **SeaPics.com:** Rodger Klein (tr). **158 Alamy Images:** Nick Hanna (tr); Martin Harvey (bl); Zute Lightfoot (br). **159 Alamy Images:** Steve Allen Travel Photography (tr); Slick Shoots (cr). **Corbis:** Cordaiy Photo Library Ltd/John Parker (cla). **SeaPics.com:** Marc Bernardi (b). **160 Alamy Images:** Aqua Image (cla). **SeaPics.com:** James D. Watt (b). **Still Pictures:** Lynn Funkhouser (cra). **161 Alamy Images:** Robert Harding Picture Library Ltd (crb); Andre Seale (clb). **Corbis:** Reuters/Handout (cra). Brian McMorrow: (tr). **SeaPics.com:** James D. Watt (br). **162–163 Oceanwide Images:** Gary Bell. **164 DK Images:** Frank Greenaway (clb). **NASA:** Image and animations provided by the SeaWiFS Project and the NASA GSFC Scientific Visualization Studio (cra). **Sue Scott:** (tl) (bc) (br). **165 Alamy Images:** Jeremy Inglis (br); Andre Seale (bl). **Image Quest Marine:** Scott Tuason (t). **166–167 Getty Images:** Stone/Kim Westerskov. **168 Science Photo Library:** Alexis Rosenfeld (bc). **169 DeepSeaPhotography.Com:** Kim Westerskov (tc). **Image Quest Marine:** Peter Parks (br). **Science Photo Library:** Susumu Nishinaga (cb). **SeaPics.com:** Ingrid Visser (c). **170 OSF/photolibrary:** (cla); Howard Hall (b). **SeaPics.com:** Peter Parks/[illegible] **171** [illegible] Batson (cr) (clb) (bl) (br). **Image Quest Marine:** Peter Herring (tr). **NOAA:** Archival Photography by

Steve Nicklas, NOS, NGS (c). **OSF/photolibrary:** Pacific Stock (tc). **172 Getty Images:** National Geographic/Paul Nicklen. **173 Corbis:** Chris McLaughlin (tr). Courtesy of **Deep Flight Submersibles:** (bc) (br). **ExploreTheAbyss.Com:** Peter Batson (cr). **Japan Agency for Marine-Earth Science and Technology (JAMSTEC):** (cb). **Woods Hole Oceanographic Instititution:** (ca). **174 Science Photo Library:** (tr). **175 DeepSeaPhotography.Com:** Kim Westerskov (br). **NOAA:** Image Courtesy of the Deep Atlantic Stepping Stones Science Party, IFE , URI-IAO, and NOAA (bc); Office of Ocean Exploration (cb). **Science Photo Library:** Dr Ken MacDonald (t). **SeaPics.com:** Mark Conlin (bl). **176 NOAA:** Commander John Bortniak, NOAA Corps (bl); Fisheries Collection (br). **Dr. P. J. Ramsay/African Coelacanth EcoSystem Programme:** (cl). **177 Alamy Images:** Ron Scott (br). **ExploreTheAbyss.Com:** Peter Batson (bc). **NASA:** Image provided by the USGS EROS Data Center Satellite Systems Branch (tr). **NOAA:** OAR/National Undersea Research Program (NURP); University of Connecticut (ca); Ocean Explorer (cb). **178 www.uwphoto.no:** Erling Svensen. **179 ExploreTheAbyss.Com:** Peter Batson (cra). **FLPA:** D. P. Wilson (c). **Jason Hall-Spencer/Marine Conservation Society:** (crb). **JM Roberts, Scottish Association for Marine Science:** (bc); AWI & Ifremer 2003 (tc) (br). **180 Alamy Images:** Travelpix (clb). **NASA:** Jacques Descloitres, MODIS Rapid Response Team, NASA/GSFC (br). **NOAA:** National Geophysical Data Center (cla). **SeaPics.com:** David Wrobel (bl). **181 Alamy Images:** Phototake Inc./Dennis Kunkel (bc). **Image Quest Marine:** Peter Parks (t). **Science Photo Library:** Steve Gschmeissner (cb). **SeaPics.com:** D.R. Schrichte (br). **182 Alamy Images:** Blickwinkel (clb). **ExploreTheAbyss.Com:** Peter Batson (cla). **SeaPics.com:** Doug Perrine (bl). **Craig Smith & Mike Degruy:** (br). **182–183 NOAA:** OAR/National Undersea Research Program (NURP) (c). **183 naturepl.com:** David Shale (bl). **Naval Historical Foundation, Washington, D.C.:** (br). **Science Photo Library:** US Geological Survey (tr). **184 Alamy Images:** Fabrice Bettex (tr). **Corbis:** The Oregonian/Doug Beghtel (bl). **184–185 Corbis:** Yann Arthus-Bertrand (b). **185 Alamy Images:** David Tipling (c). **Corbis:** Cordaiy Photo Library Ltd/John Farmar (ca); Ralph White (tl). **Planetary Visions:** Lamont-Doherty Earth Observatory (br). **186 Planetary Visions. 187 European Space Agency:** Denmann Production (clb). **NASA:** Canadian Space Agency/National Snow and Ice Data Centre (crb); GSFC (tr) (ca) (cra); JPL (c). **Science Photo Library:** David Vaughan (br). **University College London:** (cb). **188** Image courtesy of **Karen L. Von Damm. Image obtained from the DSV Alvin, with funding provided by the U.S. National Science Foundation:** (tr). **Science Photo Library:** Southampton Oceanography Centre/B. Murton (bl). **188–189 Woods Hole Oceanographic Instititution:** (c). **189 ExploreTheAbyss.Com:** Peter Batson (tc) (c) (br). **Richard T. Lutz:** (cr). **NOAA:** Ocean Explorer (bl). **190–191 FLPA:** Minden Pictures/Norbert Wu. **192 Bridgeman Art Library:** Royal Geographical Society, London, UK (tr). **Corbis:** Ecoscene/Graham Neden (cb). **193 Alamy Images:** Rosemary Calvert (b); John Digby (tl). **SeaPics.com:** Franco Banfi (tr). **194 Alamy Images:** Blickwinkel (br); Eric Ghost (fbr); K-Photos (tc). **M.A. Felton:** (bc). **NOAA:** Michael Van Woert, NOAA NESDIS, ORA (crb). **194–195 Getty Images:** Photographer's Choice/Siegfried Layda (c). **195 Alamy Images:** Giles Angel (crb); Nordicphotos/Kristjan Fridriksson (tr). **196 Corbis:** Sygma. **197 Corbis:** Bettmann (ca) (cra); Hulton-Deutsch Collection (tr); Ralph White (crb) (cb) (br). **Henning Pfeifer:** (clb). **Rex Features:** ITV (ITV/TPC) (c). **198 Alamy Images:** Bryan & Cherry Alexander Photography (fbl) (br). **NOAA:** Michael Van Woert, NOAA NESDIS, ORA (bl). Courtesy of **Don Perovich:** (fbr). **198–199 Bryan and Cherry Alexander Photography:** (t). **199 Corbis:** Bettmann (cra). **DK Images:** Harry Taylor (crb). **NOAA:** Michael Van Woert, NOAA NESDIS, ORA (bc) (br). **SeaPics.com:** John KB Ford/Ursus (c). **200 NASA:** Jacques Descloitres, MODIS Land Rapid Response Team, NASA/GSFC (cl). **200–201 Alamy Images:** Brandon Cole Marine Photography (b). **201 Alamy Images:** Popperfoto (br). **Corbis:** Paul A. Souders (tc). **SeaPics.com:** Bryan & Cherry Alexander (c); iq3-d/Peter Parks (cra). **202–203 Getty Images:** Image Bank/Mike Kelly. **204–205 DeepSeaPhotography.Com:** Kim Westerskov. **206 Alamy Images:** Bruce Coleman/Tom Brakefield (cl/*Kingdom*); Norma Jospeh (c);Visual&Written SL/Kike Calvo (cl/*Genus*). **Corbis:** Brandon D. Cole (b/*Hagfish*). **DK Images:** (cl/*Domain*); Martin Camm (cl/*Phyllum*) (cl/*Order*) (cl/*Family*) (cl/*Species*); Geoff Dann (b/*Lamprey*) (b/*Ray-Finned Fish*); Frank Greenaway (b/*Cartilaginous Fish*); David Peart (cl/*Class*). **SeaPics.com:** Mark V. Erdmann (b/*Lobe-Finned Fish*). **207 DK Images:** (*Fungi*); Neil Fletcher (*Plants*); Dave King (*Red Seaweeds*); Jane Miller (*Animals*); Karl Shone (*Brown Seaweeds*). **SeaPics.com:** iq3-d/Peter Parks (*Protists*). **208 DK Images:** (*Echinoderms*); Frank Greenaway (*Molluscs*); Dave King (*Arthropods*). **209 DK Images:** Jerry Young (*Chordates*). **210 FLPA:** Minden Pictures/Chris Newbert. **211 Conservation International:** Robert Thacker (ca); Jeffrey T. Williams/Smithsonian Institution (c). **FLPA:** Minden Pictures/Norbert Wu (crb). **Dr J. Frederick Grassle, Rutgers University:** (br). **Sue Scott:** (tr) (cra). **SeaPics.com:** Phillip Colla (cb). **212 Still Pictures:** Steven Kazlowski (cr). **213 iStockphoto.com:** Dan Schmitt (cla). **SeaPics.com:** Doug Perrine (b). **Still Pictures:** Bob Evans (tc). **214 DeepSea-Photography.Com:** Kim Westerskov (clb). **FLPA:** D. P. Wilson (cla). **OSF/photolibrary:** Mark Jones (br). **Sue Scott:** (cra) (cr) (bc). **215 M. Boyer/edge-of-reef.com:** (bc). **OSF/photolibrary:** Richard Herrmann (t). **SeaPics.com:** Masa Ushioda (br). **216 iStockphoto.com:** Dan Schmitt (bl). **Sue Scott:** (tr) (cl) (ca). **216–217 SeaPics.com:** Espen Rekdal (b). **217 Dive Gallery/Jeffrey Jeffords (www.divegallery.com):** (br). **Marine Wildlife:** Paul Kay (ca) (cr). **Sue Scott:** (tr). **218 FLPA:** Linda Lewis (cra). **Still Pictures:** Secret Sea Visions (br); Gunter Ziesler (bl). **218–219 FLPA:** Minden Pictures/Norbert Wu (c). **219 Alamy Images:** Danita Delimont (bl). **220 Alamy Images:** SCPhotos/Tom & Pat Leeson (bl); Bruce Coleman/Patrice Ricard (fbr). **iStockphoto.com:** Steffen Foerster (cl). **SeaPics.com:** Mark Conlin (cra) (clb); Chris Huss (fbl) (br). **221 Getty Images:** National Geographic/Brian J. Skerry (t). **OSF/photolibrary:** Doug Allan (bl). **222 Alamy Images:** Brandon Cole Marine Photography (bc). **FLPA:** Minden Pictures/Norbert Wu (c). **223 ExploreTheAbyss.Com:** Peter Batson (ca) (bl). **Charles G. Messing/Nova Southeastern University, Florida:** (br). **NOAA:** OAR/National Undersea Research Program (NURP); Univ. of Hawaii (cra). **OSF/photolibrary:** Norbert Wu (cla). **224 DK Images:** (bl). **Image Quest Marine:** (c); Y. Kito (br). **224–225 ExploreTheAbyss.Com:** Peter Batson (t). **225 ExploreTheAbyss.Com:** Peter Batson (cl). **OSF/photolibrary:** (br). **SeaPics.com:** iq3d/Peter Parks (cr). **226 DK Images:** Colin Keates (bl/*above*) (bl). **Science Photo Library:** Ria Novosti (cra); Sinclair Stammers (cl). **226–227 FLPA:** Minden Pictures/Fred Bavendam (b). **227 The Academy of Natural Sciences:** Ted Daeschler (cl). **DK Images:** Colin Keates (cla) (tr) (cra); Harry Taylor/Courtesy of the Royal Museum of Scotland, Edinburgh (c). **SeaPics.com:** Doug Perrine (tl). **228 Alamy Images:** Natural Visions/Heather Angel (tc). **Bridgeman Art Library:** Private Collection (cra). **DK Images:** Harry Taylor/Courtesy of the Hunterian Museum (University of Glasgow) (c); Harry Taylor/Courtesy of the Natural History Museum, London (bc). **OSF/photolibrary:** Karen Gowlett-Holmes (br). **Science Photo Library:** David Parker (clb). **229 Alamy Images:** David Fleetham (tc); Stephen Frink Collection/James D. Watt (cra). **DK Images:** (c/*Terrestrial Mammal*) (c/*Jawless Fish*); Bedrock Studios (c/*Armoured Fish*) (c/*Turtle*); Robin Carter (c/*Placodont*); Neil Fletcher (c/*Penguin*); Giuliano Fornari (c/*Ichthyosaurus*) (c/*Plesiosaur*); Jon Hughes (c/*Whale*); Colin Keates (bl) (c/*Cambrian*) (c/*Ediacaran*) (c/*Ammonite*); Harry Taylor/Courtesy of the Natural History Museum, London (c/*Shark*); Harry Taylor/Courtesy of the Royal Museum of Scotland, Edinburgh (c/*Lobe-Finned*). **Getty Images:** Science Faction/G. Brad Lewis (br). **230–231 Getty Images:** Taxi/Gary Bell. **232 MicroScope/Woods Hole:** D. J. Patterson (cb). **NOAA:** OAR/National Undersea Research Program (NURP); Lousiana Univ. Marine Consortium (cr). **Oceanwide Images:** Gary Bell (b). **OSF/photolibrary:** Phototake Inc/Dennis Kunkel (ca). **University of Illinois at Urbana-Champaign:** (cra). **233 DK Images:** M.I. Walker (tc). **Image Quest Marine:** Peter Parks (cl). **Oceanwide Images:** Rudie Kuiter (bc). **Still Pictures:** Tom E. Adams (cr). **Laura K. Sycuro, Fred Hutchinson Cancer Research Center, Seattle:** (bl). **234 Image Quest Marine:** Peter Parks (bc). **NASA:** Jacques Descloitres, MODIS Rapid Response Team, NASA/GSFC (cb). **Still Pictures:** Manfred Kage (t); Darlyne A. Murawski (bl). **235 FLPA:** D. P. Wilson (cla) (b). **Science Photo Library:** Jan Hinsch (cb). **Visuals Unlimited Inc:** Wim van Egmond (cra) (bl). **236 FLPA:** D. P. Wilson (b). **MicroScope/Woods Hole:** Alfred Wegener Institute/Mona Hoppenrath, Alex Kraberg, Tanya Morozova and David Patterson (tr). **237 Karl Embleton/Continuous Plankton Recorder Survey:** (c). **NASA:** LANDSAT (bl). **The Natural History Museum, London:** (cra). **OSF/photolibrary:** Michael Pitts/Survival Anglia (tl). **238 Alamy Images:** Karsten Wrobel (cr). **Algaebase.org:** M.D. Guiry (cl). **Sue Scott:** (crb). **238–239 Sue Scott:** (c). **239 Algaebase.org:** M.D. Guiry (crb). **Sue Scott:** (tr) (c). **240 Algaebase.org:** M.D. Guiry (cr) (br); John Huisman (cl). **Sue Scott:** (bl). **241 Alamy Images:** Bob Gibbons (t). **Algaebase.org:** M.D. Guiry (bl). **Natural Visions:** Heather Angel (br). **Sue Scott:** (cr). **242–243 Corbis:** Ralph A. Clevenger. **244 Algaebase.org:** Rob Anderson (crb); Ignacio Bárbara (cla); O. De Clerck (c); M.D. Guiry (bl). **Sue Scott:** (tr). **245 Algaebase.org:** M.D. Guiry (cra). **Sue Scott:** (cla) (bl) (br). **246 Sue Scott:** (cra). **246–247 Image Quest Marine:** Roger Steene (c). **247 Rob Houston:** (bc). **Sue Scott:** (tc) (cr). **248 Algaebase.org:** Ignacio Bárbara (cra); Coastal Imageworks/Colin Bates (bc); M.D. Guiry (tr). **Sue Scott:** (cl) (cr) (br). **249 Alamy Images:** Olivier Digoit (cra); Sami Sarkis (br); Kevin Schafer (tl). **Algaebase.org:** John Huisman (clb). **SeaPics.com:** Andrew J. Martinez (crb). **250 Natural Visions:** Heather Angel (cra) (bc). **Charles J. OKelly:** (tc). **Science Photo Library:** Alexis Rosenfeld (c). **251 Natural Visions:** Heather Angel (cl). **Jonathan Sleath:** (bl) (crb); Dr David Holyoak (cra). **252 Alamy Images:** Andrew Woodley (cl). **Sue Scott:** (bl) (cr) (br). **SeaPics.com:** Jeremy Stafford-Deitsch (ca). **253 Alamy Images:** Nature Picture Library/Jose B. Ruiz (cl); Wildscape/Jason Smalley (cr). **Sue Scott:** (cla) (b). **254 Alamy Images:** Roger Eritja (tr); Marilyn Shenton (tl). **OSF/photolibrary:** Kathie Atkinson (br). **Sue Scott:** (bl). **US Geological Survey:** Forest & Kim Starr (cl). **255 Corbis:** FLPA/Peter Reynolds (tc). **DK Images:** Richard Watson (bl). **FLPA:** Minden Pictures/Tui De Roy (tl). **OSF/photolibrary:** (br). **256 Getty Images:** Lonely Planet Images/Grant Dixon (r). **MicroScope/Woods Hole:** David Patterson, Linda Amaral Zettler, Mike Peglar and Tom Nerad (fcl). **Natural Visions:** Heather Angel (bc). **OSF/photolibrary:** Phototake Inc. (cl). **257 MicroScope/Woods Hole:** David Patterson & Aimlee Laderman (br). **Einar Timdal/University of Oslo:** (cl) (tr) (cb) (bl). **258 Alamy Images:** Brandon Cole Marine Photography (cr); Robert Fried (cla). **FLPA:** Minden Pictures/Fred Bavendam (bc). **Andy Murch/Elasmodiver.com:** (br). **OSF/photolibrary:** (clb). **SeaPics.com:** Mark Conlin (crb); Doug Perrine (tr). **259 Alamy Images:** Dave and Sigrun Tollerton (cla). **FLPA:** Minden Pictures/Birgitte Wilms (b). **SeaPics.com:** Phillip Colla (tc). **260 Alamy Images:** Andre Seale (cr). **Dr. Frances Dipper:** (fcl) (cl) (br). **Natural Resources Canda:** The Sponge Reef Project (clb). **261 Alamy Images:** Wolfgang Pölzer (bl). **Dr. Frances Dipper:** (t) (ca) (cb). **Keith Hiscock:** (tl). **Prof. Dr. Joachim Reitner/Universität Göttingen:** (br). **262 Alamy Images:** Tribaleye Images/J. Marshall (cra). **OSF/photolibrary:** Pacific Stock/David Fleetham (bl). **SeaPics.com:** Mark Conlin (c); Doug Perrine (cla). **262–263 Getty Images:** National Geographic/Paul Nicklen (b). **263 Dr. Frances Dipper:** (bc). **DK Images:** Frank Greenaway (cla). **OSF/photolibrary:** (cr). **SeaPics.com:** David Wrobel (c). **264 Alamy Images:** Reinhard Dirscherl (crb). **Dr. Frances Dipper:** (cb). **Sue Scott:** (bl). **SeaPics.com:** iq3-d/Chris Parks (t); David Wrobel (ca). **265 Richard L. Lord. 266 Marine Wildlife:** Paul Kay (br). **SeaPics.com:** Jeremy Stafford-Deitsch (cl); Steven Wolper (bc). **Still Pictures:** Kelvin Aitken (ca). **267 Alamy Images:** Andre Seale (clb). **Dr. Frances Dipper:** (br). **NOAA:** Mr. Mohammed Al Momany, Aqaba, Jordan (t). **SeaPics.com:** Doug Perrine (bc). **268 Alamy Images:** Michael Patrick O'Neill (cr) (bl); Wolfgang Pölzer (tc). **Sue Scott:** (c) (crb). **269 Sue Scott:** (cl) (tc) (c) (br). **SeaPics.com:** Franco Banfi (cr). **270 Alamy Images:** Aqua Image (cl). **OSF/photolibrary:** Tobias Bernhard (b). **SeaPics.com:** Masa Ushioda (cra). **Dr. Charlie Veron/Australian Institute of Marine Science:** Photo by Mary Stafford-Smith (c). **271 Alamy Images:** Mark Morgan (cl). **Dr. Frances Dipper:** (bl). **Marine Wildlife:** Paul Kay (crb). **SeaPics.com:** Doug Perrine (tr). **272 Alamy Images:** Nature Picture Library/Jose B. Ruiz (b). **Dr. Frances Dipper:** (ca). **Sue Scott:** (t). **SeaPics.com:** Doug Perrine (cr). **273 Alamy Images:** FLPA (c). **M. Boyer/edge-of-reef.com:** (cr) (bl). **OSF/photolibrary:** Tobias Bernhard (tr). **Sue Scott:** (br). **274 M. Boyer/edge-of-reef.com:** (cl) (cla) (bl). Courtesy of **John J. Holleman:** (crb). **Michael D. Miller:** (tr) (cr). **275 Image Quest Marine:** Peter Parks (cr); Roger Steene (cl). **Marine Wildlife:** Paul Kay (bl). **Kåre Telnes/Seawater.no:** (br). **276 Corbis:** Lawson Wood (tr). **Keith Hiscock:** (cb). **Image Quest Marine:** Roger Steene (ca). **Marine Wildlife:** Paul Kay (crb). **SeaPics.com:** Larry Madrigal (cra). **277 Dive Gallery/Jeffrey Jeffords (www.divegallery.com):** (c). **DK Images:** Steve Gorton (cla). **ExploreTheAbyss.Com:** Peter Batson (br). **Dr. Dieter Fiege:** (tc) (cra). **Image Quest Marine:** Jim Greenfield (bl). **278 DK Images:** Matthew Ward (c) (clb). **SeaPics.com:** Clay Bryce (ca). **278–279 Getty Images:** Image Bank/Mike Severns (c). **279 Alamy Images:** Robert Harding Picture Library Ltd/Sylvain Grandadam (tr). **Dive Gallery/Jeffrey Jeffords (www.divegallery.com):** (cr) (br). **DK Images:** Andreas Von Einsiedel (cra). **OSF/photolibrary:** Karen Gowlett-Holmes (cla) (ca). **SeaPics.com:** Doc White (c). **280 Alamy Images:** Daniel L. Geiger/SNAP (tc); Wildscape/Jason Smalley (tr). **SeaPics.com:** Mark Strickland (b). **281 Alamy Images:** Nature Picture Library/Jose B. Ruiz (ca). **Keith Hiscock:** (tr). **Image Quest Marine:** Peter Parks (cra). Scott Tuason (cr). **SeaPics.com:** John C. Lewis (tl). **282 Marine Wildlife:** Paul Kay (t). **SeaPics.com:** Marilyn & Maris Kazmers (bc); Espen Rekdal (cl). **283 DK Images:** Andreas von Einsiedel (ca). **FLPA:** Minden Pictures/AUSCA/D. Parer & E. Parer-Cook (crb). **Jon Moore/Coastal Assessment Liaison & Monitoring, Pembroke:** (tl). **SeaPics.com:** D. R. Schrichte (b). **284–285 Getty Images:** Taxi/Pete Atkinson. **286 Alamy Images:** Natural Visions/Heather Angel (cl). **DK Images:** Dave King (tc); Frank Greenaway/Courtesy of the Natural History Museum, London (bc). **FLPA:** Minden Pictures/Norbert Wu (cra). **Image Quest Marine:** Roger Steene (crb) (br). **287 Alamy Images:** Liquid-Light Underwater Photography (br). **Oceanwide Images:** Gary Bell (t). **SeaPics.com:** James D. Watt (clb). **288 Alamy Images:** Carol Buchanan (tl). **Dive Gallery/Jeffrey Jeffords (www.divegallery.com):** (c). **Marine Wildlife:** Lucy Kay (cra). **NOAA:** National Estuarine Research Reserve Collection (br). **289 Alamy Images:** Andre Seale (br). **FLPA:** Minden Pictures/Norbert Wu (t). **290 Corbis:** Jeffrey L. Rotman (b). **Image Quest Marine:** Peter Batson (ca). **Oceanwide Images:** Gary Bell (cra). **SeaPics.com:** Doug Perrine (tl); Jeff Rotman (crb). **291 Alamy Images:** f1 online (bl). **Image Quest Marine:** Peter Batson (bc). **Oceanwide Images:** Gary Bell (t). **SeaPics.com:** Marc Chamberlain (br). **292 Alamy Images:** Daniel L. Geiger/SNAP (cla). **DK Images:** Colin Keates/Courtesy of the Natural History Museum, London (cra). **NHPA:** Ken Griffiths (bl). **OSF/photolibrary:** Barrie Watts (clb). **SeaPics.com:** Espen Rekdal (tr). **293 Dive Gallery/Jeffrey Jeffords (www.divegallery.com):** (b). **Image Quest Marine:** Jez Tryner (cla) (ca) (cra). **NOAA:** Jamie Hall (tr). **294 DK Images:** Frank Greenaway/Courtesy of the Natural History Museum, London (cra). **Image Quest Marine:** Roger Steene (tl). **naturepl.com:** Christophe Courteau (cl). **NOAA:** Dr. Bradley Stevens (crb). **Still Pictures:** Fred Bavendam (tr). **295 FLPA:** Minden Pictures/Fred Bavendam (r). **NOAA:** Hopcroft (bl). **SeaPics.com:** Franco Banfi (cl). **296 Photo Biopix.dk:** (bl). **iStockphoto.com:** Ian Campbell (br). **SeaPics.com:** Marli Wakeling (tr). **297 Dive Gallery/Jeffrey Jeffords (www.divegallery.com):** (cr). **Ifremer (www.ifremer.fr):** A. Le Magueresse (clb). **Natural Visions:** Heather Angel (cl). **Still Pictures:** Everson (br). **298 Dive Gallery/Jeffrey Jeffords (www.divegallery.com):** (b). **DK Images:** Frank Greenaway (cr). **OSF/photolibrary:** (cl). **Still Pictures:** Lynda Richardson (tr). **299 DK Images:** Jane Burton (c); Andreas von Einsiedel (bc); Dave King (cra). **Image Quest Marine:** Masa Ushioda (crb). **David Kusner:** (tc). **OSF/photolibrary:** Green Cape Pty Ltd (bl). **300–301 Steve Smithson. 302 DK Images:** Jane Burton (bl). **Marine Wildlife:** Paul Kay (cl) (br). **SeaPics.com:** Masa Ushioda (cr). **303 Alamy Images:** Maximilian Weinzierl (tr). **DK Images:** Kim Taylor & Jane Burton (cr). **Oceanwide Images:** Gary Bell (tl). **SeaPics.com:** David B. Fleetham (b). **304 naturepl.com:** Jurgen Freund. **305 Corbis:** Roger Garwood & Trish Ainslie (tr). **SeaPics.com:** Ralf Kiefner (c) (crb). **306 Laurent Dabouineau/University U.C.O. Bretagne Nord, France:** (cl). **FLPA:** Foto Natura/Jef Meul (ca). **Nature Portfolio (www.natureportfolio.co.uk):** Bob Ford (br). **307 Alamy Images:** SNAP/Daniel L. Geiger (cl). **Karen Gowlett-Holmes:** (crb). **Marine Wildlife:** Lucy Kay (cra). **Sue Scott:** (bl) (bc). **Kåre Telnes/Seawater.no:** (tr). **308 SeaPics.com:** Phillip Colla (cl). **308–309 Getty Images:** Lonely Planet Images/Michael Aw (c). **309 Alamy Images:** David Fleetham (tr). **Corbis:** FLPA/Douglas P. Wilson (tc). **Sue Scott:** (b). **SeaPics.com:** Marli Wakeling (bl). **Still Pictures:** P. Danna (cr). **310 Alamy Images:** David Fleetham (clb). **M. Boyer/edge-of-reef.com:** (tr). **Marine Wildlife:** Paul Kay (crb). **Oceanwide Images:** Gary Bell (br). **Sue Scott:** (cla). **SeaPics.com:** D.R. Schrichte (bl). **311 Dive Gallery/Jeffrey Jeffords (www.divegallery.com):** (t). **OSF/photolibrary:** Tobias Bernhard (bl). **Sue Scott:** (br). **312 M. Boyer/edge-of-reef.com:** (tr) (cl). **Marine Wildlife:** Paul Kay (cr). **SeaPics.com:** David Wrobel (bl). **313 Dr. Frances Dipper:** (bl). **DK Images:** Frank Greenaway (cl); Colin Keates (cr). **Richard Ling:** (crb). **Charles G. Messing/Nova Southeastern University, Florida:** (tr). **314 Alamy Images:** F. Jack Jackson (clb). **Australian Institute of Marine Science:** (t). **Dr. Jacob Dafni:** Dr. A. Diamant (bc). **Image Quest Marine:** Peter Herring (br); Roger Steene (cra). **315 Alamy Images:** Lawrence Stepanowicz (cra). **ExploreTheAbyss.Com:** Peter Batson (br). **OSF/photolibrary:** Stephen Foote (bl). **SeaPics.com:** Andrew J. Martinez (cl). **316 Corbis:** Lawson Wood (clb). **FLPA:** D. P. Wilson (c). **www.uwphoto.no:** Erling Svensen (tl) (cra) (br). **317 ExploreTheAbyss.Com:** Peter Batson (ca). **NOAA:** OAR/National Undersea Research Program (NURP); College of William & Mary (tr). **SeaPics.com:** Scott Leslie (b). **318 FLPA:** Foto Natura/Jan Van Arkel (t). **Peter Funch, University of Aarhus:** (bl). **M. Antonio Todaro, University of Modena e Reggio Emilia:** (br). **319 Alamy Images:** David Fleetham (br). **M. Boyer/edge-of-reef.com:** (c). **ExploreTheAbyss.Com:** Peter Batson (tr). **Image Quest Marine:** Peter Parks (tc). **Lyubomir Klissurov:** (bl). **Still Pictures:** Roland Birke (cla). **320 M. Boyer/edge-of-reef.com:** (c). **ExploreTheAbyss.Com:** Peter Batson (cl). **Sue Scott:** (bc). **SeaPics.com:** Espen Rekdal (br). **321 M. Boyer/edge-of-reef.com:** (cla). **Dr. Frances**

Dipper: (cra). **Photo by Per R. Flood © Bathybiologica.no:** (bc). **Natural Visions:** Heather Angel (br). **Sue Scott:** (ca). **SeaPics.com:** David Wrobel (clb). **322 Corbis:** Brandon D. Cole (crb). **ExploreTheAbyss.Com:** Peter Batson (tr) (bl). **SeaPics.com:** Jonathan Bird (br). **323 Corbis:** Brandon D. Cole (crb). **OSF/photolibrary:** (tr); Zig Leszcynski (cla). **www.uwphoto.no:** Erling Svensen (bl). **324 Alamy Images:** Brandon Cole Marine Photography (cra). **DK Images:** Frank Greenaway (cl); James Stevenson (ca). **324–325 Corbis:** Denis Scott (b). **325 DK Images:** Frank Greenaway (br); Dave King (tc). **SeaPics.com:** Doug Perrine (cl) (bc). **326 Janna Nichols:** (clb). **SeaPics.com:** Doug Perrine (tl) (ca). **www.uwphoto.no:** Erling Svensen (br). **327 Andy Murch/Elasmodiver.com:** (cra). **OSF/photolibrary:** Paul Kay (bl). **Still Pictures:** Kelvin Aitken (cla) (crb). **328 Alamy Images:** Stephen Frink Collection/Marty Snyderman (br). **OSF/photolibrary:** Gerard Soury (bl). **SeaPics.com:** Saul Gonor (cra); Espen Rekdal (cla). **329 Alan Chow:** (bc). **Dr. Frances Dipper:** (br). **DK Images:** Frank Greenaway (bl). **Andy Murch/Elasmodiver.com:** (t). **330 Andy Murch/Elasmodiver.com:** (t) (br) (bc). **naturepl.com:** Bruce Rasner/Jeff Rotman (cl). **331 Alamy Images:** Jeff Rotman (cla). **DK Images:** Harry Taylor/Courtesy of the Natural History Museum, London (cl). **John A. Scarlett:** (tr). **SeaPics.com:** Scott Michael (br); David Shen (clb). **332–333 Steve Bloom Images. 334 OSF/photolibrary:** Pacific Stock (br). **Powder River Photography/Todd Mintz:** (cl). **SeaPics.com:** Richard Herrmann (tl). **335 Alamy Images:** David Fleetham (br); Michael Patrick O'Neill (tl). **DK Images:** Frank Greenaway (bl). **naturepl.com:** Jeff Rotman (cra). **SeaPics.com:** Doug Perrine (crb). **336 Image Quest Marine:** Carlos Villoch (bl). **Andy Murch/Elasmodiver.com:** (br). **SeaPics.com:** Randy Morse (c). **336–337 Marine Wildlife:** Alexander Mustard (t). **337 Alamy Images:** M.Timothy O'Keefe (cra). **SeaPics.com:** Doug Perrine (clb); Tim Rock (br). **338 Alamy Images:** WorldFoto (cra). **DK Images:** Colin Keates/Courtesy of the Natural History Museum, London (clb). **338–339 Alamy Images:** Reinhard Dirscherl (c). **339 Alamy Images:** Mark Boulton (br). **Dive Gallery/Jeffrey Jeffords (www.divegallery.com):** (cb). **OSF/photolibrary:** David Fleetham (tr). **340 Alamy Images:** Reinhard Dirscherl (bl); Stephen Frink Collection (cl); Images&Stories (tr). **DK Images:** Frank Greenaway (cla). **Naoko Kouchi:** (cla/*Background*). **SeaPics.com:** Doug Perrine (br). **341 Alamy Images:** Blickwinkel (br). **FLPA:** Minden Pictures/Fred Bavendam (cl). **OSF/photolibrary:** Dr. F. Ehrenstrom & L. Beyer (ca) (crb). **342 DK Images:** Steve Gorton (clb); Colin Keates/Courtesy of the Natural History Museum, London (tc). **Getty Images:** Taxi/Peter Scoones (cl). **Image Quest Marine:** Masa Ushioda (crb). **SeaPics.com:** Mark V. Erdmann (cr). **Andreas Svensson/Norwegian University of Science and Technology:** (bl). **343 Marine Wildlife:** Alexander Mustard (t). **Robert A. Patzner, University of Salzburg, Austria:** (br). **344 Ardea:** Pat Morris (cb). **Rick J. Coleman:** (bl). **DK Images:** (cla). **Marine Wildlife:** Alexander Mustard (tr). **Dr. Volker Neumann:** (br). **345 DK Images:** (clb). **FLPA:** Minden Pictures/Norbert Wu (crb). **Marine Wildlife:** Alexander Mustard (t). **346 Corbis:** Paul A. Souders (bl). **OSF/photolibrary:** Sue Scott (tl). **SeaPics.com:** Mark Conlin (c); Jeff Jaskolski (br). **347 Alamy Images:** FLPA/S. Jonasson (bl); Andre Seale (tl). **SeaPics.com:** Shedd Aquar/Ceisel (cr). **348 Alamy Images:** Wolfgang Pölzer (b). **Getty Images:** National Geographic/Paul Nicklen (t). **349 FLPA:** Minden Pictures/Norbert Wu (cla). **Getty Images:** National Geographic/Wolcott Henry (bc). **Image Quest Marine:** Peter Herring (tr). **OSF/photolibrary:** Paulo De Oliveira (bl). **350 DK Images:** Frank Greenaway (cr). **Keith Hiscock:** (cra). **OSF/photolibrary:** Doug Allan (b). **SeaPics.com:** Hideyuki Utsunomiya (ca). **351 Peter Ajtai:** (clb). **SeaPics.com:** Marilyn & Maris Kazmers (br). **www.uwphoto.no:** Erling Svensen (cla) (tc). **352 Image Quest Marine:** Peter Herring (t); Justin Marshall (br). **SeaPics.com:** James D. Watt (clb). **353 FLPA:** Minden Pictures/Norbert Wu (cr). **marinethemes.com:** Kelvin Aitken (tc). **Natural Visions:** Peter David (clb). **OSF/photolibrary:** Neil Bromhall (br); Rodger Jackman (ca). **354 Dr. Frances Dipper:** (clb). **FLPA:** D. P. Wilson (cla). **Oceanwide Images:** Gary Bell (bc); Rudie Kuiter (crb). **OSF/photolibrary:** Richard Herrmann (tr). **355 DK Images:** Dave King (br). **FLPA:** Minden Pictures/Norbert Wu (l). **New Zealand Seafood Industry Council Ltd:** (cr). **356 Magnum Photos:** Harry Gruyaert . **357 Alamy Images:** Charles Bowman (bl); Jeff Rotman (crb). **FLPA:** Minden Pictures/Norbert Wu (ca). **naturepl.com:** Michael Pitts (cr). **OSF/photolibrary:** Sue Scott (cb). **Sue Scott:** (cra). **SeaPics.com:** Richard Herrmann (br). **358 Dive Gallery/Jeffrey Jeffords (www.divegallery.com):** (br). **NHPA:** A.N.T. Photo Library (t). **Robert A. Patzner, University of Salzburg, Austria:** (cb). **359 Alamy Images:** Papilio/Steve Jones (crb); Wolfgang Pölzer (tl). **Dive Gallery/Jeffrey Jeffords (www.divegallery.com):** [illegible] **FLPA:** R. Dirscherl. **362 Alamy Images:** Blickwinkel (cr); Reinhard Dirscherl (b). **Marine Wildlife:** Paul Kay (cla). **SeaPics.com:** Doug Perrine (cl). **363 Dr. Frances Dipper:** (bc). **DK Images:** Jerry Young (tr). **OSF/photolibrary:** David Fleetham (tl); Pacific Stock (cr). **Robert A. Patzner, University of Salzburg, Austria:** (clb). **SeaPics.com:** V&W/Hal Beral (cra). **364 Dr. Frances Dipper:** (cla) (br). **Dive Gallery/Jeffrey Jeffords (www.divegallery.com):** (cb). **DK Images:** Frank Greenaway (cra). **SeaPics.com:** Masa Ushioda (bl). **365 DK Images:** Jerry Young (c). **Marine Wildlife:** Paul Kay (tc). **OSF/photo library:** Doug Allan (br). **SeaPics.com:** Jonathan Bird (cl); Jeremy Stafford-Deitsch (cra). **366–367 FLPA:** Minden Pictures/Chris Newbert. **368 Alamy Images:** Blickwinkel (br); Wolfgang Pölzer (ca) (bl). **DK Images:** Jane Burton (cb). **SeaPics.com:** Doug Perrine (cla). **369 Alamy Images:** Reinhard Dirscherl (tl). **DK Images:** Geoff Dann (clb); Colin Keates/Courtesy of the Natural History Museum, London (cr). **OSF/photolibrary:** Richard Herrmann (bc); Pacific Stock (br). **370 Alamy Images:** Reinhard Dirscherl (tc); Sami Sarkis (c). **Dive Gallery/Jeffrey Jeffords (www.divegallery.com):** (crb). **Marine Wildlife:** Alexander Mustard (cra). **SeaPics.com:** Doug Perrine (bl). **371 DK Images:** Dave King (bl). **OSF/photolibrary:** Richard Herrmann (br). **SeaPics.com:** Jez Tryner (tl). **372 Corbis:** Staffan Widstrand (cra). **DK Images:** Dave King (ca). **FLPA:** Minden Pictures/J. H. Editorial/Cyril Ruoso (cl); Minden Pictures/Tui De Roy (crb). **Oceanwide Images:** Gary Bell (bc). **373 Getty Images:** Image Bank/Tobias Bernhard (b). **OSF/photolibrary:** Olivier Grunewald (cla). **374 Getty Images:** Image Bank/Pete Atkinson (t). Brook Mathews, Sydney: (br). **Ilan Ben Tov, Israel:** (bl). **375 FLPA:** Peter Reynolds (cra); S. A. Team/Foto Natura (tc). **Getty Images:** National Geographic/Bill Curtsinger (bl). **SeaPics.com:** Doug Perrine (br). **376–377 Harald Slauschek/UnderwaterVisions.net. 378 Dick Bartlett:** (cra). **naturepl.com:** Constantinos Petrinos (l). **Queensland Museum, Australia (www.Qmuseum.qld.gov.au):** (br). **379 Getty Images:** Taxi/Gary Bell (r). **SeaPics.com:** Gary Bell (cla); Steve Drogin (clb). **380 Alamy Images:** Pep Roig (t). **Kraig Haver Photography:** (bl). **Still Pictures:** Michael Fairchild (br). **381 FLPA:** Minden Pictures/Mike Parry (c). **Adam Slavický:** (bl). **Scott Solar/Amazon Reptile Center:** (tl). **Dr. Adam P. Summers:** (br). **Frank Bambang Yuwono:** (tr). **382 DK Images:** Ken Findlay (clb/*Albatross*); Chris Gomersall (clb/*Curlew*); Frank Greenaway/Courtesy of The National Birds of Prey Centre, Gloucestershire (clb/*Sea Eagle*); Rob Reichenfeld (clb/*Pelican*). **iStockphoto.com:** Hans F. Meier (cla). **OSF/photolibrary:** Survival Anglia (crb). **Still Pictures:** Woodfall Wild Images/Everson (bc). **383 Alamy Images:** PhotoStockFile/Paul Wayne Wilson (bc). **OSF/photolibrary:** Doug Allan (t). **SeaPics.com:** Richard Herrmann (br). **384 Alamy Images:** Petr Svarc (cla); WorldFoto (cra). **DK Images:** (crb); Frank Greenaway (bl). **385 Alamy Images:** Malcolm Schuyl (br); David Tipling (bl). **DK Images:** Steve Gorton (clb); Dave King (cr). **Neil Fletcher:** Tomi Muukonen (cla). **Dr. Paul Hofmann:** (c). **386 Alamy Images:** Bryan & Cherry Alexander Photography (l). **DK Images:** Neil Fletcher (br). **387 Alamy Images:** Kim Westerskov (bl). **Neil Fletcher:** Barry Hughes (cra). **OSF/photolibrary:** Konrad Wothe (clb). **SeaPics.com:** Hiroya Minakuchi (ca); Kevin Schafer (br). **388–389 FLPA:** Fritz Polking. **390 Alamy Images:** INFOCUS Photos/Malie Rich-Griffith (bc). **naturepl.com:** Peter Reese (cr). **OSF/photolibrary:** Daniel Cox (tr) (tl). **391 Alamy Images:** ImageState/Pete Oxford (bl). **Neil Fletcher:** Hanne & Jens Eriksen (tr). **392 Alamy Images:** George McCallum Photography (ca); INFOCUS Photos/Malie Rich-Griffith (cb). **Neil Fletcher:** Hanne & Jens Eriksen (bl); Jonathan Grey (cla); Just Birds (br). **393 Alamy Images:** Nature Photographers Ltd/Paul Sterry (crb). **Neil Fletcher:** George Reszeter (t). **SeaPics.com:** Doug Perrine (cb). **394 Alamy Images:** Barry Bland (tr); Chris Mercer (cl). **DK Images:** Kim Taylor (fcl). **Neil Fletcher:** Just Birds (clb). **SeaPics.com:** Robert Shallenberger (br). **395 Alamy Images:** Robert E. Barber (cr); f1 online/Pölzer (bc). **Neil Fletcher:** Joe Fuhrman (tl); Mike Read (tr). **SeaPics.com:** Phillip Colla (br). **396 Alamy Images:** Blickwinkel (b). **FLPA:** Minden Pictures/Tui De Roy (ca) (tr). **397 Alamy Images:** WoodyStock/Ingo Schulz (cra). **Neil Fletcher:** Ian Montgomery (tc); Mike Read (cla). **Still Pictures:** Fritz Polking (br). **398 Alamy Images:** Mike Lane (bl); PhotoStockFile/Paul Wayne Wilson (br). **DK Images:** Cyril Laubscher (cr); Frank Greenaway/Courtesy of the Natural History Museum, London (ca). **Neil Fletcher:** Barry Hughes (cla). **399 Alamy Images:** Bryan & Cherry Alexander Photography (cr); R. & M. Thomas (b). **SeaPics.com:** Richard Herrmann (tl). **Still Pictures:** Steve Kaufman (cb); Tom Vezo (tr). **400 Alamy Images:** George McCallum Photography (br); The Photolibrary Wales (bl). **Neil Fletcher:** Just Birds (cl). **SeaPics.com:** Scott Leslie (tr). **401 Alamy Images:** Robert F. [illegible] **DK Images:** Cyril Laubscher (br). **Neil Fletcher:** Joe Fuhrman (bl); George Reszeter (cra). **402 Alamy Images:** Blickwinkel (br). **DK Images:** Harry Taylor/Courtesy of the Natural History Museum, London (bc). **Neil Fletcher:** Dudley Edmonson (cra); Barry Hughes (tc). **Getty Images:** Image Bank/Roine Magnusson (c). **403 Alamy Images:** Kevin Schafer (bl). **DK Images:** Irv Beckman (crb). **OSF/photolibrary:** David Tipling (t). **Still Pictures:** Mark Edwards (cra). **404 Alamy Images:** Brandon Cole Marine Photography (cl). **DK Images:** Philip Dowell (cr). **Marine Wildlife:** Doug Allan (bl). **Still Pictures:** Steven Kazlowski (clb). **404–405 OSF/photolibrary:** Mark Jones (c). **405 Alamy Images:** Steven J. Kazlowski (ca). **DK Images:** James Stevenson & Tina Chambers/Courtesy of the National Maritime Museum, London (crb). **OSF/photolibrary:** Pacific Stock (br). **406 DK Images:** Jerry Young (tl). **FLPA:** Foto Natura/Wil Meinderts (br). **Howard Hall Productions:** (bl). **Still Pictures:** Norbert Wu (cb). **407 FLPA:** Minden Pictures/Tui De Roy (bc). **Getty Images:** Image Bank/Joseph Van Os (tr). **Brian Lockett (www.air-and-space.com):** (bl). **Marine Wildlife:** Paul Kay (cla). **NOAA:** Captain Budd Christman, NOAA Corps (crb). **408 DK Images:** Frank Greenaway (cl). **FLPA:** Panda Photo (crb). **Still Pictures:** Fred Bruemmer (bl); Woodfall Wild Images/Tapani Rasanen (tr). **www.harwoodphotography.co.uk:** (cr). **409 NOAA:** Jan Roletto/Gulf of the Farallones National Marine Sanctuary (tl). **SeaPics.com:** Franco Banfi (br); Kevin Schafer (cr). **410–411 Steve Smithson. 412 Ardea:** Francois Gohier (tr). **Corbis:** The Mariners' Museum (cl). **FLPA:** Minden Pictures/Flip Nicklin (bl). **SeaPics.com:** Howard Hall (bc). **413 Alamy Images:** Brandon Cole Marine Photography (t); Stephen Frink Collection/James D. Watt (b). **414–415 Marine Wildlife:** Sue Flood. **416 DK Images:** Frank Greenaway (tr). **naturepl.com:** Doc White (ca). **SeaPics.com:** Doug Perrine (b). **417 Alamy Images:** Andre Seale (cla). **FLPA:** Minden Pictures/Flip Nicklin (crb). **Getty Images:** National Geographic/Brian Skerry (tr). **SeaPics.com:** John K. B. Ford/Ursus (bl). **418 FLPA:** Minden Pictures/Flip Nicklin (tr). **Getty Images:** Stone/Kim Westerskov (bc). **SeaPics.com:** Thomas Jefferson (cl); Robert L. Pitman (br). **419 Alamy Images:** Stock Connection Blue/Tom Brakefield. **420 Marine Wildlife:** Sue Flood. **421 FLPA:** Minden Pictures/Michio Hoshino (ca); Minden Pictures/Flip Nicklin (cra); Minden Pictures/Norbert Wu (bc). **naturepl.com:** Todd Pusser (crb). **Mike Scott:** (cb). **SeaPics.com:** Phillip Colla (c). **422 FLPA:** Minden Pictures/Flip Nicklin (t). **Image Quest Marine:** Masa Ushioda (br). **SeaPics.com:** Florian Graner (cl). **423 DK Images:** Peter Visscher (crb). **FLPA:** Minden Pictures/Chris Newbert (tr). **Getty Images:** Photographer's Choice/Pete Atkinson (b). **Still Pictures:** Douglas Faulkner (ca). **424–425 NASA:** Jacques Descloitres, MODIS Rapid Response Team, NASA/GSFC. **428 Alamy Images:** Bryan & Cherry Alexander Photography (cla); LOOK Die Bildagentur der Fotografen GmbH (cb). **430 Alamy Images:** Jack Stephens (cla). **Corbis:** Lowell Georgia (cra). **431 Alamy Images:** Nordicphotos/Kristjan Fridriksson (br). **432 Alamy Images:** Greenshoots Communications (c); David Sanger Photography (cla). **Corbis:** Ralph White (bl). **434 Alamy Images:** FLPA (br). **435 DK Images:** David Lyons (cl). **NASA:** Jacques Descloitres, MODIS Land Rapid Response Team, NASA/GSFC (ca). **436 Alamy Images:** Ace Stock Ltd (bc); allOver photography (clb). **DK Images:** Linda Whitwam (ca). **437 Alamy Images:** Nick Hanna (br). **438 Mads Eskesen. 439 Alamy Images:** Mike Lane (br). **Mads Eskesen:** (tr) (cra). **Horns Rev Havmøllepark (www.hornsrev.dk):** Medvind Fotografi/Bent Sørensen (bc). **Nysted Havmøllepark (www.nystedhavmoellepark.dk):** (crb). **441 Alamy Images:** Wild Places Photography/Chris Howes (br). **Corbis:** Sygma/Bernard Annebicque (cr). **442 DK Images:** Christopher & Sally Gable (tl); John Heseltine (br). **443 Alamy Images:** Vehbi Koca (crb); Rob Rayworth (bl). **NASA:** Image courtesy NASA/GSFC/MITI/ERSDAC/JAROS, and U.S./Japan ASTER Science Team (c). **444 Corbis:** (tr) (bl); Sygma/Harford Chloe (cb). **NASA:** Jacques Descloitres, MODIS Land Rapid Response Team, NASA/GSFC (br). **446 Corbis:** Cordaiy Photo Library Ltd/John Farmar (tr). **NASA:** GSFC/JPL, MISR Team (bl). **Science Photo Library:** Southampton Oceanography Centre/B. Murton (cla). **447 Alamy Images:** Wild Places Photography/Chris Howes (cra). **Corbis:** (ca). **448 FLPA:** Colin Monteath (br). **iStockphoto.com:** Patrick Roherty (bc). **449 Alamy Images:** Bryan & Cherry Alexander Photography (bl). **NASA:** Jesse Allen, NASA Earth Observatory and the HIGP Thermal Alerts Team (cr). **NOAA:** Lieutenant Philip Hall, NOAA Corps (br). **450 Corbis:** Yann Arthus-Bertrand (bl); Reuters/Supri (cla). **Still Pictures:** Friedrich Stark (c). **452 Alamy Images:** Blickwinkel (bc); Tor Eigeland (tr). **Corbis:** Jonathan Blair (cl). **453 Alamy Images:** Images of Africa Photobank/Peter Williams (tr). **454 Alamy Images:** Neil McAllister (br). **455 Alamy Images:** Julio Etchart (cra). **456 Getty Images:** AFP. **457 Corbis:** Reuters/U.S. Navy/Philip A. McDaniel (cra); Reuters/Tarmizy Harva (br). **Images acquired and processed by CRISP, National University of Singapore:** (tc) (ca). **Getty Images:** AFP/Sena Vidanagama (crb). **NOAA:** (cb). **459 iStockphoto.com:** Wesley Drake (cra). **OSF/photolibrary:** Michael Brooke (br). **460 Alamy Images:** Danita Delimont (cla). **Corbis:** Ralph White (bc). **NOAA:** Commander Richard Behn, NOAA Corps (cl). **462 NASA:** George Riggs, NASA GSFC (cra). **US Fish and Wildlife Service National Image Library:** Alaska Maritime National Wildlife Refuge/Kevin Bell (cla). **463 Corbis:** Neil Rabinowitz (tc). **464 Alamy Images:** FocusRussia (cra); Iain Masterton (cl). **Corbis:** Michael S. Yamashita (crb). **466 Alamy Images:** Chris Willson (br). **467 Still Pictures:** Henning Christoph (cra). **469 Alamy Images:** Andre Seale (br). **Corbis:** Douglas Faulkner (cr); Reuters/Alex De La Rosa (tr). **470 Still Pictures:** Richard J. Wainscoat (cra). **471 Alamy Images:** Dennis Hallinan (bc). **472 OSF/photolibrary:** Tammy Peluso (tr). **473 iStockphoto.com:** Angela Bell (cla). **SeaPics.com:** Gary Bell (tc). **474 Getty Images:** AFP/Tarik Tinazay. **475 Alamy Images:** Nick Hanna (bl); Images&Stories (cra). **Corbis:** Stephen Frink (br); Jeffrey L. Rotman (bc); Lawson Wood (crb). Getty Images: Image Bank/Zac Macaulay (tr). **OSF/photolibrary:** David Fleetham (ca). **477 Alamy Images:** Danita Delimont (cra); INTERFOTO Pressebildagentur (cb). **478 OSF/photolibrary:** Mike Hill (br). **479 iStockphoto.com:** Michal Wozniak (tr). **480 Alamy Images:** LOOK Die Bildagentur der Fotografen GmbH (cl); Bruce Percy (br). **Corbis:** Paul A. Souders (cra). **482 Alamy Images:** Bryan & Cherry Alexander (cl). **Still Pictures:** Norbert Wu (cla) (br). **484 Corbis:** Eye Ubiquitous/C. M. Leask (cra). **NOAA:** Commander John Bortniak, NOAA Corps (cla). **485 Alamy Images:** Blickwinkel (bl); Kim Westerskov (br). **486 Ardea:** Edwin Mickleburgh. **487 Alamy Images:** Graphic Science (tr); Steve Morgan (ca). **NASA:** MODIS Land Science Team (cra); Jacques Descloitres, MODIS Land Science Team (br); NASA/GSFC/LaRC/JPL, MISR Team (crb).

JACKET IMAGES: *Front:* **Getty Images:** Visuals Unlimited/David Wrobel. *Back:* **Alamy Images:** Fabrice Bettex (ftl); **DeepSeaPhotography.Com:** Kim Westerskov (ftr); **Getty Images:** Image Bank/Don & Liysa King (b); Lonely Planet Images/Peter Hendrie (tr); Science Faction/Flip Nicklin (tl). *Spine:* **Dive Gallery/Jeffrey Jeffords (www.divegallery.com)**. *Front Flap:* **Corbis:** Jeffrey L. Rotman. *Back Flap:* **Getty Images:** Photonica/Anna Grossman.

Data for the bathymetric maps in the Atlas of the Oceans chapter provided by Planetary Visions based on ETOPO2 global relief data, SRTM30 land elevation data, and the Generalised Bathymetric Chart of the Ocean. ETOPO2 published by the U.S. Department of Commerce, National Oceanic and Atmospheric Administration, National Geophysical Data Center, 2001. SRTM30 published by NASA and the National Geospatial Intelligence Agency, 2005, distributed by the U.S. Geological Survey. GEBCO One Minute Grid reproduced from the GEBCO Digital Atlas published by the British Oceanographic Data Centre on behalf of the Intergovernmental Oceanographic Commission of UNESCO and the International Hydrographic Organisation, 2003.

All other images © Dorling Kindersley
For further information see: www.dkimages.com